Advances in Science, Technology & Innovation

IEREK Interdisciplinary Series for Sustainable Development

Advances in Science, Technology & Innovation (ASTI) is a series of peer-reviewed books based on important emerging research that redefines the current disciplinary boundaries in science, technology and innovation (STI) in order to develop integrated concepts for sustainable development. It not only discusses the progress made towards securing more resources, allocating smarter solutions, and rebalancing the relationship between nature and people, but also provides in-depth insights from comprehensive research that addresses the **17 sustainable development goals (SDGs)** as set out by the UN for 2030.

The series draws on the best research papers from various IEREK and other international conferences to promote the creation and development of viable solutions for a **sustainable future and a positive societal** transformation with the help of integrated and innovative science-based approaches. Including interdisciplinary contributions, it presents innovative approaches and highlights how they can best support both economic and sustainable development, through better use of data, more effective institutions, and global, local and individual action, for the welfare of all societies.

The series particularly features conceptual and empirical contributions from various interrelated fields of science, technology and innovation, with an emphasis on digital transformation, that focus on providing practical solutions to **ensure food, water and energy security to achieve the SDGs.** It also presents new case studies offering concrete examples of how to resolve sustainable urbanization and environmental issues in different regions of the world.

The series is intended for professionals in research and teaching, consultancies and industry, and government and international organizations. Published in collaboration with IEREK, the Springer ASTI series will acquaint readers with essential new studies in STI for sustainable development.

ASTI series has now been accepted for Scopus (September 2020). All content published in this series will start appearing on the Scopus site in early 2021.

More information about this series at https://link.springer.com/bookseries/15883

Inamuddin · Rajender Boddula ·
Mohd Imran Ahamed · Anish Khan
Editors

Carbon Dioxide Utilization to Sustainable Energy and Fuels

Springer

Editors
Inamuddin
Department of Applied Chemistry
Faculty of Engineering and Technology
Zakir Husain College of Engineering
and Technology
Aligarh Muslim University
Aligarh, India

Mohd Imran Ahamed
Department of Chemistry
Faculty of Science
Aligarh Muslim University
Aligarh, India

Rajender Boddula
CAS Key Laboratory of Nanosystems
and Hierarchical Fabrication
Beijing, China

Anish Khan
Department of Chemistry
Faculty of Science
King Abdulaziz University
Jeddah, Saudi Arabia

ISSN 2522-8714 ISSN 2522-8722 (electronic)
Advances in Science, Technology & Innovation
IEREK Interdisciplinary Series for Sustainable Development
ISBN 978-3-030-72876-2 ISBN 978-3-030-72877-9 (eBook)
https://doi.org/10.1007/978-3-030-72877-9

This Springer imprint is published by the registered company Springer Nature Switzerland AG
The registered company address is: Gewerbestrasse 11, 6330 Cham, Switzerland

Contents

Chemical Valorization of CO_2

Esperanza Ruiz Martínez and José María Sánchez Hervás

Abstract

The growing CO_2 atmospheric concentrations, linked with detrimental global warming, and the declining fossil resources derived from power generation, transportation and industries are of great environmental and sustainability concerns. Chemical valorization of CO_2 into fuels, chemicals, polymers and construction materials is a promising tool to reduce fossil fuel depletion and greenhouse gas emissions, contribute to decarbonization of energy, transport and industrial sectors and to fulfill climate goals, while storing renewable energy and generating revenue. Chemical valorization of CO_2 has the potential to generate products with similar or improved quality and with smaller carbon and water footprints, energy consumption and production costs than traditionally produced counterparts. However, the deployment of these CO_2-based products is limited to about 10 million tons of CO_2 per year, due to the early stage of development of some valorization routes, the need for low-carbon, cheap and renewable energy and hydrogen and the fulfillment of quality standards and regulations. This chapter focuses on valorization technologies approaching commercialization, consuming huge amounts of CO_2 and integrating renewable energy, and reviews the maturity, the scaling-up potential, the climatic, economic and environmental benefits and the bottlenecks for commercialization of the technologies and the market size and price of the synthesized products.

Keywords

CO_2 recycling • Renewable energy storage • CO_2 to chemicals • CO_2 to fuels • CO_2 to polymers • CO_2 to building materials • Climate change • Greenhouse gas mitigation • Chemical processes • CO_2 utilization

Abbreviations

AT	Austria
BE	Belgium
Bio	Biological
BP	British Petroleum
CA	Canada
Cat	Catalytic
CH	Switzerland
CN	China
coSOEC	Co-electrolysis of water and CO_2 in Solid Oxide Electrolyte Cells
CRT	Chemical Reaction Engineering
d	Day
DE	Germany
DEA	Danish Energy Agency
DGC	Danish Gas Technology Centre
DK	Denmark
DME	Dimethyl ether
DNV	Det Norske Veritas
DTU	Technical University of Denmark
ECN	Energy research Centre of the Netherlands
ES	Spain
EU	European Union
FI	Finland
FR	France
FT	Fischer–Tropsch
HU	Hungary
IN	India
IS	Iceland
IEA	International Energy Agency
IET	Institut fur Energietechnik

E. Ruiz Martínez (✉) · J. M. Sánchez Hervás
Unit for Sustainable Thermochemical Valorisation, CIEMAT, Avenida Complutense, 40, 28040 Madrid, Spain
e-mail: esperanza.ruiz@ciemat.es

J. M. Sánchez Hervás
e-mail: josemaria.sanchez@ciemat.es

Inamuddin et al. (eds.), *Carbon Dioxide Utilization to Sustainable Energy and Fuels*, Advances in Science, Technology & Innovation, https://doi.org/10.1007/978-3-030-72877-9_1

IT	Italy
JP	Japan
KEI	Knowledge Energy Institute
k€	Thousands of Euros
MCi	Mineral Carbonation International
MSW	Municipal Solid Waste
Mt	Millions of tons
nda	No data available
NO	Norway
NL	Netherlands
NL-DE	Dutch-German collaboration
NZ	New Zealand
P2G	Power-to-Gas
PFI	Prüf- und Forschungsinstitut
PL	Poland
QAFAC	Qatar Fuel Additives Company Limited
RFA	Renewable Fuel Association
RWGS	Reverse Water Gas Shift
SE	Sweden
SLF	Swiss Liquid Future
t	Ton
Tec	Technology
TRL	Technology Readiness Level
TW	Taiwan
US	United States of America
WBT	White Biotech
y	Year
ZA	South Africa
ZSW	Zentrum für Sonnenenergie- und Wasserstoff-Forschung Baden-Württemberg

1 Introduction

Carbon dioxide is a waste gas resulting mainly from fossil fuel combustion for transportation and power generation and from other industrial processes. Carbon dioxide has a high greenhouse gas potential and is regarded as one of the foremost contributors to climate change and global warming (Ganesh 2014; Goeppert et al. 2014).

The production of energy, fuels, chemicals and materials from fossil sources continues to play a crucial role (Goeppert et al. 2014). On account of continuous increase in the atmospheric concentration of carbon dioxide and progressive depletion of fossil fuel reserves resulting from extensive power generation, chemical valorization of emitted carbon dioxide to fuels, chemicals and materials is considered as a method complementary to carbon dioxide sequestration and storing for reducing greenhouse gas emissions, enabling carbon dioxide recycling and a more sustainable exploitation of resources, while reducing costs associated with carbon dioxide capture or allows obtaining an economic revenue (Centi 2009; Zhu 2019).

There are different potential technological routes, based mainly on catalytic, biochemical, electrochemical and mineralization processes, for chemical conversion of carbon dioxide into a variety of added value products, such as methane, methanol, dimethyl ether (DME), gasoline, diesel, jet fuel, ethanol, formic acid, polymers and building materials (Zhu 2019; Chauvy et al. 2019).

Chemical valorization of wasted carbon dioxide in the form of materials, chemicals and fuels may result in a series of environmental benefits, as long as renewable power sources are utilized for the conversion of carbon dioxide. In this manner, CO_2 can become from a detrimental greenhouse gas producing global warming, into a worthwhile, renewable, endless and carbon neutral source of chemicals, materials and combustibles for the future (Ganesh 2014; Goeppert et al. 2014; Centi 2009).

The climate benefits linked with carbon dioxide utilization mainly come from substituting a fossil-based product, such as conventional chemicals, fuels or polymers, by the corresponding alternative product derived from carbon dioxide and having longer life cycle of carbon dioxide emissions. Key aspects to be consider in evaluating the climate benefits of carbon dioxide valorization include: the origin of carbon dioxide, derived from fossil fuels or biomass or captured from air, the conventional product being replaced by the carbon dioxide based product, the amount and type of energy required for carbon dioxide conversion, the retention time of carbon dioxide in the product and the scale of the potential of carbon dioxide utilization (International Energy Agency (IEA) 2019).

Carbon dioxide obtained from biomass or directly from the air can provide neutral or even negative carbon emissions for carbon dioxide valorization routes that involve the long-term permanent storage of carbon in the final product, such as in construction materials (International Energy Agency (IEA) 2019).

The carbon retention time for carbon dioxide valorization applications ranges from barely one year for fuels, until a ten of years for the majority of chemicals, to centuries and millions of years for polymers and construction materials, respectively (International Energy Agency (IEA) 2019).

The potential contribution of carbon dioxide utilization to the fulfillment of climate objectives will be determined by the timeframe and speediness of the scaling-up of these chemical valorization pathways (Chauvy et al. 2019; International Energy Agency (IEA) 2019).

There are two types of carbon dioxide-based products depending on the market size and value: products with high added value and low market size, such as formic acid, and products with low added value and large market volume, such as synthetic fuels and building materials. Fuels have the largest potential for carbon dioxide utilization due to their huge market volume, while building materials show the greatest potential of climate change mitigation mainly because of their low energy requirements and permanent retention of carbon in the product, but face other barriers to prompt deployment, such as regulations and construction standards (Chauvy et al. 2019; International Energy Agency (IEA) 2019; Aresta et al. 2014). In summary, the deployment of these CO_2-based products is limited to about 10 million tons of CO_2 per year, due to the early stage of development of some valorization routes, the need for low-carbon, cheap and renewable energy and hydrogen and the fulfillment of quality standards and regulations (Chauvy et al. 2019; International Energy Agency (IEA) 2019; Aresta et al. 2014).

Chemical valorization of carbon dioxide provides also a method for storing the excess of discontinuous renewable energy, unable to be discharged into the electricity network, in the form of carbon neutral or negative fuels, chemicals and materials, which storing, transport and utilization technologies are fully established, contributing to the decarbonization of power generation, transportation and other industrial sectors (Centi 2009).

The development and potential technology transfer of processes for the chemical valorization of carbon dioxide to fuels, chemicals and materials would help to overcome the limitations to renewable energy deployment, ensure power supplying and lessen the reliance on fossil reserves.

Chemical valorization of CO_2 toward fuels, chemicals and materials is a suitable way to provide social wellbeing and economic growing, while allowing some savings in greenhouse gas emission, contributing also to the achievement of the environmental, energy and social goals settled at global and local levels (Ganesh 2014; Goeppert et al. 2014; Centi 2009).

Presently, only a small number of processes for the synthesis of valuable products, such as urea, salicylic acid, polycarbonates and cyclic carbonates, from carbon dioxide are available at commercial scale (Alper and Yuksel Orhan 2017; Artz 2018; Fukuoka et al. 2007; Patricio et al. 2017; Quadrelli et al. 2011; Sakakura et al. 2007). Therefore, these technologies will not be reviewed further in this chapter.

There are also some innovative technologies for the conversion of carbon dioxide to various commercial products, such as chemicals, synthetic fuels, polymers and construction materials, currently at different levels of development and commercialization. The potential application of these technologies for the valorization of carbon dioxide can result in savings in carbon dioxide emissions, by improving efficiency, by partially substituting fossil-based raw materials and by using renewable energy, thus contributing to the decarbonization of power generation and other industrial sectors and to the creation of economic benefits through the production of saleable products (Zhu 2019).

A high-technology readiness level (TRL) is an indispensable requirement for market penetration of the carbon dioxide-based products (O'Connell et al. 2019). This chapter focuses on technologies of chemical valorization of carbon dioxide which are at demonstration level or close to commercialization, with technology readiness levels higher than 5. Moreover, processes of chemical valorization of carbon dioxide which still remain in an early stage of research and development are not considered in this chapter. This chapter specifically reviews valorization processes consuming large volume of carbon dioxide and demonstration projects/plants at pilot or pre-commercial scale. This chapter also highlights the capability of some chemical routes of carbon dioxide valorization to integrate renewable energy into the chemical and energy infrastructures available for the transport, storage and utilization of the chemicals and fuels synthesized from carbon dioxide.

The present chapter reviews, for each route of chemical valorization of carbon dioxide, the maturity, the potential for scaling-up, the climatic, economic and environmental benefits and the bottlenecks for further development or commercialization of the technology and the amount of carbon dioxide chemically converted, namely the carbon dioxide abatement potential, the reduction in carbon and water footprints and in energy consumption as compared with conventional production routes, as well as the market size and price of the synthesized products.

2 CO_2-Derived Fuels and Chemicals

Carbon dioxide represents an alternative to fossil fuels as a source of carbon for the production of platform chemicals, transportation fuels or energy carriers, such as methane, methanol, ethanol, dimethyl ether, formic acid, carbon monoxide or syngas, gasoline, diesel and jet fuels. The process consists of the reaction of carbon dioxide with hydrogen to produce, directly, or indirectly, from the syngas generated via the reverse water gas shift reaction, a carbon-containing compound with higher easiness of handling, transport, storage and utilization than pure hydrogen. Given that hydrogen production is a highly energy-consuming process, low-carbon hydrogen has to be produced by biomass gasification or reforming coupled with carbon dioxide capture and storage, or via electrolysis of water powered with renewable or low-carbon electricity (International Energy Agency (IEA) 2019). In general, the cost of hydrogen production dominates the operating cost of the full chemical synthesis process. In summary, carbon dioxide is a potential raw material for the production of fuels and chemicals, but its transformation to valuable products implies an energy consumption and production costs. Depending on the carbon dioxide-derived product, the production technology has different maturity and potential of carbon dioxide uptake (Chauvy et al. 2019; International Energy Agency (IEA) 2019; Agarwal et al. 2011; Álvarez et al. 2017; Götz et al. 2016; HYSYTECH 2020; Jarvis and Samsatli 2018; Kiss et al. 2016; Made in China 2020; Panzone et al. 2020; Pérez-Fortes et al. 2016a; Pérez-Fortes et al. 2016b; Pérez-Fortes and Tzimas 2016; Schmidt et al. 2016; Rego de Vasconcelos and Lavoie 2019).

Table 1 shows a non-exhaustive list of the different key performance indicators, such as TRL of the carbon dioxide conversion process, production cost and price of the specific CO_2-based product, in kilo euros per ton of synthesized product, electricity consumption per ton of synthesized product, and net carbon dioxide utilization potential, in tons of CO_2 utilized per ton of product synthesized, for the most mature technologies for the production of fuels and platform chemicals from CO_2.

2.1 Methane

Methane is used to produce heat, electricity and value-added chemicals, being an important energy vector. At the industrial level, methane is obtained, at relatively low cost, from fossil natural gas. High-purity methane, also known as substitute natural gas, can be more sustainably produced by methanation of CO_2 with renewable hydrogen, generated by electrolysis of water powered by renewable energy, via the so-called Power-to-Methane process (O'Connell et al. 2019; Rego de Vasconcelos and Lavoie 2019).

The substitute natural gas produced can be injected directly into the natural gas grid (Götz et al. 2016). In this

Table 1 Key performance indicators for the most mature technologies for the production of fuels and platform chemicals from CO_2

Product	CO_2 conversion process	TRL	Production cost (k€/t)	Price (k€/t)	Electricity (Mwh/t)	CO_2 utilization (tCO_2/t)	References
Methane	Catalytic hydrogenation	8–9	1.5–13.5	0.92	15.2	2.75	(Chauvy et al. 2019; O'Connell et al. 2019; Götz et al. 2016; Jarvis and Samsatli 2018; Panzone et al. 2020; Behrens et al. 2019; YCharts 2020)
Methanol	Catalytic hydrogenation	9	0.5–1.9	0.26	0.55	1.4	(Chauvy et al. 2019; O'Connell et al. 2019; Jarvis and Samsatli 2018; Kiss et al. 2016; Panzone et al. 2020; Pérez-Fortes et al. 2016b; Methanex 2020; Zhang et al. 2019)
Dimethyl ether	Catalytic hydrogenation	6	1.8–2.3	1.6	nda	1.9	(Chauvy et al. 2019; China Petroleum and Chemical Industry Federation 2020; Michailos et al. 2019; Wulf et al. 2018)
Formic acid	Catalytic hydrogenation	3–5	1.6	0.45–0.6	4.1	0.7	(Álvarez 2017; Jarvis and Samsatli 2018; Made in China 2020; Panzone et al. 2020; Pérez-Fortes et al. 2016a; Rego de Vasconcelos and Lavoie 2019)
	Electrochemical reduction	5–6	0.85	0.45–0.6	1.1	0.5	(Chauvy et al. 2019; Agarwal et al. 2011; HYSYTECH 2020; Pérez-Fortes and Tzimas 2016; Rego de Vasconcelos and Lavoie 2019)
Ethanol	Syngas fermentation	9	0.8	0.98	nda	1.9	(Chauvy et al. 2019; O'Connell et al. 2019; Global Petrol Prices 2020; Handler et al. 2016; Medeiros et al. 2020)
Liquid fuels: gasoline, diesel, kerosene	Reverse Water Gas Shift + Fischer Tropsch	8–9	0.5–2	0.27–0.32	6.8	2.6	(O'Connell et al. 2019; Jarvis and Samsatli 2018; Schmidt et al. 2016)

This table summarizes the different key performance indicators, such as TRL of the carbon dioxide conversion process, production cost and price of the specific CO_2-based product and electricity consumption and net carbon dioxide utilization potential in product formation, for the most mature technologies for the production of fuels and platform chemicals from carbon dioxide. Comparison of key performance indicators of different alternative technologies gives an idea of the most advantageous and advanced technology for the production of a given carbon dioxide-based product.

TRL Technology Readiness Level, nda no data available.

way, substitute natural gas can be exploited immediately using the currently available natural gas storage, transport and utilization infrastructures, minimizing the investment needed for massive deployment of CO_2 methanation technology (Navarro et al. 2018). Power-to-Methane technology constitutes a promising pathway for large-scale and long-term renewable energy storage while reducing greenhouse gas emissions. Power-to-Methane technology allows interconnecting the gas and electrical networks, resulting in a great flexibility to the balance of the energy grid (Bailera et al. 2017).

CO_2 methanation can be catalytic or biological. The biological methanation is carried out by fermentation of CO_2 in the biogas plant itself or in a detached bioreactor (O'Connell et al. 2019). Catalytic CO_2 methanation is most usually carried out over nickel-based catalysts due to their high activity and low cost (Behrens et al. 2019; Ghaib and Ben-Fares 2018) at temperatures between 200 and 550 °C, to avoid potential formation of highly toxic nickel carbonyls at 200 °C and deactivation of catalyst by sintering and coking at temperatures above 550 °C (Schaaf et al. 2014) and at moderate pressures (1–100 bar) (Götz et al. 2016; Rego de Vasconcelos and Lavoie 2019).

Since methanation reaction is exothermal, the increment of temperature in the reactor may result in equilibrium limitations in conversion and hydrothermal deactivation of the catalyst (Götz et al. 2016). Therefore, different configurations of reactor, such as three-phase, heat exchanger, fixed bed, fluidized bed and structured, are used to improve the performance of the methanation process (Rego de Vasconcelos and Lavoie 2019). Basically, there are two different methanation reactor concepts: fixed bed reactor with intermediate cooling steps, developed by TREMP/Haldor-Topsoe, Hicom/British Gas Corp, RMP/Ralph, Lurgi and Sasol/Lurgi GmbH reactors, and fluidized bed reactor, developed by Comflux/Thyssengas GmbH and Bi-Gas/Bituminous Coal Res reactors (Schaaf et al. 2014). Other reactor concepts have also been brought to market, such as catalytic wall tubular reactor and parallel plate reactor (Schaaf et al. 2014). New reaction systems have also been investigated to enable better temperature control, such as suspended reactor, also known as "slurry-phase bubble-column reactor", magnetic fluidized bed reactor, dielectric barrier discharge plasma reactor and micro-channel reactor, which can inspire new developments of industrial methanation reactors (Gao et al. 2015). There are also some commercial Ni-based catalysts for substitute natural gas production with high efficiency, lifetime, poison resistance and thermal stability, developed, among others, by Johnson Mathey (CRG), Haldor-Topsoe (PK-7R) and CLARIANT (METH130/134/135).

Different countries, especially in Europe, driven by the need to reduce greenhouse gas emissions and intensify the share of renewable energy, initiated research and development projects focused on Power-to-Methane technology (Rego de Vasconcelos and Lavoie 2019). The Power-to-Methane technology is currently in an advanced stage of development leading to a technology readiness level of up to 8–9 (O'Connell et al. 2019; Behrens et al. 2019).

Several Power-to-Methane pilot and demonstration plants, both using catalytic and biological methanation, are already in operation in different countries, and others are still being developed. Most of the Power-to-Methane plants are installed in Europe. A non-exhaustive list of plants or projects for Power-to-Methane is provided in Table 2 (Zhu 2019; O'Connell et al. 2019; Götz et al. 2016; Rego de Vasconcelos and Lavoie 2019; Wulf et al. 2018; Bailera et al. 2017; Thema et al. 2019). This table combines and updates earlier reviews by Zhu (2019), O'Connell et al. (2019), Götz et al. (2016), Rego de Vasconcelos and Lavoie (2019), Wulf et al. (2018), Bailera et al. (2017) or Thema et al. (2019), adding previously un-reviewed projects or plants and including details such as plant operational status, city and country of location of the plant, type of conversion technology of carbon dioxide, technology readiness level, CO_2 source, plant capacity, expressed in terms of total methane output in Nm^3/h and installed power in kW calculated from the lower heating value of methane product gas, as well as main references.

In most of the plants the CO_2 is sourced from biogas upgrading or from CO_2 fraction in biogas streams. In some plants the CO_2 is obtained from flue gas or industrial emissions. A few plants use CO_2 directly captured from air (O'Connell et al. 2019).

In early 2019, substitute natural gas production capacity was about 6–7 MW of methane, according to the capacity of the plants of Power-to-Methane being in operation mostly in Europe (O'Connell et al. 2019; Thema et al. 2019). For some plants, the capacity is unknown. In addition, some plants are announced, planned, projected or under construction. Therefore, the estimated substitute natural gas production capacity would be actually higher than 16 MW of CH_4 (O'Connell et al. 2019). The European natural gas grid consists of a more than 2 million km transport and distribution network and approximately 800 TWh of underground storage through Europe (Behrens et al. 2019), available to absorb the potential substitute natural gas production.

Power-to-Methane technology can contribute to fulfill the requirements of long-term and massive storage of fluctuating renewable energy, resulting in a significant demand of Power-to-Methane plants in the future (O'Connell et al. 2019). Extrapolation to the electricity demand of the European Union leads to about 25 GW of methane (O'Connell et al. 2019). Furthermore, renewable transportation fuels are also required in the European Union to achieve the target reduction of greenhouse gas emissions. The final methane

Table 2 Plants or projects for Power-to-Methane in Europe

Plant/project	Status	Location	Capacity Nm3/h (kW) CH4	Tec	TRL	CO_2 source	References
Audi e-gas	In operation	Werlte (DE)	320 (3184)	Cat	7	Biogas upgrading	(Zhu 2019; O'Connell et al. 2019; Götz et al. 2016; Rego de Vasconcelos and Lavoie 2019; Wulf et al. 2018; Bailera et al. 2017; Thema et al. 2019; Kondratenko et al. 2013)
ETOGAS ZSW	In operation	Stuttgart (DE)	12.5 (124)	Cat	6	nda	(Zhu 2019; O'Connell et al. 2019; Rego de Vasconcelos and Lavoie 2019; Wulf et al. 2018; Bailera et al. 2017; Ghaib and Ben-Fares 2018; Zentrum für Sonnenenergie-und Wasserstoff-Forschung Baden-Württemberg (ZSW) 2020)
CoSin	In operation	Barcelona (ES)	10 (41)	Cat	6	Biogas upgrading	(O'Connell et al. 2019; Naturgy 2020; Guilera et al. 2020)
P2G Rozenburg	In operation	Rozenburg (NL)	2 (2.75)	Cat	6	nda	(O'Connell et al. 2019; Wulf et al. 2018; Bailera et al. 2017; Thema et al. 2019; Vlap et al. 2015)
Stromspeicherung Wunsiedel	Planned/announced	Wunsiedel (DE)	7960	Cat		Flue gas	(O'Connell et al. 2019)
BioPower2Gas	In operation	Allendorf (DE)	60–220 (149)	Bio	6	Biogas upgrading	(O'Connell et al. 2019; Wulf et al. 2018; Bailera et al. 2017; Thema et al. 2019; BioPower2Gas 2016; Müller et al. 2017)
P2G BioCat	In operation	Avedore (DK)	200 (1990)	Bio	7	CO_2 fraction in biogas	(O'Connell et al. 2019; Götz et al. 2016; Rego de Vasconcelos and Lavoie 2019; Wulf et al. 2018; Bailera et al. 2017; Thema et al. 2019; BioCat Project 2016)
DemoSNG	In operation	Köping (SE)	Nda	Cat	6	Biomass gasification syngas /flue gas	(O'Connell et al. 2019; Rego de Vasconcelos and Lavoie 2019; Bailera et al. 2017; Thema et al. 2019)
Jupiter 1000	Under construction	Fos sur mer (FR)	25 (249)	Cat	7	Industrial emission from an ASCOMETAL factory	(O'Connell et al. 2019; Wulf et al. 2018; Thema et al. 2019; Jupiter 1000 2019)
EXYTRON	In operation	Rostock (DE)	1 (10)	Cat	7	Flue gas	(O'Connell et al. 2019; Wulf et al. 2018; Thema et al. 2019; Müller et al. 2017)

(continued)

Table 2 (continued)

Plant/project	Status	Location	Capacity Nm3/h (kW) CH4	Tec	TRL	CO_2 source	References
Eichhof	In operation	Bad Hersfeld (DE)	20 (199)	Cat	6	CO_2 fraction in biogas	(O'Connell et al. 2019; Wulf et al. 2018; Thema et al. 2019)
RAG underground-sun-conversion	Planned	Pilsbach (AT)	(10,000)	Bio	6	nda	(O'Connell et al. 2019; Underground Sun Conversion 2020)
STORE & GO	In operation	Falkenhagen (DE)	57 (567)	Cat	7	Bioethanol plant (biogas)	(O'Connell et al. 2019; Rego de Vasconcelos and Lavoie 2019; Wulf et al. 2018; Bailera et al. 2017; Thema et al. 2019; Store & Go 2020)
STORE & GO	In operation	Solothurn (CH)	11.5 (114)	Bio	7	Waste water treatment plant	(O'Connell et al. 2019; Rego de Vasconcelos and Lavoie 2019; Wulf et al. 2018; Bailera et al. 2017; Thema et al. 2019; Store & Go 2020; Hafenbradl 2017)
STORE & GO	In operation	Puglia–Troia (IT)	11 (100)	Cat	7	Air capture	(O'Connell et al. 2019; Rego de Vasconcelos and Lavoie 2019; Wulf et al. 2018; Bailera et al. 2017; Thema et al. 2019; Store & Go 2020)
Energiepark	In operation	Pirmasens-Winzeln (DE)	50 (498)	Bio	7	CO_2 fraction in biogas	(O'Connell et al. 2019; Wulf et al. 2018; Thema et al. 2019)
Energiepark extension	In operation	Pirmasens-Winzeln (DE)	(996)	Bio	7	CO_2 fraction in biogas	(O'Connell et al. 2019; Thema et al. 2019; Prüf-und Forschungsinstitut (PFI) Germany 2020)
Hybridge (AMPRION, OGE)	Announced	Southern Emsland (DE)	(100,000) (electrolyzer)	nda	nda	nda	(O'Connell et al. 2019; Thema et al. 2019; Wiede and Land 2018)
CO_2-SNG	In operation	Laziska Görne (PL)	4.5 (45)	Cat	6	Flue gas	(Wulf et al. 2018; Bailera et al. 2017; Chwoła 2020)
Power to gas Hungary	Projected	nda (HU)	(10,000) (electrolyzer)	Bio	7–8	nda	(Wulf et al. 2018; Thema et al. 2019; Bertalan and Hein 2016)
Symbio	In operation	Lyngby (DK)	nda	Bio	6	nda	(Wulf et al. 2018; Bailera et al. 2017; Thema et al. 2019; Technical University of Denmark (DTU) 2013)
Swisspower Hybridkraftwerk	Planned/Under construction	Dietikon (CH)	~ 200,000	Bio	nda	Waste water treatment plant	(Wulf et al. 2018; Thema et al. 2019; SolarServer 2020; Swisspower 2020)
Greenlab Skive	Planned	Skive Kåstrup (DK)	(10 MW methanol)	Cat	6	nda	(Wulf et al. 2018; Winther Mortensen et al. 2019; Danish Energy Agency (DEA) 2020; State of Green 2019)
Exytron Zero-Emission-Wohnpark	Projected	Alzey (DE)	2.5	Cat	7–8	nda	(Wulf et al. 2018; Thema et al. 2019)

(continued)

Table 2 (continued)

Plant/project	Status	Location	Capacity Nm3/h (kW) CH4	Tec	TRL	CO_2 source	References
Power-to-Flex	Planned	(NL-DE)	nda	Bio	6–7	nda	(Wulf et al. 2018; Thema et al. 2019)
MeGa-stoRE 2	In operation	Heden (DK)	60	Bio	6	nda	(Wulf et al. 2018; Thema et al. 2019; Danish Gas Technology Centre (DGC) 2017)
Power to Gas Biogasbooster	nda	Straubing (DE)	0.4 (40)	Bio	6	nda	(Wulf et al. 2018; Thema et al. 2019)
IET Rapperswil	nda	Rapperswil (CH)	Pilot	Cat	6	Air	(Wulf et al. 2018; Thema et al. 2019; Institut fur Energietechnik (IET) 2017)
Infinity 1	Under construction (starting up 2020)	Pfaffenhofen (DE)	~ 70 (700)	Bio	7	Waste water treatment plant	(Wulf et al. 2018; Thema et al. 2019; Nagengast 2017)
MeGa-stoRE com 1	Planned	nda (DK)	10,000,000 (electrolyzer)	Cat	nda	nda	(Thema et al. 2019; Biogas 2015)
MeGa-stoRE com 2	Planned	nda (DK)	10,000,000 (electrolyzer)	Cat	nda	nda	(Thema et al. 2019; Biogas 2015)

This table compiles Power-to-Methane pilot and demonstration plants or projects currently in operation, announced, planned, projected or under construction in Europe, as reported in literature. This table includes details about plant operational status, city and country of location of the plant, type of methanation technology, technology readiness level, CO_2 source and plant capacity, which give an idea of the level of development, deployment and sustainability of methanation technologies and the substitute natural gas production capacity thorough Europe.

Tec Technology, TRL Technology Readiness Level, nda no data available, DE Germany, Catcatalytic, ZSW Zentrum für Sonnenenergie- und Wasserstoff-Forschung Baden-Württemberg, nda no data available, ES Spain, P2G Power-to-Gas, NL Netherlands, Bio biological, DK Denmark, SE Sweden, FR France, AT Austria, CH Switzerland, IT Italy, PL Poland, HU Hungary, NL-DE Dutch-German collaboration, IET Institut fur Energietechnik

demand for road, rail, aviation and navigation vehicles will strongly depend on the level of electrification and on the fuel mix of the future scenario of vehicles (O'Connell et al. 2019).

More specifically, the techno-economic and environmental viability, industrial applicability and profitability of the CO_2 catalytic methanation technology are limited by the lack of availability of cheap and renewable hydrogen and electricity (Jarvis and Samsatli 2018). The greenhouse gas mitigation capacity of CO_2 catalytic methanation technology is conditioned by the fact that the required hydrogen and electricity come from renewable and carbon neutral or negative sources (Jarvis and Samsatli 2018). According to Artz et al. (2018), in a best-case scenario, carbon dioxide emissions can be reduced by up to 87% for CO_2 methanation as compared to conventional production routes (Artz et al. 2018).

The estimated net CO_2 utilization for the catalytic methanation process is about 2.75 tCO_2/tCH_4. Considering a global methane production capacity of about 1100–1500 $MtCH_4/y$, the potential for CO_2 uptake ranges between 3000 and 4000 $MtCO_2/y$ (Chauvy et al. 2019).

The estimated electricity consumption of the catalytic methanation process is about 15.2 MWh/tCH_4 (Götz et al. 2016; Jarvis and Samsatli 2018). Hydrogen production from water electrolysis dominates both the capital expenditure and operating cost for CO_2 catalytic methanation, which ranges between 35.8 and 38.8 M€ and 0.6–0.9 €/kWh, respectively, for a substitute gas natural production rate of nearly 311 m^3/h and a plant operational lifetime of 20 years (Götz et al. 2016; Jarvis and Samsatli 2018). Therefore, considering a low heating value of methane of about 9.95 kWh/m^3, a net capital expenditure of about 918–994 €/tCH_4 and a net operational cost around 8338–12,507 €/tCH_4 can be estimated. Other techno-economic studies estimated a methane production cost between 1500 and 6800 €/tCH_4 depending on the type of Power-to-Methane technology and carbon dioxide source utilized (Panzone et al. 2020). The product price for substitute natural gas is assumed to be equal to the current European Union natural gas import price of about 924 €/tCH_4 (YCharts 2020). Therefore, the financial viability of the Power-to-Methane technology is still lacking under the present conditions.

Power-to-Methane technology could play an important role in the future energy system, although there are still a number of technical and economic barriers, such as low efficiency and high costs, which need to be successfully resolved prior to its commercialization. In the case of catalytic methanation of CO_2, pending research areas include:

addressing catalyst robustness under intermittent and unsteady-state operation; exploring compact, efficient and intensified methanation reactors suitable for fluctuating operating conditions, such as plate reactors, micro-reactors and milli-reactors and sorption-enhanced and membrane reactors; acquiring more information on the operation dynamics of the methanation reactors to define the size of the hydrogen storage; optimization, control and management of the heat release during methanation reaction to minimize carbon deposition, prevent formation of toxic nickel carbonyls and increase the energy efficiency of the overall Power-to-Methane process; and development of new catalysts with high activity and poison tolerance at lower temperatures (200–300 °C) and with high mechanical strength and hydrothermal stability at high temperature (above 550 °C), given the high exothermicity of the reaction (Götz et al. 2016; Rego de Vasconcelos and Lavoie 2019; Behrens et al. 2019; Schaaf et al. 2014; Gao et al. 2015; Centi et al. 2013; Hu and Urakawa 2018; Mutz et al. 2015).

The European demonstration projects currently running will supply the basis for planning full-size installations by providing information on long-term performance, operational limitations and flexibility of methanation systems (Behrens et al. 2019; Bailera et al. 2017).

2.2 Methanol

Methanol has a relatively high energy density, of about 4.4 kWh/dm^3, and octane rating, of around 110, storage stability, easiness of handle and transport and established usage (Rego de Vasconcelos and Lavoie 2019; Onishi et al. 2018). Methanol has a big potential as an energy, fuel or hydrogen liquid carrier (Behrens et al. 2019). Methanol can be utilized for power generation directly, in methanol fuel cells, or indirectly, after conversion to hydrogen or gasoline, respectively, in fuel cells or in internal combustion engines (Ganesh 2014; Goeppert et al. 2014; Al-Saydeh and Zaidi 2018).

Methanol is also an important feedstock material for the fabrication of different chemical intermediates and industrial products, such as olefins, dimethyl ether, aromatics, liquid fuels, paint and polymers, becoming a clean and cost-competitive alternative to fossil fuels (Ganesh 2014; Goeppert et al. 2014; Behrens et al. 2019; Dang et al. 2019).

On an industrial scale, methanol is currently synthesized from CO_2-containing syngas, obtained mainly by natural gas reforming, using catalysts consisting of copper, zinc oxide and alumina (Kondratenko et al. 2013; Olah et al. 2009). In recent years, different companies, such as Methanex in Canada and Qatar Fuel Additives Company Limited (QAFAC) in Qatar, have implemented technologies that reduce the carbon footprint of methanol synthesized from natural gas, called "low-carbon methanol", by injecting waste CO_2 into the methanol production process (Hobson and Márquez 2018). There are also several research and development projects which focus on low-carbon methanol synthesis from renewable energy and CO_2-containing syngas, such as Carbon2Chem in Germany and FRESME in Sweden (Hobson and Márquez 2018).

Methanol can also be produced, more sustainably, by hydrogenation of recycled carbon dioxide with hydrogen produced by water electrolysis using renewable electricity, via the Power-to-Methanol process (Centi 2009; Rego de Vasconcelos and Lavoie 2019; Ali et al. 2015; Saeidi et al. 2014; Sankaranarayanan and Srinivasan 2012; Ham et al. 2012).

Renewable methanol production, via the Power-to-Methanol process, can avoid investment for energy storage and distribution, absorb and transform the excess of electricity and hydrogen into valuable chemicals and fuels in a competitive and efficient way and reduce carbon dioxide emissions associated with industry and transportation (Bergins et al. 2015).

Methanol can be synthesized directly from CO_2, or in two steps, indirectly from intermediate syngas via the reverse water gas shift reaction.

The CAMERE process for carbon dioxide hydrogenation via the reverse water gas shift reaction was developed to a pilot plant level of 100 kg CH_3OH/d by the Korean Institute of Science and Technology in Korea (Joo et al. 1999).

The hydrogenation of carbon dioxide directly to methanol occurs at 250–300 °C and 50–100 bars over same copper-based industrial catalyst modified with different additives and promoters (Behrens et al. 2019; Centi et al. 2013; Saeidi et al. 2014; Jadhav et al. 2014; Ma et al. 2009; Wang et al. 2011; Yan et al. 2014).

Direct CO_2 hydrogenation technology for methanol production can be considered almost ready to be commercialized (Centi 2009; O'Connell et al. 2019; Behrens et al. 2019; Wilson et al. 2015). According to O'Connell et al. (2019), Power-to-Methanol technology has already been developed to a technology readiness level of nearly 9 (O'Connell et al. 2019). The technical feasibility of the technology was firstly confirmed, at the pilot plant level of 100 tCH_3OH/y, by Mitsui Chemicals Osaka Works in Japan, using an own developed catalyst (Sankaranarayanan and Srinivasan 2012; Huang 2014). At the industrial level, the first plant for production of methanol from carbon dioxide was the George Olah plant, implemented by Carbon Recycling International in Iceland (Kondratenko et al. 2013; Sankaranarayanan and Srinivasan 2012; Huang 2014). However, compared to commercial technology to obtain methanol from CO, direct CO_2 hydrogenation suffers from higher consumption of hydrogen and lower productivity (Kondratenko et al. 2013). Today, some companies, such as Süd-Chemie Mitsubishi

Gas Chemical, Sinetix, Ceramatec and Haldor-Topsoe, are commercializing highly active and stable catalysts for methanol production by CO_2 hydrogenation (Kondratenko et al. 2013; Al-Saydeh and Zaidi 2018).

Until now, only a few Power-to-Methanol plants are currently operational and projects targeting at constructing pilot and demonstration plants are in progress, especially in Europe, driven by the need to reduce greenhouse gas emissions and intensify the share of renewable energy (O'Connell et al. 2019; Rego de Vasconcelos and Lavoie 2019; Al-Saydeh and Zaidi 2018). A non-exhaustive list of Power-to-Methanol plants and projects is provided in Table 3 (Zhu 2019; O'Connell et al. 2019; Rego de Vasconcelos and Lavoie 2019; Behrens et al. 2019; Wulf et al. 2018; Bailera et al. 2017; Al-Saydeh and Zaidi 2018; Hobson and Márquez 2018). This table combines and updates earlier reviews by Zhu (2019), O'Connell et al. (2019), Rego de Vasconcelos and Lavoie (2019), Behrens et al. (2019), Wulf et al. (2018), Bailera et al. (2017); Al-Saydeh and Zaidi (2018) or Hobson and Márquez (2018) adding previously un-reviewed projects and including details such as plant operational status, city and country of location of the plant, technology readiness level, CO_2 source, plant capacity, expressed in terms of total methanol production, in t/y, and installed power in kW, calculated from the lower heating value of methanol product, as well as main references.

In most of the plants, the CO_2 is sourced from waste gases of power plants and industrial processes. A few plants use CO_2 captured directly from air.

According to the size of Power-to-Methanol plants currently in operation in Europe (see Table 3), methanol production capacity is about 4300 tCH_3OH/y or nearly 3 MW of methanol, according to the methanol low heating value (O'Connell et al. 2019). The capacity of some plants is unknown. In addition, some plants are, announced, planned, projected or under construction. Therefore, the estimated methanol production capacity could be actually higher.

Methanol is one of the most important materials in the worldwide chemicals and energy markets with a continuously growing demand (Jadhav et al. 2014). Global methanol annual demand is approaching 100 million tons in 2020 (Methanol Institute 2020). Methanol synthesized from CO_2 can be used to produce almost all industrial products that are currently derived from fossil fuels, leading to the same associated demand.

The methanol demand will potentially expand even further in the future, as the use of methanol as energy, hydrogen and fuel vector increases. In fact, Power-to-Methanol technology would contribute to fulfill the requirements of long-term and massive renewable energy storage, resulting in a significant demand of Power-to-Methanol plants in the future (O'Connell et al. 2019). Furthermore, renewable transportation fuels are also needed in the EU to achieve the target reduction of greenhouse gas emissions. The final methanol demand for road, aviation, rail and navigation vehicles will strongly depend on the level of electrification and transportation fuel mix of the future scenario of vehicles (O'Connell et al. 2019).

The estimated net CO_2 utilization of Power-to-Methanol process is about 1.4 tCO_2/tCH_3OH (Chauvy et al. 2019; Jarvis and Samsatli 2018; Kiss et al. 2016). Considering an installed global production capacity of methanol of about 110 million ton of methanol per year (Methanol Institute 2020), the potential of CO_2 utilization is about 154 million tons of carbon dioxide per year. Direct CO_2 hydrogenation to methanol emits about 0.226 tCO_2/t methanol, while conventional methanol synthesis releases 0.768 tCO_2/tCH_3OH (Pérez-Fortes et al. 2016b). Therefore, the net amount of CO_2 avoided by the Power-to-Methanol process is about 2 tCO_2/tCH_3OH (Jarvis and Samsatli 2018). According to Artz et al. (2018), in a best-case scenario, carbon dioxide emissions can be reduced by up to 93% for Power-to-Methanol process as compared to conventional production routes (Artz et al. 2018).

The electricity needed for methanol synthesis by direct CO_2 hydrogenation is about 170 kWh/tCH_3OH, but can reach up to 550 kWh/tCH_3OH if hydrogen production by water electrolysis is integrated (Kiss et al. 2016). Therefore, powering the CO_2 hydrogenation with electricity from renewable sources is indispensable to keep a neutral or negative carbon balance for the process (Jarvis and Samsatli 2018).

Methanol production by CO_2 hydrogenation results also in a low water footprint compared to conventional methanol production from syngas, each with a consumption of water of about 26.4 tH_2O/tCH_3OH and 90 tH_2O/tCH_3OH, correspondingly (Jarvis and Samsatli 2018).

According to a study by Pérez-Fortes et al. (2016b), the capital expenditure and annual operating and maintenance cost for a carbon dioxide hydrogenation to methanol plant were about 200 and 293 million euros, respectively, considering an annual plant production capacity of 440,000 tons of renewable methanol and a plant operational lifetime of 20 years (Jarvis and Samsatli 2018; Pérez-Fortes et al. 2016b). A net capital cost of about 23 €/tCH_3OH and a net operational cost of around 666 €/tCH_3OH were assessed, each slightly lower and considerably higher than for a conventional syngas-based plant, respectively (Pérez-Fortes et al. 2016b). In accordance with a recent study by Zhang et al. (2019), the total cost associated with methanol production by CO_2 hydrogenation with integrated solid-oxide electrolyzer was about 500 €/tCH_3OH, considering an annual plant production capacity of 100,000 tons of renewable methanol (Zhang et al. 2019). Other techno-economic studies estimated a methanol production cost between about 500 and 1900 €/tCH_3OH, depending on

Table 3 Plants or projects for Power-to-Methanol

Plant/project	Status	Location	Capacity t/y (kW) CH3OH	TRL	CO_2 source	References
Greenfuel	In operation	Essen (DE)	0.91	4–5	Air captured	(O'Connell et al. 2019; Kanacher and Innogy 2017)
George Olah Plant	In operation	Svartsengi (IS)	4000 (~2500)	7–8	Flue gas geothermal power plant	(Zhu 2019; (O'Connell et al. 2019; Rego de Vasconcelos and Lavoie 2019; Wulf et al. 2018; Al-Saydeh and Zaidi 2018)
MefCO_2	Under construction	Niederaussem (DE)	1 t/d (~230)	6	Flue gas of power plant	(O'Connell et al. 2019; Rego de Vasconcelos and Lavoie 2019; Behrens et al. 2019; Wulf et al. 2018; MefCO_2 2020; Europe 2020a; Spire 2030 2020; Cuesta Pardo and Everis 2018; Kourkoumpas et al. 2016)
FResMe	Planned (~2019)	Luleå (SE)	2.5 t/d (~575)	6	Steel plant gases	(O'Connell et al. 2019; Cuesta Pardo and Everis 2018; Fresme project 2017; Bonalumi et al. 2018; Europe 2020b; Energy research Centre of the Netherlands (ECN) 2017)
E2fuels	Planned	Haßfurt (DE)	(507)	nda	nda	(O'Connell et al. 2019; Albert J, Chemical Reaction Engineering (CRT) University of Erlangen-Nuremberg 2020)
E-CO_2MET	Planned	Leuna (DE)	MW scale	nda	Industrial concentrated CO_2 from refinery	(Knowledge Energy Institute (KEI) 2019; TOTAL 2019; News Green Hydrogen 2019)
Power-to-Methanol Antwerp BV	Announced	Lillo (BE)	8000 (~5000)	nda	Captured CO_2	(O'Reilly 2020; Taylor 2020; Friis M, Blue World Technologies 2019)
Power2Met phase 1- Aalborg university	Planned (~2019)	Aalborg (DK)	~ 198 (~124)	nda	nda	(Friis M, Blue World Technologies 2019; Søren 2020; Hydrogen Valley 2020)
Power2Met phase 2	Planned (~2020-)	nda (DK)	~ 791–1,582 (~495–991)	nda	Biogas plant	(Friis M, Blue World Technologies 2019; Søren 2020; Hydrogen Valley 2020)
Power2Met phase 3	Planned (~2022–)	North Jutland main power plant (DK)	~ 1.1 t/h (~6000)	nda	nda	(Friis M, Blue World Technologies 2019; Søren 2020; Hydrogen Valley 2020)
BASF/BSE	Projected	nda (DE)	1.55 t/h (8587)	7	Flue gas	(Zhu 2019; Wulf et al. 2018; Bartmann and Wranik 2017; Schweitzer C- bse Engineering 2017)
Silicon Fire (SLF)	In operation	Altenrhein, Kanton St. Gallen (CH)	33 kg/h (~183)	6	nda	(O'Connell et al. 2019)
Silicon Fire (SLF 15)	Planned	nda (CH)	12 t/d (2770)	9	Biogenic or waste-CO_2	(O'Connell et al. 2019; Swiss Liquid Future 2016)
Carbon2Chem	In operation	Duisburg (DE)	nda	6–7	Steel plant gases	(Hobson and Márquez 2018; Bender and Thyssenkrupp 2018; Fona 2016)
Haldor-Topsoe	Announced	Foulum (DK)	10 kg/h (55)	nda	Biogas	(Ravn and Haldor-Topsoe 2019)

This table compiles Power-to-Methanol pilot and demonstration plants or projects currently in operation, announced, planned or under construction in Europe, as reported in literature. This table includes details about plant operational status, city and country of location of the plant, technology readiness level, CO_2 source and plant capacity, which give an idea of the level of development, deployment and sustainability of Power-to-Methanol technology and the renewable methanol production capacity thorough Europe.

TRL Technology Readiness Level, DE Germany, IS Iceland, SE Sweden, nda no data available, BE Belgium, DK Denmark, SLF Swiss Liquid Future, CH Switzerland.

the type of electrolysis technology and carbon dioxide source utilized (Panzone et al. 2020). According to Methanex Corporation, the market price for methanol is currently 260 euros per ton of methanol (Methanex 2020). Therefore, direct CO_2 hydrogenation is not yet commercially competitive to industrial syngas route for methanol production, due to the low price of methanol.

Power-to-Methanol technology could play a crucial role in chemical industry and energy system in the future. The current technological challenges include increasing energy efficiency while reducing the cost of the carbon dioxide hydrogenation process. Clearly, the development of novel catalysts and process designs is needed to accelerate widespread penetration of greener CO_2 to methanol processes.

Some commercial catalysts are already available for the carbon dioxide hydrogenation process, but there are still a number of opportunities to improvement (Centi 2009; Kondratenko et al. 2013; Sankaranarayanan and Srinivasan 2012; Ma et al. 2009; Yan et al. 2014), such as: increase conversion to methanol at low temperature by using the $Cu/ZnO/Al_2O_3$ catalyst promoted with ZrO_2, La_2O_3, Ga_2O_3 and SiO_2 (Kondratenko et al. 2013; Ali et al. 2015; Saeidi et al. 2014; Sankaranarayanan and Srinivasan 2012; Ham et al. 2012; Ma et al. 2009; Wang et al. 2011; Yan et al. 2014; Huang 2014), the use of catalysts based on Pd or Ag supported on ZnO, ZrO_2, MCM-41 and SBA-15, to improve tolerance to inhibition by water, with alkali promoters, to increase the conversion, or on carbon nanotubes, to improve selectivity (Ali et al. 2015; Ham et al. 2012; Ma et al. 2009; Wang et al. 2011; Yan et al. 2014), or on metal carbides, such as TiC, Mo_2C, Fe_3C, SiC and TaC, to improve activity, selectivity and resistance to deactivation (Ma et al. 2009; Yan et al. 2014).

Water generated in the carbon dioxide hydrogenation reaction inhibits the formation of methanol and cause hydrothermal deactivation of catalyst incremented by the exothermicity of the reaction (Centi et al. 2013; Ali et al. 2015; Sankaranarayanan and Srinivasan 2012; Ma et al. 2009; Wang et al. 2011).

There are also opportunities for optimizing reactor design, such as the development of fluidized bed membrane reactors with water-cooled walls and two-stage reactors with intermediate cooling to reduce hydrothermal deactivation and extending lifetime of the catalysts and recirculation of the feed for getting higher conversions while improving heat transfer in the reactor (Saeidi et al. 2014). Another possibility to improve reaction system is the development of membrane reactors for the separation of water from the reaction medium to shift the equilibrium to the product side, with the consequent increase in methanol production while reducing reagent consumption, facilitating operation at lower pressures and higher temperatures, preventing catalyst inhibition and deactivation by water and promoting reaction kinetics with reduced residence time and reactor size (Ham et al. 2012; Jadhav et al. 2014; Wang et al. 2011). A two-stage fixed bed reactor design also enables a smoother temperature profile along the length of the tube and increases catalyst lifetime and activity (Jadhav et al. 2014; Ma et al. 2009). Another system studied is a reactor consisting of two stages with two types of membranes: one selective to hydrogen and one to water, the so-called dual-type perm-selective membrane reactor. The dual-type perm-selective membrane reactor allows overcoming the limitation of thermodynamic equilibrium, improving kinetically limited reactions and controlling the hydrogen to carbon dioxide ratio in the reactor feed (Wang et al. 2011).

2.3 Dimethyl Ether

Dimethyl ether is a gas at ambient temperature and pressure, but liquefies by slight compression or cooling, making dimethyl ether ease to transport and storage. Dimethyl ether is a potential substitute for liquefied petroleum gas in domestic, transportation and industrial applications. Due to its good ignition features and relatively high cetane number, around 55–60 versus 40–55 for diesel, dimethyl ether can be burned, with negligible modifications, in compression ignition engines as an alternative to conventional diesel fuel for transportation, power generation and household applications. In addition, the combustion of dimethyl ether leads to lower emission of pollutants, such as nitrogen oxides, sulfur compounds, soot and carbon monoxide, than conventional fuels. Dimethyl ether can also be transformed into olefins and hydrocarbons (Álvarez et al. 2017; Li et al. 2018; Wang et al. 2009).

Commercially, dimethyl ether is obtained by catalytic dehydration of methanol. One-step synthesis of dimethyl ether from synthesis gas reached the demonstration level at a 100 t/d plant in JFE Steel Corporation in Japan (Olah et al. 2009; Yotaro et al. 2006).

Korea Gas Co. (KOGAS) has developed a proprietary catalyst and a process for the synthesis of dimethyl ether from syngas derived from the tri-reforming of methane, CO_2, and H_2. KOGAS demonstrated the indirect synthesis of dimethyl ether on a scale of 10 tDME/y. KOGAS has still to commercialize its dimethyl ether production process due to a lack of economic viability (Chung et al. 2012).

Dimethyl ether can be also produced by direct hydrogenation of carbon dioxide with renewable hydrogen over bi-functional catalysts combining an active catalyst for the synthesis of CH_3OH from CO_2, such as $CuO\text{-}ZnO\text{-}Al_2O_3$ or $CuO\text{-}TiO_2$, and an acid catalyst for dehydration of CH_3OH to dimethyl ether, such as Al_2O_3 or zeolites, in the same fixed bed or suspended reactor (Álvarez et al. 2017; Centi et al. 2013; Wang et al. 2011, 2009; Li et al. 2018; Naik et al. 2011). Optimization of catalysts for CO_2 hydrogenation to dimethyl ether in a one step includes the addition of promoters to increase the production of dimethyl ether, such as Pd and MoO_3, the selectivity to dimethyl ether, such as Ga_2O_3 and Cr_2O_3, and the lifetime of the catalyst, such as NaHZSM5 (Wang et al. 2011). The single-step technology for dimethyl ether synthesis directly from carbon dioxide and hydrogen enables to produce dimethyl ether with higher efficiency, by overpassing thermodynamic equilibrium limitations for CO_2 conversion by in-situ dehydration of the produced methanol, and with lower capital and operating costs, because both methanol synthesis and dehydration takes places in a single process, with no need for methanol purification (Olah et al. 2009; An et al. 2008).

In recent years, the production of dimethyl ether from carbon dioxide and hydrogen in a single step has advanced considerably toward demonstration, reaching a technical readiness level of 6 (Wulf et al. 2018). A new demonstration plant was developed and built over the past two years as part of the ALIGN-CCUS project funded by the European Commission and Member States. The demonstration plant is located at RWE's Innovation Centre at Niederaussem power plant in Germany. The commissioning of the ALIGN-CCUS plant started at the end of 2019. The dimethyl ether synthesis unit was supplied by Mitsubishi Hitachi Power Systems Europe and can produce up to 50 kg per day of dimethyl ether from 180 kg of CO_2 and 22 kg of hydrogen using an innovative catalyst and reactor system. CO_2 will be captured from the flue gas of the coal power station in a post-combustion amine scrubbing plant available on-site. The hydrogen will be obtained from water electrolysis in an alkaline electrolyzer with a capacity of about 120 kW supplied by Asahi Kasei Europe (O'Connell et al. 2019; RWE Power 2019).

The easiness of transport and storage, the clean combustion performance, and the relatively high energy density of dimethyl ether, makes the direct conversion of recycled CO_2 and renewable H_2 to dimethyl ether very attractive for long-term chemical storage of renewable energy and as a source of clean, renewable and climate-friendly alternative fuel.

Almost all of the dimethyl ether is currently produced by catalytic dehydration of coal-derived methanol. In China, 20 million tons of dimethyl ether production capacity were expected by 2020 (Methanol Institute 2016) at a market price close to 1600 €/t (China Petroleum and Chemical Industry Federation 2020).

According to the analysis made by Michailos et al. (2019), the total cost associated with dimethyl ether production by direct hydrogenation of captured CO_2 with integrated electrolysis ranges between 1828 and 2322 €/tDME depending on the price of renewable electricity (Michailos et al. 2019). Therefore, dimethyl ether produced by direct CO_2 hydrogenation is currently not competitive as compared with dimethyl ether derived from fossil.

The estimated net CO_2 utilization of the direct CO_2 hydrogenation to dimethyl ether process is about 1.9 tCO_2/tDME (Chauvy et al. 2019). Therefore, considering a global dimethyl ether production capacity of about 20 million tons of dimethyl ether per year (Methanol Institute 2020), the potential of CO_2 utilization is about 38 million tons of carbon dioxide per year.

The market size estimated for dimethyl ether by 2020 was for 8 billion € (Fleisch et al. 2012). Therefore, assuming a market price of about 1600 €/tDME, the global demand in 2020 is roughly 5 million tons of dimethyl ether. China consume more than 90% of the global dimethyl ether demand and mainly for heating and cooking applications (Prabowo et al. 2017). The demand of dimethyl ether will increase exponentially in the future if dimethyl ether is adopted as alternative fuel for transportation and power generation.

The demand of dimethyl ether as alternative fuel for transportation expects to increase, especially for heavy vehicles, such as trucks and ships, which cannot be easily electrified. The demand of dimethyl ether as blended fuel for vehicles driven by liquefied petroleum gases will increase, because infrastructures for distribution and end-use are already available. One the other hand, the demand of dimethyl ether as a cleaner substitute for diesel fuel in transportation applications is foreseen to increase, because the use of dimethyl ether will help vehicle manufacturers and end-users comply with the severer regulations (Methanol institute 2016; Fleisch et al. 2012; Prabowo et al. 2017).

The use of dimethyl ether as an alternative fuel to natural gas for efficient power generation could play a key role in the future, in special for power plants at isolated locations where the transportation of natural gas is challenging (Methanol institute 2016; Fleisch et al. 2012; Prabowo et al. 2017).

2.4 Formic Acid

Formic acid is broadly used in the industry, not only as insecticide, preservative, and antibacterial and reducing agent, but also as C1 building block for the production of various chemicals, as well as fuel for direct energy generation in fuel cells (Álvarez et al. 2017; Panzone et al. 2020; Rego de Vasconcelos and Lavoie 2019; Huang et al. 2018).

Despite its relatively low hydrogen content, formic acid decomposes reversibly and selectively releasing hydrogen at relatively low temperature, enabling its potential application as liquid carrier of energy and hydrogen (Álvarez et al. 2017; Panzone et al. 2020; Rego de Vasconcelos and Lavoie 2019; Singh et al. 2016).

The most common process for industrial production of formic acid consists of the synthesis of methyl formate from carbon monoxide and methanol and subsequent hydrolysis of methyl formate to formic acid at high pressure and in the presence of a huge excess of water. Formic acid can be produced in greener way by direct carbon dioxide reduction, enabling carbon dioxide utilization and chemical storage of renewable energy and hydrogen (Álvarez et al. 2017; Panzone et al. 2020; Pérez-Fortes et al. 2016a; Rego de Vasconcelos and Lavoie 2019; Huang et al. 2018).

Carbon dioxide hydrogenation to formic acid usually occurs over homogeneous catalysts in aqueous medium. Catalysts based on noble metals, especially on Ru, showed good catalytic performance. However, the homogeneous catalysts are difficult to separate from the reaction products and the yield to formic acid is still low, hindering the development of the process beyond a technology readiness level of 5 (Álvarez et al. 2017; Jarvis and Samsatli 2018; Panzone et al. 2020; Pérez-Fortes et al. 2016a; Rego de Vasconcelos and Lavoie 2019). Homogeneous catalysts require further enhancements in selectivity to formic acid and stability to become potentially applicable (Panzone et al. 2020). Some companies like British Petroleum (BP) and BASF acquired patents on the carbon dioxide hydrogenation to formic acid over homogeneous catalysts (Pérez-Fortes et al. 2016a; Pérez-Fortes and Tzimas 2016). Carbon dioxide hydrogenation to formic acid over immobilized homogeneous catalysts or heterogeneous catalysts was recently studied to deal with the problem of final separation of reaction products (Panzone et al. 2020; Rego de Vasconcelos and Lavoie 2019).

The global production of formic acid was estimated to surpass 760 million tons of formic acid by 2019 (Jarvis and Samsatli 2018; Rego de Vasconcelos and Lavoie 2019; Huang et al. 2018). The demand for formic acid is presently led by its use as chemical feedstock, but the potential application of formic acid as energy and hydrogen vector would increase its demand in the future (Jarvis and Samsatli 2018; Singh et al. 2016).

A techno-economic study by Pérez-Fortes et al. (2016a) utilizes a market price of formic acid of 650 €/t (Pérez-Fortes et al. 2016a), in line with the current market price of formic acid around 400–620 €/t (Made in China 2020). The cost estimated for the production of formic acid by carbon dioxide hydrogenation using homogeneous catalysts and considering a plant with a production capacity of 12 kt/y and a lifetime of 20 years was about 1591 €/t, corresponding to the sum of capital costs, of about 67 €/t, and operational costs, of about 1524 €/t (Pérez-Fortes et al. 2016a). Therefore, the production of formic acid by carbon dioxide hydrogenation is not economically viable yet.

According also to the study by Pérez-Fortes et al. (2016a), formic acid production by carbon dioxide hydrogenation leads to lower consumption of steam and heavy fuel oil, lower water footprint and to higher electricity consumption, of about 4.1 Mwh per ton of formic acid, as compared to the conventional method of formic acid production (Jarvis and Samsatli 2018; Pérez-Fortes et al. 2016a), resulting in a tenfold potential reduction in the greenhouse gas emissions associated with formic acid production (Gunasekar et al. 2016). Therefore, formic acid production by carbon dioxide hydrogenation is advantageous, in terms of environmental impact, compared to its conventional counterpart, as long as the electricity needed for the carbon dioxide hydrogenation and hydrogen production by water electrolysis comes from a renewable source (Jarvis and Samsatli 2018; Rego de Vasconcelos and Lavoie 2019). In addition to lower associated emissions of carbon dioxide, the process of carbon dioxide hydrogenation to formic acid has a net CO_2 utilization of 0.7 tons of CO_2 per ton of formic acid, resulting in a net amount of avoided CO_2 emissions of 2 tons of CO_2 per ton of formic acid and in a potential of CO_2 utilization in the range of 4–21 million metric tons of carbon dioxide per year only in Europe (Jarvis and Samsatli 2018; Pérez-Fortes et al. 2016a).

The electrochemical reduction of carbon dioxide to formic acid involves the conversion of CO_2, protons and electrons to formic acid at the cathode of an electrolytic cell. Oxygen is obtained as by-product at the anode of the electrochemical system (Jarvis and Samsatli 2018). The electrochemical reduction of carbon dioxide to formic acid typically operates at ambient temperature and pressure, with current efficiencies up to 100% (Jarvis and Samsatli 2018). The maturity of electrochemical reduction of CO_2 to formic acid is currently limited to a technology readiness level range of 5–6 (Chauvy et al. 2019; Pérez-Fortes and Tzimas 2016), mainly due to the high over-potentials and low selectivity in formic acid formation (Rego de Vasconcelos and Lavoie 2019).

Various life-cycle and techno-economic studies identified electro-reduction of CO_2 to formic acid as a promising technology for further scale-up and commercialization (Agarwal et al. 2011; Jarvis and Samsatli 2018; Pérez-Fortes et al. 2016a).

There are some projects aiming at demonstrating the process of carbon dioxide electro-reduction to formic acid in a relevant scale and environment, especially in Europe (Pérez-Fortes and Tzimas 2016).

Det Norske Veritas (DNV) (2007) and Mantra Venture Group (Mantra 2014, 2015) are exploring the electro-reduction of carbon dioxide to formic acid at pilot scale (Pérez-Fortes et al. 2016a; Pérez-Fortes and Tzimas 2016).

Det Norske Veritas developed the ECFORM process for the direct electro-reduction of CO_2 to formic acid and formate salts, with oxygen as by-product (Zhu 2019). DNV tested the ECFORM process in a small-scale demonstration plant, of 350 kg of formic acid per year (Chauvy et al. 2019; Pérez-Fortes et al. 2016a; Pérez-Fortes and Tzimas 2016; Det Norske Veritas (DNV) 2007; Robledo-Díez 2012).

Mantra Venture Group has finished the engineering work on a pilot plant to be installed and operated at Lafarge Cement in Canada (Mantra 2014, 2015). The pilot plant can produce 35 tons of formic acid per year by electrochemical reduction of carbon dioxide (Pérez-Fortes et al. 2016a; Pérez-Fortes and Tzimas 2016; Mantra 2014).

A new pilot plant was developed and built as part of the CELBICON project funded by the European Commission. The pilot plant includes an electrolytic cell for testing the electrochemical reduction of carbon dioxide captured from air to formic acid in a relevant environment, that is to say at a technical readiness level of 5 (HYSYTECH 2020; Jarvis and Samsatli 2018).

Recently, Twence, a waste incineration company, launched the development of a pilot plant for the electrochemical conversion of carbon dioxide captured from flue gases into formic acid. The pilot plant is based on an electrochemical reactor previously developed and tested in cooperation between Coval Energy, VoltaChem/The Netherlands Organization for Applied Scientific Research (TNO), and Technical University of Delft (Voltachem 2019).

The electrochemical reduction of carbon dioxide is ready for scaling-up, but substantial technological developments are required before applying this process to the production of formic acid at large scale. Outstanding challenges for electrochemical reduction of carbon dioxide to formic acid comprise increasing the selectivity, stability and durability of catalyst-electrodes and the development of less energy-intensive approaches to separate formic acid from the aqueous stream of products and of process in continuous for synthesis of formic acid by carbon dioxide electro-reduction (Zhu 2019; Agarwal et al. 2011; Pérez-Fortes et al. 2016a; Pérez-Fortes and Tzimas 2016; Rego de Vasconcelos and Lavoie 2019).

Based on a study by Agarwal et al. (2011), a net capital cost of about 94 € per ton of formic acid and a net operational cost of around 760 € per ton of formic acid were assessed (Agarwal et al. 2011; Jarvis and Samsatli 2018), resulting in total production cost of 854 € per ton of formic acid, considerably higher than market price for formic acid indicated above, of approximately 650 € per ton of formic acid. Therefore, electrochemical reduction of CO_2 to formic acid is not yet economically viable.

According also to Agarwal et al. (2011), the economic viability of the process of electrochemical reduction of carbon dioxide to formic acid may increase by utilizing waste water at the anode and by separating and selling by-products of the reaction such as oxygen and hydrogen (Agarwal et al. 2011).

The production of formic acid by electrochemical reduction of carbon dioxide is environmentally beneficial because of its operation at ambient temperatures and pressures resulting in a less energy and carbon intensive process.

Based on the study by Agarwal et al. (2011), a net CO_2 utilization of about 0.5 tons of CO_2 per ton of formic acid and an electricity requirement of about 1.1 MWh per ton of formic acid were assessed (Agarwal et al. 2011; Jarvis and Samsatli 2018). However, the use of renewable electricity for the production of formic acid is decisive to avoid compromising the mitigation capacity of carbon dioxide of the electrochemical reduction process.

2.5 Ethanol

Ethanol is commonly used as fuel additive and platform chemical and, in some countries is blended, mainly with gasoline, as transportation fuel. Ethanol has a relatively high energy density of about 7.4 kWh/kg (Handler et al. 2016). Ethanol is currently produced by direct or indirect ethylene hydration or biomass fermentation (Panzone et al. 2020; Bordiga and Lercher 2019).

Ethanol can be also obtained by direct hydrogenation, electrochemical reduction or biochemical conversion of CO_2 (Panzone et al. 2020; Rego de Vasconcelos and Lavoie 2019). Among the different technologies currently under development for the production of ethanol from CO_2, biochemical conversion is the only one that has reached a degree of maturity sufficient as to consider its possible commercialization (O'Connell et al. 2019).

A promising process for biochemical conversion of carbon dioxide to ethanol at industrial level is gas fermentation. Feedstocks for gas fermentation include syngas from gasification of biomass and municipal solid waste, CO-rich gas from industrial exhausts and even reformed biogas and mixtures of captured carbon dioxide and electrolytic hydrogen (Zhu 2019; O'Connell et al. 2019; Anggraini et al. 2018).

The gas fermentation process consists of reception, compression and fermentation of the syngas and separation and/or purification of ethanol. Fermentation takes place at nearly atmospheric pressure and ambient temperature, resulting in a less energy-intensive route for ethanol production with reduced carbon footprint and operating cost (Zhu 2019; O'Connell et al. 2019; Bordiga and Lercher 2019; Liew et al. 2016).

The advantages of gas fermentation in terms of efficiency, diversity of feedstocks and product selectivity resulted in the scaling-up of fermentation process for ethanol production at commercial scale (Liew et al. 2016).

LanzaTech is the leading company in process development for ethanol production by gas fermentation. LanzaTech demonstrated the gas fermentation process at pilot and pre-commercial scales. LanzaTech is currently scaling up its process, and some commercial plants are under construction or planned (Zhu 2019; O'Connell et al. 2019; Liew et al. 2016). Accordingly, the ethanol production by gas fermentation is a mature technology with a technology readiness level of 9 (O'Connell et al. 2019).

Table 4 shows pilot, pre-commercial and commercial plants based on the LanzaTech process. This table combines and updates earlier reviews by Zhu (2019), O'Connell et al. (2019) and Liew et al. (2016), including details such as plant operational status, city and country of location of the plant, CO_2 source, plant scale and capacity, expressed in terms of total ethanol production, in t/y, and installed power in kW, calculated from the lower heating value of ethanol product, as well as main references.

The global annual production of ethanol is currently about 87 million tons (Renewable Fuel Association (RFA) 2020). The global annual demand of ethanol is presently around 47.5 million tons (Renewable Fuel Association (RFA) 2020).

A techno-economic study by Medeiros et al. (2020) estimated a minimum ethanol selling price of about 811 €/t for the production of ethanol via syngas fermentation to become economically feasible (Medeiros et al. 2020). The current market price of ethanol is around 976 €/t (Global petrol prices 2020). Therefore, the production of ethanol by fermentation of carbon dioxide-rich streams can be economically competitive depending on the source of syngas, as the minimum ethanol selling price depends on the type of gas utilized (Medeiros et al. 2020).

According to a life-cycle analysis by Ou et al. (2013), ethanol production by gas fermentation is advantageous in terms of energy savings and reduction of greenhouse gas emissions compared to most bioethanol production routes (Ou et al. 2013).

Table 4 Pilot, pre-commercial and commercial plants based on the LanzaTech process for ethanol production by gas fermentation

Plant	Status	Location	Capacity t/y (kW) C_2H_5OH	Scale	CO_2 source	References
LanzaTech Blue Scope	In operation	Auckland (NZ)	45 (38)	Pilot	Steel mill off gas	(Zhu 2019; O'Connell et al. 2019; Lane 2015)
LanzaTech BaoSteel	In operation	Shanghai (CN)	300 (255)	Pre-commercial	Steel mill off gas	(Zhu 2019; O'Connell et al. 2019; Liew et al. 2016; Lane 2015)
LanzaTech Shougang	In operation	Caofeidian (CN)	300 (255)	Pre-commercial	Steel mill off-gas	(Zhu 2019; O'Connell et al. 2019; Liew et al. 2016; Lane 2015)
LanzaTech WBT	In operation	Kaohsiung (TW)	36 (31)	Pilot	Steel mill off gas	(Zhu 2019; O'Connell et al. 2019; Liew et al. 2016)
LanzaTech Sekisui	In operation	North Japan (JP)	15 (13)	Pilot	Syngas MSW gasification	(Zhu 2019; Sekisui Chemical Co., LTD. and LanzaTech 2017)
LanzaTech Freedom Pines	nda	Georgia (US)	30 (26)	Pilot	Syngas Biomass gasification	(Zhu 2019; Lane 2015)
LanzaTech Shougang	In operation	Caofeidian (CN)	46,000 (40,900)	Commercial	Steel mill off-gas	(Zhu 2019; O'Connell et al. 2019; Liew et al. 2016)
LanzaTech ArcelorMittal	Under construction	Gent (BE)	62,000 (53,600)	Commercial	Steel mill off gas	(O'Connell et al. 2019; Steleanol Project 2019)
LanzaTech Indian Oil	Under construction	Panipat (IN)	33,000 (28,100)	Commercial	Refinery off-gas	(Zhu 2019; O'Connell et al. 2019)
LanzaTech Aemetis	Under engineering	California (US)	36,000 (30,600)	Commercial	Syngas Biomass gasification	Zhu 2019; O'Connell et al. 2019)
LanzaTech Swayana	Under construction	South Africa (ZA)	57,000 (43,400)	Commercial	Ferroalloy off-gas and Titania smelter	(Zhu 2019; O'Connell et al. 2019)

This table compiles pilot, pre-commercial and commercial plants based on the LanzaTech process for ethanol production by CO_2-containing gas fermentation worldwide. This table includes details about plant operational status, city and country of location of the plant, CO_2 source, plant scale and capacity, which give an idea of the level of development, deployment and sustainability of ethanol production by gas fermentation around the world.
NZ New Zealand, CH Switzerland, TW Taiwan, JP Japan, MSW Municipal Solid Waste, nda no data available, US United States of America, BE Belgium, IN India, ZA South Africa

The production of ethanol via gas fermentation has a net CO_2 utilization of 1.9 tons of CO_2 per ton of ethanol, resulting in a potential of CO_2 uptake of about 165 million tons of carbon dioxide per year worldwide (Chauvy et al. 2019; Handler et al. 2016).

2.6 CO_2-Fischer–Tropsch Liquid Fuels

Liquid fuels, such as diesel, gasoline, and jet fuel or kerosene, are extensively utilized for transport. Much of the CO_2 emitted into the atmosphere results from the combustion of fossil fuels in vehicles, making necessary the development of methods, such as the Power-to-Liquids technology, for the production of liquid transportation fuels from CO_2, water and renewable electricity. Power-to-Liquids technology enables to decrease the greenhouse gas emissions associated with the transportation sector while obtaining an economic benefit and storing renewable energy (O'Connell et al. 2019; Schmidt et al. 2016).

Fuels derived from carbon dioxide take on special importance in the aviation sector, where the use of other low-carbon energy carriers, such as electricity or hydrogen, is particularly challenging (International Energy Agency (IEA) 2019).

There are two main ways for the production of gasoline, diesel and jet fuel or kerosene via Power-to-Liquids technology: methanol synthesis and conversion and Fischer–Tropsch synthesis and subsequent upgrading. Both Fischer–Tropsch and methanol pathway consist of three steps: hydrogen production by water electrolysis powered by renewable electricity, carbon dioxide conversion to the intermediate product, carbon monoxide or methanol, and synthesis of liquid hydrocarbons and subsequent upgrading or conversion to the target fuel, respectively (O'Connell et al. 2019; Panzone et al. 2020; Schmidt et al. 2016).

In case of the Fischer–Tropsch route, CO_2 is firstly converted to CO, by reverse water gas shift reaction or by co-electrolysis of CO_2 and water, and then, to long-chain hydrocarbons by the Fischer–Tropsch reaction in the presence of cobalt or iron catalyst (Jarvis and Samsatli 2018). The obtained synthetic crude requires subsequent conventional upgrading processes, by hydrocracking, isomerization and distillation, to gasoline, diesel or kerosene (Schmidt et al. 2016). In case of the methanol route, methanol is converted to gasoline, diesel and kerosene by a series of reactions including dimethyl ether synthesis, olefin synthesis, oligomerization and hydrotreating (O'Connell et al. 2019; Panzone et al. 2020; Schmidt et al. 2016; Rego de Vasconcelos and Lavoie 2019).

Both Power-to-Liquids routes, via Fischer–Tropsch or methanol, has a high level of technology readiness. The methanol-to-liquid fuel process is a mature and commercially available technology with a technology readiness level of 9 (O'Connell et al. 2019). The Power-to-Liquids technology, via syngas production by reverse water gas shift and subsequent Fischer–Tropsch synthesis, is almost ready for commercialization with a technology readiness level between 8 and 9 (O'Connell et al. 2019; Schmidt et al. 2016).

Integration of the individual processes involved in Power-to-Liquids technology at large scale is still under development (Schmidt et al. 2016). By the time being, a small number of Power-to-Liquids plants are currently operational and some demonstration projects are in progress, especially in Europe, driven by the need to decrease greenhouse gas emissions and increase the share of renewable energy. There are also some plants announced and planned in Switzerland, Norway and Canada (Zhu 2019; O'Connell et al. 2019; Panzone et al. 2020; Rego de Vasconcelos and Lavoie 2019; Wulf et al. 2018).

A non-exhaustive list of Power-to-Liquids plants or projects for the production of gasoline, diesel and kerosene is provided in Table 5 (Zhu 2019; O'Connell et al. 2019; Panzone et al. 2020; Rego de Vasconcelos and Lavoie 2019; Wulf et al. 2018). This table combines and updates earlier reviews by Zhu (2019), O'Connell et al. (2019), Panzone et al. (2020), Rego de Vasconcelos and Lavoie (2019) and Wulf et al. (2018), including details such as CO_2 conversion route, plant operational status, city and country of location of the plant, target product, CO_2 source, plant capacity, expressed in terms of total syncrude production, in t/y, and installed power in kW, calculated from the average lower heating value of syncrude, as well as main references.

According to the Power-to-Liquids plants currently in operation in Europe (see Table 5), the installed production capacity of renewable transportation fuels is still low, nearly 160 tons of syncrude per year or about 200 KWh, considering a low heating value of 11.94 kWh/kg for the syncrude. For some plants, the capacity is unknown. In addition, there are many commercial-scale plants announced or planned. Therefore, the production capacity of renewable fuels can considerably increase in the short term.

Renewable transportation fuels with almost zero net greenhouse gas emissions are essential in reducing greenhouse gas emissions in order to meet the Paris agreement (Schmidt et al. 2016). The final demand of renewable fuels for road vehicles, locomotive, aviation and navigation will strongly depend on the level of electrification of the transportation sector (O'Connell et al. 2019). According to a study by O'Connell et al. (2019), even for highly electrified vehicle scenarios, roughly 347 GW or, based on low heating value of syncrude and assuming a full load period of the Power-to-Liquids plant of 4000 h/year, 116 thousands of tons per year of Power-to-Liquids have to be installed in the European Union by 2050.

Table 5 Plants or projects for Power-to-Liquids

Plant/project	Route	Status	Location	Capacity t/y (kW) Fuel	Target Product	CO_2 source	References
Sunfire PtL	RWGS/coSOEC-FT	In operation	Dresden (DE)	~ 46 (~63)	Diesel	Air captured CO_2	(Zhu 2019; O'Connell et al. 2019; Panzone et al. 2020; Rego de Vasconcelos and Lavoie 2019)
Audi PtL	RWGS-FT	Announced	Laufenburg (CH)	~ 320 (~435)	Diesel	Biomass plant	(O'Connell et al. 2019; Höpfner 2017)
GreenPower2Jet	RWGS-FT	Announced	Stade/Lingen (DE)	10,000 (~14,000)	Kerosene	Air captured CO_2	(O'Connell et al. 2019; Bio-based News 2019; Neuling and Kaltschmitt 2019)
KEROSyN100	methanol	Announced	Heide (DE)	nda	Kerosene	Waste CO_2 from refinery	(O'Connell et al. 2019)
Nordic Blue Crude	coSOEC-FT	Planned	Herøya (NO)	8,000 (~10,000)	Diesel	Air captured CO_2	(Zhu 2019; O'Connell et al. 2019; Panzone et al. 2020; Sherrard 2017)
Powerfuel	RWGS-FT	In operation	Karlsruhe (DE)	~ 90 (~120)	Kerosene	Air captured CO_2	(O'Connell et al. 2019; Zitscher et al. 2019)
SOLETAIR	RWGS-FT	In operation	Lappeenranta (FI)	21 (~30)	Gasoline	Air captured CO_2	(O'Connell et al. 2019; Panzone et al. 2020; Soletair 2020)
Sundance Clean Methanol/Blue fuel Energy	Methanol	Announced	British Columbia, (CA)	693,500 (~950,000)	Gasoline + diesel	Flue gas	(O'Connell et al. 2019; Blue Fuel Energy 2020)
Carbon Engineering—Air to fuel system	RWGS + FT	Planned	nda	~ 80,000 (~100,000)	Syncrude	Air captured CO_2	(Zhu 2019; Rego de Vasconcelos and Lavoie 2019; Design Engineering 2019)

This table compiles Power-to-Liquids plants or demonstration projects for the production of gasoline, diesel and kerosene currently in operation, announced or planned mainly in Europe, as reported in literature. This table includes details about plant operational status, city and country of location of the plant, target product, CO_2 conversion route, CO_2 source and plant capacity, which give an idea of the level of development, deployment and sustainability of Power-to-Liquids technology and the potential for producing alternative gasoline, diesel or kerosene transportation fuels from carbon dioxide mostly at European level

RWGS Reverse Water Gas Shift, coSOEC Co-electrolysis of water and CO_2 in Solid Oxide Electrolyte Cells, FT Fischer–Tropsch, DE Germany, CH Switzerland, nda no data available, NO Norway, FI Finland, CA Canada

The estimated net CO_2 utilization of the Power-to-Liquids process via the Fischer–Tropsch route is about 2.6 t CO_2/t liquid fuel (Jarvis and Samsatli 2018). Therefore, considering a production capacity of Power-to-Liquids of about 116 thousands of tons per year, only in Europe, the potential of CO_2 utilization would be about 302 thousands of tons of carbon dioxide per year by 2050.

The energy consumption of the Power-to-Liquids process via the Fischer-Tropsch route is about 6.8 kWh per ton of liquid fuel produced (Jarvis and Samsatli 2018).Therefore, the Power-to-Liquids process has neutral or negative carbon balance only if the energy required for the production of the liquid fuel comes from renewable sources and if carbon dioxide is captured directly from air or from renewable sources (Jarvis and Samsatli 2018; Panzone et al. 2020).

The production of fuels through the Fischer–Tropsch route has the lowest water demand, land requirements and greenhouse gas emissions and the highest energy efficiency, as compared to other production technologies, such as biofuels, biomass-to-liquid, carbon-to-liquid and gas-to-liquid (Panzone et al. 2020; Schmidt et al. 2016).

Different techno-economic studies estimated a production cost for liquid fuels synthetized via the Fischer–Tropsch route ranging between about 536 and 1992 € per ton of liquid fuel depending mainly on the type of electrolysis process and carbon dioxide source utilized, while the production cost of the same liquid fuels from fossil sources ranges between roughly 272 and 320 € per ton of liquid fuel, depending on the type of fuel (Panzone et al. 2020). Therefore, transportation fuels produced via Power-to-Liquids are not yet cost-competitive as substitute for conventional fossil fuels.

CO_2 can be also converted directly into liquid fuels, such as gasoline (Schmidt et al. 2016; Wei et al. 2018, 2017; Choi et al. 2017; Liu et al. 2019) via the CO_2 hydrogenation reaction over iron based catalysts. However, the product distribution can be wide depending on the structure and composition of the catalysts. Hydrocarbons products mainly consisted in low molecular weight hydrocarbons, such as paraffins and olefins, instead of heavier products with higher added value as liquid transportation fuels. Different types of co-catalysts, such as copper and cobalt, promoters, such as potassium, and metal oxide supports, such as alumina, silica, titania and zeolites, are employed to tune the product distribution toward heavier hydrocarbons (Choi et al. 2017). According to literature, the combination of iron, potassium and zeolites, makes the catalyst highly active and selective for CO_2 hydrogenation to long-chain hydrocarbons (Ye et al. 2019).

The CO_2 hydrogenation to liquid fuels is considered as a modification of the traditional Fischer–Tropsch process, which uses syngas, a mixture of CO and H_2, to produce liquid fuels (Panzone et al. 2020). However, the use of a single reactor for direct CO_2 hydrogenation is advantageous compared to the classical Fischer–Tropsch route due to its ease of operation and lower CO_2 conversion cost (Ye et al. 2019).

One of the main challenges for advancing CO_2 hydrogenation to liquid fuels is increasing the energy efficiency and reducing the investment and operating costs of the process and the integration of renewable energy sources in the process (Whipple and Kenis 2010). The tendency is to maximize the yield to desired product with a minimum renewable energy input. However, the synthesis of hydrocarbons from CO_2 over conventional heterogeneous catalysts is favored at high pressures (high compression power) and low temperatures (exothermic) and are slow processes, which implies the use of high residence times (high reactor volume) or feed recycling ratios (high power consumption) to obtain significant conversions. In addition, it is necessary to improve activity, selectivity and stability of the catalyst and to minimize carbon deposition and undesirable methane formation.

Different processes have been developed in fixed bed reactors operating at high pressure, between 10 and 40 bar, and low temperature, between 200 and 350 °C, using catalysts based on iron over different supports, such as alumina, silica, titania and zeolite, or a combination of them, and improved with promoters, such as potassium, rubidium, zirconium, magnesium, copper, zinc, chromium, vanadium and lanthanum, or a combination of them (Schaaf et al. 2014; Kondratenko et al. 2013; Centi et al. 2013; Saeidi et al. 2014; Wang et al. 2011; Rodemerck et al. 2013).

Some reactor configurations, such as fluidized bed and slurry reactors, facilitate the removal of the heat generated by the carbon dioxide hydrogenation, leading to improved conversion of carbon dioxide to hydrocarbons. Other reaction configurations, such as serial or recirculating reactors, allow for increased residence time, and thus higher production of hydrocarbons in the reactor (Saeidi et al. 2014). Another way to improve the conversion of carbon dioxide to hydrocarbons is by removing in situ water using a perm-selective silica membrane integrated into a membrane catalytic reactor (Centi et al. 2013).

2.7 Carbon Monoxide—Syngas

Both syngas, a mixture of carbon monoxide and hydrogen, and pure carbon monoxide are crucial building blocks for industrial production of several fuels and chemicals (Keim 1989). Syngas is conventionally produced from fossil sources by partial oxidation of hydrocarbons or coal or by catalytic steam reforming of natural gas. Pure CO is obtained from syngas by gas separation (Jarvis and Samsatli 2018; Lavoie 2014).

Given that syngas is a feedstock for conventional production of methanol and higher hydrocarbons by mature synthesis technologies (Behrens et al. 2019; Prieto 2017), the development of technologies for carbon dioxide conversion to syngas represents a more climate-friendly route for sustainable production of fuels and chemicals.

One method to produce syngas from carbon dioxide and methane is the catalytic dry reforming of methane. Dry reforming of methane has many environmental, technical and economic advantages compared to conventional technologies for syngas production. The dry reforming process operates at milder temperatures and pressures than steam and autothermal reforming processes (Mondal et al. 2016) and uses recycled carbon dioxide instead of water for reforming, avoiding energy consumption associated with steam generation and decreasing the carbon footprint associated with the syngas production (Jarvis and Samsatli 2018; Er-rbib et al. 2012; Schwab et al. 2015).

Dry reforming of methane produces a syngas with a hydrogen to carbon monoxide ratio more compatible with some downstream synthesis processes such as Fischer–Tropsch synthesis (Jarvis and Samsatli 2018; Lavoie 2014; Schwab et al. 2015). According to a study by Mondal et al. (2016), the cost associated with methanol synthesis from dry reforming of methane is lower than the equivalent cost of methanol production from steam reforming of methane

(Jarvis and Samsatli 2018; Mondal et al. 2016). Dry reforming is also a more cost-effective alternative to partial oxidation for syngas production in plants with medium and small size (Schwab et al. 2015).

The industrial application of dry reforming of methane for valorization of carbon dioxide is still limited (Er-rbib et al. 2012), having a technology readiness level of up to 6. The Linde Group built a dry reforming pilot plant at Pullach in Germany (The Linde Group 2015). Nowadays, after completion of the testing at pilot scale, the Linde Group is aiming at upscaling the catalyst on the way for commercial application (The Linde Group 2020).

Catalysts utilized in carbon dioxide reforming of methane comprise nickel, noble metals, cobalt and metal carbides in fixed bed reactors (Lavoie 2014). The main technological challenge for dry reforming of methane is the readiness of cheap and widely available catalysts resistant to deactivation by coking and with high selectivity to syngas (Jarvis and Samsatli 2018; Behrens et al. 2019; Kim 2017).

Other way to obtain carbon monoxide from carbon dioxide is the reverse water gas shift reaction. In the reverse water gas shift reaction, carbon dioxide reacts with renewable hydrogen to form carbon monoxide and water, but subsequent removal of water is needed before the downstream Fischer–Tropsch or methanol synthesis in order to avoid catalyst inhibition and equilibrium limitation by water, enabling milder operating conditions for the synthesis process (Behrens et al. 2019; Wang et al. 2011). The reverse water gas shift reaction is not currently available at industrial scale (Behrens et al. 2019), being at a technical readiness level of about 6 (Jarvis and Samsatli 2018; Schmidt et al. 2016). The technical viability of the CAMERE process concept, consisting of three consecutive steps: reverse water gas shift reaction, water removal and conventional methanol synthesis, was demonstrated, but additional research and development is needed for commercial application of the process, due to a lack of economic feasibility (Joo et al. 1999).

Strategies to increase the yield of carbon monoxide via the reverse water gas shift reaction include removing products, mainly water, in situ from the reaction medium and using high temperatures, given the endothermicity of the reaction. The main challenge in the development of catalyst for reverse water gas shift reaction is the improvement of stability and selectivity of the catalyst at high temperature in order to avoid hydrothermal deactivation of the catalyst and undesirable methane formation, respectively (Behrens et al. 2019; Wang et al. 2011).

Other method for carbon monoxide production from carbon dioxide is the electrolysis of carbon dioxide directly to carbon monoxide and oxygen. Electrolysis of carbon dioxide operates either at high temperature in solid oxide electrolysis cells or at low temperature in gas diffusion or solution-phase electrolysis cell (Qiao et al. 2014).

Haldor Topsoe has developed and commercialized a modular technology called eCOs™ for onsite production of pure carbon monoxide from carbon dioxide via solid oxide electrolysis, with plant capacities up to a few hundred of normal cubic meter of carbon monoxide per hour (Zhu 2019; Andersen 2017; Haldor-Topsoe 2020). The first commercial system, having a capacity of 12 Nm^3/h, commenced operating in 2016 at Gas Innovations in Texas (Mittal et al. 2017).

Major challenges for increasing the deployment of solid oxide electrolysis of carbon dioxide include improving the stability of the electrodes and the ionic conductivity of the electrolytes, and enhancing the tolerance to impurities (Qiao et al. 2014).

The electrolysis of carbon dioxide at low temperature is still under development.

Siemens tested carbon dioxide electrolysis on larger-scale gas diffusion electrodes, of 10 and 100 cm^2, in flow cell configurations (Jeanty et al. 2018). Next, Siemens collaborated with Evonik in the development of a hybrid system for the production of chemicals by connecting their small electrolyzer with a fermentation process (Haas et al. 2018).

Dioxide Materials explored the electrolysis of carbon dioxide and the co-electrolysis of water and carbon dioxide in an anion-conductive membrane-based electrolyzer, resulting in the production of carbon monoxide and syngas, respectively at realistic rates (Kutz et al. 2017; Liu, et al. 2016). Remaining challenges for the low-temperature electrolysis of carbon dioxide to become feasible include improving stability and durability of electrodes, minimizing loss of carbon dioxide to the liquid electrolyte and reducing overall energy requirement under operating conditions at practical production rates of carbon monoxide (Liu et al. 2016).

3 CO_2 Chemically Derived Materials

CO_2 can also be chemically utilized for the production of materials with different applications.

3.1 Polymers

Polymers, such as polyurethanes and polycarbonates, are used in the manufacturing of plastics, resins and foams, which have a wide range of applications, such as in buildings and cars (International Energy Agency (IEA) 2019; Qin et al. 2015). Petroleum is the traditional feedstock for the production of polymers. The carbon dioxide can partially replace petrochemicals as raw material for the synthesis of polymers, such as in the copolymerization of epoxides with carbon dioxide for the production of polycarbonates (Zhu 2019; Chauvy et al. 2019; International Energy Agency (IEA) 2019; Ampelli et al. 2015).

Replacing petrochemical feedstocks with CO_2 in polymer production is potentially beneficial from an economic and environmental point of view, owing to the relatively low energy requirements and reduced impact on climate change associated with processing polymers with CO_2, and enables the production of biodegradable polymers with innovative features (Zhu 2019; Chauvy et al. 2019; International Energy Agency (IEA) 2019; Ampelli et al. 2015; Alberici 2017; Assen 2015).

Polycarbonates are classically synthesized by reaction of 1,2-diols with phosgene. Phosgene is toxic and corrosive. An alternative route for the selective synthesis of polycarbonates is the copolymerization of epoxides, such as ethylene oxide, propylene oxide, cyclohexene oxide, vinyl oxide and styrene oxide, and CO_2 over homogeneous or heterogeneous catalysts based on transition metals (Qin et al. 2015; Lu and Darensbourg 2012; Poland and Darensbourg 2017). In addition, according to Fukuoka et al. (2007), the carbon dioxide copolymerization process has lower expenditure cost as compared to conventional process using phosgene (Zhu 2019; Fukuoka et al. 2007).

There are many companies and institutions developing and producing polymers based on carbon dioxide.

The Asahi Kasei Corporation commercialized the first process for producing phosgene-free polycarbonate from ethylene oxide and carbon dioxide with high selectivity and yield to polycarbonates, avoiding the need for separation and purification of the reaction products and the treatment or disposal of residues (Zhu 2019; Chauvy et al. 2019; International Energy Agency (IEA) 2019; Fukuoka et al. 2007). Based on the Asahi Kasei copolymerization process, a commercial facility in China produces about 150,000 tons of polycarbonates per year from carbon dioxide (International Energy Agency (IEA) 2019; Fukuoka et al. 2007). Asahi Kasei Corporation licensed its technology to various companies, leading to the construction of several plants in South Korea, Russia and Saudi Arabia with a total polycarbonate production capacity of 1.07 Mt/y by 2019, enabling the incorporation of 0.185 $MtCO_2$/y into the polycarbonates produced (Zhu 2019).

Empower Materials currently sells polyethylene carbonate made from ethylene oxide and carbon dioxide (Quadrelli et al. 2011).

Econic Technologies in UK developed a catalytic process for the production of polycarbonate polyols without significant by-product formation and using up to 50 percent by weight of carbon dioxide captured in a coal-fired power plant (Quadrelli et al. 2011; Broadwith 2015; Econic Technologies 2018). Econic Technologies has released a plant for demonstrating the full polycarbonates production process from copolymerization reactor to the product treatment system (Econic Technologies 2018).

Covestro developed a plant in Dormagen, Germany, for the copolymerization of propylene oxide with carbon dioxide to produce 5000 t/y of polyether-polycarbonate polyols suitable for polyurethane foam applications, without undesirable by-products formation. The polymer, branded as cardyon®, contains up to 20 Wt.% of carbon dioxide sourced from waste gases of a nearby ammonia production plant (Zhu 2019; Chauvy et al. 2019; Covestro 2020; Fernández-Dacosta, et al. 2017; Langanke et al. 2015).

Novomer Inc. developed a process for the copolymerization of carbon dioxide and propylene oxide to polypropylene carbonate polyols, branded as Converge®. A commercial-scale facility in Texas, USA, annually produces several thousands of Converge® polyols incorporating up to 50 wt.% of carbon dioxide. The Converge® polyols are applicable to the formulation of polyurethane-based sealants, coatings, adhesives and foams with improved performance, weatherability and strength and with lower content of petrochemical feedstock and polyols production cost (Zhu 2019; Korosec 2016; Novomer 2020).

There are also some producers of polymers based on carbon dioxide in China. In 2015, Nanyang Zhongju Tianguan Low-Carbon Technology Company started to produce 25,000 t/y of biodegradable polypropylene carbonate by copolymerization of propylene oxide, derived from its ethanol production process, and carbon dioxide and was constructing a new copolymerization plant with a production capacity of 100,000 t/y. Jinlong Green Chemical Company manufactures aliphatic polycarbonate polyols and biodegradable polyurethane foam based on carbon dioxide. Inner Mongolia Mengxi High-Tech Group Company fabricates 3000 t/y aliphatic polycarbonate by copolymerization of waste carbon dioxide from an own cement kiln (Zhu 2019). Jinioong-Cas Chemail Co. produces polyols with a capacity of 10 kt/y (Aresta et al. 2014). Sinopec SABIC Tianjin Petrochemical Co. is currently building a plant in Tianjin for the production of 260,000 t/y polycarbonate. The plant will be in operation by 2020 (Aresta et al. 2014; Volkova 2019).

Newlight Technologies in USA developed a biocatalytic process to produce a biodegradable thermo-polymer, trademarked as AirCarbon, from captured CH_4 or CO_2 emissions. AirCarbon is claimed as a carbon-negative material cheaper than its oil-derived equivalents. Commercial production of AirCarbon started in 2013, and Newlight signed several Production License Agreements with different companies to sell several thousand tons of AirCarbon polymer by year from different manufacturing plants (Newlight Technologies 2020).

The synthesis of polycarbonates by copolymerization of epoxides and CO_2 is currently ready for industrial application at large scale (Zhu 2019) with a technology readiness level of 8–9 (Chauvy et al. 2019). However, despite the relative maturity of this technology, there are still opportunities for progressing. The carbon dioxide content in polycarbonates is currently limited to a maximum of 50% CO_2 by weight. Therefore, the copolymerization process needs further research and development to increase the incorporation of CO_2 into the copolymers (International Energy Agency (IEA) 2019; Quadrelli et al. 2011; Alberici 2017; Assen 2015). In some processes, the scale of production of polyether carbonates derived from CO_2 is too small to replace existing conventional polyol plants. Therefore, the carbon dioxide polymerization technology requires further development to produce polyols on a larger scale (Chauvy et al. 2019).

Current research on copolymerization of CO_2 focuses on the development of catalyst for producing polymers with tailor-made features and from renewable epoxides, such as limonene oxide, cyclohexadiene oxide and α-pinene oxide (Poland and Darensbourg 2017).

The estimated production and demand for polycarbonates worldwide were approximately 5–6 Mt/y and 4–5 Mt/y, respectively for 2016. The market price of polycarbonates varied between 3120 and 3830 € per ton during the last few years, and it probably will decrease in the future as the production process moves from the conventional method to the copolymerization of carbon dioxide (Aresta et al. 2014).

Regarding the climate benefits of the synthesis of polycarbonates from carbon dioxide copolymerization in comparison with the conventional methods based on phosgene, the estimated net CO_2 utilization for the copolymerization process is about 0.173 tons of CO_2 per ton of polycarbonate (Chauvy et al. 2019). Therefore, considering a global installed capacity of 5 million tons of polycarbonates per year (Aresta et al. 2014), the potential of CO_2 uptake is about 1 million tons of carbon dioxide per year.

3.2 *CO_2-Derived Building Materials*

CO_2 can be used in the production of building materials to partially replace water as curing agent in concrete manufacturing or as a raw material in the fabrication of concrete constituents, such as cement and construction aggregates (International Energy Agency (IEA) 2019). Building materials can be obtained from CO_2 via carbonation processes, involving the reaction of a calcium or magnesium source with carbon dioxide in the presence of water to produce carbonates suitable for concrete applications (International Energy Agency (IEA) 2019).

Alternative Ca or Mg sources for CO_2 carbonation include a variety of alkaline industrial residues from power plants or industrial processes, leading to further economic and environmental benefits, as the carbonation process converts both solid wastes and captured CO_2 into marketable products and enables the permanent storage of carbon dioxide and the stabilization of the toxic compounds contained in wastes within the carbonated products (Zhu 2019; Aresta et al. 2014).

The curing of concrete with CO_2 is a mature and encouraging technology for the utilization of CO_2, while the incorporation of CO_2 in the manufacture of the cement itself is at a lower level of development (International Energy Agency (IEA) 2019).

A crucial challenge for the deployment of CO_2 carbonation at large scale is to speed up the carbonation reaction by increasing temperature and pressure of the process and by pre-treatment of the mineral or solid waste using mechanical and chemical methods (Zhu 2019; Aresta et al. 2014).

(a) Cement-Concrete

The curing of concrete with CO_2 entails converting CO_2, calcium silicate in cement and water into hydrated calcium silicate and calcium carbonate via carbonation during the mixing of concrete (International Energy Agency (IEA) 2019). As carbonation is exothermal, the heat released during the reaction accelerates the curing process, decreasing the consumption of fossil fuels or steam for heating and resulting in energy savings and emission reduction (Zhu 2019; Aresta et al. 2014).

CarbonCure Technologies and Solidia Technologies are leading the development and commercialization of the technology for curing concrete with CO_2 (Zhu 2019; International Energy Agency (IEA) 2019; CarbonCure 2020; Solidia 2020).

The CarbonCure process for concrete curing with CO_2 shows a series of advantages versus conventional methods, such as operation under atmospheric pressure, simplifying the design of the curing chambers and using shorter curing times, resulting in cost savings, while obtaining a concrete with the same or improved quality. However, the use of liquid CO_2 penalizes the economy of the curing process. The net CO_2 emission avoided by the CarbonCure process is about 18 tons of carbon dioxide per cubic meter of concrete, as a result of the reduced use of cement and the incorporation of carbon dioxide in the concrete in comparison with concrete produced by traditional methods (Zhu 2019; Monkman and MacDonald 2017; Monkman et al. 2016).

The CarbonCure process is easily integrated into conventional systems for concrete production without disturbing normal operation of the plants. The CarbonCure process is currently implemented in many facilities for ready-mix concrete production, mostly in North America, and the retrofit of more plants is planned. The application of the CarbonCure technology for the conversion of CO_2 waste from cement industry into concrete suitable for use in construction was first demonstrated in 2018 in USA (Zhu 2019; Chauvy et al. 2019; Sanna et al. 2014).

Solidia Technologies in USA commercializes processes for the manufacture of cement and concrete using CO_2 (Zhu 2019). The Solidia technology promises to simplify and reduce the costs and the water and carbon dioxide footprints of the carbonation process compared to traditional methods of production, while obtaining cement and concrete with improved performance. As an example, the Solidia concrete requires 24 h to reach complete strength compared to the up to 28 days needed for the curing of the concrete produced by classical methods. In addition, the Solidia process for curing concrete is easy to implement in existing manufacturing facilities and is adjustable to different production methods, concrete formulations and standards. The maximum capacity of CO_2 uptake of the Solidia process is about 0.3 tons of CO_2 per ton of cement used to make the concrete (DeCristofaro and Opfermann 2015).

Solidia and CarbonCure reported a reduction in carbon footprint of concrete about 70 and 80%, respectively (CarbonCure 2020; DeCristofaro and Opfermann 2015).

The market of concrete was about 32,000 Mt/y by 2014 and was foreseen to increase at an annual rate of 0.8–1.2%/y until 2030 (Aresta et al. 2014). Therefore, the CO_2 mitigation potential of the production of building materials by carbon dioxide carbonation is substantial (Zhu 2019).

In summary, concrete cured with CO_2 may have superior performance, reduced manufacturing costs and lower water and CO_2 footprints as compared with concrete produced by conventional methods (International Energy Agency (IEA) 2019). However, compliance with existing quality standards and regulations may delay the deployment of carbon dioxide curing of concrete to the manufacture of ready-mixed and precast concrete until long-term trials demonstrate its environmental friendly performance (International Energy Agency (IEA) 2019). Meanwhile, non-structural uses of concrete, such as in the construction of floors, ditches and roads, represent the target for early application of concrete cured with CO_2 (International Energy Agency (IEA) 2019).

(b) Aggregates

Construction aggregates, small particles utilized in the fabrication of building materials, can be obtained via reaction of carbon dioxide with wastes rich in magnesium or calcium, such as steel and blast furnace slags, bauxite waste, cement kiln dust, municipal waste incineration ash, air pollution control wastes, mining residues, fly ashes and asbestos, coming from power generation and other industrial sectors, such as steel making, cement, paper and mining industries, otherwise stored in landfills (International Energy Agency (IEA) 2019; Quadrelli et al. 2011; Carbon8 2020).

Different companies are scaling up the technology of waste carbonation leading to the consumption of about 75 kilotons of carbon dioxide per year (International Energy Agency (IEA) 2019).

The British company Carbon8 Systems developed and commercialized a technology to produce an aggregate from captured CO_2 and wastes (Zhu 2019; Chauvy et al. 2019; Patricio et al. 2017; Sanna et al. 2014). The Carbon8 process stored more carbon dioxide within the aggregate than the CO_2 emitted during the carbonation process, leading to a carbon-negative balance for the process (Carbon8 2020). In addition, the Carbon8 process resulted in lower water and energy consumption, as uses rainwater for manufacturing of the aggregate and electricity only to transport the materials through the plant, without producing any waste by-product (Zhu 2019; Carbon Upcycling 2020).

The first commercial plant based on carbon8 technology is currently producing about 65,000 t/y of lightweight aggregate using wastes from the air pollution control of an incineration facility. Another plant for the production of 100,000 t/y was commissioned in 2016, and three additional plants with the same production capacity or higher would likely to be operational in 2018 (Zhu 2019; Carbon8 2020; Carbon Upcycling 2020).

Skyonic developed the SkyMine process (Carbonfree Chemicals 2020) for the capture and carbonation of carbon dioxide to produce, among others, construction aggregates, while also removing heavy metals and nitrogen and sulfur oxides from exhaust gases of industrial and power plants (Zhu 2019; Chauvy et al. 2019; Quadrelli et al. 2011). The SkyMine technology was first integrated at Capitol Aggregates cement plant in Texas, USA, in 2015. The plant can save about 83,000 tons of CO_2 by year (Zhu 2019; Chauvy et al. 2019; Sanna et al. 2014; Carbonfree Chemicals 2020).

Mineral Carbonation International (MCi) developed a technology for waste carbonation by combining CO_2 from industrial sources, such as flue gases of power generation, cement and steel industries, with low-rank minerals or wastes to manufacture of building products. MCI built a pilot plant for scaling up and testing of the waste carbonation technology in New South Wales, Australia, and is currently moving ahead toward first industrial demonstration plants for processing about 20,000 tons of carbon dioxide per year (Chauvy et al. 2019; Mineral Carbonation International (MCi) 2020).

In summary, the manufacturing of construction aggregates from wastes and carbon dioxide enables reduction of carbon dioxide emissions and disposal of industrial wastes while potentially producing marketable products. The CO_2 mineralization technologies have the potential to take advantage of the high volumes of CO_2 emissions common in the cement industry, leading to a huge potential of CO_2 mitigation. However, the overall environmental and climate benefits associated with the manufacturing of construction aggregates from wastes and CO_2 depend on the energy requirements of the specific production process.

The use of CO_2 for manufacturing of building materials offers products with superior performance and lower cost as compared with conventional methods of production of building materials (Zhu 2019; International Energy Agency (IEA) 2019).

Materials of construction based on CO_2 and wastes can be competitive versus commercial cement and aggregates as they offset the costs associated with conventional waste disposal. However, in terms of market penetration, technologies of waste carbonation are held back by the low costs of alternative raw materials, such as limestone, adherence to quality standards in the building sector and compliance with end of waste regulations. In addition, the cement industry is very conservative, and new construction materials are difficult to bring to market unless long-term trials demonstrate their safe and environmental friendly performance (Chauvy et al. 2019; International Energy Agency (IEA) 2019; Quadrelli et al. 2011).

4 Conclusions

The chemical valorization of CO_2 in the form of a wide range of fuels, chemicals and materials through a number of technologies is challenging but opens new research, industrial and business opportunities. Different technological routes for chemical valorization are being developed and approach quickly to commercialization.

Different techno-economic and life-cycle studies of different chemical valorization processes showed that carbon dioxide-derived products may have economic and environmental benefits as provide energy saving, greenhouse gas emission reduction and products with similar or improved quality and with smaller carbon and water footprints and production costs than traditionally produced counterparts.

Some technologies for the catalytic, biochemical and electrochemical conversion of carbon dioxide into chemicals or fuels are already in operation at pre-commercial scale. The most technologically mature pathway for chemical valorization of CO_2 to fuels and chemicals depends on the target product, being direct or indirect carbon dioxide hydrogenation for methanol, methane and dimethyl ether or Fischer–Tropsch synthesis, respectively and gas fermentation, electrochemical reduction and solid oxide electrolysis for ethanol, formic acid and carbon monoxide/syngas production, respectively. However, there are still some key barriers for further development and commercialization of the chemical valorization technologies and products.

In the case of catalytic alternatives to react CO_2, or derived CO, and H_2 for the synthesis of carbon based products, the current technological challenges are increasing energy efficiency while reducing the cost of the conversion process.

Clearly, the development of novel catalysts/electrocatalyst/biocatalysts and process designs are needed to accelerate widespread penetration of greener processes for CO_2 conversion to fuels and chemicals by improving catalyst efficiency, selectivity, stability and durability under long-term, intermittent and unsteady operation, by optimizing catalyst-reactor designs for better thermal control, renewable energy integration and product separation, to surpass equilibrium limitations and to avoid catalyst deactivation, as well as to develop technologically and economically viable processes for the conversion of CO_2 into fuels and chemicals on a commercial scale. Another bottleneck for the market competitiveness of CO_2-based fuels and chemicals is the cheap production of carbon-free hydrogen and the availability of low-cost renewable electricity and feedstock CO_2.

CO_2-derived fuels and chemical can provide greenhouse gas emission savings, but the utilization of low-carbon energy and hydrogen for their production is critical. In addition, widespread testing is required for these products to be accepted by outstanding quality standards.

The synthesis of polymers by copolymerization of epoxides and CO_2 is currently ready for industrial application at large scale. However, despite the relative maturity of this technology, there are still opportunities for progress. The copolymerization process needs further research and development to increase the incorporation of CO_2 into the copolymers and the scale of polymers production and for the development of catalyst for producing polymers with tailor-made features and from renewable epoxides.

The technologies for producing building materials by CO_2 carbonation such as CO_2-cured concrete and carbonated aggregates are at higher level of development and some processes are being commercialized. These technologies are easy to set up or retrofit in the existing production facilities and generate products with similar or better quality than conventionally produced counterparts with reduced manufacturing costs and lower water and CO_2 footprints, but the fulfillment of quality standards and end of waste regulations is still retarding the widespread deployment of these technologies.

Chemical valorization of carbon dioxide will continue advancing in the short-to-medium term, especially in most mature applications, such as the production of CO_2-based polymers, methane and methanol and the CO_2 carbonation. Chemical valorization of carbon dioxide will play a crucial role in a future circular economy for sustainable production of low-carbon fuels, energy and hydrogen vectors, chemicals and materials.

References

Agarwal AS, Zhai Y, Hill D, Sridhar N (2011) The electrochemical reduction of carbon dioxide to formate/formic acid: engineering and economic feasibility. Chemsuschem 4:1301–1310. https://doi.org/10.1002/cssc.201100220

Alberici S et al. (2017) Assessing the potential of CO_2 utilisation in the UK. Available from https://assets.publishing.service.gov.uk/government/uploads/system/uploads/attachment_data/file/799293/SISUK17099AssessingCO2_utilisationUK_ReportFinal_260517v2__1_.pdf. Accessed 30 June 2020

Albert J, Chemical Reaction Engineering (CRT) University of Erlangen-Nuremberg (2020) E2Fuels—development of a single-stage reaction concept for methanol-synthesis from CO_2 and renewable hydrogen via in-situ sorption. https://www.crt.tf.fau.eu/forschung/arbeitsgruppen/komplexe-katalysatorsysteme-und-kontinuierliche-verfahren/biomasse-und-nachhaltige-erzeugung-von-plattformchemikalien/e2fuels-development-of-a-single-stage-reaction-concept-for-methanol-synthesis-from-co2-and-renewable-hydrogen-via-in-situ-sorption/. Accessed 20 May 2020

Ali KA, Abdullah AZ, Mohamed AR (2015) Recent development in catalytic technologies for methanol synthesis from renewable sources: a critical review. Renew Sustain Energ Rev 44:508–518. https://doi.org/10.1016/j.rser.2015.01.010

Alper E, Yuksel Orhan O (2017) CO_2 utilization: developments in conversion processes. Petroleum 3:109–126. https://doi.org/10.1016/j.petlm.2016.11.003

Al-Saydeh S, Zaidi J (2018) Carbon dioxide conversion to methanol: opportunities and fundamental challenges.https://doi.org/10.5772/intechopen.74779

Álvarez A et al (2017) Challenges in the greener production of formates/formic acid, methanol, and DME by heterogeneously catalyzed CO_2 hydrogenation processes. Chem Rev 117:9804–9838. https://doi.org/10.1021/acs.chemrev.6b00816

Ampelli C, Perathoner S, Centi G (2015) CO_2 utilization: an enabling element to move to a resource-and energy-efficient chemical and fuel production. Philos Transact A Math Phys Eng Sci 373:1–35. https://doi.org/10.1098/rsta.2014.0177

An X, Zuo Y-Z, Zhang Q, Wang D-z, Wang J-F (2008) Dimethyl ether synthesis from CO_2 hydrogenation on $aCuO{-}ZnO{-}Al_2O_3{-}ZrO_2$/HZSM-5 bifunctional catalyst. Ind Eng Chem Res 47:6547–6554. https://doi.org/10.1021/ie800777t

Andersen M (2017) Hydrogen from 'Reverse Fuel' cells. http://www.dtu.dk/english/news/2017/03/dynamo-theme4-hydrogen-from-reverse-fuel-cells?id=e804ab15-4822-4f1c-92be-09a3e5bece1e. Accessed 30 June 2020

Anggraini I, Kresnowati M-T-A-P, Purwadi R, Setiadi T (2018) Bioethanol production via syngas fermentation. MATEC Web Conf 156:03025. https://doi.org/10.1051/matecconf/201815603025

Aresta M, Dibenedetto A, Angelini A (2014) Catalysis for the valorization of exhaust carbon: from CO_2 to chemicals, materials, and fuels. Technological use of CO_2. Chem Rev 114 (3):1709−1742. https://doi.org/10.1021/cr4002758

Artz J et al (2018) Sustainable conversion of carbon dioxide: an integrated review of catalysis and life cycle assessment. Chem Rev 118:434–504. https://doi.org/10.1021/acs.chemrev.7b00435

von der Assen NV (2015) From life-cycle assessment towards life-cycle design of carbon dioxide capture and utilization. Ph.D. thesis, University of Aachen. Available from http://publications.rwthaachen.de/record/570980/files/570980.pdf. Accessed 30 June 2020

Bailera M, Lisbona P, Romeo LM, Espatolero S (2017) Power to gas projects review: lab, pilot and demo plants for storing renewable energy and CO_2. Renew Sustain Energ Rev 69:292–312. https://doi.org/10.1016/j.rser.2016.11.130

Bartmann M, Wranik S (2017). BASF and BSE Engineering sign development agreement to transform CO_2 and excess current into methanol. In: BASF and BSE engineering join press release. https://www.basf.com/global/en/media/news-releases/2017/08/p-17-293.html. Accessed 24 May 2020

Behrens M, Bowker M, Hutchings G (2019) Thermal CO_2 reduction. In: Latimer A, Dickens C (eds) Research needs towards sustainable production of fuels and chemicals. Energy-X, pp 28–38. https://www.energy-x.eu/wp-content/uploads/2019/10/Energy-X-Research-needs-report.pdf. Accesed 20 May 2020

Bender M, Thyssenkrupp (2018) Milestone for climate protection: Carbon2Chem pilot plant opened. https://engineered.thyssenkrupp.com/en/milestone-for-climate-protection-carbon2chem-pilot-plant-opened/. Accessed 24 May 2020

Bergins C, Tran K, Koytsoumpa EI, Kakaras E, Buddenberg T, Sigurbjörnsson Ó (2015) Power to methanol solutions for flexible and sustainable operations in power and process industries. PowerGen. Available from http://www.mefco2.eu/pdf/1._Power_to_Methanol_Solutions_for_Flexible_and_Sustainable_Operations_in_Power_and_Process_Industries.pdf. Accesed 14 May 2020

Bertalan Z, Hein M (2016) Munich-based clean-tech startup Electrochaea and Hungarian utility MVM establish power-to-gas joint venture. http://news.bio-based.eu/munich-based-clean-tech-startup-electrochaea-and-hungarian-utility-mvm-establish-power-to-gas-joint-venture/. Accessed 15 May 2020

BioCat project (2016) BioCat project. https://biocat-project.com/. Accessed 15 May 2020

Lemvig Biogas (2015) MeGa-stoRE project no. 12006. Final report. Available from https://www.lemvigbiogas.com/MeGa-stoREfinalreport.pdf. Accessed 15 May 2020

BioPower2Gas (2016) BioPower2Gas project. http://www.biopower2gas.de. Accessed 15 May 2020

Blue Fuel Energy (2020) Blue fuel energy project. http://bluefuelenergy.com/. Accessed 30 June 2020

Bonalumi D, Manzolini G, Vente J, van Dijk H-A-J, Hooey L, Sarron E (2018) From residual steel gases to methanol: the FReSMe project. In: 14th international conference on greenhouse gas control technologies, GHGT-14 21st–25th October 2018, Melbourne, Australia. Available from http://www.fresme.eu/events/pdf/Bonalumi_Fresme_2019_GHGT-SSRN.pdf. Accessed 26 May 2020

Bordiga S, Lercher JA (2019) Thermal processes for syngas to fuels and chemicals. In: Latimer A, Dickens C (eds) Research needs towards sustainable production of fuels and chemicals. Energy-X 39–48. Available from https://www.energy-x.eu/wp-content/uploads/2019/10/Energy-X-Research-needs-report.pdf. Accessed 30 June 2020

Broadwith P (2015) Catalytic carbon dioxide convertors. https://www.chemistryworld.com/business/catalytic-carbon-dioxideconvertors/8308.article. Accessed 30 June 2020

Carbon8 (2020) Carbon8 aggregates. www.c8s.co.uk. Accessed 30 June 2020

CarbonCure (2020) Carbon Sense Solutions, Inc. www.novascotia.ca/nse/cleantech/docs/CarbonSense.pdf. Accessed 30 June 2020

Centi PS (2009) Opportunities and prospects in the chemical recycling of carbon dioxide to fuels. Catal Today 148:191–205. https://doi.org/10.1016/j.cattod.2009.07.075

Centi G, Quadrelli EA, Perathoner S (2013) Catalysis for CO_2 conversion: a key technology for rapid introduction of renewable energy in the value chain of chemical industries. Energ Environ Sci 6:1711–1731. https://doi.org/10.1039/C3EE00056G

Chauvy R, Meunier N, Thomas D, De Weireld G (2019) Selecting emerging CO_2 utilization products for short-to mid-term deployment. Appl Energ 236:662–680. https://doi.org/10.1016/j.apenergy.2018.11.096

Carbonfree Chemicals (2020) Capture harmful pollutants with SkyMine. http://www.carbonfreechem.com/technologies/skymine. Accessed 30 June 2020

China Petroleum and Chemical Industry Federation (2020) China's CN: market price: monthly avg: organic chemical material: dimethyl ether: 99.0% or above in Apr 2020. https://www.ceicdata.com/en/china/china-petroleum–chemical-industry-association-petrochemical-price-organic-chemical-material/cn-market-price-monthly-avg-organic-chemical-material-dimethyl-ether-990-or-above. Accessed 30 June 2020

Choi YH, Jang YJ, Park H, Kim WY, Lee YH, Choi SH, Lee JS (2017) Carbon dioxide fischer-Tropsch synthesis: a new path to carbon-neutral fuels. Appl Catal B 202:605–610. https://doi.org/10.1016/j.apcatb.2016.09.072

Chung J, Cho W, Baek Y, Lee C (2012) Optimization of KOGAS DME process from demonstration long-term test. Trans Korean Hydrogen New Energ Soc 23:559–571. https://doi.org/10.7316/KHNES.2012.23.5.559

Chwoła T et al (2020) Pilot plant initial results for the methanation process using CO_2 from amine scrubbing at the Łaziska power plant in Poland. Fuel 263:116804. https://doi.org/10.1016/j.fuel.2019.116804

Covestro (2020) Cardyon®—brighter use of CO_2. https://www.covestro.com/en/cardyon/overview. Accessed 30 June 2020

Cuesta D, Pacios R (2018) MefCO_2-Methanol fuel from CO_2. Synthesis of methanol from captured carbon dioxide using surplus electricity. Available from https://etipwind.eu/wp-content/uploads/MefCO2-slides.pdf. Accessed 20 May 2020

Dang S, Yang H, Gao P, Wang H, Li X, Wei W, Sun Y (2019) A review of research progress on heterogeneous catalysts for methanol synthesis from carbon dioxide hydrogenation. Catal Today 330:61–75. https://doi.org/10.1016/j.cattod.2018.04.021

Danish Energy Agency (DEA) (2020) Funding for two large-scale Power-to-X projects in Denmark. Fuel Cells Bull 1:9.https://doi.org/10.1016/S1464-2859(20)30024-9

Danish Gas Technology Centre (DGC) (2017) DGC-notat. Power to gas status primo 2017. Available from https://www.dgc.dk/sites/default/files/filer/publikationer/N1701_power_to_gas.pdf. Accessed 15 May 2020

DeCristofaro N, Opfermann A (2015) New CO_2-curing technology for concrete. CryoGas Int 53:28–29

Design engineering (2019) Canada invests $25M in carbon engineering. Squamish, B.C.-based company to expand CO_2 capture and air to fuels technologies to commercial scale. https://www.design-engineering.com/canada-invests-25m-in-carbon-engineering-1004033396/. Accessed 30 June 2020

Energy research Centre of the Netherlands (ECN) (2017) FReSMe—from residual steel gases to methanol. In: Trondheim 14th June 2017. Available from https://www.sintef.no/globalassets/project/tccs-9/presentasjoner/d5/7—20170614-fresme-h2020-project—dissemination.pdf. Accessed 26 May 2020

Er-rbib H, Bouallou C, Werkoff F (2012) Production of synthetic gasoline and diesel fuel from dry reforming of methane. Energ Procedia 29:156–165. https://doi.org/10.1016/j.egypro.2012.09.020

CORDIS Europe (2020) MefCO_2 project. https://cordis.europa.eu/project/rcn/193453_en.html. Accessed 20 May 2020

CORDIS Europe (2020) FResMe. https://cordis.europa.eu/project/id/727504. Accessed 26 May 2020

Fernández-Dacosta C et al (2017) Prospective techno-economic and environmental assessment of carbon capture at a refinery and CO_2 utilisation in polyol synthesis. J CO_2 Utilization 21:405–422. https://doi.org/10.1016/j.jcou.2017.08.005

Fleisch TH, Basu A, Sills RA (2012) Introduction and advancement of a new clean global fuel: the status of DME developments in China and beyond. J Nat Gas Sci Eng 9:94–107. https://doi.org/10.1016/j.jngse.2012.05.012

Fona (2016) Carbon2Chem. https://www.fona.de/en/measures/funding-measures/carbon2chem-project.php. Accessed 24 May 2020

Fresme project (2017) From residual steel gases to methanol. http://www.fresme.eu/. Accessed 25 May 2020

Friis M, Blue World Technologies (2019) e-fuels, methanol and fuel cells. Available from https://brintbranchen.dk/wp-content/uploads/2019/04/E-fuels-methanol-and-fuel-cells-Mads-Friis-Jensen.pdf. Accessed 24 May 2020

Fukuoka S, Tojo M, Hachiya H, Aminaka M, Hasegawa K (2007) Green and sustainable chemistry in practice: development and industrialization of a novel process for polycarbonate production from CO_2 without using phosgene. Polym J 39:91–114. https://doi.org/10.1295/polymj.PJ2006140

Ganesh I (2014) Conversion of carbon dioxide into methanol—a potential liquid fuel: fundamental challenges and opportunities (a review). Renew Sustain Energ Rev 31:221–257. https://doi.org/10.1016/j.rser.2013.11.045

Gao J, Liu Q, Gu F, Liu B, Zhong Z, Su F (2015) Recent advances in methanation catalysts for the production of synthetic natural gas. RSC Adv 5:22759–22776. https://doi.org/10.1039/C4RA16114A

Ghaib K, Ben-Fares F-Z (2018) Power-to-methane: a state-of-the-art review. Renew Sustain Energ Rev 81:433–446. https://doi.org/10.1016/j.rser.2017.08.004

Global petrol prices (2020) Ethanol prices, liter. https://www.globalpetrolprices.com/ethanol_prices/. Accessed 15 June 2020

Store & Go (2020) Store & Go project. https://www.storeandgo.info/about-the-project/. Accessed 15 May 2020

Goeppert A, Czaun M, Jones JP, Surya Prakash GK, Olah GA (2014) Recycling of carbon dioxide to methanol and derived products—closing the loop. Chem Soc Rev 43:7995–8048. https://doi.org/10.1039/c4cs00122b

Götz M et al (2016) Renewable power-to-gas: a technological and economic review. Renew Energ 85:1371–1390. https://doi.org/10.1016/j.renene.2015.07.066

Guilera J, Andreu T, Basset N, Boeltken T, Timm F, Mallol I, Morante JR (2020) Synthetic natural gas production from biogas in a waste water treatment plant. Renew Energ 146:1301–1308. https://doi.org/10.1016/j.renene.2019.07.044

Gunasekar GH, Park K, Jung K-D, Yoon S (2016) Recent developments in the catalytic hydrogenation of CO_2 to formic acid/formate using heterogeneous catalysts. Inorg Chem Frontiers 3:882–895. https://doi.org/10.1039/C5QI00231A

Haas T, Krause R, Weber R, Demler M, Schmid G (2018) Technical photosynthesis involving CO_2 electrolysis and fermentation. Nat Catal 1:32–39. https://doi.org/10.1038/s41929-017-0005-1

Hafenbradl D (2017) Decarbonize and store renewable power in the gas grid: demonstrated by electrochaea's power-to-methane project In:

Electrochaea press release. Available from http://www.electrochaea.com/wp-content/uploads/2020/02/PR_Electrochaeas-power-to-methane-technology-delivers-excellent-project-results.pdf. Accessed 15 May 2020

Haldor-Topsoe (2020) Produce your own carbon monoxide. https://www.topsoe.com/processes/carbon-monoxide. Accessed 16 June 2020

van der Ham LGJ, van der Berg H, Benneker A, Simmelink G, Timmer J, van Weerden S (2012) Hydrogenation of carbon dioxide for methanol production. Chem Eng Trans 29:181–186. https://doi.org/10.3303/CET1229031

Handler RM, Shonnard DR, Griffing EM, Lai A, Palou-Rivera I (2016) Life cycle assessments of ethanol production via gas fermentation: anticipated greenhouse gas emissions for cellulosic and waste gas feedstocks. Ind Eng Chem Res 55:3253–3261. https://doi.org/10.1021/acs.iecr.5b03215

Hobson C, Márquez C (2018) Renewable methanol report. Available from https://www.methanol.org/wp-content/uploads/2019/01/MethanolReport.pdf. Accessed 24 May 2020

Höpfner S (2017) Audi steps up research into synthetic fuels. In: Audi media center https://www.audi-mediacenter.com/en/press-releases/audi-steps-up-research-into-synthetic-fuels-9546. Accessed 30 May 2020

Hu L, Urakawa A (2018) Continuous CO_2 capture and reduction in one process: CO_2 methanation over unpromoted and promoted Ni/ZrO_2. J CO_2 Utilization 25:323−329. https://doi.org/10.1016/j.jcou.2018.03.013

Huang C-H (2014) A review: CO_2 utilization. Aerosol Air Qual Res 14:480−499.https://doi.org/10.4209/aaqr.2013.10.0326

Huang Y, Deng Y, Handoko AD, Goh GKL, Yeo BS (2018) Rational design of sulfur-doped copper catalysts for the selective electroreduction of carbon dioxide to formate. Chemsuschem 11:320–326. https://doi.org/10.1002/cssc.201701314

HYSYTECH (2020). CELBICON—the CO_2 reduction unit successfully closes the project. https://hysytech.com/News/celbicon-chiusura-progetto-eng. Accessed 17 June 2020

Institut fur Energietechnik (IET) (2017) Pilot and demonstration plant power-to-methane. https://www.iet.hsr.ch/index.php?id=13510&L=4. Accessed 15 May 2020

Methanol Institute (2020) https://www.methanol.org. Accessed 30 May 2020

International Energy Agency (IEA) (2019) Putting CO_2 to Use Creating value from emissions. IEA Technology Report. Available from https://ccsknowledge.com/pub/Publications/2019Sep_IEA_Putting_CO2_To_Use.pdf. Accessed 30 June 2020

Jadhav SG, Vaidya PD, Bhanage BM, Joshi JB (2014) Catalytic carbon dioxide hydrogenation to methanol: a review of recent studies. Chem Eng Res Des 92:2557–2567. https://doi.org/10.1016/j.cherd.2014.03.005

Jarvis SM, Samsatli S (2018) Technologies and infrastructures underpinning future CO_2 value chains: a comprehensive review and comparative analysis. Renew Sustain Energ Rev 85:46–68. https://doi.org/10.1016/j.rser.2018.01.007

Jeanty P, Scherer C, Magori E, Wiesner-Fleischer K, Hinrichsen O, Fleischer M (2018) Upscaling and continuous operation of electrochemical CO_2 to CO conversion in aqueous solutions on silver gas diffusion electrodes. J CO_2 Utilization 24:454−462. https://doi.org/10.1016/j.jcou.2018.01.011

Joo O-S, Jung K-D, Moon I, Rozovskii AY, Lin GI, Han S-H, Uhm S-J (1999) Carbon dioxide hydrogenation to form methanol via a reverse-water-gas-shift reaction (the CAMERE Process). Ind Eng Chem Res 38:1808–1812. https://doi.org/10.1021/ie9806848

Jupiter 1000 (2019) Jupiter 1000 project. https://www.jupiter1000.eu/english. Accessed 15 May 2020

Kanacher J. Innogy SE (2017) Greenfuel—global solution for sustainable national energy systems. In: Hydrogen net annual meeting. Available from https://brintbranchen.dk/wp-content/uploads/2017/05/Jens-Kanacher.pdf. Accessed 20 May 2020

Keim W (1989) Carbon monoxide: feestock for chemicals, present and future. J Organomet Chem 372:15–23. https://doi.org/10.1016/0022-328X(89)87071-8

Kim SM et al (2017) Cooperativity and dynamics increase the performance of NiFe dry reforming catalysts. J Am Chem Soc 139:1937–1949. https://doi.org/10.1021/jacs.6b11487

Kiss AA, Pragt JJ, Vos HJ, Bargeman G, de Groot MT (2016) Novel efficient process for methanol synthesis by CO_2 hydrogenation. Chem Eng J 284:260–269. https://doi.org/10.1016/j.cej.2015.08.101

Knowledge Energy Institute (KEI) (2019) Total plans hydrogen technology partnership with Sunfire. https://knowledge.energyinst.org/search/record?id=112830. Accessed 20 May 2020

Kondratenko EV, Mul G, Baltrusaitis J, Larrazábal GO, Pérez-Ramírez J (2013) Status and perspectives of CO_2 conversion into fuels and chemicals by catalytic, photocatalytic and electrocatalytic processes. Energ Environ Sci 6:3112–3135. https://doi.org/10.1039/C3EE41272E

Korosec K (2016) Ford is using factory emissions to make car parts. http://fortune.com/2016/05/16/bill-ford-carbon-emissions/. Accessed 30 June 2020

Kourkoumpas DS, Papadimou E, Atsonios K, Karellas S, Grammelis P, Kakaras E (2016) Implementation of the power to methanol concept by using CO_2 from lignite power plants: techno-economic investigation. Int J Hydrogen Energ 41:16674–16687. https://doi.org/10.1016/j.ijhydene.2016.07.100

Kutz RB, Chen Q, Yang H, Sajjad SD, Liu Z, Masel IR (2017) Sustainion imidazolium-functionalized polymers for carbon dioxide electrolysis. Energ Technol 5:929–936. https://doi.org/10.1002/ente.201600636

Lane J (2015) LanzaTech: biofuels digest's 2015 5-minute guide 2015. In: Biofuels Digest's. http://www.biofuelsdigest.com/bdigest/2015/01/13/lanzatech-biofuels-digests-2015-5-minute-guide/. Accessed 30 May 2020

Langanke J, Wolf A, Peters M (2015) Chapter 5—Polymers from CO_2—an industrial perspective. In: Styring P, Quadrelli EA, Armstrong K (eds) Carbon dioxide utilisation. Elsevier, Amsterdam, pp 59−71. https://doi.org/10.1016/B978-0-444-62746-9.00005-0

Lavoie J-M (2014) Review on dry reforming of methane, a potentially more environmentally-friendly approach to the increasing natural gas exploitation. Front Chem 2(81):1–17. https://doi.org/10.3389/fchem.2014.00081

Li W, Wang H, Jiang X, Zhu J, Liu Z, Guo X, Song C (2018) A short review of recent advances in CO_2 hydrogenation to hydrocarbons over heterogeneous catalysts. RSC Adv 8:7651–7669. https://doi.org/10.1039/C7RA13546G

Liew F, Martin ME, Tappel RC, Heijstra BD, Mihalcea C, Köpke M (2016) Gas fermentation-A flexible platform for commercial scale production of low-carbon-fuels and chemicals from waste and renewable feedstocks. Front Microbiol 7:694. https://doi.org/10.3389/fmicb.2016.00694

The Linde Group (2020) Linde-engineering. Innovative dry reforming process. Successful catalyst upscaling. https://www.linde-engineering.com/en/innovations/innovate-dry-reforming/index.html. Accessed 15 June 2020

Liu M, Yi Y, Wang L, Guo H, Bogaerts A (2019) Hydrogenation of carbon dioxide to value-added chemicals by heterogeneous catalysis and plasma catalysis. Catalysts 9:275. https://doi.org/10.3390/catal9030275

Liu Z et al (2016) Electrochemical generation of syngas from water and carbon dioxide at industrially important rates. J CO_2 Utilization 15:50−56. https://doi.org/10.1016/j.jcou.2016.04.011

Lu X-B, Darensbourg DJ (2012) Cobalt catalysts for the coupling of CO_2 and epoxides to provide polycarbonates and cyclic carbonates. Chem Soc Rev 41:1462–1484. https://doi.org/10.1039/C1CS15142H

Ma J, Sun N, Zhang X, Zhao N, Xiao F, Sun Y (2009) A short review of catalysis for CO_2 conversion. Catal Today 148:221–231. https://doi.org/10.1016/j.cattod.2009.08.015

Made in China (2020) 99% liquid market formic acid price. https://hnwinchem8866.en.made-in-china.com/product/cSoxHBkznqph/China-99-Liquid-Market-Formic-Acid-Price.html. Accessed 17 June 2020

Mantra (2014) Mantra officially launches pilot plant project. In: Globe News Wire. https://www.globenewswire.com/news-release/2014/03/25/1395013/0/en/Mantra-Officially-Launches-Pilot-Plant-Project.html. Accessed 17 June 2020

Mantra (2015) Mantra releases update on demonstration projects. In: Globe News Wire. https://www.globenewswire.com/news-release/2015/02/03/1395085/0/en/Mantra-Releases-Update-on-Demonstration-Projects.html. Accessed 17 June 2020

de Medeiros EM, Noorman H, Maciel Filho R, Posada JA (2020) Production of ethanol fuel via syngas fermentation: optimization of economic performance and energy efficiency. Chem Eng Sci X 5:100056. https://doi.org/10.1016/j.cesx.2020.100056

MefCO_2 (2020) MefCO_2 project. http://www.mefco2.eu/mefco2.php. Accessed 20 May 2020

Methanex (2020) Methanex posts regional contract methanol prices for North America, Europe and Asia. Methanex European Posted Contract Price. Posted 27 March 2020. https://www.methanex.com/our-business/pricing. Accessed 29 May 2020

Methanol Institute (2016) Methanol facts. DME: an emerging global fuel. Available from http://www.methanol.org/wp-content/uploads/2016/06/DME-An-Emerging-Global-Guel-FS.pdf. Accessed 30 June 2020

Michailos S, McCord S, Sick V, Stokes G, Styring P (2019) Dimethyl ether synthesis via captured CO_2 hydrogenation within the power to liquids concept: a techno-economic assessment. Energ Convers Manag 184:262–276. https://doi.org/10.1016/j.enconman.2019.01.046

Mineral Carbonation International (MCi) (2020) Mineral Carbonation International technology. http://mineralcarbonation.com/. Accessed June 2020

Mittal C, Hadsbjerg C, Blennow P (2017) Small-scale CO from CO_2 using electrolysis. Chem Eng World 52:44–46

Mondal K, Sasmal S, Badgandi S, Chowdhury DR, Nair V (2016) Dry reforming of methane to syngas: a potential alternative process for value added chemicals—a techno-economic perspective. Environ Sci Pollut Res 23:22267–22273. https://doi.org/10.1007/s11356-016-6310-4

Monkman S, MacDonald M (2017) On carbon dioxide utilization as a means to improve the sustainability of ready-mixed concrete. J Clean Prod 167:365–375. https://doi.org/10.1016/j.jclepro.2017.08.194

Monkman S, MacDonald M, Hooton RD, Sandberg P (2016) Properties and durability of concrete produced using CO_2 as an accelerating admixture. Cement Concr Compos 74:218–224. https://doi.org/10.1016/j.cemconcomp.2016.10.007

Müller KR, Rachow F, Israel J, Charlafti E, Schwiertz C, Scmeisser D (2017) Direct methanation of flue gas at a lignite power plant. Int J Environ Sci 2:425−437 ISSN 2367-8941

Mutz B, Carvalho HWP, Mangold S, Kleist W, Grunwaldt J-D (2015) Methanation of CO_2: structural response of a Ni-based catalyst under fluctuating reaction conditions unraveled by operando spectroscopy. J Catal 327:48–53. https://doi.org/10.1016/j.jcat.2015.04.006

Nagengast J (2017) Power-to-Gas-Anlagen von Electrochaea. In: Energy load. https://energyload.eu/stromspeicher/power-to-gas/power-to-gas-electrochaea/. Accessed 15 May 2020

Naik SP, Ryu T, Bui V, Miller JD, Drinnan NB, Zmierczak W (2011) Synthesis of DME from CO_2/H_2 gas mixture. Chem Eng J 167:362–368. https://doi.org/10.1016/j.cej.2010.12.087

Naturgy (2020) RIS3CAT CoSin: synthetic fuels. https://www.naturgy.com/en/get_to_know_us/innovation_and_the_future/technology_innovation_plan/ris3cat_cosin_synthetic_fuels. Accessed 14 May 2020

Navarro JC, Centeno MA, Laguna OH, Odriozola JA (2018) Policies and motivations for the CO_2 valorization through the sabatier reaction using structured catalysts. A review of the most recent advances. Catalysts 8(12):578. https://doi.org/10.3390/catal8120578

Neuling U, Kaltschmitt M (2019) GreenPower2Jet—PtL-demonstration in Northern Germany. https://www.tuhh.de/alt/iue/research/projects/greenpower2jet.html. Accessed 10 June 2020

Bio-based News (2019) Industrial-scale power-to-liquid plant planned. http://news.bio-based.eu/industrial-scale-power-to-liquid-plant-planned/. Accessed 10 June 2020

News Green Hydrogen (2019) Total, Sunfire partner on hydrogen tech for methanol production. Fuel Cells Bull 11:12.https://doi.org/10.1016/S1464-2859(19)30480-8

Novomer (2020) What's new at Novomer. https://www.novomer.com/news. Accessed 30 June 2020

O'Connell A, Konti A, Padella M, Prussi M, Lonza L (2019) Advanced alternative fuels technology market report.https://doi.org/10.2760/894775

Olah GA, Goeppert A, Prakash GKS (2009) Chemical recycling of carbon dioxide to methanol and dimethyl ether: from greenhouse gas to renewable, environmentally carbon neutral fuels and synthetic hydrocarbons. J Org Chem 74:487–498. https://doi.org/10.1021/jo801260f

Onishi N, Laurenczy G, Beller M, Himeda Y (2018) Recent progress for reversible homogeneous catalytic hydrogen storage in formic acid and in methanol. Coord Chem Rev 373:317–332. https://doi.org/10.1016/j.ccr.2017.11.021

O'Reilly C (2020) Power to methanol project launched. Hydrocarbon Engineering. https://www.hydrocarbonengineering.com/clean-fuels/11052020/power-to-methanol-project-launched/. Accessed 11 May 2020

Ou X, Zhang X, Zhang Q, Zhang X (2013) Life-cycle analysis of energy use and greenhouse gas emissions of gas-to-liquid fuel pathway from steel mill off-gas in China by the LanzaTech process. Frontiers Energ 7:263–270. https://doi.org/10.1007/s11708-013-0263-9

Panzone C, Philippe R, Chappaz A, Fongarland P, Bengaouer A (2020) Power-to-liquid catalytic CO_2 valorization into fuels and chemicals: focus on the Fischer-Tropsch route. J CO_2 Utilization 38:314−347. https://doi.org/10.1016/j.jcou.2020.02.009

Patricio J, Angelis-Dimakis A, Castillo-Castillo A, Kalmykova Y, Rosado L (2017) Method to identify opportunities for CCU at regional level—matching sources and receivers. J CO_2 Utilization 22:330−345. https://doi.org/10.1016/j.jcou.2017.10.009

Pérez-Fortes M, Tzimas E (2016) Techno-economic and environmental evaluation of carbon dioxide utilisation for fuel production. Synthesis of methanol and formic acid. Technical report, European Commission: JRC Science for Policy Report, 2016. EUR 27629 EN. https://doi.org/10.2790/981669

Pérez-Fortes M, Schöneberger JC, Boulamanti A, Harrison G, Tzimas E (2016a) Formic acid synthesis using CO_2 as raw material: techno-economic and environmental evaluation and market potential. Int J Hydrogen Energ 41:16444–16462. https://doi.org/10.1016/j.ijhydene.2016.05.199

Pérez-Fortes M, Schöneberger JC, Boulamanti A, Tzimas E (2016b) Methanol synthesis using captured CO_2 as raw material: techno-economic and environmental assessment. Appl Energ 161:718–732. https://doi.org/10.1016/j.apenergy.2015.07.067

Poland SJ, Darensbourg DJ (2017) A quest for polycarbonates provided via sustainable epoxide/CO_2 copolymerization processes. Green Chem 19:4990–5011. https://doi.org/10.1039/C7GC02560B

RWE Power (2019) Making fuels from CO_2: RWE unveils new synthesis pilot plant in Germany. https://www.alignccus.eu/news/making-fuels-co2-rwe-unveils-new-synthesis-pilot-plant-germany. Accessed 30 June 2020

Prabowo B, Yan M, Syamsiro M, Hendroko R, Biddinika M (2017) State of the art of global dimethyl ether production and it's potentional application in Indonesia. Proc Pakistan Acad Sci 54:29–39

Prieto G (2017) Carbon dioxide hydrogenation into higher hydrocarbons and oxygenates: thermodynamic and kinetic bounds and progress with heterogeneous and homogeneous catalysis. Chemsuschem 10:1056–1070. https://doi.org/10.1002/cssc.201601591

Prüf-und Forschungsinstitut (PFI) Germany (2020) Biorefinery at Pirmasens-Winzeln Energy Park. https://www.pfi-germany.de/en/research/biorefinery-at-pirmasens-winzeln-energy-park/. Accessed 15 May 2020

Qiao J, Liu Y, Hong F, Zhang J (2014) A review of catalysts for the electroreduction of carbon dioxide to produce low-carbon fuels. Chem Soc Rev 43:631–675. https://doi.org/10.1039/C3CS60323G

Qin Y, Sheng X, Liu S, Ren G, Wang X, Wang F (2015) Recent advances in carbon dioxide based copolymers. J CO_2 Utilization 11:3−9. https://doi.org/10.1016/j.jcou.2014.10.003

Quadrelli EA, Centi G, Duplan J-L, Perathoner S (2011) Carbon dioxide recycling: emerging large-scale technologies with industrial potential. Chemsuschem 4:1194–1215. https://doi.org/10.1002/cssc.201100473

Ravn S, Haldor-Topsoe (2019) Topsoe to build demonstration plant to produce cost competitive CO_2-neutral methanol from biogas and green electricity. https://blog.topsoe.com/topsoe-to-build-demonstration-plant-to-produce-cost-competitive-co2-neutral-methanol-from-biogas-and-green-electricity. Accessed 24 May 2020

Rego de Vasconcelos B, Lavoie J-M (2019) Recent advances in power-to-X technology for the production of fuels and chemicals. Front Chem 7:392. https://doi.org/10.3389/fchem.2019.00392

Renewable Fuel Association (RFA) (2020). Ethanol markets and statistics. https://ethanolrfa.org/statistics/. Accessed 15 June 2020

Robledo-Díez A (2012). Life cycle assessment on the conversion of CO_2 to formic acid. Master thesis. Norwegian University of Science and Technology. http://brage.bibsys.no/xmlui/handle/11250/234844. Accessed 17 June 2020

Rodemerck U, Holeňa M, Wagner E, Smejkal Q, Barkschat A, Baerns M (2013) Catalyst development for CO_2 hydrogenation to fuels. ChemCatChem 5:1948–1955. https://doi.org/10.1002/cctc.201200879

Saeidi S, Amin NAS, Rahimpour MR (2014) Hydrogenation of CO_2 to value-added products—a review and potential future developments. J CO_2 Utilization 5:66−81. https://doi.org/10.1016/j.jcou.2013.12.005

Sakakura T, Choi J-C, Yasuda H (2007) Transformation of carbon dioxide. Chem Rev 107(6):2365–2387. https://doi.org/10.1021/cr068357u

Sankaranarayanan S, Srinivasan K (2012) Carbon dioxide—a potential raw material for the production of fuel, fuel additives and bio-derived chemicals. Indian J Chem Sect A-Inorg Bio-Inorg Phys Theor Anal Chem 51:1252–1262

Sanna A, Uibu M, Caramanna G, Kuusik R, Maroto-Valer MM (2014) A review of mineral carbonation technologies to sequester CO_2. Chem Soc Rev 43:8049–8080. https://doi.org/10.1039/C4CS00035H

Schaaf T, Grünig J, Schuster M, Rothenfluh T, Orth A (2014) Methanation of CO_2—storage of renewable energy in a gas distribution system. Energy Sustain Soc 4:1–14. https://doi.org/10.1186/s13705-014-0029-1

Schmidt P, Weindorf W, Roth A, Batteiger V, Riegel F (2016) Power to-liquids–potentials and perspectives for the future supply of renewable aviation fuel. German Environment Agency, Dessau-Roßlau. http://www.lbst.de/news/2016_docs/161005_uba_hintergrund_ptl_barrierrefrei.pdf. Accessed 30 June 2020

Schwab E, Milanov A, Schunk SA, Behrens A, Schödel N (2015) Dry reforming and reverse water gas shift: alternatives for syngas production? Chem Ing Tec 87:347–353. https://doi.org/10.1002/cite.201400111

Schweitzer C, BSE Engineering (2017) Small scale methanol plants: a chance for re-industrialisation. In: The international methanol conference, Copenhagen, Demark, 8–10 May 2017. Available from http://www.methanol.org/wp-content/uploads/2018/04/Small-Scale-Methanol-Plants-Christian-Schweitzer_bse-Engineering.pdf. Accessed 24 May 2020

Sekisui Chemical Co., LTD, LanzaTech (2017) From trash to tank: upcycling of landfill to fuel demonstrated in Japan. Sekisui Chemical Co., LTD. and LanzaTech announce breakthrough in conversion of municipal solid waste to ethanol. Available from http://carbonrecycling.net/wp-content/uploads/2018/05/sekisui-lanzatech-release-dec62017.pdf. Accessed 30 May 2020

Sherrard A (2017) Sunfire to build 8000 tonne-per-annum power-to liquid facility in Norway. In: Bioenergy international. https://bioenergyinternational.com/biofuels-oils/sunfire-build-8-000-tonne-per-annum-power-liquid-facility-norway. Accessed 20 June 2020

Singh AK, Singh S, Kumar A (2016) Hydrogen energy future with formic acid: a renewable chemical hydrogen storage system. Catal Sci Technol 6:12–40. https://doi.org/10.1039/C5CY01276G

SolarServer (2020) Industrielle Power-to-Gas-Anlage in der Schweiz. https://www.solarserver.de/2020/02/05/industrielle-power-to-gas-anlage-in-der-schweiz/. Accessed 15 May 2020

Soletair (2020) Soletair project. https://soletair.fi/. Accessed 20 June 2020

Solidia (2020) Solidia technology. https://solidiatech.com. Accessed 30 June 2020

Søren K et al (2020) Power2Met—renewable energy to green methanol. https://vbn.aau.dk/en/projects/power2met-renewable-energy-to-green-methanol. Accessed 24 May 2020

Spire 2030 (2020) MefCO_2 (Methanol fuel from CO_2)—synthesis of methanol from captured carbon dioxide using surplus electricity. https://www.spire2030.eu/mefco2. Accessed 20 May 2020

State of Green (2019) Denmark funds new Power-to-X flag-ship projects. https://stateofgreen.com/en/partners/state-of-green/news/denmark-funds-new-power-to-x-flagship-projects/. Accessed 15 May 2020

Steleanol Project (2019) Steelanol fueling a sustainable future. http://www.steelanol.eu/en. Accessed 30 May 2020

Swiss Liquid Future AG (2016) Technical data of a standardised SLF 15 Swiss Liquid Future®-system. https://www.swiss-liquid-future.ch/technology/?lang=en. Accessed 24 May 2020

Swisspower (2020) Swisspower factsheet. Available from https://www.swisspower.ch/content/files/publications/Factsheets-PDF/factsheet_limeco_online.pdf. Accessed 15 May 2020

Taylor I (2020) Europe: consortium set to build 'Power to Methanol' plant in Antwerp. In: Europe News. Bunkerspot. https://www.bunkerspot.com/europe/50472-europe-consortium-set-to-build-power-to-methanol-plant-in-antwerp. Accessed 26 May 2020

Technical University of Denmark (DTU) (2013) SYMBIO—Biogasupgade. http://www.biogasupgrade.dk/. Accessed 15 May 2020
Econic Technologies (2018). Press release: UK's first carbon capture utilisation demonstration plant opens its doors. http://econic-technologies.com/news/uk-first-ccu-demo-plant/. Accessed 30 June 2020
Newlight Technologies (2020) Technology/AirCarbon™/News. https://www.newlight.com/. Accessed 30 June 2020
The Linde Group (2015) Linde develops a new production process for synthesis gas. http://www.linde-engineering.com/en/news_and_media/press_releases/news_20151015.html. Accessed 15 June 2020
Thema M, Bauer F, Sterner M (2019) Power-to-gas: electrolysis and methanation status review. Renew Sustain Energ Rev 112:775–787. https://doi.org/10.1016/j.rser.2019.06.030
TOTAL (2019) Integrating climate into our strategy. Available from https://www.total.com/sites/g/files/nytnzq111/files/atoms/files/total_rapport_climat_2019_en.pdf . Accessed 26 May 2020
Underground Sun Conversion (2020) Underground sun conversion project description. https://www.underground-sun-conversion.at/en/project/project-description.html. Accessed 15 May 2020
Carbon Upcycling (2020) Turning carbon dioxide into CO2NCRETE™. http://www.co2upcycling.com/. Accessed 30 June 2020
Hydrogen Valley (2020) Power2Met Project. http://hydrogenvalley.dk/power2met-en/. Accessed 24 May 2020
Det Norske Veritas (DNV) (2007) Carbon dioxide utilization: electrochemical conversion of CO_2—opportunities and challenges. Position Paper 07. https://issuu.com/dnv.com/docs/dnv-position_paper_co2_utilization. Accessed 17 June 2020
Vlap H, van der Steen A, Knijp J, Holstein J, Grond L (2015) Power-to-gas project in Rozenburg, The Netherlands. In: Technical assumptions, technology demonstration and results P2G project. DNV GL Oil & Gas. Available from https://projecten.topsectorenergie.nl/storage/app/uploads/public/5b7/54a/624/5b754a624582f744864490.pdf. Accessed 15 May 2020
Volkova M (2019) Sinopec SABIC Tianjin Petrochem begins expansion of ethylene capacity in China. http://www.mrcplast.com/news-news_open-360523.html. Accessed 30 June 2020
Voltachem (2019) Pilot for synthesis of formic acid from CO_2 at Twence waste incineration site. www.voltachem.nl/news/pilot-for-synthesis-of-formic-acid-from-co2-at-twence-waste-incineration-site. Accessed 17 June 2020
Wang S, Mao D, Guo X, Wu G, Lu G (2009) Dimethyl ether synthesis via CO_2 hydrogenation over $CuO–TiO_2–ZrO_2HZSM$-5 bifunctional catalysts. Catal Commun 10:1367–1370. https://doi.org/10.1016/j.catcom.2009.02.001
Wang W, Wang S, Ma X, Gong J (2011) Recent advances in catalytic hydrogenation of carbon dioxide. Chem Soc Rev 40:3703–3727. https://doi.org/10.1039/C1CS15008A
Wei J et al (2017) Directly converting CO_2 into a gasoline fuel. Nat Commun 8:15174. https://doi.org/10.1038/ncomms15174
Wei J et al (2018) Catalytic hydrogenation of CO_2 to isoparaffins over Fe-based multifunctional catalysts. ACS Catal 8:9958–9967. https://doi.org/10.1021/acscatal.8b02267
Whipple DT, Kenis PJA (2010) Prospects of CO_2 utilization via direct heterogeneous electrochemical reduction. J Phys Chem Lett 1:3451–3458. https://doi.org/10.1021/jz1012627
Wiede T, Land A (2018) Sektorenkopplung: amprion und Open Grid Europe geben Power-to-Gas in Deutschland einen Schub. Berlin. https://www.amprion.net/Presse/Presse-Detailseite_14983.html#:~:text=20.06.2018-,Sektorenkopplung%3A%20Amprion%20und%20Open%20Grid%20Europe%20geben%20Power%2Dto%2D,Tempo%20in%20der%20Energiewende%20machen. Accessed 15 May 2020
Wilson G, Travaly Y, Brun T, Knippels H, Armstrong K, Styring P, Krämer D, Saussez G, Bolscher H (2015) A vision for smart CO_2 transformation in Europe. Technical report, the smart CO_2 transformation (SCOT) Project. Available from http://www.scotproject.org/images/SCOT%20Vision.pdf. Accessed 28 May 2020
Winther Mortensen A, Wenzel H, Dalgas Rasmussen K, Sandermann Justesen S, Wormslev E, Porsgaard M (2019) Nordic GTL—a pre-feasibility study on sustainable aviation fuel from biogas, hydrogen and CO_2. Available from https://www.nordicenergy.org/wp-content/uploads/2019/10/Nordic-aviation-fuel-production-28-10-2019-final.pdf. Accessed 15 May 2020
Wulf C, Linßen J, Zapp P (2018) Review of power-to-gas projects in Europe. Energ Procedia 155:367–378. https://doi.org/10.1016/j.egypro.2018.11.041
Yan R, Guo H, Yang D, Qiu S, Yao X (2014) Catalytic hydrogenation of carbon dioxide to fuels. Curr Org Chem 18:1335–1345. https://doi.org/10.2174/1385272819666140424212948
YCharts (2020) European Union natural gas import price. https://ycharts.com/indicators/europe_natural_gas_price. Accessed 22 May 2020
Ye R-P et al (2019) CO_2 hydrogenation to high-value products via heterogeneous catalysis. Nat Commun 10:5698. https://doi.org/10.1038/s41467-019-13638-9
Yotaro O, Masahiro Y, Tsutomu S, Osamu I, Takashi O, Norio I (2006) New direct synthesis technology for DME (Dimethyl Ether) and its application technology. JFE technical report No. 8. Available from https://www.jfe-steel.co.jp/en/research/report/008/pdf/008-06.pdf. Accessed 30 June 2020
Zentrum für Sonnenenergie-und Wasserstoff-Forschung Baden-Württemberg (ZSW) (2020) P2G 250. https://www.zsw-bw.de/en/projects/regenerative-kraftstoffe/ptg-250-p2gr.html. Accessed 14 May 2020
Zhang H, Wang L, Van Herle J, Maréchal F, Desideri U (2019) Techno-economic optimization of CO_2-to-methanol with solid-oxide electrolyzer. Energies 12:3742. https://doi.org/10.3390/en12193742
Zhu Q (2019) Developments on CO_2-utilization technologies. Clean Energ 3:85–100. https://doi.org/10.1093/ce/zkz008
Zitscher T, Drünert S, Kaltschmitt M (2019) PowerFuel—demonstration and potential analysis of new technologies for sector coupling for the production of synthetic fuel from carbon dioxide. https://www.tuhh.de/alt/iue/research/projects/powerfuel.html. Accessed 20 June 2020

Progress in Catalysts for CO_2 Reforming

Maria do Carmo Rangel

Abstract

Since the first catalyst proposed for CO_2 reforming as early as 1928 by Fischer and Tropsch, the publications on the reaction and on catalysts as well as on new applications and technologies have increased continuously. Over the years, new catalysts with several formulations even more complex have been developed, for different reaction conditions and for several purposes and applications. Moreover, the most promising formulations have been improved by using new supports and promoters. Also, the knowledge on kinetics and mechanism as well as on the action of supports and promoters has opened opportunities for designing new catalysts for the reaction. In this chapter, the progress on the catalysts for dry reforming was discussed over the years, focusing the advances on formulations and on the action of the catalysts at different conditions for several applications.

1 Introduction

In the last decades, mankind has faced even more negative impacts of pollution on the environment as well as on the biota. In fact, the frequent discharges of pollutants both in the water bodies and in the air have greatly contributed to the water scarcity and to the poor quality of air in several urban centers (Rohi et al. 2020; Britto and Rangel 2008; Shah and Shah 2020; Rangel and Carvalho 2003). Over the years, the continuous pollution achieved dangerous levels causing alarming consequences such as global warming and climate changes (Bong et al. 2017; Töbelmann and Wendler 2020).

These pollutants are closely related to anthropogenic activities, mainly those involving the ongoing combustion of fossil fuels including natural gas and coal as well as light/heavy crude oils during industrial, agricultural, and domestic activities. Since pollution causes a severe threat to the Earth's sustainability, many policies and legislations have been adopted by governmental agencies to avoid or to decrease emissions. During the Climate Change Conference 2015 hold in Paris, for instance, it had been agreed that the greenhouse gas emissions should decrease to levels enough to limit the increase of Earth temperature to values lower than 2 °C of the temperature before industrialization (UNFCCC, COP21 2015). However, according to the BP Energy Outlook (BP Energy Outlook 2019) the carbon dioxide emissions from combustion will continuously increase in the next years achieving around 10% by 2040. Based on this projection, the Intergovernmental Panel on Climate Change recently recommended to limit the global temperature increase to less than 1.5 °C (Global Warming and of 1.5 °C Report).

Although the greenhouse effect has been assigned mainly to methane and carbon dioxide (Kischke et al. 2013), it is known (Baird 1995) that methane is a greenhouse gas much more powerful than carbon dioxide. However, carbon dioxide can remain in the atmosphere for a longer time than methane, and then its amount becomes much higher. It has been reported (Bong et al. 2017) that the current amount of carbon dioxide can remain up to 10,000 years in the atmosphere causing climate changes. In addition, carbon dioxide emissions increase each year and reached 33.1 Gt (an increase of 1.7%) in 2018, hitting a new record because of higher energy consumption, as reported by International Energy Agency (IEA International Energy Agency 2018). Therefore, carbon dioxide has been considered the most important greenhouse gas, being responsible for the highest percentage of global warming (60%). This pushes all the countries to replace fossil with renewable energy and to look for technologies for the capture, storage, and use of carbon

M. do Carmo Rangel (✉)
Instituto de Química, Universidade Federal do Rio Grande do Sul, Av. Bento Gonçalves, 9500, Agronomia, Porto Alegre, Rio Grande do Sul 90650-001, Brazil
e-mail: maria.rangel@ufrgs.br

Inamuddin et al. (eds.), *Carbon Dioxide Utilization to Sustainable Energy and Fuels*, Advances in Science, Technology & Innovation, https://doi.org/10.1007/978-3-030-72877-9_2

dioxide (Mustafa et al. 2020; Saadabadi et al. 2019; Araújo et al. 2008; Ahmed et al. 2020; Galadima and Muraza 2019). Regarding electrical power, the main sources are still fossil fuels whose use is expected to increase in the next thirty years, achieving triple of the current values before 2100. Therefore, the scientific community has been addressed to find out new technologies to use the abundant carbon dioxide coming from fossil fuels (Mustafa et al. 2020; Galadima and Muraza 2019; Zhang et al. 2020). However, this largely depends on the technologies for capturing and stocking carbon dioxide.

2 Technologies for Capturing and Storing Carbon Dioxide

The technologies for carbon dioxide capture and storage (CCS) involve the trapping of this gas from emission units (stationary and mobile), transportation, and disposal into geological formations or other places (Ahmed et al. 2020; Miranda-Barbosa et al. 2017). Different methods were proposed for removing carbon dioxide gas from outlet gases at power stations, including absorption and adsorption (Keller et al. 2018). The use of amine scrubbing for absorbing carbon dioxide from flue gas is a common method frequently used in powerhouses. Nevertheless, this technology is limited by the corroding of pieces, the discharge of toxics, and by the high energy consumption during regeneration (Tian et al. 2018). To overcome these drawbacks, several technologies have been proposed such as starburst, oxyfuel combustion, and after-combustion combined with techniques of separation including absorption, adsorption as well as membranes. Each one has advantages and disadvantages for commercial applications. The absorption based in solvents, for instance, is recognized as an efficient method but requires high amounts of energy, and several dangerous gases are discharged during the solvent recuperation. This has motivated the research for adsorbents in solid phase which require less energy, are eco-friendly, and absorb faster than liquid absorbents. In addition, they remain unchanged in severe conditions (Babu et al. 2013). Cost-effective and environment friendly adsorbents such as porous carbons, metal-organic frameworks (MOF), and zeolites, among others, have also been considered. They usually have high specific surface areas and are resistant against water and chemicals. The renewable carbon-based materials are especially useful for capturing carbon dioxide due to the high specific surface area and their hydrophilic and polarity, as well as to the functional groups on the surface. The adsorptive capacity of carbonaceous materials can be easily tailored by modifying the functional groups on the surface (phenolic carboxylic and amino, others) or by doping their surface with other elements. Despite these advantages, the commercial application of carbonaceous materials still depends on overcoming some challenges (Ahmed et al. 2020).

In line with the development of different methods for capture and storage of carbon dioxide, the use of this compound is essential for keeping sustainable environment. Therefore, incentives should be provided to industries for the implantation of the Capture, Storage and Use of Carbon Dioxide (CCSU) technologies to convert carbon dioxide into renewable energy/fuel/chemicals by reusing the industrial emissions (Ahmed et al. 2020). Besides capturing carbon dioxide from industrial processes and punctual sources, one should consider other techniques such as the Direct Air Capture (DAC). This technique uses composite materials able to sequester carbon dioxide at very low concentrations to reduce global warming (Ahmed et al. 2020; Song et al. 2019).

However, the high cost for capturing, transporting, and storing carbon dioxide largely restrains this technology usage and then the growth of carbon dioxide market. To overcome this difficulty, the governments have recently established regulations and directives on CCS technologies for making them safer and more environmentally friendly. This has motivated several companies to develop more economically advanced CCS technologies following these directives. In addition, an increased number of power plants and chemical industries which emit carbon dioxide are installing CCS plants for industrial use (Watch).

3 Technologies for Using Carbon Dioxide

Although carbon dioxide is an abundant and low cost and a nontoxic feed, its industrial usage is still difficulty. This compound is thermodynamically stable, requiring high energy that can be provided by co-reagents or electroreductive processes. This is probably the main reason for the most preferred use of carbon monoxide instead of carbon dioxide in industrial operations. Indeed, several organic syntheses using carbon dioxide have been studied but only some of them are used in industry, such as the production of urea and its derivatives, the manufactory of organic carbamates, and the electrochemical Kolbe-Schmitt for salicylic acid production (Ma et al. 2009; Alper and Orhan 2017).

The market of carbon dioxide usage in 2013 was about 200 $MtCO_2$ per year (Aresta et al. 2013), which can be considered small when compared to around 14,000 $MtCO_2$ per year emitted from big sources (Aresta et al. 2018). Currently, this use tends to increase mainly due to industrial applications of carbon dioxide, such as refrigeration, food and beverages, chemical wholesaling, and pharmaceuticals. It is expected a growth of 3.9% in the carbon monoxide market, achieving USD 9.16 billion by 2026 (Watch). In addition, carbon dioxide can be used in oil recovery (Dai et al. 2014) or as solvent in physical applications such as in supercritical plants (Aresta and Forti 1987).

However, it has been recognized that the most efficient approach is to valorize carbon dioxide by its conversion to fuels (Jiang et al. 2010) or chemicals (Aresta 2010) just after its capture, using different technologies (Mustafa et al. 2020; Alper and Orhan 2017), as shown in Fig. 1. The first group includes traditional processes whose energy demands are provided by fossil sources. They include pure decomposition of carbon dioxide, which requires high energy and an efficient catalyst, as well as optimized reaction conditions since the reaction is thermodynamically unfavorable (Mustafa et al. 2020). Even though, the feasibility of the reaction was demonstrated by some researchers (Yutaka and Bernard 1986; Itoh et al. 1993) using semipermeable membranes for oxygen extraction. However, the high values of temperature, the low conversions, and the risk of explosion by the carbon monoxide and oxygen mixture made this process not promising for commercial applications (Mustafa et al. 2020).

Other processes involve reactants with higher values of Gibbs free energy (co-reactant) than carbon dioxide, such as methane or hydrogen, which will provide the thermal energy for the reaction. To produce methanol and syngas from carbon dioxide, for instance, most of the conventional reactions involve hydrogen or methane. There are other hydrogen donors which can be used such as water and glycerol (Mustafa et al. 2020). The dry reforming of methane and of other carbon-containing compounds is included in this category and will be discussed later. The hydrogenation of carbon dioxide is another kind of reaction in which carbon dioxide reacts with hydrogen to produce methanol (Olah et al. 2008). This traditional approach is the most efficient one in industrial conversion of carbon dioxide to valuable chemicals (Mustafa et al. 2020). Other chemicals like ethene, ethanol, and formaldehyde can also be obtained by these processes, but they are not efficient and then other approaches are adopted instead.

The utilization of carbon dioxide as solvent and/or as reactant can also be convenient, providing carbon and/or oxygen to produce chemicals by several processes. Various works have demonstrated the advantages of using soft oxidants such as carbon dioxide in other reactions like ethane (Saito et al. 2019), propane (Kantserova et al. 2019),

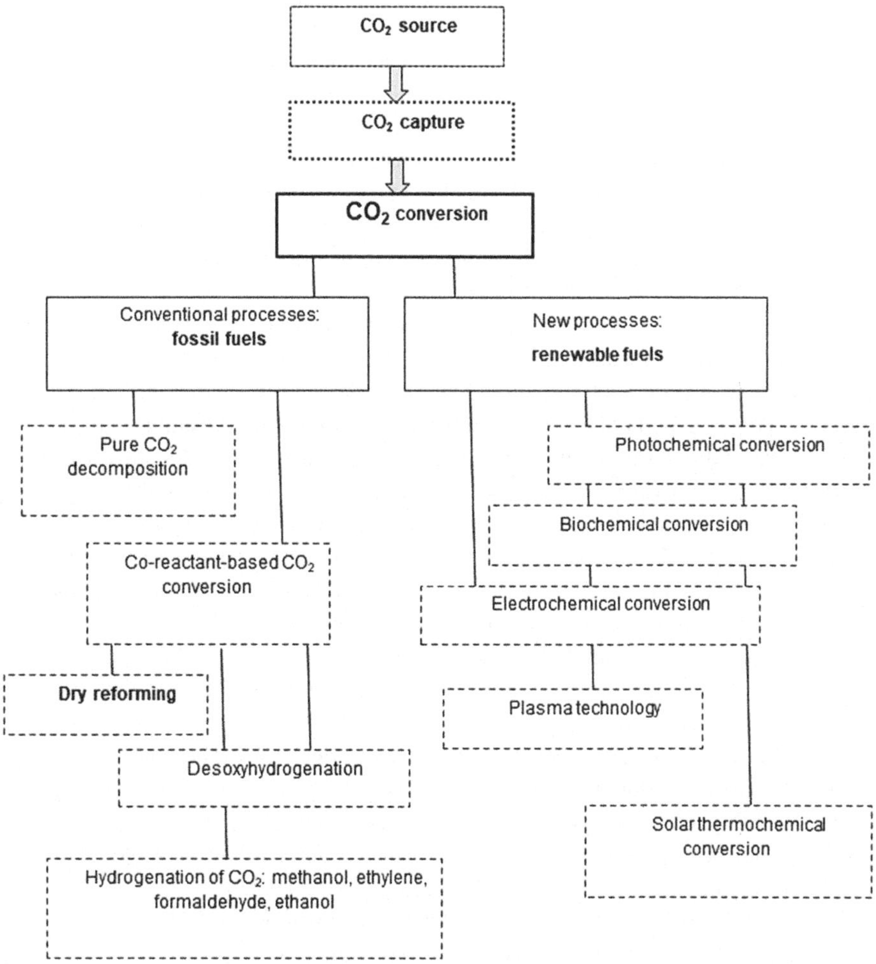

Fig. 1 Traditional and new technologies for carbon dioxide conversion (based on Mustafa et al. 2020)

isobutane (Kraemer et al. 2016) and ethylbenzene dehydrogenation (Oliveira et al. 2008; Rangel et al. 2012), along with the dry reforming of methane (Araújo et al. 2008; Alper and Orhan 2017) and methane oxidative coupling (Mo et al. 2019). These works and others have demonstrated that carbon dioxide is particularly attractive in the dehydrogenation of hydrocarbons, which are frequently reversible and limited by equilibrium. The process is carried out at high temperatures to achieve high conversions, but other reactions, such as hydrocarbons cracking, are favored decreasing the process efficiency. The use of carbon dioxide (instead of water) and a suitable catalyst makes the reaction exothermic allowing its performance at low temperatures, minimizing the cracking reaction. In this case, conversion is no more limited by thermodynamic factors (Bhasin et al. 2001).

As shown in Fig. 1, new processes for the conversion of carbon dioxide based on renewable sources have also been considered, like solar thermochemical, electrochemical, biochemical, photochemical, and plasma-chemical. All of them have limitations in practical applications. The photochemical process is an emerging approach whose challenge is to decrease the number of steps, by the selection of appropriated semiconductors. Regarding biological technology, the use of microbes is continuously expanding for converting carbon dioxide into fuels and chemicals, several microbial species being identified and efficiently contributing to the reaction. The electrochemical reduction technology is considered a promising process in which the catalyst is responsible for the efficiency of reaction. Plasma-chemical technology has been used for carbon dioxide conversion through cold and thermal plasma, the last one being considered the best option for dry reforming. The production of oxygenates by the process of plasma-chemical looks encouraging but needs to be more investigated. The process of solar thermochemical technology seems to be promising for using concentrated solar energy, which is commercially used for power generation and for the reaction of carbon dioxide and hydrogen to produce syngas, which can further produce liquid hydrocarbons (Mustafa et al. 2020).

The potential of homogeneous catalysis to produce high-value chemicals has been demonstrated. Carbonates, carbamates, urethanes, lactones, and pyrones, as well as formic acid and derivatives, can be obtained using homogeneous catalysts. On the other hand, heterogeneous catalysis can be technically more advantageous like the ease reactor, design and easier carrying on, recovery and reutilization of the catalyst (Ma et al. 2009). However, these approaches demand carbon dioxide activation by heterogeneous catalysts.

The viability of carbon dioxide as a feed for chemical processes has been pointed out some years ago by the production of C1 building block chemicals (Bertau et al. 2010; Ola et al. 2013; Aresta and Dibenedetto 2007), such as urea (commodity), formaldehyde, methanol, ethylene carbonate, formic acid, dimethyl carbonate, cyclic carbonates, copolymers, and other compounds of fine chemicals (Alper and Orhan 2017). For the industrial production of salicylic acid, polycarbonates, urea, and cyclic carbonates 130 Mt per year of carbon dioxide are used, the urea process being responsible for the most consumption (Alper and Orhan 2017; Aresta 2010).

4 Methane Dry Reforming Process

Because of the latest findings of abundant reserves of shale gas and the further improvements in hydraulic fracturing technologies (Coonrod et al. 2020; Klewiah et al. 2020), the use of methane has been greatly increased worldwide, mainly in the USA. In addition, biomethane has been successfully obtained from biogas and landfill gas produced through the anaerobic decomposition of wastes from different kinds (Wasajja et al. 2020). This scenario, associated with the advantages of the abundant and cheap carbon dioxide, made the use of methane dry reforming extremely favorable. Besides all these advantages, the process consumes the most important greenhouse gases (Baird 1995).

Through the reaction between methane and carbon dioxide, synthesis gas or syngas (gaseous hydrogen and carbon monoxide) can be obtained. The syngas is considered a sustainable and alternative fuel to fossil sources, as well as an important feedstock to many processes to produce methanol, dimethyl ether (Saravanan et al. 2017), aromatics (Yang et al. 2020), olefins (Zhao et al. 2018), ethanol (Xu et al. 2017a), higher alcohols and other high-value chemicals by Fischer–Tropsch synthesis (Panzone et al. 2020). The last reaction is the most used process worldwide to convert syngas to liquid fuels to avoid the inconvenience of storing and transporting gaseous fuels. Each chemical can require feedstocks with different hydrogen to carbon monoxide molar ratios that can be adjusted by the water gas shift reaction (Rangel et al. 2017).

As shown in Eq. 1, methane dry reforming is endothermic and is frequently performed at temperatures above700 °C to obtain suitable conversions. According to stoichiometry, hydrogen to carbon dioxide (H_2/CO) of 1.0 is expected but other parallel reactions also occur (Eq. 2–6). The decomposition of methane (Eq. 2), the disproportionation of carbon monoxide (Eq. 3) and the hydrogenation of carbon dioxide (Eq. 4) and of carbon monoxide (Eq. 5) are undesirable reactions that lead to coke formation on the catalyst. The reverse water gas shift reaction, RWGSR (Eq. 6), causes the lowering of hydrogen to carbon dioxide molar ratio to lesser than 1.0, since it consumes carbon dioxide (Wang et al. 1996; Zhang et al. 2018a).

$$CH_4 + CO_2 \rightarrow 2CO + 2H_2 \quad \Delta H_{298K} = 248\,\text{kJ mol}^{-1} \quad (1)$$

$$2CO \leftrightarrow C + CO_2 \quad \Delta H^{\circ}_{298K} = -172.6\,kJ\,mol^{-1} \qquad (2)$$

$$CH_4 \rightarrow C + 2H_2 \quad \Delta H_{298K} = 75\ kJ\ mol^{-1} \qquad (3)$$

$$CO_2 + 2H_2 \leftrightarrow C + 2H_2O \quad \Delta H_{298K} = -90\,kJ\,mol^{-1} \qquad (4)$$

$$H_2 + CO \leftrightarrow H_2O + C \quad \Delta H_{298K} = -5a1.3\,kJ\,mol^{-} \qquad (5)$$

$$CO_2 + H_2 \leftrightarrow CO + H_2O \quad \Delta H_{298K} = 41.2\,kJ\,mol^{-1} \qquad (6)$$

These reactions can be controlled by the temperature of dry reforming since each reaction has lower and upper limits to occur. For instance, at values above 640 °C methane cracking begins to occur while carbon monoxide disproportionation (Eq. 2) and the RWGS (Eq. 6) are inhibited above 820 °C. In the 557–700 °C range, methane cracking and carbon monoxide disproportionation are favorable and then carbon deposition occurs. In addition, methane dry reforming is limited by equilibrium and is favored at temperatures above727 °C. In these conditions, the disproportionation of carbon monoxide (Eq. 2) is inhibited whereas methane cracking (Eq. 3) will occur. At this condition, methane cracking is the major responsible for deactivating the catalysts by coke deposition (Wang et al. 1996; Zhang et al. 2018a).

Several studies have been performed on thermodynamic simulations for different conditions of temperature (Wang et al. 1996; Jang et al. 2016; Nikoo and Amin 2011), CO_2/CH_4 ratio (Nikoo and Amin 2011), pressure, and additional oxidant (Wang et al. 1996; Jang et al. 2016), taking the contribution of the reactions responsible for coke formation into account (Nikoo and Amin 2011). It was concluded that the best operation conditions for methane dry reforming are temperatures higher than 850 °C and low pressures to achieve higher conversions (Wang et al. 1996; Jang et al. 2016; Nikoo and Amin 2011). In addition, carbon formation should be avoided to operate the process stably (Wang et al. 1996; Jang et al. 2016; Nikoo and Amin 2011). Moreover, these works have demonstrated that one cannot predict the conversion or coke formation by thermodynamic calculation, because the rate reactions strongly depend on the selectivity of the catalyst.

Besides the thermodynamic requirements, the occurrence and the extension with which the Eqs. 1–6 occur largely depend on the selectivity of the catalyst. Therefore, a lot of work has been carried out over the years, aiming to find active, selective, and stable catalysts, which cannot go on deactivation.

4.1 Progress in Catalysts for Methane Dry Reforming (1928–1989)

The first detailed investigation dealing with dry reforming was reported in 1928 by Fischer and Tropsch (1928), who have found severe carbon formation and then the deactivation of nickel and cobalt-based catalysts. Twenty years later, Reitmeier et al. (1948) have demonstrated the advantages of using carbon dioxide to control the hydrogen to carbon monoxide ratio in the synthesis gas obtained by steam reforming of methane (Eq. 7), without producing coke. By equilibrium considerations, as well as by laboratory and pilot plant investigations over a nickel catalyst supported on a perforated disk, they have found the reaction conditions in which the syngas composition could be adjusted by changing the feedstock composition. This was the first work that pointed out the suitable conditions to obtain H_2/CO ranging from 3 to 0.5 to be used in Fischer–Tropsch synthesis for different purposes. Depending on these values, the excess carbon dioxide in the effluent gas could be recovered and recycled to the reactor. These results were used to build an industrial plant for the Bureau of Mines at Bruceton, Pennsylvania, USA.

$$CH_4 + H_2O \rightarrow CO + 3H_2 \quad \Delta H_{298K} = 206\,kJ\,mol^{-1} \qquad (7)$$

Table 1 shows the reports on dry reforming in the period from 1928 to 1989. It can be noted that few works have been carried out, most of them showing simple formulations of catalysts. In this period, the research has been addressed to find catalytically active metals or compounds. Only the commercial catalysts for steam reforming have included mixed supports or dopants. In fact, until 1990, most of the research was addressed to steam reforming, the most important industrial route to produce syngas and hydrogen since that time. However, in the following years, the interest for dry reforming has been increasingly motivated by the incentives to consume carbon dioxide for decreasing global warming and to develop technologies to produce clean energy. In addition, it has been recognized the advantages of producing syngas with low hydrogen to carbon dioxide ratio, proper to Fischer–Tropsch synthesis, to obtain hydrocarbons or oxygenates.

4.2 Progress in Catalysts for Methane Dry Reforming (1990–1999)

The first works in the 1990s have been motivated by the need of using dry reforming in the Chemical Energy Transmission Systems (CETS) for energy storage and transmission. Richardson and Paripatyadar (1990), for instance, have evaluated a noble-based catalyst (0.5 wt.% $Rh/\gamma\text{-}Al_2O_3$) in a pilot plant and have proposed a model for methane conversion, incorporating the rate and the internal and external diffusion equations. The catalyst has shown no coke deposition during dry reforming and has been successfully tested in large scale experiments in solar receivers. The choice of rhodium was based on the experience with

Table 1 Progress in catalyst development (1928–1989)

Active phase	Support	Reference
Ni, Co		Fischer and Tropsch (1928)
Ni, CuO, Fe_2O_3	SiO_2	Sodesawa et al. (1979)
Rh > Pd > Pt ≫ Ru	Al_2O_3	Sakai et al. (1984)
Ni > Co ≫ Fe	Al_2O_3	Tokunaga and Ogasawara (1989)

noble metals in steam reforming, which showed that rhodium and ruthenium were ten times more active than nickel, platinum, and palladium (Rostrup-Nielsen 1983). Although nickel has been the most economical and used catalyst in commercial processes of steam reforming (Ridler et al. 1989), it requires a lot of steam (to control coke deposition) which decreases the efficiency of the CETS (Richardson and Paripatyadar 1990).

Table 2 shows the progress on the catalyst development from 1990 to 1999. During this decade, the studies have been addressed to investigate new supports (pure or mixed) that could interact with the metal and improve its catalytic properties. In addition, the first studies have suggested possible other roles of the support besides providing the dispersion of metal particles. Moreover, the dopants have begun to be used mainly in the supports.

Therefore, other supports have been evaluated, such as europium oxide, Eu_2O_3 (Perera et al. 1991), for ruthenium or iridium, the catalysts proving to be active and producing no significant coke after several cycles. They have shown similar performances as compared to alumina-based catalysts (Ashcroft et al. 1991), suggesting that the kind of support did not affect the catalytic mechanism. Mark and Maier (1996) have come to the same conclusion studying rhodium, ruthenium, and iridium supported on alumina, titania, silica, and zirconia. However, silica, titania, and alumina have been also evaluated by Nakamura et al. (1994), who have noted a significant decrease in activity from rhodium supported on alumina to silica, titania showing an intermediate value. In another work, Bitter et al. (1996) have found the following order of stability of platinum on different supports: $Pt/\gamma\text{-}Al_2O_3 < Pt/TiO_2 < Pt/ZrO_2$. This tendency was related to coke deposition blocking the active sites. In addition, it has been noted that for platinum-based catalysts the active sites are located in the interface of metal and support. This was confirmed in another study (Bitter et al. 1998) by steady-state transient kinetics measurements and other techniques. These contradictory results are probably associated with different reaction conditions as well as different catalyst preparations. In a further report, Bitter et al. (1999) have pointed out that the rates of carbon formation and of deactivation depend on the stability of carbonates coming from carbon dioxide on the supports. The replacement of zirconia by alumina decreased the stability and favored the growth of carbon particles.

Other researchers have reported strong effects of acidity/basicity of support on the catalysts' performance. Gronchi et al. (1995), for instance, have proposed that the basic sites of lanthana were able to activate carbon dioxide, inhibiting carbon deposition on rhodium-based catalysts. Lercher et al. (1996) have also considered the role of the acidity of the support on the catalyst efficiency and have found a proportionality between the rate of coke production and the concentration of Lewis sites on the support. In line with these findings, Zhang et al. (1996) have considered that the acidity of the support could affect the performance of rhodium-based catalysts as well. They have found that the initial intrinsic activity largely depended on the support acidity, as suggested by the decreasing sequence: yttria-stabilized zirconia (YSZ); Al_2O_3; TiO_2; SiO_2; La_2O_3 and MgO. Moreover, as the metal nanoparticle size increased the catalyst activity and velocity of deactivation diminished. The catalysts based on titania and magnesia have shown the highest deactivation rates while those based on YSZ and silica showed no deactivation during 50 h on stream. The catalyst deactivation has been caused by coke formation, sintering of the metal, and rhodium poisoning by some species of the support. These effects on activity and on deactivation were found to be dependent on the support, suggesting that they are related to its interaction between metal and support. In a further paper (Efstathiou et al. 1996), the same group has investigated the dry reforming mechanism over YSZ and alumina-supported rhodium. This was the first report of experimental results for dry reforming derived from steady-state tracing, allowing to prove the origin of carbon species accumulated on rhodium surface catalysts. It has been concluded that the YSZ-based catalyst was the most resistant against carbon deposition because the ability of the support for providing lattice oxygen species, which react with carbonaceous species. These species could spill over rhodium, favoring the formation of carbon monoxide and water.

The role of the support in affecting the metal activity has been also investigated at molecular level (Basini and Sanfilippo 1995). The experiments were performed over rhodium, ruthenium, and iridium as a monolayer deposited on α-alumina, magnesia, ceria, lantana, and titania with low surface areas. It has been found that carbon dioxide could be partially dissociated producing oxidic and carbonyl species on metal surfaces. This reactivity was facilitated by surface

Table 2 Progress in catalyst development for dry reforming (1990–1999)

Active phase	Support	Promoter	Reference
0.5 wt.% Rh	γ-Al_2O_3		Richardson and Paripatyadar (1990)
Ru, Ir	Eu_2O_3		Perera et al. (1991)
Ru, Ir	Al_2O_3		Ashcroft et al. (1991)
Ir, Rh, Ni > Pd > Ru	Al_2O_3		Vernon et al. (1992)
MnO_2			Mirzabekova et al. (1992)
CeO_2			Otsuka et al. (1993)
Ru, Rh > Ir > Ni, Pt, Pd	Al_2O_3-stabilized MgO		Rostrup-Nielsen and Hansen (1993)
Rh > Ru > Ir > Pt > Pd	MgO		Qin and Lapszewicz (1994)
Re	γ-Al_2O_3		Claridge et al. (1994)
Rh	Al_2O_3 > TiO_2 > SiO_2	Al_2O_3, TiO_2, MgO	Nakamura et al. (1994)
Ni	Al_2O_3	La	Blom et al. (1994)
Ni	Al_2O_3	Pt, Rh-, CeO_2, La_2O_3.	Inui et al. (1994)
Ni, Pt	ZrO_2 > γ-Al_2O_3 > α-Al_2O_3	K	Seshan et al. (1994)
Ni	SiO_2, La_2O_3, MgO, ZrO_2, TiO_2, Al_2O_3-SiO_2	Cu, K	Swaan et al. (1994)
Co	SiO_2 > AC	MgO	Guerrero-Ruiz et al. (1994)
Ni	MgO, SiO_2	SiO_2, MgO	Takayasu et al. (1995)
Ni	Al_2O_3, SiO_2	MgO, CaO	Tang et al. (1995)
Ni	MgO, CaO, BaO, SrO		Ruckenstein and Hu (1995)
Rh, Ru, Ir	Al_2O_3, TiO_2, SiO_2, ZrO_2		Mark and Maier (1996)
Pt	Al_2O_3, TiO_2, ZrO_2		Bitter et al. (1996)
Rh	YSZ > Al_2O_3 > TiO_2 > SiO_2 > La_2O_3 > MgO		Zhang et al. (1996)
Pt, Ni	TiO_2, ZrO_2, Al_2O_3		Lercher et al. (1996)
Ni, Cu, Co, Fe, Ru, Rh, Pd, Ir, Pt	MgO, TiO_2, SiO_2, Activated Carbon, Cr_2O_3		Bradford and Vannice (1996a), Bradford and Vannice (1998), Bradford and Vannice (1999)
Ni	La_2O_3, γ-Al_2O_3, CaO		Zhaolong and Verykios (1996)
Ni	Al_2O_3	MgO, CaO, La_2O_3, CeO	Cheng et al. (1996)
Ni	α-Al_2O_3	Fe, Co, Cu, Ce	Halliche et al. (1996)
Ni	Al_2O_3	Na, K, Mg, Ca	Horiuchi et al. (1996)
Ni	Pentasil-type zeolite	K, Ca	Chang et al. (1996)
MoS_2, WS_2			Osaki et al. (1997)
$Ni_{0.03}Mg_{0.97}$		Pt, Pd, Rh	Chen et al. (1997a)
Rh	NaY > HY		Bhat and Sachtler (1997)
(Rh > Pt > Ir, Pd > Ru)	ZrO_2		Keulen et al. (1997)
Pt	SiO_2, ZrO_2	Sn	Stagg and Resasco (1997)
Ni	Al_2O_3	La–Nd–Pr	Slagtern et al. (1997)
Ni–Mg–O		Cr_2O_3, La_2O_3	Chen et al. (1998)
Ru	Activated carbon	MgO	Ferreira-Aparicio et al. (1998)
Pt	ZrO_2	Ce, La	Stagg and Resasco (1998)
Ni	CeO_2–Al_2O_3, CeO_2		Wang and Lu (1998)
Ni	Al_2O_3	MoO_3 or WO_3	York et al. (1998)
Rh	SiO_2	V	Bradford and Vannice (1999)

hydrogen atoms produced by dissociative hydrogen or methane chemisorption. The reactivities of the hydridocarbonyl and oxidic species were considered as the center of the catalytic cycle, during which one species was periodically converted into the other one. The carbon growth reactions have been inhibited by the high oxidative properties of the noble metal-oxygen adducts.

In addition, Bradford and Vanice (1996a,1998) have studied the metal-support interaction in nickel deposited on magnesia, titania, silica, and on activated carbon (AC). Less coking was noted for titania and magnesia-supported nickel, the last one being active and stable for 44 h on stream. The strong interaction of nickel and titania provided TiO_x species that decorated the surface of metal particles and then destroyed the big ensembles of nickel, responsible for coke formation. On the other hand, the nickel oxide-magnesium oxide solid solution stabilized the reduced nickel, making them resistant against coke formation. Moreover, the authors proposed that the activity of the metal could be affected by the support, changing the ability of reduced nickel for donating electrons. In a further work (Bradford and Vannice 1999), they have elucidated the role of the support by studying several metals (copper, cobalt, iron, nickel, ruthenium, rhodium, paladium, iridium, and platinum) deposited on silica or titania. It was found that both the support and the percentage of *d* character of each metal affected the activity, different trends being observed for metals on each support. For silica, platinum showed the highest activity while ruthenium exhibited the maximum activity for titania.

At that time, rhodium has seemed to be the most suitable catalyst to dry reforming, but the high cost and low availability have pushed the investigation to improve the nickel-based catalysts and to look for other active phases. In line with this need, Swaan et al. (1994) have considered the use of different supports and dopants, by studying nickel supported on silica, lantana, magnesia, zirconia, titania, alumina-silica, and potassium, and copper as dopants for silica. The low activity of the catalysts based on magnesia, titania, and potassium-doped silica was related to the metal-support interactions which have made nickel reduction more difficult. This led to the conclusion that the activity of nickel was more related to nickel phase on each support.

Other dopants, such as magnesium oxide and/or calcium oxide, have also been investigated for alumina and silica-supported nickel catalysts (Tang et al. 1995). The dopants improved silica-supported nickel, a fact that was related to their basic character, depressing carbon deposition. Horiuchi et al. (1996) have come to the same conclusion by studying the role of basic metal (sodium, potassium, magnesium, and calcium) oxides on deactivation of alumina-supported nickel. It has been proposed that the promoters decreased nickel activity in methane dehydrogenation. This effect has also been investigated by Ruckenstein and Hu (1995), doping nickel-based catalysts with alkaline earth metal (magnesium, calcium, barium, and strontium) oxide. The NiO/MgO catalyst was superior to the others, a fact that was related to a nickel oxide and magnesium oxide solid solution. In addition, Stagg and Resasco (1997,1998) have noted that tin, cerium, and lanthanum were efficient promoters for silica or zirconia-supported platinum for preventing deactivation. In another work (Slagtern et al. 1997), it has been found that different loadings of a rare-earth mixture (La, Nd, Pr) or of pure lanthanum could efficiently increase the catalytic activity of alumina-supported nickel as well as its stability. Moreover, York et al. (1998) have found that by adding molybdenum oxide (MoO_3) or tungsten (VI) oxide (WO_3) to nickel supported on alumina, carbon formation was strongly decreased, the tungsten-containing samples showing no deactivation.

Besides promoting the support, other studies have been carried out promoting the active phase. Halliche et al. (1996), for instance, have found that the activity of nickel supported on α-alumna changed in the order: nickel ~ nickel-cobalt, nickel-cerium > nickel-copper ≫ nickel-iron. The action of these additives has been assigned to the interaction of nickel with the other metals. Sometime later, Chen et al. (1997a) have evaluated the effects of doping both metal and support on the catalysts' performance by using different amounts of Pt, Pd, and Rh on the catalysts based on the $Ni_{0.03}Mg_{0.97}$ solid solution. An appreciable improvement in activity and stability was noted, the last one being related to the increased nickel reducibility. Moreover, Chen et al. (1997b) have found that small amounts of trivalent-metal oxides (Cr_2O_3, La_2O_3) have improved the resistance against coke on Ni–Mg–O catalysts. This finding was related to the ability of the dopants to create Schottky defects that increased the oxygen anions mobility in the lattice, fastening the oxidation of carbonaceous species. In addition, part of the Schottky defects (cationic vacancies) may go to the surface and favor nickel reducibility.

Another strategy adopted to improve the dry reforming catalysts was the physical mixture of different metal oxides and the catalysts, such as the study of Takayasu et al. (1995). They have found that the addition of silica to magnesia-supported nickel increased the activity and the resistance against coke deposition, while no effect was noted for the addition of magnesia, at 800 °C. It means that magnesium oxide improved the activity as a support but not in a mixture. The physically mixed silica decreased the amount of deposited coke, this being related to its ability of storing hydrogen and of spillover the CH_x adsorbed species, preventing carbon deposition.

The role of promoters has also been studied on cobalt-based catalysts (Horiuchi et al. 1996) using silica and activated carbon as supports in the range of 500–700 °C. The role of the promoter was assigned to strongly adsorbed carbon dioxide species during reaction, which could react with coke, preventing the catalyst deactivation. However, magnesium oxide was not able to avoid the cobalt particles sintering. Chang et al. (1996) have studied the role of promoters (potassium and calcium oxides) as well but using a pentasil-type zeolite-supported. They have noted that the promoters decreased coke deposition, due to carbonate species formed on alkaline promoters near nickel sites.

Other supports have also been investigated in the last years of this decade. Bhat and Sachtler (1997), for instance, have demonstrated the potential of zeolites to produce well dispersed metals, such as rhodium on Y zeolite. The catalysts showed high activity, selectivity, and stability, no deactivation or coke being noted after 30 h on stream or after repeated thermal cycles. NaY was superior to HY.

Other active phases were proposed in the last period as well. New active phase, such as MoS_2 and WS_2 has been studied by Osaki et al. (1997). No catalyst deactivation has been noted on disulfides but the rate was smaller than on Ni/SiO_2. This was related to the poor ability of disulfides for decomposing methane.

The effect of several parameters on deactivation of silica-supported nickel catalyst has been studied by Kroll et al. (1996). It has been proposed that at the beginning of the reaction, carbide-like surface was formed rapidly and then the activity changed according to the morphology of the particles and to nickel reduction degree. During reaction, nickel particles went on sintering and various forms of coke were produced. In contrast with the activity in reforming reaction, the rate of formation and the type of coke seemed to be structure sensitive, this determining the aging of the catalyst. Different types of aging can occur during reaction: (i) the faceted and flat particles produced on stream do not form coke or form low amounts of filamentary coke (toxic); (ii) during the prereduction step, under hydrogen flow, the spherical and small particles formed are able to produce encapsulating coke (toxic) and (iii) at temperatures below 700 °C, few particles sinter and the rates of carbon deposition increase, in accordance with thermodynamics. At this stage, the rate of deactivation no more depends on the treatment and on the morphology of particles. Therefore, the better condition for improving the stability of the catalyst is 700 °C in a reactor with minimized temperature gradients, without prereducing the catalyst.

At the end of this period, some workers pointed out the importance of the preparation methods for nickel supported on alkaline earth metal oxides (Ruckenstein and Hu 1997), early transition metal carbides (Lemonidou et al. 1998), rhodium and nickel-containing catalysts prepared from hydrotalcites (Basile et al. 1998), nickel-based catalysts obtained by crystallization (Suzuki et al. 1998) and nickel calcium aluminate catalysts (Lemonidou et al. 1998). However, only few works have compared different preparation methods (Chang et al. 1996; Lemonidou et al. 1998). The impregnation of the metal precursor on support previously prepared or coprecipitation methods have been used in most of the works.

On the other hand, a lot of studies have been carried out dealing with kinetic models and mechanicist studies for different catalysts with several active phases and supports (Bradford and Vannice 1996b; Kroll et al. 1997; Schuurman et al. 1997; Wang and Au 1997; Erdőhelyi et al. 1997; Zhang and Verykios 1997; Cant et al. 1997; Hei et al. 1998; O'Connor et al. 1998; Gronchi et al. 1998; Osaki et al. 1998). These works have shown that the kinetics and the mechanisms depend on the catalyst and on the support but generally, the following steps are involved: (1) methane adsorption followed by successive dehydrogenation steps to produce hydrogen and carbon on the surface; (2) dissociative adsorption of carbon dioxide to produce surface oxygen and carbon monoxide and (3) desorption of CO and hydrogen (Wang et al. 1996). By studying nickel supported on alumina and on lantana by isotopic tracing techniques Tsipouriari and Verykios (1999) have demonstrated that different reaction pathways may occur on different supports. On lantana-supported nickel, the carbonaceous intermediates species able to produce carbon monoxide are present in larger amounts than those that form carbon dioxide. The amount which produces CO is much less on alumina-supported nickel. In the reaction condition, the compound $La_2O_2CO_3$ (produced by the reaction of lantana with CO_2) can be decomposed to form carbon monoxide and then supply oxygen to oxidize the coke on the surface. Aiming to explain the role of the alkaline metals on decreasing coke on Ni/Al_2O_3, Mori et al. (1999) have proposed that these dopants can decrease the activity to methane decomposition and then decrease the concentration of hydrogen-deficient hydrocarbon species ($CH_{x,ad}$) on a nickel surface. This provides a high concentration of adsorbed oxygen atoms or CO_x species on the surface preventing coke. On the undoped catalysts, methane was readily decomposed to carbon, while carbon dioxide was not, resulting in the production of carbon.

In addition, the carbonaceous species have been identified by several techniques and their formation and decomposition have been discussed (Chen et al. 1997b). Several mechanisms for coke formation were proposed, depending on the catalyst. Briefly, the process can occur both by carbon dioxide molecular and methane molecular routes (Erdőhelyi et al. 1997; Tsipouriari et al. 1994). Three types of carbonaceous species with different reactivities can be produced on nickel catalysts: α-C, β-C, and γ-C species. The first form is supposed to be responsible for CO formation while the others cause the catalyst deactivation (Zhang and Verykios 1994). Just the stable form (coming from the disproportionation of carbon monoxide) can poison nickel. Although the forms produced by the activation of methane accumulate promptly on the surface, they are less toxic. It is believed that coke is first produced as carbon atoms that is an important intermediate because of their high reactivity.

During the period from 1990 to 1999, the main progresses in dry reforming catalysts were mostly related to the knowledge of the role of support on affecting the active phase and on the kinetic and mechanism, largely based on the evaluation of different supports and on several techniques of catalyst characterization. The use of dopants had begun in this period and a lot of compounds were evaluated but few works have explained their role. From the experiments and modeling studies, it has been concluded that the kind of support strongly affects the activity due to the interaction with the active phase. During dry reforming, carbon dioxide is adsorbed and dissociated, the basicity of the support improving these processes. This was confirmed by using different basic supports (La_2O_3, CaO) or basic promoters which have proved to prevent coke formation.

4.3 Progress in Catalysts for Methane Dry Reforming (2000–2009)

In this decade, most of the works have been addressed to investigate more deeply the most promising catalysts for methane dry reforming, aiming to improve these solids. In addition, several new combinations of active phase, supports, and promoters have been proposed, resulting in new and more complex formulations, as shown in Table 3. Moreover, the knowledge that the methods of preparing the catalysts could affect the catalysts' efficiency has addressed the investigation to the comparison of catalysts prepared by different methods.

Xu et al. (2000), for instance, have proposed nickel-containing hexaaluminates, $LaNi_yAl_{5-y}O_{19-\delta}$ (y = 0.3, 0.6, 0.9, 1.0) as an optional catalyst for methane dry reforming. The catalysts have been more active and stable alumina or silica-supported nickel. Alumina intercalated laponite (Al-laponite) has been also proposed as new support for nickel catalysts (Hwang et al. 2001). The samples have been prepared by different methods, the pore structure being tailored through the surfactant. The catalyst has shown well-developed porosity and high activity and stability. In another work (Chen et al. 2001), it was found that platinum supported on yttria-stabilized zirconia (YSZ) with different amounts of yttrium led to higher conversions and stabilities than platinum supported on alumina and silica. This finding has been assigned to the increase of adsorption, activation, and dissociation of CO_2 on YSZ.

Aiming to create new interactions between the metal and the support, the effect of previous deposition of a metal oxide over the support has also been studied. Chen et al. (2002) have prepared zirconia on silica and lanthana on alumina with different loadings of zirconia, before deposition of platinum. It has been found that zirconia spreads homogeneously on silica to form a stable surface-phase oxide whereas lantana could not be well dispersed on alumina and formed an amorphous compound with alumina. The new catalysts were three to four times more active and stable than the conventional silica and alumina-supported platinum. These findings have been related to the activation of carbon dioxide adsorbed on the basic sites of the catalysts.

Several kinds of zeolites have also been evaluated as supports for different metals (Pt, Ni, Mo, Ni), such as 5A, ZSM-5 X, Y, and K, as well as mesoporous materials (SBA-15, MCM-41). Due to the high specific surface areas, they usually provide high metal dispersion and low coke deposition (Lacheen and Iglesia 2005).

Some unusual supports such as SiC monolithic foam have been investigated, demonstrating to be promising for dry reforming catalysts. An active and stable catalyst (7 wt % Ni) has been obtained, which worked at 750 °C for 100 h on stream. Mesoporous zirconia with 3% M_2O_x (M = Ce, La, and K) have also shown potential for support of nickel-based catalysts, the sample containing 5 wt.% Ni being active and stable 50 h on stream. The catalysts were obtained by using Pluronic P53 surfactant (Rezaei et al. 2007). In addition, Zhang et al. (2008) have shown the potential of nickel nanoparticle-embedded mesoporous titania/silica catalysts prepared by impregnation of nickel in a titanium-silica gel, no organic additives or templates being used. The NiO (10 wt%) and TiO_2/SiO_2 (50:50 w/w) were the most active catalysts.

The promoters both in the support and in the active phase have also been studied. Montoya et al. (2000), for instance, have evaluated the effect of promoters (CeO or MgO) on the activity and stability of zirconia-supported nickel during the dry reforming of methane. They have found that the additives prevented the t-ZrO_2 transformation to m-ZrO_2 increasing the thermal stability and activity. The doped catalyst was stable for 200 h on stream, this being related to the metal reducibility, crystal size, and dispersion, as well as to metal-support interactions. Richardson et al. (2001) have evaluated the traditional bimetallic catalyst (Pt-Re/γ-Al_2O_3) used in industrial process of naphtha reforming to produce aromatics. They have noted that low amounts of rhenium increased the deactivation rate, but higher amounts lead to increased conversion and stability, the catalysts being stable up to 750 h. It has been proposed that rhenium was active in carbon dioxide dissociation releasing oxygen to remove the carbon and to avoid carbon clusters. Table 3 shows some examples of the different promoters proposed for the catalysts of dry reforming in this decade.

By studying the influence of different preparation methods on catalyst performance, Tang et al. (2000) have

Table 3 Progress on catalysts for dry reforming (1999–2009)

Active phase	Support	Promoter	Reference
$LaNi_yAl_{12-y}O_{19-\delta}$			Xu et al. (2000)
Ni	ZrO_2	CeO_2, MgO	Montoya et al. (2000)
Pd	CeO_2		Sharma et al. (2000)
Ni, Ru	$LaMnO_3$		Pietri et al. (2000)
Ni–La_2O_3	5A zeolite		Luo et al. (2000)
Ni	Al_2O_3	Mn	Seok et al. (2001)
Mo_2O			Sehested et al. (2001)
Pt	YSZ		Chen et al. (2001)
Ni Ru, Mn	$LaMO_3$		Pietri et al. (2001)
Pt	γ-Al_2O_3	Re	Richardson et al. (2001)
Ni	Al-laponite		Hwang et al. (2001)
Pt	ZrO_2/SiO_2, La_2O_3/Al_2O_3		Chen et al. (2002)
Mo_2C, WC	ZrO_2		Naito et al. (2002a, b)
Ni	Al_2O_3	Mo	Quincoces et al. (2002)
Ni	SiO_2, H-ZSM-5	Ru	Crisafulli et al. (2002)
Co	γ-Al_2O_3		Ruckenstein and Wang (2002)
Ru, Ni	$Ln_{1-x}Ca_xO_3$ (La, Sm, Nd)		Goldwasser et al. (2003)
Fe, Co, Ni	CeO_2		Asami et al. (2003)
Ni	Al_2O_3	V	Valentini et al. (2003)
$LaN_xCo_{1-x}Al_{11}O_{19+\delta}$			Wang et al. (2003)
Ni	CeO_2, CeO_2–ZrO_2, ZrO_2		Roh et al. (2004)
Ni	MgO–γ–Al_2O_4, $MgAl_2O_4$		Guo et al. (2004)
Mo_2C	$SiO_2 < ZrO_2 < Al_2O_3$		Treacy and Ross (2004)
Ni	CeO_2	Y_2O_3	Wang et al. (2004)
Ni	α-Al_2O_3	Sn	Hou et al. 2004)
Ni	Al_2O_3–ZrO_2		Li and Wang (2004)
Ir	$Ir/Ce_{0.9}Gd_{0.1}O_{2-x}$		Wisniewski et al. (2005)
Pt	Al_2O_3,Na–Al_2O_3, K–Al_2O_3, ZrO_2		Ballarini et al. (2005)
Ni	γ-Al_2O_3	K, CeO_2, Mn	Nandini et al. (2005)
Mo	H-ZSM-5		Lacheen and Iglesia (2005)
Ni, Co	TiO_2		Takanabe et al. (2005)
Ni	SBA-15		Meili et al. (2006)
Pt	MgO		Yang and Papp (2006)
Ru, Rh, Ir, Pt, Pd	Al_2O_3–MgO		Rezaei et al. (2006)
Ni–Co (Fe, Cu, Mn)-Al–Mg			Zhang et al. (2007a)
Ni–Ti–Al composite			Sun et al. (2007)
Pd	La_2O_3		Fernandes Júnior et al. (2007)
Rh	CeO_2–Al_2O_3, Al_2O_3		Wang et al. (2007a)
Co	MgO, ZrO_2, CeO_2/SiO_2–Al_2O_3		Mondal et al. (2007)
Ni	KH zeolite		Kaengsilalai et al. (2007)
Ni	Sm_2O_3–CaO		Zhang et al. (2007b)
Ni	MCM-41	La	Wang (2007)
Ni	ZrO2–3%M_2Ox; M: Ce, La, K		Rezaei et al. (2007)
Ni	ZSM-5	Pt	Pawelec et al. (2007)

(continued)

Table 3 (continued)

Active phase	Support	Promoter	Reference
Ni/Al/Al/Ce			Daza et al. (2008)
Ni	Al_2O_3	ZrO_2	Therdthianwong et al. (2007)
Ni	$Ce_{0.75}Zr_{0.25}O_2$	Y^{3+}, Pr^{4+}/Pr^{3+}	Wu and Zhang (2008)
Ni	SBA-15	MgO	Huang et al. (2008)
Ni	SiO_2	Gd_2O_3	Guo et al. (2008)
$LaNi_{1-y}Mg(Co)ByO_3 \pm \delta$			Gallego et al. (2008)
Ni	SiC foam		Liu et al. (2008)
Ni	La_2O_3	SiO_2	Gao et al. (2008)
Ni	TiO_2–SiO_2		Zhang et al. (2008)
Ni	$BaTiO_3$		Xiancai et al. (2008)
Ni	$La_{1-x}Pr_xNiAl_{11}O_{19}$ (hexaluminates)		Zhang et al. (2009)
Ni–Ru	BN		Wu and Chou (2009)
Ru	CeO_2–Al_2O_3		Safariamin et al. (2009)

compared Ni/γ-Al_2O_3 catalysts obtained by nickel impregnation on a commercial and on a sol-gel made γ-Al_2O_3, using organometallic precursors. They showed similar activities but the commercial support produced a catalyst more susceptible to the deposition of heavy coke leading to the interruption of reaction after 3.5 h. On the other hand, the other catalyst showed almost no coke after 80 h on stream, this being related to smaller size of initial nickel particles. The relationship between the preparation method and the activity and stability have also been studied on carbides of molybdenum and tungsten (Naito et al. 2002a). Mo_2C did not deactivate when obtained by molybdenum oxides nitridation followed by carburization. Tungsten carbides, prepared by carburization or nitridation followed by carburization, showed close activities in the beginning but that obtained from direct carbonization was more stable than the other one. In another work (Naito et al. 2002b), it has been noted that by supporting molybdenum carbide on zirconia the deactivation was suppressed, this being related to the decrease of the rate of methane dissociation.

Several other works have been addressed the effects coming from the changing of the sequence of impregnation and coimpregnation on the catalytic properties. By studying the dependence of activity and thioresistance as well as coke deposition on the impregnation sequence, for molybdenum and tungsten carbide catalysts, Quincoces et al. (2002) have found that molybdenum addition strongly affected the carbon deposition and thioresistance, regardless the impregnation sequence. On the other hand, Özkara-Aydınoğlu et al. (2009) have compared the coimpregnation and sequential impregnation in the preparation of platinum supported on zirconia-doped cerium, noting that the effect of cerium on activity and selectivity depended on cerium loading and on the preparation method. The addition of 1 wt.% Ce by impregnation improves the activity and stability of the catalysts. Dias and Assaf (2003) have also investigated the effect of changing the order of impregnation of nickel and calcium precursors. They have found out that calcium interacts with the support, competing with nickel, and favors the production of nickel reducible species mainly when nickel is added after calcium. In addition, Therdthianwong et al. (2007) have compared the coimpregnation and successive impregnation methods. They have noted that zirconia inhibited coke deposition by increasing the dissociation of carbon dioxide and then producing oxygen intermediates near the ZrO_2–Ni interface, where coke was gasified. The co-impregnated catalyst was more active but the resistance to coke was comparable for short-running period.

Besides impregnation methods, the sol-gel methods have been used and compared with other methods in several works. Xiancai et al. (2008), for instance, have compared the preparation of nickel-containing catalysts obtained by sol-gel and impregnation routes. They have found that the sol-gel method produced catalysts with higher activity than those prepared by impregnation. On the other hand, Kim et al. (2000) have compared the conventional alumina-supported nickel with aerogel catalysts obtained by sol-gel route and then supercritical drying. Very small nickel particles evenly distributed over alumina were obtained for the aerogel catalysts. Their activity increased with metal loading and all aerogel samples produced low amounts of coke, a fact that was related to the well dispersed metallic particles. The filamentous carbon production was affected by nickel particles size and was formed preferentially on particles bigger than 7 nm. In further work, Sun et al. (2007) have found that solids showing high specific surface area and with narrow pore size distribution could be obtained by sol-gel route, optimizing the hydrolysis ratio and the acid to alkoxide ratio. However, the high-temperature calcination led to a large decrease of specific surface area, because of pore structure collapse and sintering of the particles. Adding aluminum-containing precursor the specific surface area increased even after calcining the solids at 973 K. The catalyst was highly active and stable in dry reforming of methane.

The effect of calcination or reduction temperatures on the catalyst performance has been studied by several researchers. Nguyen et al. (2009) have investigated the influence of gaseous mixture used in the pretreatment on the efficiency of β-SiC extrudate-supported nickel. The pretreatment of the catalyst by carbon dioxide and oxygen, instead of direct activation under carbon dioxide and methane mixture, improved the activity. The oxygen pretreatment was supposed to stabilize the metallic nickel instead of the $NiSi_2$ phase. By studying Ni–Mg-Al catalysts prepared by continuous coprecipitation, Perez-Lopez et al. (2006) have found that the reduction temperature strongly affected the activity and selectivity while the calcination temperature did not. On the other hand, Wang and Ruckenstein (2001) have found an influence of calcination temperature, by studying the addition of cobalt to magnesia-supported nickel catalysts in dry reforming. The catalyst with 8–36 wt.% Co and calcined at 500 or 800 °C were highly active and stable, this being related to the CoO–MgO solid solution. The effect of calcination temperature has also been studied by Wang and Ruckenstein (2001), on alumina-supported cobalt. They have noted a strong dependence of the stability on cobalt amount. Stable catalysts were obtained for some loadings and calcination temperatures (6 wt% for 500 °C and 9 wt% for 1000 °C) but the other catalyst was deactivated. Two different deactivation mechanisms have been noted: coke deposition and metal oxidation, whose balance determined the stable activity.

The influence of metal amount, calcination temperature, kind of support, and alkali additives on the performance of nickel-based catalysts have been investigated by Xu et al. (2001a). They have found that these variables can determine the metal-support interactions and then the kind of active sites. In addition, it has been concluded that nickel species in weak interaction with the support were mobile and went on sintering and coking.

Catalysts prepared from perovskites have been extensively studied in this period, the precursors being obtained by different methods. Pietri et al. (2001), for instance, have investigated perovskites oxides such as $LaMO_3$ (M = Ru, Ni, Mn) prepared by sol-gel method using citrate, which was compared to solids obtained by wet impregnation. The $LaRu_{0.8}Ni_{0.2}O_3$ perovskite produced the catalyst which showed the highest activity and selectivity, being stable after 150 h on stream. On the other hand, Araújo et al. (2008) have prepared perovskites $LaRu_xNi_{-x}O_3$ ($0.0 < x < 1.0$) and have found that nickel showed higher activity and selectivity than ruthenium but the later led to more stable catalysts, decreasing coke production. All catalysts produced filamentous coke which was not toxic to the catalysts. The $LaNi_{0.8}Ru_{0.2}O_3$ catalyst was the most resistant against coke deposition. The apparent contradiction between the two works can be related to different conditions of preparation and evaluation of the catalysts. In another work (Batiot-Dupeyrat et al. 2003), it has been demonstrated that after perovskite reduction the catalyst was made of supported-lantana metallic nickel. Other perovskites have been used to prepare catalysts for dry reforming such as $Ln_{1-x}Ca_xRu_{0.8}Ni_{0.2}O_3$ (Ln = La, Sm, Nd). The highest performance was showed by $La_{0.8}Ca_{0.2}Ru_{0.8}Ni_{0.2}O_3$ and $La_{0.5}Ca_{0.5}Ru_{0.8}Ni_{0.2}O_3$ catalysts. It has been noted that the kind of A-site cations of perovskite is related to the stability and selectivity of the catalysts (Goldwasser et al. 2003). Later, Batiot-Dupeyrat et al. (2005) have demonstrated the efficiency of the auto-ignition method in the preparation of $LaNiO_3$ catalysts, which were highly active, stable, and resistant to coke deposition.

The potential of nanoparticles for preparing catalyst for dry reforming has also been demonstrated. Xu et al. (2001b) have found that nanoparticles (MgO) produced more active and stable nickel-based catalysts than the conventional magnesium oxide. In another work, Chen et al. (2005) have compared xerogel and aerogel catalysts ($Ni/CeO_2–Al_2O_3$) prepared by sol–gel combined with conventional drying and supercritical drying, respectively. The aerogel catalyst was more active, stable, and more resistant to sintering than the other one. These and other studies (Kim et al. 2000; Xu et al. 2001c) have shown that nanoparticles were able to minimize carbon deposition in fixed bed reactors, but their application on industrial scale was not feasible because the nanometric size of the particles would increase too much pressure drop in the catalytic bed. Then, the fluidized bed has been considered as more suitable option for using nanoparticles, because the pressure drop will not depend on particles diameter. However, some challenges still have to overcome such as the difficulty of fluidizing nanoparticles, because of their strong cohesive forces. Other authors have proposed soft alternatives to fluidize some nanopowders (aerogels) (Yao et al. 2002). In addition, external excitations, por instance, an acoustic field or vibration and a magnetic field, among others, are needed to achieve soft fluidization for nanopowders (Xu and Zhu 2006).

Several kinetics and mechanistic studies (Cui et al. 2006; Wang et al. 2007b) of the most efficient catalysts have been carried out in that period. Using steady-state tracing and transient techniques and in situ FTIR spectroscopy, Verykios (2003) has investigated the mechanism of dry reforming over alumina-supported rhodium. He concluded that the number of carbonaceous species was higher (0.2) than the amount of active oxygen (<0.02), both related to the formation of carbon monoxide. During reaction, three kinds of carbon-containing species (CxHy, $y \cong 0$) on the catalyst have been detected, the amounts of each one depending on temperature. It has been found that coke comes mostly from carbon dioxide than from methane whose contribution is small. Over nickel-based catalysts, Wei and Iglesia (2004) have proposed a simple and comprehensive mechanism for dry and steam reforming, decomposition of methane and water gas shift reaction, based on isotopic studies in which transport and thermodynamic artifacts were excluded. It has been found that both steam and dry reforming follow the same mechanism confirmed by the same forward methane rates, kinetic constant, kinetic isotopic effect, and activation energies for these reactions. These reactions have been shown to be limited only by C–H bond activation, which was the unique kinetically relevant reversible step. It has also been found out that activation of hydrogen, carbon dioxide, and water was reversible and achieved a quasi-equilibrium during reaction. The elementary reactions have allowed us to propose a predictive model for the growth of coke filament. It was noted that the carbon thermodynamic activity depended on several kinetic and thermodynamic features of elementary steps. These findings on nickel surfaces were similar to those previously found for supported rhodium, platinum, iridium, and ruthenium catalysts.

At the end of that decade, it was well-established that the metal and promoter loadings, the kind of support, and the preparation methods largely affect the catalysts' activity, and stability. Therefore, the selection of preparation variables combined with the kind of support and of promoters could allow the tailoring of the metal-support interaction, the metal size particles, and other characteristics leading to the possibility of designing more efficient catalysts for dry reforming.

4.4 Progress in Catalysts for Methane Dry Reforming (2010–2019)

In the last decade, the works have been developed by taking the role of the support in creating the active sites and the metal particles size into account, as well as the need of decreasing coke by using basic supports. Several excellent reviews have been devoted to the catalysts for methane dry reforming (Aramouni et al. 2018; Zhang et al. 2018b; Aziz et al. 2019; Abdulrasheed et al. 2019; Jang et al. 2019). In these papers, aspects as the role of active phase, supports, and dopants on the performance of the catalysts for dry reforming, as well as the thermodynamics, kinetics, and the mechanism of reaction have been discussed. Aramouni et al. (2018) have reported a truly clear analysis of the dry reforming catalysts addressing the requirements for designing efficient catalysts. In addition, Praserthdam and Balbuena (2018) have reported a method for evaluating catalysts for dry reforming through computational screening. Moreover, Jiang et al. (2019) have developed a fast screening of ternary rare earth for dry reforming catalysts over CeLa and CeZr oxides, by measuring the conversion and coke rates at low pressures and long times on stream. The catalysts with nickel and cobalt were the most stable, nickel strongly interacting with mixed oxides, even in the reduced state.

The number of paper have strongly increased in this period, motivated by new applications for the reaction as well as by combining dry reforming with steam reforming and/or partial reforming or autothermal reforming.

New catalysts have also been proposed in this decade, Table 4 displaying some examples. Both new and conventional catalysts have been prepared by several methods and the differences in performances were discussed. For instance, a new type of catalyst based on Ni_3Al intermetallic compound has proposed by Arkatova et al. (2011), by self-propagating high-temperature method combined with ion implantation. They have found that the addition of low amounts of platinum (<0.1 wt%) decreased coke formation and improved stability. These findings have been assigned to the ability of platinum in avoiding nickel sintering and in hindering coke formation by limiting bulk nickel carbide and then the formation of carbon filament. On other hand, Gaur et al. (2011) have investigated pyrochlore catalysts, such as lanthanum rhodium (LRZ), lanthanum nickel (LNZ), lanthanum calcium rhodium (LCRZ), and lanthanum strontium (LSZ) zirconium. LCRZ showed the highest performance in methane dry reforming, being the most active, stable, and resistant against coke deposition. The activity decreased in the order calcium rhodium > rhodium > nickel > strontium (no activity), this order being related to rhodium. The nickel-based catalyst showed the highest amount of coke while the calcium rhodium sample showed the lowest one. Moreover, Chen et al. (2012) have proposed a microfibrous entrapped composite catalyst (Cu-MFE-Ni/Al_2O_3), using a sintered-locked microfibrous carrier (~2 vol% of 8 μm diameter copper microfibers) to entrap ~27 vol% of 0.15–0.18 mm diameter Ni/Al_2O_3 particles. This arrangement has the advantages of providing high heat and mass transfer besides high mechanical resistance. These catalysts showed higher activity and stability than alumina-supported nickel, showing higher resistance to coke deposition. In addition, these catalysts allow to combine small particles (0.15–0.18 mm) with an open structure with void volume (71.3 vol %) improving the mass transfer and thus preventing pressure drop in the catalytic bed. In another study, Liu et al. (2016) have prepared alumina nanofiber-supported nickel using electrospinning technique to obtain nanostructured catalysts for dry reforming. Metallic nickel nanoparticles were highly

Table 4 Progress in catalyst for dry reforming (2010–2019)

Active phase	Support	Promoter	Reference
Ni–Mg-Al–Ce			Daza et al. (2010a)
La_2NiO_4/α–Al_2O_3			Barros et al. (2010)
Ni	Phyllosilicates		Sivaiah et al. (2010)
Ni	Al_2O_3	Ba	García-Diéguez et al. (2010)
Ni	MCM-41	Rh	Arbag et al. (2010)
Co	ZrO_2	La, Ce, Mn, Mg, K	Özkara-Aydınoğlu and Aksoylu (2010)
Ru	$Ce_{0.75}Zr_{0.25}O_2$		Chen et al. (2010)
Ce–Zr–Ni-Co(Fe, Rh)			Koubaissy et al. (2010)
Pt	MCM-41	Co, Ni	Liu et al. (2010)
Pt	Ni_3Al		Arkatova et al. (2011)
Ni	Al_2O_3–ZrO_2	CeO_2	Li et al. (2011)
Ni	La_2O_3	Mg, Ca, Sr	Sutthiumporn and Kawi (2011)
Ni	SiO_2	La, Mg, Co, Zn,	Zhu et al. (2011)
Ni	Ce–Zr-oxide	Ru, Rh	Horváth et al. (2011)
CoMgAl oxide			Gennequin et al. (2011)
Rh	Al_2O_3	TiO_2, V_2O_5	Sarusi et al. (2011)
Co	Al_2O_3	K, Sr	San José-Alonso et al. (2011)
Ni	$CeO_2ZrO_2MgAl_2O$		Corthals et al. 2011)
Ni	Si_3N_4		Shang et al. (2011)
Ni	CeZr	Rh, Ru	Pietraszek et al. (2011)
Co-Ni	Al_2O_3	Ce, Pr Sm	Foo et al. (2011)
$La_{0.8}Sr_{0.2}Ni_{0.8}M_{0.2}O_3$ perovskite	M = Bi, Co, Cr, Cu, Fe		Sutthiumporn et al. (2012)
Ni	γ-Al_2O_3	Rare Earth	Zeng et al. (2012)
Pd	Al_2O_3	K, Ca, Y, Mn, Cu	Shia and Zhang (2012)
Ni	γ-Al_2O_3	Yb	Amin et al. (2012)
Ba-aluminates			Gardner et al. (2013)
Ni	SiO_2	Mn, Zr	Yao et al. (2013)
Ni	SiC	Yb_2O_3	Guo et al. (2014)
Co	Activated carbon		Zhang et al. (2014)
Ni	SBA-15	Y_2O_3	Li et al. (2014)
Co	γ-Al_2O_3	Sr	Fakeeha et al. (2014)
Ni-Co	γ-Al_2O_3	Sr	Al-Fatesh (2015)
Ni-Co	β-SiC/$CeZrO_2$, γ-Al_2O_3/$CeZrO_2$		Aw et al. (2015)
Ni	Diatomite		Jabbour et al. (2015)
$La_{1-x}Ce_xNi_{1-y}Zn_y\ O_3$			Marmarshahi et al. (2015)
$La_{0.9}M_{0.1}Ni_{0.5}Fe_{0.5}O_3$		M = Sr, Ca	Yang et al. (2015)
Co	Activated carbon	Zr	Zhang et al. (2015)
Co	Nd_2O_3		Ayodele et al. (2016)
Ni	SiC	Sm	Bing et al. (2016)
Ni	MgO–ZnO		Singha et al. (2016)
Ni	$Ce_{1-x}Pr_xO_{2-\delta}$		Vasiliades et al. (2016)
Co	Activated carbon	Ca	Zhang et al. (2017a)
C–Ni/MgO–Al_2O_3			Jin et al. (2017)
Activated carbon	$KMnO_4$		Zhang et al. (2017b)

(continued)

Table 4 (continued)

Active phase	Support	Promoter	Reference
Ni	$ZnAl_2O_4$	Ce	Movasati et al. (2017)
Ni	SBA-16	N	Huo et al. (2017)
$SmCoO_3$			Osazuwa et al. (2017)
Ni-Co–CaO–ZrO_2			Gurav et al. (2017)
Ni	CeO_2	Gd	Stroud et al. (2017)
Ni	CeO_2–Al_2O_3	Sn	Kwon et al. (2017)
$Sr_{0.92}Y_{0.08}TiO_3$		Ni	Ramezani et al. (2018)
Ni	γ-Al_2O_3	Mn, Mg	Park et al. 2018)
Co	ZrO_2	Zn	Al-Fatesh et al. (2018)
Ni–Co	ZrO_2–Al_2O_3	Rh	Al-Fatesh et al. (2019)
Ni	MCM-41	Gd	Shah et al. (2019)
Ni	TiO_2–Al_2O_3		Li et al. (2019)
Ni	MoCeZr/$MgAl_2O_4$–MgO		Yentekakis et al. (2019)
Rh	CeO_2–ZrO		Hossain et al. (2019)
Ni	$CaFe_2O_4$		Jiang et al. (2019)
Ternary rare earth	CeLa, CeZr oxides		

dispersed (3.4 nm), leading to higher activity and resistance against coke than a sample obtained by incipient impregnation using commercial alumina. Aiming to find low cost supports, Talkhoncheh et al. (2017) have studied Al_2O_3-clinoptilolite-CeO_2 with different Al_2O_3 and clinoptilolite contents for supporting nickel nanoparticles, aiming to replace alumina with clinoptilolite. It was found that nickel crystallites decreased, and the specific surface area increased with increasing aluminum content. The performance of the catalyst with 60% of alumina was close to the catalyst with 80% of alumina which was the most efficient sample.

Several nanostructured materials have been prepared to aim to obtain high metal dispersion. García-Diéguez et al. (2010) have prepared nickel and platinum deposited on nanofibrous alumina for methane dry reforming. Well dispersed metal particles were obtained and NiO was formed instead of $NiAl_2O_4$ facilitating nickel reduction. Platinum and the support improved the activity and stability and decreased the operating temperature as compared to nickel supported on commercial alumina. In another work, Arkatova et al. (2011) have used the surfactant-assisted precipitation method to obtain nickel supported on magnesia nanoparticles. The well dispersed of reduced nickel and the basic support was believed to be responsible for the high efficiency, regarding activity and stability) of the catalyst. On the other hand, Rahemi et al. (2013) have prepared nickel supported on alumina and on alumina–zirconia nanocatalysts through impregnation and treatment with non-thermal plasma. This treatment led to well dispersed nanoparticles strongly interacting with the support. The Ni/Al_2O_3–ZrO_2 nanocatalyst showed the smallest nickel particles and the narrowest particles size distribution, as well as the highest activity and stability. Singh and Madras (2016) have used the sonochemical method to obtain well dispersed platinum and ruthenium supported on titania. The catalysts were highly active for methane reforming reaction, a fact that was related to the high metal dispersion and high oxygen storage capacity.

Frontera et al. (2012) have studied Ni-based silicalite catalyst for drying reforming, aiming to state the action of silanol groups on the efficiency of the catalysts. The amounts of defect groups on silicalite were changed by the age of gel, followed by heating, ionic exchange, and silylation procedures. The last one favored the production of smaller and more reducible nickel particles that improved conversion and decreased coke deposition. In a further report (Frontera et al. 2013), they have described the effect of silcalite, MCM-41 and, ITQ-6 on the catalytic properties of nickel-based solids for dry reforming and have found that nickel species and then the catalytic activity depended on the heterogeneity of the support surface.

By comparing the combustion and precipitation methods, Daza et al. (2010a) have studied cerium-promoted nickel catalysts by the reconstruction of Ni–Mg-Al periclase-type mixed oxide. The combustion route improved the specific surface areas and the basicity as compared to the precipitation one, giving rise to high active and selective catalysts with reduced coke. In a further work Daza et al. (2010b) have studied the same catalysts but prepared them through the reconstruction technique using the $[Ce(EDTA)]^-$

complex. In this case, the reconstruction of the mixed oxide was only partial and occurred in the external edges of the particles. They have also noted that the conversion increased, and coke decreased as cerium loading increased, this being related to the effects of $[Ce(EDTA)]^-$ on the basic character and on the metal reducibility. In addition, Barros et al. (2010) have compared the microwave assisted self-combustion method with the incipient wetness impregnation. Glycine or urea (fuel) and metal nitrates (oxidizers) were used to obtain $La_2NiO_4/\alpha\text{-}Al_2O_3$ catalysts precursors. All samples prepared by combustion were active and stable whereas the catalyst prepared by impregnation showed high coke deposition, in the form of carbon nanotubes (multi-walled). Moreover, Sivaiah et al. (2010) have used hydrothermal synthesis to prepare phyllosilicates (PS) with nickel to be used as catalyst precursors. The Ni-based 2:1 phyllosilicate catalyst showed higher thermal stability and activity than the Ni 1:1 PS compounds than a conventional catalyst. Nickel was partially reduced because of the thermally stable PS, producing highly dispersed nickel oxide nanoparticles.

Zeolites and mesoporous materials have continued to be studied in that period. Ballarini et al. (2019), for instance, have compared K-L zeolite with other basic solids (potassium-doped alumina, potassium and magnesium-doped Al mixed oxide, and magnesium oxide) for platinum-based catalysts. They have found that the amount of basic sites decreased from magnesia, potassium-doped Mg-Al, potassium-doped alumina, and potassium-doped L zeolite, which was the same order as the activity. Magnesia-supported platinum showed the highest performance because of its highest amount of strong basic sites. On the other hand, Shen et al. (2011) have prepared ordered mesoporous nickel-magnesium–aluminum oxides through evaporation induce self-assembly, utilizing Pluronic P53 (soft template). These catalysts were more active in methane dry reforming than the respective conventional solids. The ordered mesoporous structure made nickel reduction easier, allowing the decrease of nickel reduction temperature for 50–100 °C. The mesoporous structure made nickel reduction easier and prevented sintering, besides improving the resistance against coke deposition. Mesoporous ZSM-5-supported nickel nanoparticles have also been prepared (Sarkar et al. 2012), being active and stable up to 5 h during methane dry reforming. The hard template methods have also been used for preparing catalysts for dry reforming, such as ceria using KIT-6 SiO_2 as template (Djinović et al. 2012), which was further impregnated with a solution of rhodium precursor. Well dispersed rhodium particles (2.8 nm) on ceria were obtained. The catalyst was active and stable showing 11% of decrease in activity after 70 h in stream. Aiming to improve the nickel dispersion over the support, Liu et al. (2014) have prepared nickel supported on MCM-41 using glucose (Glu) and β-cyclodextrin (βCD). The modified catalysts showed smaller nickel particles size and higher activities and stability than Ni/MCM-41. In addition, Xu et al. (2016) have prepared cobalt-based ordered mesoporous materials containing magnesium and calcium by facile evaporation induced self-assembly (EISA) method. The in situ incorporated cobalt active centers were stabilized by the mesoporous framework, preventing cobalt sintering during reaction. Magnesium and barium increased the surface basicity and then carbon dioxide chemisorption, favoring coke elimination. In another work, mesoporous nickel-lanthanum-silicon mixed oxides with different lanthanum amounts have been prepared by Chen et al. (2019) using polyethylene glycol, tetraethoxysilane, and inorganic salts, in a solution of nitric acid in water and ethylene glycol. The solids showed narrow pore size distributions, high specific surface areas, and nickel nanocrystallites (5–7 nm). The catalysts were active and resistant to coke deposition, lanthanum increasing the metal dispersion and stability. The catalyst was able to inhibit the reverse water gas shift reaction and showed near 100% of selectivity to hydrogen over the catalyst with 3.0 wt% La.

The core/shell type catalysts obtained by different methods were also evaluated in dry reforming. Kang et al. (2011) have prepared nickel supported on alumina and on magnesia-alumina, using multi-bubble sonoluminescence (MBSL). All catalysts led to conversions close to equilibrium and showed high stability, the sample 10% Ni being stable for 150 h. On the other hand, Xu et al. (2017b) have encapsulated nickel oxide-magnesium oxide nanoparticles into silica matrices o obtain the core-shell structure. The catalyst was active at low temperatures (670 °C) leading to conversions above 80 and 70% of methane and carbon dioxide, respectively. It has been concluded that the confined structure improved the resistance against coke and preventing the sintering of active sites as compared to conventional catalysts. By using the microemulsion method, Zhang et al. (2019) have obtained a core–shell structure for nickel nanoparticles supported on silica, with different particles sizes and metal-support interaction after calcination at different temperatures. They showed different performances, the catalyst calcined at 600 °C showing the highest activity and stability. This was related to the smallest nickel nanoparticles and to the lowest amount of deposited coke.

Hydrotalcites have also been used as catalysts precursors in this decade. Yu et al. (2012) have prepared lanthanum-doped nickel-magnesium–aluminum catalysts, obtained from hydrotalcite prepared by precipitation. They have found that the catalysts were slightly more active than other lanthanum-containing solids. A combined support obtained from hydrotalcite and SBA-15 have been proposed by Zuo et al. (2013), by adding SBA-15 calcined at 550 °C to the hydrotalcite suspension and then calcined at 550 and

700 °C. Other catalysts were obtained by mixing the suspensions of hydrotalcites and SBA-15. Another catalyst was obtained by adding hydrotalcite calcined at 700 °C to a hydrotalcite suspension. All catalysts were active and stable at 800 °C for 50 h on stream. However, only the sample prepared by adding hydrotalcite (calcined at 700 °C) to SBA-15 suspension was stable for 500 h on stream. This was related to the strong interaction between the metal and the support and to the porosity of SBA-15. Hydrotalcites-based catalysts have also been prepared using cold plasma. Long et al. (2013) have prepared Ni-Co bimetallic catalysts by cold plasma jet decomposition, followed by reduction of hydrotalcites based on nickel, cobalt, magnesium, and aluminum. The hydrotalcite-like precursors were totally decomposed and partly reduced by cold plasma jet. The catalyst with Ni/Co = 8/2 showed the highest activity and stability (100 h). This has been associated with the well dispersed metal, leading to the smallest metal particles size and to the improved interaction between nickel and cobalt, provided by the preparation method.

Carbonaceous materials have been more studied in this decade than in the past. Xu et al. (2014) have compared the catalytic performance of raw activated carbon or modified with HNO_3 and $NaNO_3$ in dry reforming. It has been found that the activities were different at low temperatures but similar at high temperatures. $NaNO_3$ was more efficient to produce mesopores, surface oxygenated groups, and a catalyst more resistant against coke formation. On the other hand, HNO_3 produced a catalyst active in methane cracking. Carbon nanotubes have also been evaluated as catalytic support for drying reforming. Aiming to state the effect of the position of active sites, Ma et al. (2013) have deposited nickel nanoparticles on the external and internal sides of carbon nanotubes using the wet chemical method. The catalyst with inside nanoparticles was more active and stable than the others. This finding was related to a different electric densities between inside and outside nanotubes as well as to the confinement effect. In another work, Jin et al. (2017) have used carbon obtained from sucrose to prepare a carbon-magnesia-supported nickel-alumina composite from nickel-magnesium–aluminum layered double hydroxide (LDH) as nickel precursor. It has been found that the pretreated temperature of LDH, the carbonization temperature of composite, and the sucrose/layered double oxide (LDO) mass ratio affected the structure and the catalytic performances. Carbon caused an increase of specific surface area, prevented coke, and inhibited nickel sintering, resulting in catalytic stability. The highest activity and stability were shown by LDH pretreated at 350 °C, carbonized at 800 °C, and obtained using a sucrose/LDO ratio of 1.2. In addition, Li et al. (2018) have prepared nanoparticles of biochar (BC)-supported molybdenum carbide by carburization of molybdate salts. The catalysts were prepared by incipient wetness impregnation followed by carburization and have shown to be promising for dry reforming.

The solid solution has also shown potential for drying reforming catalysts. Lu et al. (2018) have found that cerium, zirconium, or co-doped flower-like NiO–MgO solid solution catalysts, obtained by solvothermal synthesis, were promising catalysts for dry reforming. When cerium and zirconium were added to $Ni_{0.1}Mg_{0.90}$, the activity and the coke resistance of (111) facets were increased, due to smaller nickel particles, resulting from the stronger metal-support interactions. In addition, the amount of surface oxygen increased improving CO_2 adsorption. The Ce/Zr(molar = 0.01:0.001) was the most active and stable catalyst.

Silica foam has been also considered in the formulation of the catalysts for dry reforming. Daoura et al. (2017) have prepared nickel supported on mesocellular silica foam (MCF) catalysts and used different methods for adding nickel: impregnation, two-solvents post-synthesis, and direct one-pot synthesis introduction. The last one produced the most promising catalyst since it increased the metal-support interaction and then produced the smallest nanoparticles.

4.5 Current Status in the Catalysts for Methane Dry Reforming

Figure 2 shows a summary of the progress in the catalysts for methane drying reforming over the years. We can see that the greatest progress in the knowledge of the reaction and of the catalyst action has occurred from 1990 to 1999. Using this understanding, in the following decades, the research has been addressed to new preparation methods to achieve suitable interactions among the catalyst components as well as to produce small metal particles. During all decades, new materials have been proposed and prepared by different methods. Since the end of 2000–2009, the understanding of the role of each component of the catalysts, as well as of the kinetics and mechanism, have allowed the researcher to design the catalysts for drying reforming (Aramouni et al. 2018) by exploring different preparation methods to achieve the desirable properties. From 1928 to April 2020, the number of research and review papers has increased over the years, as shown in Fig. 3. The strong increase noted in the last decade is mostly related to the use of steam or oxygen in dry reforming and dry reforming of biogas, as well as the interest in the reaction for other applications.

Up to April 2020, the same tendency has been noted. The preparation methods have continued to be developed and used for obtaining known and new catalysts. Some examples included the preparation of nickel-based catalyst by steam

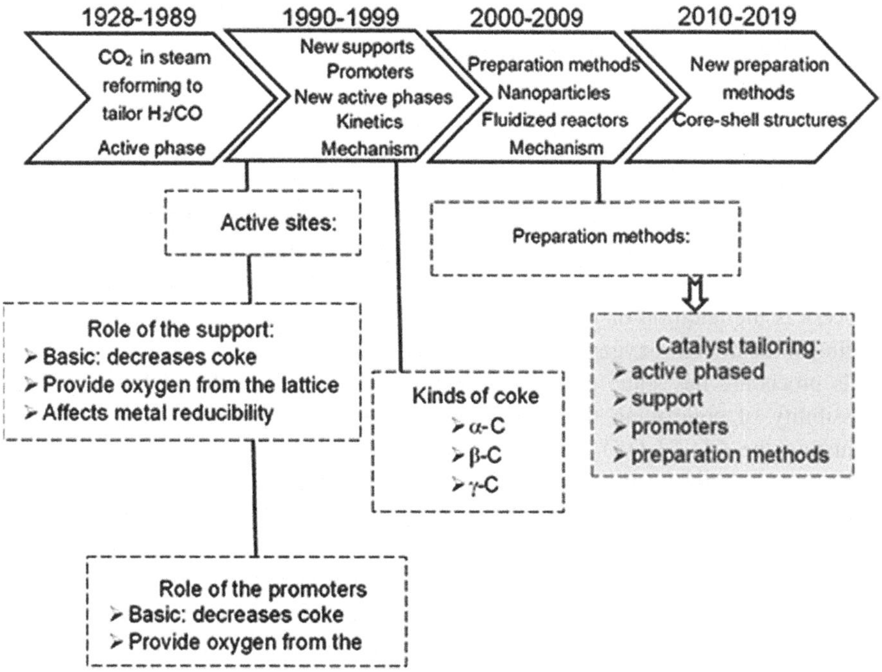

Fig. 2 Progress in catalysts for methane dry reforming over the years

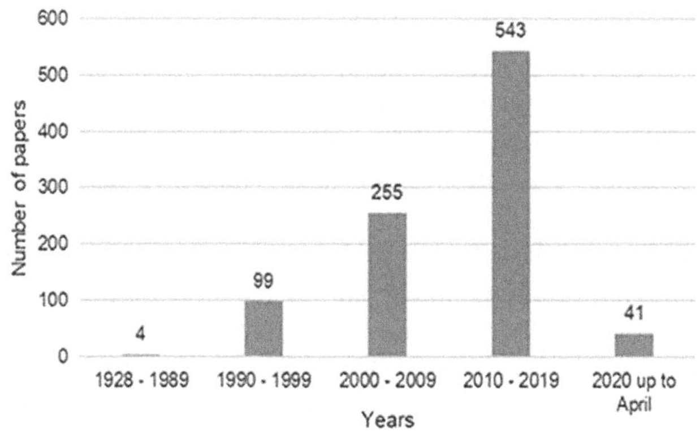

Fig. 3 Number research and review papers over the years

gasification and impregnation-pyrolysis-reduction (Zhang et al. 2020), perovskites-derived catalysts (Anil et al. 2020), hydrotalcites-derived catalysts (Zhan et al. 2020), hydroxyapatite-supported nickel catalysts (Li et al. 2020) and Ni-CeO_2 nanocrystallites decorating Al_2O_3 nanoparticle clusters (Liang et al. 2020), among others. In addition, Shah et al. (2020) have carried out the design and optimization of nanocrystalline mesoporous nickel-based catalyst using a statistically designed experiment. A significant improvement of the catalysts has been demonstrated, because of the optimization, for platinum supported on titania-alumina mixed oxide with magnesium as promoter. The optimized catalyst showed high stability on stream for 100 h and no coke deposition or metal sintering was noted.

5 Dry Reforming of Other Compounds

Besides methane, other hydrocarbons have been evaluated in dry reforming. As early as 1984, Sakai et al. (1984) have investigated the potential of alumina-supported rhodium in dry reforming of toluene, heptane, cyclohexane, and methylcyclohexane. At temperatures around 550 °C, they were all reactive, although lesser than methane. The reactivity of cyclohexane and methylcyclohexane was much higher than of toluene and heptane, this being assigned to simultaneous dehydrogenation of naphthene to its aromatic compound. Then, their intrinsic reactivity to form carbon monoxide is expected to be comparable to that of toluene.

Over the years, several catalysts have been evaluated in dry reforming of propane (Sudhakaran et al. 2018), light alkanes, C_2–C_4 (Olafsen et al. 2006), n-heptane (Puolakka and Krause 2004), toluene (Chen et al. 2013), dimethyl ether (Ma et al. 2012), ethanol, glycerol, and others (Aziz et al. 2019). In a paper in which the benefits of using carbon dioxide in hydrocarbons reforming were discussed, Adesina (2012) has emphasized the additional benefits of carbon dioxide such as the possibility of tailoring the H_2:CO ratio to provide the required ratio to produce olefins or oxygenates improved product yield, and catalyst longevity. Recently, Aziz et al. (2019) have published an excellent review discussing several aspects of the reaction such as thermodynamics, kinetics, and the requirements of the catalysis for the reaction.

6 Use of Steam or Oxygen in Dry Reforming of Methane and Other Compounds

Over the years, several works have demonstrated that the main limitation for the commercial use of drying reforming is the fast carbon deposition, which deactivates the catalyst. In order to overcome this drawback, several catalysts have been improved or new catalysts have been proposed. Other alternative to make dry reforming a viable commercial process is the addition of water or oxygen to the feed providing an additional oxygen source for the removal of coke. This procedure has some practical advantages such as the possibility of controlling the H_2/CO ratios in syngas, by adjusting the $CO_2/H_2O/O_2$ feed ratio. In addition, as dry reforming is highly endothermic, its combination with the exothermic partial oxidation would minimize the energy requirement and then decrease the process costs.

Several papers have been published on these combined processes so-called oxy-reforming. Choudhary et al. (1998), for example, have evaluated the simultaneous utilization of steam and carbon dioxide on catalysts based on magnesia-supported nickel and have obtained syngas with $H_2/CO \approx 2$. They also confirmed that the syngas composition could be adjusted by varying the amount of steam and carbon dioxide in the feed, as previously reported by Richardson and Paripatyadar (1990). In another work, (Huang et al. 2008) the influence of magnesia on the properties of Ni/SBA-15 has been studied. The catalysts were evaluated in combined steam and carbon dioxide reforming of methane. The simultaneous use of these gases has changed the performance of the undoped catalyst while the doped catalysts operated for 620 h showing a drop of conversion. The main reasons for deactivation were coke deposition and nickel sintering. Magnesia doping improved nickel dispersion through nickel-magnesia interactions and this also increased the carbon dioxide adsorption, depressing coke formation. Moreover, Guo et al. (2008) have evaluated Ni/SiO_2 catalysts doped with different amounts of Gd_2O_3 to obtain a catalyst with higher ability to adsorb and activate carbon dioxide because of carbonate species produced on the surface. A strong interaction among nickel, Gd_2O_3 and SiO_2 was detected, improving the nickel dispersion. The doped catalysts showed activity and stability in combined oxy-CO_2 reforming of methane.

In the last years, tri-reforming of methane also has received attention because of the possibility of obtaining H_2/CO ratio for methanol production and Fischer–Tropsch synthesis. This process combines steam reforming, partial oxidation, dry reforming, and the water–gas shift reaction (Abdullah et al. 2017).

7 Solid Oxide Fuel Cells Fueled with Biogas

Aiming to meet the global demands for efficient and green technologies for generating energy, the fuel cell fed with biogas emerges as one of the most promising options. A solid oxide fuel cell (SOFC) is an electrochemical device in which chemical energy of a fuel and an oxidant gas (air) is converted into electricity without irreversible oxidation. Because it does not have limitations by Carnot cycle, this device can achieve higher electrical efficiencies than the internal combustion engines. When fueled with hydrogen, the SOFC produces only water, avoiding pollution. Therefore, many works have been devoted to speed up the commercialization of SOFCs. Moreover, it has been recognized that the generation of hydrogen on board, from other fuels, is more feasible for practical applications.

From the environmental viewpoint, the utilization of biogas to fuel the SOFCs is extremely attractive, proving the generation of clean electric energy with high efficiency. Biogas is constituted by methane (50–60%), carbon dioxide (40–50%), and some impurities, produced by the anaerobic fermentation of biomass. Due to its composition, the most suitable process for generation of hydrogen on board a SOFC is the dry reforming of methane. Therefore, several papers have been devoted both to the development of SOFCs and to dry reforming of biogas. The viability of this technology has already been demonstrated by Barelli and Ottaviano (2019). Recently, Saadabadi et al. (2019) have pointed out the potential and the constraints of this technology. Regarding the internal reforming of biogas on the anode of SOFCs, they have pointed out some limitations for commercial operations. One problem is that only a few studies have been addresses to kinetics of biogas reaction and to the conditions of anode, besides the lack of research on the influence of contaminants (H_2S) found in biogas. Therefore, the developing of sulfur tolerant anodes can contribute to the development of the biogas-SOFC. Moreover, the conventional Ni-based anodes might decrease carbon deposition.

8 Commercialization of Dry Reforming Process

Despite the economic and environmental incentives, the dry reforming process is not a mature technology for commercial applications. Besides the inherent advantages of consuming two greenhouse gases, this process can operate at lower temperatures and pressures, such as 850 °C and 1 bar, achieving 100% of conversion (Mondal et al. 2016). In addition, dry reforming produces hydrogen to carbon monoxide ratio limited to 1 while the steam methane reforming and the autothermic reforming produces rates of around 3 and 2.5, respectively (Lavoie 2014; Martins et al. 2016). This makes dry reforming extremely advantageous to Fischer–Tropsch synthesis to obtain long-chain hydrocarbons and several other chemicals, such as methanol (Peña et al. 1996). Moreover, methane steam reforming is a mature technology used in commercial processes to produce pure hydrogen (Balthasar 1984).

Regarding industrial applications, in 1994 Rostrup-Nielsen (Jarvis and Samsatli 2018) has demonstrated that dry reforming of natural gas was a feasible technology, the replacement of steam by carbon dioxide in the reforming reaction having no significant impact on the reaction mechanism. However, this replacement increased the carbon formation, this being solved by using noble metal catalysts or sulfur-passivated nickel catalysst. It was noted the same sequence obtained in steam reforming: ruthenium, rhodium > iridium > nickel, palladium, platinum but the use of carbon dioxide instead of steam decreased the activity depending on the metal. No coke was formed over the noble metals a fact that was assigned to their ability in not dissolving surface carbon as does nickel. However, the availability of these metals is limited and then they may be important only for specific applications. The sulfur-passivated nickel catalyst represents a less expensive solution for large scale applications. In this case, sulfur blocks the arrangement of nickel sites active in nucleation of carbon avoiding coke formation, while enough active nickel clusters are kept for catalyzing dry reforming. In these studies, it has also been pointed out that the addition of carbon dioxide in catalytic autothermal reforming is beneficial for the reaction, such as in industrial steam reforming. Therefore, dry reforming may contribute to large scale conversion when carbon dioxide-rich natural gas or carbon dioxide-rich streams are available. However, the feasibility of the reaction depends on several factors including the availability of cheap oxygen for the competing partial oxidation processes.

Therefore, this technology still has many difficulties to be overcome, the main one related to catalyst development with a long life on stream at a price enough low for providing profit-oriented commercialization (Aramouni et al. 2018). The conventional catalysts quickly deactivate because of coke deposition and sintering (Muraza and Galadima 2015). Also, the industrial application for CCU in dry reforming is still limited (Er-Rbib et al. 2012), although it is currently developed on a commercial scale for the steel industry. Another limitation (Jangam et al. 2019) is the endothermicity of reaction which, however, can be currently overcame using renewable energy sources such as solar energy.

Even though, in 2015 the Linde Group has installed the first pilot facility of dry reforming at Pullach near Munich (The Linde Group. 2015), as a result of a research effort in partnership with BASF and others. The process used two catalysts based on nickel and on cobalt, aiming to get data at longer-term and longer-scale to study and optimize different approaches which could be used in the design of a commercial plant. This initiative greatly motivates the research efforts for the commercial application of dry reforming to produce syngas (Jarvis and Samsatli 2018). The other two industrial pilot plants successfully implemented are the SPRAG and CALCOR projects (Er-Rbib et al. 2012).

The studies involving the industrial scale of dry reforming have been performed by Mortensen and Dybkjær (2015), leading to the conclusions that this reaction is very similar to steam reforming and then the kinetic models, rate controlling steps, and reaction mechanism can be handled similarly across the two reactions. The tendencies for catalyst activity are also very close, nickel being the most used catalyst because of its activity (close to noble metals) and low price. This justifies the efforts on improving nickel catalysts by suitable support and promoters. These industrial scale studies also have shown that carbon deposition is indeed the biggest challenge of dry reforming since the stoichiometry of reaction implies favorable conditions for carbon formation. This can be controlled by suitable operating conditions or by the nickel particles size, but water cannot be completely excluded from the reaction. It means that dry reforming is difficult to operate on a large scale with realistic feedstocks and some water should be added.

The current design of the catalysts for dry reforming involves the use of intrinsic properties of catalyst components and their synergistic interactions for optimum performance and longevity. Over the years, different catalyst configurations and morphologies have been evaluated to study how the synergistic interactions among the components affect the catalyst properties such as metal dispersion, basicity, redox property, oxygen mobility, particle size, size distribution, reducibility, and mass transfer limitations of catalysts. The balance between the catalyst cost and its performance is the main challenge for industrial scale of dry reforming.

9 Conclusions

Over the years, the research on dry reforming has continuously increased, leading to the development of new and more efficient catalysts allowing the use of the reaction for several purposes and applications. The greatest progress in the catalyst development occurred from 1990 to 1999. Currently, the design of the catalysts takes the intrinsic properties of catalyst components and of their synergistic interactions into account to obtain optimum performance and longevity. Over the years, different catalyst configurations and morphologies have been evaluated to study how the synergistic interactions among the components affect the catalyst properties such as metal dispersion, basicity, redox property, oxygen mobility, particles size, particles size distribution, reducibility, and mass transfer limitations of catalysts. The balance between the catalyst cost and its performance is the main challenge for industrial scale of dry reforming. In spite of the inherent advantages of consuming two greenhouse gases and the operation at low temperatures and pressures (850 °C, 1 bar) the process is not a mature technology for commercial applications. The main difficulty is related to catalyst development with a long life on stream at a price enough low for commercialization. Another limitation is the endothermicity of reaction which can be currently overcome using renewable energy sources.

References

Abdullah B, Ghani NAA, Vo DVN (2017) Recent advances in dry reforming of methane over Ni-based catalysts. J Clean Prod 162:170–185. https://doi.org/10.1016/j.jclepro.2017.05.176

Abdulrasheed A, Jalil AA, Gambo Y, Ibrahim M, Hambali HU, Hamid MYS (2019). A review on catalyst development for dry reforming of methane to syngas: recent advances. Renew Sust Energ Rev 108:175–193. https://doi.org/10.1016/j.rser.2019.03.054

Adesina AA (2012) The role of CO_2 in hydrocarbon reforming catalysis: friend or foe? Curr Opin Chem Eng 3:272–280. https://doi.org/10.1016/j.coche.2012.03.001

Ahmed R, Liu G, Yousaf B, Abbas Q, Ullah H (2020) Recent advances in carbon-based renewable adsorbent for selective carbon dioxide capture and separation—a review. J Clean Prod 242:18409. https://doi.org/10.1016/j.jclepro.2019.118409

Al-Fatesh A (2015) Suppression of carbon formation in CH_4–CO_2 reforming by addition of Sr into bimetallic Ni–Co/γ-Al_2O_3 catalyst. J King Saud Univ Sci 27:101–107. https://doi.org/10.1016/j.jksues.2013.09.006

Al-Fatesh A, Singh SK, Kanade GS, Atia H, Fakeeha AH, Ibrahim AA, El-Toni AM, Labhasetwar NK (2018) Rh promoted and ZrO_2/Al_2O_3 supported Ni/Co based catalysts: high activity for CO_2 reforming, steam–CO_2 reforming and oxy–CO_2 reforming of CH_4. Int J Hydrog Energy 43:12069–12080. https://doi.org/10.1016/j.ijhydene.2018.04.152

Al-Fatesh AS, Atia H, Ibrahim AA, Fakeeha AH, Sing SK, Labhsetwar NK, Shaikh H, Qasim SO (2019) CO_2 reforming of CH_4: effect of Gd as promoter for Ni supported over MCM-41 as catalyst. Renew Energy 140:658–667. https://doi.org/10.1016/j.renene.2019.03.082

Alper E, Orhan OY (2017) CO_2 utilization: developments in conversion processes. Petroleum 3:109–126. https://doi.org/10.1016/j.petlm.2016.11.003

Amin MH, Tardio J, Bhargava SK (2012) An investigation on the role of ytterbium in ytterbium promoted γ-alumina-supported nickel catalysts for dry reforming of methane. Int J Hydrog Energy 38:14223–14231. https://doi.org/10.1016/j.ijhydene.2013.08.040

Anil C, Modak JM, Madras G (2020) Syngas production via CO_2 reforming of methane over noble metal (Ru, Pt, and Pd) doped $LaAlO_3$ perovskite catalyst. Mol Catal 484:110805. https://doi.org/10.1016/j.mcat.2020.110805

Aramouni NAK, Toum JG, Tarboush BA, Zeaiter J, Ahmad MN (2018) Catalyst design for dry reforming of methane: analysis review. Renew Sust Energ Rev 82:2570–2585. https://doi.org/10.1016/j.rser.2017.09.076

Araújo GC, Lima SM, Assaf JM, Peña MA, Fierro JLG, Rangel MC (2008) Catalytic evaluation of perovskite-type oxide $LaNi_{(1-x)}$-Ru_xO_3 in methane dry reforming. Catal Today 133–135:129–135. https://doi.org/10.1016/j.cattod.2007.12.049

Arbag H, Yasyerli S, Yasyerli N, Dogu G (2010) Activity and stability enhancement of Ni-MCM-41 catalysts by Rh incorporation for hydrogen from dry reforming of methane. Int J Hydrog Energy 35:2296–2304. https://doi.org/10.1016/j.ijhydene.2009.12.109

Aresta MF, Forti G (eds) (1987) Carbon dioxide as a source of carbon: biochemical and chemical uses. Nato Science Series C Springer, Netherlands

Aresta MF (2010) Carbon dioxide as chemical feedstock. Wiley-VCH

Aresta MM, Dibenedetto A (2007) Utilisation of CO_2 as a chemical feedstock opportunities and challenges. Dalton Trans 2975–2992. https://doi.org/10.1039/B700658F

Aresta MM, Dibenedetto A, Angelini A (2013) The changing paradigm in CO_2 utilization. J CO2 Utilization 3–4:65–73. https://doi.org/10.1016/j.jcou.2013.08.001

Aresta M, Nocito F, Dibenedetto A (2018) Chapter two—what catalysis can do for boosting CO_2 utilization. Adv Catal 62:49–111. https://doi.org/10.1016/bs.acat.2018.08.002

Arkatova LA, Pakhnutov OV, Shmakov AN, Naiborodenko YS, Kasatsky NG (2011) Pt-implanted intermetallides as the catalysts for CH_4–CO_2 reforming. Catal Today 171:156–167. https://doi.org/10.1016/j.cattod.2011.03.014

Asami K, Li X, Fujimoto K, Koyama Y, Sakurama A, Kometani N, Yonezawa Y (2003) CO_2 reforming of CH_4 over ceria-supported metal catalysts. Catal Today 84:27–31. https://doi.org/10.1016/S0920-5861(03)00297-9

Ashcroft AT, Cheetham AK, Green MLH, Vernon PDF (1991) Partial oxidation of methane to synthesis gas using carbon dioxide. Nature 352:225–226. https://doi.org/10.1038/352225a0

Aw MS, Zorko M, Djinović P, Pintar A (2015) Insights into durable NiCo catalysts on β-SiC/$CeZrO_2$ and γ-Al_2O_3/$CeZrO_2$ advanced supports prepared from facile methods for CH_4–CO_2 dry reforming. Appl Catal B Environm 164:100–112. https://doi.org/10.1016/j.apcatb.2014.09.012

Ayodele BV, Hossain SS, Lam SS, Osazuwa OU, Khan MR, Cheng CK (2016) Syngas production from CO_2 reforming of methane over neodymium sesquioxide supported cobalt catalyst. J Nat Gas Sci Eng 34:873–885. https://doi.org/10.1016/j.jngse.2016.07.059

Aziz MAA, Setiabudi HD, Tehd LP, Annuar NHR, Jalil AA (2019) A review of heterogeneous catalysts for syngas production via dry reforming. J Taiwan Inst Chem E 101:139–158. https://doi.org/10.1016/j.jtice.2019.04.047

Babu P, Kumar R, Praveen L (2013) A new porous material to enhance the kinetics of clathrate process: application to precombustion carbon dioxide capture. Environ Sci Technol 47:13191–13198. https://doi.org/10.1021/es403516f

Baird C (1995) Environmental chemistry. W. H. Freeman and Company, New York

Ballarini AD, Miguel SR, Jablonski EL, Scelza OA, Castro AA (2005) Reforming of CH_4 with CO_2 on Pt-supported catalysts: effect of the support on the catalytic behaviour. Catal Today 107–108:481–486. https://doi.org/10.1016/j.cattod.2005.07.058

Ballarini AD, Virgens CF, Rangel MC, Miguel SR, Grau JM (2019) Characterization and behaviour of Pt catalysts supported on basic materials in dry reforming of methane. Braz J Chem Eng 36:275–284. https://doi.org/10.1590/0104-6632.20190361s20170158

Balthasar W (1984) Hydrogen production and technology: today, tomorrow and beyond. Int J Hydrogen Energ 9:649–668. https://doi.org/10.1016/0360-3199(84)90263-5

Barelli L, Ottaviano A (2019) Solid oxide fuel cell technology coupled with methane dry reforming: a viable option for high efficiency plant with reduced CO_2 emissions. Energy 71:118–129. https://doi.org/10.1016/j.energy.2014.04.070

Barros BS, Melo DMA, Libs S, Kiennemann A (2010) CO_2 reforming of methane over $La_2NiO_4/\alpha\text{-}Al_2O_3$ prepared by microwave assisted self-combustion method. Appl Catal A Gen 378:69–75. https://doi.org/10.1016/j.apcata.2010.02.001

Basile F, Fornasari G, Poluzzi E, Vaccari A (1998) Catalytic partial oxidation and CO_2-reforming on Rh- and Ni-based catalysts obtained from hydrotalcite-type precursors. Appl Clay Sci 13:329–345. https://doi.org/10.1016/S0169-1317(98)00031-3

Basini L, Sanfilippo D (1995) Molecular aspects in syn-gas production: the CO_2-reforming reaction case. J Catal 157:162–178. https://doi.org/10.1006/jcat.1995.1277

Batiot-Dupeyrat C, Valderrama G, Meneses A, Martinez F, Barrault J, Tatibouët JM (2003) Pulse study of CO_2 reforming of methane over $LaNiO_3$. Appl Catal A Gen 248:143–151. https://doi.org/10.1016/S0926-860X(03)00155-8

Batiot-Dupeyrat C, Gallego GAS, Mondragon F, Barrault J, Tatibouët JM (2005) CO_2 reforming of methane over $LaNiO_3$ as precursor material. Catal Today 107–108:474–480. https://doi.org/10.1016/j.cattod.2005.07.014

Bertau M, Offermanns EH, Menges G, Keim W, Effenberger FX (2010) Methanol needs more attention as a fuel and raw material for the future. Chem Ing Tech 82:2055–2058. https://doi.org/10.1002/cite.201000159

Bhasin MM, McCain JH, Vora BV, Imai T, Pujadó PR (2001) Dehydrogenation and oxydehydrogenation of paraffins to olefins. Appl Catal a: Gen 221:397–419. https://doi.org/10.1016/S0926-860X(01)00816-X

Bhat RN, Sachtler WMH (1997) Potential of zeolite supported rhodium catalysts for the CO_2 reforming of CH4. Appl Catal a: Gen 150:279–296. https://doi.org/10.1016/S0926-860X(96)00277-3

Bing W, Guo GJ, Ying GY, Xiang YG (2016) Ni-Sm/SiC catalysts prepared by hydrothermal method for carbon dioxide reforming of methane. J Fuel Chem Technol 44:1473–1478

Bitter JH, Hally W, Seshan K, van Ommen JG, Lercher JA (1996) The role of the oxidic support on the deactivation of Pt catalysts during the CO_2 reforming of methane. Catal Today 29:349–353. https://doi.org/10.1016/0920-5861(95)00303-7

Bitter JH, Seshan K, Lercher JA (1998) Mono and bifunctional pathways of CO_2/CH_4 reforming over Pt and Rh based catalysts. J Catal 176:93–101. https://doi.org/10.1006/jcat.1998.2022

Bitter JH, Seshan K, Lercher JA (1999) Deactivation and coke accumulation during CO_2/CH_4 reforming over Pt catalysts. J Catal 183:336–343. https://doi.org/10.1006/jcat.1999.2402

Blom R, Dahl IM, Slagtem A, Sortland B, Spjelkavik A, Tangstad E (1994) Carbon dioxide reforming of methane over lanthanum-modified catalysts in a fluidized-bed reactor. Catal Today 21:535–543. https://doi.org/10.1016/0920-5861(94)80177-0

Bong CPC, Lim YL, Ho WS, Lim JS, Klemeš JJ, Towprayoon S, Ho CS, Lee CT (2017) A review on the global warming potential of cleaner composting and mitigation strategies. J Clean Prod 146(Supplement C):149–157. https://doi.org/10.1016/j.jclepro.2016.07.066

BP Energy Outlook 2019 edition. https://www.bp.com/content/dam/bp/business-sites/en/global/corporate/pdfs/energy-economics/energy-outlook/bp-energy-outlook-2019.pdf

Bradford MCJ, Vannice MA (1996a) Catalytic reforming of methane with carbon dioxide over nickel catalysts I. Catalyst characterization and activity. Appl Catal a: Gen 142:73–96. https://doi.org/10.1016/0926-860X(96)00065-8

Bradford MCJ, Vannice MA (1996) Catalytic reforming of methane with carbon dioxide over nickel catalysts II. Reaction kinetics. Appl Catal A: Gen 142:97–122. https://doi.org/10.1016/0926-860X(96)00066-X

Bradford MCJ, Vannice MA (1998) CO_2 reforming of CH_4 over supported Pt catalysts. J Catal 173:157–171. https://doi.org/10.1006/jcat.1997.1910

Bradford MCJ, Vannice MA (1999) The role of metal–support interactions in CO_2 reforming of CH4. Catal Today 50:87–96. https://doi.org/10.1016/S0920-5861(98)00465-9

Britto JM, Rangel MC (2008) Advanced oxidation process of phenolic compounds in industrial wastewater. Quím Nova 31:114–122. https://doi.org/10.1590/S0100-40422008000100023

Cant NW, Dümpelmann R, Maitra AM (1997) A comparison of nickel and rhodium catalysts for the reforming of methane by carbon dioxide. Stud Surf Sci Catal 107:491–496. https://doi.org/10.1016/S0167-2991(97)80380-1

Chang JS, Park SE, Chon H (1996) Catalytic activity and coke resistance in the carbon dioxide reforming of methane to synthesis gas over zeolite-supported Ni catalysts. Appl Catal a: Gen 145:111–124. https://doi.org/10.1016/0926-860X(96)00150-0

Chen Y, Tomishige K, Yokoyama K, Fujimoto K (1997a) Promoting effect of Pt, Pd and Rh noble metals to the $Ni_{0.03}Mg_{0.97}O$ solid solution catalysts for the reforming of CH_4 with CO_2. Appl Catal a: Gen 165:335–347. https://doi.org/10.1016/S0926-860X(97)00216-0

Chen Y, Tomishige K, Fujimoto K (1997b) Formation and characteristic properties of carbonaceous species on nickel-magnesia solid solution catalysts during CH_4CO_2 reforming reaction. Appl Catal a: Gen 161:L11–L17. https://doi.org/10.1016/S0926-860X(97)00106-3

Chen P, Zhang H, Lin G, Tsai K (1998) Development of coking-resistant Ni-based catalyst for partial oxidation and CO_2-reforming of methane to syngas. Appl Catal a: Gen 166:343–350. https://doi.org/10.1016/S0926-860X(97)00291-3

Chen YZ, Liaw BJ, Kao CF, Kuo JC (2001) Yttria-stabilized zirconia supported platinum catalysts (Pt/YSZs) for CH_4/CO_2 reforming. Appl Catal A Gen 217:23–31. https://doi.org/10.1016/S0926-860X(01)00592-0

Chen YZ, Liaw BJ, Lai WH (2002) ZrO_2/SiO_2- and La_2O_3/Al_2O_3-supported platinum catalysts for CH_4/CO_2 reforming. Appl Catal A Gen 230:73–83. https://doi.org/10.1016/S0926-860X(01)00996-6

Chen J, Wang R, Zhang J, He F, Han S (2005) Effects of preparation methods on properties of $Ni/CeO_2\text{–}Al_2O_3$ catalysts for methane reforming with carbon dioxide. J Mol Catal A Chem 235:302–310. https://doi.org/10.1016/j.molcata.2005.04.023

Chen J, Yao C, Zhao Y, Jia P (2010) Synthesis gas production from dry reforming of methane over $Ce_{0.75}Zr_{0.25}O_2$-supported Ru catalysts. Int J Hydrog Energy 35:1630–1642. https://doi.org/10.1016/j.ijhydene.2009.12.043

Chen W, Sheng W, Cao F, Lu Y (2012) Microfibrous entrapment of Ni/Al_2O_3 for dry reforming of methane: heat/mass transfer enhancement towards carbon resistance and conversion promotion. Int J Hydrog Energy 37:18021–18030. https://doi.org/10.1016/j.ijhydene.2012.09.080

Chen T, Liu H, Shi P, Chen D, Song L, He H, Frost RL (2013) CO_2 Ni/Palygorskite. Fuel 107:699–705. https://doi.org/10.1016/j.fuel.2012.12.036

Chen C, Wang X, Huang H, Zou X, Gu F, Su F, Lu X (2019) Synthesis of mesoporous Ni–La–Si mixed oxides for CO_2 reforming of CH_4 with a high H_2 selectivity. Fuel Process Technol 185:56–67. https://doi.org/10.1016/j.fuproc.2018.11.017

Cheng Z, Wu Q, Li J, Zhu Q (1996) Effects of promoters and preparation procedures on reforming of methane with carbon dioxide over Ni/Al_2O_3 catalyst. Catal Today 30:147–155. https://doi.org/10.1016/0920-5861(95)00005-4

Choudhary VR, Uphade BS, Mamman AS (1998) Simultaneous steam and CO_2 reforming of methane to syngas over NiO/MgO/SA-5205 in presence and absence of oxygen. Appl Catal a: Gen 168:33–46. https://doi.org/10.1016/S0926-860X(97)00331-1

Claridge JB, Green MLH, Tsang SC (1994) Methane conversion to synthesis gas by partial oxidation and dry reforming over rhenium catalysts. Catal Today 12:455–460. https://doi.org/10.1016/0920-5861(94)80167-3

Coonrod CL, Yin YB, Hanna T, Atkinson AJ, Alvarez PJJ, Tekavec TN, Reynolds MA, Wong MS (2020) Fit-for-purpose treatment goals for produced waters in shale oil and gas fields. Water Res 173:115467. https://doi.org/10.1016/j.watres.2020.115467

Corthals S, Nederkassel JV, Winne HD, Geboers J, Jacobs P, Sels B (2011) Design of active and stable $NiCeO_2ZrO_2MgAl_2O_4$ dry reforming catalysts. Appl Catal B Environm 105:263–275. https://doi.org/10.1016/j.apcatb.2011.04.008

Crisafulli C, Scirè S, Minicò S, Solarino L (2002) Ni–Ru bimetallic catalysts for the CO_2 reforming of methane. Appl Catal A Gen 225:1–9. https://doi.org/10.1016/S0926-860X(01)00585-3

Cui Y, Xu H, Ge Q, Li W (2006) Kinetic study on the CH4/CO2 reforming reaction: Ni–H in Ni/α-Al_2O_3 catalysts greatly improves the initial activity. J Mol Catal A Chem 243:226–232. https://doi.org/10.1016/j.molcata.2005.08.030

Dai ZX, Middleton R, Viswanathan H, Fessenden-Rahn J, Bauman J, Pawar R, Lee SX, McPherson B (2014) An integrated framework for optimizing CO_2 sequestration and enhanced oil recovery. Environ Sci Technol Lett 1:49–54. https://doi.org/10.1021/ez4001033

Daoura O, Kaydouh MN, El-Hassan N, Massiani P, Launay F, Boutros M (2017) Mesocellular silica foam-based Ni catalysts for dry reforming of CH_4 (by CO_2). J CO_2 Util 24:112–119. https://doi.org/10.1016/j.jcou.2017.12.010

Daza CE, Gallego J, Moreno JÁ, Mondragón F, Moreno S, Molina R (2008) CO_2 reforming of methane over Ni/Mg/Al/Ce mixed oxides. Catal Today 133–135:357–366. https://doi.org/10.1016/j.cattod.2007.12.081

Daza CE, Moreno S, Molina R (2010) Ce-incorporation in mixed oxides obtained by the self-combustion method for the preparation of high performance catalysts for the CO_2 reforming of methane. Catal Commun 12:173–179. https://doi.org/10.1016/j.catcom.2010.09.012

Daza CE, Cabrera CR, Moreno S, Molina R (2010) Syngas production from CO_2 reforming of methane using Ce-doped Ni-catalysts obtained from hydrotalcites by reconstruction method. Appl Catal A Gen 378:125–133. https://doi.org/10.1016/j.apcata.2010.01.037

Dias JAC, Assaf JM (2003) Influence of calcium content in Ni/CaO/γ-Al_2O_3 catalysts for CO_2-reforming of methane. Catal Today 85:59–68. https://doi.org/10.1016/S0920-5861(03)00194-9

Djinović P, Batista J, Pintar A (2012) Efficient catalytic abatement of greenhouse gases: methane reforming with CO_2 using a novel and thermally stable Rh–CeO_2 catalyst. Int J Hydrog Energy 37:2699–2707. https://doi.org/10.1016/j.ijhydene.2011.10.107

Efstathiou AM, Kladi A, Tsipouriari VA, Verykios XE (1996) Reforming of methane with carbon dioxide to synthesis gas over supported rhodium catalysts: II. A steady-state tracing analysis: mechanistic aspects of the carbon and oxygen reaction pathways to form CO. J Catal 158:64–75. https://doi.org/10.1006/jcat.1996.0006

Erdőhelyi A, Fodor K, Solymosi F (1997) Reaction of CH_4 with CO_2 and H_2O over supported Ir catalyst. Stud Surf Sci Catal 107:525–530. https://doi.org/10.1016/S0167-2991(97)80385-0

Er-Rbib H, Bouallou C, Werkoff F (2012) Dry reforming of methane—review of feasibility studies. Chem Eng Trans 29:163–168. https://doi.org/10.3303/CET1229028

Fakeeha AH, Naeem MA, Khan WU, Al-Fatesh AS (2014) Syngas production via CO_2 reforming of methane using Co-Sr-Al catalyst. J Ind Eng Chem 20:549–557. https://doi.org/10.1016/j.jiec.2013.05.013

Fernandes Júnior LCP, Miguel S, Fierro JLG, Rangel MC (2007) Evaluation of Pd/La_2O_3 catalysts for dry reforming of methane. Stud Surf Sci Catal 167:499–504. https://doi.org/10.1016/S0167-2991(07)80181-9

Ferreira-Aparicio P, Bachiller-Baeza B, Guerrero-Ruiz A, Rodríguez--Ramos (1998) Utilization of CO_2 in the reforming of natural gas on carbon supported ruthenium catalysts. Influence of MgO addition. Stud Surf Sci Catal 114:399–402. https://doi.org/10.1016/S0167-2991(98)80780-5

Fischer F, Tropsch H (1928) Conversion of methane into hydrogen and carbon monoxide. Brennstoff Chem 3:39–46

Foo SY, Cheng CK, Nguyen TH, Adesina AA (2011) Evaluation of lanthanide-group promoters on Co–Ni/Al_2O_3 catalysts for CH_4 dry reforming. J Mol Catal A Chem 344:28–36. https://doi.org/10.1016/j.molcata.2011.04.018

Frontera P, Macario A, Aloise A, Crea F, Antonucci PL, Nagy JB, Frusteric F, Giordano G (2012) Catalytic dry-reforming on Ni–zeolite supported catalyst. Catal Today 179:52–60. https://doi.org/10.1016/j.cattod.2011.07.039

Frontera P, Macario A, Aloise A, Antonuccia PL, Giordano G, Nagy JB (2013) Effect of support surface on methane dry-reforming catalyst preparation. Catal Today 218–219:18–29. https://doi.org/10.1016/j.cattod.2013.04.029

Galadima A, Muraza O (2019) Catalytic thermal conversion of CO_2 into fuels: Perspective and challenges. Renew Sust Energ Rev 115:109333. https://doi.org/10.1016/j.rser.2019.109333

Gallego GS, Batiot-Dupeyrat C, Barrault J, Florez E, Mondragóna F (2008) Dry reforming of methane over $LaNi_{1-y}B_yO_{3\pm\delta}$ (B = Mg, Co) perovskites used as catalyst precursor. Appl Catal A Gen 334:251–258. https://doi.org/10.1016/j.apcata.2007.10.010

Gao J, Hou Z, Guo J, Zhu Y, Zhen X (2008) Catalytic conversion of methane and CO_2 to synthesis gas over a La_2O_3-modified SiO_2 supported Ni catalyst in fluidized-bed reactor. Catal Today 131:278–284. https://doi.org/10.1016/j.cattod.2007.10.019

García-Diéguez M, Pieta LS, Herrera MC, Larrubia MA, Alemany LJ (2010) Nanostructured Pt- and Ni-based catalysts for CO_2-reforming of methane. J Catal 270:136–145. https://doi.org/10.1016/j.jcat.2009.12.010

Gardner TH, Spiveyb JJ, Kugler EL, Pakhare D (2013) CH_4–CO_2 reforming over Ni-substituted barium hexaaluminate catalysts. Appl Catal A Gen 455:129–136. https://doi.org/10.1016/j.apcata.2013.01.029

Gaur S, Haynes DJ, Spivey JJ (2011) Rh, Ni, and Ca substituted pyrochlore catalysts for dry reforming of methane. Appl Catal A Gen 403:142–151. https://doi.org/10.1016/j.apcata.2011.06.025

Gennequin C, Safariamin M, Siffert S, Aboukaïs A, Abi-Aad E (2011) CO_2 reforming of CH_4 over Co–Mg–Al mixed oxides prepared via hydrotalcite like precursors. Catal Today 176:139–143. https://doi.org/10.1016/j.cattod.2011.01.029

Goldwasser MR, Rivas ME, Pietri E, Pérez-Zurita MJ, Cubeiro ML, Gingembre L, Leclerc L, Leclercq G (2003) Perovskites as catalysts precursors: CO_2 reforming of CH_4 on Ln1−xCaxRu0.8Ni0.2O3 (Ln = La, Sm, Nd). Appl Catal A Gen 255:45–57. https://doi.org/10.1016/S0926-860X(03)00643-4

Gronchi P, Mazzocchia C, Del Rosso R (1995) Carbon dioxide reaction with methane on Al_2O_3 supported Rh catalysts. Energy Convers Mgmt 36:605–608. https://doi.org/10.1016/0196-8904(95)00078-R

Gronchi P, Centola P, Kaddouri A, Del Rosso R (1998) Transient reactions in CO_2 reforming of methane. Stud Surf Sci Catal 119:735–740

Guerrero-Ruiz A, Sepúlveda-Escribano A, Rodríguez-Ramos I (1994) Cooperative action of cobalt and MgO for the catalysed reforming of CH_4 with CO_2. Catal Today 21:545–550. https://doi.org/10.1016/0920-5861(94)80178-9

Guo J, Lou H, Zhao H, Chai D, Zheng X (2004) Dry reforming of methane over nickel catalysts supported on magnesium aluminate spinels. Appl Catal A Gen 273:75–82. https://doi.org/10.1016/j.apcata.2004.06.014

Guo J, Hou Z, Gao J, Zheng X (2008) Syngas production via combined oxy-CO_2 reforming of methane over Gd_2O_3-modified Ni/SiO_2 catalysts in a fluidized-bed reactor. Fuel 87:1348–1354. https://doi.org/10.1016/j.fuel.2007.06.018

Guo P, Jin G, Guo C, Wang Y, Tiong X, Guo X (2014) Effects of Yb_2O_3 promotor on the performance of Ni/SiC catalysts in CO_2 reforming of CH_4. J Fuel Chem Technol 42:719–726. https://doi.org/10.1016/S1872-5813(14)60033-5

Gurav HR, Dama S, Samuel V, Chilukuri S (2017) Influence of preparation method on activity and stability of Ni catalysts supported on Gd doped ceria in dry reforming of methane. J CO_2 Util 20:357–367. https://doi.org/10.1016/j.jcou.2017.06.014

Halliche D, Bouarab R, Cherifi O, Bettahar MM (1996) Carbon dioxide reforming of methane on modified $Ni/\alpha\text{-}Al_2O_3$ catalysts. Catal Today 29:373–377. https://doi.org/10.1016/0920-5861(95)00307-X

Hei MJ, Chen HB, Yi J, Lin YJ, Lin YZ, Wei G, Liao DW (1998) CO_2-reforming of methane on transition metal surfaces. Surf Sci 417:82–96. https://doi.org/10.1016/S0039-6028(98)00663-3

Horiuchi T, Sakuma K, Fukui T, Kubo Y, Osaki T, Mori T (1996) Suppression of carbon deposition in the CO_2-reforming of CH4 by adding basic metal oxides to a Ni/Al_2O_3 catalyst. Appl Catal a: Gen 144:111–120. https://doi.org/10.1016/0926-860X(96)00100-7

Horváth A, Stefler G, Geszti O, Kienneman A, Pietrasz A, Guczi L (2011) Methane dry reforming with CO_2 on CeZr-oxide supported Ni, NiRh and NiCo catalysts prepared by sol–gel technique: relationship between activity and coke formation. Catal Today 169:102–111. https://doi.org/10.1016/j.cattod.2010.08.004

Hossain MA, Ayodele BV, Cheng CK, Khan MR (2019) Optimization of renewable hydrogen-rich syngas production from catalytic reforming of greenhouse gases (CH_4 and CO_2) over calcium iron oxide supported nickel catalyst. J Energy Inst 92:177–194. https://doi.org/10.1016/j.joei.2017.10.010

Hou Z, Yokota O, Tanaka T, Yashima T (2004) Surface properties of a coke-free Sn doped nickel catalyst for the CO_2 reforming of methane. Appl Surf Sci 233:58–68. https://doi.org/10.1016/j.apsusc.2004.03.223

Huang B, Li X, Ji S, Lang B, Habimana F, Li C (2008) Effect of MgO promoter on Ni-based SBA-15 catalysts for combined steam and carbon dioxide reforming of methane. J Nat Gas Chem 17:225–231. https://doi.org/10.1016/S1003-9953(08)60055-9

Huo M, Li L, Zhao X, Zhang Y, Li J (2017) Synthesis of Ni-based catalysts supported on nitrogen-incorporated SBA-16 and their catalytic performance in the reforming of methane with carbon dioxide. J Fuel Chem Technol 45:172–181. https://doi.org/10.1016/S1872-5813(17)30012-9

Hwang KS, Zhu HY, Lu GQ (2001) New nickel catalysts supported on highly porous alumina intercalated laponite for methane reforming with CO_2. Catal Today 68:183–190. https://doi.org/10.1016/S0920-5861(01)00299-1

IEA International Energy Agency, 2018. Global energy and CO_2 status report—2019. https://www.iea.org/reports/global-energy-co2-status-report-2019

Inui T, Fujioka K, Fujii Y, Takeguchi T, Nishiyama H, Inoue MH, Tanakulrungsank W (1994) Highly active catalysts for syngas production from natural gas and its consecutive conversion to methanol and more valuable hydrocarbons. In: Curry-Hyde HE, Howe RF (eds) Natural gas conversion II

IPCC Global Warming of 1.5 °C Report. The Intergovernmental Panel on Climate Change https://www.ipcc.ch

Itoh N, Sancheza MA, Xua WC, Haraya K, Hong M (1993) Application of a membrane reactor system to thermal decomposition of CO_2. J Membrane Sci 77:245–253. https://doi.org/10.1016/0376-7388(93)85073-6

Jabbour K, Hassana NE, Davidson A, Massiani P, Casale S (2015) Characterizations and performances of Ni/diatomite catalysts for dry reforming of methane. Chem Eng J 264:351–358. https://doi.org/10.1016/j.cej.2014.11.109

Jang WJ, Jeong DW, Shim JO, Kim HM, Roh HS, Son IH, Lee SJ (2016) Combined steam and carbon dioxide reforming of methane and side reactions: thermodynamic equilibrium analysis and experimental application. Appl Energy 173:80–91. https://doi.org/10.1016/j.apenergy.2016.04.006

Jang WJ, Shim JO, Kim HM, Yoo SY, Roh HS (2019) A review on dry reforming of methane in aspect of catalytic properties. Catal Today 324:15–26. https://doi.org/10.1016/j.cattod.2018.07.032

Jangam A, Das S, Dewangan N, Hongmanorom P, Hui WM, Kawi S (2019) Conversion of CO_2 to C1 chemicals: catalyst design, kinetics and mechanism aspects of the reactions. Catal Today. https://doi.org/10.1016/j.cattod.2019.08.049 (in Press)

Jarvis SM, Samsatli S (2018) Technologies and infrastructures underpinning future CO_2 value chains: a comprehensive review and comparative analysis. Renew Sustain Energy Rev 85:46–68. https://doi.org/10.1016/j.rser.2018.01.007

Jiang Z, Xiao T, Kuznetsov VL, Edwards PP (2010) Turning carbon dioxide into fuel. Phil Trans R Soc A 368:3343–3364. https://doi.org/10.1098/rsta.2010.0119

Jiang C, Akkullu MR, Li B, Davila JC, Janik MJ, Dooley KM (2019) Rapid screening of ternary rare-earth—transition metal catalysts for dry reforming of methane and characterization of final structures. J Catal 377:332–342. https://doi.org/10.1016/j.jcat.2019.07.020

Jin L, Li Y, Feng Y, Hu H, Zhu H (2017) Integrated process of coal pyrolysis with CO_2 reforming of methane by spark discharge plasma. J Anal Appl Pyrolysis 126:194–200. https://doi.org/10.1016/j.jaap.2017.06.008

Kaengsilalai A, Luengnaruemitchai A, Jitkarnka S, Wongkasemjit S (2007) Potential of Ni supported on KH zeolite catalysts for carbon dioxide reforming of methane. J Power Sources 165:347–352. https://doi.org/10.1016/j.jpowsour.2006.12.005

Kang KM, Kim HW, Shim IW, Kwak HY (2011) Catalytic test of supported Ni catalysts with core/shell structure for dry reforming of methane. Fuel Process Technol 92:1236–1243. https://doi.org/10.1016/j.fuproc.2011.02.007

Kantserova MR, Vlasenko NV, Orlyk SM, Veltruska K, Matolinova I (2019) Effect of acid-base characteristics of In_2O_3-Al_2O_3 (ZrO_2) compositions on their catalytic properties in the oxidative dehydrogenation of propane to propylene with CO_2. Theor Exp Chem 55:207–214. https://doi.org/10.1007/s11237-019-09610-9

Keller L, Ohs B, Abduly L, Blanke P, Wessling M (2018) High capacity polyethylenimine impregnated microtubes made of carbon nanotubes for CO_2 capture. Carbon 126:338–345. https://doi.org/10.1016/j.carbon.2017.10.023

Keulen NJ, Hegarty MES, Ross JHR, Oosterkamp PF (1997) The development of platinum-zirconia catalysts for the CO_2 reforming of methane. Stud Surf Sci Catal 107:537–546. https://doi.org/10.1016/S0167-2991(97)80387-4

Kim JH, Suh DJ, Park TJ, Kim KL (2000) Effect of metal particle size on coking during CO_2 reforming of CH_4 over Ni-alumina aerogel catalysts. Appl Catal A Gen 197:191–200. https://doi.org/10.1016/S0926-860X(99)00487-1

Kischke S, Bousquet P, Ciais P, Sounois M, Canadell JG, Dlugokencky EJ, Bergamaschi P, Bergmann D, Blake DR, Bruhwiler L

(2013) Three decades of global methane sources and sinks. Nat Geos 6:813–823. https://doi.org/10.1038/NGEO1955

Klewiah I, Berawala DS, Walkera HCA, Andersen WP, Nadeau PH (2020) Review of experimental sorption studies of CO_2 and CH_4 in shales. J Nat Gas Sci Eng 73:103045. https://doi.org/10.1016/j.jngse.2019.103045

Koubaissy B, Pietraszek A, Roger AC, Kiennemann A (2010) CO2 reforming of methane over Ce–Zr–Ni–Me mixed catalysts. Catal Today 157:436–439. https://doi.org/10.1016/j.cattod.2010.01.050

Kraemer S, Rondinone A J, Tsai YT, Schwartz V, Overbury SH, Idrobo JC, Wu Z (2016) Oxidative dehydrogenation of isobutane over vanadia catalysts supported by titania nanoshapes. Catal Today 263:84–90. https://doi.org/10.1016/j.cattod.2015.09.049

Kroll VCH, Swaan HM, Mirodatos C (1996) Methane reforming reaction with carbon dioxide over Ni/SiO_2 catalyst: I. Deactivation studies. J Catal 161:409–422. https://doi.org/10.1006/jcat.1996.0199

Kroll VCH, Swaan HM, Lacombe S, Mirodatos C (1997) Methane reforming reaction with carbon dioxide over Ni/SiO_2 catalyst: II. A mechanistic study. J Catal 164:387–398. https://doi.org/10.1006/jcat.1996.0395

Kwon BW, Oh JH, Kim GS, Yoon SP, Han J, Nam SW, Ham HC (2017) The novel perovskite-type Ni-doped Sr0.92Y0.08TiO3 catalyst as a reforming biogas (CH_4 + CO_2) for H_2 production. Appl Energy 227:213–219. https://doi.org/10.1016/j.apenergy.2017.07.105

Lacheen HS, Iglesia H (2005) Stability, structure, and oxidation state of Mo/H-ZSM-5 catalysts during reactions of CH_4 and CH_4–CO_2 mixtures. J Catal 230:173–185. https://doi.org/10.1016/j.jcat.2004.11.037

Lavoie JM (2014) Review on dry reforming of methane, a potentially more environmentally friendly approach to the increasing natural gas exploitation. Front Chem 2:1–17. https://doi.org/10.3389/fchem.2014.00081

Lemonidou AA, Goula MA, Vasalos IA (1998) Carbon dioxide reforming of methane over 5 wt.% nickel calcium aluminate catalysts—effect of preparation method. Catal Today 46:175–183. https://doi.org/10.1016/S0920-5861(98)00339-3

Lercher JA, Bitter JH, Hally H, Niessen W, Seshan K (1996) Design of stable catalysts for methane-carbon dioxide reforming. Stud Surf Sci Catal 101:463–472. https://doi.org/10.1016/S0167-2991(96)80257-6

Li H, Wang J (2004) Study on CO_2 reforming of methane to syngas over Al_2O_3–ZrO^2 supported Ni catalysts prepared via a direct sol–gel process. Chem Eng Sci 59:4861–4867. https://doi.org/10.1016/j.ces.2004.07.076

Li H, Xu H, Wang J (2011) Methane reforming with CO_2 to syngas over CeO_2-promoted Ni/Al2O3-ZrO2 catalysts prepared via a direct sol-gel process. J Nat Gas Chem 20:1–8. https://doi.org/10.1016/S1003-9953(10)60156-9

Li JF, Xi C, Au CT, Liu BS (2014) Y_2O_3-promoted NiO/SBA-15 catalysts highly active for CO_2/CH_4 reforming. Int J Hydrog Energy 39:10927–10940. https://doi.org/10.1016/j.ijhydene.2014.05.021

Li R, Shahbazi A, Wang L, Zhang B, Chung CC, Dayton D, Yan Q (2018) Nanostructured molybdenum carbide on biochar for CO_2 reforming of CH_4. Fuel 225:403–410. https://doi.org/10.1016/j.fuel.2018.03.179

Li X, Huang Y, Zhang Q, Luan C, Vinokurov VA, Huang W (2019) Highly stable and anti-coking $Ni/MoCeZr/MgAl_2O_4$-MgO complex support catalysts for CO_2 reforming of CH_4: effect of the calcination temperature. Energy Convers Manag 179:166–177. https://doi.org/10.1016/j.enconman.2018.10.067

Li B, Yuan X, Li B, Wang X (2020) Impact of pore structure on hydroxyapatite supported nickel catalysts (Ni/HAP) for dry reforming of methane. Fuel Process Technol 202:106359. https://doi.org/10.1016/j.fuproc.2020.106359

Liang TY, Chen HH, Tsai DH (2020) Nickel hybrid nanoparticle decorating on alumina nanoparticle cluster for synergistic catalysis of methane dry reforming. Fuel Process Technol 201:106335. https://doi.org/10.1016/j.fuproc.2020.106335

Liu H, Li S, Zhang S, Wang J, Zhou G, Chen L, Wang X (2008) Catalytic performance of novel Ni catalysts supported on SiC monolithic foam in carbon dioxide reforming of methane to synthesis gas. Catal Commun 9:51–54. https://doi.org/10.1016/j.catcom.2007.05.002

Liu D, Cheo WNE, Wen Y, Lim YWY, Borgna A, Lau R, Yang Y (2010) A comparative study on catalyst deactivation of nickel and cobalt incorporated MCM-41 catalysts modified by platinum in methane reforming with carbon dioxide. Catal Today 154:229–236. https://doi.org/10.1016/j.cattod.2010.03.054

Liu H, Li Y, Wu H, Yang W, He D (2014) Effects of Nd, Ce, and La modification on catalytic performance of Ni/SBA-15 catalyst in CO_2 reforming of CH_4. Chinese J Catal 35:1520–1528. https://doi.org/10.1016/S1872-2067(14)60095-4

Liu L, Wang S, Guo Y, Wang B, Rukundo P, Wen S, Wang ZJ (2016) Synthesis of a highly dispersed Ni/Al_2O_3 catalyst with enhanced catalytic performance for CO_2 reforming of methane by an electrospinning method. Int J Hydrog Energy 41:17361–17369. https://doi.org/10.1016/j.ijhydene.2016.07.151

Long H, Xu Y, Zhang X, Hu S, Shang S, Yin Y, Dai X (2013) Ni-Co/Mg-Al catalyst derived from hydrotalcite-like compound prepared by plasma for dry reforming of methane. J Energy Chem 22:733–739. https://doi.org/10.1016/S2095-4956(13)60097-2

Lu Y, Jiang S, Wang S, Zhao Y, Ma X (2018) Effect of the addition of Ce and Zr over a flower-like NiO-MgO (111) solid solution for CO_2 reforming of methane. J CO_2 Util 26:123–132. https://doi.org/10.1016/j.jcou.2018.05.007

Luo JZ, Yu ZL, Ng CF, Au CT (2000) CO_2/CH_4 reforming over Ni–La_2O_3/5A: an investigation on carbon deposition and reaction steps. J Catal 194:198–210. https://doi.org/10.1006/jcat.2000.2941

Ma J, Sun N, Zhang X, Zhao N, Xiao F, Wei W, Sun Y (2009) A short review of catalysis for CO_2 conversion. Catal Today 148:221–231. https://doi.org/10.1016/j.cattod.2009.08.015

Ma Z, Jiang QJ, Wang X, Zhang WG, Ma ZF (2012) CO_2 reforming of dimethyl ether over Ni/γ-Al_2O_3 catalyst. Catal Commun 17:49–53. https://doi.org/10.1016/j.catcom.2011.10.014

Ma Q, Wang D, Wu M, Zhao T, Yoneyama Y, Tsubaki N (2013) Effect of catalytic site position: nickel nanocatalyst selectively loaded inside or outside carbon nanotubes for methane dry reforming. Fuel 108:430–438. https://doi.org/10.1016/j.fuel.2012.12.028

Mark MF, Maier WF (1996) CO_2-reforming of methane on supported Rh and Ir catalysts. J Catal 164:122–130. https://doi.org/10.1006/jcat.1996.0368

Market Watch https://www.marketwatch.com/press-release/39-growth-for-carbon-dioxide-co2-market-size-raising-to-usd-916-billion-by-2026-2019-10-14

Marmarshahi S, Niaei A, Salari D, Abedini F, Abbasi M, Kalantari N (2015) Evaluating the catalytic performance of $La_{1-x}Ce_xNi_{1-y}Zn_yO_3$ nanostructure perovskites in the carbon dioxide reforming of methane. Procedia Mater Sci 11:616–621. https://doi.org/10.1016/j.mspro.2015.11.095

Martins AR, Carvalho LS, Reyes P, Grau JM, Rangel MC (2016) Hydrogen production on alumina-supported platinum catalysts. J Mol Catal A-Chem 429:1–9. https://doi.org/10.1016/j.molcata.2016.11.040

Meili Z, Shengfu JI, Linhua H, Fengxiang Y, Chengyue L, Hui L (2006) Structural characterization of highly stable Ni/SBA-15 catalyst and its catalytic performance for methane reforming with

CO_2. Chinese J Catal 27:777–781. https://doi.org/10.1016/S1872-2067(06)60043-0

Miranda-Barbosa E, Sigfússon B, Carlsson J, Tzimas E (2017) Advantages from combining CCS with geothermal energy. Energy Procedia 114(Supplement C):6666–6676. https://doi.org/10.1016/j.egypro.2017.03.1794

Mirzabekova S, Mamedov A, Aliev V, Krylov O (1992) Conversion of C1–C2 alkanes over manganese catalysts reoxidized by carbon dioxide and oxygen. React Kinet Catal Lett 47:159–166. https://doi.org/10.1007/BF0213764

Mo W, Ma F, Ma Y, Fan X (2019) The optimization of Ni–Al_2O_3 catalyst with the addition of La_2O_3 for CO_2–CH_4 reforming to produce syngas. Int J Hydrogen Energy 44:24510–24524. https://doi.org/10.1016/j.ijhydene.2019.07.204

Mondal KC, Vasant VR, Choudhary R, Joshi UA (2007) CO_2 reforming of methane to syngas over highly active and stable supported CoO_x (accompanied with MgO, ZrO_2 or CeO_2) catalysts. Appl Catal A: Gen 316:47–52. https://doi.org/10.1016/j.apcata.2006.09.016

Mondal K, Sasmal S, Badgandi S, Chowdhury DR, Nair V (2016) Dry reforming of methane to syngas: a potential alternative process for value added chemicals-a techno-economic perspective. Environ Sci Pollut Res 23:22267–22273. https://doi.org/10.1007/s11356-016-6310-4

Montoya JA, Romero E, Monzón A, Guimon C (2000) Methane reforming with CO_2 over Ni/ZrO_2-CeO_2 and Ni/ZrO_2-MgO catalysts synthesized by sol-gel method. Stud Surf Sci Catal 130:3669–3674. https://doi.org/10.1016/S0167-2991(00)80593

Mori T, Osaki T, Horiuchi T, Sugiyama T, Suzuki K (1999) Suppression of carbon deposition during the CO_2-reforming of CH_4 by the enhancement of CO_2 adsorption. Stud Surf Sci Catal 126:365–372. https://doi.org/10.1016/S0167-2991(99)80487-X

Mortensen PM, Dybkjær I (2015) Industrial scale experience on steam reforming of CO_2-rich gas. Appl Catal A Gen 495:141–151. https://doi.org/10.1016/j.apcata.2015.02.022

Movasati A, Alavi SM, Mazloom G (2017) CO_2 reforming of methane over Ni/$ZnAl_2O_4$ catalysts: influence of Ce addition on activity and stability. Int J Hydrog Energy 42:16436–16448. https://doi.org/10.1016/j.ijhydene.2017.05.199

Muraza O, Galadima A (2015) A review on coke management during dry reforming of methane. Int J Energy Res 39:1196–1216. https://doi.org/10.1002/er.3295

Mustafa A, Lougou BG, Shuai Y, Wang Z, Tan H (2020) Current technology development for CO_2 utilization into solar fuels and chemicals: a review. J Energy Chem 49:96–123. https://doi.org/10.1016/j.jechem.2020.01.023

Naito S, Tsuji M, Sakamoto Y, Miyao T (2002) Marked difference of catalytic behavior by preparation methods in CH_4 reforming with CO_2 over Mo2C and WC catalysts. Stud Surf Sci Catal 415–423

Naito S, Tsuji M, Sakamoto Y, Miyao T (2002b) Mechanistic difference of the CO_2 reforming of CH_4 over unsupported and zirconia supported molybdenum carbide catalysts. Catal Today 77:161–165. https://doi.org/10.1016/S0920-5861(02)00242-0

Nakamura J, Aikawa K, Sato K, Uchijima T (1994) The Role of support in methane reforming with CO_2 over rhodium catalysts. Stud Surf Sci Catal 90:495–500. https://doi.org/10.1016/S0167-2991(08)61865-0

Nandini A, Pant KK, Dhingra SC (2005) K-, CeO_2-, and Mn-promoted Ni/Al_2O_3 catalysts for stable CO_2 reforming of methane. Appl Catal A Gen 290:66–174. https://doi.org/10.1016/j.apcata.2005.05.016

Nguyen DL, Leroi P, Ledoux MJ, Pham-Huub C (2009) Influence of the oxygen pretreatment on the CO2 reforming of methane on Ni/β-SiC catalyst. Catal Today 141:393–396. https://doi.org/10.1016/j.cattod.2008.10.019

Nikoo ML, Amin NAS (2011) Thermodynamic analysis of carbon dioxide reforming of methane in view of solid carbon formation. Fuel Process Technol 92:678–691. https://doi.org/10.1016/j.fuproc.2010.11.027

O'Connor AM, Meunier FC, Ross JRH (1998) An in-situ DRIFTS study of the mechanism of the CO_2 reforming of CH_4 over a Pt/ZrO_2 catalyst. Stud Surf Sci Catal 119:819–824. https://doi.org/10.1016/S0167-2991(98)80533-8

Ola O, Maroto-Valer MM, Mackintosh S (2013) Turning CO2 into valuable chemicals. Energy Proced 37:6704–6709. https://doi.org/10.1016/j.egypro.2013.06.603

Olafsen A, Daniel C, Schuurman Y, Råberg LB, Olsbye U, Mirodatos C (2006) Light alkanes CO_2 reforming to synthesis gas over Ni based catalysts. Catal Today 115:179–185. https://doi.org/10.1016/j.cattod.2006.02.053

Olah GA, Goeppert A, Prakash GKS (2008) Chemical recycling of carbon dioxide to methanol and dimethyl ether: from greenhouse gas to renewable, environmentally carbon neutral fuels and synthetic hydrocarbons. J Org Chem 74:487–498. https://doi.org/10.1021/jo801260f

Oliveira SB, Barbosa DP, Monteiro APM, Rabelo D (2008) Evaluation of copper supported on polymeric spherical activated carbon in the ethylbenzene dehydrogenation. Catal Today 133–135:92–98. https://doi.org/10.1016/j.cattod.2007.12.040

Osaki T, Horiuchi T, Suzuki K, Mori T (1997) Catalyst performance of MoS_2 and WS_2 for the CO_2-reforming of CH_4 suppression of carbon deposition. Appl Catal a: Gen 155:229–238. https://doi.org/10.1016/S0926-860X(96)00391-2

Osaki T, Fukaya H, Horiuchi T, Suzuki K, Mori T (1998) Isotope effect and rate-determining step of the CO_2-reforming of methane over supported Ni catalyst. J Catal 180:106–109. https://doi.org/10.1006/jcat.1998.2265

Osazuwa OU, Setiabudi HD, Rasid RA, Cheng CK (2017) Syngas production via methane dry reforming: a novel application of $SmCoO_3$ perovskite catalyst. J Nat Gas Sci Eng 37:435–448. https://doi.org/10.1016/j.jngse.2016.11.060

Otsuka K, Ushiyama T, Yamanaka I (1993) Partial oxidation of methane using the redox of cerium oxide. Chem Lett 22:1517–1520. https://doi.org/10.1246/cl.1993.1517

Özkara-Aydınoğlu S, Aksoylu AE (2010) Carbon dioxide reforming of methane over Co-X/ZrO_2 catalysts (X = La, Ce, Mn, Mg, K). Catal Commun 11:1165–1170. https://doi.org/10.1016/j.catcom.2010.07.001

Özkara-Aydınoğlu S, Özensoy E, Aksoylu E (2009) The effect of impregnation strategy on methane dry reforming activity of Ce promoted Pt/ZrO_2. Int J Hydrogen Energ 34:9711–9722. https://doi.org/10.1016/j.ijhydene.2009.09.005

Panzone C, Philippe R, Chappaz A, Fongarland P, Bengaouer A (2020) Power-to-Liquid catalytic CO_2 valorization into fuels and chemicals: focus on the Fischer-Tropsch route. J CO_2 Util 38:314–347. https://doi.org/10.1016/j.jcou.2020.02.009

Park JH, Yeo S, Kang TJ, Shin HR, Heo I, Chang TS (2018) Effect of Zn promoter on catalytic activity and stability of Co/ZrO_2 catalyst for dry reforming of CH_4. J CO_2 Util 23:10–19. https://doi.org/10.1016/j.jcou.2017.11.002

Pawelec B, Damyanova S, Arishtirova K, Fierro JLG, Petrov L (2007) Structural and surface features of PtNi catalysts for reforming of methane with CO_2. Appl Catal A Gen 323:188–201. https://doi.org/10.1016/j.apcata.2007.02.017

Peña MA, Gómez JP, Fierro JLG (1996) New catalytic routes for syngas and hydrogen production. Appl Catal a: Gen 144:7–57. https://doi.org/10.1016/0926-860X(96)00108-1

Perera JSHQ, Couves JW, Sankar G, Thoma JM (1991) The catalytic activity of Ru and Ir supported on Eu203 for the reaction,

CO_2 + CH_4 → 2H2 + 2CO: a viable solar-thermal energy system. Catal Lett 11:219–226. https://doi.org/10.1007/BF00764088

Perez-Lopez OW, Senger A, Marcilio NR, Lansarin MA (2006) Effect of composition and thermal pretreatment on properties of Ni–Mg–Al catalysts for CO_2 reforming of methane. Appl Catal A Gen 303:234–244. https://doi.org/10.1016/j.apcata.2006.02.024

Pietraszek A, Koubaissy B, Roger AC, Kiennemann A (2011) The influence of the support modification over Ni-based catalysts for dry reforming of methane reaction. Catal Today 176:267–271. https://doi.org/10.1016/j.cattod.2010.12.015

Pietri E, Barrios A, Goldwasser MR, Pérez-Zurita MJ, Cubeiro ML, Goldwasser J, Leclercq L, Leclercq G, Gingembre L (2000) Optimization of Ni and Ru catalysts supported on $LaMnO_3$ for the carbon dioxide reforming of methane. Stud Surf Sci Catal 130:3657–3662. https://doi.org/10.1016/S0167-2991(00)80591-1

Pietri E, Barrios A, Gorlzalez O, Goldwasser MR, Pérez-Zurita MJ, Cubeiro ML, Goldwasser J, Leclercq L, Leclercq G, Gingembre L (2001) Perovskites as catalysts precursors for methane reforming: Ru based catalysts. Stud Surf Sci Catal 136:381–386

Praserthdam S, Balbuena PB (2018) Evaluation of dry reforming reaction catalysts via computational screening. Catal Today 312:23–34. https://doi.org/10.1016/j.cattod.2018.04.017

Puolakka J, Krause OI (2004) CO_2 reforming of n-heptane on a Ni/AkCb catalyst. Stud Surf Sci Catal 153:137–140

Qin D, Lapszewicz J (1994) Study of mixed steam and CO_2 reforming of CH_4 to syngas on MgO-supported metals. Catal Today 21:551–560. https://doi.org/10.1016/0920-5861(94)80179-7

Quincoces CE, de Vargas SP, Grange P, González MG (2002) Role of Mo in CO_2 reforming of CH_4 over Mo promoted Ni/Al_2O_3 catalysts. Mater Lett 56:698–704. https://doi.org/10.1016/S0167-577X(02)00598-0

Rahemi N, Haghighi M, Babaluo AA, Jafari MF, Estifaee P (2013) Synthesis and physicochemical characterizations of Ni/Al_2O_3–ZrO_2 nanocatalyst prepared via impregnation method and treated with non-thermal plasma for CO_2 reforming of CH_4. J Ind Eng Chem 19:1566–1576. https://doi.org/10.1016/j.jiec.2013.01.024

Ramezani Y, Meshkani F, Rezaei M (2018) Promotional effect of Mg in trimetallic nickel-manganese-magnesium nanocrystalline catalysts in CO_2 reforming of methane. Int J Hydrog Energy 43:22347–22356. https://doi.org/10.1016/j.ijhydene.2018.09.222

Rangel MC, Carvalho MFA (2003) Impact of automotive catalysts in the control of air quality. Quím Nova 26:265–277. https://doi.org/10.1590/S0100-40422003000200021

Rangel MC, Monteiro APM, Oportus M, Reyes P, Ramos MS, Lima SB (2012) Ethylbenzene dehydrogenation in the presence of carbon dioxide over metal oxides. In: Liu GC (ed) Capturing, utilization and reduction greenhouse gases. InTech, USA

Rangel MC, Querino PS, Borges SMS, Marchetti SG, Assaf JM, Vásquez DPR, Rodella CB, Silva TF, Silva AHM, Ramon AP (2017) Hydrogen purification over lanthanum-doped iron oxides by WGSR. Catal Today 296:262–271. https://doi.org/10.1016/j.cattod.2017.05.058

Reitmeier RE, Atwood K, Bennett HA, Baugh HM (1948) Production of synthesis gas by reacting light hydrocarbons with steam and carbon dioxide. Ind Eng Chem 40:620–626

Rezaei M, Alavi SM, Sahebdelfar S, Yan ZF (2006) Syngas production by methane reforming with carbon dioxide on noble metal catalysts. J Nat Gas Chem 15:327–334. https://doi.org/10.1016/S1003-9953(07)60014-0

Rezaei M, Alavi SM, Sahebdelfar S, Yan ZF (2007) Mesoporous nanocrystalline zirconia powders: a promising support for nickel catalyst in CH_4 reforming with CO_2. Mater Lett 61:2628–2631. https://doi.org/10.1016/j.matlet.2006.10.053

Richardson JT, Paripatyadar SA (1990) Carbon dioxide reforming of methane with supported rhodium. Appl Catal 61:293–309. https://doi.org/10.1016/S0166-9834(00)82152-1

Richardson JT, Hung JK, Zhao J (2001) CO_2-CH_4 reforming with Pt-Re/T-Al_20_3 catalysts. Stud Surf Sci Catal 136:203–208

Ridler DE, Twigg MV (eds) (1989) Catalyst handbook, 2nd edn. Wolfe, London

Roh HS, Potdar HS, Jun KW (2004) Carbon dioxide reforming of methane over co-precipitated Ni–CeO_2, Ni–ZrO_2 and Ni–Ce–ZrO_2 catalysts. Catal Today 93–95:39–44. https://doi.org/10.1016/j.cattod.2004.05.012

Rohi G, Ejofodomi O, Ofualagba G (2020) Autonomous monitoring, analysis, and countering of air pollution using environmental drones. Heliyon 6:e3252. https://doi.org/10.1016/j.heliyon.2020.e03252

Rostrup-Nielsen JR (1983) Catalytic steam reforming. In: Anderson JR, Boudart M (eds) Catalysis, science and technology, vol 5. Springer-Verlag, Berlin

Rostrup-Nielsen JR, Hansen JHB (1993) CO_2-reforming of methane over transition metals. J Catal 144:38–49. https://doi.org/10.1006/jcat.1993.1312

Ruckenstein E, Hu YH (1995) Carbon dioxide reforming of methane over nickel/alkaline earth metal oxide catalysts. Appl Catal a: Gen 133:149–161. https://doi.org/10.1016/0926-860X(95)00201-4

Ruckenstein E, Hu YH (1997) The effect of precursor and preparation conditions of MgO on the CO_2 reforming of CH_4 over NiO/MgO catalysts. Appl Catal a: Gen 154:185–205. https://doi.org/10.1016/S0926-860X(96)00372-9

Ruckenstein Y, Wang HY (2002) Carbon deposition and catalytic deactivation during CO_2 reforming of CH_4 over Co/γ-Al_2O_3 catalysts. J Catal 205:289–293. https://doi.org/10.1006/jcat.2001.3458

Saadabadi SA, Thattai AT, Fan L, Lindeboom REF, Spanjers H, Aravind PV (2019) Solid oxide fuel cells fueled with biogas: potential and constraints. Renew Energy 134:194e214. https://doi.org/10.1016/j.renene.2018.11.028

Safariamin M, Tidahy LH, Abi-Aad E, Siffert S, Aboukaïs A (2009) Dry reforming of methane in the presence of ruthenium-based catalysts. C R Chimie 12:748–753. https://doi.org/10.1016/j.crci.2008.10.021

Saito H, Seki H, Hosono Y, Higo T, Seo JG, Maeda S, Hashimoto K, Ogo S, Sekine Y (2019) Dehydrogenation of ethane via the Mars–van Krevelen mechanism over $La_{0.8}Ba_{0.2}MnO_{3-\delta}$ perovskites under anaerobic conditions. J Phys Chem C 123:26272–26281. https://doi.org/10.1021/acs.jpcc.9b06475

Sakai Y, Saito H, Sodesawa T, Nozaki F (1984) Catalytic reactions of hydrocarbon with carbon dioxide over metallic catalysts. React Kinet Catal Lett 24:253–257. https://doi.org/10.1007/BF02093437

San José-Alonso D, Illán-Gómez MJ, Román-Martínez MC (2011) K and Sr promoted Co alumina supported catalysts for the CO_2 reforming of methane. Catal Today 176:187–190. https://doi.org/10.1016/j.cattod.2010.11.093

Saravanan K, Ham H, Tsubaki N, Bae JW (2017) Recent progress for direct synthesis of dimethyl ether from syngas on the heterogeneous bifunctional hybrid catalysts. Appl Catal B Environ 217:494–522. https://doi.org/10.1016/j.apcatb.2017.05.085

Sarkar B, Tiwari R, Singha RK, Suman S, Ghosh S, Acharyya SS, Mantri K, Konathala LNS, Pendem C, Bal R (2012) Reforming of methane with CO_2 over Ni nanoparticle supported on mesoporous ZSM-5. Catal Today 198:209–214. https://doi.org/10.1016/j.cattod.2012.04.029

Sarusi I, Fodor K, Baán K, Oszkó A, Pótári G, Erdőhelyi A (2011) CO_2 reforming of CH_4 on doped Rh/Al2O3 catalysts. Catal Today 171:132–139. https://doi.org/10.1016/j.cattod.2011.03.075

Schuurman Y, Kroll VCH, Ferreira-Aparicio P, Mirodatos C (1997) Use of transient kinetics techniques for studying the methane reforming by carbon dioxide. Catal Today 38:129–135. https://doi.org/10.1016/S0920-5861(97)00046-1

Sehested J, Jacobsen CJH, Rokni S, Rostrup-Nielsen JR (2001) Activity and stability of molybdenum carbide as a catalyst for CO_2 reforming. J Catal 201:206–212. https://doi.org/10.1006/jcat.2001.3255

Seok SH, Han SH, Lee JS (2001) The role of MnO in Ni/MnO-Al_2O_3 catalysts for carbon dioxide reforming of methane. Appl Catal A Gen 215:31–38. https://doi.org/10.1016/S0926-860X(01)00528-2

Seshan K, ten Barge HW, Halty W, van Keulen ANJ, Ross JRH (1994) Carbon dioxide reforming of methane in the presence of nickel and platinum catalysts supported on ZrO_2. Stud Surf Sci Catal 81:285–290. https://doi.org/10.1016/S0167-2991(08)63882-3

Shah A, Shah M (2020) Characterisation and bioremediation of wastewater: a review exploring bioremediation as a sustainable technique for pharmaceutical wastewater. Groundw Sustain Dev 11:100383. https://doi.org/10.1016/j.gsd.2020.100383

Shah M, Bordoloi A, Nayak AK, Mondal P (2019) Effect of Ti/Al ratio on the performance of Ni/TiO_2-Al_2O_3 catalyst for methane reforming with CO_2. Fuel Process Technol 192:21–35. https://doi.org/10.1016/j.fuproc.2019.04.010

Shah M, Mondal P, Nayak AK, Bordoloi A (2020) Advanced titania composites for efficient CO_2 reforming with methane: statistical method versus. experiment. J CO_2 Util 39:101160. https://doi.org/10.1016/j.jcou.2020.101160

Shang R, Guo X, Mu S, Wang Y, Jin G, Kosslick H, Schulz A, Guo X (2011) Carbon dioxide reforming of methane to synthesis gas over Ni/Si_3N_4 catalysts. Int J Hydrog Energy 36:4900–4907. https://doi.org/10.1016/j.ijhydene.2011.01.034

Sharma S, Hilaire S, Gorte RG (2000) A study of CH_4 reforming by CO_2 and H_2O on ceria-supported Pd. Stud Surf Sci Catal 130:677–682. https://doi.org/10.1016/S0167-2991(00)81036-8

Shen W, Momoi H, Komatsubara K, Saito T, Yoshida A, Naito S (2011) Marked role of mesopores for the prevention of sintering and carbon deposition in dry reforming of methane over ordered mesoporous Ni–Mg–Al oxides. Catal Today 171:150–155. https://doi.org/10.1016/j.cattod.2011.04.003

Shia C, Zhang P (2012) Effect of a second metal (Y, K, Ca, Mn or Cu) addition on the carbon dioxide reforming of methane over nanostructured palladium catalysts. Int J Hydrog Energy 115–116:190–200. https://doi.org/10.1016/j.apcatb.2011.12.002

Singh SA, Madras G (2016) Sonochemical synthesis of Pt, Ru doped TiO_2 for methane reforming. Appl Catal A: Gen 518:102–114. https://doi.org/10.1016/j.apcata.2015.10.047

Singha RK, Yadav A, Agrawal A, Shukla A, Adak S, Sasaki T, Bal R (2016) Synthesis of highly coke resistant Ni nanoparticles supported MgO/ZnO catalyst for reforming of methane with carbon dioxide. Appl Catal B-Environ 191:165–178. https://doi.org/10.1016/j.apcatb.2016.03.029

Sivaiah MV, Petit S, Barrault J, Batiot-Dupeyrat C, Valange S (2010) CO_2 reforming of CH_4 over Ni-containing phyllosilicates as catalyst precursors. Catal Today 157:397–403. https://doi.org/10.1016/j.cattod.2010.04.042

Slagtern A, Olsbye U, Blom R, Dahl IM, Fjellvåg H (1997) Characterization of Ni on La modified Al_2O_3 catalysts during CO_2 reforming of methane. Appl Catal a: Gen 165:379–390. https://doi.org/10.1016/S0926-860X(97)00220-2

Sodesawa T, Dobashi A, Nozaki F (1979) Catalytic reaction of methane with carbon dioxide. React Kinet Catal Lett 12:107–111. https://doi.org/10.1007/BF02071433

Song C, Liu Q, Deng S, Lic H, Kitamura Y (2019) Cryogenic-based CO_2 capture technologies: state-of-the-art developments and current challenges. Renew Sust Energ Rev 101:265–278. https://doi.org/10.1016/j.rser.2018.11.018

Stagg SM, Resasco DE (1997) Effects of promoters and supports on coke formation on Pt catalysts during CH_4 reforming with CO_2. Stud Surf Sci Catal 111:543–550. https://doi.org/10.1016/S0167-2991(97)80198-X

Stagg SM, Resasco DE (1998) Effect of promoters on supported Pt catalysts for CO_2 reforming of CH_4. Stud Surf Sci Catal 119:813–818

Stroud T, Smith TJ, Saché EL, Santos JL, Centeno MAH, Arellano-Garcia H, Odriozola JA, Reina TR (2017) Chemical CO_2 recycling via dry and bi reforming of methane using Ni-Sn/Al_2O_3 and Ni-Sn/CeO_2-Al_2O_3 catalysts. Appl Catal B-Environ 224:125–135. https://doi.org/10.1016/j.apcatb.2017.10.047

Sudhakaran MSP, Sultana L, Hossain MM, Pawlat J, Diatczyk J, Brüser V, Reuter S, Mok YS (2018) Iron–ceria spinel ($FeCe_2O_4$) catalyst for dry reforming of propane to inhibit carbon formation. J Ind Eng Chem 61:142–151. https://doi.org/10.1016/j.jiec.2017.12.011

Sun H, Wang H, Zhang J (2007) Preparation and characterization of nickel–titanium composite xerogel catalyst for CO_2 reforming of CH_4. Appl Catal B-Environ 73:158–165. https://doi.org/10.1016/j.apcatb.2006.07.019

Sutthiumporn K, Kawi S (2011) Promotional effect of alkaline earth over Ni–La_2O_3 catalyst for CO_2 reforming of CH_4: role of surface oxygen species on H_2 production and carbon suppression. Int J Hydrog Energy 36:14435–14446. https://doi.org/10.1016/j.ijhydene.2011.08.022

Sutthiumporn K, Maneerung T, Kathiraser Y, Kawi S (2012) CO2 dry-reforming of methane over La0.8Sr0.2Ni0.8M0.2O3 perovskite (M = Bi Co, Cr, Cu, Fe): Roles of lattice oxygen on C-H activation and carbon suppression. Int J Hydrog Energy 37:11195–11207. https://doi.org/10.1016/j.ijhydene.2012.04.059

Suzuki S, Hayakawa T, Hamakawa S, Suzuki K, Shishido T, Takehira K (1998) Sustainable Ni catalysts prepared by SPC method for CO_2 reforming of CH_4. Stud Surf Sci Catal 119:783–788

Swaan HM, Kroll VCH, Martin GA, Mirodatos C (1994) Deactivation of supported nickel catalysts during reforming of methane by carbon dioxide. Catal Today 21:571–578. https://doi.org/10.1016/0920-5861(94)80181-9

Takanabe K, Nagaoka K, Nariai K, Aika K (2005) Titania-supported cobalt and nickel bimetallic catalysts for carbon dioxide reforming of methane. J Catal 232:268–275. https://doi.org/10.1016/j.jcat.2005.03.011

Takayasu O, Takegahara Y, Matsuura I (1995) Thermogravimetric study in connection with the CO_2 reforming reaction of CH_4. Energy Convers Mgmt 36:597–600. https://doi.org/10.1016/0196-8904(95)00076-P

Talkhoncheh SK, Haghighi M, Jodeiri N, Aghamohammadi S (2017) Hydrogen production over ternary supported Ni/Al2O3-clinoptilolite-CeO2 nanocatalyst via CH_4/CO_2 reforming: influence of support composition. J Nat Gas Sci Eng 46:699–709. https://doi.org/10.1016/j.jngse.2017.08.027

Tang SB, Qiu FL, Lu SJ (1995) Effect of supports on the carbon deposition of nickel catalysts for methane reforming with CO_2. Catal Today 24:253–255. https://doi.org/10.1016/0920-5861(95)00036-F

Tang S, Ji L, Lin J, Zeng HC, Tan KL, Li K (2000) CO_2 reforming of methane to synthesis gas over sol–gel-made Ni/γ-Al_2O_3 catalysts from organometallic precursors. J Catal 194:424–430. https://doi.org/10.1006/jcat.2000.2957

The Linde Group (2015) Linde develops a new production process for synthesis gas. https://www.linde-engineering.com/en/news_and_media/press_releases/news_20151015.html

Therdthianwong S, Siangchin C, Therdthianwong A (2007) Improvement of coke resistance of Ni/Al_2O_3 catalyst in CH_4/CO_2 reforming by ZrO_2 addition. Fuel Process Technol 89:160–168. https://doi.org/10.1016/j.fuproc.2007.09.003

Tian Z, Huang J, Shao G, He Q, Cao S, Yuan S (2018) Ultra-microporous N-doped carbon from polycondensed framework precursor for CO_2-adsorption. Microporous Mesoporous Mater 257:19–26. https://doi.org/10.1016/j.micromeso.2017.08.012

Töbelmann D, Wendler T (2020) The impact of environmental innovation on carbon dioxide emissions. J Clean Prod 244:118787.https://doi.org/10.1016/j.jclepro.2019.118787

Tokunaga O, Ogasawara S (1989). Reduction of carbon dioxide with methane over Ni-catalyst. React Kinet Catal Lett 39:69–74. https://doi.org/10.1007/BF02061857

Treacy D, Ross JRH (2004) Carbon dioxide reforming of methane over supported molybdenum carbide catalysts. Stud Surf Sci Catal 147:193–198. https://doi.org/10.1016/S0167-2991(04)80050-8

Tsipouriari VA, Verykios XE (1999) Carbon and oxygen reaction pathways of CO_2 reforming of methane over Ni/La2O3 and Ni/Al_2O_3 catalysts studied by isotopic tracing techniques. J Catal 187:85–94. https://doi.org/10.1006/jcat.1999.2565

Tsipouriari VA, Efstathiou AM, Zhang ZL, Verykios XE (1994) Reforming of methane with carbon dioxide to synthesis gas over supported Rh catalysts. Catal Today 21:579–587. https://doi.org/10.1016/0920-5861(94)80182-7

UNFCCC COP21. Report of the conference of the parties on its twenty-first session, held in Paris from 30 Nov–13 Dec 2015.https://unfccc.int/resource/docs/2015/cop21/eng/10a02.pdf

Valentini A, Carreño NLV, Probst LFD, Lisboa-Filho PN, Schreiner WH, Leite ER, Longo E (2003) Role of vanadium in $Ni:Al_2O_3$ catalysts for carbon dioxide reforming of methane. Appl Catal A Gen 255:211–220. https://doi.org/10.1016/S0926-860X(03)00560-X

Vasiliades MA, Makri MM, Djinović P, Erjavec B, Pintar A, Efstathioua AM (2016) Dry reforming of methane over 5 wt% $Ni/Ce_{1-x}Pr_xO_{2-\delta}$ catalysts: performance and characterisation of active and inactive carbon by transient isotopic techniques. Appl Catal B-Environ 197:168–183. https://doi.org/10.1016/j.apcatb.2016.03.012

Vernon PDF, Green MLH, Cheetham AK, Ashcroft AT (1992) Partial oxidation of methane to synthesis gas, and carbon dioxide as an oxidising agent for methane conversion. Catal Today 13. https://doi.org/10.1016/0920-5861(92)80167-L

Verykios XE (2003) Mechanistic aspects of the reaction of CO_2 reforming of methane over Rh/Al_2O_3 catalyst. Appl Catal A Gen 255:101–111. https://doi.org/10.1016/S0926-860X(03)00648-3

Wang S (2007) Mesoporous silica supported Ni catalysts for CO_2 reforming of methane. Stud Surf Sci Catal 165:795–798. https://doi.org/10.1016/S0167-2991(07)80439-3

Wang HY, Au CT (1997) Carbon dioxide reforming of methane to syngas over SiO_2-supported rhodium catalysts. Appl Catal A: Gen 155:239–252. https://doi.org/10.1016/S0926-860X(96)00398-5

Wang S, Lu GQM (1998) Role of CeO_2 in $Ni/CeO_2–Al_2O_3$ catalysts for carbon dioxide reforming of methane. Appl Catal B-Environ 19:267–277. https://doi.org/10.1016/S0926-3373(98)00081-2

Wang HY, Ruckenstein E (2001) CO_2 reforming of CH_4 over Co/MgO solid solution catalysts—effect of calcination temperature and Co loading. Appl Catal A Gen 209:207–215. https://doi.org/10.1016/S0926-860X(00)00753-5

Wang S, Lu GQ, Millar GJ (1996) Carbon dioxide reforming of methane to produce synthesis gas over metal-supported catalysts: state of the art. Energy Fuels 10:896–904. https://doi.org/10.1021/ef950227t

Wang JX, Liu Y, Cheng T, Li WX, Bi YL, Zhen K (2003) Methane reforming with carbon dioxide to synthesis gas over Co-doped Ni-based magnetoplumbite catalysts. Appl Catal A Gen 250:13–23. https://doi.org/10.1016/S0926-860X(03)00497-6

Wang JB, Wu YX, Huang TJ (2004) Effects of carbon deposition and de-coking treatments on the activation of CH4 and CO_2 in CO_2 reforming of CH_4 over Ni/yttria-doped ceria catalysts. Appl Catal A Gen 272:289–298. https://doi.org/10.1016/j.apcata.2004.05.053

Wang R, Liu X, Ge Q, Li W, Xu H (2007a) Studies of carbon deposition over $Rh\text{-}CeO_2/Al_2O_3$ during CH_4/CO_2 reforming. Stud Surf Sci Catal 167:379–384. https://doi.org/10.1016/S0167-2991(07)80161-3

Wang SG, Liao XY, Hu J, Cao DB, Li YW, Wang J, Jiao H (2007b) Kinetic aspect of CO_2 reforming of CH_4 on Ni(1 1 1): a density functional theory calculation. Surf Sci 601:1271–1284. https://doi.org/10.1016/j.susc.2006.12.059

Wasajja H, Lindeboom REF, Lier JB, Aravind PV (2020) Techno-economic review of biogas cleaning technologies for small scale off-grid solid oxide fuel cell applications. Fuel Process Technol 197:106215. https://doi.org/10.1016/j.fuproc.2019.106215

Wei J, Iglesia E (2004) Isotopic and kinetic assessment of the mechanism of reactions of CH_4 with CO_2 or H_2O to form synthesis gas and carbon on nickel catalysts. J Catal 224:370–383. https://doi.org/10.1016/j.jcat.2004.02.032

Wisniewski M, Boréave A, Gélin P (2005) Catalytic CO_2 reforming of methane over $Ir/Ce_{0.9}Gd_{0.1}O_{2-x}$. Catal Commun 6:596–600. https://doi.org/10.1016/j.catcom.2005.05.008

Wu JCS, Chou HC (2009) Bimetallic Rh–Ni/BN catalyst for methane reforming with CO_2. Chem Eng J 148(5):Q39-545. https://doi.org/10.1016/j.cej.2009.01.011

Wu Q, Chen J, Zhang J (2008) Effect of yttrium and praseodymium on properties of $Ce_{0.75}Zr_{0.25}O_2$ solid solution for CH_4–CO_2 reforming. Fuel Process Technol 89:993–999. https://doi.org/10.1016/j.fuproc.2008.03.006

Xiancai L, Qanhong H, Yifeng Y, Juanrong C, Zhihua L (2008) Effects of sol–gel method and lanthanum addition on catalytic performances of nickel-based catalysts for methane reforming with carbon dioxide. J Rare Earth 26:864–868. https://doi.org/10.1016/S1002-0721(09)60022-3

Xu C, Zhu J (2006) Parametric study of fine particle fluidization under mechanical vibration. Powder Technol 161:135–144. https://doi.org/10.1016/j.powtec.2005.10.002

Xu Z, Zhen M, Bi Y, Zhen K (2000) Catalytic properties of Ni modified hexaaluminates $LaNi_yAl_{12-y}O_{19-\delta}$ for CO_2 reforming of methane to synthesis gas. Appl Catal a: Gen 198:267–273. https://doi.org/10.1016/S0926-860X(99)00518-9

Xu Z, Li Y, Zhang J, Chang L, Zhou R, Duan Z (2001a) Bound-state Ni species-a superior form in Ni-based catalyst for CH_4/CO_2 reforming. Appl Catal A Gen 210:45–53. https://doi.org/10.1016/S0926-860X(00)00798-5

Xu Z, Li Y, Zhang J, Chang L, Zhou R, Duan Z (2001) Ultrafine NiO–La_2O_3–Al_2O_3 aerogel: a promising catalyst for CH_4/CO_2 reforming. Appl Catal A Gen 213:65–71. https://doi.org/10.1016/S0926-860X(00)00881-4

Xu BQ, Wei JM, Wang HY, Sun KQ, Zhu QM (2001) Nano-MgO: novel preparation and application as support of Ni catalyst for CO_2 reforming of methane. Catal Today 68(1–3):217–225. https://doi.org/10.1016/S0920-5861(01)00303-0

Xu L, Liu Y, Li Y, Zhu L, Ma X, Zhang Y, Argyle MD, Fan M (2014) Catalytic CH_4 reforming with CO_2 over activated carbon based catalysts. Appl Catal A Gen 469:387–397. https://doi.org/10.1016/j.apcata.2013.10.022

Xu L, Zhang X, Chen M, Qi L, Nie D, Ma Y (2016) Facilely fabricating Mg, Ca modified Co based ordered mesoporous catalysts for CO_2 reforming of CH_4: the effects of basic modifiers. Int J Hydrog Energy 41:17348–17360. https://doi.org/10.1016/j.ijhydene.2016.08.013

Xu D, Zhang H, Ma H, Qian W, Ying W (2017) Effect of Ce promoter on Rh-Fe/TiO_2 catalysts for ethanol synthesis from syngas. Catal Commun 98:90–93. https://doi.org/10.1016/j.catcom.2017.03.019

Xu J, Xiao Q, Zhang J, Sun Y, Zhu Y (2017b) NiO-MgO nanoparticles confined inside SiO_2 frameworks to achieve highly catalytic performance for CO_2 reforming of methane. Mol Catal 432:31–36. https://doi.org/10.1016/j.mcat.2017.02.019

Yang M, Papp H (2006) CO_2 reforming of methane to syngas over highly active and stable Pt/MgO catalysts. Catal Today 115:199–204. https://doi.org/10.1016/j.cattod.2006.02.047

Yang E, Noh Y, Ramesh S, Lim SS, Moon DJ (2015) The effect of promoters in $La_{0.9}M_{0.1}Ni_{0.5}Fe_{0.5}O_3$ (M = Sr, Ca) perovskite catalysts on dry reforming of methane. Fuel Process Technol 134:404–413. https://doi.org/10.1016/j.fuproc.2015.02.023

Yang X, Su X, Chen D, Zhang T, Huang Y (2020) Direct conversion of syngas to aromatics: a review of recent studies. Chinese J Catal 41:561–573. https://doi.org/10.1016/S1872-2067(19)63346-2

Yao W, Guangsheng Gu, Fei W, Jun W (2002) Fluidization and agglomerate structure of SiO_2 nanoparticles. Powder Technol 124:152–159. https://doi.org/10.1016/S0032-5910(01)00491-0

Yao L, Zhu J, Peng X, Tong D, Hu C (2013) Comparative study on the promotion effect of Mn and Zr on the stability of Ni/SiO_2 catalyst for CO_2 reforming of methane. Int J Hydrog Energy 38:7268–7279. https://doi.org/10.1016/j.ijhydene.2013.02.126

Yentekakis IV, Goula G, Hatzisymeon M, Argyropoulou IB, Botzolaki G, Kousi K, Kondarides DI, Taylor MJ, Parlett CMA, Osatiashtiani A, Kyriakou G, Holgado JP, Lambert RM (2019) Effect of support oxygen storage capacity on the catalytic performance of Rh nanoparticles for CO_2 reforming of methane. Appl Catal B-Environ 243:490–501. https://doi.org/10.1016/j.apcatb.2018.10.048

York APE, Suhartanto T, Green MLH (1998) Influence of molybdenum and tungsten dopants on nickel catalysts for the dry reforming of methane with carbon dioxide to synthesis gas. Stud Surf Sci Catal 119:777–782

Yu X, Wang N, Chu W, Liu M (2012) Carbon dioxide reforming of methane for syngas production over La-promoted NiMgAl catalysts derived from hydrotalcites. Chem Eng J 209:623–632. https://doi.org/10.1016/j.cej.2012.08.037

Yutaka N, Bernard C (1986) Production of carbon monoxide by direct thermal splitting of carbon dioxide at high temperature. Bull Chem Soc Jpn 59:1997–2002. https://doi.org/10.1246/bcsj.59.1997

Zeng S, Zhang L, Zhang X, Wang Y, Pan H, Su H (2012) Modification effect of natural mixed rare earths on Co/γ-Al_2O_3 catalysts for CH_4/CO_2 reforming to synthesis gas. Int J Hydrog Energy 37:9994–10001. https://doi.org/10.1016/j.ijhydene.2012.04.014

Zhan Y, Song K, Shi Z, Wan C, Pan J, Li D, Au C, Jiang L (2020) Influence of reduction temperature on Ni particle size and catalytic performance of Ni/Mg(Al)O catalyst for CO_2 reforming of CH4. Int J Hydrog Energy 45:2794–2807. https://doi.org/10.1016/j.ijhydene.2019.11.181

Zhang ZL, Verykios XE (1994) Carbon dioxide reforming of methane to synthesis gas over supported Ni catalysts. Catal Today 21:589–595. https://doi.org/10.1016/0920-5861(94)80183-5

Zhang Z, Verykios X (1997) Performance of Ni/La_2O_3 catalyst in carbon dioxide reforming of methane to synthesis gas. Stud Surf Sci Catal 107:511–516. https://doi.org/10.1016/S0167-2991(97)80383-7

Zhang ZL, Tsipouriari VA, Efstathiou AM, Verykios XE (1996) Reforming of methane with carbon dioxide to synthesis gas over supported rhodium catalysts: I. Effects of support and metal crystallite size on reaction activity and deactivation characteristics. J Catal 158:51–63. https://doi.org/10.1006/jcat.1996.0005

Zhang J, Wang H, Dalai AK (2007a) Development of stable bimetallic catalysts for carbon dioxide reforming of methane. J Catal 249:300–310. https://doi.org/10.1016/j.jcat.2007.05.004

Zhang WD, Liu BS, Tian YL (2007b) CO_2 reforming of methane over Ni/Sm_2O_3–CaO catalyst prepared by a sol–gel technique. Catal Commun 8:661–667. https://doi.org/10.1016/j.catcom.2006.08.020

Zhang S, Wang J, Liu H, Wang X (2008) One-pot synthesis of Ni-nanoparticle-embedded mesoporous titania/silica catalyst and its application for CO_2-reforming of methane. Catal Commun 9:995–1000. https://doi.org/10.1016/j.catcom.2007.09.033

Zhang K, Zhou G, Li J, Cheng T (2009) The electronic effects of Pr on $La_{1-x}Pr_xNiAl_{11}O_{19}$ for CO_2 reforming of methane. Catal Commun 10:1816–1820. https://doi.org/10.1016/j.catcom.2009.06.007

Zhang G, Su A, Du Y, Qu J, Xu Y (2014) Catalytic performance of activated carbon supported cobalt catalyst for CO_2 reforming of CH_4. J Colloid Interface Sci 433:149–155. https://doi.org/10.1016/j.jcis.2014.06.038

Zhang G, Hao L, Jia Y, Du Y, Zhang Y (2015) CO_2 reforming of CH_4 over efficient bimetallic Co–Zr/AC catalyst for H_2 production. Int J Hydrogen Energy 40:12868–12879. https://doi.org/10.1016/j.ijhydene.2015.07.011

Zhang G, Zhao P, Xu Y, Qu J (2017) Characterization of Ca-promoted Co/AC catalyst for CO_2-CH_4 reforming to syngas production. J CO_2 Util 18:326–334. https://doi.org/10.1016/j.jcou.2017.02.013

Zhang G, Sun Y, Zhao P, Xu Y, Su A, Qu J (2017) Characteristics of activated carbon modified with alkaline $KMnO_4$ and its performance in catalytic reforming of greenhouse gases CO_2/CH_4. J CO_2 Util 20:129–140. https://doi.org/10.1016/j.jcou.2017.05.013

Zhang G, Liu J, Xu Y, Sun Y (2018a) A review of CH_4-CO_2 reforming to synthesis gas over Ni-based catalysts in recent years (2010–2017). Int J Hydrogen Energ 43:15030–15054. https://doi.org/10.1016/j.ijhydene.2018.06.091

Zhang Z, Liu L, Shen B, Wu C (2018b) Preparation, modification and development of Ni-based catalysts for catalytic reforming of tar produced from biomass gasification. Renew Sustain Energy Rev 94:1086–1109. https://doi.org/10.1016/j.rser.2018.07.010

Zhang L, Wang F, Zhu J, Han B, Fan W, Zhao L, Cai W, Li Z, Xu L, Yu H, Shi W (2019) CO_2 reforming with methane reaction over Ni@SiO_2 catalysts coupled by size effect and metal-support interaction. Fuel 25:15954. https://doi.org/10.1016/j.fuel.2019.115954

Zhang J, Ren M, Li X, Hao Q, Chen H, Ma X (2020) Ni-based catalysts prepared for CO_2 reforming and decomposition of methane. Energy Convers Manag 205:112419. https://doi.org/10.1016/j.enconman.2019.112419

Zhao M, Cui Y, Sun, J, Zhang Q (2018) Modified iron catalyst for direct synthesis of light olefin from syngas. Catal Today 316:142–148. https://doi.org/10.1016/j.cattod.2018.05.018

Zhaolong Z, Verykios XE (1996) Carbon dioxide reforming of methane to synthesis gas over Ni/La_2O_3 catalysts. Appl Catal a: Gen 138:109–133. https://doi.org/10.1016/0926-860X(95)00238-3

Zhu J, Peng X, Yao L, Shen J, Tong D, Hu C (2011) The promoting effect of La, Mg, Co and Zn on the activity and stability of Ni/SiO_2 catalyst for CO_2 reforming of methane. Int J Hydrog Energy 36:7094–7104. https://doi.org/10.1016/j.ijhydene.2011.02.133

Zuo ZJ, Shen CF, Tan PJ, Huang W (2013) Ni based on dual-support Mg-Al mixed oxides and SBA-15 catalysts for dry reforming of methane. Catal Commun 41:132–135. https://doi.org/10.1016/j.catcom.2013.07.007

Fuel Generation from CO_2

Mariany C. Deprá, Ihana A. Severo, Rafaela B. Sartori, Patrícia Arrojo, Leila Q. Zepka, and Eduardo Jacob-Lopes

Abstract

Aware of the uncontrolled increase in greenhouse gas emissions, and designated as the main contributor, carbon dioxide (CO_2), until now, has been the hottest point in research on gaseous waste management. Thus, the use of CO_2 is attracting more and more interest in renewable fuel generation processes. In this sense, this chapter aims to compile a wide range of applications for the use of CO_2 as an essential raw material in the conversion of fuels. Beyond briefly discussing the technologies for direct chemical conversion of CO_2 for the generation of alternative fuels, our approach focused on examining in more detail the biological use of CO_2, through microalgae-based processes. Finally, the chapter presents some recommendations for the future exploration of CO_2 generation technologies for fuels.

Keywords

Biofuels • Microalgae • Renewable energy • CO_2 recycling • Carbon capture and utilization

1 Introduction

Since the beginning of the industrial revolution, economic growth has skyrocketed at the expense of environmental quality. This is because the increased requirements for fossil fuels have resulted in direct consequences in terms of greenhouse gas emissions (Ike et al. 2020). Recently, carbon dioxide (CO_2) emissions showed exponential behaviors, reaching a historical high above 33 Mt CO_2 in 2019 (IEA 2020). As a result, the considerable dependence on fossil fuels and a rapid increase in energy demand has resulted in high environmental burdens, threatening global sustainability in the short and medium-term worldwide (O'Ryan et al. 2020).

Current sources of fossil fuel are insufficient to meet global energy consumption, as they will double in the next thirty years and it is estimated to triple before the 2100s (Mustafa et al. 2020). Faced with this scenario, two distinct aspects emerge. The first is the importance of highlighting bioenergy systems: What will be the fraction of the use of renewable fuels in the coming years? Subsequently, to encourage the technology choices for the generation of clean fuels. Indeed, do public policies address the continued utilization of fossil fuels as a global emergency issue? (Mantulet et al. 2020).

On the other hand, aware of the imminent environmental collapse, significant research and development efforts focused on alternatives for reducing and controlling carbon (Sun et al. 2020). Among the alternatives, employing negative emission technologies have been developed, and the concept of CO_2 capture, storage, and use have become the main target of modern industrial progress (Zhu et al. 2020). In this way, the relentless search for the promise of reducing environmental impacts, through the recycling of CO_2, enabling the generation of fuels represented a set of complex systems to be elucidated.

As widespread, CO_2 conversion is scientifically a challenging task, but it has expressive advantages. Theoretically, some techniques, such as the direct conversion of CO_2 into fuels such as syngas and methanol, have been proposed. However, inherent characteristics of the reaction, as the low applicability and lack of thermodynamic improvement, are still determining factors for the non-consolidation of these technologies (Kumaravel et al. 2020).

In contrast, it is believed that the processes of capture and biological use of carbon are by far the most promising alternative (Severo et al. 2018). The technology-mediated by processes based on microalgae, use the photosynthetic

M. C. Deprá · I. A. Severo · R. B. Sartori · P. Arrojo · L. Q. Zepka · E. Jacob-Lopes (✉)
Bioprocess Intensification Group, Federal University of Santa Maria (UFSM), Santa Maria, RS 97105-900, Brazil

Inamuddin et al. (eds.), *Carbon Dioxide Utilization to Sustainable Energy and Fuels*, Advances in Science, Technology & Innovation, https://doi.org/10.1007/978-3-030-72877-9_3

apparatus to convert CO_2 in the presence of light and water, resulting in biomass available for the most diverse sources of biofuels (Andrade et al. 2020). However, despite its notorious potential, its large-scale application is still not practiced commercially in a major portion due to the unfavorable economics of the process (Vadlamani et al. 2019).

Given that, this chapter objectives to highlights and describe the main current technologies for the direct conversion of CO_2 into fuels, aiming at the capture and mitigation of greenhouse gases. Besides, a more in-depth approach to the importance of biological carbon capture and storage systems for the generation of biofuels from microalgae was also noteworthy.

2 Approaches for Directly Converting CO_2 to Fuels

2.1 Pure CO_2 Decomposition Technology

Originating in the mid-1980s, driven mainly by increased industrial activity and the oil energy crisis, prospective studies have resulted in considerable progress in investigating the potential of carbon dioxide as a reagent in energy synthesis, more precisely as a final product of biofuel, hydrocarbons (Freund and Roberts 1996).

Under stoichiometric aspect, the partition of the CO_2 molecule can be approached through Eq. 1:

$$CO_2(g) \rightarrow CO(g) + \frac{1}{2}O_2 \quad \Delta H^{\circ} = +293\,\text{kJ/mol} \tag{1}$$

However, a priori, the partition of the molecule provides such a high value of H°. To date, pure CO_2 decomposition is ineffective due to the high stability of the molecule. So is because the free energy of Gibbs formation has values in the order of −397 kJ/mol, for the bond between the carbon and oxygen molecules. As a result, the thermal demand for energy is extremely high, in addition to requiring a highly active catalyst and requiring ideal conditions for the chemical reaction to occur (Mustafa et al. 2020). Furthermore, thermodynamically, both enthalpy and entropy are unfavorable for conversion.

Therefore, even though initial attempts have described direct CO_2 decomposition as potential alternatives, in fact, only from a theoretical point of view, this technology has shown promise. In this way, we portray a simplified approach just to provide a complete perspective on the topic.

2.2 Reagent-Based CO_2 Conversion Technology

Based on previous studies, it can be observed that the direct division of CO_2, demands high energy values, and, therefore, an alternative to circumvent this process bottleneck, presents itself in the practical approach of the conversion of CO_2 associated with the use of a co-reagent.

Given the relationship between energy exchanges and heat availability of a chemical reaction, it is strongly favorable to convert CO_2 in the presence of a reagent whose Gibbs energy values are higher. Among the main ones, we can mention methane and hydrogen. In short, both the chemical energy of the hydrogen in the methane molecule and the hydrogen in the hydrogen gas serve as support in the conversion of CO_2.

Currently, most of the conventional processes for converting CO_2 into energy sources, more precisely in methanol and syngas, use methane and hydrogen as co-reagents. Under the thermodynamic aspects, the combination of water and carbon dioxide is of interest for exploration, since it does not have any conventional approach and results in high value-added chemicals.

Among the main CO_2 conversions with their respective co-reagents, two main methods stand out, namely the dry deformation of methane and the hydrogenation of CO_2. In this sense, more details about these technological advances are discussed below.

2.2.1 Dry Deformation of Methane Technology

Fisher and Tropsch were the first to investigate dry methane reform (DRM) in the mid-1928s (Arora and Prasad 2016). Conceptually, this process is known to utilize the major greenhouse gases, that is, residual carbon dioxide and methane, for the chemical synthesis of natural gas (Gangadharan et al. 2012).

As noted in Eq. 2, the dry reform reaction is known to be an endothermic reaction that uses catalysts favorable to accelerate the reaction. Among some catalysts used, it is known that noble metals such as ruthenium and rhodium demonstrated superior catalytic performance and resistance to coke, however, most of the processes under test make use of level and copper-based catalysts (Gao et al. 2020).

$$\begin{gathered} CH_4(g) + CO_2(g) \rightarrow 2CO(g) + 2H_2(g) \\ \Delta H^{\circ}_{298k} = +247\,\text{kJ/mol} \end{gathered} \tag{2}$$

This promising technology has long received broad interest due to environmental concerns and commercial relevance. Due to its numerous advantages, the dry methane reform ensures the reduction of greenhouse gases, concurrent with a fuel generation process, in addition to eliminating the separation of gases from the final products, resulting in more available process costs (Abdullah et al. 2017). In principle, the products derived can be applied as energy resources for the chemical industries, power plants, as well as for existing vehicles (Song et al. 2020). However, to date, this process is not widely used on an industrial scale, as some bottlenecks still require more in-depth clarifications such as the formation of coke in the catalyst, beyond the substantial heat demand necessary for the reaction to take place (Egelske et al. 2020).

2.2.2 Catalytic Hydrogenation of CO_2

A consistent approach for recycling CO_2 released into the ecosystem can be associated with the processes of hydrogenation of CO_2. In theory, the production of valuable chemicals, such as methanol, has been exploited mainly for presenting promising characteristics of efficient technology in reducing greenhouse gases (Huff and Sanford 2011).

Nowadays, it is estimated that the demand for methanol represents about 78 million tons per year in the global market (Kothandaraman and Heldebrant 2020). Among the most diverse applications, methanol can be used as a final product, but also as an intermediate product for the production of unsaturated hydrocarbons, fuel mixtures, the formation of acetic acid, beyond formaldehyde.

Chemically, the hydrogenation of CO_2 is given by the Eq. 3:

$$CO_2(g) + 3H_2(g) \rightarrow CH_3OH(g) + H_2O \quad \Delta H^{\circ}_{298k} = -49.9\,\text{kJ/mol} \tag{3}$$

Under the stoichiometric aspects, the reaction has an exothermic character, competing with the methanation reaction. Because of this obstacle, there is a substantial need to use catalysts with high adsorption and CO_2 activation, without breaking the chemical bond of the molecule. Thus, the synthesis of methanol has equilibrium characteristics limited by high temperatures, and therefore the need for catalysts operating effectively at temperatures below 450 K (Al-Saydeh and Zaidi 2018).

Given this scenario, operating restrictions make the process complex in performance. Therefore, the most diverse catalysts were evaluated. In the current state, industries opt for raw materials of metal nanoparticles supported on an oxide substrate. Likewise, copper-based catalysts have been widely marketed for more than 50 years. Technically, these catalysts do not allow the cleavage of molecular bonds, preventing the formation of methane. However, lower hydrogenation activities are observed at temperatures below 500 K (Rodriguez et al. 2015).

Although much progress has been achieved, and its importance in terms of mitigating CO_2 emissions have been recognized, to date, no large-scale adoption has yet been made (Li et al. 2020). Therefore, more theoretical and experimental investigations are needed at the laboratory level and pilot scales, to improve the effective applicability of catalysts to increase the performance of the process and thus consolidate this technology.

3 Biological CO_2 Fixation for Fuels

As previously seen, the current technologies for converting CO_2 into fuels are far from being consolidated because they present significant technical challenges, as well as economic and environmental hurdles around the world. Thus, the development of biological technologies that aim to capture and use CO_2 for biofuels is of great importance (Singh et al. 2020).

Currently, the biological capture of CO_2 based on microalgae is considered a promising method to mitigate CO_2 emissions, given its vital role in determining CO_2 capture rates and levels of biomass production (Cheng et al. 2019). This can be attributed to its metabolic versatility since microalgae can assimilate inorganic carbon via photosynthesis, after a sequence of carboxylation, reduction, and regeneration reactions. Given these proprieties associated with photosynthesis, microalgae can fix variable carbon sources, including CO_2 from the atmosphere, industrial exhaust gases, and soluble carbon builders (Table 1) (Sydney et al. 2010).

As a result of the photosynthetic process, the CO_2 intake is transformed into lipid carbohydrates, and proteins, which can later be bioconverted into various potential renewable biofuels (Table 2). However, the differences in the

Table 1 Data of biomass productivity and CO_2 fixation rate from microalgae

Microalgae strain	Biomass ($kg/m^3/d$)	CO_2 fixation rate ($kg/m^3/d$)	References
Spirulina platensis	0.15	0.32	Sydney et al. (2010)
Chlorella Vulgaris	0.13	0.25	Sydney et al. (2010)
Synechocystis aquatilis	0.03	0.05	Zhang et al. (2001)
Anabaena sp.	0.31	1.45	López et al. (2009)
Botryococcus braunii	0.21	0.50	Sydney et al. (2010)
Dunaliella tertiolecta	0.14	0.27	Sydney et al. (2010)
Aphanothece microscopica Nageli	0.30	0.56	Jacob-Lopes et al. (2009)

Adapted from Sydney et al. (2019).

Table 2 Biochemical composition of microalgae expressed on a dry matter basis (% dry weight)

Microalgal species	Protein (%)	Carbohydrates (%)	Lipid (%)
Scenedesmus obliquus	50–56	10–17	12–14
Scenedesmus dimorphus	8–18	21–52	16–40
Chlorella vulgaris	51–58	12–17	14–22
Chlorella pyrenoidosa	57	26	2
Dunaliella bioculata	49	4	8
Dunaliella salina	57	32	6
Euglena gracilis	39–61	14–18	14–20
Porphyridium cruentum	28–39	40–57	9–14
Spirulina platensis	46–63	8–14	4–9
Spirulina maxima	60–71	13–16	6–7
Synechoccus sp.	63	15	11
Anabaena cylindrica	43–56	25–30	4–7

Adapted from Becker (1994), Suganya et al. (2016).

composition of the microalgal biomass can be attributed to several factors, which are specific to the species, as well as external factors that influence the metabolic apparatus such as temperature, light intensity, pH, carbon availability (Andrade et al. 2019).

In this sense, microalgae are gaining strength on the bridge between carbon reduction and bioenergy production, since microalgae biomass is considered a CO_2 neutral substitute for fossil fuels, using the net transport of atmospheric CO_2 to biomass (Choi et al. 2019a). Besides, as microalgae can proliferate throughout the year, they can produce target products during all seasons, becoming promising candidates for the application of fuel generation (Choi et al. 2019b; Tang et al. 2020).

However, for microalgal biomass to be used as a feedstock for renewable fuels, it is necessary to understand the type of biomass conversion technology that is adequate, given its peculiar characteristics (Choudhary et al. 2020). To that end, more details will be discussed below in the thermochemical and biochemical conversion technologies conversion of biomass microalgae (Choo et al. 2020).

3.1 Thermochemical Conversion

The conversion of microalgae biomass into useful energy-rich products is gaining prominent significance, and there has been a growing focus in this area of research over the past few years. Thermochemical conversion provides a more straightforward route to produce biofuels as they involve controlled heating and/or oxidation of biomass to produce heat or energy (Chen et al. 2015; Fan et al. 2020).

Based on the literature, thermochemical technologies associated with microalgae biomass show great promise for the production of fuels and renewable electricity, as it helps to comply with the production standards established in many places (EIA 2012). Besides, in some cases, the thermochemical production renewable of electricity, fuels, and its derivative products is the most effective use of biomass for displacing fossil energy (Tanger et al. 2013).

Among several processes, the most fundamental thermochemical conversion technologies involve torrefaction, pyrolysis, liquefaction, gasification, and direct combustion (Ong et al. 2019). Figure 1 summarizes the different types of thermochemical conversion processes, their classification, process conditions, and the types of biofuels formed. Also, Table 3 shows numerous studies that can be found in the literature on microalgae, specifically to obtain renewable fuel.

3.1.1 Torrefaction

Torrefaction is a promising pretreatment technique or also known as pre-pyrolysis, as it operates at milder temperatures when compared to other processes (between 200 and 300 °C). In torrefaction, biomass is thermally degraded to remove total moisture and low molecular weight organic volatile content. This technique was implemented in order to overcome the disadvantage of the low calorific values of microalgae by increasing the density and their energy content by 30% when compared to conventional biomass (Koo et al. 2014).

Recently, roasting has been recognized as an effective method to produce hydrophobic solid fuels with low operating costs, since this technology can improve its homogeneity, heating value, grinding, and resistance to biodegradation. The main products generated are biochar, bio-oil, and biogas, in which biochar is the main product produced during this process (Dai et al. 2019; Gan et al. 2019).

During the torrefaction process, the most important parameter is the temperature, followed by the duration time

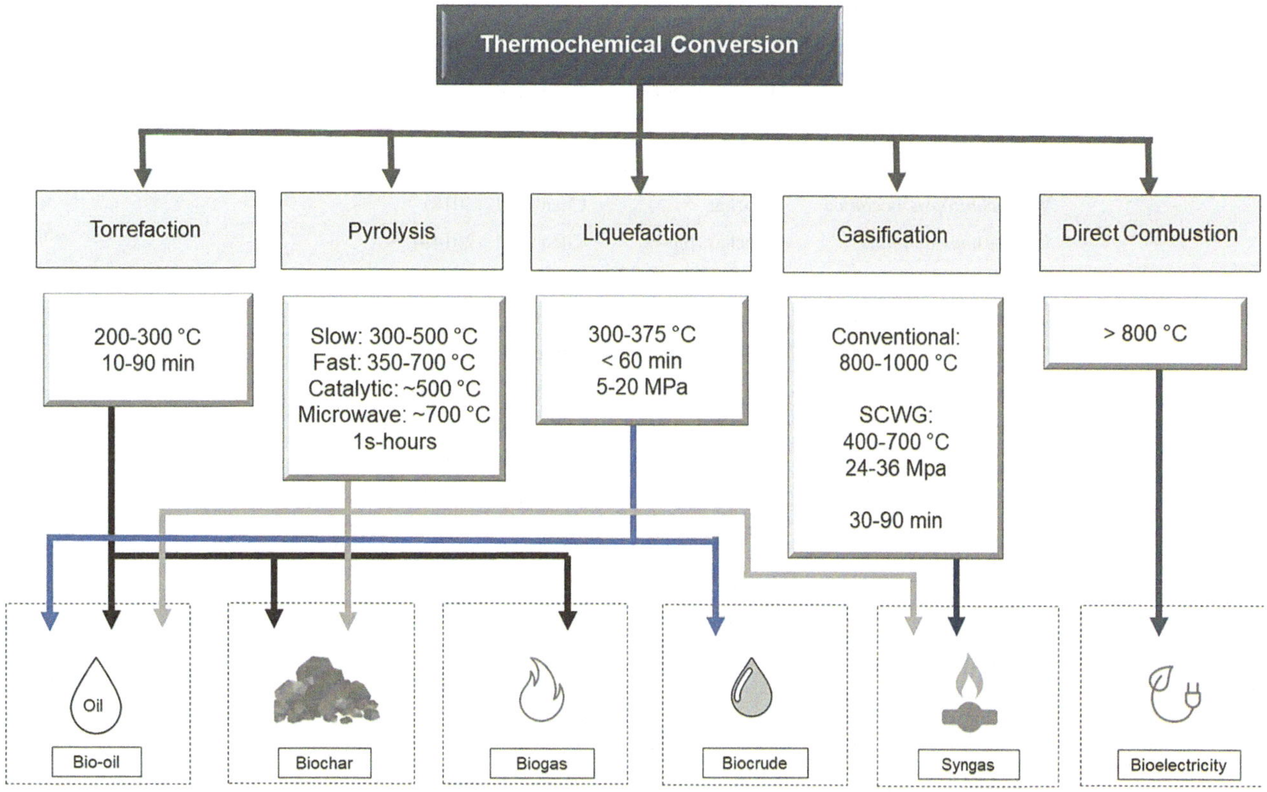

Fig. 1 Biomass thermochemical conversion. Adapted from Chen et al. (2015)

(Kumar et al. 2017). In fact, some authors report that an increase in roasting temperature and residence time increased the content of carbon, ash, and fixed carbon, as well as the grinding index of HHV (higher heating values) in microalgae, such as *Spirulina platensis* and *Scenedesmus obliquus* (Wu et al. 2012). Chen et al. (2014a) also demonstrated that higher temperatures showed better thermal degradation of *Chlorella sorokiniana* and *Chlamydomonas* sp. However, despite increasing the quality of fuel, more severe conditions consume more energy and lead to lower energy efficiency. Consequently, the optimization of the process (in terms of time and temperature) becomes essential in the selection of the ideal operating conditions in terms of fuel quality and energy consumption.

3.1.2 Pyrolysis

Pyrolysis is the thermal decomposition of biomass at high levels of temperatures in the absence of oxygen. Products formed after thermal decomposition include biochar, bio-oil, and gases (such as syngas), in which their values and qualities are mainly influenced by the type of reaction, operating conditions, and properties of the biomass used (Mathimani et al. 2019). In this case, pyrolysis from microalgae biomass is more advantageous because it has more stable properties and achieves higher bio-oil yields than lignocellulosic biomass, for example (Fan et al. 2020). Furthermore, the properties of the biochar produced from pyrolysis by microalgae biomass can be transformed into excellent raw material for other processes, such as adsorption cleansing in fuel processing, or renewable holder for advanced batteries (Baldinelli et al. 2018).

The pyrolysis process is based on moisture dehydration, decomposition of organic structures, and disintegration of residual solids, divided into different reactions depending on the rate of heating. First, slow pyrolysis is optimized to produce solid fuel (coal) at a relatively low temperature (300–500 °C) over a longer residence time, ranging from minutes to hours (Kumar et al. 2017). At this stage, fixed-bed reactors are most commonly employed.

Fast or instant pyrolysis, with waiting times ranging from seconds to minutes, is optimized for the production of high-quality bio-oil and gas production by secondary decomposition at elevated temperatures (Ong et al. 2019). This process generally occurs at temperatures ranging from 350 to 700 °C. It is considered the most viable technique for the future replacement of fossil fuels by the high bio-oil yield, reaching a conversion rate of 95% from biomass to liquid (Brennam and Owend 2010). Here, the use of fluidized-bed reactors also provides some more advantages, mainly due to high heating rates, rapid devolatilization, and better process control (Grierson et al. 2009).

Table 3 The main products generated by microalgae in different thermochemical technologies

Technology	Microalgae	Fuel	References
Torrefaction	*Chlamydomonas* sp	Biochar, bio-oil	Chen et al. (2016, 2018)
	Chlorella sorokiniana	Biochar	Chen et al. (2014a)
	Chlorella vulgaris	Biochar	Bach and Chen (2017)
	Nannochloropsis oceanica	Biochar	Chen et al. (2018)
	Scenedesmus obliquus	Biochar, bio-oil	Chen et al. (2014b)
	Spirulina platensis	Biochar	Wu et al. (2012)
)Pyrolysis	*Chlorella protothecoides*	Bio-oil	Maliutina et al. (2018)
	Chlorella sp.	Bio-oil	Borges et al. (2014)
	Chlorella vulgaris	Bio-oil	Belotti et al. (2014), Chang et al. (2015)
	M. aeruginosa	Bio-oil, biochar	Chen et al. (2016)
	Nannochloropsis sp.	Bio-oil	Chen et al. (2016), Borges et al. (2014)
	Pavlova sp.	Bio-oil	Aysu et al. (2017)
	Scenedesmus sp.	Bio-oil, biochar	Chaiwong et al. (2013), Kim et al. (2014)
	Schizochytrium limacinum	Syngas	Hong et al. (2017)
	Spirulina sp.	Bio-oil	Anand et al. (2017)
Liquefaction	*Chlorella pyrenoidosa*	Bio-oil	Zhang et al. (2013)
	Chlorella sp.	Bio-oil	Jazrawi et al. (2013), Biller et al. (2015)
	D. tertiolecta	Bio-oil	Chen et al. (2015)
	Desmodesmus sp.	Bio-oil	Torri et al. (2012), Alba et al. (2012)
	Nannochloropsis sp.	Biocrude, Bio-oil	Brown et al. (2010), Faeth and Savage (2016), Saber et al. (2016a)
	Spirulina platensis	Bio-oil	Jena et al. (2012), Elliott et al. (2015)
	Scenedesmus dimorphous	Biocrude	Biller et al. (2012)
Gasification	*Chlorella vulgaris*	Syngas	Bach and Chen (2017), Liu et al. (2017)
	Chlorella pyrenoidosa	Syngas	Duan et al. (2018)
	Nannochloropsis gaditana	Syngas	Soreanu et al. (2017), Caputo et al. (2016)
	Nannochloropsis oculata	Syngas	Adnan et al. (2017)
	Nannochloropsis sp.	Syngas	Koo et al. (2014)
	Spirulina sp.	Syngas	Yang et al. (2013)
	Tetraselmis sp.	Syngas	Alghurabie et al. (2013)
Direct combustion	*Arthrospira platensis*	Energy	Markou et al. (2015)
	Chlamydomonas reinhardtii	Energy	Choi et al. (2019a)
	Chlorella sp.	Energy	Rizzo et al. (2013)
	Chlorella prothotecoides	Energy	Peng et al. (2001)
	Chlorella vulgaris	Energy	Markou et al. (2015)
	Neochloris oleoabundans	Energy	Choi et al. (2019a)
	Spirulina platensis	Energy	Rizzo et al. (2013)
	Tetraselmis suecica	Energy	Giostri et al. (2016)

Pyrolysis of catalytic biomass, which is another way to improve product quality, is also a process of interest. In this process, in which the temperature range is around 500 °C, catalysts are introduced to improve the quantity and quality mainly of gaseous products and make it more useful for obtaining energy through combustion (Goyal et al. 2008). Furthermore, microwave-assisted pyrolysis ($\sim$700 °C) has advantages that include internal heating consistent to particle size, simplicity of control, and much more uniformity. However, until the moment few studies of microalgae pyrolysis with microwave-assisted heat were reported (Chen et al. 2015; Fan et al. 2020).

Despite presenting several types of pyrolytic conversion, research studies have focused mainly on rapid pyrolysis processes, since the disadvantages of slow pyrolysis involve secondary cracks, condensation, and polymerization of the product, which reduces HHV and the high demand for energy (Kumar et al. 2017). In addition, the yield and HHV of bio-oil from the microalgae biomass may vary from 18 to 72 wt% and 24–41 MJ/kg, respectively (Kim et al. 2014; Miao et al. 2004; Wang et al. 2013).

3.1.3 Thermochemical Liquefaction

Among the thermochemical processes, thermochemical liquefaction, or also called hydrothermal, is usually applied to the liquefaction of raw material to produce liquid fuel through the decomposition of their structures, where the water acts as an organic solvent and hydrogen donor (Fan et al. 2020). In general, liquefaction is carried out for less than 60 min at moderate temperatures below 400 °C and pressures up to 20 MPa to keep the solvent in a liquid state (Galadima and Muraza 2018; Saber et al. 2016b).

Hydrothermal liquefaction (HTL) is a substitute for pyrolysis. It attracts much attention due to its advantage of converting wet biomass into energy, which eliminates the use of expensive pretreatments, such as biomass drying methods (Chen et al. 2015; Brennam and Owend 2010). In this case, microalgae are excellent raw materials for liquefaction due to their high mass fraction of water in about 80–90%. Furthermore, the small microalgae particles end up facilitating the thermal transfer and improve the process conditions. Indeed, these factors imply a more straightforward implementation and reduced costs (Mathimani et al. 2019).

Briefly, the mechanistic pathway of microalgae biomass under HTL conditions reveals that macromolecules are hydrolyzed and decomposed to produce some fragments of fatty acids, amino acids, and monosaccharides through a series of reactions such as dehydration, deoxygenation, and decarboxylation (Kumar et al. 2017). Subsequently, the small, highly reactive, and unstable compounds are polymerized and separated into different products depending mainly on operational conditions and the end-use of the application (Chen et al. 2015; Fan et al. 2020).

The main product obtained from this process is bio-oil or biocrude, a liquid synthetic fuel similar to crude oil, which can be converted into hydrocarbon fuels or co-refined for use in various other applications (Galadima and Muraza 2018; Saber et al. 2016b). Biocrude is defined as a complex liquid mixture of hydrocarbons, nitrogenous and oxygenated cyclic compounds, fatty acids, esters, among others. These fuels are similar to heavy oil, although they have a higher nitrogen and oxygen content, coming up to 7% and 19%, respectively (Barreiro et al. 2013; Guo et al. 2015).

Although still at a very early stage of development, HTL bio-oils are thought to tend to be more stable than pyrolysis bio-oils due to the lower levels of reactive oxygenated species (Jena and Das 2011; Palomino et al. 2020). However, little research has examined the effects of physicochemical properties during the storage of biocrude from microalgae in the HTL process. What is known so far is that similar fuels report an increase in their viscosity, density, and molecular weight, during transport and long-term storage (Palomino et al. 2020; Kosinkova et al. 2016). Given this scenario, the use of microalgae biomass is still a challenging issue that HTL will need to overcome to become a real alternative to biofuel production.

3.1.4 Gasification

Gasification is another type of thermochemical conversion process used to convert carbonaceous polymers from biomass at high temperatures (800–1000 °C) in the presence of gasifying agents, such as steam, air, and oxygen. Gasification is an improved technique as it provides versatile synthetic gas (syngas) and other valuable gases as co-products. On average, syngas is a gas mixture (CO, H_2, CO_2, N, and CH_4) with a typical calorific value of 4–6 MJ/m^3 that can be converted into liquid biofuel via the Fischer–Tropsch process or burned directly for power generation (Du et al. 2016; Ramos and Rouboa 2020).

The entire biomass gasification mechanism involves several reactions depending on the temperature variation (Raheem et al. 2016). These reactions are broadly characterized in four phases: (i) drying, to remove residual moisture, around 150–200 °C; (ii) pyrolysis or devolatilization, to generate the breakdown of biomass into less complex molecules, between 250 and 500 °C; (iii) combustion or oxidation, referring to the formation of volatiles and solid carbon formed in the second stage of the pyrolysis reaction (500–700 °C); and finally, (iv) gasification, in which small molecules of high energy value and small amounts of residual char and tar are formed as final products (Basu 2010).

Supercritical water gasification (SCWG) is an alternative to conventional gasification routes directed to wet microalgae biomass. Similar to hydrothermal liquefaction, supercritical gasification occurs in an aqueous medium at high temperatures and a pressure range between 24 and 36 MPa. Here, biomass in supercritical water quickly breaks down, resulting in elevated emission of fuel gases rich in H_2 and CH_4 (Mathimani et al. 2019).

Finally, comprise the path of gasification is somewhat challenging due to the complex structures of microalgae biomass. Unlike the conventional gasification process, SCWG is still a commercial technique under development. In addition, the high energy demand for the separation and purification of gases and also for the preheating of the water used in the process seems to be an apparent disadvantage (Raheem et al. 2016).

3.1.5 Direct Combustion

With the increased incentive to use renewable raw materials in the generation of safe fuels, different thermochemical conversion routes are presented. Among them, the direct combustion of biomass is still the dominant path of bioenergy, contributing about 11% of the world's energy (Tanger et al. 2013; Du et al. 2016).

Direct combustion occurs in simple equipment, such as an oven, boiler, or steam turbine at temperatures above 800 °C. The biomass is burned directly in the presence of oxygen, converting the stored chemical energy into heat, mechanical force, or electricity. In this process, the heat produced must be used immediately, since storage is not yet a viable option (Clark and Deswarte 2008).

Due to the high moisture content of microalgae biomass, the need for processes such as centrifugation, drying, and crushing, is still an obstacle to financial costs and energy expenditure. According to Kumar et al. (2017), additions of pretreatment methods increase the final cost of processing microalgae biomass by USD 1 kg^{-1}. In this case, direct combustion is mainly directed to strains with low moisture content (<50%) (Goyal et al. 2008). Furthermore, the co-firing of the biomass of microalgae in coal plants can also be an alternative way to reduce greenhouse gas and pollution emissions. Finally, additional engineering improvements and technological incentives in the conversion paths can potentially increase the adhesion of microalgae to the combustion process (Raheem et al. 2015).

3.2 Biochemical Conversion

In addition to the thermochemical conversion, it is possible to produce a variety of liquid and gaseous biofuels, as well as bioenergies from microalgae through technological biochemical conversion routes. Biochemical conversion uses biological and biochemical processes to convert mainly biomass into energy by-products, through methods such as transesterification, fermentation, hydrolysis, anaerobic digestion, or direct conversion. Multiple steps are involved during this microalgae-to-fuels procedure, including unit operations of the upstream and downstream processing phases (Deprá et al. 2018). The main biochemical conversion pathways and subclasses are illustrated in Fig. 2.

3.2.1 Biodiesel

The biochemically converted microalgae energy product most researched by the scientific community is biodiesel. It has been considered as a potential direct substitute or blending component for petroleum derived diesel. The higher heating value of biodiesel of about 40 MJ/kg is comparable with petrodiesel, whose value is approximately 43 MJ/kg. As characteristics, biodiesel has physical and chemical properties similar to those of fossil diesel, which favors its use in combustion engines, discarding any type of modification in the ignition system or fuel injector (Daroch et al. 2013). These attributes are due to the quantity and quality of the single-cell microalgae oil, determined by the lipid content and fatty acid profile.

From an environmental perspective, the benefits of biodiesel are already well known: renewability and sustainability, due to reduced emissions of CO_2, CO, SO_x, NO_x, and unburned hydrocarbons, biodegradability, and non-toxicity (Deprá et al. 2018).

In terms of general processing, after cultivation, harvesting, and drying, the oil is extracted, and then it is converted into biodiesel through transesterification. This procedure is the most commonly used to convert oil into biodiesel. It consists of a chemical reaction between triglycerides and alcohol to form fatty acid methyl esters (FAME) and glycerol. Usually, the reaction occurs in the presence of a catalyst, which can be acidic, basic, or enzymatic via heterogeneous/homogeneous process or in-situ transesterification. Currently, numerous studies discuss new techniques to transesterify microalgae lipids, as well as merits and demerits (Pragya et al. 2013).

Oleaginous microalgae produce considerable levels of lipids ranging from 20 to 50% in dry weight or, in a broader estimate, from 5000 to 100,000 L/ha/day (Behera et al. 2019). Species like *Botryococcus braunii* and *Chlorella vulgaris* have been identified as the most suitable for biodiesel production. However, in the case of *B. braunii*, this microalgae synthesizes more than 80% dry weight of hydrocarbon content, it has a low cell growth rate, causing a deficit in lipid productivity, ranging from 0.10 to 0.4 g/L/day (Tasić et al. 2016). In addition, some studies have shown the microalgae cultivation under CO_2 input for biodiesel production. For example, *S. obliquus* has been suitable for CO_2 capture and use from industrial flue gases and biodiesel generation (Lam et al. 2012). An important observation is that the lipid content and productivity depend on the growth conditions, and these parameters can be further improved with changes in genetic and metabolic engineering.

The biochemical conversion of biomass into biodiesel requires downstream processing steps that are energy-intensive (e.g., harvesting and drying), resulting in high manufacturing costs. Also, the upstream pretreatment steps need to be improved, that is, boosting productivity strains and cultivation systems. Although the environmental impacts are relatively minor compared to other feedstock, bulk production of microalgae biodiesel is not yet

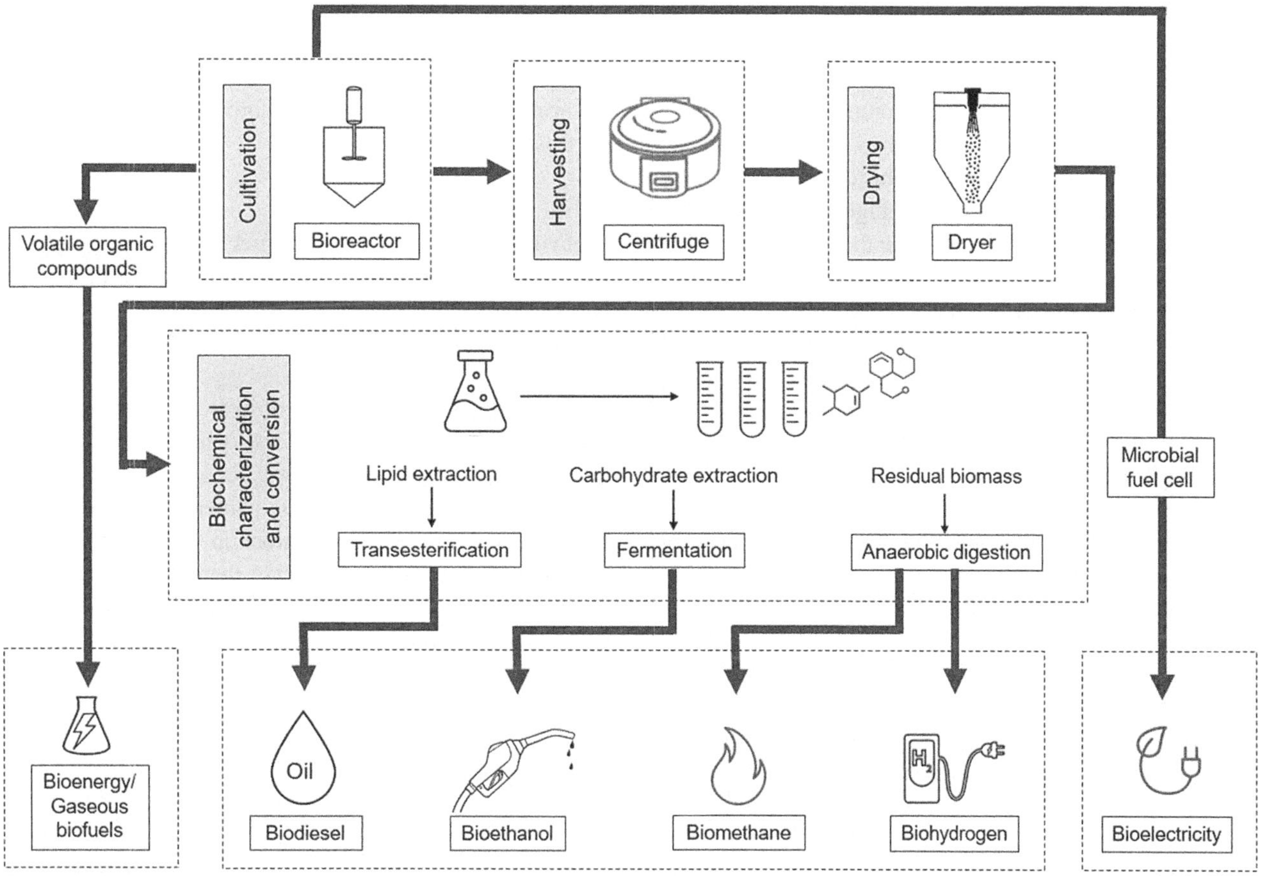

Fig. 2 Biochemical conversion routes for microalgae energy-based products

commercially competitive in technical and economic terms (Severo et al. 2019).

3.2.2 Bioethanol

The microalgae bioethanol production is performed through the biochemical conversion process of carbohydrates, such as starch, cellulose, and intracellular glycogen, via fermentation and hydrolysis. Bioethanol is the second most attractive biofuel for production from a third-generation feedstock. In addition to being considered a clean fuel, its global manufacture has shown rapid growth and the expectation today is that it will reach the mark of 160 billion liters produced.

Bioethanol processing occurs due to the presence of polysaccharides with low or no lignin and hemicellulose content in the microalgae cell walls, which facilitates the hydrolysis and fermentation of sugars. From there, the final product, ethanol, can be generated after distillation. Besides, to break down the carbohydrate molecule or starch polymers into monomers, and convert it into bioethanol, it is necessary to perform the rupture of cell wall structure via pretreatment before fermentation (Silva and Bertucco 2016). There are many methods for this purpose, including physical (e.g., microwave, ultrasound, autoclave, and high-pressure homogenization), chemical (acid, basic, and supercritical CO_2), enzymatic (alpha- and glucoamylase), and combined (dark fermentation and photofermentation). This step is crucial because it minimizes the crystallinity degree of the cellulose matrix, increases surface area, and favors the assimilation of the substrate by the enzymes to hydrolyze microalgal biomass (Phwan et al. 2018).

Some genus of microalgae such as *Chlorella*, *Scenedesmus*, *Dunaliella*, and *Tetraselmis* have considerable carbohydrate content, around 40% of dry weight (Behera et al. 2019). The species *C. vulgaris*, for example, can reach levels of approximately 55%. This percentage indicates that microalgae can be an excellent feedstock for ethanol fermentation. Evidently, this factor depends on the cultivation method and environmental parameters. Depending on the species and pretreatment technique, the bioethanol yield has been reported to be in the range of 0.08–0.12 g per g of dry biomass (Bhattacharya and Goswami 2020).

Additionally, microalgae have been applied for environmental mitigation and parallel bioethanol production. By way of exemplification, the study carried out by Ho et al. (2013), evaluated the development of a two-stage process for

CO_2 fixation, carbohydrate productivity, and bioethanol generation using *S. obliquus* CNW-N via hydrolysis and fermentation.

While this matrix is attractive, the production of stand-alone microalgae bioethanol is postulated as a costly process in energy, with relatively high operating costs, due to ethanol purification (distillation stage), and high capital investment. These limitations make it unfeasible for the scale manufacturing of this biofuel under current economic conditions. However, in Brazil, research suggests integrated production in the biorefinery context (Klein et al. 2018). This strategy has the potential for profit and to generate revenues with the appreciation of other products of marketing interest.

3.2.3 Biomethane

Microalgae biomethane—often called biogas—is a gaseous fuel produced from anaerobic digestion and dark fermentation. It has some benefits concerning other biofuels, such as the potential use as a source of heat and bioelectricity, the lower implementation cost of the process compared to emerging biofuel/bioenergy technologies (i.e., gasification and pyrolysis), and nutrient cycling to make biofertilizers. This technological route for biochemical conversion of biomass into biofuel is technically feasible and can be widely applied for the treatment of various types of waste (Zhu et al. 2014).

Concerning processing, the biomass or residual sludge is treated as follows: (i) hydrolysis, (ii) acidogenesis, (iii) acetogenesis, (iv) methanogenesis, and, finally, (v) biogenic gas production. This, in turn, consists mainly of biomethane and CO_2 and a minority of hydrogen, ammonia, and sulfides (Behera et al. 2019). Also, the wet biomass, after the harvesting stage, can be digested to generate biomethane after lipid extraction (Mehrabadi et al. 2015).

The theoretical yield established for methane ranges from 0.48 to 0.80 L per g of volatile solids. In contrast, the experimental yield from different species of microalgae is quite lower, reducing it to less than half in some cases. For example, species *Ulva* sp. and *Gracilaria* sp., and the genus *Macrocystis* present methane yields generally in the order of 0.2, 0.28–0.4, and 0.39–0.41 $m^3\ kg^{-1}$, respectively (Singh and Gu 2010). It is worth mentioning that, due to the cell wall structure of species, cell wall structure of species, and the performance of anaerobic digestion is very dependent on the type of strain, and the same applies to the biomethane yield (Uggetti et al. 2017).

However, some problems are inherent to microalgae biogas production, such as heating the digesters, which has been a concern since considerable energy input is required. In addition, the low carbon/nitrogen (C/N) ratio of biomass, generally 5:1–10:1. This proportion reduces methanization. Co-digestion has been suggested to increase the C/N ratio, adding agricultural wastes (Zhu et al. 2014). Also, integrated strategies to improve the microalgae biogas production process have been investigated. The biomethane generation from CO_2 capture and use has the potential for economic and environmental updating in an optimized system, as proposed by Bose et al. (2019).

3.2.4 Biohydrogen

Biohydrogen is an attractive biofuel because its combustion does not generate carbon-based by-products and has a high energy density (142 kJ/g) compared to other fuels, such as methane (56 kJ/g), natural gas (54 kJ/g), and gasoline (47 kJ/g).

The biochemical conversion of hydrogen from microalgae can be classified in two ways, one dependent and the other independent of light. In the first case, also called water biophotolysis, microalgae synthesize particular enzymes (hydrogenases), which accept H^+ electrons in the final phase of the photosynthetic process and release them and oxygen (Jiménez-Llanos et al. 2020). The general reaction of water photolysis can be described according to Eqs. 4 and 5. The electrons pass from photosystems I and II to enzymatic activity.

$$H_2O(g) \xrightarrow{\text{Hidrogenase}} 4H^+ + 4e^- + \frac{1}{2}O_2 \quad (4)$$

$$4H^+ + 4e^- \xrightarrow{\text{Hidrogenase}} 2H_2 \quad (5)$$

In the second case, dark fermentation, organic compounds are fermented with anaerobic bacteria. Usually, hydrogen generation is attributed to nitrogen biofixation, with a reduction in photosynthetically produced H^+ electrons (Show et al. 2019).

The competence to synthesize H_2 by microalgae is attributed to growth conditions, including light intensity, wavelength, and percentage of carbon in the biomass. In terms of species, *Chlamydomonas reinhardtii* has been further investigated for biohydrogen production, followed by *Chlorella pyrenoidosa*, *C. vulgaris*, and *S. platensis*. But this biofuel is currently undergoing research because of low yield in dark fermentation and high energy cost in light-dependent fermentation (Rashid et al. 2013). Current trends involve integrated processes to improve microalgae biohydrogen production and metabolic engineering tools.

3.2.5 Bioelectricity

A way to produce electricity by microalgae is through microbial fuel cells (MFCs). This technological route of direct biochemical conversion consists of integrating both wastewater and CO_2 from exhaust gases and generating bioelectricity from metabolic reactions. Microalgae are degraded at the anode, releasing electrons. They flow through an external system integrated into the chamber to the cathode. In the cathode chamber, electrons reduce O_2, which is combined with protons to form water. Thus, in this part of

Table 4 Production of volatile organic compounds from microalgae and their energy content

Volatile organic compounds	Energy potential (kJ/kg)
2-ethyl-1-hexanol	5420
2-propyl-1-heptanol	6720
2-methylbutanal	3240
Hexanal	3880
2,4-heptadienal	4050
2,4-decadienal, (E, Z)	6000
2-methoxy-2-methylpropane	3480
3,3-dimethylhexane	5420
2,4-dimethylheptane	6070
4,7-dimethyl-undecane	8660
2-propanone	1940
2,4-dimethyl-3-pentanone	4530
4-octen-3-one	5180
6-methyl-5-hepten-2-one	4940
Acetophenone	4220
β-ionone	7700
2-phenylpropene	4870
Total	86,320

Adapted from: Severo et al. (2018).

the system, microalgae can donate and receive electrons at the same time to generate bioenergy (Dias et al. 2020).

C. vulgaris, *C. pyrenoidosa*, and *C. reinhardtii* have been reported as potential species of biocathode in MFCs chambers, with a maximum power density of 13–27.5, 30.15, and 78 mWm^{-2}, respectively (Baicha et al. 2016). But, to have optimal performance, the operational parameters (pH, temperature, light, and substrate loading rate) must be improved to keep the electrons flow in balance and, consequently, the electric current. Yet, the technical challenges regarding the microalgae MFCs process are many. The devices currently researched and developed are not practical, in addition to the costs associated with the processing of cell culture in anodic and cathodic chambers (Bakonyi et al. 2020).

3.2.6 Volatile Organic Compounds

An emerging method of direct biochemical conversion to generate microalgae-based energy is through volatile organic compounds (VOCs) fraction. These biomolecules are secondary metabolites naturally released by microalgae cells in the environment. They belong to different chemical classes, such as alcohols, aldehydes, esters, hydrocarbons, ketones, terpenes, sulfur compounds, and carboxylic acids, which have considerable energy content (Deprá et al. 2018; Severo et al. 2020a).

The VOCs of microalgae have been identified as potential advanced sources of bioenergy and/or biofuels. Some species of microalgae produce fuel precursor compounds derived from petroleum, such as short-chain or higher alcohols, alkanes, alkenes, and aliphatic hydrocarbons with desirable combustibility properties (Severo et al. 2020b).

In terms of application, the microalgae VOCs were evaluated in a study developed by Severo et al. (2018). The bioprocess consists of the CO_2 capture and use and conversion into O_2, CO_2 unconverted and mainly VOCs, which were released into the photobioreactor exhaust gases. The total energy potential obtained was 86.320 kJ/kg (Table 4). In addition, the estimated energy generation rate was around 15,247.78 $kJ/m^3/d$.

By way of comparison, the energy potentials of gasoline, liquefied petroleum gas, diesel oil, and natural gas are close to 47,300, 46,100, 44,800, and 39,360 kJ/kg, respectively. Regardless of these potentialities, the two main limitations of this bioenergy technology are based on the low productivity of VOCs, due to the working capacity of the cultivation systems, and because there is currently no efficient recovery technique to fractionate and collect these biomolecules.

4 Conclusion and Future Perspectives

Although CO_2 capture technologies are relatively mature, the use of captured CO_2 remains a critical challenge that requires a lot of future research.

In a real-world application, to further promote the production of biofuels by microalgae-based processes, problems associated with the selectivity of the microorganism, production cost, harvesting stages, and biomass nutrient value must be addressed.

However, regardless of the chemical or biological conversion process of generating fuels from CO_2, in general, the next advances to promote the high performance of technologies, as well as the progress of industrial viability, must be deeply explored. Only then, the immense potential attributed to these technologies can be used.

Finally, in the near future, CO_2 recycling technologies could become resources required by different sectors of the global economy and affect the regulation and policy of the CO_2 based fuel market.

References

Abdullah B, Ghani NAA, Vo DVN (2017) Recent advances in dry reforming of methane over Ni-based catalysts. J Clean Prod 162:170–185. https://doi.org/10.1016/j.jclepro.2017.05.176

Adnan MA, Susanto H, Binous H, Muraza O, Hossain MM (2017) Feed compositions and gasification potential of several biomasses including a microalgae: a thermodynamic modeling approach. Int J Hydrogen Energy 42:17009–17019. https://doi.org/10.1016/j.ijhydene.2017.05.187

Alba GL, Cristian T, Chiara S, Jaapjan VS, Daniele F, Kersten SRA, Brilman DWF (2012) Hydrothermal treatment (HTT) of microalgae: evaluation of the process as conversion method in an algae biorefinery concept. Energy Fuels 26:642–657. https://doi.org/10.1021/ef201415s

Alghurabie IK, Hasan BO, Jackson B, Kosminski A, Ashman PJ (2013) Fluidized bed gasification of Kingston coal and marine microalgae in a spouted bed reactor. Chem Eng Res Des 91:1614–1624. https://doi.org/10.1016/j.cherd.2013.04.024

Al-Saydeh SA, Zaidi SJ (2018) Carbon dioxide conversion to methanol: opportunities and fundamental challenges. In: Carbon dioxide chemistry, capture and oil recovery, vol 41. https://doi.org/10.5772/intechopen.74779

Anand V, Gautam R, Vinu R (2017) Non-catalytic and catalytic fast pyrolysis of *Schizochytrium limacinum* microalga. Fuel 205:1–10. https://doi.org/10.1016/j.fuel.2017.05.049

Andrade DS, Telles TS, de Castro GHL (2019) The Brazilian microalgae production chain and alternatives for its consolidation. J Cleaner Prod 119526. https://doi.org/10.1016/j.jclepro.2019.119526

Andrade DS, Telles TS, Castro GHL (2020) The Brazilian microalgae production chain and alternatives for its consolidation. J Clean Prod 250:119526. https://doi.org/10.1016/j.jclepro.2019.119526

Arora S, Prasad RJRA (2016) An overview on dry reforming of methane: strategies to reduce carbonaceous deactivation of catalysts. RSC Adv 6(110):108668–108688. https://doi.org/10.1039/C6RA20450C

Aysu T, Ola O, Maroto-Valer MM, Sanna A (2017) Effects of titania based catalysts on in-situ pyrolysis of Pavlova microalgae. Fuel Process Technol 166:291–298. https://doi.org/10.1016/j.fuproc.2017.05.001

Bach QV, Chen WH (2017) Pyrolysis characteristics and kinetics of microalgae via dynamics and severity of microalgae residues in torrefaction. Biores Technol 169:258–264. https://doi.org/10.1016/j.biortech.2014.06.086

Baicha Z, Salar-García MJ, Ortiz-Martínez VM, Hernández-Fernández FJ, DelosRíos AP, Labjar N, Elmahi M (2016) A critical review on microalgae as an alternative source for bioenergy production: a promising low cost substrate for microbial fuel cells. Fuel Process Technol 154:104–116. https://doi.org/10.1016/j.fuproc.2016.08.017

Bakonyi P, Peter J, Koter S, Mateos R, Kumar G, Koók L, Nemestóthy N (2020) Possibilities for the biologically-assisted utilization of CO_2-rich gaseous waste streams generated during membrane technological separation of biohydrogen. J CO_2 Utilization 36:231–243. https://doi.org/10.1016/j.jcou.2019.11.008

Baldinelli A, Dou X, Buchholz D, Marinaro M, Passerini S, Barelli L (2018) Addressing the energy sustainability of biowaste-derived hard carbon materials for battery electrodes. Green Chem 20:1527–1537. https://doi.org/10.1039/C8GC00085A

Barreiro DL, Prins W, Ronsse F, Brilman W (2013) Hydrothermal liquefaction (HTL) of microalgae for biofuel production: state of the art review and future prospects. Biomass Bioenerg 53:113–127. https://doi.org/10.1016/j.biombioe.2012.12.029

Basu P (2010) Biomass gasification and pyrolysis: practical design and theory. Academic Press. https://doi.org/10.1016/C2009-0-020099-7

Becker EW (1994) Microalgae: biotechnology and microbiology, vol 10. Cambridge University Press. ISBN:0-521-35020-4

Behera B, Acharya A, Gargey IA, Aly N, Balasubramanian P (2019) Bioprocess engineering principles of microalgal cultivation for sustainable biofuel production. Bioresour Technol Rep 5:297–316. https://doi.org/10.1016/j.biteb.2018.08.001

Belotti G, Caprariis BD, Filippis PD, Scarsella M, Verdone N (2014) Effect of Chlorella vulgaris growing conditions on bio-oil production via fast pyrolysis. Biomass Bioenergy 61:187–195. https://doi.org/10.1016/j.biombioe.2013.12.011

Bhattacharya M, Goswami S (2020) Microalgae—a green multi-product biorefinery for future industrial prospects. Biocatal Agric Biotechnol 25:101580. https://doi.org/10.1016/j.bcab.2020.101580

Biller P, Ross AB, Skill SC, Lea-Langton A, Balasundaram B, Hall C et al (2012) Nutrient recycling of aqueous phase for microalgae cultivation from the hydrothermal liquefaction process. Algal Res 1:70–76. https://doi.org/10.1016/j.algal.2012.02.002

Biller P, Sharma BK, Kunwar B, Ross A (2015) Hydroprocessing of bio-crude from continuous hydrothermal liquefaction of microalgae. Fuel 159:197–205. https://doi.org/10.1016/j.fuel.2015.06.077

Borges FC, Xie Q, Min M, Muniz LAR, Farenzena M, Trierweiler JO et al (2014) Fast microwave-assisted pyrolysis of microalgae using microwave absorbent and HZSM-5 catalyst. Biores Technol 166:518–526. https://doi.org/10.1016/j.biortech.2014.05.100

Bose A, Lin R, Rajendran K, O'Shea R, Xia A, Murphy JD (2019) How to optimise photosynthetic biogas upgrading: a perspective on system design and microalgae selection. Biotechnol Adv 107444. https://doi.org/10.1016/j.biotechadv.2019.107444

Brennam L, Owend P (2010) Biofuels from microalgae—a review of technologies for production, processing, and extractions of biofuels and co-products. Renew Sustain Energy Rev 14:557–577. https://doi.org/10.1016/j.rser.2009.10.009

Brown TM, Duan P, Savage PE (2010) Hydrothermal liquefaction and gasification of *Nannochloropsis sp*. Energy Fuels 24(6):3639–3646. https://doi.org/10.1021/ef100203u

Caputo G, Dispenza M, Rubio P, Scargiali F, Marotta G, Brucato A (2016) Supercritical water gasification of microalgae and their constituents in a continuous reactor. J Supercrit Fluids 118:163–170. https://doi.org/10.1016/j.supflu.2016.08.007

Chaiwong K, Kiatsiriroat T, Vorayos N, Thararax C (2013) Study of bio-oil and bio-char production from algae by slow pyrolysis. Biomass Bioenerg 56:600–606. https://doi.org/10.1016/j.biombioe.2013.05.035

Chang YM, Tsai WT, Li MH (2015) Chemical characterization of char derived from slow pyrolysis of microalgal residue. J Anal Appl Pyrol 111:88–93. https://doi.org/10.1016/j.jaap.2014.12.004

Chen WH, Huang MY, Chang JS, Chen CY (2014a) Thermal decomposition dynamics and severity of microalgae residues in torrefaction. Biores Technol 169:258–264. https://doi.org/10.1016/j.biortech.2014.06.086

Chen WH, Wu ZY, Chang JS (2014b) Isothermal and non-isothermal torrefaction characteristics and kinetics of microalga Scenedesmus obliquus CNW-N. Biores Technol 155:245–251. https://doi.org/10.1016/j.biortech.2013.12.116

Chen HW, Lin BJ, Huang MY, Chang JS (2015) Thermochemical conversion of microalgal biomass into biofuels: a review. Biores Technol 184:314–327. https://doi.org/10.1016/j.biortech.2014.11.050

Chen YC, Chen WH, Lin BJ, Chang JS, Ong HC (2016) Impact of torrefaction on the composition, structure and reactivity of a microalga residue. Appl Energy 181:110–119. https://doi.org/10.1016/j.apenergy.2016.07.130

Chen WH, Chu YS, Liu JL, Chang JS (2018) Thermal degradation of carbohydrates, proteins and lipids in microalgae analyzed by evolutionary computation. Energy Convers Manage 160(209):219. https://doi.org/10.1016/j.enconman.2018.01.036

Cheng J, Zhu Y, Zhang Z, Yang W (2019) Modification and improvement of microalgae strains for strengthening CO_2 fixation from coal-fired flue gas in power plants. Bioresour Technol 121850. https://doi.org/10.1016/j.biortech.2019.121850

Choi YY, Patel AK, Hong ME, Chang WS, Sim SJ (2019a) Microalgae bioenergy with carbon capture and storage (BECCS): an emerging sustainable bioprocess for reduced CO_2 emission and biofuel production. Bioresource Technol Rep 7:100270. https://doi.org/10.1016/j.biteb.2019.100270

Choi HI, Hwang SW, Sim SJ (2019b) Comprehensive approach to improving life-cycle CO_2 reduction efficiency of microalgal biorefineries: a review. Bioresour Technol 121879. https://doi.org/10.1016/j.biortech.2019.121879

Choo MY, Oi LE, Ling TC, Ng EP, Lee HV, Juan JC (2020) Conversion of microalgae biomass to biofuels. In: Microalgae cultivation for biofuels production. Academic Press, pp 149–161. https://doi.org/10.1016/B978-0-12-817536-1.00010-2

Choudhary P, Assemany PP, Naaz F, Bhattacharya A, de Siqueira Castro J, do Couto EDA, Malik A (2020) A review of biochemical and thermochemical energy conversion routes of wastewater grown algal biomass. Sci Total Environ 137961. https://doi.org/10.1016/j.scitotenv.2020.137961

Clark J, Deswarte F (2008) Introduction to chemicals from biomass. In: Stevens CV (ed) Wiley series in renewable resources. Wiley

Dai L, Wang Y, Liu Y, Ruan R, He C, Yu Z et al (2019) Integrated process of lignocellulosic biomass torrefaction and pyrolysis for upgrading bio-oil production: a state-of-the-art review. Renew Sustain Energy Review 107:20–36. https://doi.org/10.1016/j.rser.2019.02.015

Daroch M, Geng S, Wang G (2013) Recent advances in liquid biofuel production from algal feedstocks. Appl Energy 102:1371–1381. https://doi.org/10.1016/j.apenergy.2012.07.031

de Silva CEF, Bertucco A (2016) Bioethanol from microalgae and cyanobacteria: a review and technological outlook. Process Biochem 51:1833–1842. https://doi.org/10.1016/j.procbio.2016.02.016

Deprá MC, dos Santos AM, Severo IA, Santos AB, Zepka LQ, Jacob-Lopes E (2018) Microalgal biorefineries for bioenergy production: can we move from concept to industrial reality? Bioenergy Res. https://doi.org/10.1007/s12155-018-9934-z

Dias RR, Sartori RB, Severo IA, Deprá MC, Zepka LQ, Jacob-Lopes E (2020) Microalgae-based systems applied to bioelectrocatalysis. In: Inamuddin, Boddula R, Lichtfouse E (eds) Methods for electrocatalysis: advanced materials and allied applications. Springer Nature Switzerland, pp 241–262

Du X, Zhao C, Liu D, Lin CSK, Wilson K, Luque K, Clark J (2016) Introduction: an overview of biofuels and production technologies. In: Luque R, Lin CSK, Wilson K, Clark J (eds) Handbook of biofuels production 2nd edn. Processes and Technologies, pp 3–12. https://doi.org/10.1016/B978-0-08-100455-5.00001-1

Duan PG, Li SC, Jiao JL, Wang F, Xu YP (2018) Supercritical water gasification of microalgae over a two-component catalyst mixture. Sci Total Environ 630:243–253. https://doi.org/10.1016/j.scitotenv.2018.02.226

Egelske BT, Keels JM, Monnier JR, Regalbuto JR (2020) An analysis of electroless deposition derived Ni-Pt catalysts for the dry reforming of methane. J Catal 381:374–384. https://doi.org/10.1016/j.jcat.2019.10.035

EIA (Energy Information Administration) (2012) Most states have renewable portfolio standards. Available online at: http://www.eia.gov/todayinenergy/detail.cfm?id=4850. Accessed 20 April 2020

Elliott DC, Biller P, Ross AB, Schmidt AJ, Jones SB (2015) Hydrothermal liquefaction of biomass: developments from batch to continuous process. Biores Technol 178:147–156. https://doi.org/10.1016/j.biortech.2014.09.132

Faeth JL, Savage PE (2016) Effects of processing conditions on biocrude yields from fast hydrothermal liquefaction of microalgae. Biores Technol 206:290–293. https://doi.org/10.1016/j.biortech.2016.01.115

Fan L, Zhang H, Li J, Wang Y, Leng L, Li J, Yao A, Lu Q, Yuan W, Zhou W (2020) Algal biorefinery to value-added products by using combined processes based on thermochemical conversion: a review. Algal Res 47:101819. https://doi.org/10.1016/j.algal.2020.101819

Freund HJ, Roberts MW (1996) Surface chemistry of carbon dioxide. Surf Sci Rep 25(8):225–273

Galadima A, Muraza O (2018) Hydrothermal liquefaction of algae and bio-oil upgrading into liquid fuels: role of heterogeneous catalysts. Renew Sustain Energy Rev 81:1037–1048. https://doi.org/10.1016/j.rser.2017.07.034

Gan YY, Ong HC, Ling TC, Chen WH, Chong CT (2019) Torrefaction of de-oiled Jatropha seed kernel biomass for solid fuel production. Energy 170:367–374. https://doi.org/10.1016/j.energy.2018.12.026

Gangadharan P, Kanchi KC, Lou HH (2012) Evaluation of the economic and environmental impact of combining dry reforming with steam reforming of methane. Chem Eng Res Des 90(11):1956–1968. https://doi.org/10.1016/j.cherd.2012.04.008

Gao Y, Jiang J, Meng Y, Aihemaiti A, Ju T, Chen X, Yan F (2020) A novel nickel catalyst supported on activated coal fly ash for syngas production via biogas dry reforming. Renewable Energy 149:786–793. https://doi.org/10.1016/j.renene.2019.12.096

Giostri A, Binotti M, Macchi E (2016) Microalgae cofiring in coal power plants: innovative system layout and energy analysis. Renewable Energy 95:449–464. https://doi.org/10.1016/j.renene.2016.04.033

Goyal HB, Seal D, Saxena RC (2008) Bio-fuels from thermochemical conversion of renewable resources: a review. Renew Sustain Energy Rev 12:504–517. https://doi.org/10.1016/j.rser.2006.07.014

Grierson S, Strezov V, Ellem G, Mcgregor R, Herbertson J (2009) Thermal characterisation of microalgae under slow pyrolysis conditions. J Anal Appl Pyrol 85:118–123. https://doi.org/10.1016/j.jaap.2008.10.003

Guo Y, Yeh T, Song W, Xu D, Wang S (2015) A review of bio-oil production from hydrothermal liquefaction of algae. Renew Sustain Energy Rev 48:776–790. https://doi.org/10.1016/j.rser.2015.04.049

Ho S-H, Li P-J, Liu C-C, Chang J-S (2013) Bioprocess development on microalgae-based CO_2 fixation and bioethanol production using *Scenedesmus obliquus* CNW-N. Biores Technol 145:142–149. https://doi.org/10.1016/j.biortech.2013.02.119

Hong Y, Chen W, Luo X, Pang C, Lester E, Wu T (2017) Microwave-enhanced pyrolysis of macroalgae and microalgae for syngas production. Biores Technol 237:47–56. https://doi.org/10.1016/j.biortech.2017.02.006

Huff CA, Sanford MS (2011) Cascade catalysis for the homogeneous hydrogenation of CO_2 to methanol. J Am Chem Soc 133 (45):18122–18125. https://doi.org/10.1021/ja208760j

IEA International Energy Agency (2020) Global energy and CO_2 status report 2019. Access at: https://www.iea.org/reports/global-energy-co2-status-report-2019/emissions

Ike GN, Usman O, Sarkodie SA (2020) Fiscal policy and CO_2 emissions from heterogeneous fuel sources in Thailand: evidence from multiple structural breaks cointegration test. Sci Total Environ 702:134711. https://doi.org/10.1016/j.scitotenv.2019.134711

Jacob-Lopes E, Scoparo CHG, Lacerda LMCF, Franco TT (2009) Effect of light cycles (night/day) on CO_2 fixation and biomass production by microalgae in photobioreactors. Chem Eng Process 48(1):306–310. https://doi.org/10.1016/j.cep.2008.04.007

Jazrawi C, Biller P, Ross AB, Montoya A, Maschmeyer T, Haynes BS (2013) Pilot plant testing of continuous hydrothermal liquefaction of microalgae. Algal Res 2:268–277. https://doi.org/10.1016/j.algal.2013.04.006

Jena U, Das KC (2011) Comparative evaluation of thermochemical liquefaction and pyrolysis for bio-oil production from microalgae. Energy Fuels 25:5472–5482. https://doi.org/10.1021/ef201373m

Jena U, Das KC, Kastner JR (2012) Comparison of the effects of Na_2CO_3, $Ca_3(PO_4)_2$, and NiO catalysts on the thermochemical liquefaction of microalga Spirulina platensis. Appl Energy 98:368–375. https://doi.org/10.1016/j.apenergy.2012.03.056

Jiménez-Llanos J, Ramírez-Carmona M, Rendón-Castrillón L, Ocampo-López C (2020) Sustainable biohydrogen production by *Chlorella* sp. microalgae: a review. Int J Hydrogen Energy 45:8310–8328. https://doi.org/10.1016/j.ijhydene.2020.01.059

Kim SW, Koo BS, Le DH (2014) A comparative study of bio-oils from pyrolysis of microalgae and oil seed waste in a fluidized bed. Biores Technol 162:96–102. https://doi.org/10.1016/j.biortech.2014.03.136

Klein BC, Bonomi A, Filho RM (2018) Integration of microalgae production with industrial biofuel facilities: a critical review. Renew Sustain Energy Rev 82:1376–1392. https://doi.org/10.1016/j.rser.2017.04.063

Koo KY, Lee SH, Jung UH, Roh HS, Yoon WL (2014) Syngas production via combined steam and carbon dioxide reforming of methane over Ni–Ce/$MgAl_2O_4$ catalysts with enhanced coke resistance. Fuel Process Technol 119:151–157. https://doi.org/10.1016/j.fuproc.2013.11.005

Kosinkova J, Ramirez JA, Ristovski ZD, Brown RJ, Rainey TJ (2016) Physical and chemical stability of bagasse bio-crude from liquefaction stored in real conditions. Energy Fuels 30(12):10499–10504. https://doi.org/10.1021/acs.energyfuels.6b02115

Kothandaraman J, Heldebrant DJ (2020) Towards environmentally benign capture and conversion: heterogeneous metal catalyzed CO_2 hydrogenation in CO_2 capture solvents. Green Chem

Kumar G, Shobana S, Chen WH, Bach QV, Kim SH, Atabani AE, Chang JS (2017) A review of thermochemical conversion of microalgal biomass for biofuels: chemistry and processes. Green Chem 19:44–67. https://doi.org/10.1039/C6GC01937D

Kumaravel V, Bartlett J, Pillai SC (2020) Photoelectrochemical conversion of carbon dioxide (CO_2) into fuels and value-added products. ACS Energy Lett. https://doi.org/10.1021/acsenergylett.9b02585

Lam MK, Lee KT, Mohamed AR (2012) Current status and challenges on microalgae-based carbon capture. Int J Greenhouse Gas Control 10:456–469. https://doi.org/10.1016/j.ijggc.2012.07.010

Li S, Guo L, Ishihara T (2020) Hydrogenation of CO_2 to methanol over Cu/AlCeO catalyst. Catal Today 339:352–361. https://doi.org/10.1016/j.cattod.2019.01.015

Liu G, Liao Y, Wu Y, Ma X, Chen L (2017) Characteristics of microalgae gasification through chemical looping in the presence of steam. Int J Hydrogen Energy 42:22730–22742. https://doi.org/10.1016/j.ijhydene.2017.07.173

López CG, Fernández FA, Sevilla JF, Fernández JS, García MC, Grima EM (2009). Utilization of the cyanobacteria *Anabaena* sp. ATCC 33047 in CO_2 removal processes. Bioresour Technol 100 (23):5904–5910. https://doi.org/10.1016/j.biortech.2009.04.070

Maliutina K, Tahmasebi A, Yu J (2018) Pressurized entrained-flow pyrolysis of microalgae: enhanced production of hydrogen and nitrogen-containing compounds. Biores Technol 256:160–169. https://doi.org/10.1016/j.biortech.2018.02.016

Mantulet G, Bidaud A, Mima S (2020) The role of biomass gasification and methanisation in the decarbonisation strategies. Energy 193:116737. https://doi.org/10.1016/j.energy.2019.116737

Markou G, Iconomou D, Sotiroudis T, Israilides C, Muylaert K (2015) Exploration of using stripped ammonia and ash from poultry litter for the cultivation of the cyanobacterium *Arthrospira platensis* and the green microalga *Chlorella vulgaris*. Biores Technol 196:459–468. https://doi.org/10.1016/j.biortech.2015.08.007

Mathimani T, Baldinelli A, Rajendran K, Prabakar D, Matheswaran M, Leeuwen RPV, Pugazhendhi A (2019) Review on cultivation and thermochemical conversion of microalgae to fuels and chemicals: process evaluation and knowledge gaps. J Cleaner Prod 208:1053–1064. https://doi.org/10.1016/j.jclepro.2018.10.096

Mehrabadi A, Craggs R, Farid MM (2015) Wastewater treatment high rate algal ponds (WWT HRAP) for low-cost biofuel production. Bioresour Technol 202–214.https://doi.org/10.1016/j.biortech.2014.11.004

Miao X, Wu Q, Yang C (2004) Fast pyrolysis of microalgae to produce renewable fuels. J Anal Appl Pyrol 71:855–863. https://doi.org/10.1016/j.jaap.2003.11.004

Mustafa A, Lougou BG, Shuai Y, Wang Z, Tan H (2020) Current technology development for CO_2 utilization into solar fuels and chemicals: a review. J Energy Chem. https://doi.org/10.1016/j.jechem.2020.01.023

O'Ryan R, Nasirov S, Álvarez-Espinosa A (2020) Renewable energy expansion in the Chilean power market: a dynamic general equilibrium modeling approach to determine CO_2 emission baselines. J Clean Prod 247:119645. https://doi.org/10.1016/j.jclepro.2019.119645

Ong HC, Chen WH, Farooq A, Gan YY, Lee KT, Ashokkumar V (2019) Catalytic thermochemical conversion of biomass for biofuel production: a comprehensive review. Renew Sustain Energy Rev 113:109266. https://doi.org/10.1016/j.rser.2019.109266

Palomino A, Godoy-Silva RD, Raikova S, Chuck C (2020) The storage stability of biocrude obtained by the hydrothermal liquefaction of microalgae. Renewable Energy 145:1720–1729. https://doi.org/10.1016/j.renene.2019.07.084

Peng W, Qingyu W, Pingguan T, Nanming Z (2001) Pyrolytic characteristics of microalgae as renewable energy source determined by thermogravimetric analysis. Biores Technol 80:1–7. https://doi.org/10.1016/s0960-8524(01)00072-4

Phwan CK, Ong HC, Chen W-H, Ling TC, Ng EP, Show PL (2018) Overview: comparison of pretreatment technologies and fermentation processes of bioethanol from microalgae. Energy Convers Manage 173:81–94. https://doi.org/10.1016/j.enconman.2018.07.054

Pragya N, Pandey KK, Sahoo PK (2013) A review on harvesting, oil extraction and biofuels production technologies from microalgae. Renew Sustain Energy Rev 24:159–171. https://doi.org/10.1016/j.rser.2013.03.034

Raheem A, Sivasangar S, Azlina WAK, Yap YT, Danquah MK, Harun R (2015) Thermogravimetric study of *Chlorella vulgaris* for syngas production. Algal Res 12:52–59. https://doi.org/10.1016/j.algal.2015.08.003

Raheem A, Memon LA, Abbasi SA, Yap YHT, Danquah MK, Harun R (2016) Potential applications of nanotechnology in thermochemical conversion of microalgal biomass. In: Rai M, Silva SS (eds) Nanotechnology for bioenergy and biofuel production. Green Chemistry and Sustainable Technology, pp 91–116. https://doi.org/10.1007/978-3-319-45459-7_5

Ramos A, Rouboa A (2020) Syngas production strategies from biomass gasification: numerical studies for operational conditions and quality indexes. Renewable Energy 155:1211–1221. https://doi.org/10.1016/j.renene.2020.03.158

Rashid N, Rehman MSU, Memon S, Rahman ZU, Lee K, Han JI (2013) Current status, barriers and developments in biohydrogen production by microalgae. Renew Sustain Energy Rev 22:571–579. https://doi.org/10.1016/j.rser.2013.01.051

Rizzo AM, Prussi M, Bettucci L, Libelli IM, Chiaramonti D (2013) Characterization of microalga Chlorella as a fuel and its thermogravimetric behavior. Appl Energy 102:24–31. https://doi.org/10.1016/j.apenergy.2012.08.039

Rodriguez JA, Liu P, Stacchiola DJ, Senanayake SD, White MG, Chen JG (2015) Hydrogenation of CO_2 to methanol: importance of metal–oxide and metal–carbide interfaces in the activation of CO_2. ACS Catal 5(11):6696–6706. https://doi.org/10.1021/acscatal.5b01755

Saber M, Golzary A, Hosseinpour M, Takahashi F, Yoshikawa K (2016a) Catalytic hydrothermal liquefaction of microalgae using nanocatalyst. Appl Energy 183:566–576. https://doi.org/10.1016/j.apenergy.2016.09.017

Saber M, Nakhshiniev B, Yoshikawa K (2016b) A review of production and upgrading of algal bio-oil. Renew Sustain Energy Rev 58:918–930. https://doi.org/10.1016/j.rser.2015.12.342

Severo IA, Deprá MC, Barin JS, Wagner R, de Menezes CR, Zepka LQ, JacobLopes E (2018) Bio-combustion of petroleum coke: the process integration with photobioreactors. Chem Eng Sci 177:422–430. https://doi.org/10.1016/j.ces.2017.12.001

Severo IA, Siqueira SF, Deprá MC, Maroneze MM, Zepka LQ, Jacob-Lopes E (2019) Biodiesel facilities: what can we address to make biorefineries commercially competitive? Renew Sustain Energy Rev 112:686–705. https://doi.org/10.1016/j.rser.2019.06.020

Severo IA, Deprá MC, Dias RR, Barin JS, de Menezes CR, Wagner R, Jacob-Lopes E (2020a) Bio-combustion of petroleum coke: the process integration with photobioreactors. Part II—sustainability metrics and bioeconomy. Chem Eng Sci 213:115412. https://doi.org/10.1016/j.ces.2019.115412

Severo IA, Pinheiro PN, Vieira KR, Zepka LQ, Jacob-Lopes E (2020b) Biological conversion of carbon dioxide into volatile organic compounds. In: Conversion of carbon dioxide into hydrocarbons, vol 2 technology. Springer, Cham, pp 45–73. https://doi.org/10.1007/978-3-030-28638-5_2

Show KY, Yan Y, Zong C, Guo N, Chang JS, Lee DJ (2019) State of the art and challenges of biohydrogen from microalgae. Bioresour Technol 121747. https://doi.org/10.1016/j.biortech.2019.121747

Singh J, Gu S (2010) Commercialization potential of microalgae for biofuels production. Renew Sustain Energy Rev 14:2596–2610. https://doi.org/10.1016/j.rser.2010.06.014

Singh A, Nanda S, Berruti F (2020) A review of thermochemical and biochemical conversion of miscanthus to biofuels. In: Biorefinery of alternative resources: targeting green fuels and platform chemicals. Springer, Singapore, pp 195–220. https://doi.org/10.1007/978-981-15-1804-1_9

Song Y, Ozdemir E, Ramesh S, Adishev A, Subramanian S, Harale A, Albuali M, Fadhel BA, Jamal A, Moon D, Choi SH (2020) Dry reforming of methane by stable Ni–Mo nanocatalysts on single-crystalline MgO. Science 367(6479):777–781. https://doi.org/10.1126/science.aav2412

Soreanu G, Tomaszewicz M, Fernandez-Lopez M, Valverde JL, Zuwała J, Sanchez-Silva L (2017) CO_2 gasification process performance for energetic valorization of microalgae. Energy 119:37–43. https://doi.org/10.1016/j.energy.2016.12.046

Suganya T, Varman M, Masjuki HH, Renganathan S (2016) Macroalgae and microalgae as a potential source for commercial applications along with biofuels production: a biorefinery approach. Renew Sustain Energy Rev 55:909–941. https://doi.org/10.1016/j.rser.2015.11.026

Sun Y, Li Y, Cai B-F, Li Q (2020) Comparing the explicit and implicit attitudes of energy stakeholders and the public towards carbon capture and storage. J Cleaner Prod 120051. https://doi.org/10.1016/j.jclepro.2020.120051

Sydney EB, Sturm W, de Carvalho JC, Thomaz-Soccol V, Larroche C, Pandey A, Soccol CR (2010) Potential carbon dioxide fixation by industrially important microalgae. Biores Technol 101(15):5892–5896. https://doi.org/10.1016/j.biortech.2010.02.088

Sydney EB, Sydney ACN, de Carvalho JC, Soccol CR (2019) Potential carbon fixation of industrially important microalgae. In: Biofuels from Algae. Elsevier, pp 67–88. https://doi.org/10.1016/B978-0-444-64192-2.00004-4

Tang DYY, Khoo KS, Chew KW, Tao Y, Ho SH, Show PL (2020) Potential utilization of bioproducts from microalgae for the quality enhancement of natural products. Bioresour Technol 122997. https://doi.org/10.1016/j.biortech.2020.122997

Tanger P, Field JL, Jahn CE, DeFoort MW, Leach JL (2013) Biomass for thermochemical conversion: targets and challenges. Front Plant Sci 218(4):1–20. https://doi.org/10.3389/fpls.2013.00218

Tasić MB, Pinto LFR, Klein BK, Veljkovićv VB, Filho RM (2016) Botryococcusbraunii for biodiesel production. Renew Sustain Energy Rev 64:260–270. https://doi.org/10.1016/j.rser.2016.06.009

Torri C, Alba LG, Samorì C, Fabbri D, Brilman DWF (2012) Hydrothermal treatment (HTT) of microalgae: detailed molecular characterization of HTT oil in view of HTT mechanism elucidation. Energy Fuels 26:658–671. https://doi.org/10.1021/ef201417e

Uggetti E, Passos F, Solé M, Garfí M, Ferrer I (2017) Recent achievements in the production of biogas from microalgae. Waste Biomass Valor 8:129–139. https://doi.org/10.1007/s12649-016-9604-3

Vadlamani A, Pendyala B, Viamajala S, Varanasi S (2019) High productivity cultivation of microalgae without concentrated CO_2 input. ACS Sustain Chem Eng 7(2):1933–1943. https://doi.org/10.1021/acssuschemeng.8b04094

Wang K, Brown RC, Homsy S, Martinez L, Sidhu SS (2013) Fast pyrolysis of microalgae remnants in a fluidized bed reactor for bio-oil and biochar production. Biores Technol 127:494–499. https://doi.org/10.1016/j.biortech.2012.08.016

Wu KT, Tsai CJ, Chen CS, Chen HW (2012) The characteristics of torrefied microalgae. Appl Energy 100:52–57. https://doi.org/10.1016/j.apenergy.2012.03.002

Yang KC, Wu KT, Hsieh MH, Hsu HT, Chen CS, Chen HW (2013) Cogasificationof woody biomass and microalgae in a fluidized bed. J Taiwan Inst Chem Eng 44:1027–1033. https://doi.org/10.1016/j.jtice.2013.06.026

Zhang K, Miyachi S, Kurano N (2001) Evaluation of a vertical flat-plate photobioreactor for outdoor biomass production and carbon dioxide bio-fixation: effects of reactor dimensions, irradiation and cell concentration on the biomass productivity and irradiation utilization efficiency. Appl Microbiol Biotechnol 55(4):428–433. https://doi.org/10.1007/s002530000550

Zhang JX, Chen WT, Zhang P, Luo ZY, Zhang YH (2013) Hydrothermal liquefaction of Chlorella pyrenoidosa in sub- and supercritical ethanol with heterogeneous catalysts. Biores Technol 133:389–397. https://doi.org/10.1016/j.biortech.2013.01.076

Zhu LD, Hiltunen E, Antila E, Zhong JJ, Yuan ZH, Wang ZM (2014) Microalgal biofuels: flexible bioenergies for sustainable development. Renew Sustain Energy Rev 30(2014):1035–1046. https://doi.org/10.1016/j.rser.2013.11.003

Zhu C, Zhai X, Xi Y, Wang J, Kong F, Zhao Y, Chi Z (2020) Efficient CO_2 capture from the air for high microalgal biomass production by a bicarbonate pool. J CO_2 Utilization 37:320–327. https://doi.org/10.1016/j.jcou.2019.12.023

Thermodynamics of CO_2 Conversion

Joshua Gorimbo and Ralph Muvhiiwa

Abstract

The work has looked at possible processes to utilise atmospheric carbon dioxide (CO_2) as well as the thermodynamic complexities involved in trying to convert CO_2 to other useful fuels. Thermodynamic data from literature, as well as simple calculations from the Attainable Region, have shown that most CO_2 conversion processes with water (H_2O) require heat and work addition as the enthalpy and Gibbs free energy are positive. The Sabatier reaction that produces methane (CH_4) and H_2O and reverse water–gas shift (rWGS) which make syngas (carbon monoxide (CO) and hydrogen (H_2)) is possible reactions in utilising CO_2, but all occur at the expense of feeding in H_2 with CO_2 in the presence of a catalyst. The outcome of this research suggests that it is important to try and stop CO_2 emissions at source. This is because thermodynamics shows that the reaction processes involving CO_2 are not sustainable when trying to make fuel as a product.

Keywords

Carbon dioxide • Sabatier reaction • Reverse water gas shift reaction • Reverse combustion • Bosch reaction

Nomenclature

GHG	Greenhouse gas
AR	Attainable Region
NASA	National Aeronautics and Space Administration
rWGS	Reverse water–gas shift
FTS	Fischer–Tropsch synthesis
G	Gibbs free energy
H	Enthalpy

J. Gorimbo (✉) · R. Muvhiiwa
Institute for the Development of Energy for African Sustainability (IDEAS) Research Unit, University of South Africa (UNISA), Florida Campus, Private Bag X6, Johannesburg, 1710, South Africa

J. Gorimbo · R. Muvhiiwa
Keqiao Green Energy Materials Joint Laboratory, Zhijiang College of Zhejiang University of Technology, Shaoxing, 312030, China

1 Introduction

Efforts towards designing systems that decrease carbon dioxide (CO_2) emissions must increase to meet the Paris agreement objectives. The Paris agreement aimed to maintain the global average temperature increase under 2 °C above the pre-industrial levels preventing them from rising beyond dangerous thresholds (Agreement 2015). Any increase in temperature beyond that point would be dangerous and cause irreversible climate change. However, many of the country representatives argued that a more stringent cap of 1.5 °C, though ambitious, would be even more efficacious. Another key issue discussed exhaustively at the conference was that countries should reduce the CO_2 gas emitted by human activities to a level that can be absorbed naturally by the environment, Paris Climate Change Conference (COP21), 2015. CO_2 gas is colourless and has no scent. In addition to the respiration of all living organisms, human beings continue to introduce CO_2 via combustion of fossil fuels and the fermentation of sugars. As energy demand increase due to global population growth, more CO_2 is emitted into the atmosphere, and ultimately contributing to the deteriorating environment causing climate change. It is therefore imperative to find ways to decrease the amount of CO_2 emitted to the atmosphere. Naturally, plants take in CO_2 during the photosynthesis process, and forests have contributed immensely in absorbing about 25% of the emitted CO_2. The other 25% is absorbed naturally by oceans and the balance continues to collect in the atmosphere.

Energy utilisation in the Anthropocene era is predominantly based on the combustion of carbonaceous fuels, inherently petroleum, fossil fuels—coal and natural gas

Inamuddin et al. (eds.), *Carbon Dioxide Utilization to Sustainable Energy and Fuels*,
Advances in Science, Technology & Innovation, https://doi.org/10.1007/978-3-030-72877-9_4

giving off CO_2. However, the produced CO_2 is a greenhouse gas (GHG), which its increase in the atmosphere has caused serious environmental and ecological damage (Vijayavenkataraman et al. 2012). Op. cit. informed evidence of the effects of climate change vis-a-vis rising sea level, increasing global temperature, melting ice sheets and acidification of ocean waters (Vijayavenkataraman et al. 2012). GHGs such as CO_2, methane (CH_4) and NO_x trap heat in the atmosphere. Global temperatures have been observed to proportionally increase with increasing GHGs concentrations 'hockey stick graph'. As a result, some experts and policy-makers are calling for an explicit cap on CO_2 concentrations, in addition to the global temperature targets of 2 °C outlined in the Paris climate accord in November 2015.

Globally, over 36 billion tonnes of CO_2 are discharged into the atmosphere per year and this is projected to continuously rise (Hannah and Max 2017). To curb this catastrophic CO_2 build-up, researchers are now looking at ways of reducing the CO_2 in the atmosphere by (i) capturing and utilising the atmospheric CO_2, (ii) research into the use of bio-neutral fuels. Trapping the CO_2 in the atmosphere and reusing it as a feed material for useful chemicals could contribute to a more sustainable way of achieving the environmental target. Recent advances in CO_2 capture and utilisation provide effective ways for the chemical conversion of CO_2 into synthetic-organic fuels (Stewart and Hessami 2005; Dowson and Styring 2017; Choi et al. 2017). Such use of CO_2 as a feed in organic synthesis could contribute positively to the fight against climate change.

This chapter is a general overview on thermodynamics of CO_2 conversion, the limitations and potential CO_2 utilisation processes. Understanding the thermodynamics of CO_2 conversion and utilisation is an integral part of trying to reduce its emission, and thus, the fight to combat climate change. Several scientific publications are available on CO_2 capture and utilisation (Mukherjee et al. 2019). Also, different pathways have been explored for CO_2 conversion to organic molecules (Dowson and Styring 2017; Liu et al. 2015).

2 Carbon Dioxide Capture

Capturing CO_2 entails physical and chemical separation of the compound from a gas mixture and either store or reuse it. However, this process is more attractive and sustainable if the concentrations are high at separation and capturing points. Though unconventional, CO_2 capture directly from ambient air has been done by spraying sodium hydroxide (Stolaroff et al. 2008). Preferably, the process has been applied amongst the anthropogenic sources of CO_2 such coal powerplants, petroleum refineries, natural gas plants and other large point sources, such as flue gas. Gas mixtures rich in CO_2 include emissions from, fossil fuel-based power plants, industrial manufacturing processes, air crafts trains and ships and other biomass combustion processes.

Currently, three capturing technologies are known for trapping CO_2 from the flue gas and these are classified as oxy-fuel combustion, pre-and post-combustion (Mukherjee et al. 2019) with post-combustion frequently used. The CO_2 is released into the atmosphere as a mixture with other constituents mainly SO_x, NO_x, N_2, H_2O, CO, O_2 and other particulate matter. Therefore, different methods have been applied to different systems depending on the applicability and effectiveness in capturing CO_2. Available methods include carbon-based renewable adsorbent (Ahmed et al. 2020), supported monoethanolamide (Sun et al. 2011), amine-based CO_2 capture (Dutcher et al. 2015), membrane separation (Lee et al. 2017), molecular sieves (Donald Carruthers et al. 2012) and cryogenic separation (Meisen and Shuai 1997).

3 Carbon Dioxide Utilisations

The art of capturing CO_2 from either the atmosphere or flue gas will only aid in minimising levels of CO_2 in the atmosphere. However, capturing CO_2 and converting it to fuels and chemicals have attracted a considerable attention to researchers. In abating global warning, the captured CO_2 could find use as a building block in organic synthesis such as demonstrated by Liu et al. (2015). Table 1 shows studied reactions that can be applied to utilise CO_2 provided the correct amount of heat and work is made available. The National Aeronautics and Space Administration (NASA) is currently exploring the use CO_2 (Sabatier reaction) as a means of producing H_2O in space exploration (Vogt et al. 2019). The methane produced via the process could be a viable option for renewable energy (Stangeland et al. 2017).

The Sabatier reaction is a high temperature and negative atmospheric pressure process (250–450 °C; 0.57–0.91 bar) where CH_4 and H_2O are formed from reacting CO_2 and H_2 (Junaedi et al. 2019; Müller et al. 2013). Other useful reactions that consume CO_2 include the WGS reaction.

In addition to the Sabatier reaction, NASA is also studying the Bosch reaction as another pathway to reduce CO_2 from the Space Station atmosphere and to generate clean H_2O (Holmes et al. 1970). The primary objective for this process is to eliminate on-board gaseous storage and overboard venting to enable the recovery of the metabolically consumed oxygen (O_2) (Wagner et al. 1988; Spina and Lee 1985). The Bosch reaction involves the reaction of CO_2 and H_2 to give H_2O and carbon (C) at high temperature and pressure with the aid of a catalyst (Wagner et al. 1988; Bunnell et al. 1991).

Reverse water–gas shift (rWGS) is another way of converting CO_2 with the use of heterogeneous iron catalysts, the

Table 1 Possible reactions for CO_2 utilisation

Process name	Reaction	Condition
Sabatier reaction	$CO_2(g) + 4H_2(g) \leftrightarrow CH_4(g) + 2H_2O(g)$	250–450 °C; 08.3–13.2 psia Nickel, ruthenium or alumina catalyst
Reverse water gas shift reaction	$CO_2(g) + H_2(g) \leftrightarrow CO(g) + H_2O(g)$	400–750 °C, Iron catalyst
Reverse combustion	$CO_2(g) \leftrightarrow C(s) + O_2(g)$	–
Reverse-Boudouard reaction	$C(s) + CO_2(g) \rightarrow 2CO(g)$	773 K, Ni catalysts
Bosch reaction	$CO_2(g) + 2H_2(g) \leftrightarrow C(s) + 2H_2O(g)$	530–730 °C; 7–15 psig Iron, cobalt, nickel or Ruthenium catalyst
CO_2 hydrogenation (CO_2-FT)	$nCO_2(g) + 3nH_2(g) \leftrightarrow C_nH_{2n} + 2nH_2O(g)$ $nCO_2(g) + (3n+1)H_2(g) \leftrightarrow C_nH_{2n+2} + 2nH_2O(g)$	320 °C; 3 MPa, Na–Fe_3O_4 catalyst

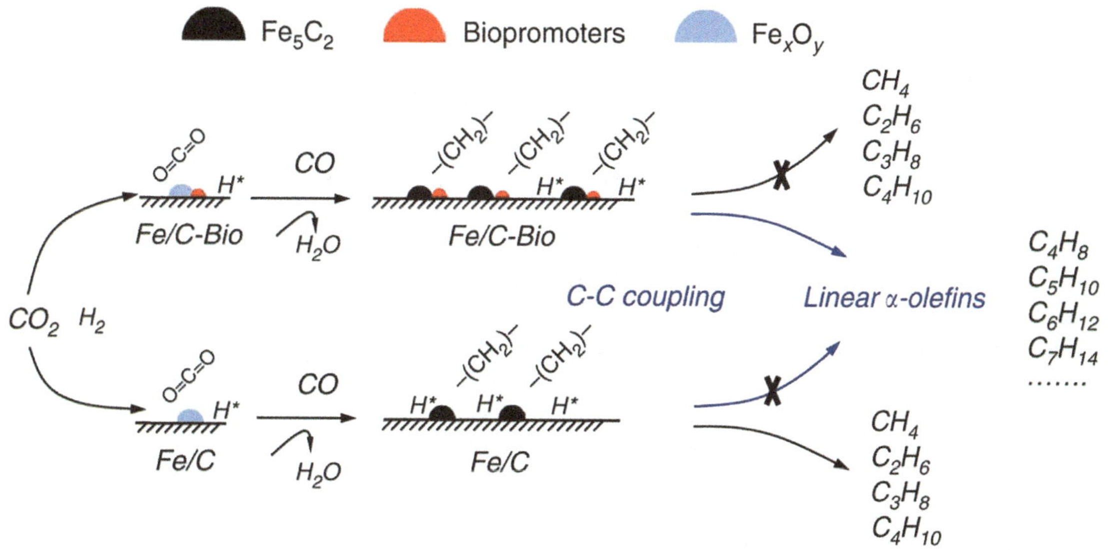

Fig. 1 Reaction scheme for CO_2 direct conversion to linear α-olefins with the use of a bio-promoted catalysts C_4–C_7 olefins and C_1–C_4 paraffins are produced (Guo et al. 2018)

process also demands high temperatures (Pastor-Pérez et al. 2017; Hla et al. 2009). Other catalysts used include copper, platinum and rhodium immobilised on a variety of supports (Daza and Kuhn 2016). Reverse water–gas shift catalysis appears to be a well-researched method for converting CO_2 to CO, which can then be used in the generation of an energy-dense transportable fuel via Fischer–Tropsch (Daza and Kuhn 2016; Gorimbo et al. 2017). Other reactions for CO_2 utilisation include direct synthesis to methanol (Marlin et al. 2018), reverse combustion (Jiang et al. 2017), reverse-Boudouard reaction which uses elemental C and CO_2 to produce CO (Osaki and Mori 2006). Reverse-Boudouard reaction finds use in CO_2 reforming of CH_4 from the viewpoint of suppressing the deposition of C (Osaki et al. 1996).

However, the fallback of these CO_2 conversion reactions in Table 1 is the need for a tremendous input of energy. This indicates the need for further research into renewable energy sources, like solar power to power these processes. Indirect production of liquid fuels from CO_2 hydrogenation via Fischer–Tropsch (CO_2-FT) is also possible (Wei et al. 2017). CO_2 hydrogenation tests in continuous flowing fixed-bed reactor indicated the feasibility of the reaction using modified Fe/C with bio promoters (Guo et al. 2018). Figure 1 shows the product distribution of the study.

Reaction scheme for CO_2 direct conversion to linear α-olefins with the use of a bio-promoted catalysts C_4–C_7 olefins and C_1–C_4 paraffins are produced (Guo et al. 2018).

4 Thermodynamic Considerations

The rWGS reaction is equilibrium limited. According to Le Chatelier's principle, CO productivity increases with temperature as the reaction is endothermic. Pressure has less influence by virtue of the balanced stoichiometric coefficients on the reaction. The same principle dictates that thermodynamic calculations at ambient pressure, showing enhanced CO_2 conversion when excess H_2 is added to the

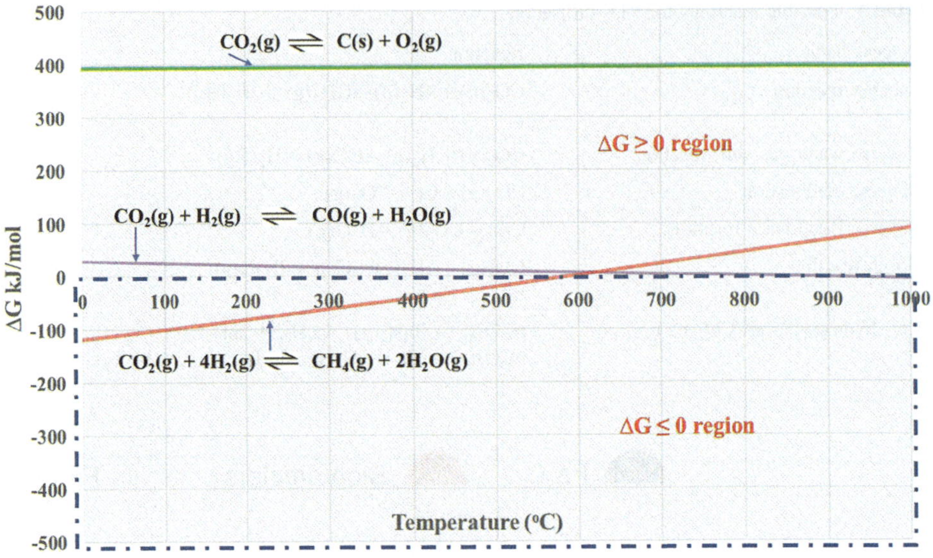

Fig. 2 Gibbs free energy for reverse combustion, CO_2 methanation and rWGS reaction

system. The CO productivity can also be increased by-product separation (Daza and Kuhn 2016). Enthalpy change (ΔH) values at standard conditions for reverse-Boudouard reaction are obtained from Osaki and Mori (2006) while that for the rWGS and CO_2 methanation were given by Pastor-Perez et al. (2017) and Gorimbo and Hildebrandt (2020).

The thermodynamics of CO_2 conversion are shown in Fig. 2, where ΔG values are plotted against temperature at ambient pressure. Thermodynamics gives an indication of the possibility of the CO_2 conversion reactions. The most significant parameter in elucidating what is possible in the conversion of CO_2 is the ΔG because it shows the thermodynamic amount of work available to the system for the reaction to occur at a given temperature and pressure. The more negative the ΔG, the more spontaneous the reaction is or the more of the reversible work that may be performed by the thermodynamic system.

Thermodynamic calculations at ambient pressure show that CO_2 conversion via the Sabatier reaction is feasible below 600 °C while the rWGS is favoured at temperatures beyond 600 °C. The ΔG for the reverse combustion shows no variation with increasing temperature and thus remains in the thermodynamically non-favoured region. Direct conversion of CO_2 and CH_4 is another possible reaction but high temperatures are needed for the conversion to happen. Standard thermodynamic data of possible CO_2 conversion routes as given in Table 2.

The Sabatier reaction has been used in space exploration while the rWGS reaction is used to adjust the CO/H_2 ratio in Fischer–Tropsch synthesis (FTS). In both reactions, CO_2 is efficiently used as feed at the expense of co-feeding with H_2. However, the strong C=O double bond which requires a lot of heat to break the bond hinders CO_2 hydrogenation. Experimental studies still need to be conducted to investigate a suitable catalyst/pathway for sustainable breakdown of CO_2. The reactions of CO_2 demand high-energy input therefore are not benign as the energy used also yields CO_2. In other words, the reverse combustion and the rWGS involve positive ΔG, and thus, they are not favoured at temperature below 600 °C.

5 Thermodynamics of CO_2

In order to cut down on harmful CO_2 emissions while continuing to use industrial technology, engineers and researchers have been investigating ways to design more efficient manufacturing processes that are less harmful to the environment and to improve the efficiency of already existing systems. Some of the most ground-breaking research, to achieve resource efficiency while achieving environmental targets, to date, has been found in the field of using renewable energies. However, most systems still emit CO_2 beyond permitted goals. Therefore, the question arising that must be addressed is: *How can we utilise CO_2 sustainably to help achieve these goals?*

5.1 The Thermodynamic Attainable Region (AR)

Scientists are starting to look into designing sustainable systems that can feed CO_2 using process synthesis techniques. This section employs a thermodynamic technique that can help us visualise the thermodynamics of reactions involving CO_2. The use of the thermodynamic Attainable

Table 2 Values at standard conditions giving an indication of the easy with which the reaction can happen

Reaction	ΔH^{θ}_{298k} (kJ/mol)	ΔS^{θ}_{298k} (J/mol.K)	ΔG^{θ}_{298k} (kJ/mol)
Well researched pathways			
$CO_2(g) + H_2(g) \leftrightarrow CO(g) + H_2O(g)$	+41.13	+42.02	+28.53
$CO_2(g) + 4H_2(g) \leftrightarrow CH_4(g) + 2H_2O(g)$	−164.75	−172.45	−113.33
Other possible pathways			
$CO_2(g) \leftrightarrow C(s) + O_2(g)$	+393.51	−2.88	+394.36
$CH_4(g) + CO_2(g) \leftrightarrow 2CO(g) + 2H_2(g)$	+247.02	+256.54	+170.53
$CH_4(g) + 2CO_2(g) \leftrightarrow 3CO(g) + H_2(g) + H_2O$	+288.16	+298.59	+199.14

Region (AR) is vital in understanding the complexity involved with the direct use of CO_2 as a feed-in process systems. The AR is a thermodynamic region that shows all possible reactions for a given set of species and restrictions applied to it. This can be used to set performance targets for systems providing valuable information which include (i) optimum operating conditions of temperature, (ii) heat and work requirements, (iii) best use of feed to make preferred product and also meet emissions targets. It is therefore important to understand the thermodynamics of CO_2 conversion using the AR in order to try and achieve these goals.

In this case, three reactions where CO_2 is feed are used to analyse the thermodynamic heat and work associated with feeding CO_2 using the AR. The system is studied at various temperatures to enable us to see whether changes in temperature can help build processes that can consume CO_2 sustainably. Three simple reactions involving the rWGS, Sabatier reactions and reverse combustion are considered, Eqs. (1–3).

$$CO_2 + 4H_2 \rightarrow CH_4 + 2H_2O \quad \text{Sabatier reaction} \tag{1}$$

$$CO_2 + H_2 \rightarrow CO + H_2O \quad \text{rWGS} \tag{2}$$

$$CO_2 \rightarrow C + O_2 \quad \text{Reverse combustion} \tag{3}$$

The seven species in the reactions above are arranged in matrix form to help find the independent mass balances using the Gaussian elimination method; in this case (#-compounds–#elements (7–3 = 4)). The species are arranged in matrix form (Table 3) using the three respective elements C, H and O and Gaussian elimination was applied to make each column have a one in only one row and other values as zero will help find the independent mass balances. However, the independent mass balances may not reflect the actual reactions occurring in the process or reactor but will help give the boundaries of the AR showing all possible reactions from the given constraints and species specified. To try and analyse a perceived sustainable system, it is assumed a feed of 1 mol CO_2 and 1 mol H_2O is used. Yet, if we feed in CO_2 with H_2 as in Eqs. (1) and (2), the AR will yield the well-researched reactions in Table 2. One of the main constraints applied is that the researchers try to avoid using H_2 as feed and are interested in the reactions associated with feeding H_2O instead. Methane is also a possible feed but in this case it is considered not sustainable to feed in this fuel to the process. Please note that if H_2 and CH_4 can be allowed as feed, the whole mass balance changes, thus changing the AR boundaries and revealing other possible mass balances associated.

Table 3 Matrix showing molar amounts of each element

C	H	O	Compounds
1	0	2	CO_2
1	0	0	C
0	2	0	H_2
0	0	2	O_2
1	0	1	CO
1	2	0	CH_4
0	2	1	H_2O

According to Yin (1989), the row/s consisting of zeros on all the columns in Table 4 corresponds to the independent reactions for the system. These are indicated with grey shade in Table 4.

The four-independent mass balances are given in Eqs. (4–7)

Table 4 Independent mass balances

C	H	O	Material balances
0	0	0	CO_2 - C - O_2
1	0	0	C
0	1	0	$1/2H_2$
0	0	0	O_2 – 2(CO-C)
0	0	1	CO-C
0	0	0	CH_4 - $2H_2$ -C
0	0	0	H_2O - H_2 -CO+C

$$CO_2 \rightarrow C + O_2 \quad \Delta H^\circ = 393.5\,kJ/mol, \Delta G^\circ = 394.3\,kJ/mol \quad \varepsilon_1 \tag{4}$$

$$2C + O_2 \rightarrow 2CO \quad \Delta H^\circ = -110.5\,kJ/mol, \Delta G^\circ = -137.1\,kJ/mol \quad \varepsilon_2 \tag{5}$$

$$2H_2 + C \rightarrow CH_4 \quad \Delta H^\circ = -74.7\,kJ/mol, \Delta G^\circ = -50.8\,kJ/mol \quad \varepsilon_3 \tag{6}$$

$$C + H_2O \rightarrow H_2 + CO \quad \Delta H^\circ = 131.2\,kJ/mol, \Delta G^\circ = 91.4\,kJ/mol \quad \varepsilon_4 \tag{7}$$

Stoichiometric mass balances for feed of 1 mol CO_2 and 1 mol H_2O

$$\begin{aligned}
&-\mathbf{N}CO_2 = \mathbf{N_o}CO_2 - \varepsilon_1 \geq 0\\
&\varepsilon_1 \leq \mathbf{1}\\
&-\mathbf{N}O_2 = \mathbf{N_o}O_2 + \varepsilon_1 - \varepsilon_2 \geq 0\\
&\varepsilon_1 - \varepsilon_2 \geq = 0\\
&-\mathbf{N}C = \mathbf{N_o}C + \varepsilon_1 - 2\varepsilon_2 - \varepsilon_3 - \varepsilon_4 \geq 0\\
&\varepsilon_1 - 2\varepsilon_2 - \varepsilon_3 - \varepsilon_4 \geq 0\\
&-\mathbf{N}H_2 = \mathbf{N_o}H_2 - 2\varepsilon_3 + \varepsilon_4 \geq 0\\
&\varepsilon_4 - 2\varepsilon_3 \geq 0\\
&-\mathbf{N}H_2O = \mathbf{N_o}H_2O - \varepsilon_4 \geq 0\\
&\varepsilon_4 \leq \mathbf{1}\\
&-\mathbf{N}CO = \mathbf{N_o}CO + 2\varepsilon_2 + \varepsilon_4 \geq 0\\
&2\varepsilon_2 + \varepsilon_4 \geq 0\\
&\mathbf{N}CH_4 = \mathbf{N_o}CH_4 + \varepsilon_3 \geq 0\\
&\varepsilon_3 \geq 0
\end{aligned}$$

The seven inequalities above are solved using MATLAB '***mupad***' function to find vertices for the AR for the system. The result gives us the vertices used to plot an AR which helps us visualise the thermodynamics of CO_2 processes in a 2-D graphical scheme.

MATLAB '***mupad***' script to find vertices when $N^\circ_{CO_2}$ and $N^\circ_{H_2O} = 1$, consider $\varepsilon_1, \varepsilon_2, \varepsilon_3, \varepsilon_4$ as a, b, c, d, respectively.

```
k:=[{a<=1,a-b>=0,a-2*b-c-d>=0,d-2*c>=0,d<=1,2*b
+d>=0,c>=0},a+b+c+d]:linopt::corners(k,[a,b,c,d])
```

Vertices obtained;

{[0, 0, 0, 0], [0, -1/2, 0, 1], [1/2, -1/2, 1/2, 1], [1, 0, 0, 0], [1, 0, 0, 1], [1, 1/2, 0, 0], [1, -1/2, 0, 1], [1, -1/2, 1/2, 1], [1, -1/4, 1/2, 1]}

The vertices correspond to $[\varepsilon_1, \varepsilon_2, \varepsilon_3, \varepsilon_4]$. After obtaining the extent vertices above, we add the respective independent mass balances Eqs. (4–7), multiplying each mass balance with the respective extent in the vertex to obtain the overall material balance for each of the AR vertices as shown in Eq. (8).

$$\text{AR vertex} = \varepsilon_1 * \text{Eq.}(4) + \varepsilon_2 * \text{Eq.}(5) + \varepsilon_3 * \text{Eq.}(6) + \varepsilon_4 * \text{Eq.}(7) \tag{8}$$

5.2 Using Hess's Law to Transform the Extents to *G-H* AR @ 25°C

According to Hess's law, heat requirements for a chemical process are the same irrespective of the pathway or steps are taken to make the products. This means that ΔH and ΔG for the reactions on each vertex can be obtained using Eqs. (9) and (10). The extents together with the enthalpy and Gibbs free energy at 25 °C, 1 bar were used to estimate the ΔH and ΔG for the combined reaction process. Similarly, the ΔH and ΔG can simply be obtained directly from the overall mass balances attained in Eq. (8) for each extent vertex (products–reactants).

$$\Delta H = \varepsilon_1 \Delta H_1 + \varepsilon_2 \Delta H_2 + \varepsilon_3 \Delta H_3 + \varepsilon_4 \Delta H_4 \tag{9}$$

$$\Delta G = \varepsilon_1 \Delta G_1 + \varepsilon_2 \Delta G_2 + \varepsilon_3 \Delta G_3 + \varepsilon_4 \Delta G_4 \tag{10}$$

The ΔH and ΔG for each reaction produced from each vertex are plotted, and the AR is obtained in terms of *G* and *H* which is giving *G-H* AR. Figure 3 shows the *G-H* AR at 25 °C, which is the region bound by joining all the outer vertices. This region represents all possible combinations of reacting species present in the independent balances when 1 mol CO_2 and 1 mol H_2O are fed. The reaction with the minimum ΔG is the one that is most likely to dominate other reactions in the AR because it has the most available work to make products.

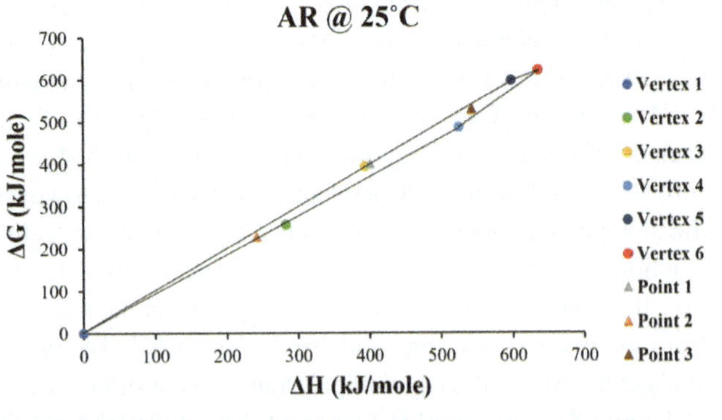

Fig. 3 Thermodynamic Attainable Region for CO_2 conversion processes at 25 °C

Important Vertices and Points to note:

- Vertex 1 = feed which correspond to 1 mol $CO_2 + 1$ mole H_2O
- Vertex 2 = $3C + 1.5O_2 \rightarrow 3CO$, $\Delta H° = 282.9$ kJ/mol, $\Delta G° = 257.1$ kJ/mol
- Vertex 3 = $\varepsilon_1 = CO_2 \rightarrow C + O_2$, $\Delta H° = 393.5$ kJ/mol, $\Delta G° = 394.3$ kJ/mol
- Vertex 4 = $CO_2 + H_2O \rightarrow CO + H_2 + O_2$, $\Delta H° =$ 524.8 kJ/mol, $\Delta G° = 485.7$ kJ/mol
- Vertex 5 = $CO_2 + H_2O \rightarrow 0.5C + 0.5CH_4 + 1.5O_2$, $\Delta H° = 597.9$ kJ/mol, $\Delta G° = 597.5$ kJ/mol
- Vertex 6 = $CO_2 + H_2O \rightarrow C + H_2 + 1.5O_2$, $\Delta H° =$ 635.3 kJ/mol, $\Delta G° = 622.9$ kJ/mol
- Point 1 = $0.5CO_2 + H_2O \rightarrow 0.5CH_4 + O_2$, $\Delta H° =$ 401.2 kJ/mol, $\Delta G° = 400.3$ kJ/mol
- Point 2 = $H_2O \rightarrow H_2 + 0.5O_2$, (minimum G) $\Delta H° =$ 241.8 kJ/mol, $\Delta G° = 228.5$ kJ/mol

temperatures but constant pressure can be obtained by using Eq. (11). Pressure deviations are considered not significant on the enthalpy Gibbs free energy for the temperatures studied.

$$\Delta H° = \Delta H_o^° + R \int_{T_o}^{T} \frac{\Delta C_p^°}{R} dT \quad (11)$$

where $\Delta H°$ and $\Delta H_o^°$ are enthalpy of formation at temperature T and T_o, respectively. The heat capacity is specified as:

$$\frac{\Delta C_p^°}{R} = A + BT + CT^2 + DT^{-2} \quad (12)$$

Integrate Eq. (11) to give Eq. (13) where $\tau = \frac{T}{T_o}$

$$\Delta H° = \Delta H_o° + R * [\Delta A * T_o * (\tau - 1) + \frac{\Delta B}{2} * T_o^2 * (\tau^2 - 1) + \frac{\Delta C}{3} * T_o^3 * (\tau^3 - 1) + \frac{\Delta D}{T_o} * (\frac{\tau - 1}{\tau})] \quad (13)$$

- Point 3 = $CO_2 + H_2O \rightarrow 0.5CO + 0.5CH_4 + 1.25O_2$, $\Delta H° = 542.7$ kJ/mol, $\Delta G° = 528.9$ kJ/mol.

At 25 °C, the whole AR lies in the positive region for both enthalpy and Gibbs free energy. This means that both heat and work must be added to make the respective products at the AR vertices when CO_2 is a feed. It is then important to find the effects of increasing temperature on the G-H AR. The electrolysis process (Point 2) exists at minimum G in the G-H AR at 25 °C but the process still requires work addition as the ΔG is positive.

ΔG at various temperatures can be calculated by using Eq. (14), (Smith et al. 2018).

$$\Delta G_T^° = \Delta H_o^° - \frac{T}{T_o}(\Delta H_o^° - \Delta G_o^°) + R \int_{T.}^{T} \frac{\Delta C_p^°}{R} dT - RT \int_{T.}^{T} \frac{\Delta C_p^°}{R} \frac{dT}{T} \quad (14)$$

where

$$\int_{T.}^{T} \frac{\Delta Cp°}{R} dT = \Delta A * T_o * (\tau - 1) + \frac{\Delta B}{2} * \frac{\Delta B}{2} * T_o^2 * (\tau^2 - 1) + \frac{\Delta C}{3} * T_o^3 * (\tau^3 - 1) + \frac{\Delta D}{To} * (\frac{\tau - 1}{\tau})$$

and

$$\int_{T.}^{T} \frac{\Delta Cp°}{R} \frac{dT}{T} = \Delta A * ln\tau + [\Delta B T_o + (\Delta C T_o^2 + \frac{\Delta D}{\tau^2 T_o^2})(\frac{\tau + 1}{2})] * (\tau - 1)$$

5.3 Increasing Temperature on *G-H* AR

Considering ideal gas behaviour, the change in enthalpy (ΔH) for each component in the mass balances at various

As temperature is increased, the G-H AR enlarges as shown in Figs. 4 and 5. ΔG for Vertex 2 becomes less positive at minimum G predicting CO as product, Eq. (15). At 900 °C, both Point 2 and Vertex 2 have the same ΔG and

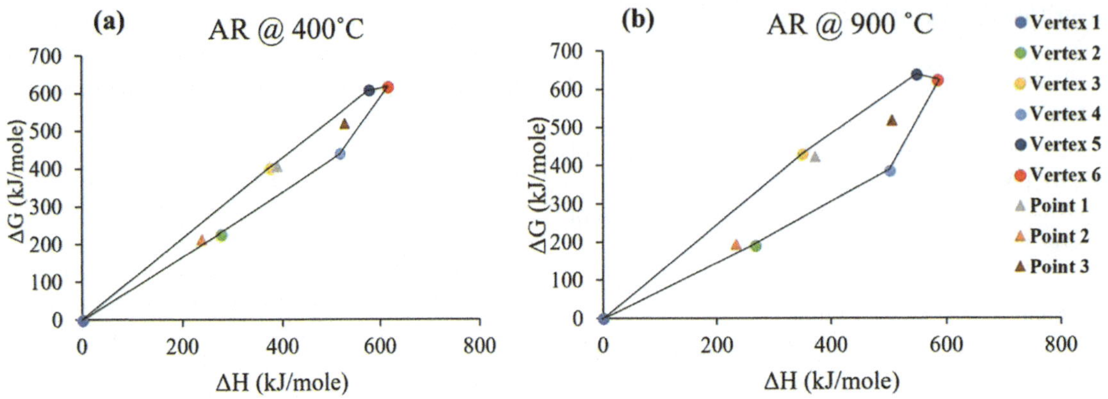

Fig. 4 Thermodynamic Attainable Region for CO_2 conversion processes at **a** 400 °C and **b** 900 °C

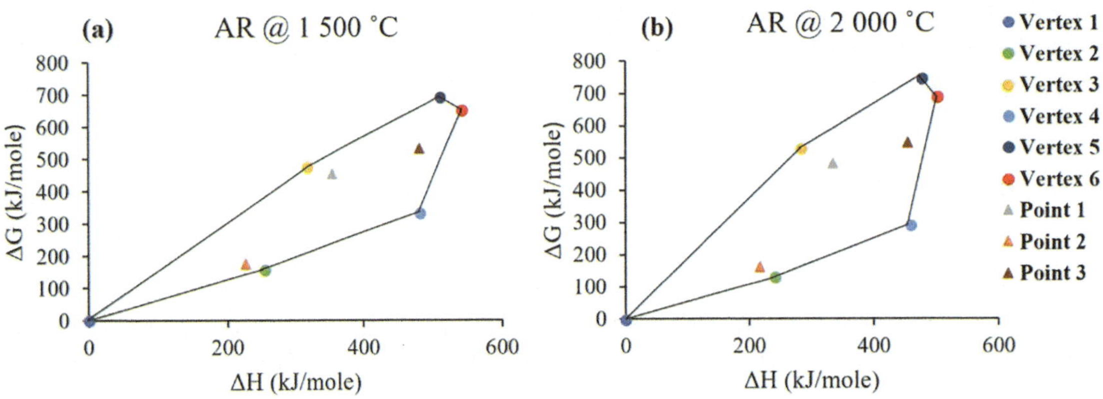

Fig. 5 Thermodynamic Attainable Region for CO_2 conversion processes at **a** 1500 °C and **b** 2000 °C

are both at minimum *G*. This means that at 900 °C H_2, O_2 or CO are all possible products. This means that mixing can be used to separate the products from the two possible reactions. However, the reaction for making CO is more endothermic hence requires heat.

$$\text{Vertex 2 @ 2000 °C} = 3C + 1.5O_2 \rightarrow 3CO, (\text{minimum } G),$$
$$\Delta H^\circ = 216.9\,\text{kJ/mol}, \Delta G^\circ = 160.8\,\text{kJ/mol} \quad (15)$$

The *G-H* AR has also shown that even at high temperatures it is not sustainable to invest in process reactions with CO_2 and H_2O as feed as these require huge work and heat addition. Nevertheless, the addition of catalysts and changing reacting pressures may help encourage conversion of CO_2 to other products but the overall cost may weigh heavily on the overall efficiency. Above 900 °C, incomplete combustion of C to CO is at minimum *G* and becomes the favoured reaction in the *G-H* AR. Ideally, the reaction at Vertex 4 would be the ideal process as it produces syngas and O_2 from feeding CO_2 and H_2O, but this process is not sustainable as this would only be feasible at enormous temperatures with work addition. The heat to achieve these high temperatures will probably make the whole process inefficient and thus not sustainable. The simple use of the thermodynamic Attainable Region has shown that it is difficult to achieve a sustainable process that feeds CO_2 and H_2O without adding numerous amounts of work and heat into the system. However, previous research has shown that co-feeding H_2 or CH_4 with CO_2 may help to make products like CH_4, CO and other hydrogenation products but it is expensive to obtain the CH_4 or H_2 gas to use as feed in such processes. This last section is meant to give a glimpse of how the AR shows the region which includes the defined well-researched processes. Here, a feed combination of 1 mol CO_2 + 1 mol H_2 + 1 mol CH_4 is used and the new stoichiometric mass balance is shown below:

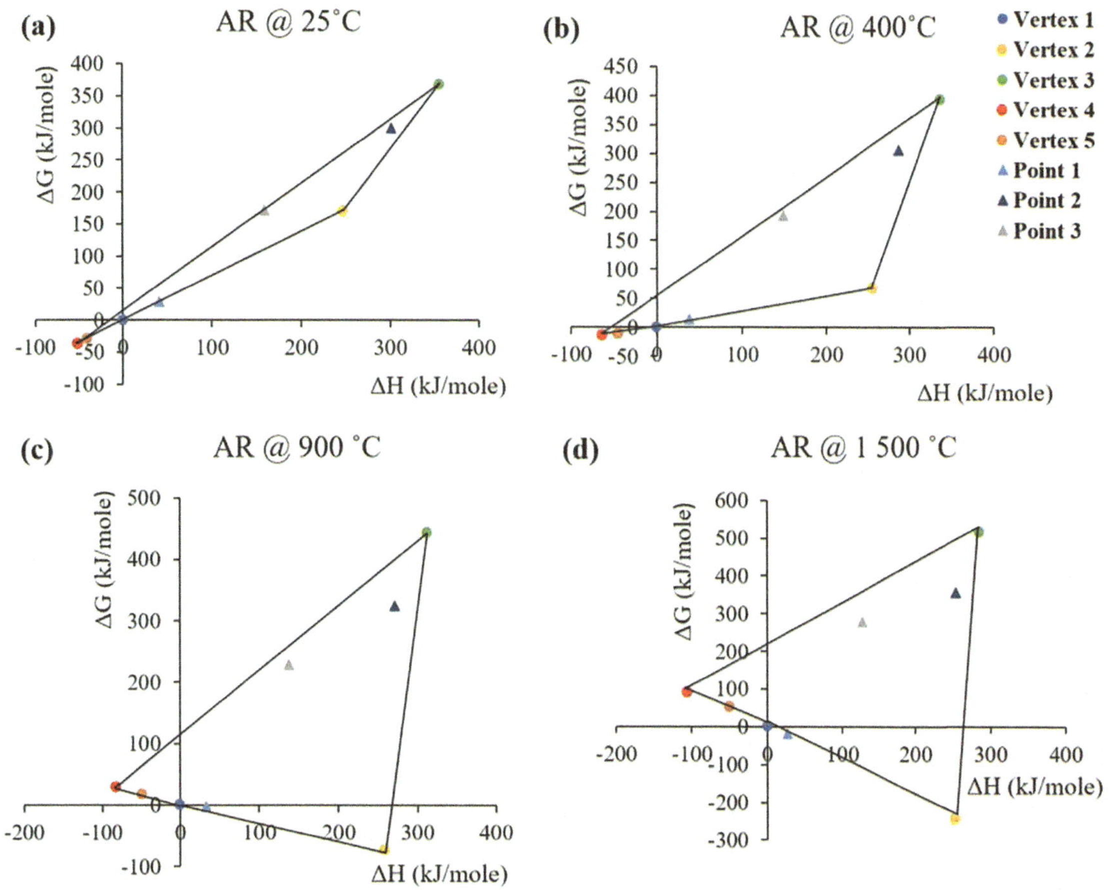

Fig. 6 Thermodynamic Attainable Region for CO_2 conversion processes with H_2 and CH_4 at **a** 25 °C, **b** 400 °C, **c** 900 °C and **d** 1500 °C

$- NCO_2 = N_oCO_2 - \varepsilon_1 \geq 0$

$\varepsilon_1 \leq 1$

$- NO_2 = N_oO_2 + \varepsilon_1 - \varepsilon_2 \geq 0$

$\varepsilon_1 - \varepsilon_2 \geq = 0$

$- NC = N_oC + \varepsilon_1 - 2\varepsilon_2 - \varepsilon_3 - \varepsilon_4 \geq 0$

$\varepsilon_1 - 2\varepsilon_2 - \varepsilon_3 - \varepsilon_4 \geq 0$

$- NH_2 = N_oH_2 - 2\varepsilon_3 + \varepsilon_4 \geq 0$

$2\varepsilon_3 - \varepsilon_4 \leq 1$

$- NH_2O = N_oH_2O - \varepsilon_4 \geq 0$

$\varepsilon_4 \geq 0$

$- NCO = N_oCO + 2\varepsilon_2 + \varepsilon_4 \geq 0$

$2\varepsilon_2 + \varepsilon_4 \geq 0$

$NCH_4 = N_oCH_4 + \varepsilon_3 \geq 0$

$\varepsilon_3 \leq 1$

As before, using '*mupad*' function in MATLAB we use the script below to obtain the vertices for the mass balance boundaries.

```
k:=[{a<=1,a-b>=0,a-2*b-c-    d>=0,2*c<=1,d<=0,2*b
+d>=0,c<=1},a+b+c+d]:linopt::corners(k,[a,b,c,d])
```

Vertices obtained;

{[0, 0, 0, 0], [1/4, 1/4, 1/4, -1/2], [1/2, 0, 1/2, 0], [1, 1, -1, 0], [1, 1, 0, -1], [1, 0, 1/2, 0], [1, 1, -1/2, -2], [1, 1/4, 1/2, 0]}

Figure 6a–d shows the AR when 1 mol of CO_2, H_2 and CH_4 each is fed at various temperatures

Important Vertices and Points to note in Fig. 6:

- Vertex 1 = feed which correspond to 1 mol CO_2 + 1 mol H_2 + 1 mol CH_4
- Vertex 2 = $3CH_4 + CO_2 \rightarrow 2CO + 2H_2$, ΔH° = 247.2 kJ/mol, ΔG° = 170.8 kJ/mol
- Vertex 3 = $CO_2 + H_2 \rightarrow 0.5C + 0.5CH_4 + O_2$, ΔH° = 356.2 kJ/mol, ΔG° = 368.9 kJ/mol
- Vertex 4 = minimum $G = G = CO_2 + 0.5CH_4 + H_2 \rightarrow 1.5C + 2H_2O$, ΔH° = −52.7 kJ/mol, ΔG° = −37.4 kJ/mol

- Vertex 5 = Sabatier reaction = $0.25CO_2 + H_2 \rightarrow 0.25CH_4 + 0.5H_2O$, $\Delta H° = -41.2$ kJ/mol, $\Delta G° = -28.4$ kJ/mol
- Point 1 = rWGS = $CO_2 + H_2 \rightarrow H_2O + CO$, $\Delta H° =$ 41.2 kJ/mol, $\Delta G° = 28.6$ kJ/mol
- Point 2 = $CO_2 + H_2 \rightarrow 0.5CH_4 + 0.5CO + 0.75O_2$, $\Delta H° = 300.9$ kJ/mol, $\Delta G° = 300.4$ kJ/mol
- Point 3 = $0.5CO_2 + H_2 \rightarrow 0.5CH_4 + 0.5O_2$, $\Delta H° =$ 159.4 kJ/mol, $\Delta G° = 171.8$ kJ/mol

Figure 6 shows the same results given in Fig. 2 but plotted on an AR showing CO_2 fed with either H_2 or CH_4. The AR also shows that Vertex 4 is at minimum G at 25 and 400 °C makes C and H_2O and the minimum G is in the region where G is negative. This reaction does not need work addition to happen. Again, we notice the Sabatier reaction which is Vertex 5 is also close to minimum G and the Gibbs free energy for this reaction is negative. The reaction at minimum G is most favoured meaning that it may be difficult to make CH_4 using the Sabatier reaction without forming C. Thermodynamics predict that the formation of C is more favoured compared to the formation of CH_4. Figure 6b, c shows that the Sabatier reaction changes from negative G to positive somewhere at temperatures between the 400 and 900 °C plot (around 600 °C as shown in Fig. 2). The same observation is also observed for Point 1 which represents the rWGS where the reaction becomes spontaneous at very high temperatures above 1000 °C. As temperatures are increased to 900 and 1500 °C, the products at Vertex 2 which makes syngas (CO and H_2) becomes feasible at minimum G where CO_2 is co-fed with CH_4. This analysis shows that the Sabatier reaction (CO_2 + H_2) and the one that makes syngas from CO_2 and CH_4 are the most attractive processes to help utilise CO_2 only if the H_2 and CH_4 are made available cheap and sustainably. We have observed using the AR that CO_2 conversion processes pose a difficulty amongst researchers in trying to come to a balance between the sustainability element; (process energy and work efficiency) and overall economics of the system. It is therefore highly advised to try and reduce CO_2 emissions at source by adopting renewable energies. Further, research is also recommended on other CO_2 conversion mechanisms especially using biological means.

6 Conclusion

Anthropic CO_2 discharge into the atmosphere must be drastically reduced, but several strategies have to be undertaken to this purpose. Contemporary research focuses on chemical means of CO_2 conversion to useful chemicals. This chapter is dedicated to understanding the thermodynamic pathways of CO_2 conversion with H_2O. A brief digression on the basic concept of CO_2 capture and utilisation is explored first. The Sabatier reaction is the only less energy-intensive process thus far and carries high potential use in large-scale conversion of carbon dioxide. Thermodynamics dictates that work must be input for the rWGS reaction at low temperature but ΔG values only become negative at temperatures greater than 600 °C. More so, the Attainable Region has shown the thermodynamic barriers of high work and energy requirements in trying to make useful products from feeding CO_2 and H_2O. Currently, thermodynamics have shown that it is difficult to achieve sustainable reaction processes that can help reduce atmospheric CO_2. It is therefore recommended to move towards renewable sources of fuel in a bid to reduce further CO_2 emissions.

Acknowledgements The authors are grateful for the financial support provided by the University of South Africa (UNISA), the University of the Witwatersrand, National Research Foundation (NRF) of South Africa and the Institute for the Development of Energy for African Sustainability (IDEAS) research unit at UNISA.

References

Ahmed R, Liu G, Yousaf B, Abbas Q, Ullah H, Ali MU (2020) J Clean Prod 242:118409
Bunnell CT, Boyda RB, Lee MG (1991) 21st International conference environment on system 253–259
Choi YH, Jang YJ, Park H, Kim WY, Lee YH, Choi SH, Lee JS (2017) Appl Catal B Environ 202:605–610
Daza YA, Kuhn JN (2016) RSC Adv 6:49675–49691
Donald Carruthers J, Petruska MA, Sturm EA, Wilson SM (2012) Microporous Mesoporous Mater 154:62–67
Dowson GRM, Styring P (2017) Front Energy Res 5:1–11
Dutcher B, Fan M, Russell AG, Appl ACS (2015) Mater Interfaces 7:2137–2148
Gorimbo J, Hildebrandt D (2020) Thermochemical conversion of carbon dioxide to carbon monoxide by reverse water-gas shift reaction over the ceria-based catalyst. In: Inamuddin et al (eds) Conversion of carbon dioxide into hydrocarbons (vol 1). Springer Nature Switzerland, pp 43–61
Gorimbo J, Lu X, Liu X, Yao Y, Hildebrandt D, Glasser D (2017) Ind Eng Chem Res 56
Guo L, Sun J, Ji X, Wei J, Wen Z, Yao R, Xu H, Ge Q (2018) Commun Chem 1:1–8
Hannah R, Max R (2017) Publ. Online OurWorldInData.Org. Retrieved from https://ourworldindata.org/Co2-and-other-greenhouse-gas-emissions Online Resour
Hla SS, Duffy GJ, Morpeth LD, Cousins A, Roberts DG, Edwards JH (2009) Catal Commun 11:272–275
Holmes RF, Keller EE, King CD (1970) Natl Aeronaut Sp Adm
Jiang L, Chen Z, Ali SMF (2017) Fuel 207:302–311
Junaedi C, Hawley K, Walsh D, Subir Roychoudhury Abney MB, Perry JL (2019) Am Inst Aeronaut Astronaut 1–10
Lee S, Binns M, Lee JH, Moon JH, Yeo JG, Yeo YK, Lee YMM, Kim JK (2017) J Memb Sci 541:224–234
Liu Q, Wu L, Jackstell R, Beller M (2015) Nat Commun 6
Marlin DS, Sarron E, Sigurbjörnsson Ó (2018) Front Chem 6:1–8
Meisen A, Shuai X (1997) Energy Convers Manag 38:37–42
Mukherjee A, Okolie JA, Abdelrasoul A, Niu C, Dalai AK (2019) J Environ Sci (china) 83:46–63

Müller K, Fleige M, Rachow F, Schmeißer D (2013) Energy Procedia 40:240–248
Osaki T, Mori T (2006) React Kinet Catal Lett 89:333–339
Osaki T, Horiuchi T, Suzuki K, Mori T (1996) J Chem Soc—Faraday Trans 92:1627–1631
Paris Agreement (2015)
Pastor-Pérez L, Baibars F, Le Sache E, Arellano-García H, Gu S, Reina TR (2017) J CO_2 Util 21:423–428
Smith JM, Van Ness HC, Abbott MM, Swihart MT (2018) Introduction to chemical engineering thermodynamics. McGraw Hill education, New York
Spina L, Lee MC (1985) SAE Tech Pap 1–10
Stangeland K, Kalai D, Li H, Yu Z (2017) Energy Procedia 105:2022–2027
Stewart C, Hessami MA (2005) Energy Convers Manag 46:403–420
Stolaroff JK, Keith DW, Lowry GV (2008) Environ Sci Technol 42:2728–2735
Sun Z, Fan M, Argyle M (2011) Ind Eng Chem Res 50:11343–11349
Vijayavenkataraman S, Iniyan S, Goic R (2012) Renew Sustain Energy Rev 16:878–897
Vogt C, Monai M, Kramer GJ, Weckhuysen BM (2019) Nat Catal 2:188–197
Wagner RC, Carrasquillo R, Edwards J, Holmes R (1988) SAE Tech Pap 1–9
Wei J, Ge Q, Yao R, Wen Z, Fang C, Guo L, Xu H, Sun J (2017) Nat Commun 8:1–8
Yin F (1989) Chem Eng Commun 83:117–127

Enzymatic CO_2 Conversion

Pravin D. Patil, Anup D. Chahande, Deepali T. Marghade, Vivek P. Bhange, and Manishkumar S. Tiwari

Abstract

The ever-growing energy and chemical demand have resulted in the absurd usage of fossil fuels, which has led to a significant increase in the CO_2 concentration in the environment. In recent times, the development of efficient methods for CO_2 capture, sequestration, and utilization is one of the key research areas under exploration. The conversion of CO_2 into chemicals and fuels via chemical, photochemical, enzymatic, and electrochemical routes is an efficient method of recycling CO_2. Among these, the enzymatic conversion of CO_2 to value-added products is an effective method due to their superior chemo-selectivity and stereospecificity. However, there is an urgent need for the exploration of new enzymes and reaction operations to advance catalytic performance. The current chapter focuses on the in vivo and in vitro biocatalyzed reaction for the reduction of CO_2 into the value-added product by employing several enzymes. The chapter explores opportunities and challenges associated with practicing such biocatalysts for converting CO_2 into value-added products. Moreover, the chapter offers an insight into the industrial application of CO_2 and its products.

Keywords

Biocatalysis • Dehydrogenases • Enzymes • CO_2 reduction • CO_2 conversion • Multi-enzyme cofactor regeneration

1 Introduction

The absurd usage of fossil fuels by humans has led to a steep rise in global carbon emissions. Carbon dioxide (CO_2) accounts for about 76% of the total greenhouse gas emissions (Marpani et al. 2017; Srikanth et al. 2017). A substantial hike in CO_2 emission in short span time has unbalanced the natural equilibrium between CO_2 emission and natural conversion of CO_2 via photosynthesis. However, in 2019, 33 gigatons reduction in global CO_2 emissions was recorded due to the use of advanced techniques in the power sector, along with the use of nuclear power and renewable sources (mainly wind and solar) as an alternative to the conventional sources (Gielen et al. 2019). Moreover, the conversion of CO_2 into value-added products has helped in the overall reduction of CO_2. It is essential to develop carbon–neutral approaches to minimize the adverse impact of CO_2 emission on the environment. In these regards, the conversion of greenhouse CO_2 to value-added products has emerged as a compelling way to dawdling down the global CO_2 accumulation rate. Scientists are exploring various pathways for the conversion of atmospheric CO_2 into valuable industrial products (such as fuels, basic chemicals, and advanced products) (Duc Long et al. 2017). Although the employment of several chemical routes and biocatalytic conversion pathways have proved their potential for the conversion of atmospheric CO_2 under optimum temperature conditions, there is still considerable scope for the advancements of the existing methods. Nevertheless, the development of a sustainable and cost-efficient conversion

P. D. Patil (✉)
Department of Basic Science and Humanities, Mukesh Patel School of Technology Management and Engineering, SVKM's NMIMS University, Mumbai, Maharashtra 400056, India
e-mail: pravin.patil@nmims.edu

A. D. Chahande · V. P. Bhange
Department of Biotechnology, Priyadarshini Institute of Engineering and Technology, Nagpur, Maharashtra 440019, India

D. T. Marghade
Department of Applied Chemistry, Priyadarshini Institute of Engineering and Technology, Nagpur, Maharashtra 440019, India

M. S. Tiwari
Department of Chemical Engineering, Mukesh Patel School of Technology Management and Engineering, SVKM's NMIMS University, Mumbai, Maharashtra 400056, India

Inamuddin et al. (eds.), *Carbon Dioxide Utilization to Sustainable Energy and Fuels*, Advances in Science, Technology & Innovation, https://doi.org/10.1007/978-3-030-72877-9_5

process with zero or even negative emissions casts a significant challenge to the scientific community.

1.1 CO_2 as a Greenhouse Gas

The continuously improving living standards of humans have excessively raised the CO2 emissions, especially after the inception of the industrial era, which ultimately accelerated the natural greenhouse effect leading to climate change. The link between elevated CO2 concentration in the atmosphere with climate shifts was first put forward by a Swedish scientist Svante Arrhenius in 1896 (Anderson et al. 2016). The adverse effect on the environment was recently verified by a research group analyzing the spectrum of the radiative flux, leaving the top of the atmosphere. The group recorded a strong band near 15 μm due to CO2 concentration at temperatures similar to the earth's surface. This elaborated on the direct relationship between the concentration of CO2 in the atmosphere with the earth's surface temperature (Zhong and Haigh 2013). Therefore, to reduce the yearly rise in global temperature, an increase in green cover and implementation of carbon capture, storage, and utilization are the best possible options in the current situation.

1.2 Carbon Capture, Storage, and Utilization

Carbon capture, storage (sequestration), and utilization, also known as CCSU, has become an emerging field that is efficiently resisting climate change in recent years. Surpassing anthropogenic CO_2 emissions in the carbon cycle has stimulated the concept of CCSU. The central idea of CCSU is to develop a more efficient artificial carbon cycle to balance CO_2 sequestration and generation of anthropogenic CO_2 emissions by trapping and converting it into valuable chemicals. The steps involved in this approach are as follows; (a) capturing of anthropogenic CO_2 during production stage (at power plants and cement manufacturing units), (b) sequestration of CO_2, which includes liquefaction of CO_2 by transportation through pipelines and injecting in storage site such as oil reservoirs to enhance oil recovery, (c) and finally converting it into value-added chemicals (Boot-Handford et al. 2014). However, the transformation of CO_2 into valuable commodity products is challenging due to the limitation of the process conferred by thermodynamic stability and kinetically inertness of CO_2.

Several approaches for carbon-capturing have been assessed in the recent past. Among many, oxy-fuel combustion and gasification in thermal power plants have proved their potential for trapping CO_2 (Helseth 2012). Other industries, such as cement, iron–steel, and natural gas, employed the alkaline absorption process for CO_2 capture. Further, trapped CO_2 was pressured and injected in the geological formation, mineralization process, and landfill process employing CO_2 sequestration (Boot-Handford et al. 2014). However, the last stage of CCSU, the utilization of CO_2 to obtain value-added chemicals, remains the most challenging. To transform CO_2 into commercially valuable products (such as phenols, alcohols, epoxides, and pyrrole) along with the extraction of supercritical CO_2, many researchers profoundly used various ways for carbon utilization, such as chemical, photochemical, electrochemical, and enzymatic conversion. (Shi et al. 2015). The CO_2 conversion is an exergonic process in which carbon atom has a +4 oxidation state. While, in the conversion of CO_2 into products through a reduction pathway, carbon atom only has +2 oxidation state making the process endogenic (Aresta and Dibenedetto 2007) and needs energy and hydrogen to carry out the conversion. A few researchers uncovered the reduction pathways for the production of several organic products and fuels (methanol, formic acid, methane, etc.). These reduction pathways require high over-potentials to overcome the thermodynamic barriers of CO_2 reduction by using a suitable catalyst (Olah et al. 2009; (Olah et al. 2009; Aresta 2010). Back and the group synthesized highly selective nanostructured CuAu alloy catalyst using an electrochemical deposition process. The selectivity of the catalyst for the conversion of CO_2 into alcohol was enhanced by 18-folds than pure Cu (Back et al. 2015). Similarly, another group investigated the electrocatalytic reduction method for the conversion of CO_2 into CO and formate (Gamba 2018). Further, the utilization of transition metal compounds as a catalyst in the reduction pathway was highlighted by (Kortlever et al. 2015) and (Feng et al. 2017). It was also observed that the output of reduction processes was hindered by the hydrogen evolution side reaction (Kortlever et al. 2015). All these methods of hydrogenation generally formed light hydrocarbons containing less than four carbon numbers. Lower olefinic products were obtained after increasing the selectivity of an iron-based catalyst by incorporating alkali promoters (Guo et al. 2018). The use of promoters generally enhanced CO_2 adsorption and raised the percentage of formation of olefinic products. Moreover, the use of bimetallic catalysts was promoted by various researchers to attain enhanced selectivity and productivity of lower olefin production. The results of the impregnation of potassium on the Al_2O_3-supported Fe-Co bimetallic catalyst instead of the homogenous catalyst were evaluated by (Satthawong et al. 2015). The boosted C2-C4 olefin formation was noted when potassium promoted Fe catalyst was utilized during the hydrogenation of anthropogenic CO_2 in a fixed-bed microreactor (Riedel et al. 2001). Gnanamani et al. synthesized and characterized the couples of Co–Fe bimetallic catalysts for the hydrogenation of anthropogenic CO_2 (Gnanamani et al. 2006).

Furthermore, researchers synthesized higher and linear α-olefins, which were finally converted into high valued commodity chemicals such as biodegradable detergents, lubricants, polymers (polyolefins), and many other expedient compounds for industries (Skupińska 1991). Zhai et al. developed Zn- and Na-amended Fe catalyst using the co-precipitation method and observed that the modified catalyst was more selective toward $C5^+$ alkenes in Fischer–Tropsch synthesis (Zhai et al. 2016). Also, researchers noted that the impregnation of Na on Fe catalyst augmented the surface basicity leading to an increase in the yield of olefins (Wei et al. 2017). Wei et al. reported an efficient and multifunctional Na–Fe_3O_4/HZSM-5 catalyst, which was found to be 78% selective to higher olefinic product (C5-C11) (Wei et al. 2017).

Even though all these conventional CO_2 conversion methods were potential and promising, extreme temperature–pressure conditions, utilization of expensive metal catalysts, and reaction agents' requirement made these less worthy (Jang et al. 2018). This fact made scientists switch over to the development of novel CO_2 conversion methods offering lower energy barriers and high outputs, along with superior selectivity that can attained under mild reaction conditions. This ultimately drove the research toward bio-based materials. The incorporation of biocatalyst instead of metallic catalyst not only upsurge the selectivity but also intensifies the production (Schlager et al. 2017). The approach of applicability of biocatalyst, such as pure enzyme and microorganisms (whole-cell system) was researched thoroughly (Srikanth et al. 2017; Duc Long et al. 2017; Fixena et al. 2016). Among enzymes and microorganisms, the enzymes were appeared more promising due to the non-requirement of nutrients (Schlager et al. 2017).

1.3 CO_2 as a Chemical Feedstock

The transformation of CO_2 gas into commodity valued products such as fuels, basic chemicals, and advanced products (nanomaterials, biodegradable polymers, etc.), is the utmost need for carbon–neutral strategies that put forward CO_2 as a chemical feedstock. In recent years, CO_2 has been conceived as a potential feedstock and preparatory material for the synthesis of derived fuels and essential industrial chemicals. The utilization of anthropogenic CO_2 remarkably decreases the requirement of fossil fuels to some extent while reducing itself in the process. The anthropogenic CO_2 can be directly used in oil recovery, microcapsules formulation, fire-extinguishers, air conditioners, and dry-cleaning. It can also be employed in beverage industries (as a fumigant) and agrochemical industry. Moreover, it can also be used as a solvent in the supercritical state with the higher dissolution capacity (Agarwal et al. 2016). Similarly, anthropogenic CO_2, through biochemical, photochemical, and electrochemical methods, can be processed into renewable fuels and other energy forms (such as syngas and methane) along with the laboratory/industrial chemicals (formic acid, ethanol, dimethyl ethanol, methanol, etc.) (Aresta and Dibenedetto 2007; Bushuyev et al. 2018). The valorization of CO_2 into different products categorized through various methods or pathways is illustrated in Fig. 1. The biocatalytic conversion method has emerged as an efficient ecologically balanced and a low-temperature method for the transformation of CO_2 into useful chemicals with zero or even negative emissions (Marpani et al. 2017). In this regard, many researchers have explored efficient and robust enzymes for the development of a single or multi-enzyme cascade system that can be employed in the

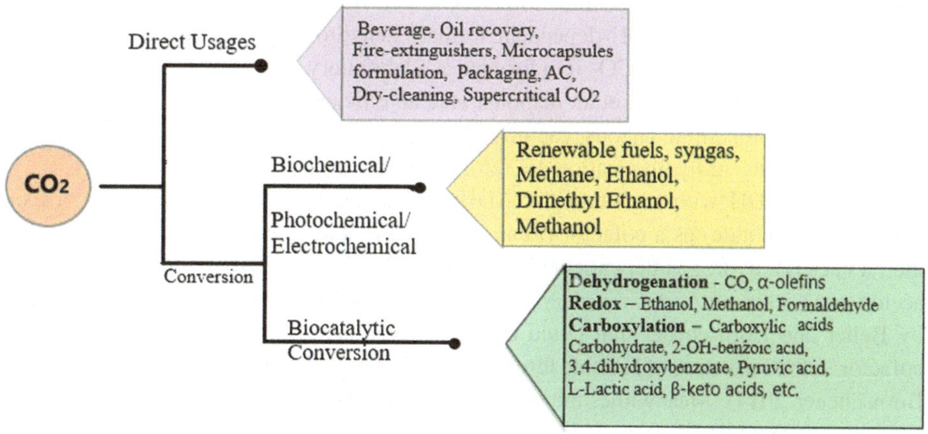

Fig. 1 Sustainable (direct or indirect) utilization of CO_2

successful conversion of CO_2 into commodity valued products. The fascinating properties, such as biodegradability and the superior selectivity of enzymes, make this method significantly promising for the conversion of CO_2.

1.4 CO_2 Conversion with Enzymes

For the CO_2 conversion, enzymatic catalyzed reduction methods have emerged as an attractive alternative to the chemical catalysts in recent years. Various researchers have investigated enzyme-induced CO_2 conversion reactions facilitated through a reduction mechanism. Aresta et al. utilized carbon monoxide dehydrogenases in the reduction of CO_2 to carbon monoxide (Aresta and Dibenedetto 2007). However, this type of conversion method has several downsides, such as the requirement of the sol–gel process, the low solubility of catalyst in water, undesired reactivity of the silica precursor, and the formation of alcohol, which adversely affects the bioactivity of the enzyme (Xu et al. 2006). These downsides can be tackled by encapsulating dehydrogenases in the alginate-silicate composite and employ in the process. Moreover, the encapsulation could accelerate enzymatic activity and stability during storage (Xu et al. 2006).

The most valued product obtained from the enzymatic conversion of CO_2 is formic acid/formate. Formate is capable of hoarding energy engendered from renewable sources such as wind, solar, and hydro, which makes it a potentially beneficial product. It is widely used as a livestock feed additive and fabric-finishing reagent. It also is being used in paper and pulp industries and low-temperature fuel cells (Duc Long et al. 2017). However, the formation of hydrogen gas as a by-product during the conversion of CO_2 into formic acid/formate makes the process inefficient due to the lower selectivity (Jang et al. 2018). Kim et al. reported a new, more efficient electro-enzymatic method with 100% selectivity mainly focused on the NADH-dependent formate dehydrogenase (FDH) conversion of CO_2 into formate at a laboratory scale (Kim et al. 2014). Zhao et al. demonstrated an efficient in situ system for simultaneous capture of CO_2 with H_2 evolution and its re-hydrogenation to reform formic acid catalyzed by FDH with regenerated NADH (nicotinamide adenine dinucleotide) as a cofactor (Zhao et al. 2019). Further, a bacterial hydrogen-dependent CO_2 reductase from the acetogenic bacterium *Acetobacterium woodii* was discovered by Beller and Bornscheuer that could directly use H_2 as a cofactor in the formation of formate through CO_2 (Beller and Bornscheuer 2014). Meanwhile, the enzymatic transition of CO_2 into carboxylic acids and carbohydrate via catabolic pathways have manifested in vitro studies for the production of valued industrial compounds (Duc Long et al. 2017).

In recent years, separate immobilization of the four enzymes on magnetite nanoparticles during cascade conversion of CO_2 into methanol was studied. This study illustrated the distinct functionalities and enzyme characteristics on immobilization of alcohol dehydrogenase, formaldehyde dehydrogenase (FaDH), FDH, and glutamate dehydrogenase on magnetite nanoparticles (Marques Netto et al. 2018). Furthermore, the development of an efficient multi-enzyme system for the transformation of CO_2 into ethanol has promoted the research activities in this domain (Schlager et al. 2017). This type of conversion process requires cofactor to act as an electron and a proton donor for the facilitation of oxidation along with the reduction reaction. The process was demonstrated by using carbon monoxide dehydrogenase, where ferredoxin serves as the electron and proton donor along with NADH as the cofactor (Schlager et al. 2017). Further, the immobilization of enzymes on flat-sheet polymeric membranes was investigated that could preserve enzyme activity (Yadav et al. 2014).

Furthermore, a new route of carbonic anhydrase mediated catalytic conversion of CO_2 into bicarbonate was explored by various researchers (Hong et al. 2015; Yu et al. 2012; Zhang et al. 2013). Carbonic anhydrase mainly balances acid–base concentration in blood by CO_2 to bicarbonate conversion. An efficient potassium carbonate-based absorption process was developed by Zhang et al. using immobilized carbonic anhydrase on nonporous silica-based nanoparticles to retain enzymatic activity (Zhang et al. 2013). Various other pathways, such as photosynthetic carbon reduction cycle into glucose using ATP (Aoshima 2007), carboxylation of epoxides (Volbeda and Fontecilla-Camps 2005), and aromatic transformation (Aresta and Dibenedetto 2002) incorporating decarboxylase enzymes were extensively investigated. Regardless of these investigations, the more improved method for enzymatic conversion of CO_2 with enhanced productivity is yet to be surfaced. Therefore, an in vitro immobilized multi-enzyme approach can be explored to accomplish the sustainable production of valuable commodity chemicals using CO_2.

2 Natural Conversion of CO_2 in Cells

Natural conversion or fixation of CO_2 into organic material is considered as a starting point of biological evolution (Berg et al. 2010; Fuchs 2011). The six primary fixation cycles (reductive citric acid cycle, Calvin cycle, 3-hydroxypropionate cycle, reductive acetyl-CoA route, dicarboxylate/4-hydroxybutyarte cycle, and 3-hydroxypropionate/4-hydroxybutyrate cycle) have been reported to represent CO_2 metabolic processes in cells (Shi et al. 2015; Berg et al. 2010). These cycles provide information about the reaction routes for the in vitro enzymatic conversion of CO_2 (Fig. 2).

The majority of photosynthetic organisms, prokaryotes, algae, and plants extensively use the Calvin cycle for

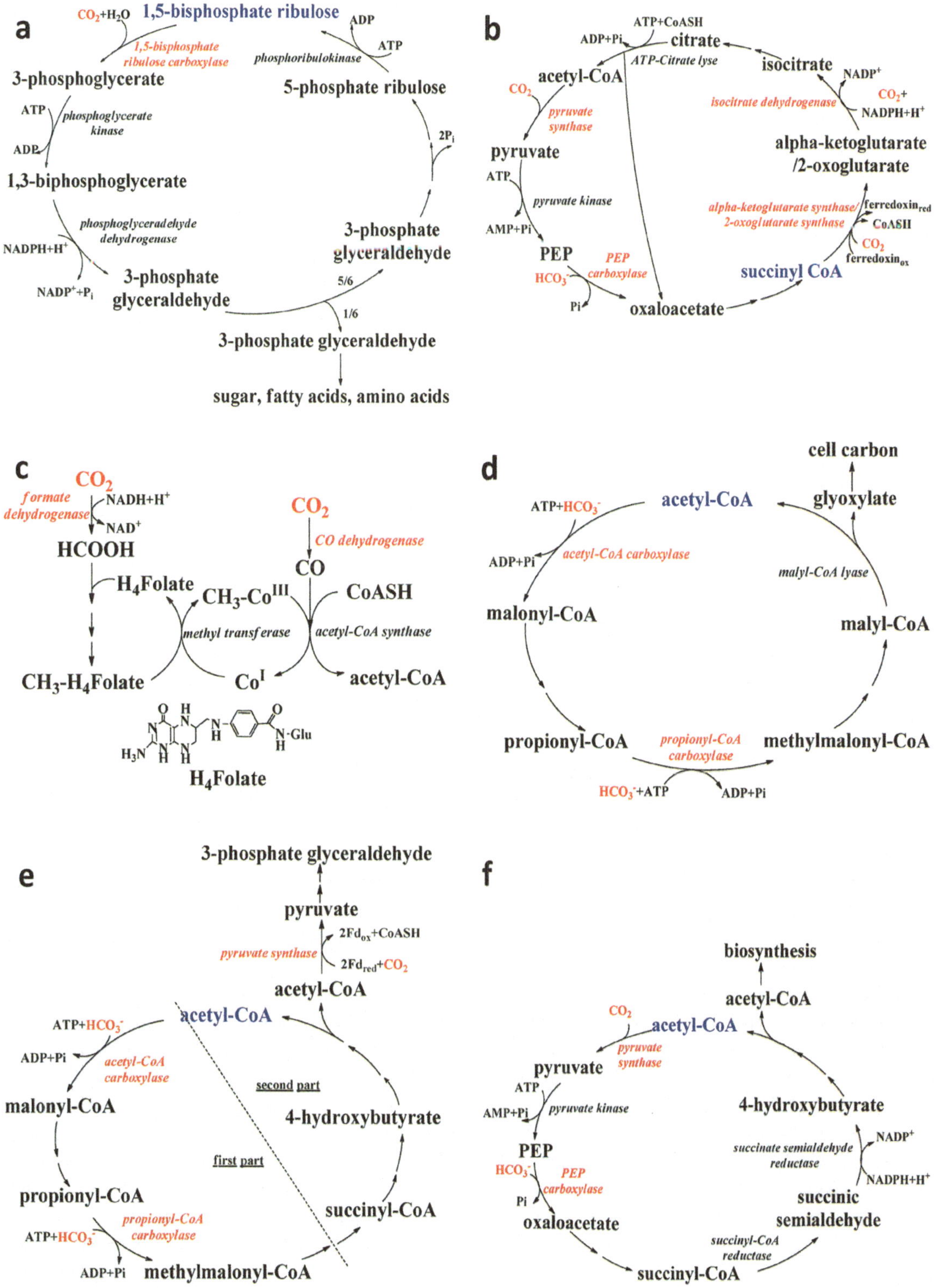

Fig. 2 General illustration of the six major routes of the CO_2 metabolic process in cells. reprinted with permission Shi et al. (2015)

biosynthesis. Calvin cycle is considered as the most prominent natural method of CO_2 conversion (Shi et al. 2015; Ducat and Silver 2012). A number of autotrophic bacteria use the Calvin cycle to fix CO_2 (Shively et al. 1998). The cycle involves three stages (CO_2 fixation, reduction, and CO_2 receptor regeneration) carried by three different enzymes (Fig. 2a). The first step consists of the conversion of CO_2 to 3-phosphoglycerate via 1,5-bisphosphate ribulose bisphosphate carboxylase (RuBisCO) catalyzed reaction between 1,5-bisphosphate ribulose and CO_2. The formed product is then reduced to 3-phosphate glyceraldehyde by phosphoglyceraldehyde dehydrogenase in the second stage. In the last stage, the part of 3-phosphate glyceraldehyde is converted to 5-phosphate ribulose, which subsequently undergoes condensation to form 1,5-bisphosphate ribulose and can be used for a new cycle. In contrast, part of the product gets converted into fatty acids, amino acids, sugar, and many other products.

The reductive citric acid cycle, also called a reverse Krebs cycle or reverse TCA (tricarboxylic acid) cycle, converts the water and CO_2 into carbon compounds (Fig. 2b) (Fuchs 2011; Buchanan and Arnon 1990). This cycle was reported in an anaerobic green phototrophic sulfur bacterium (*Chlorobium limicola),* and also found in certain bacteria that grow by employing sulfate reduction (Gregersen et al. 2011; Hügler et al. 2005). The cycle comprises four different stages, of which the first stage is α- ketoglutarate synthase/2-oxoglutarate synthase catalyzed synthesis of α-ketoglutarate/2-oxoglutarate via reductive carboxylation of succinyl-CoA with CO_2. The next step consists of the conversion of α-ketoglutarate/2-oxoglutarate and CO_2 into isocitrate at the expanse of NADPH (reduced form) and is catalyzed by isocitrate dehydrogenase. The formed isocitrate is then isomerized to citrate, which is then cleaved by adenosine triphosphate (ATP) citrate lyase into acetyl-CoA and oxaloacetate. The acetyl-CoA is then carboxylated with CO_2 by pyruvate synthase to pyruvate. The formed pyruvate is finally converted into succinyl-CoA via a series of enzymatic reactions while forming two key intermediates, *i.e.*, phosphoenolpyruvate and oxaloacetate.

The reductive acetyl-CoA cycle (also named as Wood–Ljungdhal route) is a non-cyclic route and was proposed by Wood and Ljungdahl in 1965 (Fig. 2c) (Ljungdahl and Wood 1969). This cycle is mainly carried by species of methanogenic *Euryarchaeota,* and acetogenic *Eubacteria* (Ragsdale 2008). This cycle can occur in two distinct ways. One route contains the FDH (NADH-dependent) catalyzed conversion of CO_2 into formate, which is subsequently captured by tetrahydrofolate and reduced to methyl-H_4-folate, a methyl group compound. The formed product is then converted to methylated corrinoid protein (CH_3–Co^{III}) via methyl transfer by methyltransferase to cobalt center of the heterodimeric corrinoid iron–sulfur protein (Co^I). An alternative to this route, CO dehydrogenase catalyzes the reduction of CO_2 to CO, and acetyl-CoA synthase accepts a methyl group from CH_3–Co^{III} to form acetyl-CoA via conversion of CoASH, CO, and methyl group.

The 3-hydroxypropionate route consists of two bicarbonate molecules to be fixed and is generally seen in the aerobic phototrophic bacterium, *Chloroflexaceae* sp. (Fig. 2d) (Tabita 2009). Initially, one molecule of carbonate gets converted to malonyl-CoA in the presence of ATP by acetyl-CoA carboxylase, which is then converted to propionyl-CoA via sequential reduction of the terminal carboxylate group. The formed propionyl-CoA is then converted to methylmalonyl-CoA by propionyl-CoA carboxylase via carboxylation reaction with another carbonate molecule. The methylmalonyl-CoA is then transformed to malyl-CoA via isomerization and series of redox reactions, which is then split into acetyl-CoA and glyoxylate for restocking the cycle.

The 3-hydroxypropionate/4-hydroxybutyrate cycle is seen in thermoacidophile archaea of the genera *Acidianus, Metallospharea*, and *Sulfolobus* (Berg et al. 2007; Ramos-Vera et al. 2009). The overall cycle can be divided into two parts (Fig. 2e). In the first part, malonyl-CoA is synthesized via ATP-dependent condensation of acetyl-CoA with bicarbonate by acetyl-CoA carboxylase. This malonyl-CoA is then transformed into propionyl-CoA via a five-enzyme cascade reaction, which is then used by propionyl-CoA carboxylase to form methylmalonyl-CoA in the presence of ATP and bicarbonate. The formed methylmalonyl-CoA then undergoes isomerization to yield succinyl-CoA, which is a starting molecule for the second part. In the second part, two molecules of 4-hydroxybutyrate are formed via succinyl-CoA reduction. The one molecule of 4-hydroxybutyrate is then transformed via a series of cascade reactions into two molecules of acetyl-CoA. The one molecule of acetyl-CoA is then used for the next cycle as a receptor of CO_2, while the second molecule uses CO_2 to form 3-phosphate glyceraldehyde catalyzed by pyruvate synthase.

Anaerobic members of *Desulfurococcales* and *Thermopsroteales* use the dicarboxylate/4-hydroxybutyrate cycle (Ramos-Vera et al. 2009). The cycle starts with the synthesis of pyruvate via carboxylation of acetyl-CoA with CO_2 catalyzed by pyruvate synthase (Fig. 2f). The formed pyruvate is then transformed in phosphoenolpyruvate (PEP), which is subsequently converted to oxaloacetate by carboxylation with bicarbonate. The oxaloacetate is further converted to succinyl-CoA, which later reduced to succinic semialdehyde and then into 4-hydroxybutyrate. The formed 4-hydroxybutyrate undergoes a β-oxidation reaction to give two molecules of acetyl-CoA. One molecule of acetyl-CoA serves as a receptor of CO_2 in the next cycle, while another molecule is used for biosynthesis.

As described in the above-mentioned cycles, the type of enzyme governs the use of either CO_2 or bicarbonate as a source of carbon in the CO_2 conversion or fixation reaction. The cells create a suitable physicochemical environment during the conversion reaction to prevent the enzyme denaturation. Also, the direction of reaction and the reaction rate of CO_2 conversion in the cells are influenced by the presence of a specific type of enzymes (synthases, lyases, oxidoreductases, etc.).

3 Enzymatic Conversion of CO_2 in Cells

3.1 Conversion of CO_2 by a Single Enzyme (*in vitro*)

Various biosynthetic pathways can attain conversion of CO_2 into various industrially important and useful products. Different research teams have developed the idea of performing those reactions in vitro with the help of purified enzymes instead of utilizing classic biosystems by taking inspiration from various biosynthetic routes. Enzyme-facilitated reactions are eco-friendly and cost-effective and have advantages of superior selectivity, specificity, over whole-cell systems (Schlager et al. 2017; Amao 2018; Bolivar et al. 2006). Several enzymes, including FDH, FaDH (FaldDH), carbon monoxide dehydrogenase (CODH), carbonic anhydrase (CA), RuBisCO, and carboxylase, can present CO_2 capturing and reducing capability (Fig. 3) (Schlager et al. 2017; Amao 2018). The following section covers some of the possible mechanisms for CO_2 reduction (in vitro) using these enzymes.

3.1.1 Formate Dehydrogenase

FDH is an oxidoreductive enzyme that can catalyze the reaction of conversion of formic acid to CO_2 and vice versa. To facilitate the reaction (Fig. 4), it requires various cofactors, such as NADH/NAD^+ or cytochrome or quinine (Shi et al. 2015; Amao 2018; Alissandratos and Easton 2015).

These reduction and oxidation pathways were further studied for the possible *in vitro* use of FDH to convert CO_2 into a useful product. To increase the stability, activity, specificity, and reusability of FDH, several immobilization techniques were employed. The distinct research groups have investigated FDH immobilization on various support materials and their application for CO_2 conversion. Yang Lu and the group studied the effect of novel alginate silica (ALG-SiO_2) hybrid gel for FDH immobilization. Researchers were able to maintain around 69% activity of immobilized FDH after ten consecutive cycles with 80% storage stability (up to 1 month). The study reported a 95.6% yield of formic acid at 37 °C and pH 7 (Lu and Z. yi Jiang, S. wei Xu, H. Wu, 2006). In another report, FDH from *Candida boidinii* (CbFDH) along with cofactor liposomes was studied using an external loop airlift bubble column to assess their stability. Liposomes encapsulated CbFDH/cofactor system showed a highly stable system toward the reduction of CO_2 over the free enzyme. CbFDH, in its free form, tends to rapidly deactivate in an airlift bubble column due to interactions with the gas–liquid interface. The fractional NADH oxidation of 23.6% with CO_2 gas was reported (Yoshimoto et al. 2010a,2010b). Instead of using any solid support for the immobilization of enzyme, Min Hoo Kim and the group used cross-linked aggregation (CLEAs) method. In CLEAs, enzymes are precipitated by using

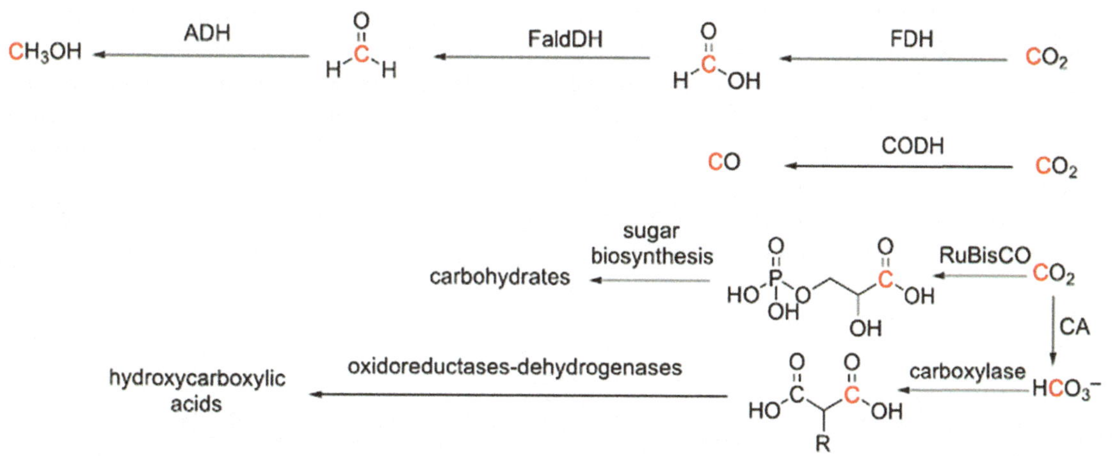

Fig. 3 Biocatalytic routes for the conversion of CO_2 into compounds with carbon in the reduced oxidation states. Creative Commons Attribution License © 2015 Alissandratos and Easton (2015)

(1) NAD-dependent FDH

$$HCO_2H + NAD(P)^+ \rightleftharpoons CO_2 + NADPH + H^+$$

(2) Cytochrome-dependent FDH

$$HCO_2H + 2Cyt\ b_1^{(III)} \rightleftharpoons CO_2 + 2Cyt\ b_1^{(II)} + 2H^+$$

$$HCO_2H + 2Cyt\ c_{553}^{(III)} \rightleftharpoons CO_2 + 2Cyt\ c_{553}^{(II)} + H^+$$

(3) FDH-N

$$HCO_2H + Quinone \rightleftharpoons CO_2 + Quinol + H^+$$

(4) FDH (NAD(+), ferredoxin)

$$2HCO_2H + NAD^+ + 2Fd_{ox} \rightleftharpoons 2CO_2 + NADH + 2Fd_{red}$$

Fig. 4 Catalytic reaction of formic acid oxidation and CO_2 reduction with FDH and redox-substrate. Cyt: cytochrome, Fd: ferredoxin. Copyright ©2018 Elsevier B.V. All rights reserved, reprinted with permission Amao (2018)

various organic solvents, salts, and precipitants such as polyethylene glycol (PEG). For the formation of FDH-CLEAs, saturated ammonium sulfate solution was used with glutaraldehyde or dextran polyaldehyde as a crosslinker. FDH-CLEAs formed by this method showed enhanced thermal and operational stability than the free form of FDH. The study unveiled a 25% CO_2 reduction in the process (Kim et al. 2013). Haiyan Song and the group developed a CO_2 electroreduction system facilitated with NAD-dependent FDH (derived from *Thiobacillus* sp. KNK65MA) for formate production with Cu-nano particles (CuNPs). CuNPs served as a support matrix for FDH immobilization. CuNPs electrodeposited on carbon felt were found to be a good immobilization substrate. Almost 8.5 mM formate production was achieved in the investigation (Song et al. 2019). Yijing Chen and the group used a zirconium-based metal–organic framework (ZrMOF) NU-106 for FDH and electron mediator encapsulation. Metal–organic frameworks have proved their potential in the enhancement of stability of enzyme (Patil and Yadav 2018; Nadar et al. 2019). A new approach of a semiartificial photosynthetic (SAP) system with efficient CO_2 reduction with recyclable enzymes was explored. In this study, a semiartificial photosynthetic system was investigated where NAD^+ was converted to cofactor NADH by a rhodium-based electron mediator. Turnover frequency of formic acid was found to be about 865/h in 24 h (Chen et al. 2020). The recent reports unveiled that the FDH contains electroactive metal at the center, which is NADH-independent, and it can carry out the reaction of CO_2 to formate in the presence of small electric potential. With this regard, Su Keun Kuk and the group reported the synthesis of the electrode, which was NADH free W-containing FDH from *Clostridium ljungdahlii* (ClFDH) immobilized on conductive polyaniline (PANi) hydrogel. PANi-ClFDH electrode showed stable CO_2-to-formate transformation. CO_2-to-formate conversion by PANi-C1FDh electrode was reported with a conversion rate of 1.42 μmol/ h at an overpotential of 40 mV (Kuk et al. 2019). Further, for the degradation of CO_2, a NAD-dependent FDH isolated from *Candida boidinii* (CbFDH) was investigated. However, research unveiled that CbFDH has a very low CO_2 reducing activity due to the interaction of the pure enzyme with the hydrophobic groups. In this regard, NAD-dependent FDH from *Thiobacillus* sp. *KNK65MA* (TsFDH) has been isolated, which enhanced activity and, ultimately, CO_2 reduction (Bolivar et al. 2006; Choe et al. 2015).

3.1.2 Carbonic Anhydrase

Carbonic anhydrase (CA) or carbonate dehydratases is a metalloenzyme capable of catalyzing the reaction between CO_2 and water while producing carbonic acid (Fig. 5). It is an essential enzyme in the living system for hydration and dehydration of CO_2 depending upon the pH of the system (Giri and Pant 2020; Park and Lee 2019). CA has gained more interest in the scientific community since the enzyme has zinc ion at the active site and can rapidly hydrate and dehydrates CO_2, almost 10^7 molecules per second (Jun et al. 2020). Although several attempts have been made to develop a process by using purified CA, a purified form of CA renders a few limitations, including high sensitivity to the environment. Since the catalytic activity of CA mostly depends on the temperature and pH of the system, a change in experimental condition can hamper the activity of the enzyme. Moreover, the activity and stability of the enzyme, along with the cost, have limited the broad adoption. In order to tackle these shortcomings, several attempts have been made to immobilize the enzyme on the solid support since

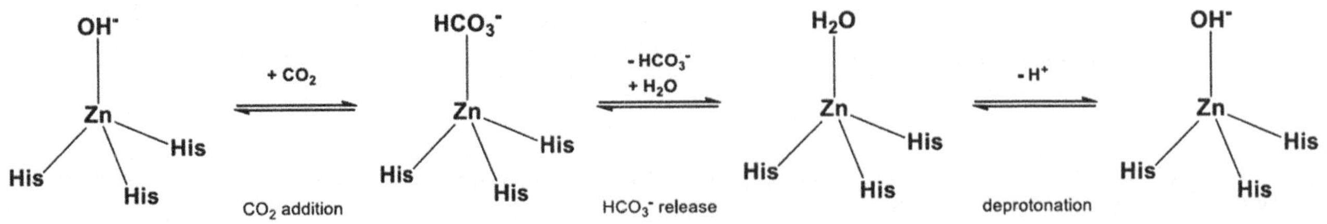

Fig. 5 Proposed mechanism for the catalytic cycle of CA. Creative Commons Attribution License © 2019 Royal Society Park and Lee (2019)

immobilization can efficiently increase the stability, activity, and reusability of enzyme (Giri and Pant 2020; Park and Lee 2019; Bose and Satyanarayana 2017).

To improve the economic productivity, reusability, and stability of CA, Moon et al. performed a whole-cell immobilization method. During this study, CA from *Hydrogenovibrio marinus* was expressed in *E. coli,* and the whole-cell lysate or purified enzyme was further subjected to immobilization in polyurethane foam (PUF). Researchers were able to increase the yield by 16-folds than the free form of enzyme. Also, the activity of enzyme was found to be increased with excellent reusability up to nine cycles. PUF immobilized CA enhanced the CO_2 capture performance of the reactor by 80% (Moon et al. 2020). It was evidently showed that CA was capable of performing a rapid conversion of CO_2 to bicarbonate; however, the stability of CA was significantly low. To increase the stability and activity of the enzyme, Seung-Hyun Jun and the group developed a method of immobilizing CA on electrospun polymer nanofibers. These nanofibers were synthesized by the polymerization of polystyrene and poly(styrene-co-maleic anhydride). CA from bovine erythrocytes (bCA) was immobilized on to nanofibers by employing enzyme precipitation, covalent attachment, and cross-linking under agitation. Researchers were able to maintain the activity of enzyme up to 63% of its initial activity for 868 days. Moreover, enhanced stability and reusability of CA were also observed, along with the enhanced conversion of CO_2 to bicarbonate (Jun et al. 2020). To improve the reusability and recovery of enzyme during the CO_2 absorption into carbonate solutions, Peirce and the group used paramagnetic nanoparticles as solid support for the immobilization of CA. Paramagnetic nanoparticles were synthesized by co-precipitation of Fe^{3+} and Fe^{2+} ions in the 7.5% of ammonium solution. CA was immobilized by covalent bonding on paramagnetic nanoparticles, and the immobilization showed fairly good stability and activity of enzyme with enhanced CO_2 absorption (Peirce et al. 2018). Jingwei Hou and the team used gas–liquid membrane contactor for CO_2 removal. They used immobilized CA on TiO_2-based nanoparticles as a solvent to capture CO_2 from flue gas. During the experimentation, it was observed that the reusability of immobilized CA improved the efficiency of the overall process. The biocatalytic activity of the enzyme remained satisfactory up to ten cycles of reuse (Hou et al. 2016). In another approach, a microbial α-carbonic anhydrase isolated from thermophilic bacteria, *Sulfurihydrogenibium yellowstonense* YO3AOP1 found in hot spring, was used for CO_2 adsorption. The αCA named SspCA reported having a high thermal stability. This SspCA could retain its bioactivity even after heating up to 70 °C. The team used SspCA by immobilizing it on PUF. The immobilized enzyme was used in a three-phase trickle bed reactor to mimic the industrial process. During the process, SspCA showed enhanced efficiency toward the conversion of bicarbonate (Migliardini et al. 2014).

3.1.3 Carbon Monoxide Dehydrogenase

One of the oldest pathways for the fixation of CO_2 is Wood–Ljungdahl pathway. This pathway exists in acetogenic bacteria and methanogenic archaea. In this pathway, CO_2 is utilized to generate methyl group and acetyl Co-A. The two-stage reaction is carried out by the bifunctional enzyme carbon monoxide dehydrogenase/acetyl-CoA synthase (CODH/ACS) complex. Pathway requires hydrogen as an electron donor (Schut et al. 2016; Xavier et al. 2018). The generalized reaction of CO_2 fixation by CODH/ACS is depicted in Fig. 6. Considering the in vivo pathway to fix CO_2 by carbon monoxide dehydrogenase, a number of research groups have carried out significant work on electrolytic and photoreduction of the reversible conversion of CO_2/CO (Shin et al. 2003; Parkin et al. 2007; Woolerton et al. 2010). Recently, Jeong Eun Hyeon and the group developed an effective and stable method by immobilizing the CO-conversion system on the cell surface of CO_2 consuming microorganisms that use CO as sole carbon source.

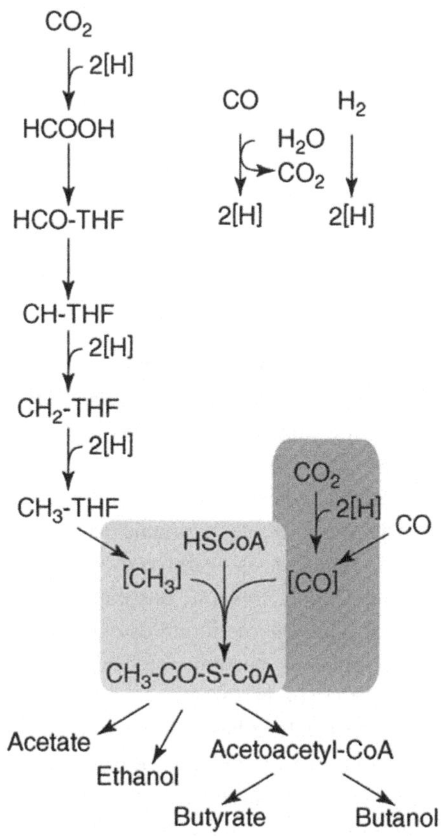

Fig. 6 Schematic representation of the reductive acetyl-CoA pathway of bacteria. Copyright © 2007 Elsevier B.V. All rights reserved, reprinted with permission (Henstra et al. 2007)

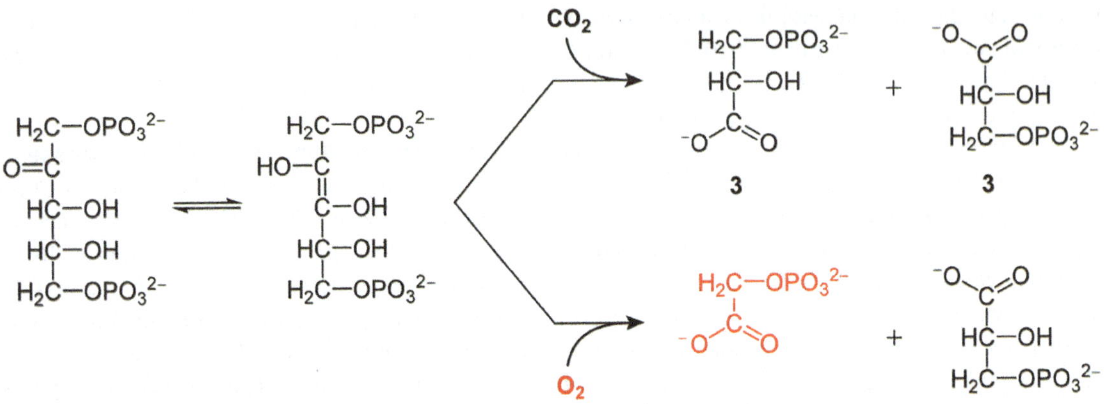

Fig. 7 RuBisCO: CO_2 fixation and oxygenation. Creative Commons Attribution License © 2011 ShareAlike License

They designed a carbon monoxide conversion system (CO-conversion system) by fusing mCbpA scaffolding protein, carbon monoxide dehydrogenase enzyme, and carbon monoxide binding protein. The entire assembly was immobilized on the surface of polyhydroxybutyrate-producing *Ralstonia eutropha*. The entire construction was used as a whole-cell biocatalyst. Carbon monoxide facilitated the interactions between carbon monoxide binding protein and carbon monoxide dehydrogenase, which converted it into CO_2, and generated CO_2 was further used as a carbon source by *Ralstonia eutropha* (Hyeon et al. 2015). In another study, Zhang and the group attempted the photoreduction of CO_2 to CO by employing an aqueous suspension of carbon monoxide dehydrogenase (CODH), silver nanoclusters stabilized by polymethacrylic acid (AgNCs-PMAA), and TiO_2 sealed in CO2 environment. The study unveiled enhanced efficiency of CODH to catalyze the reaction, probably due to the direct electron transfer from photoexcited AGNCs to the CODH (Zhang et al. 2018).

3.1.4 Ribulose-1,5-bisphosphate Carboxylase/Oxygenase (RuBisCO)

Ribulose-1,5-bisphosphate carboxylase/oxygenase or commonly abbreviated as RuBisCO is one of the oldest and abundantly found enzymes for the CO_2 fixation in nature. It is evident that most of the inorganic carbon in nature is converted to biomass is catalyzed by RuBisCO by employing the Calvin Benson Bassham cycle (Erb and Zarzycki 2018). The enzyme found its existence in a variety of life forms, such as bacteria, archaea, and eukaryotes. Nearly 50% of the soluble protein found in plant leaves is RuBisCO. It can catalyze both the carboxylation and oxygenation reaction. In carboxylation, it catalyzes the reaction of the conversion of ribulose-1,5-bisphosphate (RuBP) into two molecules of 3-phosphoglycerate (3PG). It catalyzes the oxygenation reaction to fix the O_2 while producing 2-phosphoglycolate (2PG), which is a toxic compound for the enzymes involved in carbon metabolism (Fig. 7). Despite being a major part of CO_2 fixation in plants, RuBisCO has some limitations, including a slower catalytic rate of reaction. It also requires a post-translation reaction, and the overall reaction is generally dependent on the concentration of CO_2 present in the reaction environment (Erb and Zarzycki 2018,2016; Iñiguez et al. 2020).

Hence, there is a need for improvement in the activity rate of RuBisCO for the CO_2 fixation. To achieve that, various attempts have been made by researchers in the recent past. Mitra et al. attempted to enhance the RuBisCO activity by mixing the enzyme with fungal ergosterol. In this approach, the activity of RuBisCO was assessed by in vivo and in vitro mixing of ergosterol, a metabolite produced by two fungal species, *Aspergillus niger,* and *Fusarium oxysporum.* The study unveiled an enhanced activity of enzyme with improved CO_2 fixation (Mitra et al. 2016). In order to study the cell-free system, Hery and the team embedded the enzyme RuBisCO (isolated from *Ralstonia eutropha*) into conducting polymer [polypyrrole (PPy)] doped with a bulky anion [dodecylbenzenesulfonate (DBS)] scaffold. Though the enzyme activity was found to be slightly less than the native form, the immobilization method could be further modified (Hery et al. 2016). Satagopan and the group conceptualized the immobilization of different enzymes responsible for the reaction of the entire pathway. In this study, they developed a self-assembly nanostructure for two structurally different isolated enzymes from *Rhodospirillum rubrum* and *Ralstonia eutropha.* The study showed the potential use of self-assembly nanostructures to engineer the entire pathway reprinted in the living cell (Satagopan et al. 2017). In 2019, Wunder and the group applied a different approach for the formation of CO_2 fixing liquid droplets resembling the microalgal pyrenoids. Eukaryotes compartmentalize concentrate RuBisCO in a subcellular microcompartment called pyrenoids. Their main function is to facilitate CO_2 fixation by maintaining the CO_2 rich environment around RuBisCO. This comes under an operating part of the carbon-concentrating mechanism (CCM). The research group successfully demonstrated the formation of pyrenoid like matrix by mixing RuBisCO and essential pyrenoid component 1 (EPYC1) protein by liquid–liquid phase separation process (LLPS) (Wunder et al. 2019).

3.2 Conversion of CO_2 by a Multi-Enzyme Cascade *in vitro*

CO_2 fixation is an important and major metabolic activity performed by the living cell to convert it into a useful product. This activity is performed with the help of series of enzymes in different metabolic pathways mainly using (1) Calvin–Benson–Bassham (CBB) cycle, (2) Arnon–Buchanan cycle (reductive TCA cycle), (3) Wood–Ljungdahl pathway (reductive acetyl-CoA pathway), and (4) Acyl-CoA carboxylation pathways in a variety of living cells (Alissandratos and Easton 2015). It is essential to screen new approaches for CO_2 fixation that can be handy to support limited natural CO_2 fixing mechanism. Among many, the utilization of cell-free enzyme or immobilized multiple enzymes cascade system for CO_2 fixation is an up-and-coming area. These types of cascades have several advantages over cellular conversion. In such multi-enzyme cascade systems, the reaction rate of the enzyme can be enhanced, and a large concentration of CO_2 can be processed while reusing enzymes for several cycles. Moreover, these systems offer ease in separation and purification of the end product.

One of the important products from CO_2 conversion is methanol, which has many industrial uses. In 2007, Bilal El-Zahab and the group demonstrated the production of methanol from CO_2 by employing a group of enzymes (FDH, FaDH, and glutamate dehydrogenases). The reaction catalyzed by these enzymes was NADH dependent reaction. To carry out the in vitro reaction, all the enzymes and NADH were covalently immobilized onto the polystyrene microparticle ($\sim$500 nm) (Fig. 8). The reaction was performed by bubbling the CO_2 gas in the reaction mixture, and a promising improvement in cofactor utilization was observed during the production of methanol from CO_2 (El-Zahab et al. 2008). Another approach for methanol production was investigated by constructing an immobilized multi-enzyme cascade. The support matrix for the enzyme entrapment was fabricated as a hybrid microcapsule from catechol modified gelatin (GelC). Enzyme-GelC cascade was synthesized by systematic stepwise entrapment of three enzymes (FAD, FaDH, and alcohol dehydrogenase) via physical and covalent attachment by forming silica nanoparticle layer facilitating enzyme-GelC-Si cascade (Fig. 9). This microreactor showed an enhanced methanol yield (up to 71%), along with high reusability and mechanical strength (Wang et al. 2014).

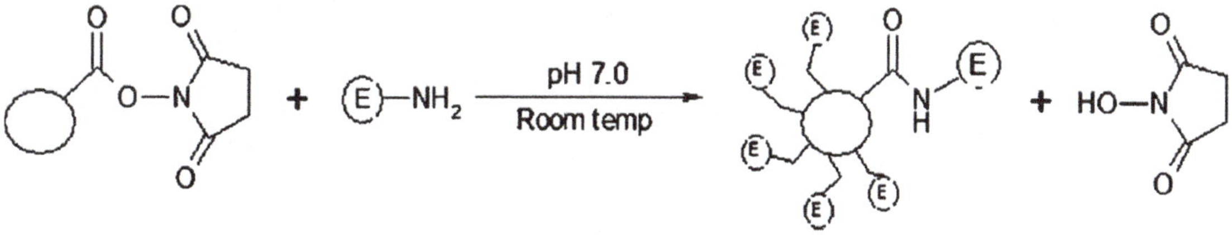

Fig. 8 Chemical route for the attachment of enzymes and cofactor onto polystyrene particles. Copyright © 2007 John Wiley and Sons. All rights reserved, reprinted with permission El-Zahab et al. (2008)

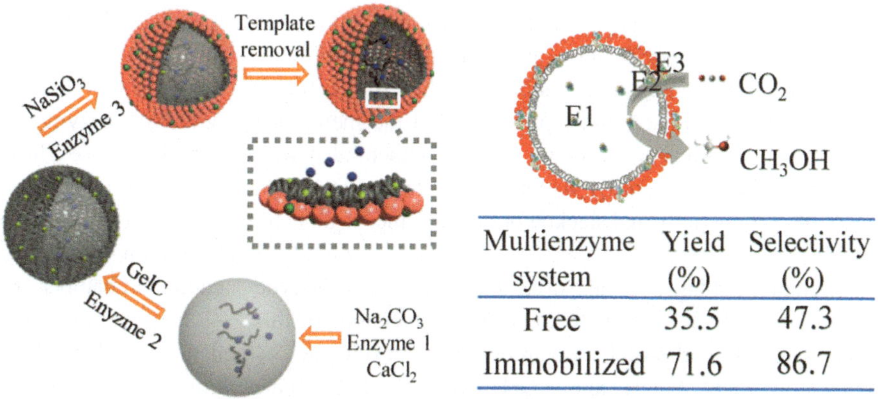

Fig. 9 GelCSi microcapsules with enzymes. Copyright © 2014 American Chemical Society. All rights reserved, reprinted with permission Wang et al. (2014)

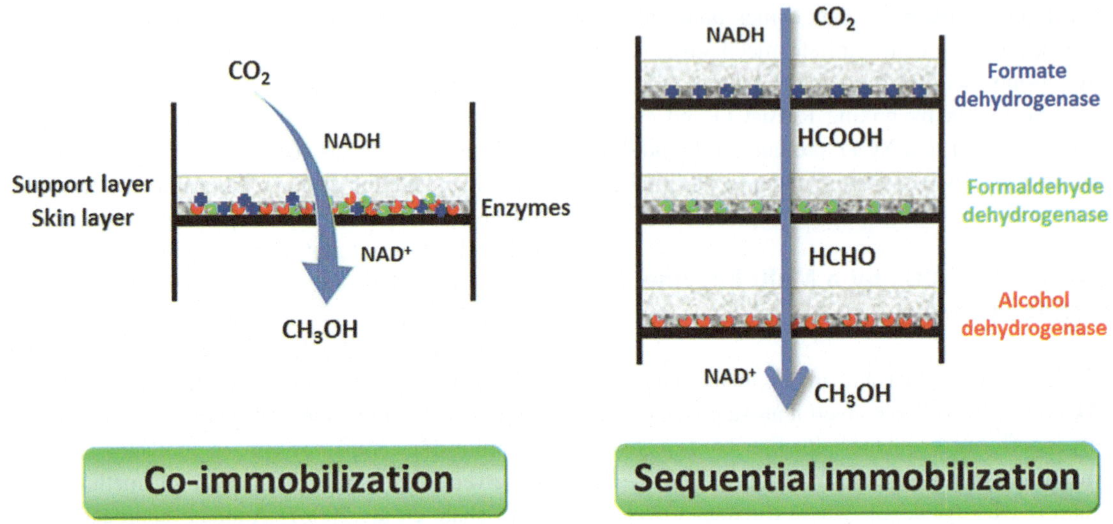

Fig. 10 Co-immobilization and sequential immobilization of enzymes in membranes for methanol production from CO_2. Copyright ©2015 Elsevier B.V. All rights reserved, reprinted with permission Luo et al. (2015)

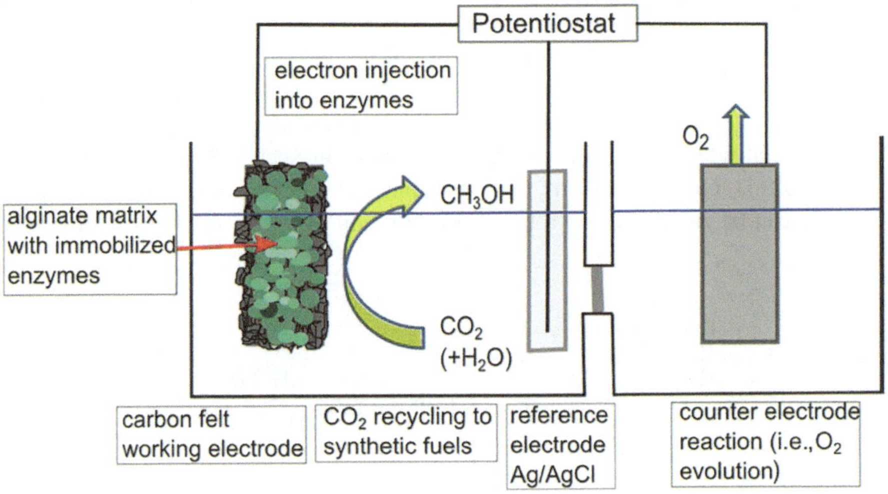

Fig. 11 Electrochemical CO_2 reduction using enzymes. Copyright © 2016 John Wiley and Sons. All rights reserved, reprinted with permission Schlager et al. (2016)

Similarly, another study was investigated by carrying out the co-immobilization of FDH, FaDH, and alcohol dehydrogenase (ADH) on the commercially available ultrafiltration (UF) flat-sheet polymeric membrane by employing a simple pressure filtration method. Two distinct approaches were used; co-immobilization and sequential immobilization. For co-immobilization, a simple method of UF membrane fouling was used under pressure until the membrane was saturated with enzymes (Fig. 10). In sequential immobilization, three different UF membranes were saturated with enzymes sequentially and stacked onto each other (Luo et al. 2015).

Stefanie Schlager et al. used the bio electrocatalytic reduction method for CO_2 to methanol conversion. In this technique, a carbon-felt electrode was developed by depositing FDH, FaDH, and ADH within alginate-silicate hybrid gel (Fig. 11). The reaction carried out in this reactor does not require NADH as cofactor as the electron is directly provided by the carbon felt electrode to the enzyme. The method could successfully produce nearly 0.15 ppm methanol (Schlager et al. 2016). Further, to mimic the natural photosynthetic pathway, Su Keun Kuk and the team developed NADH-regenerating photochemical (PEC) cell with three enzyme cascade system (FDH, FaDH, and alcohol dehydrogenase). In this PEC cascade system, cascade water was used as an electron donor, and NADH was generated, which was further used as a cofactor for the CO_2 to methanol formation employing an enzymes cascade system in the cathodic chamber (Kuk et al. 2017).

For the methanol production from CO_2, a magnetic nanoparticle multi-enzyme cascade was developed by immobilizing ADH, FaDH, and FDH, powered by NAD +

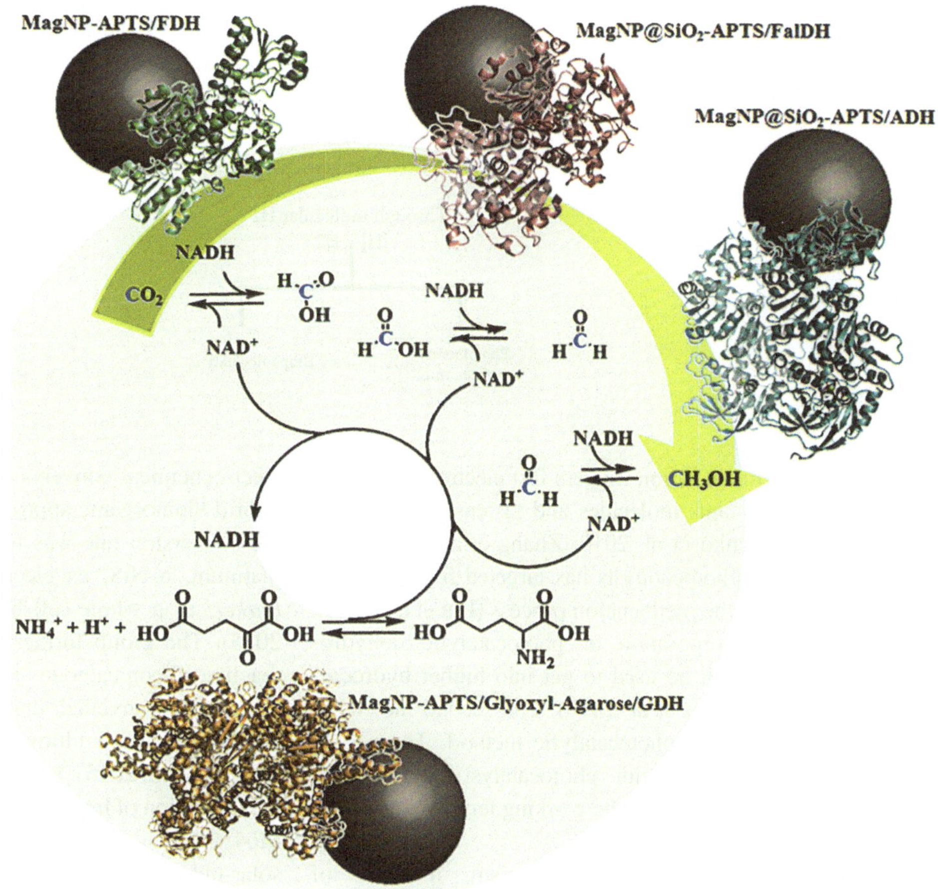

Fig. 12 CO_2-methanol cycle using immobilized FDH, FaDH, and ADH. Creative Commons Attribution License © 2017 Marques Netto et al. (2018)

/NADH and glutamate dehydrogenase (GDH) as the co-enzyme regenerating system (Fig. 12).

3.3 Other Ways *(Photocatalytic CO_2 Methanation)*

The hydrogenation of CO_2 into high calorific valued fuel, i.e., methane, portrays a potential utilization of CO_2. Methane is one such important gas that can be instilled directly into gas pipelines or used as a raw material for the synthesis of various useful chemicals. The solar methanation of CO_2 emerged as a promising catalytic route for large-scale production of beneficial chemicals. Presently, solar methanation of CO_2 is mainly carried out by two approaches. In the first approach, H_2 from renewable sources is used to reduce CO_2 into methane while making the process greener (Aziz et al. 2015). In the second approach, the methanation process is facilitated through water rupturing to produce H_2 that participates in the hydrogenation of CO_2 while converting it into CH_4 and oxygen (O_2) using photon energy (Ulmer et al. 2019).

$$CO_2 + 2H_2O \rightarrow CH_4 + 2O_2 \tag{1}$$

Further, the water-splitting methanation process is facilitated through two steps. In the first step, photocatalytic splitting of water into [H] and O_2 carried out, followed by the transformation of CO_2 into methane.

$$4H_2O \rightarrow 8[H] + 2O_2 \tag{2}$$

$$8[H] + CO_2 \rightarrow 4CH_4 + 2H_2O \tag{3}$$

In this process, a reducing agent [H] is formed which transit to either molecular H_2 or H^+/e^- pair. In the molecular pathway, the hydrogenation or reduction of CO_2 is mainly carried through molecular H_2. However, when the reduction of CO_2 is carried out through H^+/e^- pair, it is known as the proton/electron pathway (Ulmer et al. 2019). Furthermore, the photothermal/plasmonic and bio-photocatalytic methods are perceived through the molecular H_2 mechanism. The proton–electron pair mechanism is used in heterogeneous and homogeneous photoredox methanation methods (Fig. 13). The first molecular pathway is supported by the photothermal/plasmon-driven method. In photothermal and plasmon-driven methanation, the photon-induced excitation of the catalyst increases the catalyst's temperature via the photothermal effect, which then facilitates the hydrogenation process. The inclusion of plasmonic nanometals in

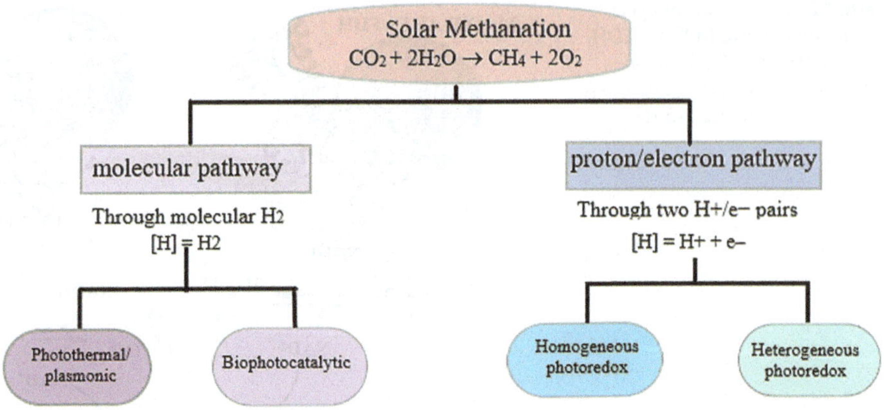

Fig. 13 Diagrammatic representation of the different techniques employed to solar-induced methanation process

plasmon-driven methanation triggers fast electron liberation, which excites reactants molecules and fastens the chemical conversion (Kryachko et al. 2010; Zhang et al. 2017). The use of plasmonic photocatalysts has targeted higher hydrocarbon products in the methanation process (Liu et al. 2018). Otherwise, the incorporation of photocatalytic dehydrogenative coupling can be used to get into higher hydrocarbons from methane (Yi et al. 2017). The second molecular pathway method is bio-photocatalytic method. The use of biomaterials (microbes) with photocatalyst in the photomethanation process lowers the working temperature while attaining increased productivity.

In a proton-electron pair mechanism, the use of photon-activated redox catalyst increased the conversion rate of CO_2 into CH_4 via light-induced redox reactions (Tada et al. 2012). In heterogeneous photoredox catalysis, photon-induced hole carried oxidation of water into O_2 and H^+ with simultaneous photoexcited electronic reduction of CO_2 in CH_4. However, homogeneous photoredox methanation system mainly comprises of photosensitizer molecules for absorbing light. It is a molecular photocatalyst that transferred energy to a sacrificial electron donor, which liberated electrons for the reduction of CO_2 into methane in an aqueous medium (Kuk et al. 2017; Sakimoto et al. 2017). The heterogeneous catalysis method is more beneficial than homogenous catalysis concerning selectivity, separation, conduction, and low production cost (Aziz et al. 2015). Researchers are continuously working to boost selectivity and efficiency of heterogeneous photoredox methanation systems and trying to switch to continuous flow mode from the batch mode system.

Out of these four methods of solar methanation depicted in Fig. 13, high selectivity productivity of the bio-photocatalytic process for the manufacturing of CH_4 and higher hydrocarbons of industry value by employing natural or engineered biomaterials is emerging as a most appealing research field. Nichols and the team successfully demonstrated electrochemical conversion of CO_2 to CH_4 by employing a hybrid bioinorganic approach. In this photo-biomethanation, the conversion rate was found to be increased (86%) using platinum, α-NiS, as electrocatalysts, and *Methanosarcina barkeri* as a whole-cell biocatalytic system (Nichols et al. 2015). The group further successfully incorporated a photoactive silicon cathode along with *Methanosarcina barkeri.* The results unveiled that the 175 mV overpotential was developed, which ultimately increased the conversion rate (Nichols et al. 2015). Wagner et al. successfully demonstrated the utilization of hybrid systems embraced with sulfur-ligated Mo4 + and formylmethanofuran dehydrogenase for solar-induced methanation of CO_2 (Wagner et al. 2016). Fixena et al. reported an efficient hybrid system for bio-photocatalytic CO_2 methanation in a single step using *Rhodopseudomonas palustris* with thiosulfate as an electron donor (Fixena et al. 2016). Van Duc Long et al. reported that the utilization of doubly substituted form of the nitrogenase MoFe protein increased the conversion rate of CO_2 into CH_4 (Duc Long et al. 2017). Sakimoto et al. elaborated on the development of photosynthetic biohybrid systems by incorporating high efficient inorganic photocatalyst with enzyme for reduction of CO_2 (Sakimoto et al. 2017). Similarly, Torella et al. modified the photosynthetic biohybrid systems using biocompatible water-splitting catalyst pair, which produced H_2 and O_2, which was further consumed by a CO_2 reducing bacterium *Cupriavidus necator* (Torella et al. 2015).

4 Industrial Applications

CO_2 valorization comprising the conversion of CO_2 into useful products (fuels/chemicals) is an effective solution for the reduction of greenhouse effects while exploring it as a promising energy storage method. This section comprises a summary of valuable chemicals/fuels, along with several bioproducts obtained from CO_2 (Table 1).

Table 1 Summery of valuable bioproducts obtained from CO_2 using microorganisms

S. no	Microorganism	Reaction condition	Product and yield	Applications	References
1	*Synechococcus* sp. PCC 7002	Artificial seawater medium as a substrate (25% v/v); Duration: 30 d (12:12 h day/night cycle)	Ethanol 0.1 g/L/d	Biofuel	(Kopka et al. 2017)
2	*Citrobacter amalonaticus*	Incorporation of CO_2 and H_2; Sucrose as a substrate	Succinic acid Production: 12 g/L, Productivity: 0.36 g/L/h, Yield: 0.48%	Precursors	(Amulya and Mohan 2019)
3	*Cupriavidus nectar*	2.84 kg CO_2/Kg PHB (Acetic acid as an indirect source of Valeric acid +CO_2)	Bioplastic PHA Production: 0.413 g/L/h	Petroleum-based biodegradable substitutes for plastics	(Garcia-Gonzalez and Wever 2018)
4	*Synechococcus elongates*	BG11, CO_2 (5%); $NaHCO_3$ 0.5 M; 30 °C, 21 d	Isoprene Production: 1.26 g/ L	Building block in synthetic rubber	(Gao et al. 2016)
5	*Anabaena* sp. PCC7120	BG11 medium; 30 °C, 14 d, 120 rpm,	Limonene Production: 3.6 μg/L/O.D./h	Cyclic hydrocarbon for jet fuels	(Halfmann et al. 2014)
6	*Synechococcus elongates* PCC7942	BG11, CO_2 (5%); 100 $\mu E/m^2/s$, 30 °C	Biodiesel (fatty acids ethyl esters) Production: 10 mg/L/O.D	Alternative to petroleum fuels	(Lee et al. 2017)
7	*C. vulgaris*	CO_2 (8%); 47 $\mu E/m^2/s$, 25 °C	Fatty acids Production: 29.5 mg/L	Chemical, food, and energy industry	(Ortiz Montoya et al. 2014)
8	*S. elongates and P. putida*	BG-11, CO_2 (2%); pH 7.5; 30 °C, 220 rpm,16 d	Bioplastic PHA Production: 23.8 mg/L/d	Biodegradable substitutes for (petroleum-based) plastics	(Löwe et al. 2017)
9	*L. delbrueckii*	Intracellular enzyme Temperature: 50 °C Time: 5 min	$CaCO_3$ Production: 183.00 mg $CaCO_3$/ mg protein	Raw material for ceramics, cement, sugar refining, iron, glass, and steel production	(Li et al. 2015)

4.1 Alcohols

The production of alcohol has been studied in order to diminish atmospheric CO_2 concentration by utilizing it as a feedstock. Alcohols are an assuring substitute for petroleum-based fuels (non-renewable) derived from microorganisms and have attained considerable attention in the recent past. Ethanol is one such promising biofuel that can be produced from CO_2 by employing several types of microorganisms. A study investigated a strain of the cyanobacterium *Synechococcus* sp. PCC 7002 grown in a medium of artificial seawater for ethanol production using CO_2 as a feedstock (Kopka et al. 2017). *n*-Propanol and isopropanol are generally used as a solvent for producing propylene as an essential monomer in the plastic industry and can be obtained from CO_2 utilizing numerous microorganisms. Using a precursor (2-ketobutyrate), *n*-propanol was obtained from threonine, and higher alcohols were produced by the addition of acetyl-CoA to this precursor. Similarly, the condensation of acetyl-CoA to acetoacetyl-CoA produced isopropanol and formed acetoacetate. Further, acetoacetate was transformed into acetone by decarboxylation while reducing it to isopropanol (Walther and François 2016). Using an integrated bio-electrochemical system, isopropanol was produced from CO_2 using *R. eutropha* at the production rate of 216 mg/L (Torella et al. 2015).

4.2 Organic Acids

Organic acids such as lactate, butyrate, 3-hydroxypropionate (3 HP), succinate, and acetate play an essential role as a carbon source for the biosynthesis of distinct products, including bioplastics. Organic acids are primarily used as precursors in pharmaceutical, polymer, electronic, food/beverage, and agriculture industries (Gong et al. 2015). Succinic acid from *Citrobacter amalonaticus* using H_2 and CO_2 was produced with a yield of 48.5% (Amulya and Mohan 2019). In another study, the conversion of CO_2 to 3-Hydroxypropionic acid using cyanobacteria was employed as a promising method for the recycling of CO_2 to valuable products. A chemical, 3-Hydroxypropionic acid is mainly used as a precursor to acrylic acid. Moreover, it is also widely employed for the synthesis of polymers. Two pathways (malonyl-CoA and β-alanine dependent) for converting CO_2 into 3 HP using cyanobacteria were investigated. Genes from *S. elongatus,* involved in malonyl-CoA pathway, were enabled. In this study, malonyl-CoA and β-alanine dependent pathways produced 3 HP at a rate of 665 mg/L and 186 mg/L, respectively (Lan et al. 2015).

4.3 Terpenoids

Terpenoids or isoprenoids are organic compounds that have wide applications in the sector of disinfectants, pharmaceuticals, fragrances, colorants, cosmetics, agrochemicals, and flavorings. Many terpenoids are beneficial to humans and widely used as drugs. Artemisinin is one such compound that is commonly utilized as an antimalarial drug (Agtmael et al. 1999). Isoprene is an essential precursor in the synthesis of terpenoids. The cyanobacterial production of isoprene was investigated using CO_2 as a feedstock. Isoprene is a common chemical that is being used in synthetic rubber, lubricants, and adhesives. About 40% of photosynthetically fixed carbon was attained by employing the isoprene biosynthetic pathway carried by *Synechococcus elongates* resulted in a significant production rate (1.26 g/L) of isoprene (Gao et al. 2016).

4.4 Fatty Acids

Fatty acids, including derivatives, are valuable products in the food additives, chemicals, and fuel industry. Conversion of fatty acids generates alcohol that can replace fossil fuel-derived products without additional chemical modifications (Yunus and Jones 2018). For instance, fatty acid ethyl ester (FAEEs) from CO_2 was produced from a model cyanobacterium, *Synechococcus elongatus* PCC 7942 (Lee et al. 2017). Similarly, essential fatty acids abundant biomass, *Chlorella vulgaris* CCAP 211 was grown in a photobioreactor supplied with air, along with CO_2 as the carbon source. Biomass *C. vulgaris* has proved its great value for human health when used as a food additive (Ortiz Montoya et al. 2014).

4.5 Polyhydroxyalkanoates

Polyhydroxyalkanoates (PHAs) are assuring bioplastics that possess some desirable features, including biodegradability. It can be produced in large fermenters by employing *Cupriavidus necator* or a recombinant *Escherichia coli* (Dawes 1988). Further, by using prokaryotic algae (cyanobacteria), PHA can be produced in an alternate way (Chen 2009). A new strain *Cupriavidus eutrophus* B-10646 was used under autotrophic growth conditions for the synthesis of PHA. PHA polymers with various significant fractions of 4-hydroxybutyrate (4HB), 3-hydroxyvalerate (3HV), 3-hydroxybutyrate (3HB), and 3-hydroxyhexanoate (3HHx) monomer units derived from mixed carbon substrate (CO_2 and PHA monomer precursors–hexanoate, valerate, c-butyrolactone) were investigated. The obtained polymers were found with a varied degree of crystallinity and temperature features (Volova et al. 2013).

4.6 Calcium Carbonate

Much attention has been gathered by precipitation of calcium carbonate ($CaCO_3$) induced by bacteria for its role in geological processes (Li et al. 2013). Zinc containing at their active site can facilitate the conversion of CO_2 to bicarbonate ions using a strong biological catalyst, carbonic anhydrase (CA). In order to attain efficient CO_2 sequestration, more than fifty different microbial CAs have been investigated to date. Carbonation of CO_2 by CAs at a low concentration while being economically viable substantially benefits CO_2 sequestration (Sharma et al. 2020). A potent CA was produced by several bacterial strains that reformed CO_2 into bicarbonate ions. In the presence of Ca^{2+} ion, it was further transformed into $CaCO_3$ (Shi et al. 2015; Thakur et al. 2018). Similarly, the CA from *Lactobacillus delbrueckii* was identified and employed to the synthesis of $CaCO_3$ at 50°C (Li et al. 2015). However, a high energy requirement, restricted market area, lack of industrial involvement, and inadequate technology advancements restrict the wide adoption of the approach (Zheng et al. 2017).

5 Summary and Future Prospects

Enzymatic conversion of CO_2 is a technology that can efficiently diminish global warming while producing value-added chemicals and fuels. However, critical scientific approaches and investigations are needed for large-scale enzymatic conversion of CO_2. Currently used enzymes are costly and possess high sensitivities toward the environment on a commercial scale. Further, enzymatic conversion of CO_2 includes reactions that are cofactor dependent, and some cofactors are expensive to express and have confined availability, which critically limits large-scale utilization. Therefore, enzymatic pathways without requiring cofactors need to be developed. Besides, a low reaction rate is another barrier to the use of enzymatic CO_2 conversion in the industry. To overcome these shortcomings, biocatalysts should be robust and more efficient. Also, there is an urgent need for the exploration of new enzymes and reaction operations to advance catalytic performance. Low productivity and the high price of expression and purification methods have limited the wide adoption of enzyme assisted CO_2 conversion. Though attempts have been made to optimize the enzymatic systems, these technological improvements are still inadequate for wider practical adoption. Also, the impurities existing in flue gases need to be considered since it significantly affects the rate of reaction in the process. There is a need for improvements in the absorbent formulation for CO_2 capture, as several chemicals utilized in the process are expensive and demands high energy for regeneration.

References

Agarwal AS, Rode E, Sridhar N, Hill D (2016). Handbook of Climate Change Mitigation and Adaptation. https://doi.org/10.1007/978-1-4614-6431-0

Alissandratos A, Easton CJ (2015) Biocatalysis for the application of CO_2 as a chemical feedstock. Beilstein J Org Chem 11:2370–2387. https://doi.org/10.3762/bjoc.11.259

Amao Y (2018) Formate dehydrogenase for CO_2 utilization and its application. J CO_2 Util 26:623–641. https://doi.org/10.1016/j.jcou.2018.06.022

Amulya K, Mohan SV (2019) Fixation of CO_2, electron donor and redox microenvironment regulate succinic acid production in Citrobacter amalonaticus. Sci Total Environ 695:133838. https://doi.org/10.1016/j.scitotenv.2019.133838

Anderson TR, Hawkins E, Jones PD (2016) CO_2, the greenhouse effect and global warming: from the pioneering work of arrhenius and callendar to today's earth system models. Endeavour 40:178–187. https://doi.org/10.1016/j.endeavour.2016.07.002

Aoshima M (2007) Novel enzyme reactions related to the tricarboxylic acid cycle: phylogenetic/functional implications and biotechnological applications. Appl Microbiol Biotechnol 75:249–255. https://doi.org/10.1007/s00253-007-0893-0

Aresta M (2010) Carbon dioxide as chemical feedstock. Wiley-VCH

Aresta M, Dibenedetto A (2007) Utilisation of CO_2 as a chemical feedstock: opportunities and challenges. J Chem Soc Dalt Trans 2975–2992 https://doi.org/10.1039/b700658f

Aresta M, Dibenedetto A (2002) Development of environmentally friendly syntheses: use of enzymes and biomimetic systems for the direct carboxylation of organic substrates. Rev Mol Biotechnol 90:113–128. https://doi.org/10.1016/S1389-0352(01)00069-1

Aziz MAA, Jalil AA, Triwahyono S, Ahmad A (2015) CO_2 methanation over heterogeneous catalysts: recent progress and future prospects. Green Chem 17:2647–2663. https://doi.org/10.1039/c5gc00119f

Back S, Kim H, Jung Y (2015) Selective heterogeneous CO_2 electroreduction to methanol. ACS Catal 5:965–971. https://doi.org/10.1021/cs501600x

Beller M, Bornscheuer UT (2014) CO_2 fixation through hydrogenation by chemical or enzymatic methods. Angew. Chemie - Int. Ed. 53:4527–4528. https://doi.org/10.1002/anie.201402963

Berg IA, Kockelkorn D, Ramos-Vera WH, Say RF, Zarzycki J, Hügler M, Alber BE, Fuchs G (2010) Autotrophic carbon fixation in archaea. Nat Rev Microbiol 8:447–460. https://doi.org/10.1038/nrmicro2365

Berg IA, Kockelkorn D, Buckel W, Fuchs G (2007) A 3-Hydroxypropionate/4-Hydroxybutyrate autotrophic carbon dioxide assimilation pathway in archaea. Science 318(80-.):1782–1786. https://doi.org/10.1126/science.1149976.

Bolivar JM, Wilson L, Ferrarotti SA, Fernandez-Lafuente R, Guisan JM, Mateo C (2006) Stabilization of a formate dehydrogenase by covalent immobilization on highly activated glyoxyl-agarose supports. Biomacromol 7:669–673. https://doi.org/10.1021/bm050947z

Boot-Handford ME, Abanades JC, Anthony EJ, Blunt MJ, Brandani S, Dowell NM, Fernández JR, Ferrari MC, Gross R, Hallett JP, Haszeldine RS, Heptonstall P, Lyngfelt A, Makuch Z, Mangano E, Porter RTJ, Pourkashanian M, Rochelle GT, Shah N, Yao JG, Fennell PS (2014) Carbon capture and storage update, Energy Environ Sci 7:130–189. https://doi.org/10.1039/c3ee42350f

Bose H, Satyanarayana T (2017) Microbial carbonic anhydrases in biomimetic carbon sequestration for mitigating global warming: prospects and perspectives. Front Microbiol 8:1615. https://doi.org/10.3389/fmicb.2017.01615

Buchanan BB, Arnon DI (1990) A reverse KREBS cycle in photosynthesis: consensus at last. Photosynth Res 24:47–53. https://doi.org/10.1007/BF00032643

Bushuyev OS, De Luna P, Dinh CT, Tao L, Saur G, Van De Lagemaat J, Kelley SO, Sargent EH (2018) What should we make with CO_2 and how can we make it? Joule 2:825–832. https://doi.org/10.1016/j.joule.2017.09.003

Chen GQ (2009) A microbial polyhydroxyalkanoates (PHA) based bio- and materials industry. Chem Soc Rev 38:2434–2446. https://doi.org/10.1039/b812677c

Chen Y, Li P, Zhou J, Buru CT, Aorević L, Li P, Zhang X, Cetin MM, Stoddart JF, Stupp SI, Wasielewski MR, Farha OK (2020) Integration of enzymes and photosensitizers in a hierarchical Mesoporous metal-organic framework for light-driven CO_2 reduction. J Am Chem Soc 142:1768–1773. https://doi.org/10.1021/jacs.9b12828

Choe H, Ha JM, Joo JC, Kim H, Yoon HJ, Kim S, Son SH, Gengan RM, Jeon ST, Chang R, Jung KD, Kim YH, Lee HH (2015) Structural insights into the efficient CO_2-reducing activity of an NAD-dependent formate dehydrogenase from Thiobacillus sp. KNK65MA, Acta Crystallogr. Sect D Biol Crystallogr 71:313–323. https://doi.org/10.1107/S1399004714025474

Dawes EA (1988) Polyhydroxybutyrate: an intriguing biopolymer. Biosci Rep 8:537–547. https://doi.org/10.1007/BF01117332

Long NVD, Lee J, Koo KK, Luis P, Lee M (2017) Recent progress and novel applications in enzymatic conversion of carbon dioxide. Energies 10:1–19. https://doi.org/10.3390/en10040473

Ducat DC, Silver PA (2012) Improving carbon fixation pathways. Curr Opin Chem Biol 16:337–344. https://doi.org/10.1016/j.cbpa.2012.05.002

El-Zahab B, Donnelly D, Wang P (2008) Particle-tethered NADH for production of methanol from CO_2 catalyzed by coimmobilized enzymes. Biotechnol Bioeng 99:508–514. https://doi.org/10.1002/bit.21584

Erb TJ, Zarzycki J (2016) Biochemical and synthetic biology approaches to improve photosynthetic CO_2-fixation. Curr Opin Chem Biol 34:72–79. https://doi.org/10.1016/j.cbpa.2016.06.026

Erb TJ, Zarzycki J (2018) A short history of RubisCO: the rise and fall (?) of Nature's predominant CO_2 fixing enzyme. Curr Opin Biotechnol 49:100–107. https://doi.org/10.1016/j.copbio.2017.07.017

D.M. Feng, Y.P. Zhu, P. Chen, T.Y. Ma, Recent advances in transition-metal-mediated electrocatalytic CO_2 reduction: from homogeneous to heterogeneous systems, Catalysts. 7 (2017). https://doi.org/10.3390/catal7120373

Fixena KR, Zhenga Y, Harrisb DF, Shawb S, Yangb ZY, Deanc DR, Seefeldtb LC, Harwooda CS (2016) Light-driven carbon dioxide reduction to methane by nitrogenase in a photosynthetic bacterium. Proc Natl Acad Sci USA 113:10163–10167. https://doi.org/10.1073/pnas.1611043113

Fuchs G (2011) Alternative pathways of carbon dioxide fixation: insights into the early evolution of life? Annu Rev Microbiol 65:631–658

Gamba I (2018) Biomimetic approach to CO_2 reduction, 2018

Gao X, Gao F, Liu D, Zhang H, Nie X, Yang C (2016) Engineering the methylerythritol phosphate pathway in cyanobacteria for photosynthetic isoprene production from CO_2. Energy Environ Sci 9:1400–1411. https://doi.org/10.1039/c5ee03102h

Garcia-Gonzalez L, De Wever H (2018) Acetic acid as an indirect sink of CO_2 for the synthesis of polyhydroxyalkanoates (PHA): comparison with PHA production processes directly using CO_2 as feedstock. Appl Sci 8:1416. https://doi.org/10.3390/app8091416

Gielen D, Boshell F, Saygin D, Bazilian MD, Wagner N, Gorini R (2019) The role of renewable energy in the global energy transformation. Energy Strateg Rev 24:38–50. https://doi.org/10.1016/j.esr.2019.01.006

Giri A, Pant D (2020) Carbonic anhydrase modification for carbon management. Environ Sci Pollut Res 27:1294–1318. https://doi.org/10.1007/s11356-019-06667-w

Gnanamani A, Jayaprakashvel M, Arulmani M, Sadulla S (2006) Effect of inducers and culturing processes on laccase synthesis in Phanerochaete chrysosporium NCIM 1197 and the constitutive expression of laccase isozymes. Enzyme Microb Technol 38:1017–1021. https://doi.org/10.1016/j.enzmictec.2006.01.004

Gong Z, Shen H, Zhou W, Wang Y, Yang X, Zhao ZK (2015) Efficient conversion of acetate into lipids by the oleaginous yeast Cryptococcus curvatus. Biotechnol Biofuels 8:189. https://doi.org/10.1186/s13068-015-0371-3

Gregersen LH, Bryant DA, Frigaard N-U (2011) Mechanisms and evolution of oxidative sulfur metabolism in green sulfur bacteria, Front Microbiol 2. https://doi.org/10.3389/fmicb.2011.00116

Guo L, Sun J, Ji X, Wei J, Wen Z, Yao R, Xu H, Ge Q (2018) Directly converting carbon dioxide to linear α-olefins on bio-promoted catalysts. Commun Chem 1:1–8. https://doi.org/10.1038/s42004-018-0012-4

Halfmann C, Gu L, Zhou R (2014) Engineering cyanobacteria for the production of a cyclic hydrocarbon fuel from CO_2 and H_2O. Green Chem 16:3175–3185. https://doi.org/10.1039/c3gc42591f

Helseth JM (2012) Biomass with CO_2 capture and storage (Bio-CCS): the way forward for europe. Ind Biotechnol 8:182–188. https://doi.org/10.1089/ind.2012.1529

Henstra AM, Sipma J, Rinzema A, Stams AJ (2007) Microbiology of synthesis gas fermentation for biofuel production. Curr Opin Biotechnol 18:200–206. https://doi.org/10.1016/j.copbio.2007.03.008

Hery TM, Satagopan S, Northcutt RG, Tabita FR, Sundaresan V-B (2016) Polypyrrole membranes as scaffolds for biomolecule immobilization. Smart Mater Struct 25:125033. https://doi.org/10.1088/0964-1726/25/12/125033

Hong SG, Jeon H, Kim HS, Jun SH, Jin E, Kim J (2015) One-pot enzymatic conversion of carbon dioxide and utilization for improved microbial growth. Environ Sci Technol 49:4466–4472. https://doi.org/10.1021/es505143f

Hou J, Zulkifli MY, Mohammad M, Zhang Y, Razmjou A, Chen V (2016) Biocatalytic gas-liquid membrane contactors for CO_2 hydration with immobilized carbonic anhydrase. J Memb Sci 520:303–313. https://doi.org/10.1016/j.memsci.2016.07.003

Hügler M, Wirsen CO, Fuchs G, Taylor CD, Sievert SM (2005) Evidence for Autotrophic CO_2 fixation via the reductive tricarboxylic acid cycle by members of the ε subdivision of proteobacteria. J Bacteriol 187:3020–3027. https://doi.org/10.1128/JB.187.9.3020-3027.2005

Hyeon JE, Kim SW, Park C, Han SO (2015) Efficient biological conversion of carbon monoxide (CO) to carbon dioxide (CO_2) and for utilization in bioplastic production by Ralstonia eutropha through the display of an enzyme complex on the cell surface. Chem Commun 51:10202–10205. https://doi.org/10.1039/c5cc00832h

Iñiguez C, Capó-Bauçà S, Niinemets Ü, Stoll H, Aguiló-Nicolau P, Galmés J (2020) Evolutionary trends in RuBisCO kinetics and their co-evolution with CO_2 concentrating mechanisms. Plant J 101:897–918. https://doi.org/10.1111/tpj.14643

Jang J, Jeon BW, Kim YH (2018) Bioelectrochemical conversion of CO_2 to value added product formate using engineered Methylobacterium extorquens. Sci Rep 8:1–7. https://doi.org/10.1038/s41598-018-23924-z

Jun SH, Yang J, Jeon H, Kim HS, Pack SP, Jin E, Kim J (2020) Stabilized and immobilized carbonic anhydrase on Electrospun Nanofibers for enzymatic CO_2 conversion and utilization in expedited microalgal growth. Environ Sci Technol 54:1223–1231. https://doi.org/10.1021/acs.est.9b05284

Kim MH, Park S, Kim YH, Won K, Lee SH (2013) Immobilization of formate dehydrogenase from Candida boidinii through cross-linked enzyme aggregates. J Mol Catal B Enzym 97:209–214. https://doi.org/10.1016/j.molcatb.2013.08.020

Kim S, Kim MK, Lee SH, Yoon S, Jung KD (2014) Conversion of CO_2 to formate in an electroenzymatic cell using Candida boidinii formate dehydrogenase. J Mol Catal B Enzym 102:9–15. https://doi.org/10.1016/j.molcatb.2014.01.007

Kopka J, Schmidt S, Dethloff F, Pade N, Berendt S, Schottkowski M, Martin N, Dühring U, Kuchmina E, Enke H, Kramer D, Wilde A, Hagemann M, Friedrich A (2017) Systems analysis of ethanol production in the genetically engineered cyanobacterium Synechococcus sp. PCC 7002. Biotechnol Biofuels 10:56. https://doi.org/10.1186/s13068-017-0741-0

Kortlever R, Shen J, Schouten KJP, Calle-Vallejo F, Koper MTM (2015) Catalysts and reaction pathways for the electrochemical reduction of carbon dioxide. J Phys Chem Lett 6:4073–4082. https://doi.org/10.1021/acs.jpclett.5b01559

Kryachko IA, Tyutyunnikov SI, Shalyapin VN (2010) Calculation of thermal process during the absorption of microwave impulse energy by a spherical nanocluster in liquid medium. Phys Part Nucl Lett 7:273–280. https://doi.org/10.1134/S1547477110040059

Kuk SK, Singh RK, Nam DH, Singh R, Lee J-K, Park CB (2017) Photoelectrochemical reduction of carbon dioxide to methanol through a highly efficient enzyme cascade. Angew. Chemie Int. Ed. 56:3827–3832. https://doi.org/10.1002/anie.201611379

Kuk SK, Gopinath K, Singh RK, Kim TD, Lee Y, Choi WS, Lee JK, Park CB (2019) NADH-Free Electroenzymatic reduction of CO_2 by conductive hydrogel-conjugated formate dehydrogenase. ACS Catal 9:5584–5589. https://doi.org/10.1021/acscatal.9b00127

Lan EI, Chuang DS, Shen CR, Lee AM, Ro SY, Liao JC (2015) Metabolic engineering of cyanobacteria for photosynthetic 3-hydroxypropionic acid production from CO_2 using Synechococcus elongatus PCC 7942. Metab Eng 31:163–170. https://doi.org/10.1016/j.ymben.2015.08.002

Lee HJ, Choi J, Lee SM, Um Y, Sim SJ, Kim Y, Woo HM (2017) Photosynthetic CO_2 conversion to fatty acid ethyl esters (FAEEs) using engineered cyanobacteria. J Agric Food Chem 65:1087–1092. https://doi.org/10.1021/acs.jafc.7b00002

Li W, Chen WS, Zhou PP, Zhu SL, Yu LJ (2013) Influence of initial calcium ion concentration on the precipitation and crystal morphology of calcium carbonate induced by bacterial carbonic anhydrase. Chem Eng J 218:65–72. https://doi.org/10.1016/j.cej.2012.12.034

Li CX, Jiang XC, Qiu YJ, Xu JH (2015) Identification of a new thermostable and alkali-tolerant α-carbonic anhydrase from Lactobacillus delbrueckii as a biocatalyst for CO_2 biomineralization. Bioresour. Bioprocess. 2:1–9. https://doi.org/10.1186/s40643-015-0074-4

Liu L, Puga AV, Cored J, Concepción P, Pérez-Dieste V, García H, Corma A (2018) Sunlight-assisted hydrogenation of CO_2 into ethanol and C^{2+} hydrocarbons by sodium-promoted Co@C nanocomposites. Appl Catal B Environ 235:186–196. https://doi.org/10.1016/j.apcatb.2018.04.060

Ljungdahl LG, Wood HG (1969) Total synthesis of acetate from CO_2 by heterotrophic bacteria. Annu Rev Microbiol 23:515–538. https://doi.org/10.1146/annurev.mi.23.100169.002503

Löwe H, Hobmeier K, Moos M, Kremling A, Pflüger-Grau K (2017) Photoautotrophic production of polyhydroxyalkanoates in a synthetic mixed culture of Synechococcus elongatus cscB and Pseudomonas putida cscAB, Biotechnol Biofuels 10:190. https://doi.org/10.1186/s13068-017-0875-0

Lu Y, Jiang ZY, Xu SW, Wu H (2006) Efficient conversion of CO_2 to formic acid by formate dehydrogenase immobilized in a novel alginate-silica hybrid gel, Catal Today 115:263–268. https://doi.org/10.1016/j.cattod.2006.02.056

Luo J, Meyer AS, Mateiu RV, Pinelo M (2015) Cascade catalysis in membranes with enzyme immobilization for multi-enzymatic conversion of CO_2 to methanol. N Biotechnol 32:319–327. https://doi.org/10.1016/j.nbt.2015.02.006

Marpani F, Pinelo M, Meyer AS (2017) Enzymatic conversion of CO_2 to CH_3OH via reverse dehydrogenase cascade biocatalysis: Quantitative comparison of efficiencies of immobilized enzyme systems. Biochem Eng J 127:217–228. https://doi.org/10.1016/j.bej.2017.08.011

Marques Netto CGC, Andrade LH, Toma HE (2018) Carbon dioxide/methanol conversion cycle based on cascade enzymatic reactions supported on superparamagnetic nanoparticles, An Acad Bras Cienc 90(2018):593–606. https://doi.org/10.1590/0001-3765201720170330

Migliardini F, De Luca V, Carginale V, Rossi M, Corbo P, Supuran CT, Capasso C (2014) Biomimetic CO_2 capture using a highly thermostable bacterial α-carbonic anhydrase immobilized on a polyurethane foam. J Enzyme Inhib Med Chem 29:146–150. https://doi.org/10.3109/14756366.2012.761608

Mitra J, Narad P, Paul PK (2016) In silico, in vitro and in vivo approach in understanding the functional relationship between ergosterol and Rubisco. Photosynthetica 54:517–523. https://doi.org/10.1007/s11099-016-0211-0

Moon H, Kim S, Jo BH, Cha HJ (2020) Immobilization of genetically engineered whole-cell biocatalysts with periplasmic carbonic anhydrase in polyurethane foam for enzymatic CO_2 capture and utilization. J CO_2 Util 39:101172. https://doi.org/10.1016/j.jcou.2020.101172

Nadar SS, Vaidya L, Maurya S, Rathod VK (2019) Polysaccharide based metal organic frameworks (polysaccharide–MOF): A review. Coord Chem Rev 396:1–21. https://doi.org/10.1016/j.ccr.2019.05.011

Nichols EM, Gallagher JJ, Liu C, Su Y, Resasco J, Yu Y, Sun Y, Yang P, Chang MCY, Chang CJ (2015) Hybrid bioinorganic approach to solar-tochemical conversion. Proc Natl Acad Sci U S A 112:11461–11466. https://doi.org/10.1073/pnas.1508075112

Olah GA, Goeppert A, Prakash GKS (2009) Beyond oil and gas : the methanol economy. Wiley-VCH

Olah GA, Goeppert A, Prakash GKS (2009) Beyond Oil and gas: the methanol economy: 2nd Edn, Wiley-VCH https://doi.org/10.1002/9783527627806

Ortiz Montoya EY, Casazza AA, Aliakbarian B, Perego P, Converti A, De Carvalho JCM (2014) Production of Chlorella vulgaris as a source of essential fatty acids in a tubular photobioreactor continuously fed with air enriched with CO_2 at different concentrations, Biotechnol Prog 30:916–922. https://doi.org/10.1002/btpr.1885

Park D, Lee MS (2019) Kinetic study of catalytic CO_2 hydration by metal-substituted biomimetic carbonic anhydrase model complexes. R Soc Open Sci 6:190407. https://doi.org/10.1098/rsos.190407

Parkin A, Seravalli J, Vincent KA, Ragsdale SW, Armstrong FA (2007) Rapid and efficient electrocatalytic CO_2/CO interconversions by Carboxydothermus hydrogenoformans CO dehydrogenase I on an electrode. J Am Chem Soc 129:10328–10329. https://doi.org/10.1021/ja073643o

Patil PD, Yadav GD (2018) Rapid In Situ Encapsulation of Laccase into metal-organic framework support (ZIF-8) under biocompatible conditions. ChemistrySelect 3:4669–4675. https://doi.org/10.1002/slct.201702852

Peirce S, Russo ME, Perfetto R, Capasso C, Rossi M, Fernandez-Lafuente R, Salatino P, Marzocchella A (2018) Kinetic characterization of carbonic anhydrase immobilized on magnetic nanoparticles as biocatalyst for CO_2 capture. Biochem Eng J 138:1–11. https://doi.org/10.1016/j.bej.2018.06.017

Ragsdale SW (2008) Enzymology of the wood-ljungdahl pathway of acetogenesis. Ann N Y Acad Sci 1125:129–136. https://doi.org/10.1196/annals.1419.015

Ramos-Vera WH, Berg IA, Fuchs G (2009) Autotrophic carbon dioxide assimilation in Thermoproteales revisited. J Bacteriol 191:4286–4297. https://doi.org/10.1128/JB.00145-09

Rao NS, Rao PS, Dinakar A, Rao PVN, Marghade D (2017) Fluoride occurrence in the groundwater in a coastal region of Andhra Pradesh, India. Appl Water Sci 7:1467–1478. https://doi.org/10.1007/s13201-015-0338-3

Riedel T, Schaub G, Jun KW, Lee KW (2001) Kinetics of CO_2 hydrogenation on a K-promoted Fe catalyst. Ind Eng Chem Res 40:1355–1363. https://doi.org/10.1021/ie000084k

Sakimoto KK, Kornienko N, Yang P (2017) Cyborgian material design for solar fuel production: the emerging photosynthetic biohybrid systems. Acc Chem Res 50:476–481. https://doi.org/10.1021/acs.accounts.6b00483

Satagopan S, Sun Y, Parquette JR, Tabita FR (2017) Synthetic CO_2-fixation enzyme cascades immobilized on self-assembled nanostructures that enhance CO_2/O_2 selectivity of RubisCO. Biotechnol Biofuels 10:175. https://doi.org/10.1186/s13068-017-0861-6

Satthawong R, Koizumi N, Song C, Prasassarakich P (2015) Light olefin synthesis from CO_2 hydrogenation over K-promoted Fe-Co bimetallic catalysts. Catal Today 251:34–40. https://doi.org/10.1016/j.cattod.2015.01.011

Schlager S, Dumitru LM, Haberbauer M, Fuchsbauer A, Neugebauer H, Hiemetsberger D, Wagner A, Portenkirchner E, Sariciftci NS (2016) Electrochemical reduction of carbon dioxide to methanol by direct injection of electrons into immobilized enzymes on a modified electrode. Chemsuschem 9:631–635. https://doi.org/10.1002/cssc.201501496

Schlager S, Dibenedetto A, Aresta M, Apaydin DH, Dumitru LM, Neugebauer H, Sariciftci NS (2017) Biocatalytic and bioelectrocatalytic approaches for the reduction of carbon dioxide using enzymes. Energy Technol 5:812–821. https://doi.org/10.1002/ente.201600610

Schut GJ, Lipscomb GL, Nguyen DMN, Kelly RM, Adams MWW (2016) Heterologous production of an energy-conserving carbon monoxide dehydrogenase complex in the Hyperthermophile Pyrococcus furiosus. Front Microbiol 7:29. https://doi.org/10.3389/fmicb.2016.00029

Sharma T, Sharma S, Kamyab H, Kumar A (2020) Energizing the CO_2 utilization by chemo-enzymatic approaches and potentiality of carbonic anhydrases: a review. J Clean Prod 247:119138. https://doi.org/10.1016/j.jclepro.2019.119138

Shi J, Jiang Y, Jiang Z, Wang X, Wang X, Zhang S, Han P, Yang C (2015) Enzymatic conversion of carbon dioxide. Chem Soc Rev 44:5981–6000. https://doi.org/10.1039/C5CS00182J

Shin W, Lee SH, Shin JW, Lee SP, Kim Y (2003) Highly Selective Electrocatalytic Conversion of CO_2 to CO at -0.57 V (NHE) by carbon monoxide dehydrogenase from Moorella thermoacetica. J Am Chem Soc 125:14688–14689. https://doi.org/10.1021/ja037370i

Shively JM, van Keulen G, Meijer WG (1998) Something from almost nothing: carbon dioxide fixation in chemoautotrophs. Annu Rev Microbiol 52:191–230. https://doi.org/10.1146/annurev.micro.52.1.191

Skupińska J (1991) Oligomerization of α-Olefins to higher oligomers. Chem Rev 91:613–648. https://doi.org/10.1021/cr00004a007

Song H, Ma C, Liu P, You C, Lin J, Zhu Z (2019) A hybrid CO_2 electroreduction system mediated by enzyme-cofactor conjugates coupled with Cu nanoparticle-catalyzed cofactor regeneration. J CO_2 Util 34:568–575. https://doi.org/10.1016/j.jcou.2019.08.007

Srikanth S, Alvarez-Gallego Y, Vanbroekhoven K, Pant D (2017) Enzymatic electrosynthesis of formic acid through carbon dioxide reduction in a bioelectrochemical system: effect of immobilization and carbonic anhydrase addition. ChemPhysChem 18:3174–3181. https://doi.org/10.1002/cphc.201700017

Tabita FR (2009) The hydroxypropionate pathway of CO_2 fixation: Fait accompli: Fig. 1. Proc Natl Acad Sci 106:21015–21016. https://doi.org/10.1073/pnas.0912486107

Tada S, Shimizu T, Kameyama H, Haneda T, Kikuchi R (2012) Ni/CeO_2 catalysts with high CO_2 methanation activity and high CH_4 selectivity at low temperatures. Int J Hydrogen Energy 37:5527–5531. https://doi.org/10.1016/j.ijhydene.2011.12.122

Thakur IS, Kumar M, Varjani SJ, Wu Y, Gnansounou E, Ravindran S (2018) Sequestration and utilization of carbon dioxide by chemical and biological methods for biofuels and biomaterials by chemoautotrophs: opportunities and challenges. Bioresour Technol 256:478–490. https://doi.org/10.1016/j.biortech.2018.02.039

Torella JP, Gagliardi CJ, Chen JS, Bediako DK, Colón B, Way JC, Silver PA, Nocera DG (2015) Efficient solar-to-fuels production from a hybrid microbial-water-splitting catalyst system. Proc Natl Acad Sci USA 112:2337–2342. https://doi.org/10.1073/pnas.1424872112

Ulmer U, Dingle T, Duchesne PN, Morris RH, Tavasoli A, Wood T, Ozin GA (2019) Fundamentals and applications of photocatalytic CO_2 methanation. Nat Commun 10:1–12. https://doi.org/10.1038/s41467-019-10996-2

Van Agtmael MA, Eggelte TA, Van Boxtel CJ (1999) Artemisinin drugs in the treatment of malaria: from medicinal herb to registered medication. Trends Pharmacol Sci 20:199–205. https://doi.org/10.1016/S0165-6147(99)01302-4

Volbeda A, Fontecilla-Camps JC (2005) Structural bases for the catalytic mechanism of Ni-containing carbon monoxide dehydrogenases, Dalt Trans 3443–3450. https://doi.org/10.1039/b508403b.

Volova TG, Kiselev EG, Shishatskaya EI, Zhila NO, Boyandin AN, Syrvacheva DA, Vinogradova ON, Kalacheva GS, Vasiliev AD, Peterson IV (2013) Cell growth and accumulation of polyhydroxyalkanoates from CO_2 and H_2 of a hydrogen-oxidizing bacterium, Cupriavidus eutrophus B-10646. Bioresour Technol 146:215–222. https://doi.org/10.1016/j.biortech.2013.07.070

Wagner T, Ermler U, Shima S (2016) The methanogenic CO_2 reducing-and-fixing enzyme is bifunctional and contains 46 [4Fe-4S] clusters. Science 354(80-.):114–117. https://doi.org/10.1126/science.aaf9284

Walther T, François JM (2016) Microbial production of propanol. Biotechnol Adv 34:984–996. https://doi.org/10.1016/j.biotechadv.2016.05.011

Wang X, Li Z, Shi J, Wu H, Jiang Z, Zhang W, Song X, Ai Q (2014) Bioinspired approach to multi-enzyme cascade system construction for efficient carbon dioxide reduction. ACS Catal 4:962–972. https://doi.org/10.1021/cs401096c

Wei J, Ge Q, Yao R, Wen Z, Fang C, Guo L, Xu H, Sun J (2017) Directly converting CO_2 into a gasoline fuel. Nat Commun 8:1–8. https://doi.org/10.1038/ncomms15174

Woolerton TW, Sheard S, Reisner E, Pierce E, Ragsdale SW, Armstrong FA (2010) Efficient and clean photoreduction of CO_2 to CO by enzyme-modified TiO_2 nanoparticles using visible light. J Am Chem Soc 132:2132–2133

Wunder T, Oh ZG, Mueller-Cajar O (2019) CO_2-fixing liquid droplets: towards a dissection of the microalgal pyrenoid. Traffic 20:380–389. https://doi.org/10.1111/tra.12650

Xavier JC, Preiner M, Martin WF (2018) Something special about CO-dependent CO_2 fixation. FEBS J 285:4181–4195. https://doi.org/10.1111/febs.14664

Xu SW, Lu Y, Li J, Jiang ZY, Wu H (2006) Efficient conversion of CO_2 to methanol catalyzed by three dehydrogenases co-encapsulated in an alginate-silica (ALG-SiO_2) hybrid gel. Ind Eng Chem Res 45:4567–4573. https://doi.org/10.1021/ie051407l

Yadav RK, Oh GH, Park NJ, Kumar A, Kong KJ, Baeg JO (2014) Highly selective solar-driven methanol from CO_2 by a photocatalyst/biocatalyst integrated system. J Am Chem Soc 136:16728–16731. https://doi.org/10.1021/ja509650r

Yi H, Niu L, Song C, Li Y, Dou B, Singh AK, Lei A (2017) Photocatalytic dehydrogenative cross-coupling of alkenes with alcohols or azoles without external oxidant. Angew Chemie Int Educ 56:1120–1124. https://doi.org/10.1002/anie.201609274

Yoshimoto M, Yamasaki R, Nakao M, Yamashita T (2010b) Stabilization of formate dehydrogenase from Candida boidinii through liposome-assisted complexation with cofactors. Enzyme Microb Technol 46:588–593. https://doi.org/10.1016/j.enzmictec.2010.02.012

Yoshimoto M, Yamashita T, Yamashiro T (2010) Stability and reactivity of liposome-encapsulated formate dehydrogenase and cofactor system in carbon dioxide gas-liquid flow, Biotechnol Prog 26. https://doi.org/10.1002/btpr.409.

Yu Y, Chen B, Qi W, Li X, Shin Y, Lei C, Liu J (2012) Enzymatic conversion of CO 2 to bicarbonate in functionalized mesoporous silica. Microporous Mesoporous Mater 153:166–170. https://doi.org/10.1016/j.micromeso.2011.12.005

Yunus IS, Jones PR (2018) Photosynthesis-dependent biosynthesis of medium chain-length fatty acids and alcohols. Metab Eng 49:59–68. https://doi.org/10.1016/j.ymben.2018.07.015

Zhai P, Xu C, Gao R, Liu X, Li M, Li W, Fu X, Jia C, Xie J, Zhao M, Wang X, Li YW, Zhang Q, Wen XD, Ma D (2016) Highly tunable selectivity for syngas-derived alkenes over zinc and sodium-modulated Fe_5C_2 Catalyst, Angew Chemie Int Educ 55:9902–9907. https://doi.org/10.1002/anie.201603556

Zhang S, Lu H, Lu Y (2013) Enhanced stability and chemical resistance of a new nanoscale biocatalyst for accelerating CO_2 absorption into a carbonate solution. Environ Sci Technol 47:13882–13888. https://doi.org/10.1021/es4031744

Zhang X, Li X, Zhang D, Su NQ, Yang W, Everitt HO, Liu J (2017) Product selectivity in plasmonic photocatalysis for carbon dioxide hydrogenation. Nat Commun 8:1–9. https://doi.org/10.1038/ncomms14542

Zhang L, Can M, Ragsdale SW, Armstrong FA (2018) Fast and selective Photoreduction of CO_2 to CO catalyzed by a complex of carbon monoxide dehydrogenase, TiO_2, and Ag Nanoclusters. ACS Catal 8:2789–2795. https://doi.org/10.1021/acscatal.7b04308

Zhao Z, Yu P, Shanbhag BK, Holt P, Zhong YL, He L (2019) Sustainable recycling of formic acid by bio-catalytic CO_2 capture and re-hydrogenation, C. 5(2019):22. https://doi.org/10.3390/c5020022

Zheng Y, Zhang W, Li Y, Chen J, Yu B, Wang J, Zhang L, Zhang J (2017) Energy related CO_2 conversion and utilization: advanced materials/nanomaterials, reaction mechanisms and technologies. Nano Energy 40:512–539. https://doi.org/10.1016/j.nanoen.2017.08.049

Zhong W, Haigh JD (2013) The greenhouse effect and carbon dioxide. Weather 68:100–105. https://doi.org/10.1002/wea.2072

Electrochemical CO_2 Conversion

I. A. Novoselova, S. V. Kuleshov, and A. A. Omel'chuk

Abstract

The chapter is a review of the present state of research concerning electrochemical conversion of carbon dioxide from conducting liquid and solid media of different chemical composition. They include solid, aqueous, non-aqueous (organic and ionic liquids), and molten salt electrolytes. It reports a comparative analysis of the effectiveness of using these electrolytes, as well as products obtained by the electrochemical decomposition of CO_2, the properties of these products, and the prospects for their use. Special emphasis has been made on the electrochemical decomposition of carbon dioxide from salt melts, several variants of decomposition have been considered, the advantages and disadvantages of each variant have been analyzed. The chapter reports an analysis of literature data on this subject as well as research results of the authors on the direct electrochemical conversion of CO_2 into valuable solid-phase chemicals with added value. They include: nanoscale solid carbon of different structure and morphology (carbon nanotubes, fibers, graphene); fine tungsten and molybdenum carbide powders; double tungsten (molybdenum) carbides with cobalt and nickel.

Keywords

Carbon dioxide • Electrochemical conversion • Molten salts • Electroreduction • Carboneous materials

I. A. Novoselova (✉) · S. V. Kuleshov · A. A. Omel'chuk
V.I. Vernadskii Institute of General and Inorganic Chemistry, National Academy of Sciences of Ukraine, 32/34 Pr. Palladina, Kiev-142, 03680, Ukraine
e-mail: iness@ionc.kiev.ua; ; inessa_nov@ukr.net

1 Introduction

The global warming, caused by the increase in the emission of greenhouse gases, such as carbon dioxide (CO_2), methane (CH_4), nitrous oxide (N_2O) and others, has been recognized as a serious environmental problem of the humanity. Today, there are large-scale extraction and consumption of fossil hydrocarbons, which results in huge emissions of carbon dioxide into the atmosphere. Its concentration, which is an indicator of energy consumption by the humanity from fossil fuels, increases at a high rate. The plants and water reservoirs do not cope any more with the natural utilization of such volumes of CO_2. At the present time, the annual increase in CO_2 is 3200–3600 million tons. According to the report of the International Group of Experts on Climate Change (IGECC), the problem of global climate changes can no longer be put off until later. It is necessary to act right now; otherwise, irreparable damage to the planet's ecosystems can be caused as early as by 2030, i.e., in 10 years.

According to the calculations of the IGECC, if CO_2 emissions continue to increase at this rate, the average annual temperature on the Earth will increase by 1.5–4.5 °C by the end of the twenty-first century (Mac Dowell et al. 2017; Chu and Majumdar 2012; Chu et al. 2016). This means a 0.3 °C increase in temperature within a decade, which exceeds by a factor of 3 the level of adaptability of natural ecosystems. Therefore, the effective utilization of carbon dioxide is a topical scientific and environmental problem of the world scientific community.

Two ways of solving this problem are possible. Way 1: the reduction of CO_2 emission mainly through the change and possible partial giving up of processes for energy generation from coal and petroleum resources. Way 2: the use of CO_2 as a renewable resource, i.e., the reuse of existing CO_2 as a source of carbon for producing fuels and chemicals with added value. Nowadays, the use of CO_2 as a new raw material for manufactured goods and chemical precursors for fine organic synthesis is intensively discussed. Carbon

Inamuddin et al. (eds.), *Carbon Dioxide Utilization to Sustainable Energy and Fuels*, Advances in Science, Technology & Innovation, https://doi.org/10.1007/978-3-030-72877-9_6

dioxide is a renewable carbon raw material for industry; it is environmentally harmless and readily available for the decomposition of organics. The concept underlying CO_2 utilization is as follows: direct power generation by the combustion of fossil resources produces CO_2, which can be converted back to energy carrier, and the carbon circle is effected thereby. This approach could be of double benefit: (1) recycling of CO_2; (2) balancing of its anthropogenic emission to the atmosphere.

The CO_2 utilization methods can be conditionally divided into biological, chemical, and physical ones. Among the chemical methods, the catalytic conversion of CO_2 and its utilization in the reactions of oxidative dehydrogenation of alkyl aromatic substances to corresponding olefines (or aldehydes and acids) are noteworthy. Of great interest is the reaction between carbon dioxide and methane as a promising method for the utilization of two greenhouse gases at the same time to form a synthesis gas ($CO + H_2$), which is the main raw material for organic synthesis:

The electrochemical method is an efficient but still poorly developed method. It involves the electrochemical decomposition of CO_2 at the cathode. The chemical composition of cathodic products can be changed dramatically by varying the electrolysis conditions (temperature, current density, applied bath voltage, bath composition, electrode materials) (Qiao Liu Zhang 2016). In recent years, the interest in the CO_2 electroreduction reaction (CO_2RR) has increased in view of a number of applied problems, namely:

- the problem of CO_2 utilization in closed systems;
- synthesis of CO_2-based organic compounds, especially formic and oxalic acids for chemical industry;
- production of liquid fuel (methanol) for future power complexes.

The use of molten salts as reaction media for CO_2RR is one of the possible ways of CO_2 conversion. The study of electrode reactions involving CO_2 is of interest for the investigation of the transport properties of gases, disclosure of the general picture of the electrochemical behavior of gases in ionic melts, and for the creation of the basis and for the control of the electrochemical synthesis of carbonaceous inorganic compounds. A peculiarity of the CO_2RR in salt melts is the deposition of a solid-state carbon on the cathode in contrast to aqueous, organic (ionic liquids) electrolytes. The investigation of this process is dealt with in a large number of original papers, reviews, and books. These works contain certain contradictions as to the compositions of electrolysis products; there are also contradictions in the interpretation of obtained data and proposed mechanisms of electrode processes.

The aim of this chapter is to consider the present state of research on the electrochemical conversion of carbon dioxide from conducting media of different chemical composition: solid, aqueous, non-aqueous (organic and ionic liquids), and molten salt electrolytes. The chapter reports a comparative analysis of the effectiveness of using these electrolytes, as well as characteristics of cathodic products obtained by carbon dioxide electrochemical splitting and prospects for their use. Special emphasis is made on the electrochemical reduction of carbon dioxide in salt melts, several variants of decomposition are shown, the advantages and disadvantages of each variant are analyzed. The chapter reports an analysis of literature data on this subject and research results of the authors on the direct electrochemical conversion of carbon dioxide.

2 Electrochemical CO_2 Conversion

2.1 Fundamentals of the Process

Carbon dioxide capture and storage (CCS) are becoming basic processes for the sustainable and stable development of the humanity. If carbon dioxide conversion from wastes to low-carbon fuels or valuable chemicals is effected at the same time, this can become a major breakthrough in the development of modern power engineering. One of the possible ways of this transformation is electrochemical CO_2 conversion. Therefore, the huge interest in developing such technologies in many laboratories in the world is quite understandable.

Carbon dioxide is a stable and non-combustible molecule since the carbon and oxygen atoms are in the highest valence state. The CO_2 ($O = C = O$) molecule is linear, and the length of the $C = O$ double bond is 0.116 nm. All of the four covalent carbon–oxygen bonds are polar; however, the molecule as a whole is non-polar owing to its linearity. The oxygen atoms in the CO_2 molecule are weak Lewis bases, i.e., have free electron pairs. Therefore, in electrochemical processes, carbon dioxide can only be reduced, and during cathodic discharge, the carbon atom is attacked by electrons, which may cause the molecule to bend.

The first works on electrochemical CO_2 reduction date from the early nineteenth century. Research in this area intensified in the 1980s after the petroleum embargo in the 1970s. In the early twenty-first century, the interest in research in this direction increased noticeably owing to the fact that the anthropogenic carbon dioxide emissions into the atmosphere reached record levels.

Electrochemical carbon dioxide conversion is the transformation of carbon dioxide into chemical substances with a high energy potential in the case of using electrical energy. Electrochemical CO_2 reduction can take place in gases, aqueous, non-aqueous (organic and ionic liquid based), solid-phase and salts melt conducting media in a wide temperature range (from room temperatures to 1000 °C). The main reduction products can be solid-phase carbon of different structure and morphology, carbon monooxide (CO), formic acid (HCOOH), methane (CH_4), methanol (CH_3OH), formaldehyde (CH_2O), oxalic acid ($H_2C_2O_4$), ethylene (C_2H_4), ethanol (CH_3CH_2OH), and others.

The device for the electrochemical splitting of CO_2 includes three essential elements: a working electrode (cathode), a counter electrode (anode), and an electrolyte of different chemical composition containing dissolved CO_2. The device can also have a non-essential element—a membrane, which is generally used in low-temperature electrolytes while obtaining products in a gas phase. CO_2RR proceeds at the cathode; an oxidation process occurs at the anode; it is generally an oxygen evolution reaction (OER). To increase the efficiency of the process, catalysts are often used both in the cathodic and in the anodic process. In the electrolyte, which should have ionic conductivity, the base electrolyte ions and dissolved CO_2 are delivered to the surface of the electrodes. The membrane separates the oxidation and reduction products with maintaining charge balance.

To effect the splitting of CO_2 (the rupture of double carbon–oxygen bonds), energy supply from outside is required. The thermodynamic expenditure for CO_2 reduction is closed to that for the hydrogen evolution reaction (HER). For example, such products as CO and C_2H_4 are thermodynamically formed at potentials of − 0.11 and + 0.07 V, respectively, relative to a RHE electrode (Ross et al. 2019; Hori et al. 2008). In practice, however, a higher energy is needed. In addition to overcoming the activation barrier to CO_2 discharge, the CO_2 reduction process involves many other steps, which can retard the CO_2RR. This is a kinetic contribution to overpotential. For instance, in the case of using aqueous electrolytes, protons participate in the cathodic reaction together with CO_2 to form H_2, which additionally complicates the process. Besides, when using high bath voltages, parallel CO_2 electroreduction paths are possible, which result in numerous products. This significantly reduces the selectivity of the process.

The main advantage of the electrochemical way of CO_2 reduction is the possibility to control the process (preparation of desired products) by means of electrolysis parameters (potentials applied to the cathode, current density, and electrolysis time). Besides the above parameters, researchers can control the process by means of a whole set of independent variables: the nature and composition of the electrolytic bath, the composition, structure and morphology of the materials of the electrodes, the operating temperature of the process and electrolyzer design. Other advantages of the electrochemical way of CO_2 reduction are also noteworthy:

- the components of electrolytic baths can be completely recycled, so that the total consumption of reagents can be minimized;
- the source of electricity for the process can be alternative energy, including solar, wind, hydroelectric, and tidal processes;
- the components of electrochemical reactions are compact and easy to use on a large scale.

However, there are serious problems in realizing sustainable processes, such as low CO_2 electroreduction rate even when using electrocatalysts and high bath voltages, low energy efficiency because of the possible occurrence of competing reactions or a base electrolyte decomposition reaction at high potentials, low selectivity in the preparation of the desired product, high operating costs, low activity of the catalyst and its insufficient life time, incomplete understanding of the mechanisms of electrode reactions and methods of controlling them.

In recent years, certain successes in solving the above problems by developing novel electrocatalysts, electrolytic media, and electrolyzer configurations have been achieved. Different approaches have been proposed, which can bring the process closer to thermodynamic limits and make it technologically viable. This chapter will review briefly research results available in literature on electrochemical CO_2 conversion in different electrolytic media and describe briefly the possibilities, disadvantages, and prospects for these developments. The main attention will be given to the consideration of carbon dioxide electroreduction from molten salt media to form solid-phase carbon since the authors of the chapter have been working in this direction for almost 40 years.

Let us consider the parameters of an electrochemical process, from which its efficiency and the prospects for developments for further industrial scaling can be assessed:

- *Overpotential* (η, *V*)—the difference between the thermodynamic voltage and the actual voltage of the working electrode (cathode) at which the CO_2RR takes place. When developing the technology, the overpotential must be minimized, which is achieved by using catalysts.
- *Total electrolytic cell (bath) voltage* (*E*, *V*)—which includes potentials both for the anodic and for the cathodic process ($E_{cell} = E_{anode} - E_{cathode}$). To achieve satisfactory rates of the process, the overpotential of cathodic and anodic processes can reach several hundred millivolts. As a result, the voltage of the bath usually

exceeds the thermodynamic potentials of the electrode reactions. Therefore, catalysts must be used not only for the cathodic, but also for the anodic process since it consumes almost a half of supplied electric energy.

- *Current efficiency or faradaic efficiency (FE, %)* is the percentage of current that is actually spent in a particular electrolytic process on obtaining the desired product. Current efficiency characterizes the selectivity of the process and is defined as the ratio of the actually obtained product to the product theoretically expected from calculation based on the Faraday law in the case of consuming the same quantity of electricity. *FE* can be represented by the formula:

$$\mathrm{FE} = m_{\mathrm{real}}/m_{\mathrm{theor}} \cdot 100\% = m_{\mathrm{real}}/k \cdot I \cdot t \cdot 100\% \quad (1)$$

where m_{real} is the amount (mass) of the product actually obtained as a result of electrolysis, m_{theor} is the amount of the product that must be obtained theoretically according to the first Faraday law, k is the electrochemical equivalent of the product, and I is the current flowing in the cell within the time *t*.

- *Energy efficiency (EE, %)* characterizes the efficiency of conversion of the precursor into the desired product at the actual applied electric potential. *EE* takes into account the energy losses due to the low selectivity (FE) and high overvoltage of the process and can be represented by the formula:

$$\mathrm{EE} = E^0 \cdot (\mathrm{FE})/(E^0 + \eta) = E^0 \cdot (FE)/E \quad (2)$$

where E^0 is the reversible cell voltage, FE—current efficiency, η is the overpotential of the cell, and $E = (E^0 + \eta)$ is the potential of the cell at the required current density.

-*Electric energy consumption (EEC, kWh/m³)* is one more measure of the efficiency of electrolyzer operation, which is used to obtain a gaseous product and characterizes the quantity of electric energy required to obtain 1 m^3 of the gaseous product at normal conditions. The formula for the calculation of *EEC* is as follows:

$$\mathrm{EEC} = E \cdot n \cdot F/(\mathrm{FE}) \cdot V_m \quad (3)$$

where E is the cell voltage, n is the number of transferred electrons, and V_m is the molar volume of an ideal gas in normal conditions.

2.2 Variants of Electrochemical Conversion of CO_2

The nature and composition of the electrolyte, cathode material, the nature and morphology of catalysts, as well as the electrolysis conditions (temperature, current density, bath voltage, electrolysis duration, and electrolyzer configuration) are the key factors that determine the CO_2RR path and mechanisms, as well as the composition of the products obtained.

Depending on the nature of the electrolytes in which the electroreduction of CO_2 takes place, four variants of its conversion are noteworthy: from aqueous, non-aqueous (organic and ionic liquids), solid, and molten salt electrolytes. Table **1** presents briefly the main characteristics of processes in these media and the compositions of the products obtained.

2.2.1 Aqueous Electrolytes

Most of the research on the CO_2RR has been carried out in aqueous electrolytes. CO_2-saturated slightly acidic, pH neutral or alkaline solutions containing anions HCO_3^-, SO_4^{2-}, ClO_4^- or Cl^-, and alkali metal cations are used most often (see Sect. 1 in Table 1). In a strongly acidic medium, the selectivity of cathodic products sharply decreases since almost all current is spent on the competing hydrogen evolution reaction. The electrochemical synthesis of hydrocarbons from CO_2 in an aqueous medium is a multistage process with adsorbed intermediates (especially with adsorbed CO). The main products are carbon monoxide, methane, ethylene, formate, and some alcohols (methanol, ethanol, and propanol).

The preparation of carbon monoxide can be represented by the following equations. CO_2 is reduced at the cathode to CO:

$$CO_2 + H_2O + 2\bar{e} = CO + 2\,OH^- \quad (4)$$

This process is usually accompanied by a competing hydrogen evolution reaction, which proceeds in alkaline media as follows:

$$2\,H_2O + 2\bar{e} = 2\,OH^- + H_2 \quad (5)$$

The co-electroreduction of carbon dioxide and hydrogen depending on the cathode material and electrolysis parameters can take place to form various hydrocarbons and alcohols by the following reactions:

$$CO_2 + H_2O + 2\bar{e} = HCOO^- + OH^- \quad (6)$$

$$CO_2 + 6\,H_2O + 8\bar{e} = CH_4 + 8\,OH^- \quad (7)$$

$$2\,CO_2 + 8\,H_2O + 12\bar{e} = C_2H_4 + 12\,OH^- \quad (8)$$

Table 1 Electrochemical CO_2 reduction in various ionic media

	Electrolytes	*T*, °C	Reactants	Reduction rate (mA/cm^2) Faraday efficiency (Jiang et al. 2019)	Products
1	**Aqueous Solutions** $KHCO_3$; $NaHCO_3$; NaOH; K_2CO_3; Na_2SO_4; K_2SO_4; $NaClO_4$ + Pyridinium; LiCl in MeOH; $LiClO_4$ in MeOH	0–100	CO_2 + H^+ + H_2O + others	0.1–100 10–90%	–CO –organic compounds (formiate and formic acid; formaldehyde, methane; methanol; acetaldehyde; ethylene; ethanol)
2	**Non-aqueous Solutions** - Organic alcohols (CH_3OH, C_2H_5OH + others) - Ionic liquids	20–200	CO_2 + H^+ + others CO_2 CO_2 + others	0.1–10 20–95%	–CO –organic compounds
3	**Solid Electrolytes** – O^{2-} ion conducting – H^+ ion conducting	400–900	CO_2	100–1500 25–98%	–CO –CH_4 in the presence of H_2 and H_2O -solid C
4	**Molten salts** Li,Na,K|CO_3; Li,Na|CO_3; MeHal-Me$_x$$CO_3$ (Me -Li, Na, K, Ca, Ba, Mg; Hal - Cl, F)	450–900	–CO_3^{2-} –CO_2 or CO_3^{2-} CO_3^{2-} + OH^- CO_2 or CO_3^{2-} + metal ions	100–1500 70–100%	–CO –solid carbon of different structure and morphology –sungas –metal carbides

$$2\,CO_2 + 9\,H_2O + 12\bar{e} = C_2H_5OH + 12\,OH^- \quad (9)$$

Oxygen is evolved at the anode in accordance with any of the two reactions:

$$2\,OH^- = \tfrac{1}{2}\,O_2 + H_2O + 2\bar{e} \quad (10)$$

$$H_2O = \tfrac{1}{2}\,O_2 + 2\,H^+ + 2\bar{e} \quad (11)$$

At high current densities, a part of CO_2 can be transformed to carbonate (or bicarbonate) anions, which can diffuse through the liquid electrolyte and be discharged at the anode. In this case, the O_2 evolution at the anode is accompanied by CO_2 evolution, which can take place by one of the two reactions:

$$CO_3^{2-} + 2\,H^+ = CO_2 + H_2O \quad (12)$$

$$HCO_3^- + H^+ = CO_2 + H_2O \quad (13)$$

The configuration of the electrolyzer (the so-called H-type electrolytic cell) is used most often. In this cell, both electrodes (anode and cathode) are immersed in electrolytic solutions separated by a membrane (anolyte and catholyte, respectively). The design of an *H*-type cell with an anion-conducting membrane is presented in Fig. 1.

The H-cell model has serious disadvantages. Because of the slow diffusion rate and low solubility of carbon dioxide in water, the mass transfer rate will be low; therefore, it is difficult to achieve current densities above 30 mA/cm^2 (Weekes et al. 2018). The use of gas diffusion electrode is a way to avoid mass transfer limitations. A schematic of a cell with a gas diffusion cathode is shown in Fig. 2. In some electrolyzer designs, both electrodes are of a gas diffusion type (Weekes et al. 2018; Verma et al. 2018).

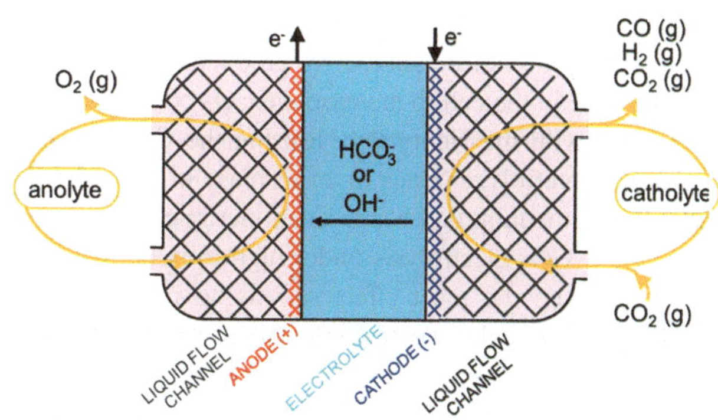

Fig. 1 Schematic illustration of an electrolysis cell in *H*-cell configuration for CO_2 conversion in aqueous electrolytes. Taken from (Küngas 2020)

Changing the temperature affects the process in a complex fashion. For instance, higher temperature leads to a decrease in CO_2 solubility, which causes a change in the solution pH, to an increase in diffusion coefficient and CO_2

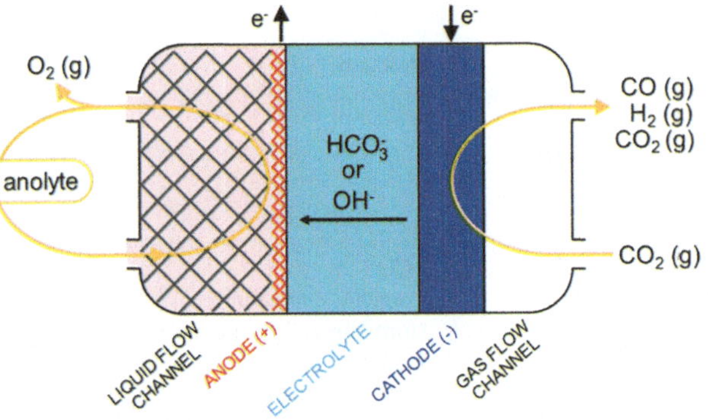

Fig. 2 Schematic illustration of an electrolysis cell with a gas diffusion cathode for CO_2 conversion in aqueous electrolytes. Taken from (Küngas 2020)

reactivity (Zhang et al. 2020; Endroda et al. 2017). Lowering the temperature suppresses the competing hydrogen evolution reaction and increases the product yield.

One of the main problems of the CO_2RR in aqueous media (low current densities) is associated with the low CO_2 solubility in water at normal conditions. The aqueous solution saturated with CO_2 at the standard temperature and pressure contains only 0.034 mol or 1.5 g/L of CO_2 (Aschenbrenner and Styring 2010).

This problem can be solved by producing an excessive CO_2 pressure in the cell to increase the CO_2 concentration in accordance with the Henry law. Raising the CO_2 gas pressure in the electrolyzer leads to an increase in gas solubility in the liquid medium and hence to an increase in the rate of the process. CO_2 electrolysis under high pressure (of up to 60 atm) can be the most effective way to realize a commercially reasonable electrochemical process. The high current density and efficiency observed in this case for different electrodes are comparable with those obtained in high-temperature solid oxide and molten salt electrolytes. However, this can greatly increase the electrolyzer construction cost.

The nature of the cathodic material has a key impact on the reaction path and the composition of products in aqueous electrolytes. The mechanism of the CO_2RR in aqueous media was studied for many years, and the main goal of the research was to understand the reason of the formation of different products on different metal cathodes (Hori 2008). The schematic of the mechanism is well represented in Ref (Jones et al. 2014) and shows different paths in which CO_2 can be reduced depending on the cathode material. The first step, the formation of a $CO_2^{\bullet-}$ anion radical, is a determining one since it limits the rate of the entire reduction process. The adsorption and electrochemical activities of the $CO_2^{\bullet-}$ anion radical determine what will be the final product of CO_2 reduction. The $CO_2^{\bullet-}$ can readily react with water or any other compound in the electrolyte. The reasons of the different behavior of metals are their different abilities to adsorb the $CO_2^{\bullet-}$ radical and subsequently reduce CO. The metals used as a cathode for CO2RR can be divided into four groups. The first group of metals, which include Pb, Hg, In, Sn, Cd, and Tl, has a very weak adsorption capacity for the $CO_2^{\bullet-}$ radical and a low ability to reduce CO. The product of CO_2 reduction on these metals is formate or formic acid. The second group of metals, which includes such metals as Au, Ag, Zn, and Ga, can adsorb the intermediate compound $CO_2^{\bullet-}$ well, but cannot reduce CO. The main product at the cathodes made of the second group metals is carbon monoxide. The third group of metals includes copper and its alloys with other metals. This group is characterized by high adsorption to the $CO_2^{\bullet-}$ radical and reduces CO_2 to polyatomic alcohols and hydrocarbons. Therefore, it is Cu and its alloys that have been investigated most widely and fully for use as the cathode in the CO_2RR. The latest results, obtained on this metal, on increasing the efficiency and selectivity of the process are promising and are fairly fully described in numerous recent reviews (Ren et al. 2016; Hoang et al. 2018; Huang et al. 2019; Nitopy et al. 2019). There is still a lot of work to be done before the industrial implementation of the technology can be fulfilled. It is necessary to deepen understanding the mechanisms of CO_2RR, to improve the practical characteristics of efficiency. Nanostructuring of catalysts, the use of bimetallic catalysts and alkaline electrolytes can reduce the overvoltage in the bath and increase the selectivity for particular products. For instance, in recent years, a tremendous progress has been made in developing gas diffusion electrodes (GDEs) and devices with continuous electrolyte flow (Endroda et al. 2017). For example, in the Ref (Dinh et al. 2018) a selectivity for ethylene (C_2H_4) of 70% during 150 h of work with a current density of 100 mA/cm^2 was achieved, and the energy efficiency of full cell for CO_2 conversion to C_2H_4 was 34%.

The fourth group of metals, which includes such metals as Ni, Fe, Pt, and Ti, has so strong adsorption ability for hydrogen that they practically rule out CO_2 electrochemical reduction in aqueous media. In view of the fact that reaction proceeds in aqueous media, it is necessary take into account the hydrogen evolution reaction (HER), which competes with CO_2 reduction and is easier for most metals. In practice, many CO_2 recovery catalysts are often chosen not because of their ability to catalyze CO_2 reduction, but because of their high over potential in the HER.

2.2.2 Non-Aqueous Electrolytes

Much less research compared with aqueous media has been carried out in non-aqueous electrolytes, namely in organic and ionic liquids. The changeover to this type of solvents is due to the fact that the solubility of CO_2 in them is noticeably higher. Therefore, much effort of researchers was directed to studying the solubility of CO_2 in non-aqueous media. It was shown in (Chang 1992; Lail et al. 2014) that the solubility of carbon dioxide can be increased by four, five, eight, or even twenty times compared with water, if acetonitrile, methanol, propylene carbonate, or DMF are used as the solvent, respectively. Therefore, non-aqueous solvent can be regarded as an alternative to producing a high pressure in the electrolyzer with aqueous electrolytes.

References (Saeki et al. 1995; Ohta et al. 1998) report a study of CO_2 reduction on different metals in methyl alcohol where high current densities at a copper electrode in the CO evolution reaction are recorded. The CO_2RR in methanol electrolyte was also intensively studied in (Kaneco et al. 2002; Ohya et al. 2009) on different cathodic materials. The electrolyzer designs and electrode materials that are used to carry out the CO_2RR in non-aqueous media are similar to those used in aqueous electrolytes.

An interesting approach—to use a mixture of tetrafluoroethane and carbon dioxide in supercritical state as a medium for CO_2 reduction—was proposed in Abbott and Eardley (2000). This approach can make it possible in future to solve any problems with carbon dioxide concentration.

In recent 10 years, publications have appeared dealing with the use of a relatively new class of electrolytes—ionic liquids—as a medium for effecting the CO_2RR (Buzzeo et al. 2004; Silvester and Compton 2006; Rees and Compton 2011; Alvarez-Guerra et al. 2015; Faggion et al. 2019). This is due to the fact that ionic liquids (ILs) possess a whole set of valuable (in applied terms) physicochemical properties. For instance, ILs have a high selective ability to absorb CO_2 and to stabilize charged $CO_2^{\bullet-}$ anion radicals, have a wide electrochemical potential window at low temperatures (Hayyan et al. 2013), a high thermal and chemical stability, and a low volatility. All this makes it possible to use them as electron transfer mediators for redox catalysis (Balasubramanian et al. 2006; Rosen et al. 2011). This area of research is at the initial stage of developments, and high characteristics of the process (current density, current efficiency, energy efficiency) have not yet been achieved (Faggion et al. 2019). An important hindrance to the commercialization of the use of organic and ionic liquids for the CO_2RR is their high cost.

2.2.3 Solid Oxide Electrolytes

Research into the electroreduction of CO_2 in solid electrolytes started in the 1960s to obtain oxygen in closed systems for prospective space flights to Mars (the atmosphere of Mars contains 95% CO_2) (Chandler and Pjllara 1966). The CO_2 electrolysis system based on solid oxide electrolysis cell (SOEC) will be soon tested on Mars as part of NASA's Program (Hartvigsen et al. 2017).

In the new millennium, the researchers' interest in this subject has increased again, but it was already associated with the solution of important environmental and energy problems, namely electrochemical CO_2 conversion into new fuels and chemicals. This subject was considered in numerous publications (original articles, reviews, and books). Especially many publications have appeared in the past 10 years (Küngas 2020; Zheng et al. 2019; Song et al. 2019; Zhang et al. 2017; Uhm and Kim 2014). After half a century of research, certain successes have been achieved; however, many problems still remain.

A typical solid oxide electrochemical cell (SOEC) includes a solid electrolyte for ion transfer, an anode for oxygen evolution, and a cathode for electrochemical CO_2 reduction. In the SOEC, gaseous CO_2 directly diffuses into the porous electrode and is electrochemically reduced to gaseous products. Therefore, the cathodic reaction is fairly simple owing to the absence of side reactions, which are typical of liquid electrolyte. Besides, temperature significantly accelerates the process and increases current density in compare with aqueous and organic electrolytes at the same bath voltage.

The electrolyte in the SOEC is a solid ceramic material. At temperatures higher than about 600 °C, the electrolyte becomes conductive on the oxide anion O^{2-}, but remains non-conductive on electrons. The working temperature of the SOEC is usually within a range of 700–900 °C. The materials that are usually used for solid oxide electrolyte contain stabilized zirconium dioxides, such as yttrium-stabilized zirconium dioxide (YSZ), Y_2O_3 and ZrO_2 solid solution, scandium-stabilized zirconium dioxide (ScSZ), as well as gadolinium-doped cerium (GDC) or samarium-doped cerium (SDC). A fairly complete review of solid electrolyte materials that can be used for the CO_2RR is presented in Kharton et al. (2004).

A typical single SOEC for CO_2 electroreduction has a sandwich plate structure with the solid electrolyte located between the porous electrodes. Two types of electrolytes are used in the SOEC: oxygen ion conducting and hydrogen (proton) ion conducting electrolytes. Therefore, the reactions at the electrodes in these electrolytes will be different. For oxygen ion conducting electrolytes, the CO_2 molecules are reduced at the cathode to CO and oxygen ions according to the equation:

$$CO_2 + 2\overline{e} \rightarrow CO + O^{2-} \quad (14)$$

Oxygen ions formed at the cathode are transported to the anode and oxidized:

$$O^{2-} \rightarrow {}^1\!/_2\, O_2 + 2\overline{e} \quad (15)$$

In the proton-conducting electrolyte, CO_2RR is closely related to water electrolysis. The H_2O molecules lose electrons at the anode to form O_2 and H^+ (Eq. (16)). The formed protons are transported to the cathode, then are reduced and react with CO_2 to form CO or other valuable chemical substances (Eqs. (17–19)).

$$H_2O - 2\bar{e} \rightarrow 2\,H^+ + {}^1\!/_2\,O_2 \qquad (16)$$

$$2\,H^+ + 2\bar{e} \rightarrow H_2 \qquad (17)$$

$$CO_2 + 2\,H^+ + 2\bar{e} \rightarrow CO + H_2O \qquad (18)$$

$$CO_2 + 8H^+ + 8\bar{e} \rightarrow CH_4 + 2H_2O \qquad (19)$$

A SOEC with an oxide ion conducting electrolyte is schematically shown in Fig. 3.

CO_2 is supplied into the cathodic portion of the electrolyzer through gas channels. On the porous cathode (which is called fuel electrode), CO_2 is reduced to CO according to reaction (14). The O^{2-} ions are incorporated into the solid oxide electrolyte, pass through the porous cathode to the anode (the so-called oxygen electrode) and are oxidized there to molecular oxygen in accordance with reaction (15). The formed O_2 gas is removed from the cell through the gas channels. As long as pure CO_2 gas (or a mixture of CO-CO_2 gases) is supplied to the fuel electrode, the resulting product will not contain H_2 and H_2O—an advantage that cannot be realized in aqueous and organic electrolytes.

The cathode material in the SOEC must possess: (1) a high electronic and ionic conductivity; (2) a high catalytic activity; chemical and mechanical stability; compatibility with the electrolyte. Noble metals, fluorites, oxides with perovskite structure are usually used as cathodic materials.

The anodic material, like the cathodic one, must also possess the similar characteristics. At present, two main classes of materials are used as anodes in the SOEC: (1) noble metals (Pt and Au); (2) electron-conducting oxides, such as perovskite oxides and bilayer perovskite oxides. The noble metals cannot compete with electron-conducting oxides because of their high cost. The latter are becoming promising candidates for anode materials in industrial SOECs. The main products formed in SOEC are usually carbon monooxide and methane. Moreover, the selectivity and performance characteristics of SOEC are higher than those observed in liquid electrolytes (Spinner et al. 2012). The electrochemical CO_2 reduction to synthesis gas (CO and H_2) is an important direction for using SOEC. In recent years, there has been an increase in research in the field of co-electrolysis of CO_2 and H_2O electrolysis (Belyaev et al. 1998; Graves et al. 2011; Ebbesen et al. 2012; Li et al. 2013; Ni 2012) for the production of syngas, which can be used later for the production of various hydrocarbon fuels, such as methane, methanol, dimethyl ether, or synthetic diesel fuel. The success achieved in recent years in the development of the SOEC technology for CO_2 reduction is largely due to multiyear research to develop solid oxide fuel cells (SOFCs). SOFCs are electrochemical devices, which generate electrical energy by the oxidation of CH_4. The same materials of the cell and battery can be used interchangeably in both fuel cell and electrolysis modes. It was shown (Graves et al. 2014; Irvine et al. 2016), that regular changeover between two modes significantly reduces the rate of cell degradation.

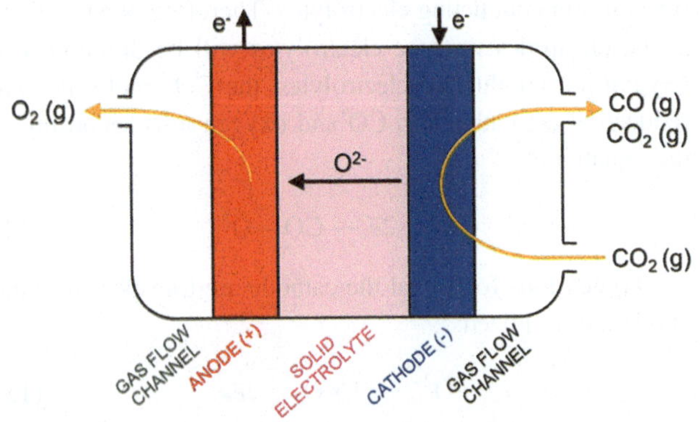

Fig. 3 Schematic illustrations of solid oxide electrolysis cell for electrochemical CO_2 conversion. Taken from (Küngas 2020)

The review (Küngas 2020) presents a parallel analysis of three competing variants of electrochemical CO_2 reduction to CO from aqueous electrolytes, molten carbonates, and solid oxide systems. Based on the overall level of the technological readiness, SOEC stands out as industry perspective for the production of carbon monoxide. The SOEC technology offers a higher efficiency independent of the chosen efficiency rating scale. The high conversion efficiency was checked on setups with industrially significant electrode areas of > 8000 cm^2. However, low-temperature electrolytes can offer interesting variants for the production of polyatomic hydrocarbons.

2.2.4 Molten Salt Electrolytes

In contrast to aqueous and organic electrolytes, the main peculiarity of CO_2 electroreduction in molten salt media is the formation of products not only in the gas phase (CO), but also in the solid phase (nanosized carbon structures of different morphology, refractory metal carbides). In molten electrolytes, the complete four-electron reduction of CO_2 to elemental carbon can take place. This is a great advantage for the convenient storage of the products of recycled carbon dioxide. Moreover, recent research has shown that the solid-phase products (carbon nanotubes, nanofibers, carbon monooxide, and amorphous carbon structures) of electrochemical CO_2 conversion are chemical substances with high added value since they possess unique properties and have good prospects for use in modern devices and technologies. Solid-phase carbon can also be produced in solid oxide electrolytes (Boehm 1978), but due to the low efficiency of the process and the difficulties in separation of the product from the reaction medium, this process is not promising.

In molten media, the direct and indirect electroreduction of carbon dioxide can be effected. In the former case, carbon dioxide dissolved in a molten salt is reduced, and which is introduced into the electrolyzer either by flow or under pressure. The creation of excessive pressure is necessary to increase the gas concentration in the electrolyte. Indirect CO_2 reduction takes place in salt (carbonate, chloride-carbonate, chloride-oxide) melts when a carbon dioxide atmosphere is created over the melt. Depending on the electrolysis conditions and modes, the discharge of the carbonate anion at the cathode can result in the formation of a carbon monoxide or a solid carbon phase and in the release of oxide anions. The oxide anion in the melt can bind carbon dioxide from the gas phase to form a carbonate anion. In this way, the carbonate anion consumed in the electrolyte is regenerated.

Molten individual and mixed alkali and alkaline-earth metal carbonates or halide-carbonate mixtures of different composition are used as electrolytes. The choice of the electrolytic bath composition can be made based on preliminary thermodynamic calculations of the decomposition voltages of alkali and alkaline-earth metal carbonates and chlorides. Such investigations have been made in many works (Bartlett and Jonson 1966; Shapoval et al. 1982; Chery et al. 2014; Yin and Wang 2018). This made it possible to purposefully develop the composition of electrolytic baths and helped a lot to propose mechanisms of electrode processes.

A simplified schematic of a molten carbonate electrolysis cell (MCEC) is shown in Fig. 4. The MCEC resembles the H-type cell which is used for aqueous electrolytes (Fig. 1). The difference is the absence of the diaphragm between the cathode and anode chambers in the electrolyzers using molten salts.

The other advantages of the CO_2RR in molten salts are: (1) the absence of competing hydrogen evolution reaction, (2) high selectivity of the C and CO products, (3) mild synthesis conditions, (4) simple hardware implementation, (5) the possibility of controlling the electrolysis conditions, (6) low energy consumption for electrolysis, (7) the use of cheap precursors, (8) easy doping during synthesis, and (9) monodispersity of the product.

An advantage of the MCEC is also the fact that the gaseous CO_2 supplied into the electrolyzer and the CO and O_2 products do not mix, allowing pure gases to be removed from the cell. Moreover, the method depends weakly on the percentage of possible impurities in the supplied gas (e.g., SO_2) (Chen et al. 2017) and can potentially use CO_2 streams from industrial flue gases.

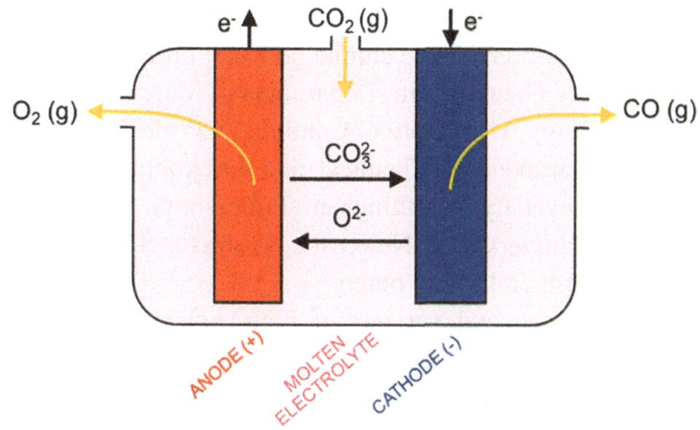

Fig. 4 Schematic illustrations of electrolysis cell design for electrochemical CO production in molten carbonate electrolysis cell. Taken from (Küngas 2020)

The investigation of this process is already dealt with in a large number of original papers and reviews. The next section of the chapter is devoted to a more detailed discussion of several variants of effecting the CO_2RR in molten electrolytes and to the present state of this line of research.

3 Electrochemical CO_2 Conversion from Molten Salts

3.1 Present State of Electrochemical Reduction of CO_2in Molten Salts for the Production of Solid-Phase Carbonaceous Nanomaterials

In recent ten years, many new works have appeared in scientific literature devoted to the electrochemical extraction of solid-phase carbon from carbonate and halide-carbonate melts. This is partially a replication of the works carried out in the 1960s (Smirnov et al. 1965; Delimarskii et al. 1968, 1970, 1971; Ingram et al. 1966) at the modern experimental level. Interest to this subject is due to the international strategy of the world scientific community for reducing atmospheric anthropogenic CO_2 emissions and search for methods for the capture, accumulation, and conversion of the greenhouse gas.

At the present time, the biggest laboratories of the world are engaged in studying the electroreduction of carbonate anions and carbon dioxide in molten salts. We mention only some of them: in the USA (Department of Chemistry, George Washington University, Professor Stuart Licht, Jiawen Ren et al. (are developing the so-called STEP (Solar Thermal Electrochemical Process)); several research centers in China (Wuhan University, Professor Dihua Wang, Professor Wei Xiao et al.; Northeastern University, Shenyang, Professor Ali Reza Kamali; University of Science and

Technology, Beijing, Professor Shuqiang Jiao et al.); France (Institut de Rechercher de chimie de Paris, Professor Michel Cassir et al.); Great Britain (Department of Materials Science and Metallurgy, University of Cambridge, Professor D. Fray et al.; Department of Chemical and Environmental Engineering, University of Nottingham, Professor G. Chen et al.); Australia (University of Newcastle, Professor Scott Wilfred Donne et al.); and many others.

There are several variants of high-temperature electrochemical synthesis (HTES) of solid-phase carbon materials (CNMs) in molten salts. Some characteristics of these variants are listed in Table 2.

There is a seeming likeness in the pointed methods: (1) the reactions take place in an ionic melt; (2) the approach of splitting of a carbon precursor is the discharge reactions at the interface electrode/molten salt electrolyte; and (3) the produced solid-phase carbon is dispersed in the reaction salt medium. But the mechanisms of product formation in these variants are dramatically different.

The first variant was widely developed by many authors in 2000–2010s (Chen and Fray 2003; Sytchev and Kaptay 2009, 2003; Gupta et al. 2014). The essence of this method is the electrowinning of alkali or alkaline-earth metals on a graphite cathode followed by the formation of nanosized solid carbon by the interaction of the metal being deposited with graphite. The authors used different compositions of the electrolytes; the most studied systems are molten LiCl and NaCl. The main disadvantage of this synthesis variant is the instability of the electrolysis—the rapid destruction of the graphite cathode.

In recent 10 years, a great contribution to the development of this synthesis variant has been made by prof. A. Kamali et al. (Kamali et al. 2011; Kamali and Fray 2014; Rezaei and Kamali 2018; Kamali 2020). It was demonstrated in these works that the variant 1 in Table 2 is a prospect method for the obtaining of nanosized carbon structures of different morphology (including high-quality 3D graphene nanosheets) economically and in a large scale. It was shown that produced carbon nanomaterials have very interesting characteristics in a variety of promising applications. In the proposed method, the electrolysis takes place on a graphite cathode in LiCl (NaCl) molten salt in dry, moist, or hydrogen-containing atmospheres, which causes the exfoliation of the graphite. The mechanisms are proposed to explain the occurring processes.

Variant 2 (electroreduction of carbonate anion) and variant 3 (electroreduction of carbon dioxide) can be considered as subvariations of the one method, when new solid carbon phase is formed from the molten salt phase at the cathode as a result of the electrochemical reduction of oxygen-containing carbon compounds (CO_3^{2-} or CO_2). To realize a stable and long-duration electrolysis, it is more appropriate to use carbon dioxide as a carbon source, since in this case there are no by-products. To raise the current density (rate) of the cathode reaction, it is necessary to increase solubility of carbon dioxide in the melt that can be reached by creation of excess gas pressure in the cell.

Variant 2 can be divided into two subvariants: (2a) concentrated molten electrolytes of individual carbonate salts, as well as binary and ternary systems of alkali and alkaline-earth metal carbonates are used as base electrolyte; (2b) halide-carbonate salt mixtures (carbonate-diluted) are used as electrolytes. In both cases, the cathodic product is a solid-phase carbon powder, whose structure and morphology are determined by the electrolysis conditions.

Variant 3 also can be divided into two subvariants: direct and indirect reduction of carbon dioxide.

The direct CO_2 electroreduction in chloride melts was studied in Kiev at the V.I. Vernadskii Institute of General and Inorganic Chemistry of NAS of Ukraine in the 70-80 s of the twentieth century (Delimarskii et al. 1970, 1971; Kushkhov et al. 1987; Novoselova 1988; Shapoval et al. 1995) for the development of a carbonate fuel cell, the estimation of the regeneration of molecular O_2 from CO_2 and electrochemical synthesis of tungsten and molybdenum carbides (Novoselova et al. 2001). The use of carbon dioxide as a precursor opens up a real prospect for the creation of a waste-free continuous cycle technology. Since there are no by-products during the synthesis, the consumable (CO_2) is recovered by the anode reaction (oxidation of the oxygen anion on a graphite anode). The electrolysis duration will be limited only by the time of complete combustion of the graphite anode. The combustion rate of the graphite anode was estimated—0.1 g/Ah (Novoselova et al. 2016). Section 3.2 will present in more detail the results of the direct electroreduction of CO_2, which were carried out in our laboratory.

The direct CO_2 electroreduction was also studied in Halman and Zukerman (1987). In this work, the authors used flow CO_2 gas cell (flow rate was 10–50 mL/min). The solid carbon product was obtained only in an equimolar mixture Al,Na|Cl at 360 °C on Cu and Al electrodes.

Interesting variant of indirect electrochemical CO_2 conversion to solid carbon is proposed in Ref (Otake et al. 2013). In this case, chloride-oxide melts (e.g., $LiCl$-Li_2O or $CaCl_2$-CaO) are used, and the electrolysis is carried out in closed cells under carbon dioxide atmosphere. Some authors who worked with these systems offer the following process mechanism. First, the electroreduction of metals (Li or Ca) from their oxides, dissolved in the chloride melt, occurs, and then the thermal reduction of gaseous CO_2 by the reduced metals Li or Ca takes place. Other authors propose a different mechanism: first, carbon dioxide dissolves in $LiCl$-Li_2O ($CaCl_2$-CaO) melts, and then interacts with the

Table 2 Variants of electrochemical synthesis of CNMs in molten salts

Electrode materials	Bath composition	T, °C	Products	Ref
1. Intercalation of alkali metal into graphite				
Cathode–graphite; anode–graphite	- LiCl - $LiCl-SnCl_2$	610–900	-*MWCNTs, amorphous carbon, Li in MWCNTs -*SWCNTs, MWCNTs -high-quality 3D graphene nanosheets Sn-filled MWCNTs	(Chen and Fray 2003; Kamali et al. 2011; Kamali and Fray 2014; Rezaei and Kamali 2018; Kamali 2020; Sytchev and Kaptay 2009; Gupta et al. 2014)
	NaCl; - $Na,K\|Cl–MgCl_2$	800–850	MWCNTs (Ø 5–400 nm); Mg in MWCNTs	(Kamali and Fray 2014; Resaei and Kamali 2018; Kamali 2020; Sytchev and Kaptay 2003)
2. Electroreduction of CO_3^{2-}: **2a. In pure carbonate melts:**				
1. cathode–*GC; anode–CG;	- $Li_2CO_3–CO_2$(5 atm.)	750	MWCNTs (characteristics are absent)	Devyatkin (2005)
2. cathode–mild steel; Pt; anode–graphite;	- $Li,K\|CO_3–CO_2$ - $Li,Na,K\|CO_3$	540–700	Amorphous carbon; various carbon nanostructures (sheet, rings, and quasi-spheres)	(Ijije et al. 2014a, b; Ijije and Chen 2016)
3. cathode–Ni, anode–SnO_2; Pt/Ti	- $Li,Na,K\|CO_3$ - $Li,Na,K\|CO_3$-Li_2SO_4 (CO_2-SO_2 atm.)	450–775	Carbon nanoflake, nanosheet and heart-shaped nanostructured cage, graphite	(Tang et al. 2013; Du et al. 2019; Deng et al. 2016a)
4. cathode–Au, C; anode–Au	- $Li,Na\|CO_3$; - $Li,K\|CO_3$; -$Li,Na,K\|CCO_3$; - $Na,K\|CO_3$	450–850	C (characteristics are absent) The emphasis is on the cathodic process mechanism	(Chery et al. 2014, 2015; Yin and Wang 2018)
5. cathode–steel, Au, Cu,Ti, C	-$Li,Na,K\|CO_3$–CO_2 atmosphere - $Li,Na\|CO_3$; - $Li,K\|CO_3$	500–700	amorphous and graphitized carbons, with plate-like and spherical carbon aggregates	(Glenn et al. 2020; Hughes et al. 2018, 2020)
2b. In chloride-carbonate melts:				
1. cathode–GC; W; anode–Ni	- $Li,Na\|Cl–Na_2CO_3$ (Ar or CO_2 atmosphere)	650–800	quasi-spherical carbon particles accompanied by small amounts of wire-like carbon	Ge et al. (2016)
2. cathode–Ni; anode–graphite	- $Li,K\|Cl–CaCO_3$ (CO_2 atmosphere)	450–650	-micron-sized hollow carbon spheres.; ultrathin sheets; quasi-spherical particles, coral–like carbon and carbon nanofibers	Deng et al. (2018)
3. cathode—W; Anode—Pt, graphite	- $LiCl–Li_2CO_3$ (CO_2 atmosphere)	700	micron-sized amorphous carbon quasi-spherical structures	Ge et al. (2015)
4. cathode–Ni; Anode—graphite	- $Li,Na,K\|CO_3–Li_2O–CO_2$ (Li_2CO_3); - $Li,K\|Cl–Li_2O/CaO–CO_2$ ($Li_2CO_3/CaCO_3$)	450 450	quasi-spherical carbon, micron-sized carbon sheets; micron-sized carbon sheets and shell-like carbon	Deng et al. (2016a)
5. cathode–graphite, Ti, Ti_xO	- $Li,K\|Cl–K_2CO_3$	450–500	C films are comprised of micron-sized "quasi-spherical" carbon particles with graphitized and amorphous phases	Song et al. (2012)

(continued)

Table 2 (continued)

Electrode materials	Bath composition	T, °C	Products	Ref
3. Electroreduction of CO_2 3a. Direct electroreduction:				
1. cathode–Pt, Au, GC; anode–GC	1. Na,K|Cl–CO_2 (10 atm.); 2. Na,K,Cs|Cl–CO_2 (15 atm.)	700–800 550	1. MWCNTs (Ø 5–200 nm), bunches of tubes, nano-Fe in MWCNTs, carbon fibers; 2. superdispersed amorphous carbon, nano-Pt in carbon	(Novoselova et al. 2008a, b; Novoselova et al. 2016)
2. cathode–Pt, Au, Cu, Ni, steel, Al, $LaRhO_3$-coated Ti plate	- Li,Na|Cl; Li,K, Na|F; - Al,Na|Cl–flow systems with Ar or CO_2	360–900	C CO	Halman and Zukerman (1987)
3b. Indirect electroreduction:				
1.cathode–steel; anode–ZrO_2	-LiCl–Li_2O–CO_2 -$CaCl_2$–CaO–CO_2	650 900	nanoscale amorphous carbon, rob-like graphite crystals	Otake et al. (2013)
2. cathode–steel, Pt; anode–Pt	-Li,K|Cl–CO_2	540–700	amorphous carbon, nanoscale carbon sheets, quasi-spheres carbon particles	Ijije et al. (2014a, b)

*GC –glassy-carbon; Ø–outer diameter of tubes; MWCNTs and SWCNTs–multiwall and singe-wall carbon nanotubes

melt with the formation of a carbonate anion. The latter can subsequently be reduced electrochemically at the cathode with production solid carbon of various structures (amorphous, graphite—like carbon tubes and fibers).

Variants (2) and (3) in Table 2 were called recently in literature "Molten Salt CO_2 Capture and Electrochemical Transformation" (MSCC–ET) (Deng et al. 2016a; Weng et al. 2019; Liu et al. 2020). The MSDD–ET process is schematically represented in Fig. 5. As a rule the MSDD–ET process consists of two sequential stages: CO_2 is first captured by O^{2-} ions of the melt, and complex anions $[C_{(n+1)}O_{(n+3)}]^{2-}$ (e.g., $CO_3{}^{2-}$ or $C_2O_5{}^{2-}$) are formed, which are then reduced to carbon at the cathode. At the same time, O_2 evolves on the anode surface.

The MSCC–ET process is a very promising line of research for industrial scaling. In this process, carbon and oxygen are the only products of the electrochemical splitting of CO_2 at the temperature below 800 °C. The applications of electrolytic carbon are determined by its physicochemical and functional properties, and O_2 evolution at the anode provides an ecologically friendly process. The investigations of the MSCC–RT process focus mainly on three aspects: (1) the disclosure of the mechanisms of electrode processes for better control of the entire process; (2) the study of the properties of electrodeposited carbon, including morphology, particle size, and specific surface areas; and (3) the study of the use of different materials for electrodes, in particular inert anode materials for oxygen evolution.

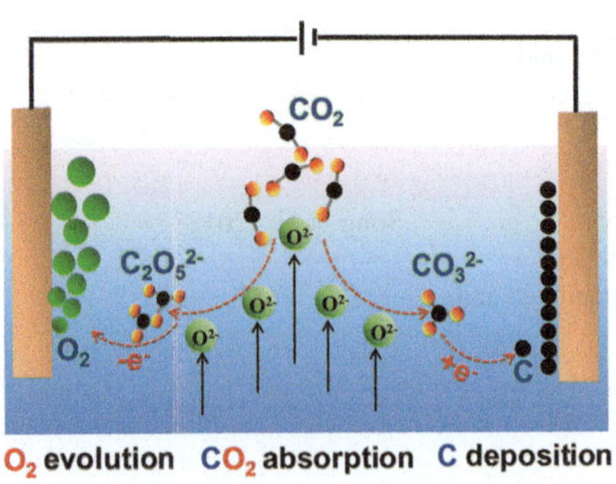

Fig. 5 Schematic illustration of the MSCC-ET process. Reprinted from (Weng et al. 2019) with permission of Elsevier

Very interesting and promising studies are carried out by Professor Stuart Licht and coworkers in the USA (Licht et al. 2010; Wu et al. 2016; Johnson et al. 2017; Wang et al. 2019; Ren et al. 2019; Liu et al. 2020). They proposed a variant of the MSCC-ET process, the so-called Solar Thermal Electrochemical Photo (STEP) process for the preparation of CNMs. In the STEP process, the visible ultraviolet part of the solar spectrum is concentrated on a photovoltaic device to produce the electrical energy, which is spent on electrolysis. At the same time, the thermal energy of the sun is concentrated on another device for heating the electrolyzer. This allows one to use the full spectrum of sunlight and realize STEP electrolysis with a very high efficiency. The STEP process is shown schematically in Fig. 6.

In Ref (Licht et al. 2010) the possibility of developing large-scale electrolyzers for the decomposition of CO_2, which combine the STEP process for the obtaining of C (CO), has been assessed. Research showed that products can be obtained with a solar efficiency of the order of 50%. The STEP is able to connect in series 3 carbonate electrolytes (molten Li_2CO_3) using only the power of one Spectrolab CPV (concentrator photovoltaic device).

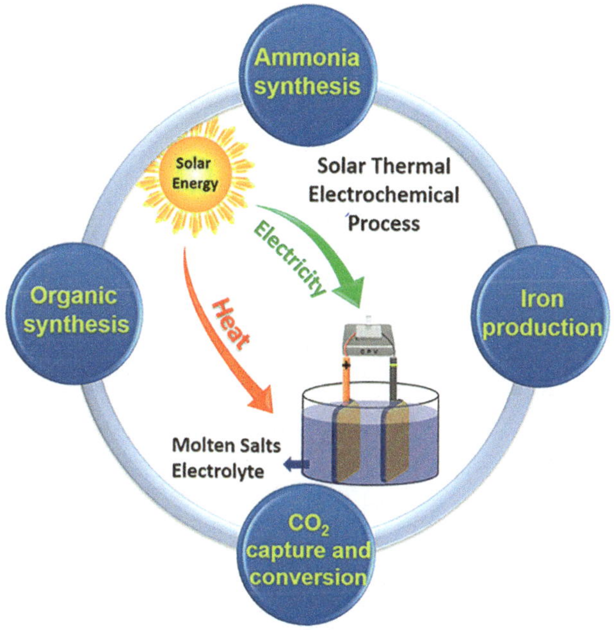

Fig. 6 Schematic representation of the solar thermal electrochemical process. Reprinted from Ren et al. (2019) with permission Copyright 2019 American Chemical Society

In spite of the large number of investigations carried out in the world to study the cathodic discharge of carbonate anion and carbon dioxide, there is no generally acknowledged mechanism in literature up to now which would reliably and fully describe the occurring electrochemical processes. Understanding the mechanisms of the processes is very important for the selective production of nanosized carbon materials with a given structure and morphology. In Sect. 3.3, our idea of the mechanisms of these processes will be presented.

3.2 Direct Electrochemical Reduction of CO_2 in Chloride Melts

The electroreduction of carbon dioxide was investigated in a three-electrode cell made of special stainless steel at temperature up to 800 °C and at an excessive gas pressure up to 15 atm (Kushkhov et al. 1987; Novoselova et al. 2007, 2008a, b, 2016). The creation of an excess pressure of CO_2 in the electrolyzer is necessary to increase its solubility in the chloride melt. The cyclic voltammetry, linear sweep voltammetry, potentiostatic and galvanostatic electrolysis were the methods of the study. The details of the experiment and the design of the electrochemical cell are described in Novoselova et al. (2016). Chloride melts of different composition were used as background electrolytes: a ternary eutectics—Na,K,Na|Cl,F and Na,K,Cs|Cl; a binary equimolar salt mixture Na,K|Cl.

It was detected (Novoselova et al. 1988) that CO_2RR occurred in one stage at cathode polarization rates of 0.2—10 V/s on Pt, Au, GC electrodes in the indicated melts (Fig. 7a, b). The voltammograms recorded one cathode wave in the potential range of −0.2–−0.5 V and one anode wave of dissolution of the cathodic product in the potential range of −0.1–0 V relative to Pt-carbonate ($Pt|CO_3^{2-}/CO_2, O_2$) reference electrode. The currents of the anode and cathode waves are almost equal. (Fig. 7b).

An analysis of the obtained voltammograms was carried out under the various conditions of the polarization scanning. It was found that with an increase of scanning rates of the electrode, a shift of the peak potentials of the cathodic and anodic waves along the potential axis is observed, and their difference exceeds that for a reversible four-electron process (Fig. 7b). This indicates that the rate determining stage of the cathode process is the charge transfer stage and its irreversible nature. The linear dependence of the cathodic wave currents on the pressure of the CO_2 gas in the cell (Fig. 7c) and dependence of the wave currents on the polarization rate (Fig. 7d) of the electrode indicates on the diffusion control of the cathode process. In general, the mixed kinetics of the electroreduction of CO_2 is observed, and the cathodic process is controlled by both the electron transfer rate and the rate of gas diffusion to the electrode. The cathodic product obtained by electrolysis of the melt NaCl-KCl-CO_2 (10 atm) at platinum electrode at temperature 750 °C and a potential of—0.8 V versus Pt– carbonate reference electrode, according to *X*-ray phase analysis, was carbon.

If scanning rate of cathode is lower than 0.1 V/s, voltammogram exhibits two cathodic waves (A) and (B) at Pt electrode: wave (A) $E_{1/2} = -0.44$ V, wave (B) $E_{1/2} = -0.78$ V vs ($Pt|CO_3^{2-}/CO_2$-O_2) reference electrode (Fig. 8a). If the reverse of the polarization scanning occurs after the limiting current of the first cathodic wave (A), only one anodic wave (C) is recorded on the cyclic voltammogram. If the reverse of the scanning occurs after the limiting current of the second cathodic wave (B), two anodic waves (C) and (D) are detected at Pt cathode. The peculiarity of the cyclic voltammograms recorded at the gold cathode (Fig. 8b) is the present of only one anodic wave (D) on the reverse curve that can be explained by the absence of interaction between carbon and gold.

Potentiostatic electrolysis was fulfilled at potentials of the first and second cathodic waves to elucidate the mechanism of the cathode process. The gray films were obtained at potential—0.6 V and consisted of metals Ca, Mg, Zn, and some others (base electrolyte impurities) (Fig. 9a). The black film was obtained at potential—1.0 V (Fig. 9b,c) and consisted of polycrystalline carbon according to XRD and electron-diffraction methods.

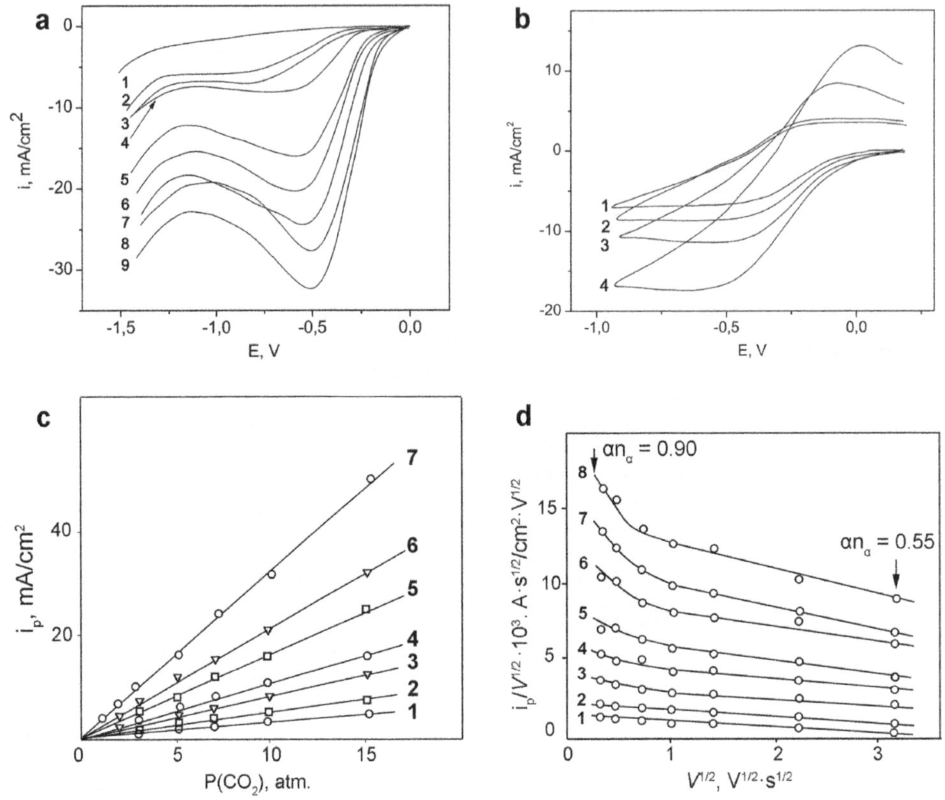

Fig. 7 Voltammograms of NaCl-KCl (1:1)–CO_2 melt at a Pt electrode at T = 700 ^{0}C: **a** at different CO_2 pressures, atm.: 0 (1); 1 (2); 2.5 (3); 5 (4); 7.5 (5); 10 (6); 12.5 (7); 15 (8); 17 (9); V = 5 Vs^{-1}; **b** at different scanning rates, Vs^{-1}: 1 (1); 2 (2); 5 (3); 10 (4); P_{CO2} = 5 atm.; **c** plots of i_p against CO_2 pressure in the melts Na,K|Cl (1 ÷ 6) and Na,K|Cl,F (7) at different scanning rates, Vs^{-1}: 0,1 (1); 0,2 (2); 0,5 (3); 1 (4); 5 (5,7); 10 (6); **d** plots $ip/v^{1/2}$ against $v^{1/2}$ at different gas pressures, atm.: 1 (1); 2 (2); 2.5 (3); 5 (4); 7.5 (5); 10 (6); 12 (7); 15 (8). Taken from (Novoselova et al. 2016) with permission from Elsevier

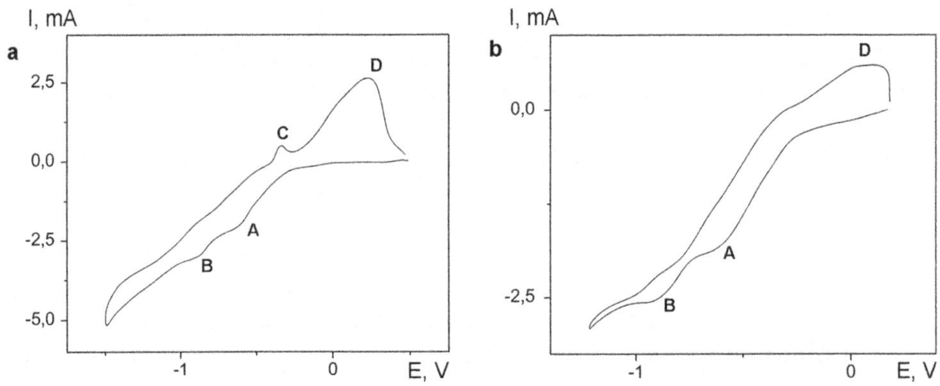

Fig. 8 Cyclic voltammograms of the system Na,K|Cl-CO_2 (10 atm), 750 ^{0}C, V = 0.1 Vs^{-1}: **a** at Pt electrode, S_{Pt} = 0.17 cm^2; **b** at Au electrode, S_{Au} = 0.2 cm^2. Taken from (Novoselova et al. 2016) with permission of Elsevier

To explain the obtained results, a three-stage ECE (electrochemical–chemical–electrochemical) mechanism of the direct electroreduction of CO_2 was proposed in Novoselova et al. (2016). The cathodic wave (A) corresponds to the first electrochemical stage of the general cathode process and represents an irreversible 2-electron reduction of carbon dioxide dissolved in the melt with the formation of an unstable CO_2^{2-} anion:

$$CO_2 + 2\bar{e} \rightarrow CO_2^{2-} \quad (20)$$

Anion CO_2^{2-} has strong reducing properties (that is indicated by the reduction products of the electrolysis carried out at the potential of the first wave) and which is an unstable short-lived particle (that is indicated by the absence of anodic current after the first cathodic wave). The formation of this ion was found in Borucka (1977).

The second chemical stage of the general process is the dissociation of CO_2^{2-} anion with the formation of carbon monoxide and oxide anions:

$$CO_2^{2-} \rightleftarrows CO + O^{2-} \quad (21)$$

The second cathodic wave (B) corresponds to the third electrochemical stage of the general process and consists in an irreversible 2-electron reduction of CO up to solid carbon:

$$CO + 2\bar{e} \rightarrow C + O^{2-} \quad (22)$$

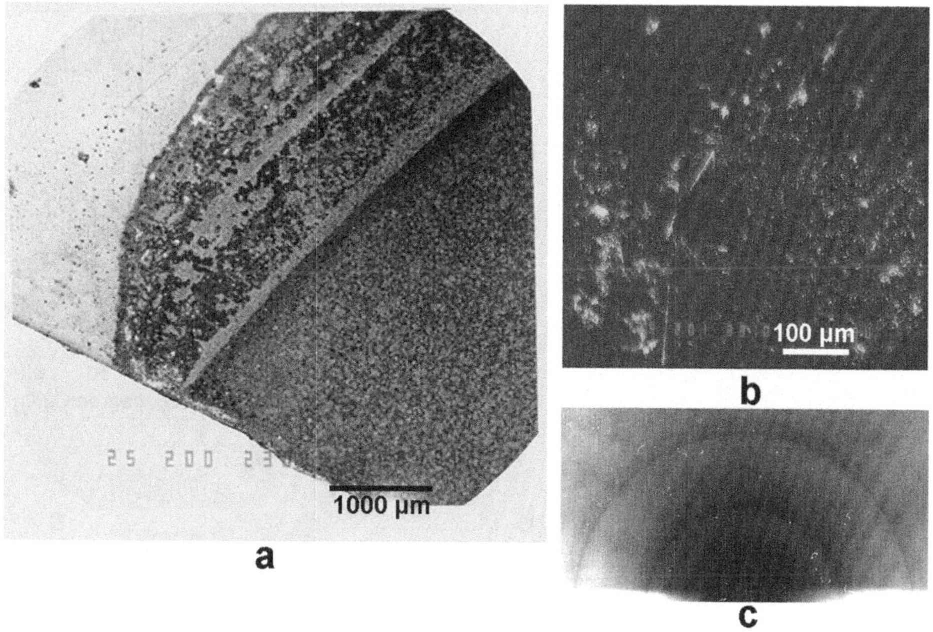

Fig. 9 SEM images (**a**, **b**) of cathodic products and electron diffraction pattern (**c**) of product (**b**) produced in the system Na,K|Cl-CO_2 (10 atm.) at Pt plate electrodes at 750 ^{0}C and potentials: **a** −0.6 V; **b** −1.0 V. Taken from (Novoselova et al. 2016) with permission of Elsevier

Carbon dioxide acts simultaneously not only as a carbon precursor, but also as an acceptor of O^{2-} anions that are released at the cathode and binds them into a carbonate anion:

$$CO_2 + O^{2-} = CO_3^{2-} \quad (23)$$

Oxide anions O^{2-} strongly inhibit the kinetics of the cathodic process. So, excess pressure of CO_2 in the cell also prevents the autoinhibition of CO_2RR.

The overall cathode reaction may be represented as:

$$3CO_2 + 4\overline{e} \rightarrow C + 2CO_3^{2-} \quad (24)$$

Carbonate-ion discharge takes place at the anode with formation of carbon dioxide and oxygen:

$$CO_3^{2-} = CO_2 + 1/2\, O_2 + 2\overline{e} \quad (25)$$

The overall electrochemical reaction is:

$$CO_2 = C + O_2 \quad (26)$$

Interesting results were obtained while studying the phase composition and morphology peculiarities of carbon deposits, which were produced in different chloride systems. Figure 10 shows the result of XRD analysis and the morphology of the carbon powder, which was obtained from a Na,K|Cl melt under excess CO_2 pressure of 10 atm on platinum cathode at temperature of 750 °C and current density of 0.1 A cm^{-2}. The XRD analysis showed that the diffraction pattern exhibited in the 2-theta regions two halos: the first with center at 26° and the second with the center at 42° (Fig. 10a). The presence of these halos indicates the existence of carbon (graphitized) planes in the product. The XRD data also showed austenite FCC lattice peaks. The lattice can consist of nickel and iron atoms. The reason of these impurities in the deposit may be the uncontrolled oxidation of the electrolyzer materials. During the long-term (up to 5 h) electrolysis, the reactor walls and other electrolyzer parts can oxidize, and the oxidation products can get into the melt. The oxidized components can dissolve in the chloride melt and then be reduced at the cathode. The relative ratio of the carbon and FCC phases was estimated in each pattern based on ratios of the intensities of the diffraction peaks of carbon {002} and austenite {111} and was between 0.5 and 2.0 in different samples. At the ratio 1:1, the carbon product contained the maximum percentage of carbon tubes in the sample according to the microscopy data (up to 40 vol%) (Fig. 10b, c). Thus, the austenite phase catalyzes the formation of a tubular carbon structure in salt melts. This observation is already purposefully used by researchers (Ren et al. 2019) to fabricate well oriented carbon nanotubes in great amount.

If refractory metal ions are added to the electrolytic bath under investigation, then the co-electroreduction of carbon and metal to form a metal carbide can be realized. The main condition for the formation of a carbide phase over a wide current density range is the closeness of the carbon and metal deposition potentials. Researches in this direction are currently under way by some groups (Stulov et al. 2017; Dubrovskiy et al. 2018; Kushkhov and Adamokova 2007) and in our laboratory (Novoselova et al. 2001, 2003, 2014, 2018, 2020a). Nanoscale powders of tungsten carbide (WC, W_2C), molybdenum carbide (Mo_2C), and

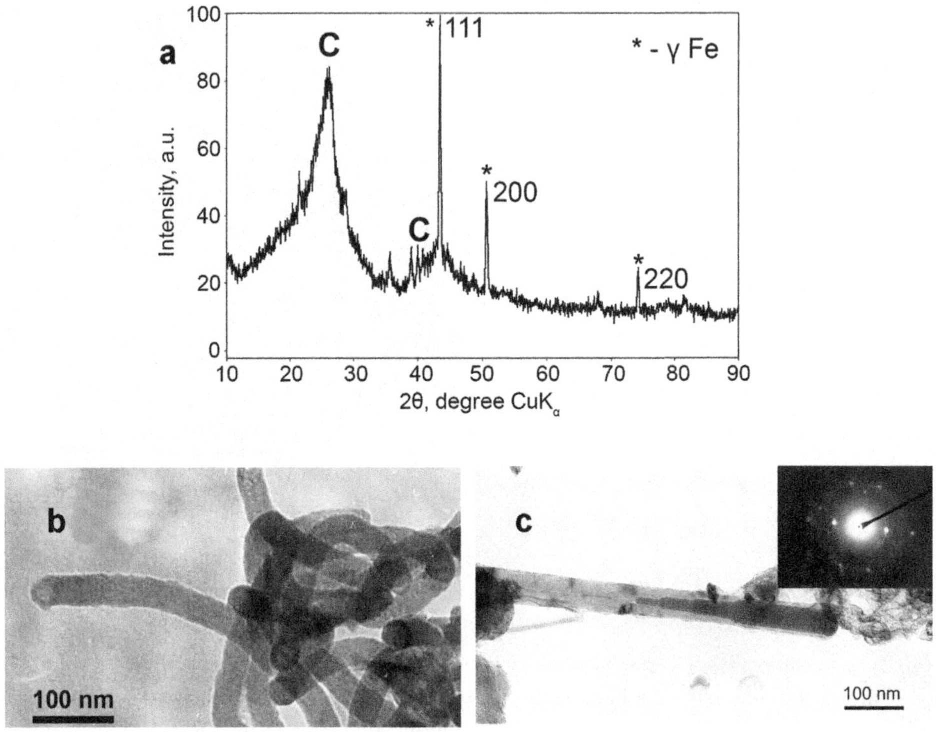

Fig. 10 a XRD pattern of cathodic product obtained in the melt Na,K|Cl-CO_2 (10 atm.) and its SEM images (**b, c**)

composite materials based on them have been synthesized (Novoselova 1988). Double tungsten (molybdenum) carbides with cobalt or nickel can also be obtained by high-temperature electrochemical synthesis (Kushkhov et al. 1990a, b). To this end, the composition of the electrolytic bath must be selected in such way that the co-electroreduction of already three components (carbon, refractory metal, and cobalt (nickel)) is realized. For this purpose, the formation of new electrochemically active species in situ by acid–base interactions in melts is used (Novoselova et al. 2020a). Single and double tungsten and molybdenum carbides synthesized electrolytically from molten salts possess a whole set of properties that are interesting in applied aspect. They have nanosized grains, high electronic conductivity values, a high chemical stability in acidic media, and a unique morphology (nanofibers, nanohoneycombs) (Fig. 11). These characteristics make it possible to use the above materials as catalysts and electrode materials in modern electrochemical devices (Novoselova et al. 2018).

3.3 Indirect Electrochemical Reduction of CO_2 in Molten Salts

Indirect CO_2RR in molten salts generally involves three main stages: (1) capture of CO_2 by molten salt through its physical or chemical dissolution; (2) electrochemical conversion of CO_2 to C or CO on the cathode and O_2 on the anode; and (3) purification and separation of products. Electrolysis should be carried out in closed or flow cells in an atmosphere of carbon dioxide. The most commonly used electrolytes are single, binary, or ternary carbonate systems, necessarily containing cations of lithium, calcium, magnesium, or barium at temperatures 400–1000 °C. These electrolytes have a wide electrochemical potential window and good dissolving ability for CO_2. Halide-carbonate and halide-oxide mixtures can also be used.

The first stage—the capture of CO_2 by molten salt—is investigated in detail by Deng et al. (Deng et al. 2016a, b). They study the thermodynamics and kinetics of CO_2 static absorption in Li_2O containing carbonate and chloride melts at various CO_2 partial pressures and temperatures.

Wang et al. (Jiang et al. 2019; Deng et al. 2018, 2016a, 2019; Ge et al. 2015; Song et al. 2012; Wu et al. 2017) developed a process for the capture and electrochemical conversion of CO_2 and called it MSCC–ET process. This scientific group conducted an extensive research on the thermodynamics and kinetics of the cathode and anode discharges of carbonate anions in carbonate and chloride-carbonate melts, to study the properties of the electrolytic carbon product, the stability and efficiency of anode materials, current efficiency, and energy consumption. Licht et al. were the first to propose to combine the processes of a CO_2 capture and electrolysis in molten salt with STEP process (Licht et al. 2010; Wu et al. 2016; Johnson et al. 2017; Wang et al. 2019; Ren et al. 2019; Liu et al. 2020). Kaplan et al. succeeded in converting CO_2 to CO with

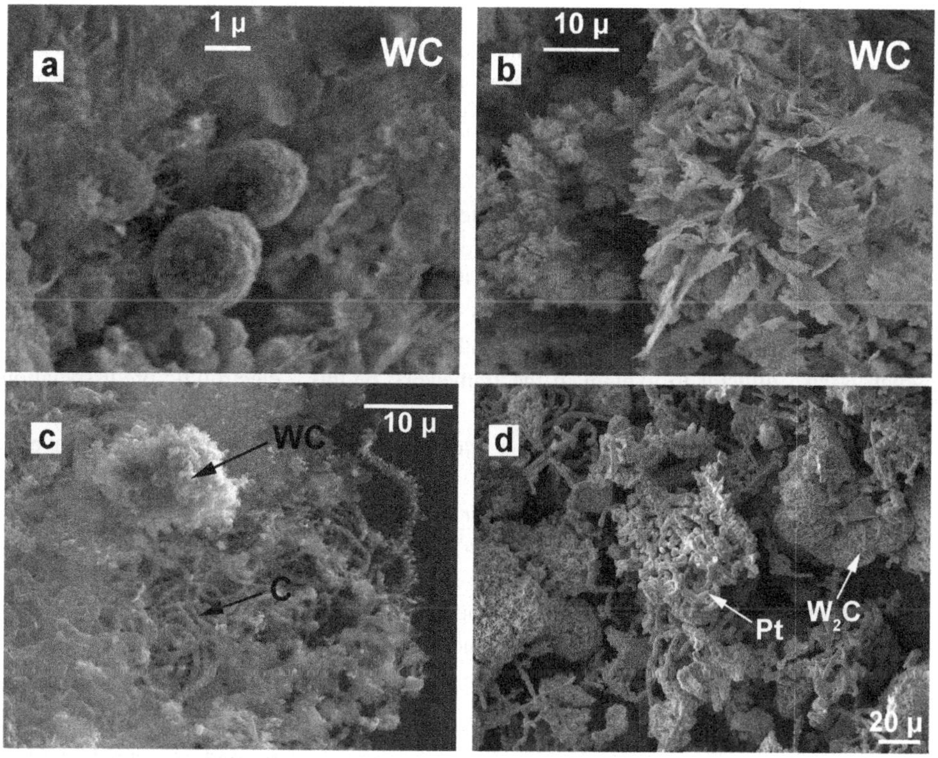

Fig. 11 SEM images of electrolytic powders of tungsten carbides: WC (**a, b**), composites WC + C (**c**), and W_2C + Pt (d), produced in the system NaCl-KCl-$Na_2W_2O_7$-CO_2

almost 100% Faraday efficiency by electrolysis of molten lithium carbonate (Kaplan et al. 2010, 2014).

The chemical composition and the form of the cathodic products in the MSCC-ET process can be quite different. Carbon powders of different morphology, structure and dispersity, carbon films, carbide coatings and powders, carbon monoxide and synthesis gas (Wu et al. 2017) were obtained under various experimental conditions. But research mostly focused on solid-phase carbon production. Graphene was obtained in molten chloride according to Refs (Liu et al. 2020; Hu et al. 2016) and hollow carbon spheres (HCS) were obtained by Deng et al. (2017). Composition and properties of carbon products depend on the conditions and modes of electrolysis (cathode potential, bath potential, current density, electrolyte composition, electrolysis duration, cathode and anode materials, temperature, etc.). For the synthesis of a product with a given composition and properties, it is necessary to elucidate the mechanisms of occurring processes.

3.4 The Mechanisms of Electrode Reactions Occurring at the Cathode and Anode

In spite the large number of publications devoted to the study of the cathodic discharge of the carbonate anion and carbon dioxide in molten salts, there is still no universally recognized mechanism that would completely describe the results obtained. Understanding the mechanisms of processes is very important for developing a sustainable, controlled, and industrially scalable method for the production of nanoscale carbon-containing materials (nanosized carbon structures and carbides of refractory metals) with given properties.

During electrolysis, it is also possible to reduce CO_3^{2-} ions to carbon monoxide (CO) according to reaction (27). However, as it has been shown by thermodynamic calculations (Chery et al. 2014; Yin and Wang 2018) and experiments carried out (Kaplan et al. 2010, 2014) reaction (27) will proceed at temperatures above 850 °C.

$$CO_3^{2-} + 2e^- \rightarrow CO + 2\,O^{2-} \tag{27}$$

To date, the literature suggests three possible mechanisms for the cathodic discharge of the carbonate anion to carbon in the solid phase.

(1) *Direct one step transfer of 4 electrons* via *reaction (28).*

The produced O^{2-} anions can react with CO_2 (from the atmosphere above the melt) and regenerate the CO_3^{2-} ions according to (29):

$$CO_3^{2-} + 4\bar{e} \rightarrow C + 3O^{2-} \tag{28}$$

$$CO_2 + O^{2-} = CO_3^{2-} \tag{29}$$

Despite the fact that this mechanism is used by many authors, there are certain contradictions in it, namely:

–effect of the cationic environment on the features of carbonate-anion discharge is not taken into account, but this effect has been recorded by many researchers (Ijije and Chen 2016; Deng et al. 2016a; Chery et al. 2015; Hughes et al. 2018; Wang et al. 2019).

–limiting currents are observed on the voltammograms of carbonate-anion discharge in pure carbonate melts, which indicates kinetic difficulties of the discharge process. This fact was not explained in the framework of this model;

–chemical composition and morphology of the carbon product strongly depends on the specified cathode potential, which is also difficult to explain in the framework of this mechanism.

(2) *Two-step process, involving the two successive stages with the transfer of two electrons in each stage and the formation of CO (reactions* (30) and (30a)*) or a hypothetical intermediate anion* CO_2^{2-} *according to reactions* (31) and (31a) (Ito et al. 1992).

The formation of CO_2^{2-} ion can take place by the combination of a CO molecule and a O^{2-} ion and was detected in Borucka (1977).

$$CO_3^{2-} + 2\bar{e} \rightarrow CO + 2O^{2-} \quad (30)$$

$$CO + 2\bar{e} \rightarrow C + O^{2-} \quad (30a)$$

$$CO_3^{2-} + 2\bar{e} \rightarrow CO_2^{2-} + O^{2-} \quad (31)$$

$$CO_2^{2-} + 2\bar{e} \rightarrow C + 2O^{2-} \quad (31a)$$

All these remarks for the first CO_2RR path can be surely related to the second mechanism.

(3) *Electrowinning of an alkali metal on the cathode, which then chemically reduces carbonate anions to carbon according reactions* (32–34).

This is indirect electrochemical CO_2 conversion.

$$M_2CO_3 \rightleftharpoons 2M^+ + CO_3^{2-} \quad (32)$$

$$M^+ + x\bar{e} \rightarrow M \quad (33)$$

$$2xM + M_xCO_3 \rightarrow C + 3M_xO \quad (34)$$

where *M* is monovalent metal.

The authors (Smirnov et al. 1965; Deanhard et al. 1986) attributed this discharge mechanism to the absence of limiting current in voltammograms before the exponential discharge of the base electrolyte in both pure carbonate and fluoride-carbonate melts. It was therefore postulated that CO_3^{2-} ions are not electrochemically active species, but are rather reduced in a chemical reaction following the electrochemical formation of an alkali metal. However, recent studies (Chery et al. 2014, 2015; Ge et al. 2016; Deng et al. 2018) have shown that limiting currents of carbon electroreduction are recorded on voltammograms both in pure concentrated carbonate and halide-carbonate melts. It depends on the experimental conditions and especially on the nature of cathodic material.

There is also a fourth mechanism which can explain the observed peculiarities of the carbonate anion discharge, and which is being developed in our laboratory. This is the so-called acid–base type of preceding electrochemical reaction.

In molten electrolytes, some discharging elements (tungsten, silicon, carbon) are in the form of stable anionic complexes, which can be reduced at the cathode without electrolytic predissociation ($[CO_3]^{2-} \rightleftarrows C^{4+} + 3\,O^{2-}$).

Yu. K. Delimarskii and V. I. Shapoval (Delimarskii et al. 1972, 1985) have extended the concept of acid–base equilibria to melts containing CO_3^{2--} oxyanions and described the effect of the following equilibria existing in ionic melts on the electroreduction kinetics of the above anion:

$$[CO_3]^{2-} \rightleftarrows CO_2 + O^{2-} \quad (35)$$

The authors established a law, according to which in the presence of such equilibria, the species with the most acidic properties, namely CO_2 species, are reduced in the first place. The acid–base equilibria of the type (35) are often called in literature Lux-Flood equilibria, and the basicity of the melt is expressed by the pO value, which is equal to the negative logarithm of oxide ion activity. The pO value is the same ionic characteristic for oxygen-containing molten systems as pH for aqueous solutions. The prelogarithmic factor of the equation of dependence of equilibrium potential for CO_3^{2-} on the basicity of the melt is:

$$E = E^o + 1.725RT/F \cdot pO \quad (36)$$

An important approach of controlling electrode processes is changing the "basicity" or "acidity" of the ionic melt. The measures of the "basicity" or "acidity" of ionic melts are the concentrations of such ions as O^{2-}. The problem of the effect of acid–base equilibria on the kinetics of electrode processes in ionic melts was considered in (Shapoval et al. 1974; Shapoval 1975). By shifting acid–base equilibria by adding oxygen ion acceptors or donors to the electrolyte, one can conduct the electrolysis in the desired direction and within definite limits.

Therefore, we proposed the following fourth discharge mechanism of thermally unstable carbonates.

(4) A preceding chemical reaction of acid–base type first takes place (reaction (35)*), and then an electrochemical discharge of carbon dioxide formed by this reaction occurs at the cathode (reaction* (37)*):*

$$CO_3^{2-} \rightleftarrows \{CO_2 + O^{2-}\} + 4\bar{e} \rightarrow C + 3O^{2-} \quad (37)$$

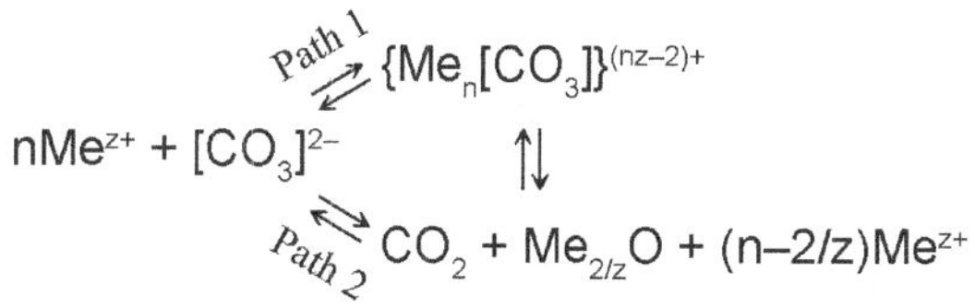

(38)

Scheme 1 Reaction (38)

Thus, the electrochemically active species is CO_2. This assumption was confirmed in effecting a direct reduction of carbon dioxide dissolved in a Na, K|Cl salts mixture under an excess pressure at 700–750 °C (Novoselova et al. 2016), which was shortly described in Sect. 3.2. The above reactions (35) and (37) are somewhat simplified since they do not take into account the possible effect of the cationic environment of discharging species. It has been shown by the quantum chemistry method (Shapoval et al. 1984; Soloviev 1985; Novoselova et al. 2020b) that by changing the cationic composition of the electrolyte, one can transform anionic carbonate complexes into a new active state, cationized carbonate complexes according to Scheme 1—reaction (38).

where *z*—is the cation charge, *n*—is the order of reaction with respect to cation or coordination number (CN) of the complex.

Thus, cation–anion interaction can result either in the formation of cationized anions (metal complexes) or in anion dissociation both under the direct influence of cations and through the intermediate stage of formation of "short-lived" metal complexes.

It has been shown by the method of voltammetry (Novoselova et al. 2020b) that there is no wave of cathodic discharge in the Na, K, Rb|Cl chloride melt, containing sodium carbonate, up to the electroreduction potentials of the base electrolyte. This means that in the chloride melt containing weakly polarizing Na^+, K^+, Rb^+ cations, the CO_3^{2-} anion shows no electrochemical activity in the temperature range of 570–700 °C. This is accounted for by the large values of the activation barriers to the two- and four-electron reduction of CO_3^{2-}.

When Li^+, Ca^{2+}, Mg^{2+}cations (cations with high specific charge) in the form of chlorides are added to this melt, waves appear in all three electrolytes in the same potential range (−0.7 to −0.9 V versus to a silver reference electrode). The catodic product of electrolysis at this potential is solid carbon. This means that the carbonate anion interacts with these cations, and a new electrochemically active species is formed. It was suggested that it can be a cationized carbonate complex. Subsequently, it decomposes, owing to its instability, into an oxide anion O^{2-} and carbon dioxide, which is then reduced at the cathode. In this case, the carbonate anion is discharged under kinetic conditions, viz. with a preceding chemical reaction of acid–base type. The reaction path will depend on the concentration of cations and their specific charge values.

Only two principal anodic reactions are possible: the oxidation of CO_3^{2-} ions to CO_2 and O_2 via reaction (39): (40).

$$2CO_3^{2-} \rightarrow 2CO_2 + O_2 + 4\bar{e} \tag{39}$$

Besides the formation of O_2 at the anode via reaction (39), O^{2-}ions produced at the cathode can migrate toward the anode where they can be oxidized to O_2 according to reaction (40): (41):

$$2O^{2-} \rightarrow O_2 + 4\bar{e} \tag{40}$$

3.5 Prospects for CO_2 Conversion in Molten Salts

The capture of CO_2 with a molten salt and its electrochemical conversion (MSCC–ET) is a promising way for the economic and effective use of this greenhouse gas. In this process, captured CO_2 (in the form of dissolved gas or carbonate ions) is selectively converted to carbonaceous materials with high added value and CO, demonstrating a high current density (of up to 1.5 A/cm^2) and current efficiency (of up to 100%). A practical application of MSCC–ET is the reuse of deposited carbon to generate electricity by reoxidation. The electrical energy that is used to produce electrolytic carbon can be stored as possible chemical energy in the carbon deposit. If necessary, this chemical energy can be converted back to electricity by the repeated electrochemical and chemical oxidation of electrolytic carbon. Storage and transportation of solid-state electrolytic carbon is an inexpensive and environmentally friendly approach. The composition of the process products can be changed by selecting the necessary electrolysis conditions and they can be carbon, carbon monoxide, or synthesis gas, as shown in Fig. 12.

The cathodic products (carbonaceous materials) obtained by CO_2 electrolysis in molten salts have a high added value. They have been tested as electrode materials for batteries (Groult et al. 2006; Ge et al. 2015; Licht et al. 2016), supercondensors (Hughes et al. 2018, 2020; Yin et al. 2013); in hydrogen production electrolyzers (Novoselova et al. 2018), and as pollutant absorbents. The results obtained indicate promising prospects of using the above materials in modern devices and technologies.

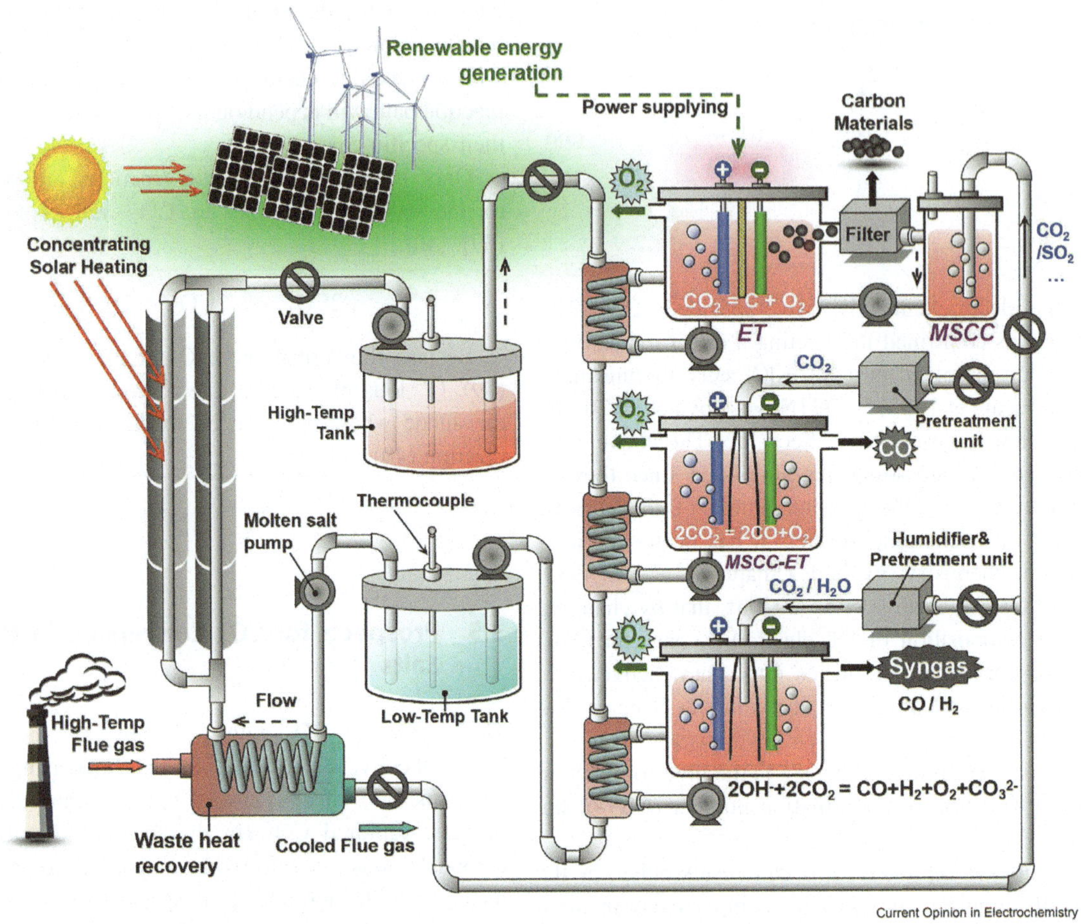

Fig. 12 Schematic of molten salt CO_2 capture and electrochemical transformation (MSCC-ET) process with heat exchange between molten salt outside heat by solar and/or high-temperature flue gas. Taken from (Jiang et al. 2019) with permission from Elsevier

Nevertheless, much unknown still remains in terms of basic research and know-hows for technological developments. The challenges facing the scientists working in this area are well covered in the review (Jiang et al. 2019). Here are some of them. Although various carbon materials have been obtained and characterized, the detailed mechanisms of CO_3^{2-} electroreduction as well as the paths of the formation of new crystal structures from reduced carbon atoms and the principles of its control are unclear up to now. This knowledge is very important for the control of the synthesis and further use of the product. Moreover, there are no optimization composition of the molten electrolyte and the behavior of the gas at the molten salt–electrode–CO_2 gas interface. The development of a reliable and readily available inert anode is not yet completed. In engineering aspect, the following issues are very topical: preparation of starting CO_2 from different sources for electrolysis; design, stability, and service life of electrolyzers and their component parts; collecting, separating, and washing the products free from the reaction medium.

4 Conclusions

In recent years, the electrolytic conversion of CO_2 has attracted great attention of researchers due to the following advantages:

(1) the process can be controlled by the parameters and conditions of electrolysis (electrode potentials, current density, bath composition, electrode materials, temperature);

(2) base electrolytes can be completely recycled, so that total reagent consumption can be minimized;

(3) the source of electricity used for the process can be obtained without generating new CO_2 using renewable energy sources, such as solar and wind energy, hydroelectric, geothermal, and thermoelectric processes;

(4) electrochemical reaction cells are compact and modular for large-scale applications.

The combination of CO_2 electroreduction with a renewable energy sources for the production of expensive fuels and chemicals is a very important and prospect strategy to create a sustainable energy economy. The electrochemical conversion of carbon dioxide can be realized from a wide range of conducting media (aqueous, non-aqueous (organic and ionic liquids), solid oxide, and molten electrolytes). Compared to aqueous and non-aqueous electrolytes, the high-temperature technologies—SOEC and MSCC-ET (STEP)—have the highest efficiency of the processes and may use the flue gases. In addition, MSCC-ET (STEP) technologies do not need complex and expensive catalysts. Based on an analysis of the readiness levels of the technologies considered, SOEC presently is the most effective and mature technology for the production of CO from CO_2.

The MSCC-ET process, especially in the view of the STEP process, is expediently used in the preparation of solid-phase carbon materials (carbon sheets, tubes, fibers, graphene). These materials can be used to generate and store energy, as protective coatings and absorbents. In turn, low-temperature aqueous and organic electrolytes can offer very promising opportunities for the production of other valuable chemicals, such as formic acid, methane, methanol, ethanol, or ethylene. None of these compounds can be obtained by the high-temperature electrolysis of molten carbonates or solid oxides.

References

Abbott AP, Eardley ChA (2000) Electrochemical reduction of CO_2 in a mixed supercritical fluid. J of Phys Chem B 104:775–779

Alvarez-Guerra M, Albo J, Alvarez-Guerra E, Irabien A (2015) Ionic liquids in the electrochemical valorisation of CO_2. Energy Environ Sci 8:2574–2599

Aschenbrenner O, Styring P (2010) Comparative study of solvent properties for carbon dioxide absorption. Energy Environ Sci 3:1106–1113

Balasubramanian R, Wang W, Murray RW (2006) Redox ionic liquid phases: ferrocenatedimidazoliums. J Am Chem Soc 128:9994–9995

Bartlett HE, Jonson KE (1966) Electrolytic reduction and Elinghem diagrams for oxy-anion systems. Can J Chem 44(18):2119–2129

Belyaev V, Galvita V, Sobyanin V (1998) Effect of anodic current on carbon dioxide reforming of methane on Pt electrode in a cell with solid oxide electrolyte. React Kinet Catal Lett 63:341–348

Boehm HP (1978) Carbon from solid-state reduction of carbonate ions. Carbon 16(1):77–78

Borucka A (1977) Evidence for the existence of stable CO_2^{2-} ion and response time of gas electrodes in molten alkali carbonates. J of Electrochem Soc 124:972–976

Buzzeo MC, Evans RG, Compton RG (2004) Non-haloaluminate room-temperature ionic liquids in electrochemistry–a review. Chem Phys Chem 5:1106–1120

Chandler HW, Pjllara FZ (1966) Oxygen generation in solid electrolyte system. AJCHE chemical engineering progress. Series Aerospace Life Support 62:29–37

Chang CJ (1992) The solubility of carbon dioxide in organic solvents at elevated pressures. Fluid Phase Equilib 74:235–242

Chen GZ, Fray DJ (2003) Recent development in electrolytic formation of carbon nanotubes in molten salts. Journal of Mining and Metallurgy, Section: B-Metallurgy 39:309–342

Chen Zh, Gu Yu, Hu L et al (2017) Synthesis of nanostructured graphite via molten salt reduction of CO_2 and SO_2 at a relatively low temperature. J. Mater Chem A 5:20603–20607

Chery D, Albin V, Lair L, Cassir M (2014) Thermodynamic and experimental approach of electrochemical reduction of CO_2 in molten carbonates. Int J Hydrogen Energy 39:12330–12339

Chery D, Lair V, Cassir M (2015) Overview on CO_2 valorization: challenge of molten carbonates. Frontiers in Energy Research 3:43

Chu S, Majumdar A (2012) Opportunities and challenges for a sustainable energy future. Nature 488:294–303

Chu S, Cui Y, Liu N (2016) The path towards sustainable energy. Nat Mater 16:16–22

Deanhardt ML, Stern KH, Kende A (1986) Thermal decomposition and reduction of carbonate ion in fluoride melts. J Electrochem Soc 133:1148–1151

Delimarskii YuK (1985) Some peculiarities of the electrolysis of ionic melts. Electrochim Acta 30:1007–1010

Delimarskii YuK, Shapoval VI (1972) Equations for cathodic polarograms taking account of the acid-base characteristics of oxygen-containing molten salts. Theor Exp Chem 8:378–382

Delimarskii YuK, Shapoval VI, Grishchenko VF et al (1968) Features of cathodic carbon evolution during the electrolysis of molten carbonates. Dokl Akad Nauk SSSR 183:1332–1334 (in Russian)

Delimarskii YuK, Shapoval VI, Vasilenko VA et al (1970) Electrolysis of fused carbonates of alkali metals under pressure. Zh Prikl Khim 43:2634–2638 (in Russian)

Delimarskii YuK, Shapoval VI, Vasilenko VA (1971) Importance of the kinetic process during the electroreduction of CO_3^{2-} in molten potassium chloride-sodium chloride. Russ J Electrochem 7:1301–1304 (in Russian)

Deng BW, Chen ZG, Gao MX et al (2016a) Molten salt CO_2 capture and electro-transformation (MSCC-ET) into capacitive carbon at medium temperature: effect of the electrolyte composition. Faraday Discuss 190:241–258

Deng B, Tang J, Mao X et al (2016b) Kinetic and thermodynamic characterization of enhanced carbon dioxide absorption process with lithium oxide-containing ternary molten carbonate. Environ Sci Technol 50:10588–10595

Deng B, Mao X, Xiao W, Wang D (2017) Microbubble effect-assisted electrolytic synthesis of hollow carbon spheres from CO_2. J Mater Chem A 5:12822–12827

Deng B, Tang J, Gao M et al (2018) Electrolytic synthesis of carbon from the captured CO_2 in molten LiCl–KCl–$CaCO_3$: Critical roles of electrode potential and temperature for hollow structure and lithium storage performance. Electrochim Acta 259:975–985

Deng B, Gao M, Ru Y, Wang D et al (2019) Critical operating conditions for enhanced energy-efficient molten salt CO_2 capture and electrolytic utilization as durable looping applications. Appl Energy 255:113862

Devyatkin SV (2005) Electrochemical synthesis of carbon nanotubes in molten carbonates. In: 7th international symposium on molten salts chemistry and technology 2005, Toulouse, France, 29 Aug.–2 Sept., 2005 Proceedings 1: 515–517

Dinh C, Burdyny T, Kibria M et al (2018) CO_2 electroreduction to ethylene via hydroxide-mediated copper catalysis at an abrupt interface. Science 360:783–787

Du K, Yu R, Gao M et al (2019) Durability of platinum coating anode in molten carbonate electrolysis cell. Corros Sci 153:12–18

Dubrovskiy A, Makarova O, Kuznetsov S (2018) Effect of the molybdenum substrate shape on Mo_2C coating electrodeposition. Coatings 8:442. https://doi.org/10.3390/coatings8120442

Ebbesen SD, Knibbe R, Mogensen M (2012) Co-electrolysis of steam and carbon dioxide in solid oxide cells. J Ectrochem Soc 159:F482–F489

Endroda B, Bencsik G, Darvas F et al (2017) Continuous-flow electroreduction of carbon dioxide. Prog Energy Combust Sci 62:133–154

Faggion DJ, Gonçalves WDG, Dupont J (2019) CO_2 electroreduction in ionic liquids. Front Chem 7:102

Fennell PS, Shah N, Maitland GC (2017) The role of CO_2 capture and utilization in mitigating climate change. Nat Clim Chang 7:243–249. https://doi.org/10.1038/nclimate3231

Ge JB, Hu LW, Wang W et al (2015) Electrochemical conversion of CO_2 into negative electrode materials for Li-ion batteries. Chem Electro Chem 2:224–230

Ge JB, Wang S, Hu LW et al (2016) Electrochemical deposition of carbon in LiCl-NaCl-Na_2CO_3 melts. Carbon 98:649–657

Glenn MJ, Allen JA, Donne SW (2020) Carbon electro-catalysis in the direct carbon fuel cell utilizing alkali metal molten carbonates: a mechanistic review. J Power Sources 453: 227662

Graves C, Ebbesen SD, Mogensen M et al (2011) Sustainable hydrocarbon fuels by recycling CO_2 and H_2O with renewable or nuclear energy. Renew Sustain Energy Rev 15:1–23

Graves C, Ebbesen SD, Jensen SH et al (2014) Eliminating degradation in solid oxide electrochemical cells by reversible operation. Nature Mater 14:239–244

Groult H, Kaplan B, Lantelme F et al (2006) A preparation of carbon nanoparticles from electrolysis of molten carbonates and use as anode materials in lithium-ion batteries. Solid State Ionics 177:869–875

Gupta R, Schwandt C, Fray D (2014) Preparation of tin-filled carbon nanotubes and nanoparticles by molten salt electrolysis. Carbon 70:142–148

Halman M, Zukerman K (1987) Electroreduction of carbon dioxide to carbon monoxide in molten LiCl + NaCl, LiF + KF + NaF, Li_2CO_3 + Na_2CO_3 + K_2CO_3 and $AlCl_3$ + NaCl. J of Electroanal Chemistry and Interfacial Electrochemistry 235:369–380

Hartvigsen J, Elangovan S, Elwell J, Larsen D (2017) Oxygen production from Mars atmosphere carbon dioxide using solid oxide electrolysis. ECS Trans 78(1):2953

Hayyan M, Mjalli FS, Hashim MA et al (2013) Investigating the electrochemical windows of ionic liquids. J Indus Eng Chem 19:106–112

Hoang Th, Verma S, Ma S et al (2018) Nanoporous copper–silver alloys by additive-controlled electrodeposition for the selective electroreduction of CO_2 to ethylene and ethanol. J Am Chem Soc 140(17):5791–5797

Hori Y (2008) Electrochemical CO_2 reduction on metal electrodes. In: Vayenas CG, White RE, Gamboa-Aldeco ME (eds) Modern aspects of electrochemistry, vol 42. Springer. New York, NY, pp 89–189

Hu L, Song Y, Jiao S et al (2016) Direct Conversion of Greenhouse Gas CO_2 into Graphene via Molten Salts Electrolysis. Chem Sus Chem 9:588–594

Huang J, Mensi M, Oveisi E et al (2019) Structural sensitivities in bimetallic catalysts for electrochemical CO_2 reduction revealed by Ag–Cu nanodimers. J Am Chem Soc 141(6):2490–3249

Hughes MA, Allen JA, Donne SW (2018) The properties of carbons derived through the electrolytic reduction of molten carbonates under varied conditions: Part I. A study based on step potential electrochemical spectroscopy. J Electrochem Soc 165:A2608–A2624

Hughes MA, Allen JA, Donne SW (2020) Optimized electrolytic carbon and electrolyte systems for electrochemical capacitors. Chem Electro Chem 7:266–282

Ijije HV, Chen GZ (2016) Electrochemical manufacturing of nanocarbons from carbon dioxide in molten alkali metal carbonate salts: roles of alkali metal cations. Advances in Manufacturing 4:23–32

Ijije HV, Sun Ch, Chen GZ (2014a) Indirect electrochemical reduction of carbon dioxide to carbon nanopowders in molten alkali carbonates: Process variables and product properties. Carbon 73:64–73

Ijije HV, Lawrence RC, Chen GZ (2014b) Carbon electrodeposition in molten salts: electrode reactions and applications. RSC Adv 4:35808–35817

Ingram MD, Baron B, Janz GJ (1966) The electrolytic deposition of carbon from fused carbonates. Electrochemica Acta 11:1629–1639

Irvine J, Neagu D, Verbraeken M et al (2016) Evolution of the electrochemical interface in high-temperature fuel cells and electrolysers. Nat Energy 1:15014

Ito Y, Shimada T, Kawamura H (1992) Electrochemical formation of thin carbon film from molten chloride system. Proc Electrochem Soc 16:574–585

Jiang R, Gao M, Mao X, Wang D (2019) Advancements and potentials of molten salt CO_2 capture and electrochemical transformation (MSCC–ET) process. Curr Opin Electrochem 17:38–46

Johnson M, Ren J, Lefler M et al (2017) Carbon nanotube wools made directly from CO_2 by molten electrolysis: Value driven pathways to carbon dioxide greenhouse gas mitigation. Mater Today Energy 5:230–236

Jones JP, Prakash SGK, Olah GA (2014) Electrochemical CO_2 reduction: recent advances and current trends. Isr J Chem 54(1):1451–1466

Kamali AR, Fray DJ (2014) Towards large scale preparation of carbon nanostructures in molten LiCl. Carbon 77:835–845

Kamali AR, Schwandt C, Fray DJ (2011) Effect of the graphite electrode material on the characteristics of molten salt electrolytically produced carbon nanomaterials. Mater Charact 62:987–994

Kamali AR (2020) Green production of carbon nanomaterials in molten salts and applications. Springer Nature Singapore Ltd.

Kaneco S, Iiba K, Yabuuchi M et al (2002) High efficiency electrochemical CO_2 – to - methane conversion method using methanol with lithium supporting electrolytes. Ind Eng Chem Res 41:(21)5165–5170

Kaplan V, Wachtel E, Gartsman K et al (2010) Conversion of CO_2 to CO by electrolysis of molten lithium carbonate. J Electrochem Soc 157:B552–B556

Kaplan V, Wachtel E, Lubomirsky I (2014) CO_2 to CO electrochemical conversion in molten Li_2CO_3 is stable with respect to sulfur contamination. J Electrochem Soc 161(1):F54–F57

Kharton VV, Marques FMB, Atkinson A (2004) Transport properties of solid oxide electrolyte ceramics: a brief review. Solid State Ionics 174:135–149

Küngas R (2020) Review–Electrochemical CO_2 reduction for CO production: comparison of low- and high-temperature electrolysis technologies. J Electrochem Soc 167(4):044508

Kushkhov KhB, Adamokova MN (2007) Electrowinning of metallic tungsten, molybdenum and their carbides from low-temperature halide-oxide melts. Russ J Electrichem 43:1049–1059 (in Russian)

Kushkhov KhB, Shapoval VI, Novoselova IA (1987) Electrochemical behavior of carbonic acid under excessive pressure in equimolecular melt of potassium and sodium chlorides. Russ J Electrochem 23:952–956 (in Russian)

Kushkhov KhB, Novoselova IA, Supatashvili DG, Shapoval VI (1990a) Joint electroreduction of fluoride oxide complexes of tungsten and carbon dioxide in a chloride-fluoride melt. Russ J Electrochem 26:48–51 (in Russian)

Kushkhov KhB, Supatashvili DG, Novoselova IA, Shapoval VI (1990b) Joint electroreduction of various ionic forms of tungsten with cations of nickel and cobalt in halide melts. Russ J Electrochem 26:720–723 (in Russian)

Lail M, Tanthana J, Coleman L (2014) Non-aqueous solvent (NAS) CO_2 capture. Process Energy Procedia 63:580–594

Li W, Wang H, Shi Y, Cai N (2013) Performance and methane production characteristics of H_2O-CO_2 co-electrolysis in solid oxide electrolysis cells. Int J Hydrogen Energy 38:11104–11109

Licht S, Wang B, Ghosh S et al (2010) A new solar carbon capture process: solar thermal electrochemical photo (STEP) carbon capture. J Phys Chem Lett 15:2363–2368

Licht S, Douglas A, Ren J et al (2016) Carbon nanotubes produced from ambient carbon dioxide for environmentally sustainable lithium-ion and sodium-ion battery anodes. ACS Central Sci 2:162–168

Liu X, Wang X, Licht G, Licht S (2020) Transformation of the greenhouse gas carbon dioxide to graphene. J CO_2 Utilization 36:288–294

Ni M (2012) An electrochemical model for syngas production by co-electrolysis of H_2O and CO_2. J Power Sources 202:209–216

Nitopi S, Bertheussen E, Scott S et al (2019) Progress and perspectives of electrochemical CO_2 reduction on copper in aqueous electrolyte. Chem Rev 119(12):7610–7672

Novoselova IA (1988) High temperature electrochemical synthesis of molybdenum and tungsten carbides under excessive carbon dioxide pressure. Thesis of PhD. Kiev: V. I. Vernadskii institute of general and inorganic chemistry, NASU (in Russian)

Novoselova IA, Kushkhov KhB, Malyshev VV, Shapoval VI (2001) Theoretical foundations and implementation of high-temperature electrochemical synthesis of tungsten carbides in ionic melts. Theor Found Chem Eng 35:175–187

Novoselova IA, Kuleshov SV, Omel'chuk AA et al (2020b) Cationic catalysis during the discharge of carbonate anions in molten salts. ECS Trans 98:317–332

Novoselova IA, Volkov SV, Oliinyk NF, Shapoval VI (2003) High-Temperature electrochemical synthesis of carbon-containing inorganic compounds under excessive carbon dioxide pressure. J Min Metall Section: B-Metall 39:281–293

Novoselova IA, Oliinyk NF, Volkov SV et al (2008b) Electrochemical synthesis of carbon nanotubes from dioxide in molten salts and their characterization. Phys E 40:2231–2237

Novoselova IA, Nakoneshnaya EP, Karpushin NA et al (2014) Electrochemical synthesis of composite materials based on nanoscale powders of tungsten carbides in molten salts. Metallofiz Noveishie Tekhnol 36(4):491–508 (in Russian)

Novoselova IA, Kuleshov SV, Volkov SV et al (2016) Electrochemical synthesis, morphological and structural characteristics of carbon nanomaterials produced in molten salts. Electrochim Acta 211:343–355

Novoselova I, Kuleshov S, Fedoryshena E et al (2018) Electrochemical synthesis of tungsten carbide in molten salts, its properties and applications. ECS Trans 86:81–94

Novoselova IA, Oliynyk NF, Volkov SV (2007) Electrolytic production of carbon nano-tubes in chloride-oxide melts under carbon dioxide pressure. In: Veziroglu TN, et al. (ed) Hydrogen materials and chemistry of carbon nanomaterials, Springer-Verlag, Berlin, pp 459–465

Novoselova IA, Oliinyk NF, Voronina AB, Volkov SV (2008a) Electrolytic generation of nano-scale carbon phases with framework structures in molten salts on metal cathodes. Zeitschrift fur Naturforschung 63a:467–474

Novoselova IA, Skryptun IN, Omelchuk AA, Soloviev VV (2020a) Cationic electrocatalysis in effecting the electrosynthesis of tungsten carbide nanopowders in molten salts. Chapter 9 in book: Methods for Electrocatalysis. Edited by Inamuddin M.Phil, Rajender Boddula, Abdullah M. Asiri. Springer Nature Switzerland

Ohta K, Kawamoto M, Mizuno T et al (1998) Electrochemical reduction of carbon dioxide in methanol at ambient temperature and pressure. J Applied Electrochemistry 28:717–724

Ohya S, Kaneco S, Katsumata H et al (2009) Electrochemical reduction of CO_2 in methanol with aid of CuO and Cu_2O. Catal Today 148:329–334

Otake K, Kinoshita H, Kikuchi T, Suzuki RO (2013) CO_2 gas decomposition to carbon by electro-reduction in molten salts. Electrochim Acta 100:293–299

Qiao J, Liu Y, Zhang J (eds) (2016) Electrochemical Reduction of Carbon Dioxide: Fundamentals and Technologies. CRC press. Taylor and Francis Group.394 p.

Rees NV, Compton RG (2011) Electrochemical CO_2 sequestration in ionic liquids; a perspective. Energy Environ Sci 4:403–408

Ren D, Ang B, Yeo B (2016) Tuning the selectivity of carbon dioxide electroreduction toward ethanol on oxide-derived Cu_xZn catalysts. ACS Catal 6(12):8239–8247

Ren J, Yu A, Peng P et al (2019) Recent advances in solar thermal electrochemical process (STEP) for carbon neutral products and high value nanocarbons. Acc Chem Res 52:3177–3187

Ren J, Johnson M, Singhal R, Licht S (2017) Transformation of the greenhouse gas CO_2 by molten electrolysis into a wide controlled selection of carbon nanotubes. J CO_2 Utilization18:335–344

Rezaei A, Kamali AR (2018) Green production of carbon nanomaterials in molten salts, mechanisms and applications. Diam Relat Mater 83:146–161

Rosen BA, Salehi-Khojin A, Thorson MR et al (2011) Ionic liquid–mediated selective conversion of CO_2 to CO at low overpotentials. Science 334:643–644

Ross MB, De Luna P, Li Y et al (2019) Designing materials for electrochemical carbon dioxide recycling. Nat Catal 2:648–658

Saeki T, Hashimoto K, Kimura N (1995) Electrochemical reduction of CO_2 with high current density in a CO_2 + methanol medium. II. CO formation promoted by tetrabutylammonium cation. J Electroanal Chem 39(15):77–82

Shapoval VI (1975) Kinetics of electrode processes with conjugate reaction of aside-base type in molten salts. Thesis of Disser for Doctor of Chem Sci Kiev: V. I. Vernadskii Institute of General and Inorganic Chemistry, NASU (in Russian)

Shapoval VI, Kushkhov KhB, Novoselova IA (1982) Thermodynamic foundation of the electrochemical synthesis of tungsten, molybdenum and boron carbides. Ukr Khim Zh 46(7):738–742 (in Russian)

Shapoval VI, Soloviev VV, Lavrinenko-Ometsinskaya ED, Kushkhov KhB (1984) Quantum-chemical study of the polarization features of NO_3^- and CO_3^{2-} anions under the influence of the cationic environment. Ukr Khim Zh 50:917–921 (in Russian)

Shapoval V, Malyshev VV, Novoselova IA, Kushkhov KhB (1995) Modern problems of HTES of transition metal compounds. Russ Chem Rev 64:125–132

Shapoval VI, Delimarskii YuK, Grishchenko VF (1974) Electrochemical processes with fast and slow acid-base reactions in molten electrolytes. In: Ionic Melts, Issue 1:222–240. Naukova Dumka, Kiev (in Russian)

Silvester DS, Compton RG (2006) Electrochemistry in room temperature ionic liquids: a review and some possible applications. Zeitschrfür Physikal Chem 220:1247–1274

Smirnov MV, Lyubimtseva IYa, Tsiovkina LA (1965) Cathodic processes on a gold electrode in the lithium oxide-lithium carbonate molten mixture. Russ J Electrochem 1:566–568 (in Russian)

Soloviev VV (1985) Interaction of anions NO_3^- and CO_3^{2-} with cations Li^+ and Be^{2+} on the background of ionic melts. PhD diss., Kiev: V. I. Vernadskii Institute of General and Inorganic Chemistry (in Russian)

Song Y, Zhang X, Xie K et al (2019) High temperature CO_2 electrolysis in solid oxide electrolysis cells: developments, challenges, and prospects. Adv Mater 31(50):1902033

Song Q, Xu Q, Shang X et al (2012) Electrochemical deposition of carbon films on titanium in molten LiCl–KCl–K_2CO_3.Thin Solid Films 520:6856–6863

Spinner NS, Vega JA, Mustain WE (2012) Recent progress in the electrochemical conversion and utilization of CO_2. Catal Sci Technol 2:19–28

Stulov YuV, Dolmatov VS, Dubrovskii AR, Kuznetsov SA (2017) Coatings by refractory metal carbides: deposition from molten salts, properties, application. Russ J Appl Chem 90(5):676–683

Sytchev BJ, Kaptay G (2003) Electrochemical study of the electrodeposition and intercalation of sodium into graphite as the first step of carbon nano-tudes formation. Journal of Mining and Metallurgy, Section: B-Metallurgy 39:369–381

Sytchev BJ, Kaptay G (2009) Influence of current density on the erosion of a graphite cathode and electrolytic formation of carbon nanotubes in molten NaCl and LiCl. Electrochim Acta 54:6725–6731

Tang D, Yin H, Mao X, et al (2013) Effects of applied voltage and temperature on the electrochemical production of carbon powders from CO_2 in molten salt with an inert anode Electrochimica Acta 114:567–573

Uhm S, Kim YD (2014) Electrochemical conversion of carbon dioxide in a solid oxide electrolysis cell. Curr Appl Phys 14(5):672–679

Verma S, Hamasaki Y, Kim C et al (2018) Insights into the low overpotential electroreduction of CO_2 to CO on a supported gold catalyst in an alkaline flow electrolyzer. ACS Energy Lett 3:193–198

Wang X, Liu X, Licht G, Licht S (2019) Exploration of alkali cation variation on the synthesis of carbon nanotubes by electrolysis of CO_2 in molten carbonates. J CO_2 Utilization 34:303–312

Weekes DM, Salvatore DA, Reyes A et al (2018) Electrolytic CO_2 reduction in a flow cell. Acc Chem Res 51(4):910–918

Weng W, Tang L, Xiao W (2019) Capture and electro-splitting of CO_2 in molten salts. Rev J Energy Chem 28:128–143

Wu H, Li Z, Ji D et al (2016) One-pot synthesis of nanostructured carbon materials from carbon dioxide via electrolysis in molten carbonate salts. Carbon 106:208–217

Wu HJ, Liu Y, Ji DQ et al (2017) Renewable and high efficient syngas production from carbon dioxide and water through solar energy assisted electrolysis in eutectic molten salts. J Power Sources 362:92–104

Yin HY, Mao XH, Tang DY et al (2013) Capture and electrochemical conversion of CO_2 to value-added carbon and oxygen by molten salt electrolysis. Energy Environ Sci 6:1538–1538

Yin H, Wang D (2018) Electrochemical valorization of carbon dioxide in molten salts. In: Li L, Wong-Ng W, Huang K, Cook LP (eds) Materials and Processes for CO_2 Capture, Conversion, and Sequestration, 1st edn. Wiley, Inc, pp 267–295

Zhang L, Hu S, Zhu X, Yang W (2017) Electrochemical reduction of CO_2 in solid oxide electrolysis cells. J of Energy Chemistry 26 (4):593–601

Zhang J, Luo W, Züttel A (2020) Crossover of liquid products from electrochemical CO_2 reduction through gas diffusion electrode and anion exchange membrane. J of Catalysis 385:140–145

Zheng Y, Yu B, Wang J et al (2019) Carbon dioxide reduction through advanced conversion and utilization technologies. Taylor & Francis, CRC Press, Series: Electrochemical energy storage and conversion, Boca Raton

Inessa A Novoselova, candidate of chemical sciences (PhD, electrochemistry) Scopus ID: 7004650001

Research interests: thermodynamics and electrochemistry of molten salts, high temperature electrochemical synthesis of refractory compounds (carbides, borides, silicides).

Serhii V Kuleshov, junior researcher Scopus ID: 57189995628

Research interests: electrochemistry of molten salts, thermodynamics, electrolysis of melts.

Anatoliy A Omel'chuk, corresponding member of National Academy of Sciences of Ukraine (Technical Chemistry), professor, doctor of chemical sciences (Dr., electrochemistry). Scopus ID: 7004008408

Research interests: electrochemistry of molten salts, solid-state electrolytes, thin-layer electrolysis of melts.

Supercritical Carbon Dioxide Mediated Organic Transformations

Bubun Banerjee and Gurpreet Kaur

Abstract

Supercritical carbon dioxide became an efficient and greener alternative of various toxic organic solvents. Non-toxicity, non-flammability, high dissolvability, inexpensive and plenty of availability make this nonconventional solvent more attractive. During last few years, various organic transformations were carried out in supercritical carbon dioxide as environmentally benign solvent. The present chapter summarizes an "up to date" developments on supercritical carbon dioxide mediated various organic transformations reported so far.

Keywords

Supercritical carbon dioxide • Efficient solvent • Green reaction medium • Sustainable developments • Organic transformation

1 Introduction

To protect our *Mother Nature* from the ever-increasing chemical pollutions, scientists are constantly trying to modify chemical processes. Among various parameters, solvent plays an important role in chemical transformations. Various organic solvents have been used for diverse chemical reactions. Though these processes afforded excellent yields still they suffered from sustainability issue as most of the organic solvents are volatile as well as toxic in nature. Under this background, search for environmentally benign alternative solvents is still remaining a valid exercise. Involvement of supercritical carbon dioxide ($scCO_2$) as a reaction medium helped a lot to fulfil this goal (Jessop and Leitner, 1999). Non-toxicity, non-flammability, high dissolvability, inexpensive and plenty of availability make $scCO_2$ attractive as a greener reaction medium (Skouta, 2009). Compared to other organic solvents, it has many advantages including higher substrate selectivity, zero surface tension, poor solvation effect, large diffusion coefficient, better reaction rate, and facile separation of reactants, catalysts and products (Jessop et al. 1995a, b). Diffusivity of solutes is much higher in $scCO_2$. After completion of the reaction, it can be easily removed from the reaction mixture. Based on the requirement, the solvent properties of $scCO_2$ can easily be controlled by adjusting pressure or temperature of the medium (Leitner, 2002). This unique property of $scCO_2$ helped a lot to design new synthetic approaches involving various organometallic catalysts (Leitner, 2004). Both homogeneous (Jessop et al. 1999) as well as heterogeneous (Baiker, 1999) catalysts are effective in $scCO_2$. Biocatalyzed reactions are also undergone smoothly in $scCO_2$ (Nakamura, 1990; Aaltonen and Markku, 1991). On the other hand, $scCO_2$ recognized as an efficient reaction medium for free-radical scavenging reactions due to the inability of radical chain transfer through the medium (Tanko and Blackert, 1994). With many advantages, it has one limitation related to the solubility of polar reactants. Only non-polar reactants become soluble up to an acceptable limit. To overcome the solubility issue of polar reactants, co-solvents or sometimes complexing agents were added additionally.

Because of these above-mentioned advantageous properties, $scCO_2$ employed as an efficient reaction medium for the diverse organic transformations which include polymerization (Cooper and DeSimone, 1996); glycerolysis of soybean oil (Jackson and King, 1997); fatty acid synthesis (Karakas et al. 1997); hydrogenation (Jessop et al. 1994; Jessop et al. 1995a, b; Jessop et al. 1996); synthesis of *N, N*-dimethylformamide (Kröcher et al. 1996); acylation of glucose (Tsitsimpikou et al. 1998); nucleophilic substitution (DeSimone et al. 2001); synthesis of β-Carotene (Zhang

B. Banerjee (✉) · G. Kaur
Department of Chemistry, Akal University, Talwandi Sabo, Bathinda, Punjab 151302, India
e-mail: bubun_chm@auts.ac.in

Inamuddin et al. (eds.), *Carbon Dioxide Utilization to Sustainable Energy and Fuels*, Advances in Science, Technology & Innovation, https://doi.org/10.1007/978-3-030-72877-9_7

et al. 2020); hydroformylation reactions (Rathke et al. 1991) etc. The following sections will describe some other valuable organic transformations which were carried out in supercritical carbon dioxide as an efficient reaction medium.

2 Applications of Supercritical Carbon Dioxide

2.1 Hydrogenation Reactions

Only one cyanide group of adiponitrile (**1**) was reduced selectively by using a mixture of rhodium/alumina (Rh/Al_2O_3) as catalyst in supercritical carbon dioxide ($scCO_2$) which produced 6-aminocapronitrile (**2**) with excellent yield (Fig. 1) (Chatterjee et al. 2010). The reduction was carried out at 80 °C. Interestingly it was reported that, hexamethylenediamine (**3**) didn't produce even as by-product. When the same reaction was carried out in ethanol, a mixture of products (with product ratio **2**:**3** = 70:30) was formed. The catalyst was recovered and recycled successfully for further runs without any significant loss in its catalytic activities. Under the same optimized conditions, terephthalonitrile (**4a**), isophthalonitrile (**4b**) and phthalonitrile (**4c**) also underwent smoothly and yielded the desired amino-nitrile products with excellent yields. In a continuous flow reactor, hydrogenation of 5-nitrobenzo [*d*] [1, 3] dioxole (**6**) afforded the corresponding benzo[*d*] [1, 3]dioxol-5-amine (**7**) in $scCO_2$ using a novel palladium catalyst at 90 °C (Fig. 2) (Hitzler and Poliakoff, 1997).

2.2 Asymmetric Hydrogenation Reactions

In 1995, asymmetric hydrogenation of several α-enamides (**8**) was achieved in $scCO_2$ using a Et-DuPHOS ligand functionalized cationic rhodium complex as catalyst at 40 °C (Fig. 3) (Burk et al. 1995). The enantiomeric excesses (ee's) of the α-emino ester derivatives (**9**) were comparable to those obtained in conventional solvents. Very next year, asymmetric hydrogenation of tiglic acid (**10**) was carried out using Ru $(OCOCH_3)_2(H_8\text{-binap})$ as catalyst in $scCO_2$

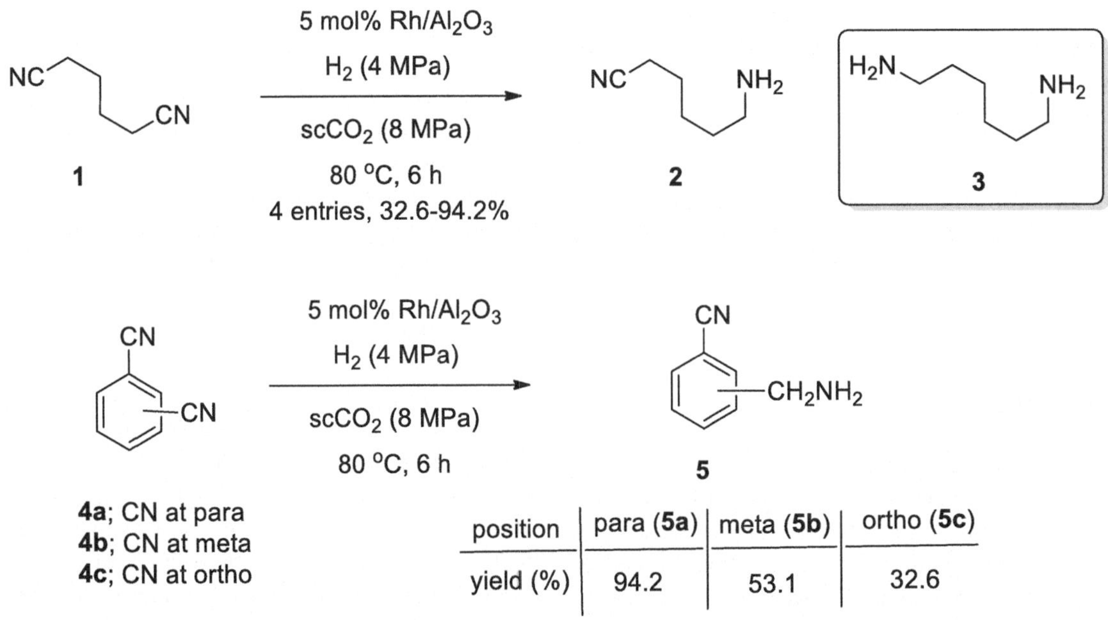

position	para (**5a**)	meta (**5b**)	ortho (**5c**)
yield (%)	94.2	53.1	32.6

Fig. 1 Supercritical carbon dioxide mediated selective reduction of cyanide

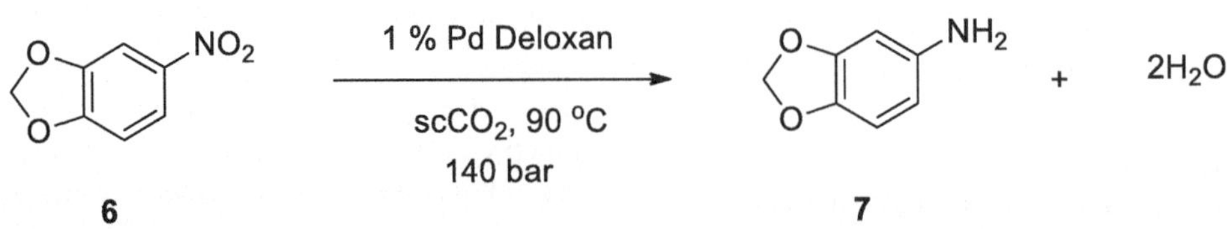

Fig. 2 Supercritical carbon dioxide mediated selective reduction of nitro to amine

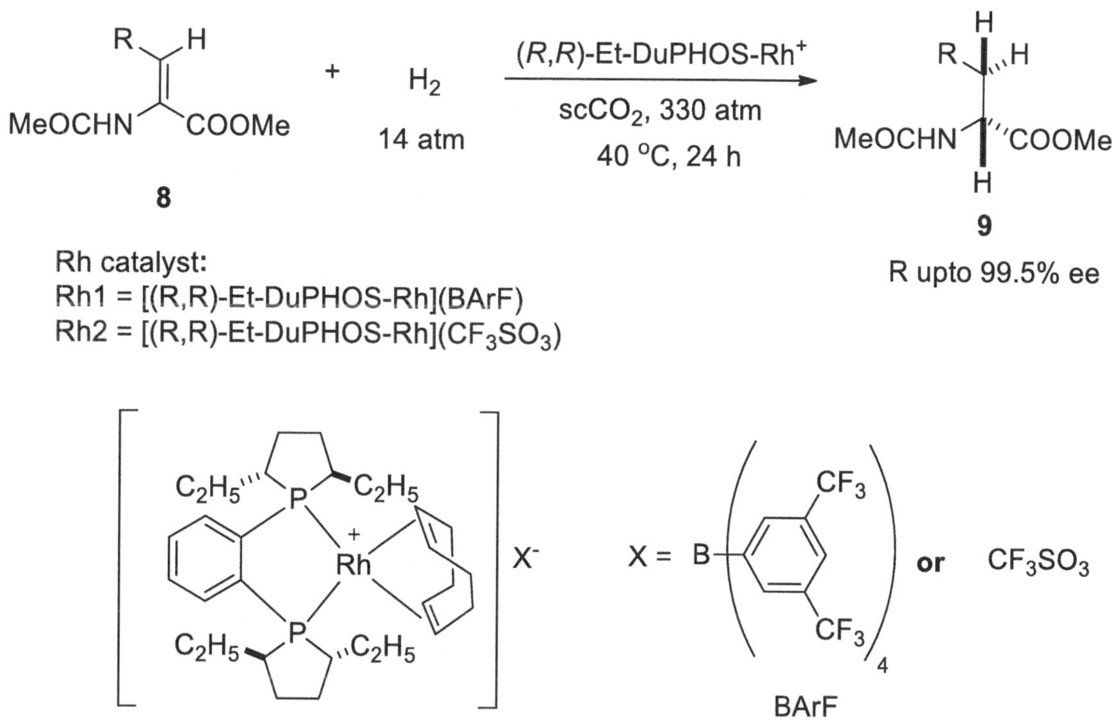

Fig. 3 Supercritical carbon dioxide mediated asymmetric hydrogenation of α-enamides

(Fig. 4) (Xiao et al. 1996). The ee of the synthesized (*S*)-2-methylbutanoic acid (**11**) was found better (81%) than that in hexane (73%) at 50 °C.

2.3 Diels–Alder Reaction

In 1987, Diels–Alder reaction between but-1, 3-diene and ethylene was carried out in supercritical carbon dioxide ($scCO_2$) (Paulaitis and Alexander, 1988). Next year, Diels–Alder reaction of methyl acrylate with cyclopentadiene was accomplished in supercritical carbon dioxide (Kim and Johnston 1988). Later on, a number of approaches were reported for the Diels–Alder reactions with different starting components in $scCO_2$ (Ikushima and Ito 1990; Ikushima and Saito 1992; Renslo et al. 1997; Clifford et al. 1998; Lin and Akgerman 1999; Fukuzawa et al. 2003). Recently, in 2017, Diels–Alder reaction between cyclopentadiene (**12**) and 1,3-butadiene (**13**) was achieved in $scCO_2$without using any polymerization inhibitor at 250 °C which afforded the corresponding 5-vinyl-2-norbornene (**14**) with 25% yield (Fig. 5) (Meng et al. 2017). A small amount of 3*a*, 4, 7, 7*a*-tetrahydro-1*H*-indene (**15**) was also formed as the by-product. Compound **9** was the major product when the same reaction carried out in traditional solvents such as tetrahydrofuran, hexane, and methanol. Synthesis of 5-methyl-3*a*, 4, 7, 7*a*-tetrahydroisobenzofuran-1, 3-dione (**17**) was achieved via the Diels–Alder reaction of isoprene (**13a**) and maleic anhydride (**16**) using anhydrous aluminium chloride as catalyst in $scCO_2$ at 33 °C (Fig. 6) (Ikushima et al. 1991).

2.4 Coupling Reaction

Bi-aryls are generally synthesized by using both Stille (Stille 1986) and Suzuki (Suzuki 1999) coupling reactions using palladium catalysts in the presence of phosphine ligand in traditional organic solvents. In 2007, palladium-catalyzed oxidative homo-coupling of arylboronic acids (**18**) was carried out smoothly using commercially available amino-functionalized resin as an efficient chelating ligand in supercritical carbon dioxide which afforded the corresponding symmetrical bi-aryls (**19**) with moderate to excellent yields (Fig. 7) (Zhou et al. 2007). During optimization it was observed that addition of small amount of water increased the yield of the desired product. Coupling of thiophene-2-boronic acid and *n*-butylboronic acid afforded 68% and 53% yield respectively whereas arylboronic acids with both electrons donating as well as withdrawing substituent produced excellent yields of the desired products. Heck coupling reaction between butyl acrylate (**20**) and bromobenzene (**21**) was also achieved in supercritical carbon dioxide which afforded the corresponding butyl cinnamate (**22**) with excellent yields (Fig. 8) (Gordon and Holmes 2002). The reaction was catalyzed by using commercially

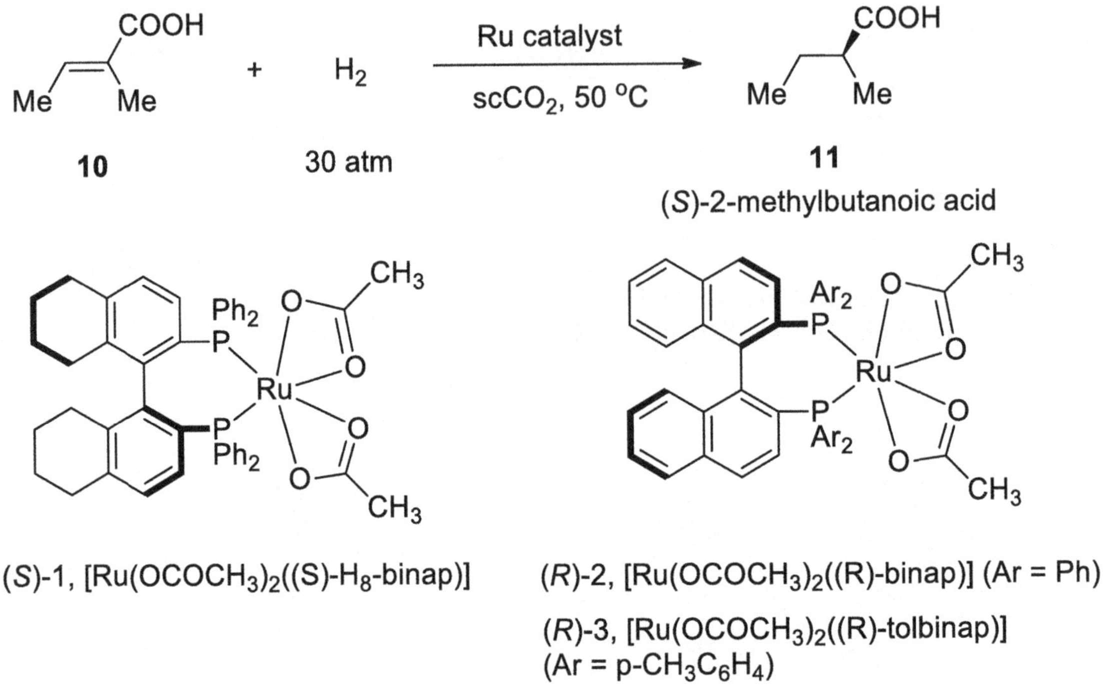

Fig. 4 Supercritical carbon dioxide mediated hydrogenation of tiglic acid

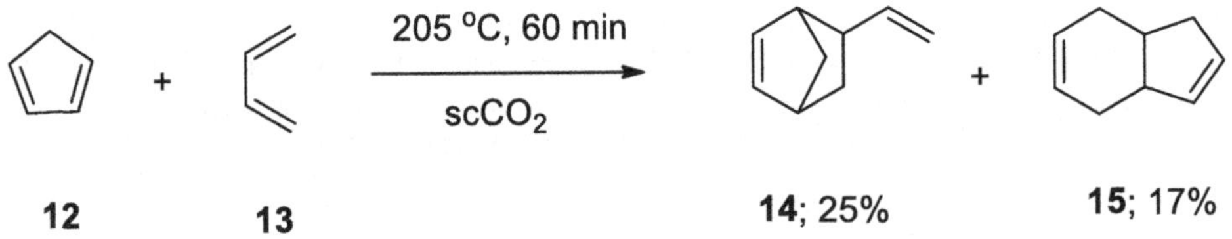

Fig. 5 Supercritical carbon dioxide mediated Diels–Alder reaction

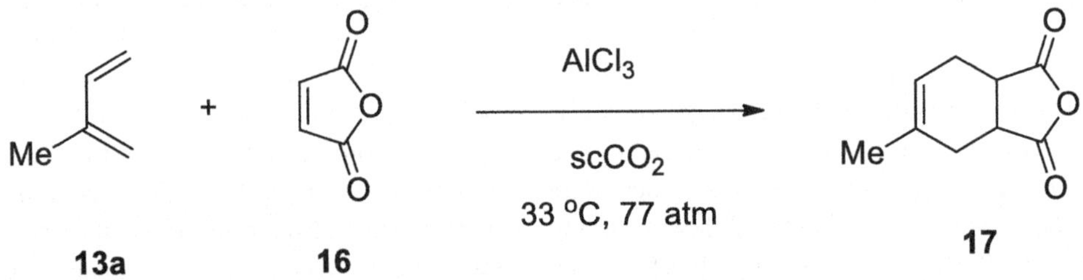

Fig. 6 Supercritical carbon dioxide mediated Diels–Alder reaction of isoprene and maleic anhydride

available phosphine functionalized polystyrene supported palladium acetate in the presence of molten tetraalkylammonium salt as base. Synthesis of methyl 3-(4-nitrophenyl) acrylate (**22a**) was accomplished by using another similar Heck reaction protocol between methyl acrylate (**20a**) and 4-nitrobromobenzene (**21a**) in supercritical carbon dioxide by using SBA-15-supported triphenyl phosphine functionalized palladium catalyst in the presence of tetrabutyl ammonium acetate as base (Fig. 9) (Feng et al. 2011). Copper catalyzed Glaser coupling reaction of phenyl acetylene (**23**) was carried out in $scCO_2$ which afforded the corresponding 1, 4-diphenylbuta-1, 3-diyne (**24**) with excellent yield using sodium acetate as a solid base instead of amines (Fig. 10) (Li and Jiang, 1999). Addition of methanol in the reaction mixture enhanced the product yield. It was assumed that the low viscosity of $scCO_2$ favoured the coupling reaction to occur at the terminal position of phenyl acetylene.

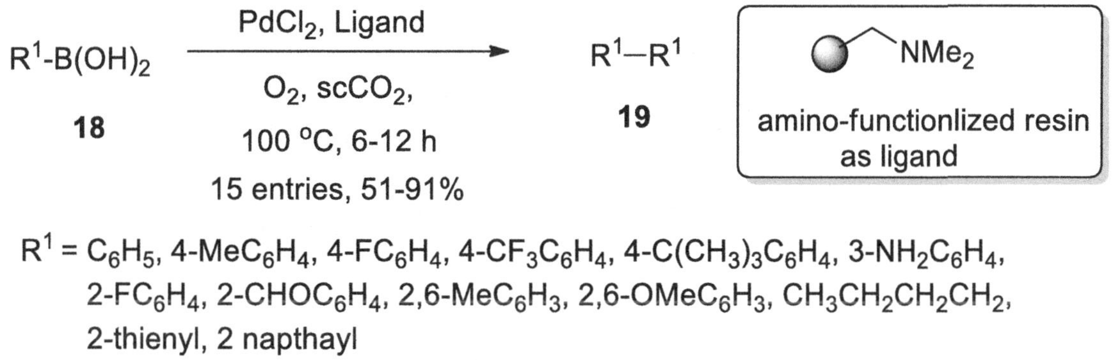

Fig. 7 Supercritical carbon dioxide mediated homo-coupling reaction

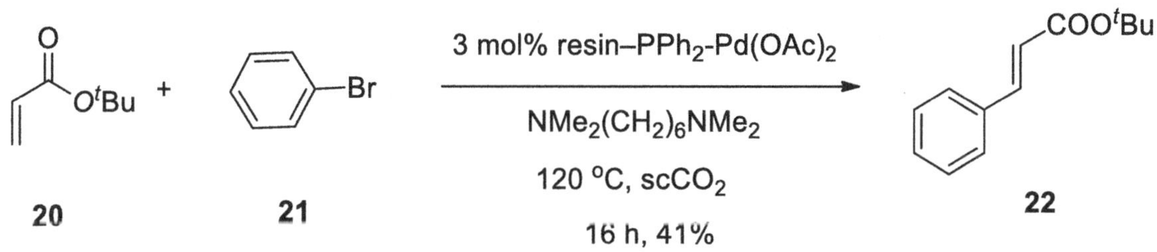

Fig. 8 Supercritical carbon dioxide mediated Heck coupling reaction

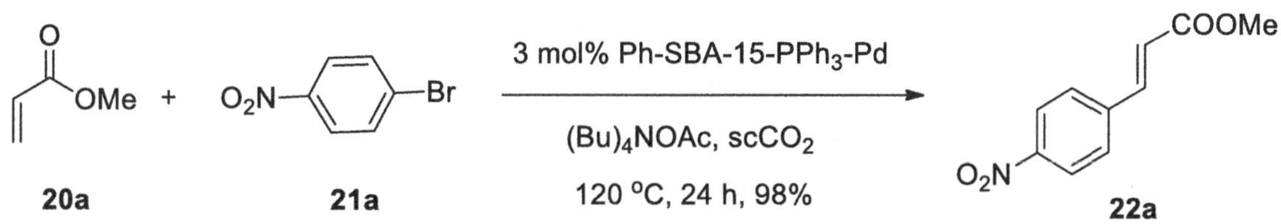

Fig. 9 Supercritical carbon dioxide mediated Heck reaction using SBA-15-supported palladium catalyst

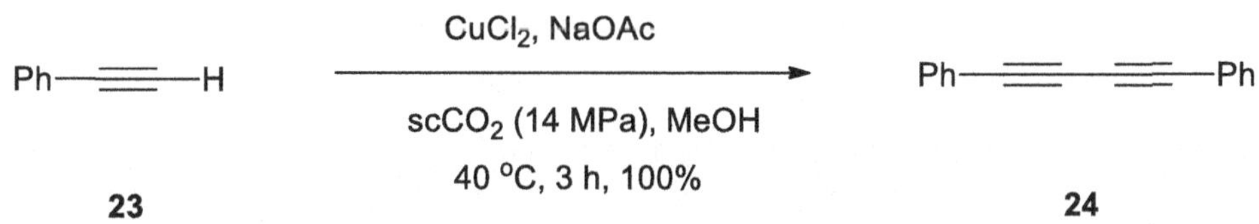

Fig. 10 Supercritical carbon dioxide mediated copper catalyzed Glaser coupling reaction

2.5 Oxidation Reaction

In 2000, Wacker oxidation reaction was first time reported in supercritical carbon dioxide (Fig. 11) (Jiang et al. 2000). Using this procedure 1-octene (**25**) was selectively oxidized to the corresponding methyl ketone (**26**) using molecular oxygen in the presence of palladium chloride as catalyst and copper chloride as co-catalyst at 40 °C. Citronellal (**28**) was synthesized as the major product from the catalytic oxidation of dihydromyrcene (**27**) with molecular oxygen using a catalytic mixture of $(MeCN)_2PdClNO_2$ and $CuCl_2$ in supercritical carbon dioxide as solvent and tertiary butanol as co-solvent at 80 °C (Fig. 12) (Ran et al. 2004). Compound **29** was produced as by-product. In absence of supercritical carbon dioxide the yield of the desired product was very low. It was reported that the chemo-selectivity of the reaction depends on the pressure of carbon dioxide, reaction temperature and the molar ratio of the catalysts. Plausible mechanism of this oxidation reaction is shown in Fig. 13.

Synthesis of epoxycyclooctane (**31**) was achieved with excellent yield via the oxidation of *cis*-cyclooctene (**30**) in supercritical carbon dioxide as the reaction medium using dioxygen as the primary oxidant and 2-methyl-

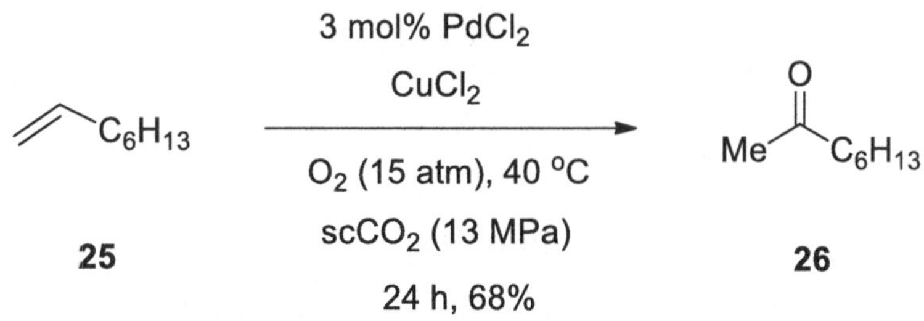

Fig. 11 Supercritical carbon dioxide mediated Wacker oxidation reaction

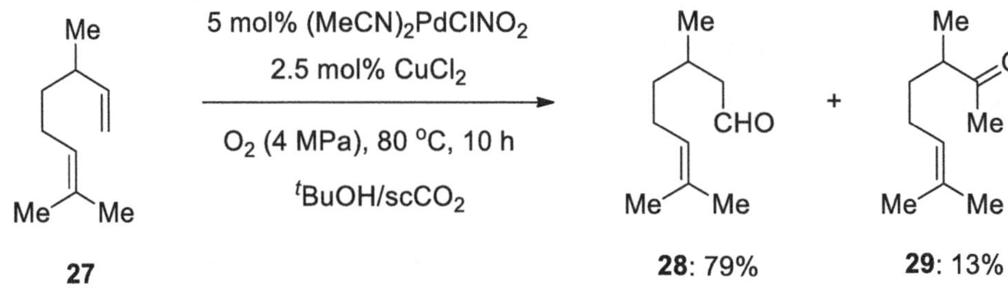

Fig. 12 Supercritical carbon dioxide mediated oxidation of dihydromyrcene

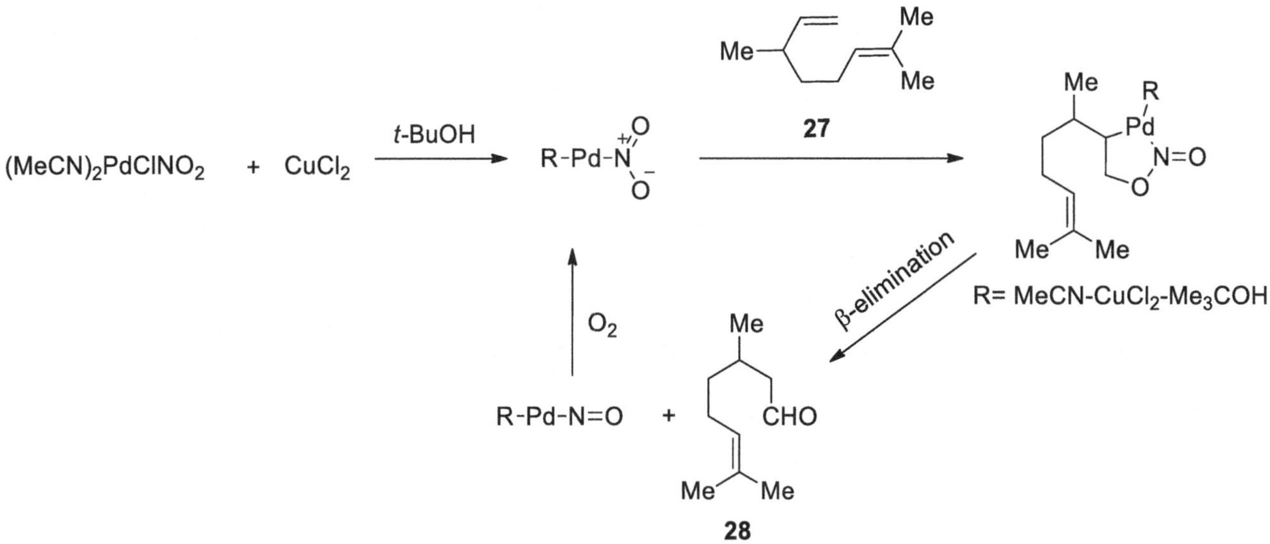

Fig. 13 Plausible mechanism for the synthesis of Citronellal in supercritical carbon dioxide

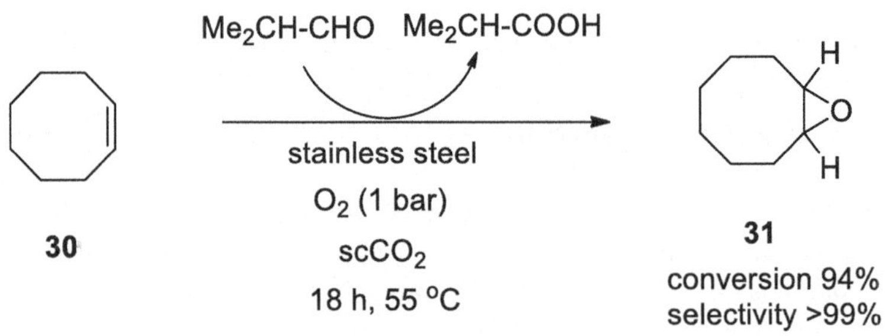

Fig. 14 Supercritical carbon dioxide mediated oxidation reaction of olefins

propionaldehyde as sacrificial co-oxidants (Fig. 14) (Loeker and Leitner 2000). It was reported that no additional catalyst was required for this reaction; the stainless steel container wall itself played the catalytic role. Proposed mechanism is shown in Fig. 15.

In 1994, oxidation of cyclohexane by molecular oxygen was carried out without using any catalyst in $scCO_2$ though the yield of the desired product was very less (Srinivas and Mukhopadhyay, 1994). Slightly improved yield was observed by using 5, 10, 15, 20-tetrakis (pentafluorophenyl)

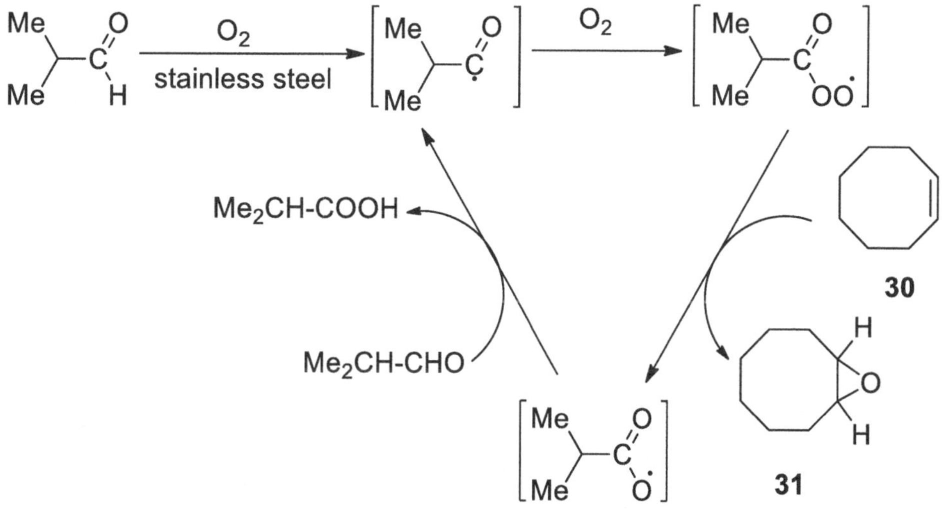

Fig. 15 Plausible mechanism for the synthesis of epoxycyclooctane in supercritical carbon dioxide

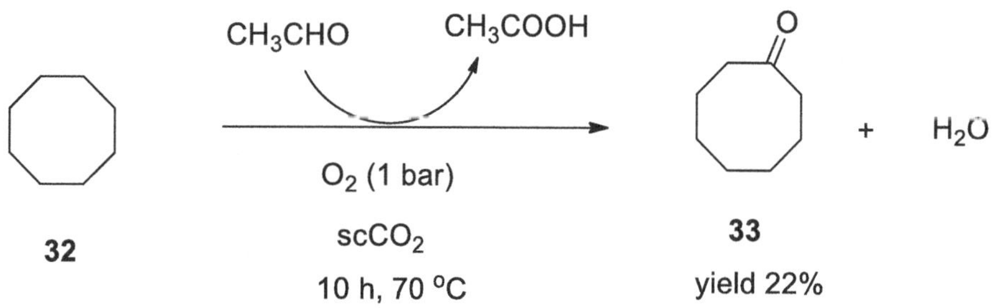

Fig. 16 Supercritical carbon dioxide mediated oxidation of cyclooctane

porphyrin (FeCl(tpfpp) as catalyst in $scCO_2$ (Wu et al. 1997). Another oxidative approach was designed by Leitner and his research group in $scCO_2$ as an inert reaction medium using dioxygen as the primary oxidant and acetaldehyde as sacrificial co-oxidants (Theyssen and Leitner 2002). Using this approach they oxidized cyclo-octane (**32**) selectively to cyclooctanone (**33**) with 20% yield (Fig. 16). It was proposed that the reaction proceeded via free radical pathway and supercritical carbon dioxide provides an inert environment which favoured the selective oxidation of cyclo-octane by using molecular oxygen-acetaldehyde mixture. It was observed that the selectivity of this transformation is dependent upon the density of compressed carbon dioxide. The same reaction yielded lower product under other inert atmosphere such as in gaseous nitrogen, argon etc.

2.6 Baeyer–Villiger Oxidation Reaction

The aerobic Baeyer–Villiger oxidation of a wide range of both cyclic and acyclic ketones (**34**) to the corresponding esters or lactones (**35**) was carried out in supercritical carbon dioxide using aldehydes as a co-reductant (Fig. 17). (Palazzi et al. 2002) The same reaction produced lower yields in conventional organic solvents such as toluene, dichloromethane and ethyl acetate.

2.7 Iodination Reaction

Di-iodoalkene is an important intermediate in organic. A simple, efficient and convenient protocol was reported for the stereoselective diiodination of both electron-rich and electron-deficient alkynes (**36**) which afforded trans-diiodoalkenes (**38**) with excellent yields in supercritical CO_2 at 40 °C (Fig. 18) (Li et al. 2003). Reaction of aqueous KI solution and Ce $(SO_4)_2$ generated molecular iodine in situ in supercritical carbon dioxide. It was proposed that the reaction proceeded through the formation of three-membered iodonium intermediate (**37**). Subsequent attack of I^- from the reverse side of the intermediate **37** affords the corresponding trans-diiodoalkenes (**38**).

2.8 Polymerization Reaction

In 1995, polymerization reaction was first carried out in supercritical carbon dioxide (Pernecker et al. 1995). Later on, in 2011, another facile procedure was reported for the

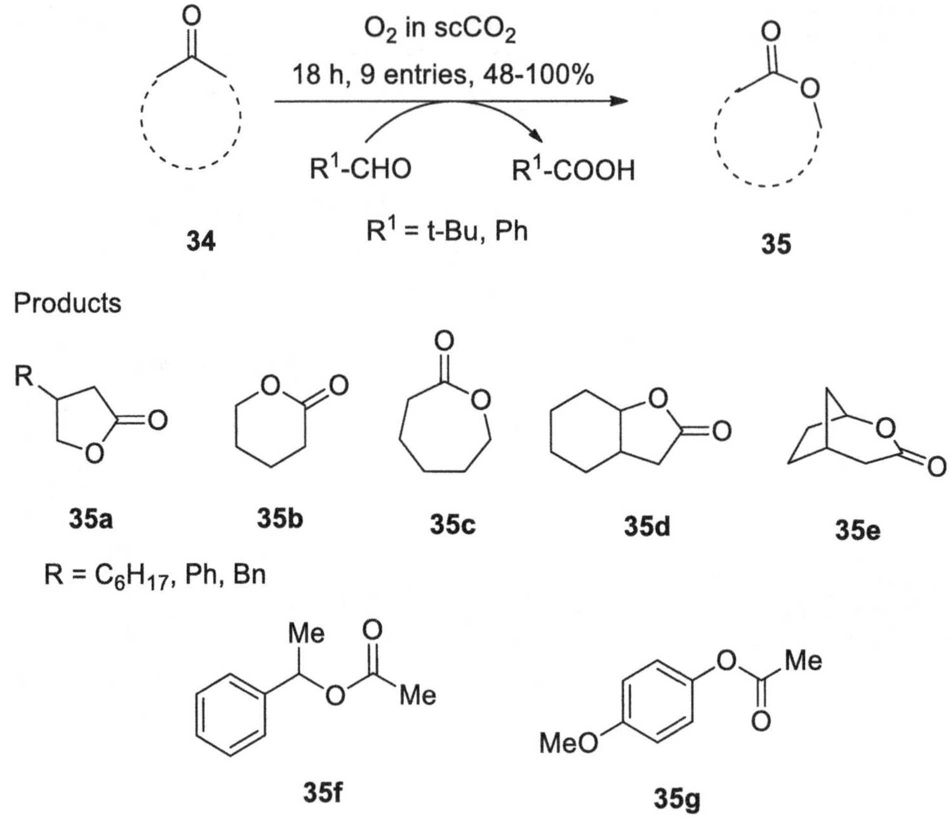

Fig. 17 Supercritical carbon dioxide mediated Baeyer–Villiger oxidation reaction

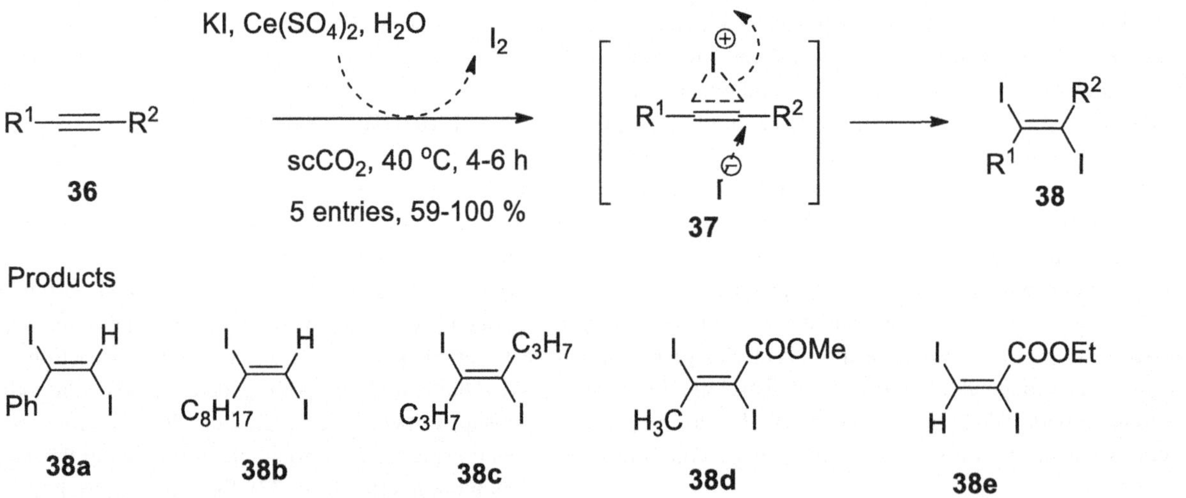

Fig. 18 Supercritical carbon dioxide mediated iodination of alkynes

ring-opening polymerization of *p*-dioxanone (**39**) using tin (II) ethyl hexanoate [$Sn(Oct)_2$] as catalyst and a catalytic amount of well-defined fluorinated triblock copolymer i.e. poly (caprolactone)-perfluropolyether-poly (caprolactone) (PCL-PFPE-PCL) as stabilizer in supercritical carbondioxide (Fig. 19) (Wang et al. 2011). It was found that the mean size of the polymer **40** depended on the stabilizer's concentration as well as on the rate of agitation. The same polymerization reaction carried out in common organic solvents produced number of by-products.

Polymerization of phenylacetylene (**23**) was carried out using a rhodium complex (acetyl acetonate) Rh (12, 5-norbornadiene) [(acac)Rh(nbd)] as catalyst in supercritical carbondioxide (Fig. 20) (Hori et al. 1999). Under this

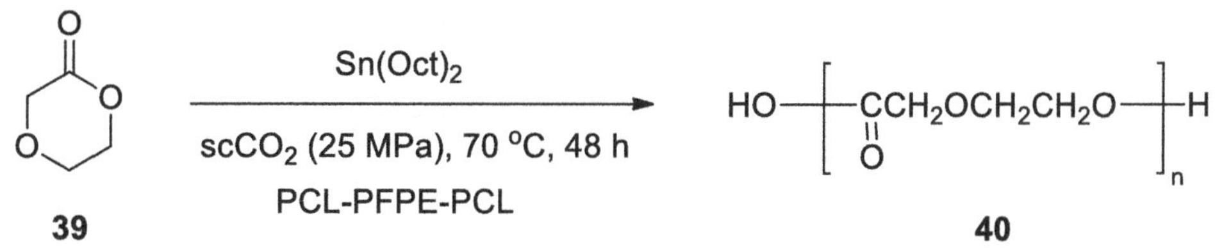

Fig. 19 Supercritical carbon dioxide mediated polymerization reaction using tin (II) ethyl hexanoate as catalyst

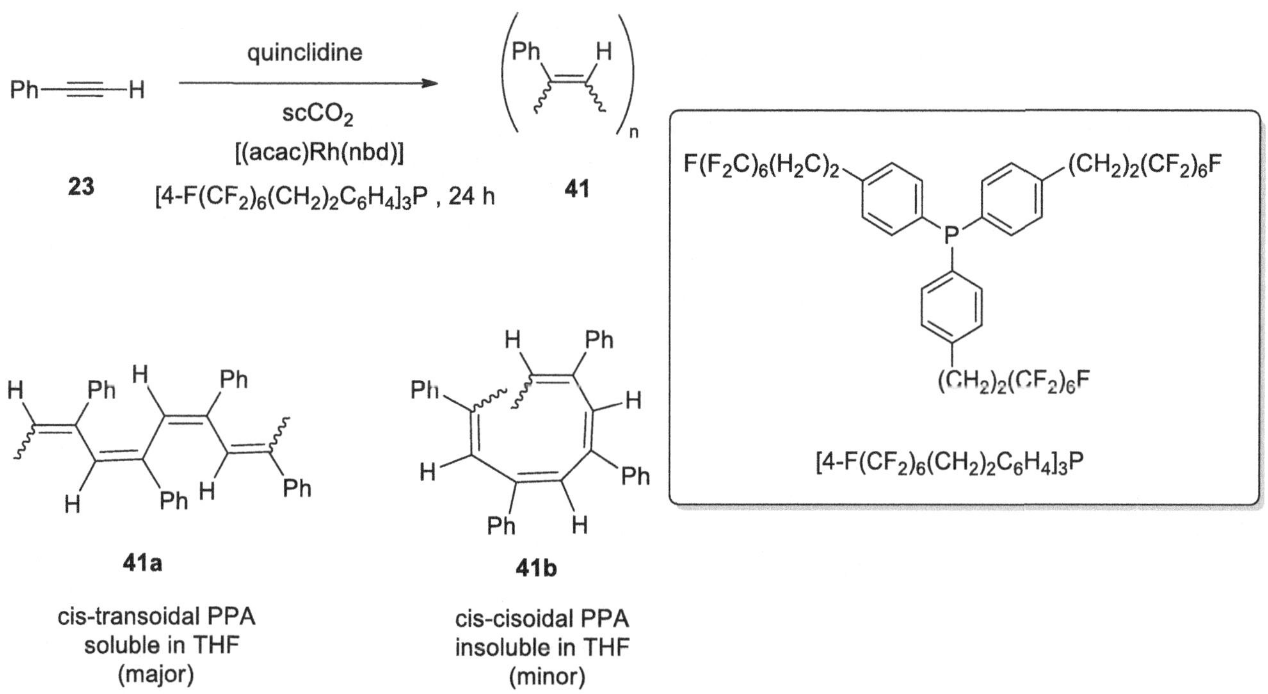

Fig. 20 Supercritical carbon dioxide mediated polymerization using rhodium complex as catalyst

condition a mixture of cis-transoidal (THF-soluble) as well as *cis-cis*oidal (THF-insoluble) polymers (**41**) was formed with higher efficiency. The same polymerization afforded lesser stereoregularity and molecular weight of the polymer in conventional solvents such as THF or hexane. Stereoselectivity of the catalyst toward the *cis*-transoidal polymer (**41a**) was increased on addition of a phosphine-based ligand ([4-F $(CF_2)_6(CH_2)_2C_6H_4]_3P$) which helped to enhance catalyst miscibility in compressed CO_2.

2.9 Carbonylation Reaction

A simple, facile and efficient procedure for the chemoselective carbonylation of amines was reported in supercritical carbon dioxide (Fig. 21) (Li et al. 2001). Various methyl *N*-alkylcarbamate derivatives (**44**) were synthesized through the carbon monoxide (**43**) insertion to amines (**42**) in the presence of methanol using $PdCl_2$ as catalyst and $CuCl_2$ as co-catalyst in supercritical carbon dioxide at 40 °C under 10 MPa

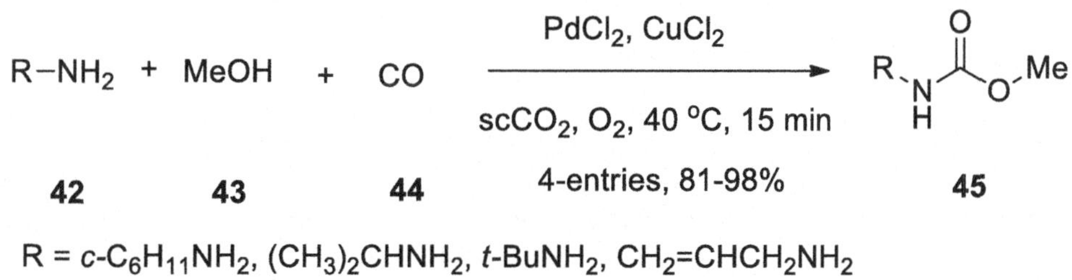

Fig. 21 Supercritical carbon dioxide mediated carbonylation reaction in the presence of palladium as well as copper chloride as catalysts

pressure into an autoclave. It was found that the selectivity was dependent on the concentration of the methanol.

Another palladium catalyzed carbonylation reaction was reported in supercritical carbon dioxide in the presence of pyridine at 60 °C (Fig. 22) (Zhang et al. 2003). Under this conditions, addition of carbon monoxide (**44**), alcohol (**43**) and carbon tetrachloride (**47**) to 1-octene (**46**) was achieved which produced [alkyl 2-(2, 2, 2-triehloroethyl)octanoate (**48**) with 60–79% yields. All three alcohols (ethanol, methanol and isopropanol) yielded the desired products under the optimized reaction conditions. Compound **49** was produced here as a by-product. It was observed that the yield of the desired product very much dependent on the reaction temperature, pressure and the amounts of alcohol and the base used. It was proposed that the reaction undergoes through the formation of intermediate **50** using in situ generated Pd (0) as catalyst (Fig. 23). $(Me_3Si)_3SiH$-promoted efficient free-radical carbonylation of 1-iodooctane (**51**) and acrylonitrile (**52**) with CO (**44**) proceeded the corresponding 1-cyano-3-undecanone (**53**) with 80% in the presence of a catalytic amount of 2, 2′-azobis(isobutyronitrile) (AIBN) as a radical initiator in supercritical carbon dioxide at 80 °C (Fig. 24) (Kishimoto and Ikariya, 2000). The same reaction when carried out in benzene afforded lower yield. Under the same optimized reaction conditions, the radical carbonylative ring-closer reaction of 6-iodohexylacrylate (**54**) was also achieved which yielded the corresponding eleven-membered oxacycloundecane-2, 5-dione (**55**) within 2 h (Fig. 25).

The carbonylation of (2-iodophenyl)methanol (**56**) with carbon monoxide (**44**) yielded the corresponding isobenzofuran-1(3*H*)-one (**57**) using triethyl phosphite functionalized Pd complex as a catalyst and triethyl amine as base in $scCO_2$ (Fig. 26) (Kayaki et al. 1999; Ikariya et al. 2000).

2.9.1 Acetalization Reaction

In 1999, acetylation of methyl acrylate (**20a**) was first achieved in supercritical carbon dioxide with high selectivity using palladium catalyst (Jia et al. 1999). Later on, another simple and facile procedure for the acetylation of methyl acrylate (**20a**) with methanol (**43**) was developed by using palladium chloride as a catalyst and polystyrene-supported hydroquinone (PS-HQ) as co-catalyst in $scCO_2$ under oxygen atmosphere which yielded the corresponding methyl 3,

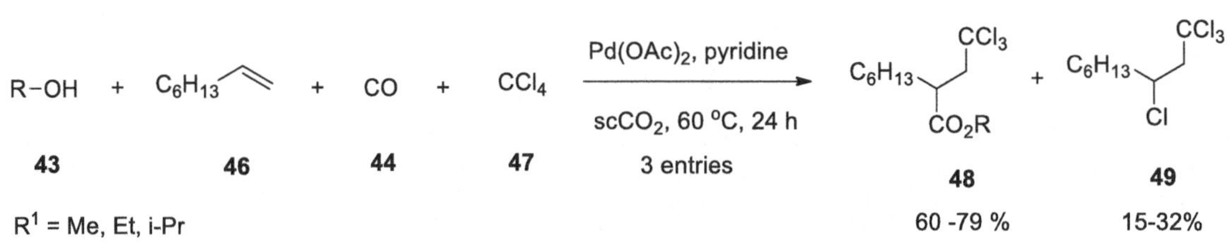

Fig. 22 Supercritical carbon dioxide mediated palladium catalyzed carbonylation reaction on 1-octene

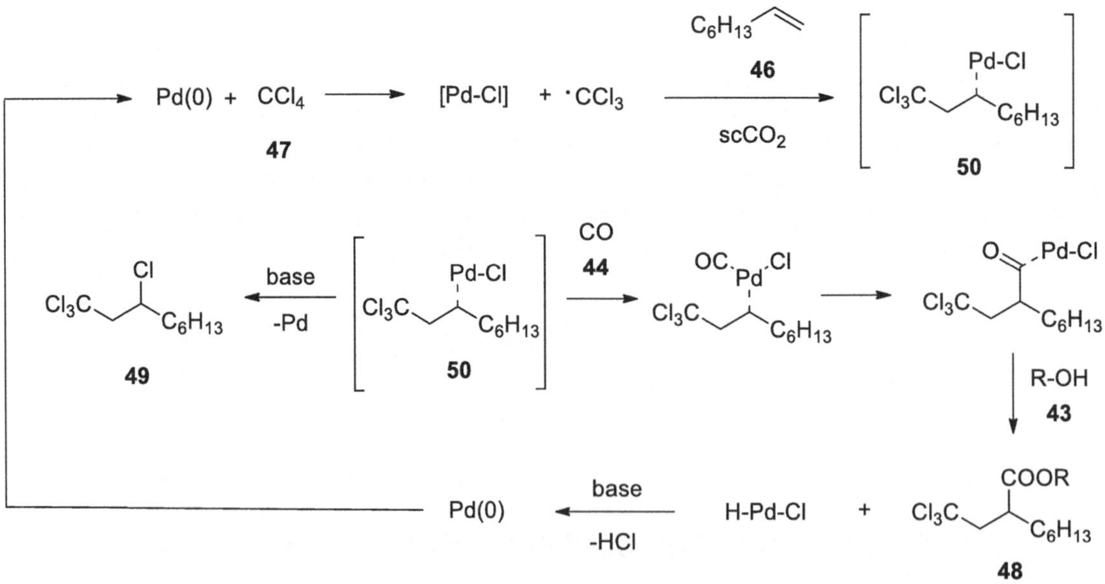

Fig. 23 Plausible mechanism of palladium catalyzed carbonylation reaction in supercritical carbon dioxide

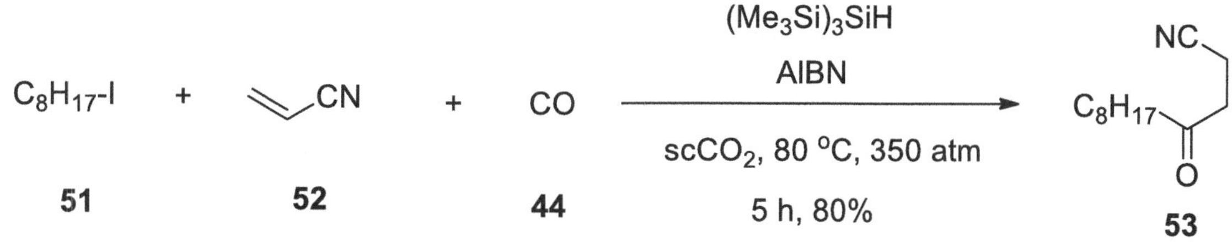

Fig. 24 Supercritical carbon dioxide mediated silane-promoted carbonylation of alkyl halides

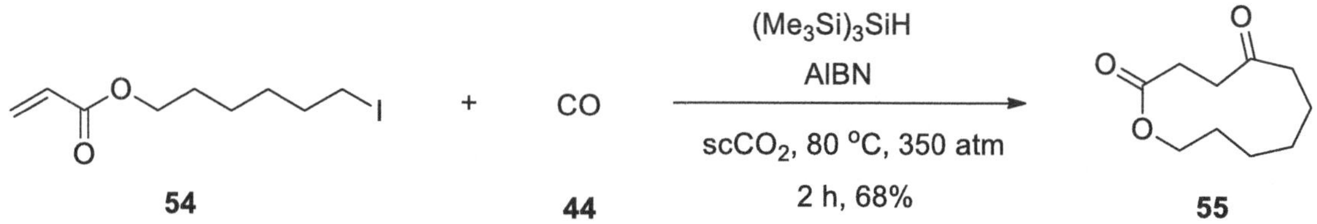

Fig. 25 Supercritical carbon dioxide mediated carbonylative ring-closing reaction of 6-iodohexylacrylate

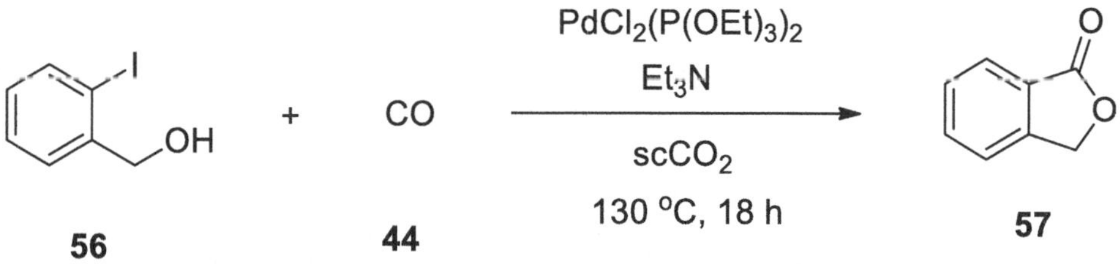

Fig. 26 Supercritical carbon dioxide mediated carbonylative ring-closing reaction of (2-iodophenyl) methanol

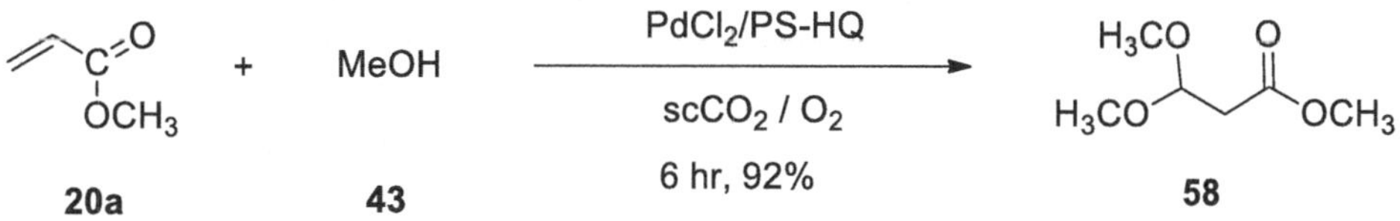

Fig. 27 Supercritical carbon dioxide mediated acetylation reaction of methyl acrylate

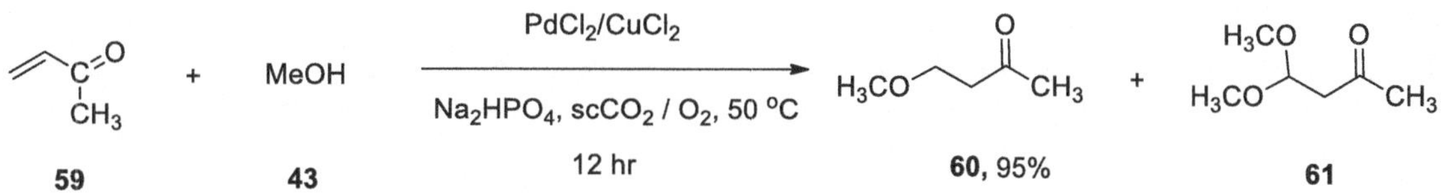

Fig. 28 Supercritical carbon dioxide mediated acetylation of methyl vinyl ketone

3-dimethoxypropanoate (**58**) with excellent yield (Fig. 27) (Wang et al. 2007). In 2006, the same batch of reaction was carried in the presence of palladium chloride as catalyst and polystyrene-supported benzoquinone (PS-BQ) as the co-catalyst in $scCO_2$ (Wang et al. 2006). It was observed that under the optimized reaction conditions, co-catalyst PS-HQ is more efficient than the PS-BQ in terms of selectivity as well as product yield. Aerobic acetylation of methyl vinyl ketone (**59**) in supercritical carbon dioxide yielded mono-acetylated product i.e. 4-methoxybutan-2-one (**60**) with high selectivity than the corresponding acetal product (**61**). The reaction was carried out in the presence of palladium as catalyst and copper chloride as co-catalyst at 50 °C (Fig. 28) (Ouyang et al. 2002).

2.9.2 Olefin Metathesis Reaction

Ring-opening metathesis polymerization of norbornene (**62**) was achieved in $scCO_2$ using [Ru $(H_2O)_6$] $(OTs)_2$ as a catalyst at 65 °C (Fig. 29) (Mistele et al. 1996; Hamilton et al. 1998). Under this condition *cis* product predominates over

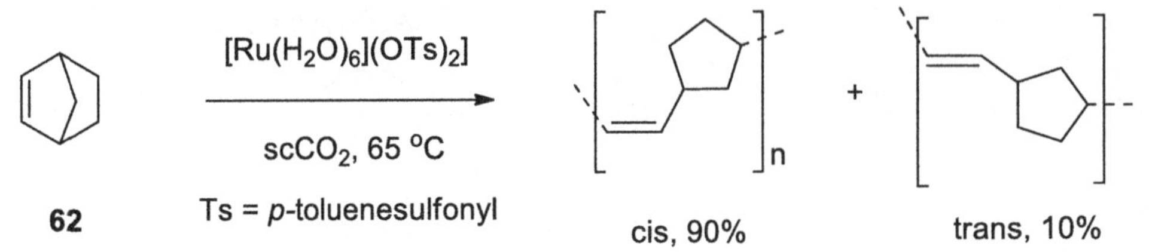

Fig. 29 Supercritical carbon dioxide mediated ring-opening metathesis polymerization

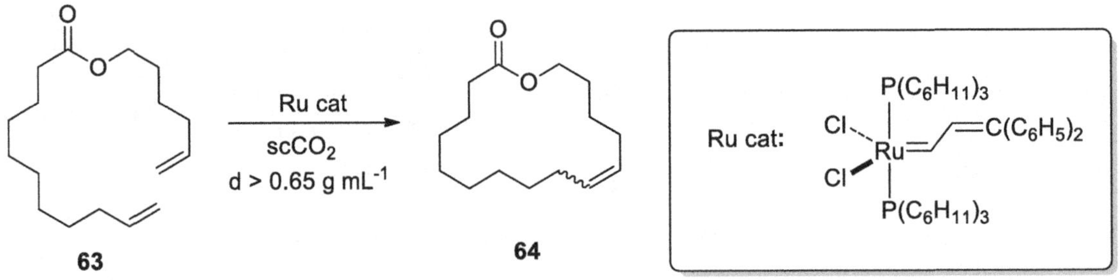

Fig. 30 Supercritical carbon dioxide mediated ring-closing metathesis reaction

trans. Later on, in 2005, the same reaction was also carried out by using Grubbs' catalyst in $scCO_2$ (Hu et al. 2005).

Ring-closing metathesis of a long-chain diene (**63**) afforded the corresponding sixteen-membered cyclic compound (**64**) using a ruthenium-based catalyst in $scCO_2$ with density more than 0:65 g/ml (Fig. 30) (Furstner et al. 1997, 2001). It was observed that at lower density (d < 0:65 g/ml) of $scCO_2$, oligomerization predominant over ring-closing metathesis process. That's why; it was assumed that the higher density favoured the intramolecular reaction.

2.9.3 Synthesis of heterocycles

Synthesis of α-alkylidene Cyclic Carbonates

Supercritical carbon dioxide mediated a mild and efficient protocol was developed for the cyclization of propargyl alcohols (**65**) with CO_2 which afforded the corresponding α-alkylidene cyclic carbonates (**66**) using polymer-supported cuprous iodide as catalyst (Fig. 31) (Jiang et al. 2008). After completion of the reaction, the catalyst was recovered by simple filtration and recycled several times without any loss of its catalytic activity.

Synthesis of 4-Methyleneoxazolidin-2-Ones

Synthesis of 4-methyleneoxazolidin-2-ones (**67**) was achieved via the CO_2 fixation with propargylic alcohols (**65**) and primary amines (**42**) without using any catalyst in $scCO_2$. Here supercritical carbon dioxide was used both as reaction medium as well as starting material. (Fig. 32) (Xu et al. 2011).

Synthesis of 5-Alkylidene-1, 3-Oxazolidin-2-Ones

A series of 5-alkylidene-1, 3-oxazolidin-2-ones (**69**) was synthesized via the CO_2 fixation with propargylic amines

Fig. 31 Supercritical carbon dioxide mediated synthesis of α-alkylidene cyclic carbonates

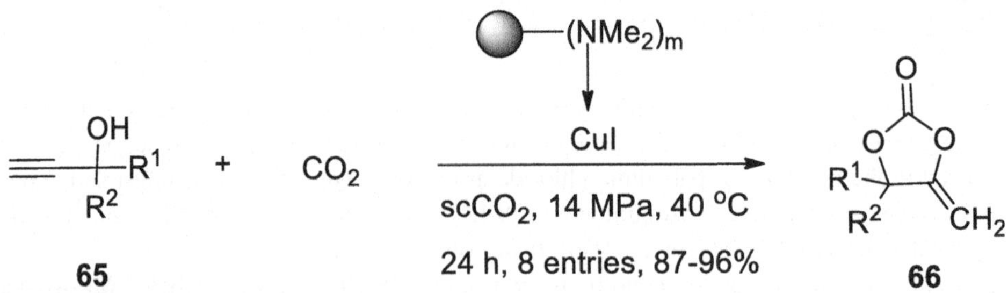

R^1= CH_3, H
R^2= $CH_2CH_2CH_3$, $C(CH_3)_2$
R^1:R^2 = cyclopentyl, cuclohexyl, benzyl

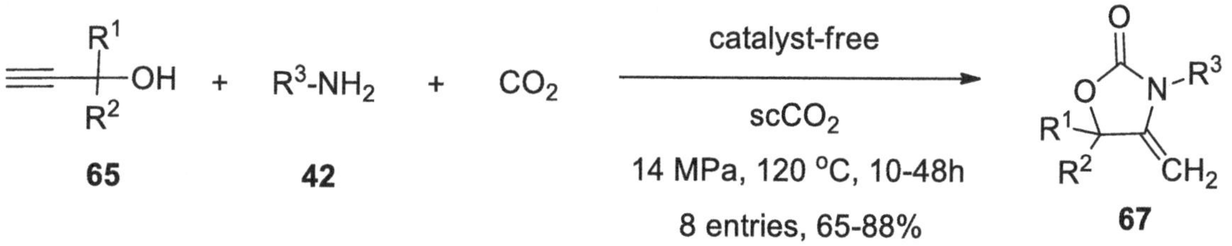

R^1 = Me, C_5H_{10}, i-Pr, n-C_6H_{13}
R^2 = H, Me
R^3 = n-Bu, allyl, Cy, Ph, sec-Bu, ^{t}Bu

Fig. 32 Chemical fixation of carbon dioxide under supercritical condition

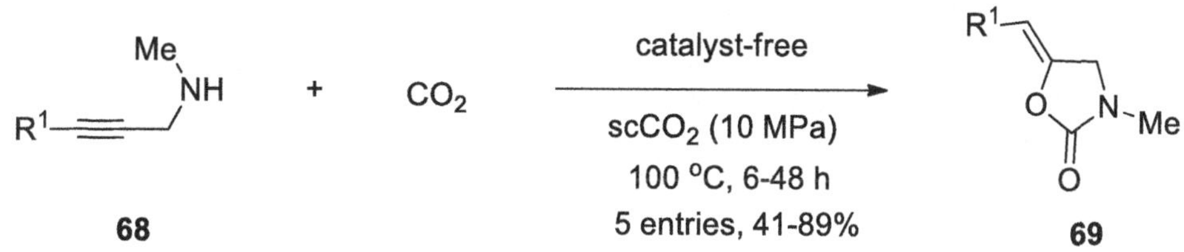

R^1= C_6H_5, 4-$CF_3C_6H_4$, 3,5-$(CF_3)_2C_6H_3$, 4-CNC_6H_4, 4-$COCH_3C_6H_4$

Fig. 33 Supercritical carbon dioxide mediated synthesis of 5-alkylidene-1, 3-oxazolidin-2-ones

(**68**) without using any catalyst under supercritical conditions (Fig. 33) (Kayaki et al 2006; Maggi et al. 2007).

Synthesis of 6-Phenyl-3a, 4-Dihydro-1H-Cyclopenta[C] furan-5(3H)-One

Synthesis of 6-phenyl-3a,4-dihydro-1*H*-cyclopenta[*c*] furan-5(3*H*)-one (**71**) was accomplished via Pauson–Khand Reaction starting from (3-(allyloxy)prop-1-yn-1-yl) benzene (**70**) and carbon monoxide (**44**) in $scCO_2$ using catalytic amount of dicobalt octacarbonyl [$Co_2(CO)_8$] as catalyst at 90 °C (Fig. 34) (Jeong et al. 1997).

Synthesis of 3, 4, 5, 6-Tetraethyl-2H-Pyran-2-One

3,4,5,6-Tetraethyl-2*H*-pyran-2-one (**42**) was synthesized with excellent selectivity by the coupling reaction between carbon dioxide and diethyl acetylene (**36a**) in $scCO_2$ using a

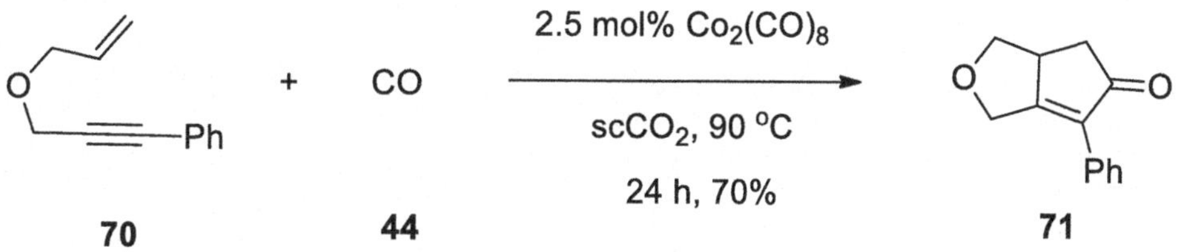

Fig. 34 Supercritical carbon dioxide mediated Pauson–Khand Reaction

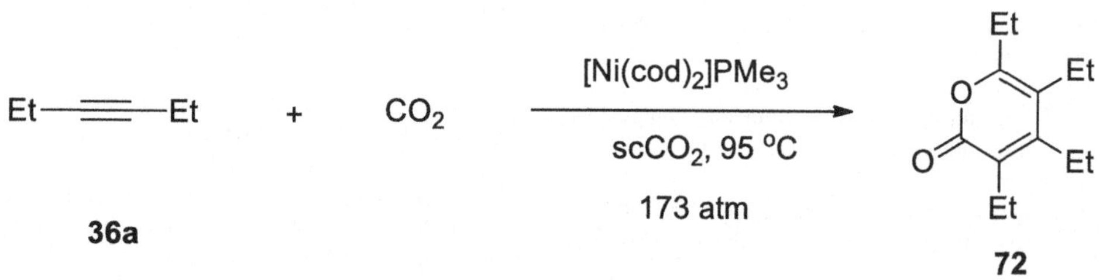

Fig. 35 Supercritical carbon dioxide mediated synthesis of 3, 4, 5, 6-tetraethyl-2*H*-pyran-2-one

catalytic amount of [Ni(cod)$_2$]/PMe$_3$ as catalyst (Fig. 35) (Jessop, 1998).

3 Conclusions

Several recent reports have shown that supercritical carbon dioxide (scCO$_2$) can be utilized as an efficient alternative of conventional organic solvents. It is non-toxic, non-flammable, inexpensive and easily available. Moreover, as a reaction medium it has many advantages which include higher substrate selectivity, zero surface tension, better reaction rate, poor solvation effect, large diffusion coefficient, high diffusivity, and facile separation of products from the reaction mixture. Because of these favourable properties, scCO$_2$ has been used as an efficient reaction medium for the diverse organic transformations. In this chapter, we have described supercritical carbon dioxide mediated various organic transformations.

Acknowledgements Authors are thankful to Prof. Gurmail Singh, Vice-Chancellor, Akal University for his wholehearted encouragement and support. BB is grateful to Akal University and Kalgidhar Trust, Baru Sahib, Himachal Pradesh, India for providing financial assistance.

References

Aaltonen O, Markku R (1991) Biocatalysis in supercritical CO_2. ChemTech 21:240–248

Baiker A (1999) Supercritical fluids in heterogeneous catalysis. Chem Rev 99:453–474

Burk MJ, Feng S, Gross MF, Tumas W (1995) Asymmetric catalytic hydrogenation reactions in supercritical carbon dioxide. J Am Chem Soc 117:8277–8278

Chatterjee M, Sato M, Kawanami H, Yokoyama T, Suzuki T, Ishizaka T (2010) An efficient hydrogenation of dinitrile to aminonitrile in supercritical carbon dioxide. Adv Synth Catal 352:2394–2398

Clifford AA, Pople K, Gaskill WJ, Bartle KD, Rayner CM (1998) Potential tuning and reaction control in the Diels-Alder reaction between cyclopentadiene and methyl acrylate in supercritical carbon dioxide. J Chem Soc Faraday Trans 94:1451–1456

Cooper AI, DeSimone JM (1996) Polymer synthesis and characterization in liquid/supercritical carbon dioxide. Curr Opin Solid State Mater Sci 1:761–766

DeSimone J, Selva M, Tundo P (2001) Nucleophilic displacements in supercritical carbon dioxide using silica-supported phase-transfer agents. J Org Chem 66:4047–4049

Feng XJ, Yan M, Zhang X, Bao M (2011) Preparation and application of SBA-15-supported palladium catalyst for Heck reaction in supercritical carbon dioxide. Chin Chem Lett 22:643–646

Fukuzawa S, Metoki K, Esumi S (2003) Asymmetric Diels-Alder reactions in supercritical carbon dioxide catalyzed by rare earth complexes. Tetrahedron 59:10445–10452

Furstner A, Koch D, Langemann K, Leitner W, Six C (1997) Olefin metathesis in compressed carbon dioxide. Angew Chem Int Ed Engl 36:2466–2469

Furstner A, Ackermann L, Beck K, Hori H, Koch D, Langemann K, Liebl M, Six C, Leitner W (2001) Olefin metathesis in supercritical carbon dioxide. J Am Chem Soc 123:9000–9006

Gordon RS, Holmes AB (2002) Palladium-mediated cross-coupling reactions with supported reagents in supercritical carbon dioxide. Chem Commun 2002:640–641

Hamilton JG, Rooney JJ, DeSimone JM, Mistele C (1998) Stereochemistry of ring-opened metathesis polymers prepared in liquid CO_2 at high pressure using $Ru(H_2O)_6(Tos)_2$ as catalyst. Macromolecules 31:4387–4389

Hitzler MG, Poliakoff M (1997) Continuous hydrogenation of organic compounds in supercritical fluids. Chem Commun 1997:1667–1668

Hori H, Six C, Leitner W (1999) Rhodium-catalyzed phenylacetylene polymerization in compressed carbon dioxide. Macromolecules 32:3178–3182

Hu X, Blanda MT, Venumbaka SR, Cassidy PE (2005) Ring-opening metathesis polymerization (ROMP) of norbornene in supercritical carbon dioxide using well-defined metal carbene catalysts. Polym Adv Technol 16:146–149

Ikariya T, Kayaki Y, Kishimoto Y, Noguchi Y (2000) Highly efficient carbonylation reactions of organic halides in supercritical carbon dioxide. Prog Nucl Energy 37:429–434

Ikushima Y, Ito S (1990) A Diels-Alder reaction in supercritical carbon dioxide medium. J Chem Eng Jpn 23:96–98

Ikushima Y, Saito N, Arai M (1991) High-pressure fourier transform infrared spectroscopy study of the Diels-Alder reaction of isoprene and maleic anhydride in supercritical carbon dioxide. Bull Chem Soc Jpn 64:282–284

Ikushima Y, Saito N, Arai M (1992) Supercritical carbon dioxide as reaction medium: examination of its solvent effects in the near-critical region. J Phys Chem 96:2293–2297

Jackson MA, King JW (1997) Lipase-catalyzed glycerolysis of soybean oil in supercritical carbon dioxide. JAOCS. 74:103–106

Jeong N, Hwang SH, Lee YW, Lim JS (1997) Catalytic Pauson-Khand reaction in super critical fluids. J Am Chem Soc 119:10549–10550

Jessop PG (1998) Homogeneously-catalyzed syntheses in supercritical fluids. Top Catal 5:95–103

Jessop PG, Ikariya T, Noyori R (1994) Homogeneous catalytic hydrogenation of supercritical carbon dioxide. Nature 368:231–233

Jessop PG, Ikariya T, Noyori R (1995) Selectivity for hydrogenation or hydroformylation of olefins by hydridopentacarbonylmanganese(I) in supercritical carbon dioxide. Organometallics 14:1510–1513

Jessop PG, Ikariya T, Noyori R (1995) Homogeneous catalysis in supercritical fluids. Science 269:1065–1069

Jessop PG, Hsiao Y, Ikariya T, Noyori R (1996) Homogeneous catalysis in supercritical fluids: hydrogenation of supercritical carbon dioxide to formic acid, alkyl formates, and formamides. J Am Chem Soc 118:344–355

Jessop PG, Ikariya T, Noyori R (1999) Homogeneous catalysis in supercritical fluids. Chem Rev 99:475–493

Jessop PG, Leitner W (1999) Chemical synthesis using supercritical fluids. Wiley-VCH

Jia L, Jiang H, Li J (1999) Palladium (II)-catalyzed oxidation of acrylate esters to acetals in supercritical carbon dioxide. Chem Commun 1999:985–986

Jiang H, Jia L, Li J (2000) Wacker reaction in supercritical carbon dioxide. Green Chem 2:161–164

Jiang H-F, Wang A-Z, Liu H-L, Qi C-R (2008) Reusable Polymer-Supported amine-copper catalyst for the formation of α-alkylidene cyclic carbonates in supercritical carbon dioxide. Eur J Org Chem 2008:2309–2312

Karakas G, Dogu T, Somer TG (1997) Reactivity of CO_2 during thermal cracking of heavy paraffins under supercritical conditions. Ind Eng Chem Res 36:4445–4451

Kayaki Y, Noguchi Y, Iwasa S, Ikariya T, Noyori R (1999) An efficient carbonylation of aryl halides catalysed by palladium complexes with phosphite ligands in supercritical carbon dioxide. Chem Commun 1999:1235–1236

Kayaki Y, Yamamoto M, Suzuki T, Ikariya T (2006) Carboxylative cyclization of propargylamines with supercritical carbon dioxide. Green Chem 8:1019–1021

Kemmere M, de Vries T, Vorstman M, Keurentjes J (2001) A novel process for the catalytic polymerization of olefins in supercritical carbon dioxide. Chem Eng Sci 56:4197–4204

Kim S, Johnston KP (1988) Adjustment of the selectivity of a Diels-Alder reaction network using supercritical fluids. Chem Eng Comm 63:49–59

Kishimoto Y, Ikariya T (2000) Supercritical carbon dioxide as a reaction medium for silane-mediated free-radical carbonylation of alkyl halides. J Org Chem 65:7656–7659

Kröcher O, Köppel RA, Baiker A (1996) Synthesis of N, N-dimethylformamide by heterogeneous catalytic hydrogenation of supercritical carbon dioxide. Process Technology Proceedings 12:91–96

Leitner W (2002) Supercritical carbon dioxide as a green reaction medium for catalysis. Review Acc Chem Res. 35:746–756

Leitner W (2004) Recent advances in catalyst immobilization using supercritical carbon dioxide. Pure Appl Chem 76:635–644

Li J-H, Jiang H-F (1999) Glaser coupling reaction in supercritical carbon dioxide. Chem Commun 1999:2369–2370

Li J-H, Jiang H-F, Chen MC (2001) Palladium-catalyzed carbonylation of primary amines in supercritical carbon dioxide. Chin J Chem 19:1153–1156

Li J-H, Xie Y-X, Yin D-L, Jiang H-F (2003) Diiodination of alkynes in supercritical carbon dioxide. Chin J Chem 21:714–716

Loeker F, Leitner W (2000) Steel-promoted oxidation of olefins in supercritical carbon dioxide using dioxygen in the presence of aldehydes. Chem Eur J 6:2011–2015

Maggi R, Bertolotti C, Orlandini E, Oro C, Sartoria G, Selva M (2007) Synthesis of oxazolidinones in supercritical CO_2 under heterogeneous catalysis. Tetrahedron Lett 48:2131–2134

Meng F-Q, Feng X-J, Wang W-H, Bao M (2017) Synthesis of 5-vinyl-2-norbornene through Diels-Alder reaction of cyclopentadiene with 1,3-butadiene in supercritical carbon dioxide. Chin Chem Lett 28:900–904

Mistele CD, DeSimone JM, Thorp HH (1996) Ring-opening metathesis polymerizations in carbon dioxide. J Macromol Sci Pure Appl Chem A33:953–960

Nakamura K (1990) Biochemical reactions in supercritical fluids. Trends Biotech 8:288–292

Ouyang X-Y, Jiang H-F, Cheng J-S, Zhang Q-J (2002) Aerobic oxidation of methyl vinyl ketone in supercritical carbon dioxide. Chin J Chem 20:1326–1329

Palazzi C, Francio G, Bolm C, Leitner W (2002) Baeyer-Villiger oxidation in compressed CO_2. Chem Commun 1588–1589

Paulaitis ME, Alexander GC (1987) Reactions in supercritical fluids. A case study of the thermodynamic solvent effects on a Diels-Alder reaction in supercritical carbon dioxide. Pure Appl Chem 59:61–68

Pernecker T, Deák G, Kenned JP (1995) Carbocationic polymerization in supercritical CO_2. IV.* the isomerization polymerization of 3-methyl-1-butene and 4-methyl-1-pentene. Macromolecular Reports, Part A 32:969–978

Ran X-G, Jiang H-F, Zhu X-H (2004) Palladium-Catalyzed oxidation of dihydromyrcene to citronellal in supercritical carbon dioxide. Chin J Chem 22:1384–1386

Rathke JW, Klingler RJ, Krause TR (1991) Propylene hydroformylation in supercritical carbon dioxide. Organometallics 10:1350–1355

Renslo AR, Weinstein RD, Tester JW, Danheiser RL (1997) Concerning the regiochemical course of the Diels-Alder reaction in supercritical carbon dioxide. J Org Chem 62:4530–4533

Skouta R (2009) Selective chemical reactions in supercritical carbon dioxide, water, and ionic liquids. Green Chem Lett Rev 2:121–156

Srinivas P, Mukhopadhyay M (1994) Oxidation of cyclohexane in supercritical carbon dioxide medium. Ind Eng Chem Res 33:3118–3124

Stille JK (1986) Palladium-katalysierte kupplungsreaktionen organischer elektrophile mit organozinn-verbindungen. Angew Chem 98:504–519

Suzuki A (1999) Recent advances in the cross-coupling reactions of organoboron derivatives with organic electrophiles, 1995–1998. J Organomet Chem 576:147–168

Tanko JM, Blackert JF (1994) Free-radical side-chain bromination of alkylaromatics in supercritical carbon dioxide. Science 263:203–205

Theyssen N, Leitner W (2002) Selective oxidation of cyclooctane to cyclooctanone with molecular oxygen in the presence of compressed carbon dioxide. Chem Commun 2002:410–411

Tsitsimpikou C, Stamatis H, Sereti V, Daflos H, Kolisis FN (1998) Acylation of glucose catalysed by lipases in supercritical carbon dioxide. J Chem Technol Biotechnol 71:309–314

Wang Z-Y, Jiang H-F, Ouyang X-Y, Qia C-R, Yang S-R (2006) Pd(II)-catalyzed acetalization of terminal olefins with electron-withdrawing groups in supercritical carbon dioxide: selective control and mechanism. Tetrahedron 62:9846–9854

Wang ZY, Jiang HF, Qi CR, Shen YX, Yang SR (2007) PS-HQ: More convenient *in situ* polymeric cocatalyst for the $PdCl_2$-catalyzed acetalization in supercritical carbon dioxide. Chin Chem Lett 18:969–972

Wang TQ, Zhao XL, Hao JY (2011) In situ generation of biodegradable poly(p-dioxanone) microparticles by polymerization in supercritical carbon dioxide. Chin Chem Lett 22:1509–1512

Wu X-W, Oshima Y, Koda S (1997) Aerobic oxidation of cyclohexane catalyzed by Fe(III)(5,10,15,20-tetrakis(pentafluorophenyl)porphyrin)Cl in sub- and super-critical CO_2. Chem Lett 1997:1045–1046

Xiao J-L, Nefkens SCA, Jessop PG, Ikariya T, Noyori R (1996) Asymmetric hydrogenation of α, βs-unsaturated carboxylic acids in supercritical carbon dioxide. Tetrahedron Lett 37:2813–2816

Xu JX, Zhao JW, Jia ZB (2011) Efficient catalyst-free chemical fixation of carbon dioxide into 2-oxazolidinones under supercritical condition. Chin Chem Lett 22:1063–1066

Zhang Q-J, Sun J-H, Jiang H-F, Ouyang X-Y, Cheng J-S (2003) Palladium- Catalyzed addition of carbon monoxide and carbon tetrachloride to 1-octene in supercritical carbon dioxide. Chin J Chem 21:1525–1527

Zhang Y, Honda M, Fukaya T, Wahyudiono, Kanda H, Goto M (2020) One-step preparation of *Z*-isomer-rich β-carotene nanosuspensions utilizing a natural catalyst, allyl isothiocyanate, *via* supercritical CO_2. Symmetry 12:777. https://doi.org/10.3390/sym12050777

Zhou L, Xu QX, Jiang HF (2007) Palladium-catalyzed homo-coupling of boronic acids with supported reagents in supercritical carbon dioxide. Chin Chem Lett 18:1043–1046

Theoretical Approaches to CO_2 Transformations

Hossein Sabet-Sarvestani, Mohammad Izadyar, Hossein Eshghi, and Nazanin Noroozi-Shad

Abstract

Nowadays, the issue of CO_2 conversion becomes an urgent necessity for human civilization, which any ignorant will be caused irreparable damage to the future of human life. Thus, all researchers in the CO_2 utilization field have been employed their facilities to overcome the CO_2 issue. In the past decade, computational chemical modeling was mainly used to describe the observed results of a carried out reaction. Nowadays, regarding the rapid evolution of computational software and hardware, computational chemical sciences have been converted to powerful tools in the description of the obtained results, studying the mechanism of a reaction and designing the novel structures. The use of computational techniques in the investigation of CO_2 transformation to value-added chemicals affords brilliant results that attract remarkable attention among scientists. Among various theoretical techniques, density functional theory (DFT) modeling is a powerful and efficient approach to explore the mechanisms of CO_2 conversion and investigate novel catalysts for more efficient CO_2 transformation. DFT-based approaches are valuable strategies to overcome the disadvantages of trial-and-error experimental processes such as tedious and time-/labor-consuming repetitions. In this chapter, a comprehensive discussion by the DFT calculations is represented about the investigation of the mechanism of the catalytic CO_2 transformation into value-added materials such as CO, CH_4, CH_3OH, HCOOH, and heterocyclic compounds in the presence of heterogeneous, homogeneous, organo-based, and photo- and electro-catalysts. Moreover, the DFT-based design of novel catalytic systems, challenges, and opportunities in CO_2 transformations is outlined.

H. Sabet-Sarvestani · M. Izadyar (✉) · H. Eshghi · N. Noroozi-Shad
Department of Chemistry, Faculty of Science, Ferdowsi University of Mashhad, Mashhad, Iran
e-mail: izadyar@um.ac.ir

Abbreviations

AFIL	Amine-functionalized ionic liquid
BDN	1,8-Bis(diethylamino) naphthalenes
PNP(Ph)	2,6-Bis(diphenylphosphino) methylpyridine
PNP(tBu)	2,6-Bis(di-tert-butylphosphino) methylpyridine
BCP	Bond critical point
CBM	Conduction band minimum
DFT	Density functional theory
DBN	Diazabicyclo[4.3.0]non-5-ene
DBU	1,8-Diazabicyclo[5.4.0]undec-7-ene
DNP	Double numeric plus polarization
ECP	Effective core potentials
ESM	Energetic span model
EGR	Enhanced gas recovery
EGS	Enhanced geothermal systems
EOR	Enhanced oil recovery
FLP	Frustrated Lewis pair
GGA	Generalized gradient approximation
GO	Graphene oxide
HOMO	High occupied molecular orbital
HER	Hydrogen evolution reaction
HFC	Hydrofluorocarbon
IEF	Integral equation formalism
IEFPCM	Integral equation formalism polarizable continuum model
IPCC	International Panel on Climate Change
N_k	Local nucleophilicity indices
LUMO	Low unoccupied molecular orbital
MOF	Metal−organic framework
MESP	Molecular electrostatic potential
MO	Molecular orbitals
P_k^-	Mulliken atomic spin density
NBO	Natural bond orbital
NHC	N-heterocyclic carbine
NHO	N-heterocyclic olefin
NHE	Normal hydrogen electrode
NMR	Nuclear magnetic resonance
PAW	Projector-augmented wave

Inamuddin et al. (eds.), *Carbon Dioxide Utilization to Sustainable Energy and Fuels*, Advances in Science, Technology & Innovation, https://doi.org/10.1007/978-3-030-72877-9_8

PBE	Perdew–Burke–Ernzerhof
PW	Perdew–Wang
PFC	Perfluorocarbon
PR	Phosphorous reagent
P-ylide	Phosphorus ylide
PCM	Polarizable continuum model
PED	Potential energy diagram
TsCl	P-Toluenesulfonyl chloride
QTAIM	Quantum theory of atoms in molecules
RRKJ	Rappe Rabe Kaxiras Joannopoulos
RDS	Rate-determining step
RWGS	Reverse water gas shift
SEM	Scanning electron microscopy
SMD	Solvation model based on density
SFLP	Surface frustrated Lewis pairs
SB	Superbase
TBAB	Tetrabutylammonium bromide
THF	Tetrahydrofuran
TMG	1,1,3,3-Tetramethylguanidine
TDI	TOF-determining intermediate
TDTS	TOF-determining transition state
TEM	Transmission electron microscopy
TBD	1,5,7-Triazabicyclo[4.4.0]dec-5-ene
TOF	Turnover frequency
VSEPR	Valence shell electron pair repulsion theory
vdW	Van der Waals
WGSR	Water gas shift reaction
XPS	X-ray photoelectron spectroscopy

1 Carbon Dioxide Properties

CO_2 is a colorless, odorless, and non-flammable gas that is heavier than air. Carbon dioxide is odorless gas at low concentrations, but at high concentrations, it has a potent acidic smell. Also, it is an inert material that does not react with many substances. Thus, its storage, liquefaction, solidification, and handling are very easy and safe. Table 1 shows the molecular properties of CO_2 (Topham et al. 2000; Nakamura et al. 2015).

As a simple triatomic molecule, the carbon atom of CO_2 is covalently double bonded to two oxygen atoms which consist of linear geometry, having short and equivalent C–O bonds (1.1602(8) Å) (Gershikov and Spiridonov 1983). This fashion of bonds causes a nonpolar molecule. The observed behaviors of CO_2 in the solid, liquid, and gas phases are the outcomes of a molecular quadrupole that is due to the shape of molecule and electron distribution. Figure 1a illustrates Lewis structure of CO_2, based upon valence shell electron pair repulsion theory (VSEPR), and its electrostatic potential (Fig. 1b), depicted on the electrostatic potential map (0.002 a.u.) at the PBE1PBE/aug-cc-pVTZ level of theory (Murphy et al. 2015). It can be concluded that the oxygen and carbon atoms are the centers of electrophilic and nucleophilic attacks, respectively. In other words, the carbon and oxygen atoms can behave as Lewis acid and base, respectively.

The molecular orbitals (MOs), which are the outcome of the molecular orbital calculations, are helpful tools to afford the basic information about the studied molecule (Nakamura et al. 2015). Figure 2 depicts the molecular orbital diagram and frontier orbitals (high occupied molecular orbital (HOMO), low unoccupied molecular orbital (LUMO)) of CO_2. The illustrated MOs in this figure are not experimentally detectable as a physical object. However, they are used frequently in organic and organometallic chemistry as an efficient concept to justify and evaluate the molecule properties. From the orbital viewpoint, the MOs are considered as the most figurative images to explanation and characterization of the molecular behavior in a typical reaction (Nakamura et al. 2015; Luther Iii 2004).

It is notable that, both HOMO and LUMO energies are degenerate. Thus, the CO_2 molecule has four degenerated electrons with similar energy levels, which can transfer in a typical reaction. Energy calculation reveals that the transformation of one of the four electrons diminishes the degeneracy, immediately. Therefore, the transformation of four electrons to or from the outside of CO_2 does not take place, simultaneously. Moreover, CO_2 possesses two degenerate orbitals including four unoccupied sites that give electron acceptation character for this molecule. An organic molecule having such properties is rare. Indeed, CO_2 is neither a complex contains transition metal, nor an electron-rich solid specie. The degenerated frontier orbitals of CO_2 are slightly similar to the topological characters of the *d* orbitals (two HOMOs) and *f* orbitals (two LUMOs) in transition metals and lanthanoids, respectively (Nakamura et al. 2015).

2 CO_2 Transformation as an Undeniable Necessity

Carbon dioxide is an important component of the earth matrix, found in the core, crust, and also the atmosphere, considerably. According to the carbon cycle, soluble carbon dioxide in water can react with other components, be solidified in carbonated stones, and freely emitted into the atmosphere (Hofmann et al. 2009). For centuries, the exchange cycle had established CO_2 equilibrium until the Industrial Revolution. However, due to global industrialization in the past years, the consumption of fossil fuels has intensified, which has been resulted in an irregular increase in the concentration of greenhouse gases in the atmosphere. Respiration procedure by living creatures on the earth and

Table 1 Molecular properties of carbon dioxide

Formula	CO_2
Molecular weight	44.0098 (g/mol)
Critical temperature	31.04 °C
Boiling point	−78.46 °C
Melting point	−56.6 °C
Density	1.977 g/L (gas at 1 atm and 0 °C) 0.914 g/L (liquid at 34.3 atm and 0 °C) 1.512 g/L (solid at −56.6 °C)
Dipole moment	0D
Point group	$D_{\infty h}$
Dielectric constant	1.000922 (gas)
Viscosity	0.0147 mN s m^{-2} (gas)
Vapor pressure	5728.9 kPa (at 20 °C)
Sublimation point	−78.92 °C
Solubility in water	835 mL/kg (at 20 °C and 101 kPa)
Enthalpy	193.90 J/g (at 20 °C)
Entropy	632.625 J g^{-1} K^{-1} (at 20 °C)

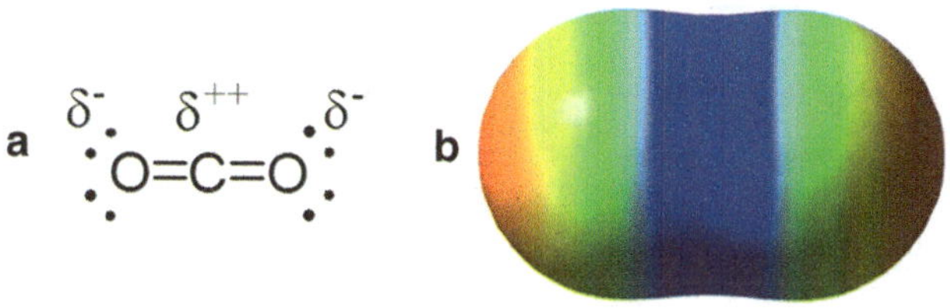

Fig. 1 Lewis structure (**a**) and electrostatic potential map (0.002 a.u.) (**b**) of CO_2 molecule (Murphy et al. 2015)

the sea causes CO_2 emission into the atmosphere (Rafiee et al. 2018). Similar to this respiration, the decay process of the corpse of buried organisms by oxygen releases CO_2 molecules. The produced CO_2 by respiration or decomposition processes is absorbed by plants and transforms into carbohydrates, via the photosynthesis process, gradually. The consumed water and CO_2 by algae, plants, and cyanobacteria via the photosynthesis process yield carbohydrates and oxygen as the main product and by-product, respectively. Both of these products are essential for other living creatures. By starting the Industrial Revolution, excessive fossil fuel extraction by human beings has disrupted the CO_2 equilibrium in the environment. This extraction causes two forms of CO_2 emission including released pure CO_2 gas to the atmosphere, due to the extraction procedure, and the by-product of the combustion process (Artz et al. 2018).

Until the Svante Arrhenius report, there was no anxiety about the irregular emission of CO_2. However, in the 1880s, he was the first investigator that anticipated global warming due to the libration of CO_2 from the combustion of fossil fuels (Maslin 2008), which received numerous criticisms, firstly. However, after an investigation of variations of climate temperature by Guy Stewart Callendar, Arrhenius's theory was attracted numerous attention (Rafiee et al. 2018).

The outcomes of irregular fossil fuel consumption are increasing the global temperature beyond the normal value (Saxena et al. 2014), socking the coastlines due to the increment of water level in the sea, and occurring unpredictable floods that affect the global economics and ecosystem, catastrophically (Mukherjee et al. 2019). Six main greenhouse gases, namely carbon dioxide (CO_2), methane (CH_4), nitrous oxide (N_2O), hydrofluorocarbons (HFCs), perfluorocarbons (PFCs), and sulfur hexafluoride (SF_6) are specified by Kyoto Protocol as the agents which remarkably affect the environment quality (Zhang and Da 2015).

Among the greenhouse gases, CO_2 is the most important gas due to the most contribution to global warming (Abeydeera et al. 2019). Because of the combustion of fossil fuel, among the global industries, about 60% of CO_2 production belong to industries, such as cement production, iron and steel factories, manufacturing petrochemical materials, gas refiner mills, generators of electrical power, and transportation sector (Yaumi et al. 2017). For example, the contribution of electricity manufacture, agriculture, and forestry is evaluated by about 25% and 24% of the total CO_2 emission, respectively (Yaumi et al. 2017). However, regarding inexpensive and abundant sources of coal, it is applied in thermal power plants, abundantly. Thus, the emitted CO_2 is evaluated up to 2249 lbs/MWh (Spigarelli and Kawatra 2013). Figure 3 depicts the produced CO_2 from different energy sections (Mukherjee et al. 2019).

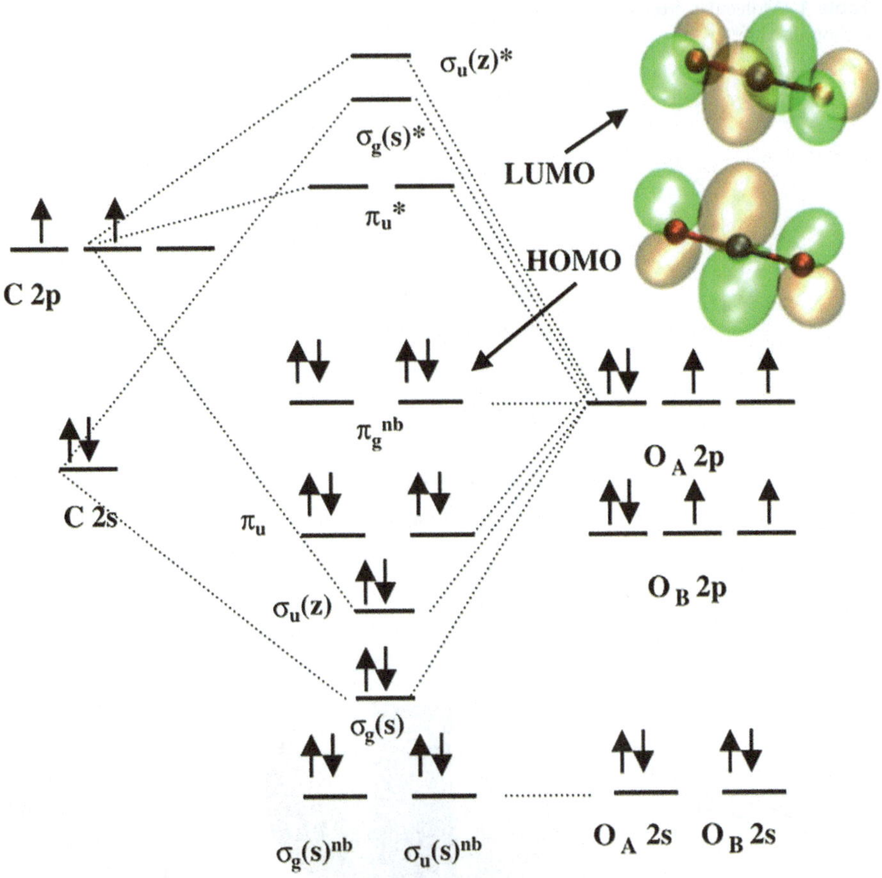

Fig. 2 Molecular orbital diagram and frontier orbitals (HOMO and LUMO) of CO_2 (Luther Iii 2004)

In 2014, the International Panel on Climate Change (IPCC) declared that the released greenhouse gas via human activities is the main reason for global warming and the related climate change. Because of industrialization and climate variation, caused by human activities including emission of CO_2, nitrous oxide (N_2O), methane (CH_4), and water vapor (H_2O) into the atmosphere, human life is put in danger (Lee and Park 2015). Investigations show that 76% of the overall volume of greenhouse gases is CO_2, as the consequence of the combustion of fossil fuels in various sections of the industry. Thus, CO_2 emission has undeniable effects on global warming, environmental changes, and ecosystem conditions (Siqueira et al. 2017). As reported by world environmental organizations, carbon dioxide emissions in ambient air have been increased from 22.15 Gt in 1990 to 36.14 Gt in 2014. The beginning of the Industrial Revolution has become the main reason for the global warming phenomenon (Rashidi and Yusup 2016).

Therefore, to reach sustainable development, consciousness about increasing environmental problems and global climate change is necessary. Hence, scientific societies need to apply their potentials toward resolving new challenges, such as mitigation of climate change, conservation of the environment, and replacement of fossil fuels by renewable energies. The first investigation on carbon emission was reported in 1981, and afterward, carbon emission researches have been increased, gradually. Figure 4 depicts the gradual growth of research publications corresponding to the carbon emission starting from 1980 to June 2019. According to this plot, about 479 research articles were reported in 2018, which reveal the growing importance and concern about this issue among the scientific communities (Abeydeera et al. 2019).

Various solutions, such as a decrease of irregular use of energy resources, changing sources of fossil fuels by renewable and sustainable energies, and utilization of CO_2, have been considered to reduce CO_2 emission and consumption of fossil fuels. All these solutions are considerable approaches to solving the greenhouse gas problem. However, most of the new sources of energy are less efficient than fossil fuels (mainly natural gas) and regarding the economic possibility viewpoint are not competitive. Based on reported studies, it is not possible to attain the target of 100% sustainable energy systems in a short or medium period. Hence, to provide energy, fossil fuel consumption will continue (OECD I 2016). Therefore, CO_2 capture and utilization are approaches that can be regarded as a solution to global warming. But, after three decades of investigation, these approaches need greater maturity.

After the industrialization and the maturation of chemical industries, CO_2 has been applied in numerous industries.

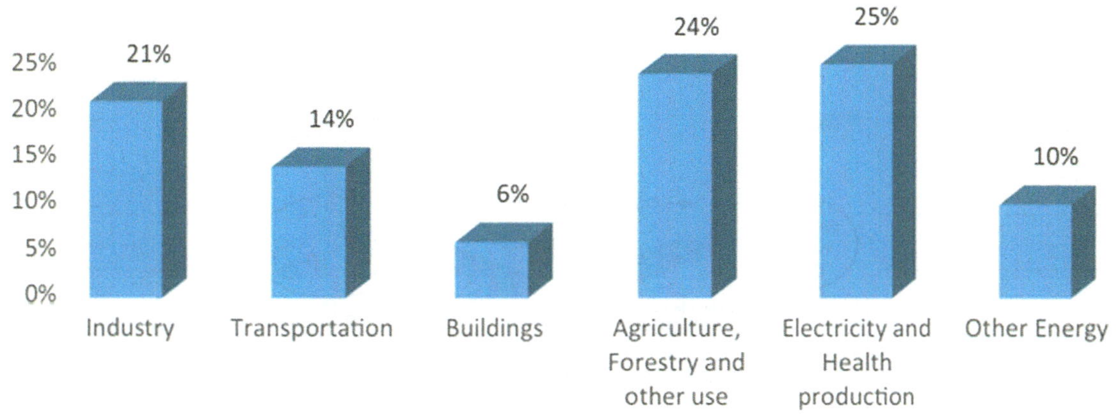

Fig. 3 Produced CO_2 from different energy sections (Mukherjee et al. 2019)

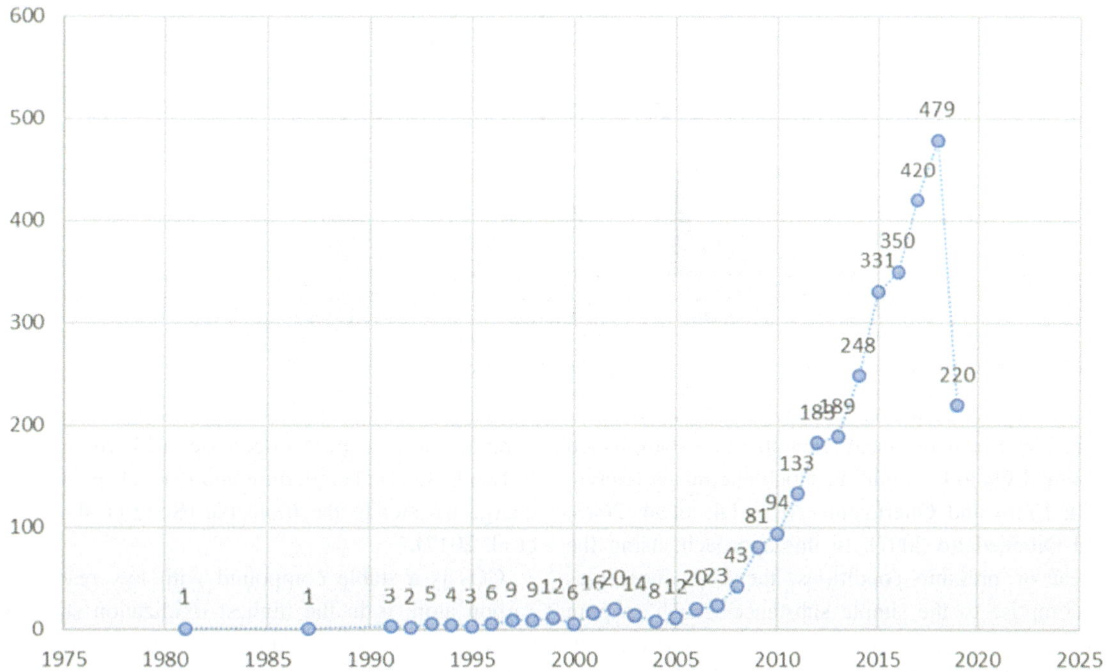

Fig. 4 Trend of research publications about carbon emission starting from 1980 to June 2019 (Abeydeera et al. 2019)

The utilization techniques can be considered into two categories. First, the direct use of CO_2 without any transformation to value-added compounds (Huang and Tan 2014). The manufacturing of carbonated drinks in beverage industries can be considered as the earliest and direct application of CO_2. Also, enhanced oil recovery (EOR) techniques by CO_2 injection remarkably improves the economic efficiency of oil extraction from the oil reservoir owning to high mutual solubility of supercritical CO_2 and oil, as a hydrophobicity material (Dai et al. 2014). As well as, increasing the CO_2 pressure leads to a substantial decrease in the viscosity of the CO_2–oil mixture. Accordingly, CO_2 injection can enhance production by about 15%. Moreover, it can also be used in enhanced gas recovery (EGR) and enhanced geothermal systems (EGS). Application of CO_2 in the production of dry ice, fire extinguisher, supercritical solvent, and the refrigerant is the next direct CO_2 utilization without any transformation. However, these direct usages for CO_2 are limited in volume and have a small consequence on the overall CO_2 concentration (Muradov 2014), as well as the CO_2 molecules in all the mentioned applications stay pure and do not convert to another chemical. The left side of Fig. 5 depicts the physical utilization of CO_2.

The second category of CO_2 utilization is its transformation to value-added material and fuels (Aresta 2010). This approach has been attracted very attention among scientific

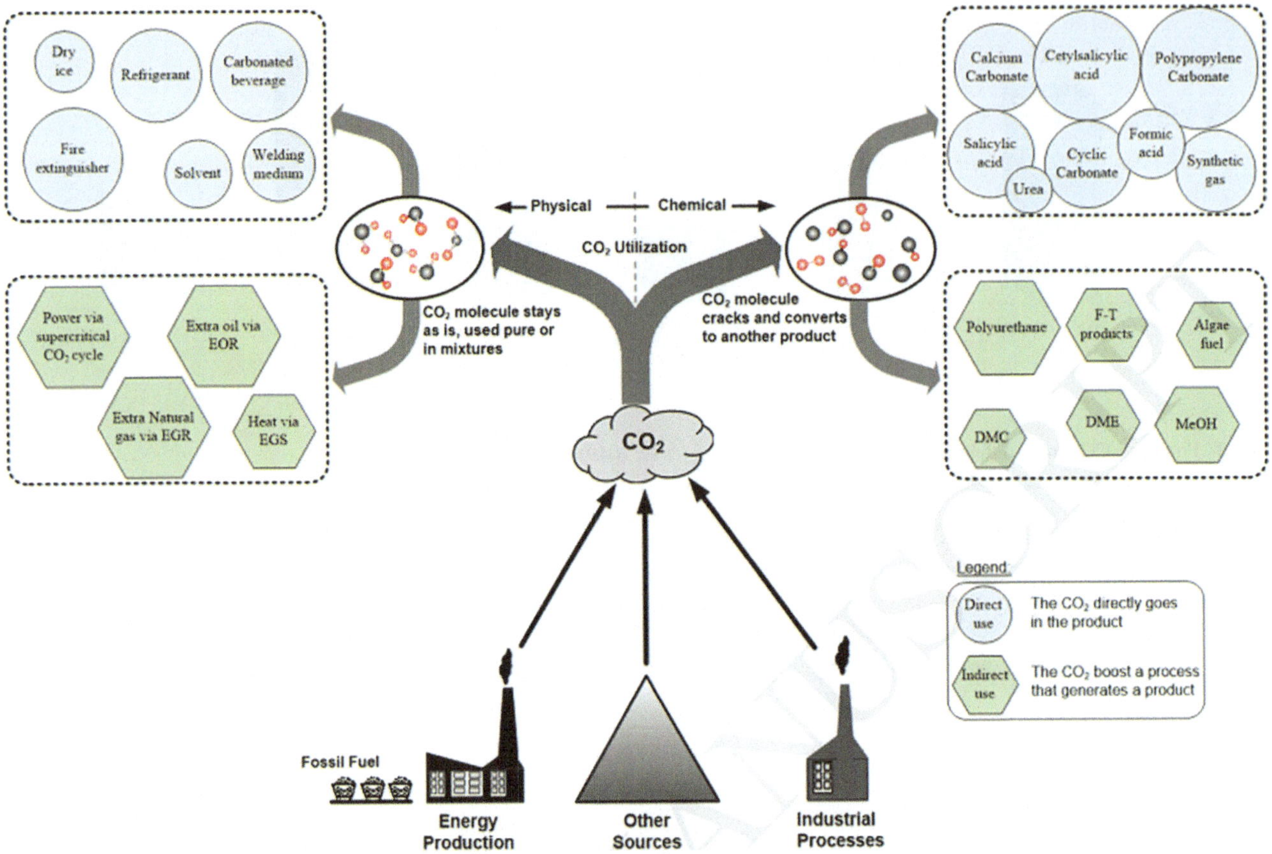

Fig. 5 Various chemical and physical routs in CO_2 utilization (Rafiee et al. 2018; Song et al. 2017)

communities. Hence, widespread attempts are accomplished for converting CO_2 to C1 building block chemicals (Bertau et al. 2010; Keim and Offermanns 2010; Ola et al. 2013; Aresta and Dibenedetto 2007). In this approach, using the specific heat or pressure conditions, the CO_2 molecules mostly decompose to the simple substances (such as pure carbon or CO) or in the presence of a catalyst react with other chemicals to produce value-added materials, such as hydrocarbons. Thus, the CO_2 molecule is a constituent of a chemical product, in which the product cannot be synthesized without CO_2. The right side of Fig. 5 shows the application of CO_2 in the manufacture of value-added chemicals, as chemical utilization.

3 CO_2 Activation

The activation of a molecule means elevation of the reactivity of the molecule. Therefore, it can be considered by making an obvious change in molecular properties in comparison with the properties of a stable ground state. Regarding the molecular structure, CO_2 activation can be considered in three methods such as (1) bending of the O–C–O angle from linear equilibrium geometry. Notably, more bending causes a greater decrease in LUMO energy. (2) One or two C–O bond stretching and (3) CO_2 polarization due to charge transfer to the molecule (Song et al. 2017; Álvarez et al. 2017).

CO_2 is a stable compound with low reactivity, and its carbon atom is in the highest oxidization state, thermodynamically. Thus, overcoming thermodynamic barrier energy is an essential step in CO_2 activation, which is an important challenge in its conversion to value-added chemicals. The greatest limitation in the industrial application of CO_2, as a starting material, is its thermodynamic stability and/or kinetically inert in a typically favorable transformation. It can be concluded that a large energy source is essential to CO_2 transformation (Sakakura et al. 2007). Generally, most of the usual reported methods in CO_2 activation are limited to the use of highly reactive chemicals (i.e., energy-rich substrates, epoxides, and aziridines) and/or vigorous situations, such as high temperature and pressure, although side reactions are unavoidable. Thus, designing and using efficient catalyst systems have remarkable advantages such as effectually activate CO_2 and/or the substrate, decrease the energy barrier of activation, modify the rate of the activation reaction, and decrease the production of by-products. It can be concluded that developing a convenient functional catalyst with an efficient active site to achieve high selectivity of the goal products is an important necessity in CO_2 transformations. Therefore, impressive functional catalysts have key roles to get an efficient and selective procedure for CO_2 transformation (Liu et al. 2015).

3.1 *Methodologies of CO_2 Activation*

The oxygen atom of CO_2, as a Lewis base, has a nucleophilicity property, while the carbon atom, as Lewis acid, behaves as an electrophile (Song et al. 2017). Thus, as shown in Fig. 6, two general methods for CO_2 activation can be considered. One method involves the catalysts having nucleophile characters such as superbase, N-heterocyclic carbine (NHC) (Kayaki et al. 2009), N-heterocyclic olefin (NHO) (Wang et al. 2013), tungstate, frustrated Lewis pair (FLP) (Bontemps 2016), and hydroxyl group-containing compounds (Yang et al. 2011) which involve with the carbon atom of CO_2. In another method, the oxygen atom is affected by the species having Lewis acid orbital such as a transition metal, yielding metal–CO_2 complex. In 1971, Aresta and coworkers reported the first almost planar metal–CO_2 complex, as $(PCy_3)_2Ni(CO_2)$ in which Ni atom is linked to CO_2 molecule and phosphine ligands (Aresta et al. 1975). This complex has two unequal C–O bonds (1.17 and 1.22 Å) and an O–C–O angle of 133° in the linked CO_2. Afterward, various transition metals such as iron, ruthenium, and palladium, with different oxidation states, have been investigated in this issue (Paparo and Okuda 2017). Moreover, using high energy and active chemicals such as unsaturated chemicals (alkene, alkyne, etc.), three-membered rings (aziridines, epoxides), and organometallic structures is another key to successful CO_2 conversion. Figure 7 depicts the energy profile of the CO_2 reaction with high energy substrates, which leads to low energy chemicals (e.g., organic carbonates and carbamates) (Paparo and Okuda 2017).

Besides the CO_2 activation, there are still various scientific and practical difficulties in CO_2 conversion. Other problems are CO_2 collection from the atmosphere by proper technologies and economical aspects of the whole process of CO_2 conversion using renewable resources of energy. Nevertheless, low reactivity and selectivity of the catalysts for the chemical transformations are the most considerable challenges, which are related to the high chemical durability of CO_2. Thus, understanding the basic chemical procedure of the transformation and investigation of efficient, cost-effective, and eco-friendly catalysts are essential solvation for this challenge.

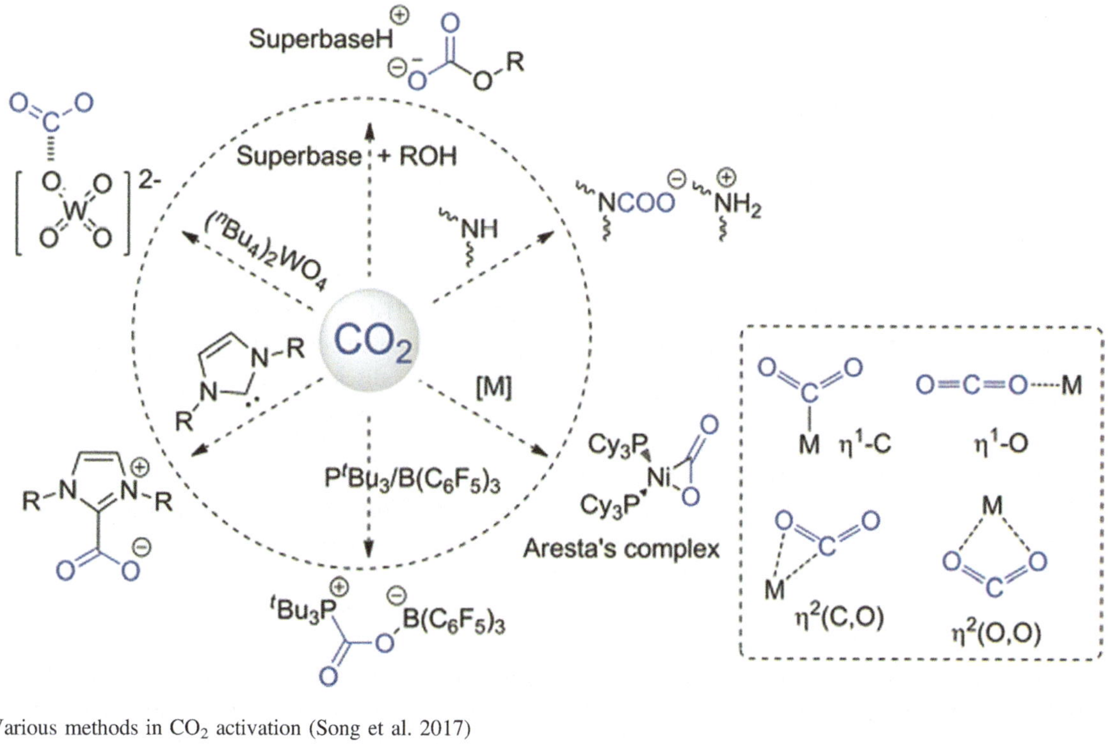

Fig. 6 Various methods in CO_2 activation (Song et al. 2017)

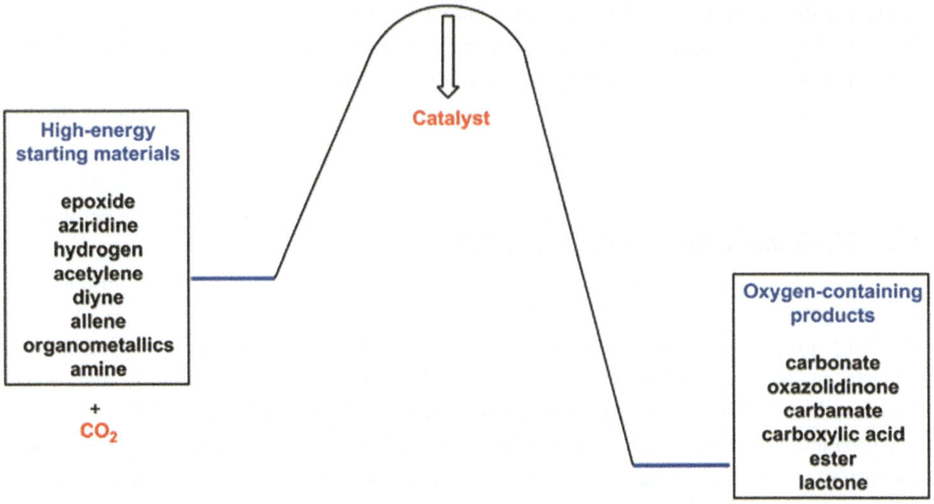

Fig. 7 Energy profile of the CO_2 reaction with high energy substrates (Song et al. 2017)

4 Theoretical Insight of CO_2 Transformation

Despite the various investigations, the determination of the mechanism of a catalytic CO_2 conversion is a challenging problem (Cheng et al. 2013). From the experimental viewpoint, the catalyst activity and complex structures, the proportion of reactant, catalyst, and substrate in the reaction, the circumstance of the interactions between the involved species and energy levels in the catalytic reactions make understanding the mechanism of CO_2 conversion as a very difficult issue. Consequently, the mechanism investigation based on a trial-and-error attitude is a very tedious and time-consuming task that includes designing synthesis–structure–property correlations. This procedure is not so effective nor fully perfect and is very labor-consuming due to the various number of conditions to be studied. Therefore, low yield and negligible selectivity are two obstacles in the catalytic transformation of CO_2, which can be improved remarkably by the logical design of catalysts, predicting key structure–activity relationships, and evaluation of specific structures (Wang et al. 2011; Centi and Perathoner 2009).

Experimental limitations give a unique opportunity to computational modeling. Due to current advances in theoretical methods such as novel methods, algorithms, and efficient computational tools, the accessibility of applicable software packages for electronic structure calculations, theoretical CO_2 transformation and simulations can play a significant role in two overall main approaches including understanding catalytic reaction mechanisms and designing new efficient catalysts to CO_2 transformation (Pople 1999; Kohn and Sham 1965; Kohn 1999; Ben-Nun and Martínez 2002; Bartlett and Musiał 2007). However, theoretical approaches can aid in a deep understanding of determining factors on the structures, binding energies, bond lengths, vibrational frequencies, intermolecular long-range interactions, dispersion forces, and reactivity of the catalysts, which are the main factors in the investigation of the reaction mechanism. Also, computational methods can be used as a predictive tool in the catalyst designing and the controlled synthesis of the target materials. Based on previous experience in the theoretical transformation of CO_2, in this chapter, we focus on the mechanism investigation and catalyst designing based on computational approaches. It is noticeable that the experimental features of CO_2 transformation are discussed in several reviews (Olah et al. 2009; Riduan and Zhang 2010; Mikkelsen et al. 2010; Ma et al. 2009; Hunt et al. 2010; Darensbourg 2010; Spinner et al. 2012; Cokoja et al. 2011; Yaashikaa et al. 2019; Schilling and Das 2018) and are not considered in this chapter. The chapter is divided into two sections. The first section deals with the theoretical aspects and possible reaction pathways of the catalytic CO_2 conversion to value-added materials such as CO, CH_4, HCOOH, CH_3OH, and heterocyclic compounds. In the second section, theoretical modeling and design of novel catalysts are discussed, and successful examples, within challenges and opportunities, are provided in this approach.

4.1 *The Theoretical Approach in CO_2 Conversion to Value-Added Chemicals*

4.1.1 Carbon Monoxide

The water gas shift reaction (WGSR) is a traditional process, which is applied for the transformation of synthesis gas to needed hydrogen for ammonia manufacture in the fertilizer industry and petroleum refineries (Bs et al. 2010). Nowadays, the produced hydrogen of this reaction is used as fuel for power production and transportation. The primary application of the reaction back to 1888 (Bs et al. 2010) and its advantage refer to the Haber ammonia synthesis process and evolution of catalyst by Bosch and Wilde in 1912 (Bs et al. 2010). The developed catalyst is based on iron and chromium, which catalyzed the reaction at 400–500 °C and reduce the existing carbon monoxide. Nowadays, reverse water gas shift (RWGS) reaction (Eq. 1), as an endothermic reaction, is a well-known procedure in CO production from CO_2, which is regarded as the elementary step for many other hydrogenation reactions, such as the Sabatier reaction (Liu et al. 2015; Wang et al. 2018a) and methanol synthesis (Hu et al. 2013). RWGS reaction is carried out in the gas phase and thermodynamically favorable at high temperatures.

$$\begin{gathered} CO_2 + H_2 \rightleftharpoons CO + H_2O \\ \Delta H^0_{298} = 41.3 \text{kJ mol}^{-1} \end{gathered} \quad (1)$$

The mechanisms of RWGS reaction can be classified into two overall classes, namely the redox mechanism and associative mechanism (Su et al. 2017). In the redox mechanism, a catalyst bed is responsible for the oxidation–reduction cycle (Bs et al. 2010), in which CO_2 and H_2 are firstly adsorbed on the active site of reducing metal or metal oxide. Equation 2 depicts steps of the reactions on an adsorbed species (star notation denotes a typical metal or metal oxide). At the end of the mechanism, the oxidized catalyst is reduced by H_2, and the active sites are reproduce again (Saeidi et al. 2017; Yan et al. 2014).

$$\begin{gathered} H_2 + * \rightarrow H_2^* \quad \text{Step 1} \\ H_2^* \rightarrow H^* + H^* \quad \text{Step 2} \\ CO_2 + * \rightarrow CO_2^* \quad \text{Step 3} \\ CO_2^* + * \rightarrow CO^* + O^* \quad \text{Step 4} \\ H^* + O^* \rightarrow OH^* + * \quad \text{Step 5} \\ OH^* + H^* \rightarrow H_2O^* \quad \text{Step 6} \\ OH^* + OH^* \rightarrow H_2O^* + O^* \quad \text{Step 7} \\ H_2O^* \rightarrow H_2O + * \quad \text{Step 8} \\ CO^* \rightarrow CO + * \quad \text{Step 9} \end{gathered} \quad (2)$$

The second mechanism named associative mechanism also named dissociative (Aresta et al. 2016; Chen et al. 2000) or formate (Cheng et al. 2013; Scibioh and Viswanathan 2018) mechanism. The critical step in this mechanism is described by an adsorption–desorption model, in which the adsorbed intermediate (carbonate, formate, carbonyl, etc.) is produced during the reaction between the absorbed species. Then, the produced product decomposes to form H_2 and a monodentate carbonate (Podrojkova et al. 2020). On the other hand, the reaction of CO_2* and dissociated H* produces bidentate formate intermediate by the adsorption of oxygen atoms on the metal surface. The reaction is followed by the formation of HCOOH via more hydrogenation which decomposes into HCO* and OH* and afterward into CO* and H*. In another pathway, CO_2 is adsorbed on the surface of the carbon atom, followed by the hydrogenation of the oxygen atoms. Due to more hydrogenation, the produced COOH* decomposes into COH* and OH* forming CO* and H*. Also, COOH* intermediate can be directly decomposed into CO* and OH*. Finally, H_2O is formed by the addition of OH* and H* intermediates. Equation 3 shows the steps of this mechanism, and Fig. 8 depicts two separate pathways for the associative mechanism.

$$\begin{gathered} H_2 + * \rightarrow H_2^* \quad \text{Step 1} \\ H_2^* \rightarrow H^* + H^* \quad \text{Step 2} \\ CO_2 + * \rightarrow CO_2^* \quad \text{Step 3} \\ CO_2^* + H^* \rightarrow HCOO^*(COOH^*) + * \quad \text{Step 4} \\ HCOO^*(COOH^*) + * \rightarrow HCO^*(COH^*) + OH^* \quad \text{Step 5} \\ HCO^*(COH^*) + * \rightarrow CO^* + H^* \quad \text{Step 6} \\ COOH^* + * \rightarrow CO^* + OH^* \quad \text{Step 7} \\ OH^* + H^* \rightarrow H_2O^* \quad \text{Step 8} \\ H_2O^* \rightarrow H_2O + * \quad \text{Step 9} \\ CO^* \rightarrow CO + * \quad \text{Step 10} \end{gathered} \quad (3)$$

Cai and coworkers were employed the DFT method to investigate the performance of the MoP (0001) surface, in comparison with the Ni_2P (0001) surface (Fig. 9) by the generalized gradient approximation (GGA) using the Perdew−Burke−Ernzerhof (PBE) functional. Higher adsorption energies and charge for MoP (0001) confirm that the intermediates are more stable on the MoP (0001) surface. However, both MoP and Ni_2P are active in the RWGS reaction. Additionally, the obtained potential energy diagram (PED) (Fig. 10) in mechanistic study illustrates that in the case of MoP (0001) the direct path is more favorable, whereas the COOH-mediated mechanism is a better path for progressing the reaction in the presence of Ni_2P (0001) as a catalyst (Guharoy et al. 2019).

In another study, the performance of the surfaces of Cu@Mo_2C (001) and Cu_4@Mo_2C (001) (Fig. 11) was explored in the RWGS reaction by DFT calculations using the PBE functional (Jing et al. 2018).

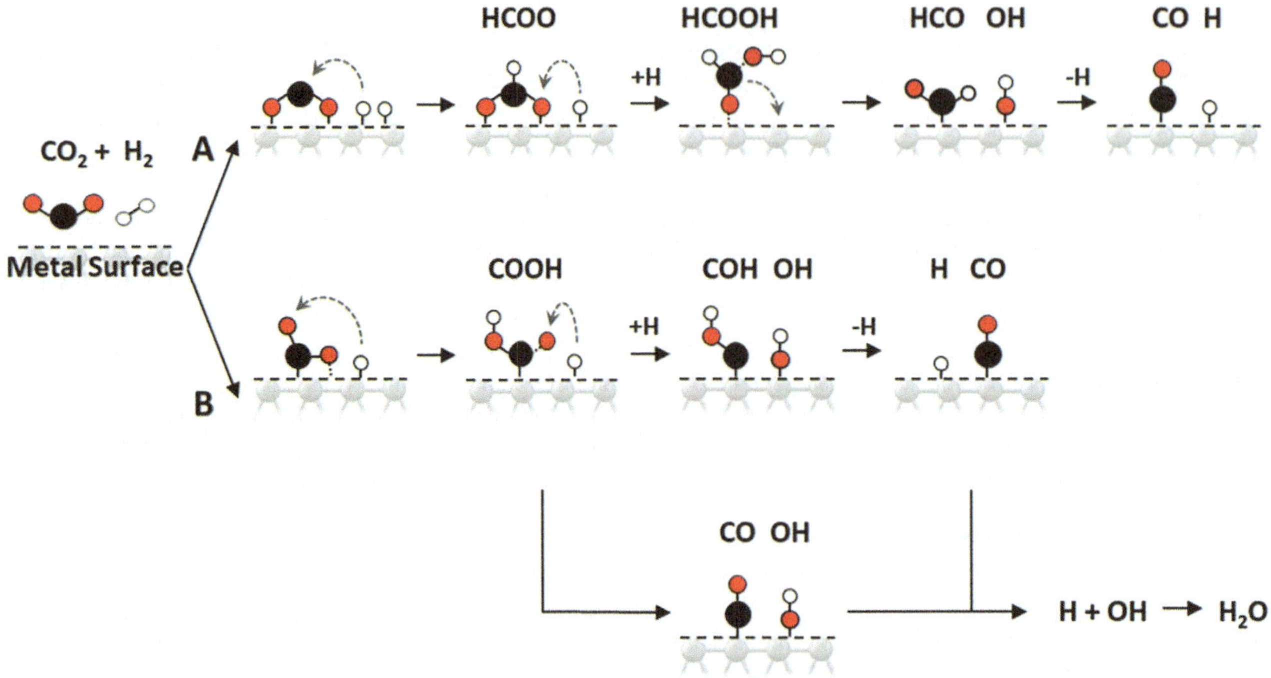

Fig. 8 Two separated pathways for the associative mechanism, **a** formation of HCOO intermediate and **b** formation of COOH intermediate (Podrojkova et al. 2020)

Fig. 9 Crystal structures of the applied MoP (0001) and Ni_2P (0001) surfaces in RWGS reaction (Guharoy et al. 2019)

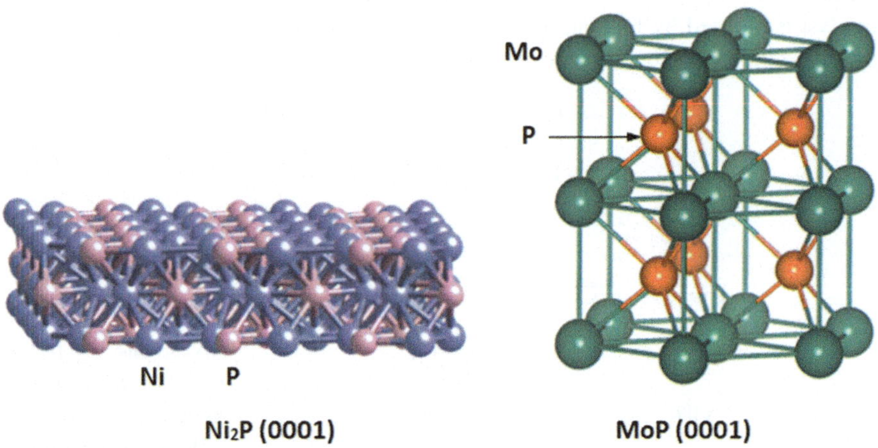

The results show that the doped β-Mo_2C (001) surface by Cu needs lower energy for desorption of CO. Therefore, the introduction of non-noble metal copper is a useful technique in CO release. Also, mechanism studies illustrate that the reaction progress by the redox mechanism is the most preferred path for both Cu@Mo_2C (001) and Cu_4@Mo_2C (001) surfaces. Also, regarding the rate-determining step (RDS) of the redox mechanism on the PEDs, corresponding to Cu@Mo_2C (001) and Cu_4@Mo_2C (001) surfaces (Fig. 12), the latter surface is a better candidate for the proceeding of the RWGS reaction (Jing et al. 2018).

As mentioned in two previous examples, DFT calculations based on the plane-wave basis sets are a privileged approach to exploring the solid-state materials as catalysts in the RWGS reaction. However, DFT calculations by Gaussian-type orbitals are another method for the investigation of this reaction. Guo and coworkers investigated Cu_{12}TM (TM = Co, Rh, Ir, Ni, Pd, Pt, Ag, Au) as the bimetallic metal catalysts in RWGS reaction, in which 6-31G(d,p) basis set and LANL2DZ pseudopotential were used to describe the H, C, O atoms and metal elements, respectively (Zhang and Guo 2018).

Figure 13 illustrates the optimized structure of Cu_{12}TM and adsorbed reactants, intermediates, and products, in the RWGS reaction. Figure 14 depicts the PED of the reaction on the Cu_{13} and Cu_{12}TM clusters. The results indicate the

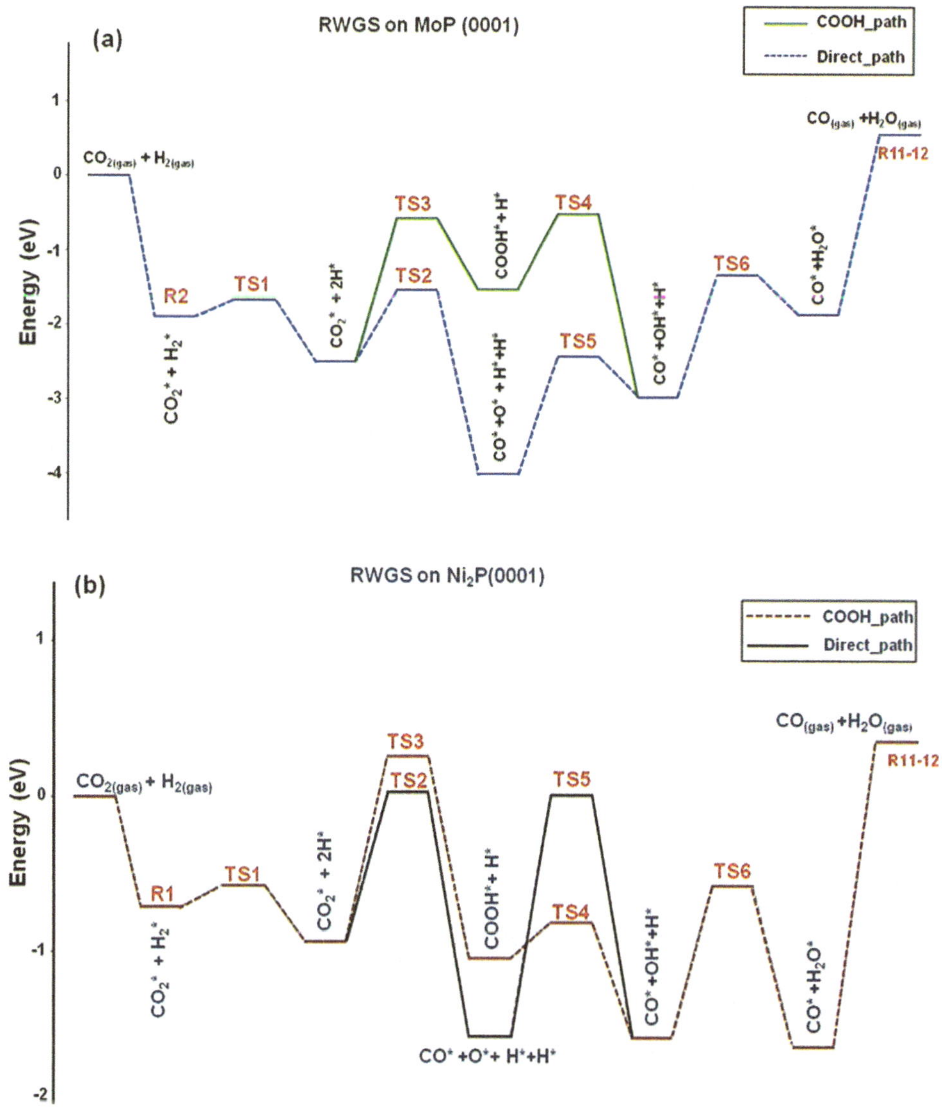

Fig. 10 Potential energy diagram (PED) for the RWGS reaction on **a** MoP (0001) and **b** Ni_2P (0001) (Guharoy et al. 2019)

redox mechanism, in which CO_2 is cleaved to CO* and O* is more favorable than both carboxyl and formate mechanisms. Also, as it is shown in PED, H_2O formation is the RDS of the reaction (Zhang and Guo 2018).

In other DFT calculations by the Gaussian-type orbitals, four mechanisms were considered to explore the thermodynamic and kinetic aspects of the RWGS reaction in the presence of the Rh–Mo_6S_8 cluster as a catalyst. Besides the general mechanisms in the RWGS reaction, the authors considered the possibility of formic acid formation for the possible paths of the reaction mechanism. Equation 4 illustrates the proposed steps of the formation and consumption of formic acid in the RWGS reaction (Cao et al. 2016).

Figure 15 depicts the optimized species in the reaction, and Fig. 16 illustrates the PEDs corresponding to (A) the redox mechanism, (B) the carboxyl mechanism, (C) the directly formic acid decomposition to CO, and (D) the formation of a CHO as the intermediate through the reaction. Based on the studied mechanisms, the reaction progress through the direct decomposition of HCOOH to CO and H_2O (mechanism C) leads to a lower value of barrier energy in comparison with other mechanisms. Also, due to the high energy barrier of RDS (76.01 kcal mol^{-1}), the decomposition of HCOOH to CHO* is unlikely to proceed (Cao et al. 2016).

$$H_2 + * \rightarrow H_2^* \quad \text{Step 1} \tag{4}$$

$$H_2^* \rightarrow H^* + H^* \quad \text{Step 2}$$

$$CO_2 + * \rightarrow CO_2^* \quad \text{Step 3}$$

$$CO_2^* + H^* \rightarrow HCOO^* + * \quad \text{Step 4}$$

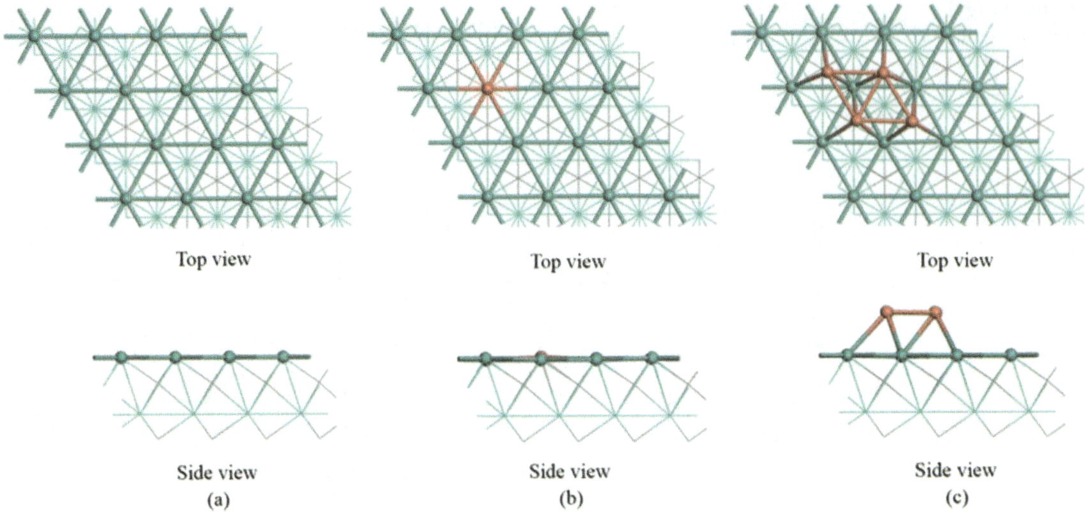

Fig. 11 Optimized crystal structures of the Mo-terminated β-Mo_2C (001) surface, Cu@Mo_2C (001), and Cu_4@Mo_2C (001) surfaces, in which Mo, C, and Cu atoms have been detected by blue, gray, and orange color, respectively (Jing et al. 2018)

$$HCOO^* + H^* \rightarrow HCOOH^* + * \quad \text{Step 5}$$

$$HCOOH^* + * \rightarrow CO^* + H_2O^* \quad \text{Step 6}$$

$$HCOOH^* + * \rightarrow HCO^* + OH^* \quad \text{Step 7}$$

$$HCO^* + * \rightarrow CO^* + H^* \quad \text{Step 8}$$

$$OH^* + H^* \rightarrow H_2O^* + * \quad \text{Step 9}$$

$$H_2O^* \rightarrow H_2O + * \quad \text{Step 10}$$

$$CO^* \rightarrow CO + * \quad \text{Step 11}$$

Despite the various studies about the RWGS reaction, the subjects, such as the effects of the metal catalysts and related mechanisms along the spontaneous, dynamic, and high-temperature reaction, need more investigation and clarification (Choi et al. 2017). It was revealed that to reach a satisfactory level of conversion, high temperature is required, thermodynamically. Therefore, remarkable attempts are performed to modify catalytic activity and selectivity for carrying out the RWGS reaction at lower temperatures (Yang et al. 2018).

4.1.2 Methane

CO_2 reduction to methane (CO_2 methanation or Sabatier reaction) shows numerous benefits among other chemical reaction because methane can be straightly driven into the natural gas pipelines, and it can be applied as fuel or starting material for the generation of other chemicals (Frontera et al. 2017). Also, CO_2 conversion to methane is a simple reaction that proceeds under atmospheric pressure and room temperature. Regarding thermodynamic property, CO_2 methanation remains the most advantageous transformation than other reactions, which form hydrocarbons or alcohols (Aziz et al. 2015). This reaction is exothermic, pressure-dependent, and thermodynamically favorable at low temperatures (Eq. 5). But, the full reduction of the carbon atom of CO_2 with the highest oxidation state has a significant limitation, kinetically. Thus, to attain an acceptable rate and selectivity, the use of a catalyst is necessary.

$$CO_2 + 4H_2 \rightleftharpoons CH_4 + 2H_2O \qquad \Delta G_{298} = -114 \text{ kJ mol}^{-1} \tag{5}$$

Two general reaction mechanisms can be considered for CO_2 methanation on the solid surfaces. The first route is passing through the CO intermediate via RWGS reaction followed by further reduction and CO conversion to CH_4 (Wei and Jinlong 2011). However, the second mechanism is a direct reduction, in which the reaction of CO_2 with H_2 forms methane without CO formation as an intermediate. Figure 17 depicts (A) direct hydrogenation mechanism and (B) possible routes for the mechanism including the CO intermediate (Lapidus et al. 2007).

Ren and colleagues report a DFT investigation about methanation of CO_2 on the Ni(111) surfaces within the GGA and Perdew–Wang (PW91) functional (Ren et al. 2015). Figure 18 depicts the most stable configurations of the reactants and intermediates involved in the investigated mechanisms.

Three mechanisms were considered for this conversion. In the first path, HCOO species are formed by the reaction between CO_2 and H. CH_4 is produced due to the dissociation reaction of HCOO to CO and OH and hydrogenation of CO.

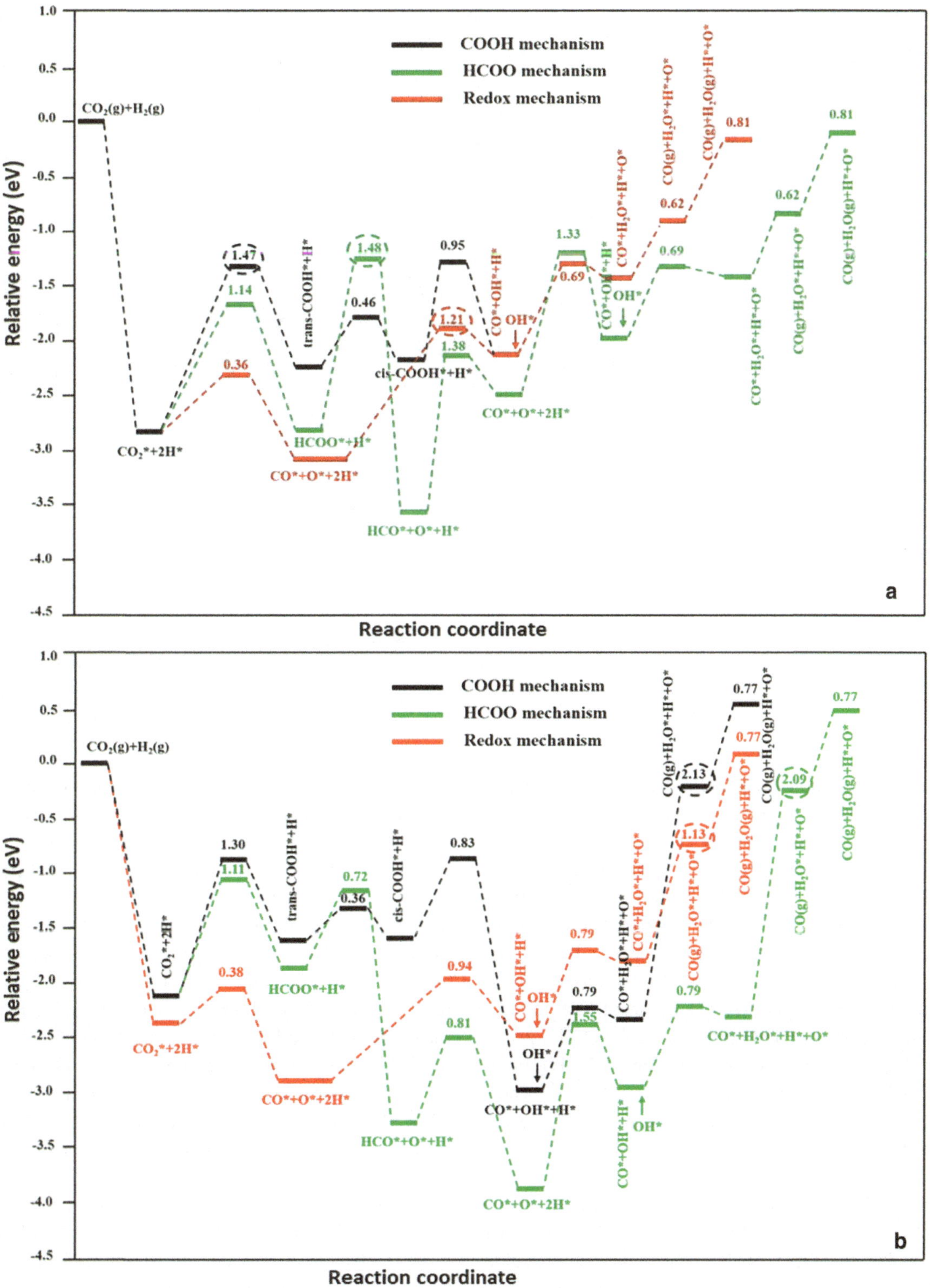

Fig. 12 PED of the RWGS reaction in the presence of **a** Cu@Mo_2C (001) and **b** Cu_4@Mo_2C (001) surfaces and corresponding energy values for RDS (Jing et al. 2018)

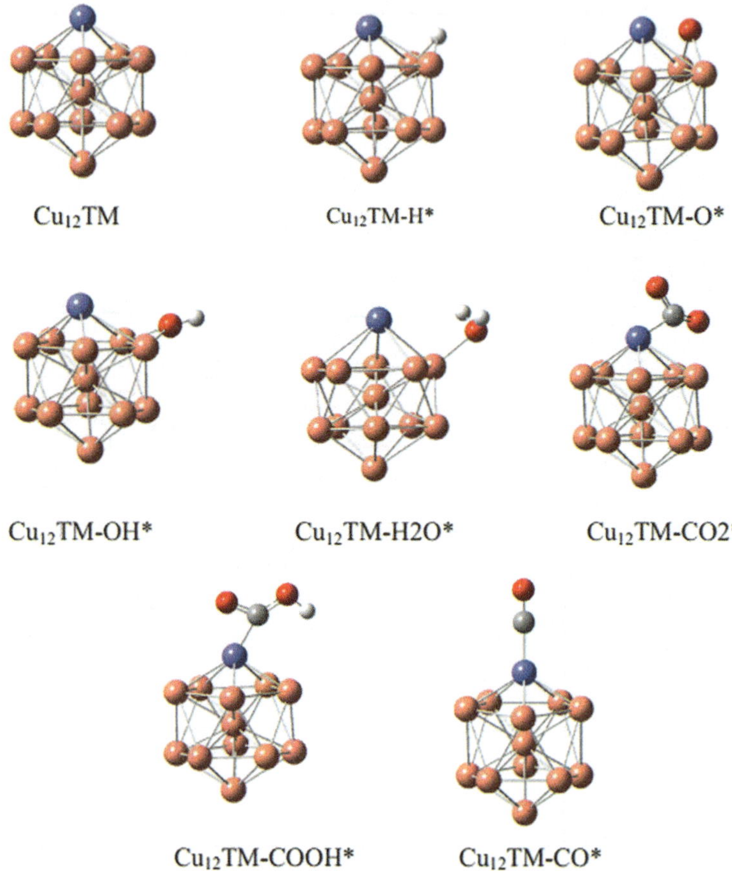

Fig. 13 Optimized structure of $Cu_{12}TM$ and adsorbed reactants, product, and intermediates in the RWGS reaction (Zhang and Guo 2018)

In this path, dissociation of HCOO shows the highest energy barrier (306.8 kJ mol^{-1}). In the second route, CO_2 directly is split into CO and O on the Ni(111) surface, without any HCOO formation. Then, CO decomposition and hydrogenation of C atoms lead to CH_4 molecules. The RDS for this mechanism is 237.4 kJ mol^{-1}, which belongs to the decomposition of CO into C and O species. In the third route, CO_2 is firstly hydrogenated to $C(OH)_2$ species, which dissociate into CH_2O and OH in the next step. Further hydrogenation leads to CH_2O dissociation into CH_2 species. RDS for this path belongs to the formation of $C(OH)_2$ on the Ni(111) surface, corresponding to 292.3 kJ mol^{-1}. Therefore, regarding all studied mechanisms, path 2 is the best candidate for reaction progress. Figure 19 depicts the PED diagrams for the studied pathways.

In other DFT calculations, CO_2 methanation is studied on Ni_4/t-ZrO_2(101), Ni4/VO-t-ZrO_2(101), and Ni_4/H-t-ZrO_2(101) surfaces within the GGA method. Geometry optimizations were obtained using the PBE functional and double numeric polarized (Han et al. 2017) basis set. This basis set is analogous in measure and feature to the 6-31G(d,p) basis set, as a Pople basis set. Figure 20 shows the optimized crystal structures of the studied surfaces.

Based on the obtained results, the reaction passes through the CO intermediate for all surfaces, and direct reduction of CO_2 to methane without any formation of CO intermediate was not observed. On the other hand, methanol production, based on the steps of Eq. 6, is a competitive reaction with methane formation. In the case of Ni_4/t-ZrO_2(101) catalyst, the difference between the highest barrier energies of CH_4 and CH_3OH formation is 261.8 kJ mol^{-1} and 197.9 kJ mol^{-1}, respectively. Therefore, CH_3OH formation reduces the productivity and selectivity of the CO hydrogenation into the CH_4 product. In the case of Ni_4/VO-t-ZrO_2(101) catalyst, the highest barrier energies for the production of CH_4 and CH_3OH from CO are 157.8 kJ mol^{-1} and 202 kJ mol^{-1}, respectively, which specify that the formation of CH_4 is more favorable, thermodynamically and kinetically. However, regarding the highest barrier energies for the conversion of CO to CH_4 (246.9 kJ mol^{-1}) and CH_3OH (274 kJ mol^{-1}) for Ni_4/H-t-ZrO_2(101) catalyst, it can be concluded that the hydroxyl groups and oxygen vacancies on Ni4/t-ZrO2 surfaces are useful to the development of CH_4. Moreover, the selectivity of CO conversion to methane is enhanced due to the oxygen vacancies. Figure 21 depicts PED diagrams of the reaction for the investigated surfaces (Han et al. 2017).

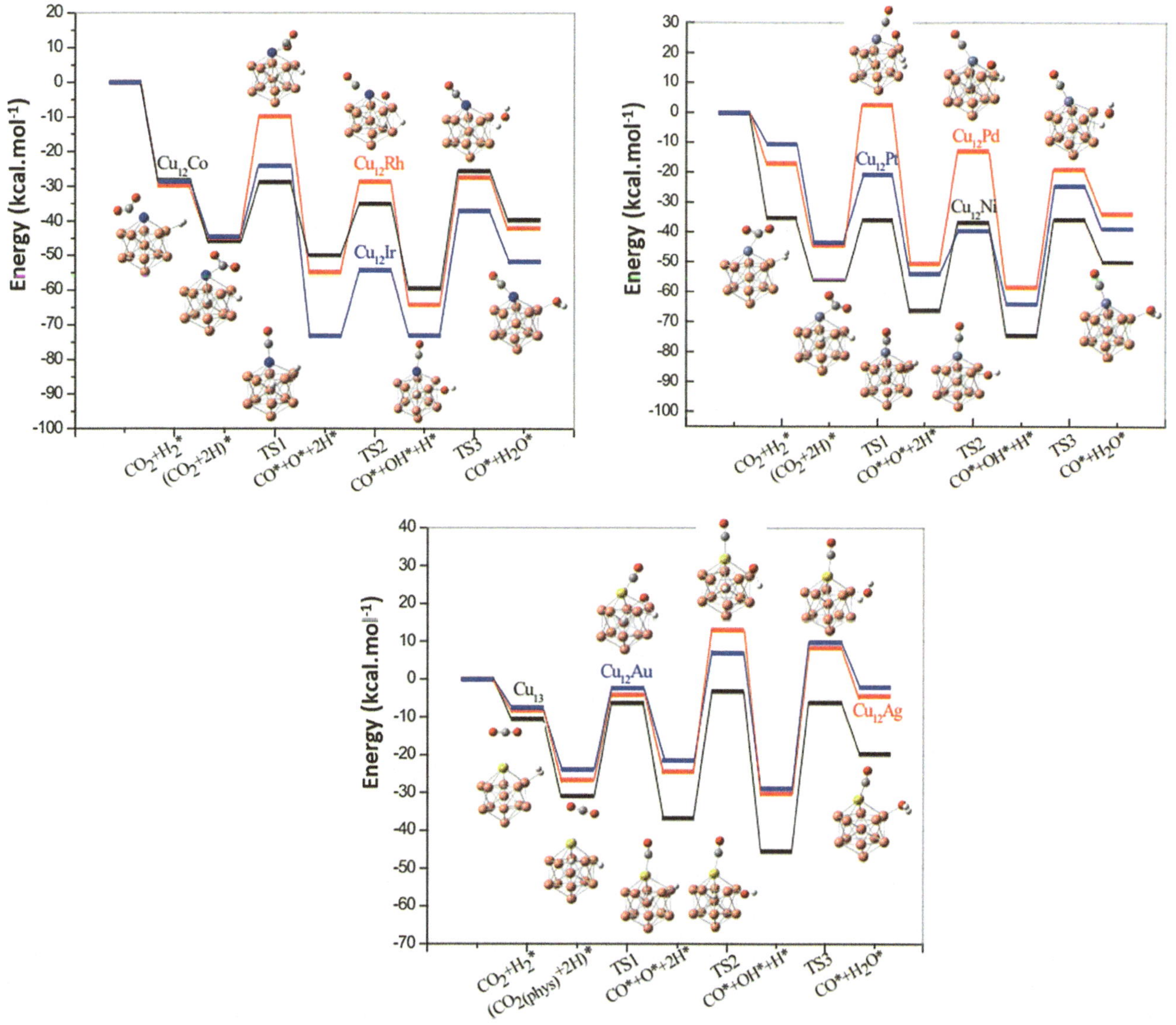

Fig. 14 PED of the RWGS reaction on Cu_{13} and $Cu_{12}TM$ clusters, within the involved species in the different steps (Cao et al. 2016)

$$CO \rightarrow C + O \quad \text{Step 1} \tag{6}$$

$$CO + H \rightarrow CHO \quad \text{Step 2}$$

$$CO + H \rightarrow COH \quad \text{Step 3}$$

$$CHO + H \rightarrow CH_2O \quad \text{Step 4}$$

$$CHO + H \rightarrow CHOH \quad \text{Step 5}$$

$$CHO \rightarrow CH + O \quad \text{Step 6}$$

$$CH_2O \rightarrow CH_2 + O \quad \text{Step 7}$$

$$CH_2O + H \rightarrow CH_2OH \quad \text{Step 8}$$

$$CH_2O + H \rightarrow CH_3O \quad \text{Step 9}$$

$$CH_3O + H \rightarrow CH_3OH \quad \text{Step 10}$$

$$CH_3O \rightarrow CH_3 + O \quad \text{Step 11}$$

$$CH_3 + H \rightarrow CH_4 \quad \text{Step 12}$$

Applied catalysts in CO_2 methanation are not limited to heterogeneous surfaces, and Wei and coworkers reported a DFT investigation by the cationic Ir-pincer complex $((POCOP)Ir(H)(acetone)^+$ (POCOP = 2,6-bis(dibutylphosphinito)phenyl)) as the catalyst and silane as the reducing agent. The investigation of a four-step mechanism was performed by the hybrid meta exchange–correlation M06 functional, in which the effective core potentials (ECPs) of Hay and Wadt with double-ζ valence basis sets (LanL2DZ) were applied to study the Ir metal atom. Also, these basis sets have been improved by polarization functions (ζ_f =

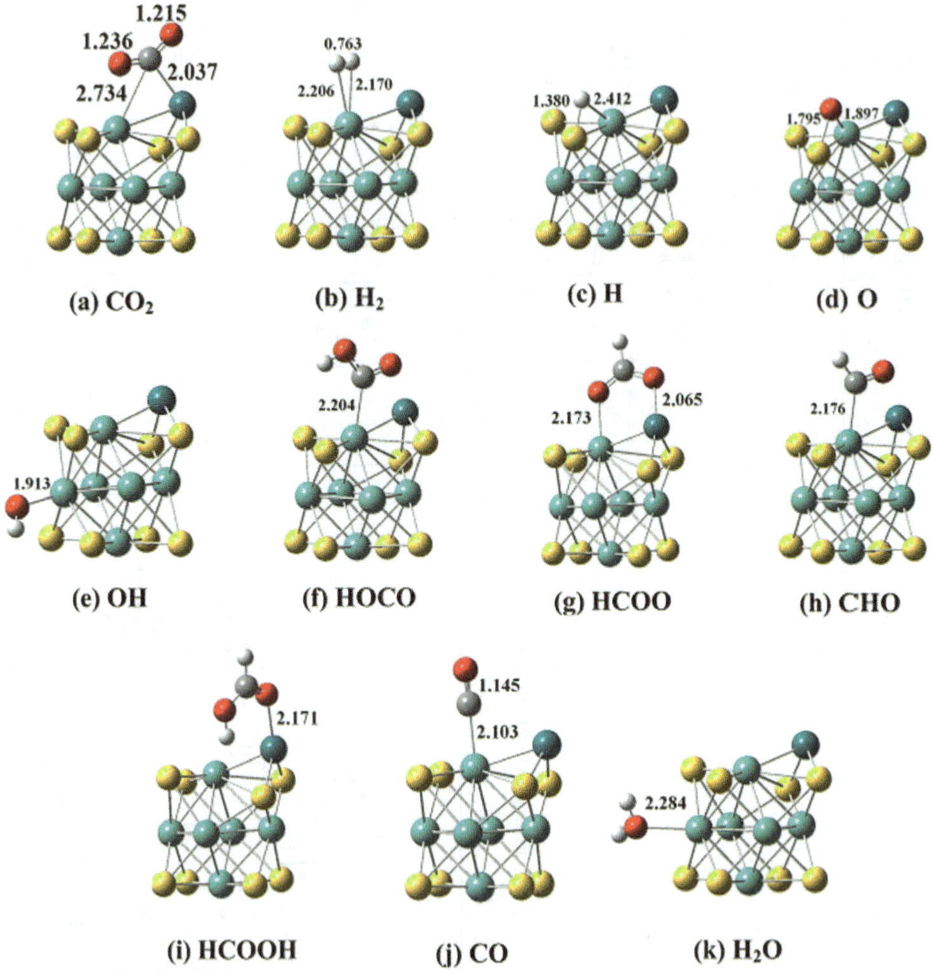

Fig. 15 Stable reactants, intermediates, and products in the catalyzed reaction by Rh–Mo_6S_8 cluster, in which Mo, S, Rh, H, O, and C atoms are colored by light blue, yellow, dark blue, white, red, and gray, respectively (Cao et al. 2016)

0.938). The 6-311G(d, p) basis set was applied to describe other atoms. Figure 22 depicts the overall reaction and the studied mechanism for the reaction (Fang et al. 2018).

The reaction progresses by sequential reduction of CO_2 molecule to silylformate, bis(silyl)acetal, methoxysilane, and methane as the final product, respectively. In stage 1, the hydrosilylation of CO_2 to silylformate was investigated via three routes. In the first step of path A, the cationic Ir-pincer complex activates CO_2, accomplished by the Ir–CO_2 moiety activation by a silane molecule. Bridged hydrogen between Ir and Si atoms is the result of Path B, which is started by the interaction between the cationic Ir-pincer complex and the silane molecule. This path is followed by the activation of a CO_2 molecule by the silane iridium adduct. CO_2 molecule inserts into the iridium-hydride bond of the cationic Ir-pincer complex in path C. Then, the produced iridium formate reacts with a free silane to produce silylformate (HCOO-$SiMe_3$). Figure 23 depicts the involved species and corresponding relative energies in stage 1.

Similar to step 1, three pathways can be considered for the conversion of silylformate to bis(silyl)acetal ($H_2C(OSiMe_3)_2$) or formaldehyde ($H_2C{=}O$) in step 2. Path A proceeds through the breaking of the next silane Si–H bond and C=O bond reduction of silylformate, path B is related to the ionic S_N2 outer-sphere mechanism, accomplished by the silylformate nucleophilically attacking the ^{1}-silane iridium complex, and pathway C corresponds to silylformate insertion into the Ir–H bond. The obtained results show that path B (Fig. 24) passes through lower barrier energy than A and C paths. The calculated RDS energy for this step is about 12.2 kcal mol^{-1}. In the third step, bis(silyl) acetal is converted to methoxysilane, in which the obtained RDS energy for this step is 16.4 kcal mol^{-1}. Finally, in the fourth step, by overcoming to activation energy about 22.9 kcal mol^{-1}, methoxysilane is reduced to methane (Fig. 25). Also, it is notable that formaldehyde production by the cationic Ir-pincer complex is not probable.

Electrochemically CO_2 reduction reaction (CO_2RR) to fuels and organic feedstock is another approach toward CO_2 utilization. CO_2RR can be scaled to facilitate the large-scale storage of chemical products (Wang et al. 2011; Gattrell et al. 2007; Costentin et al. 2013). Since the products of CO_2RR are extracted from petrochemical sources, their manufacture via CO_2RR could decrease the global demand

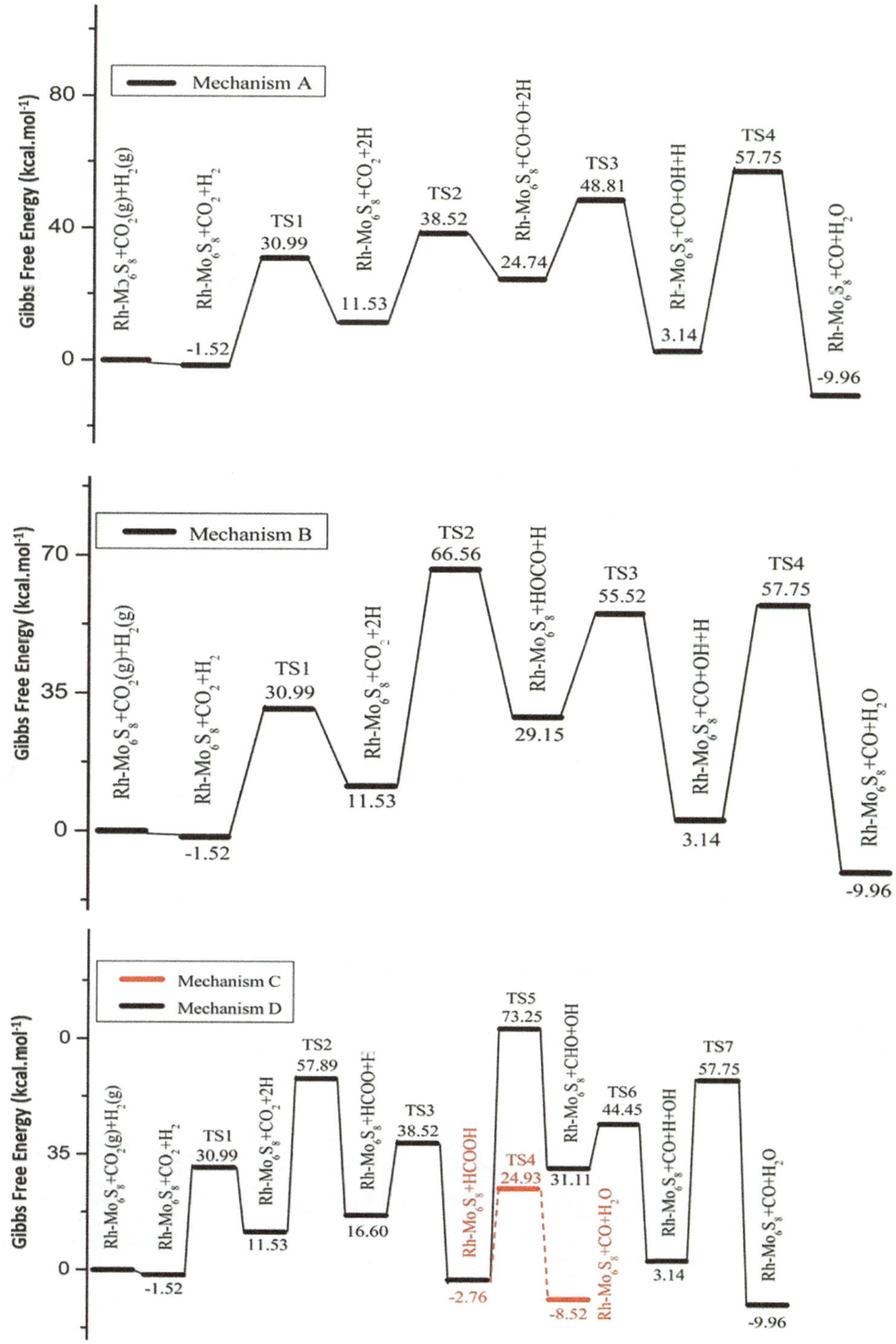

Fig. 16 PED corresponding to **a** the redox mechanism, **b** the carboxyl mechanism, **c** the directly formic acid decomposition to CO, and **d** the formation of a CHO intermediate along with the reaction (Cao et al. 2016)

for fossil fuels. Lihui and coworkers reported the CO_2 reduction into CH_4 on the Cu (111) via electrochemical reaction (Ou et al. 2019). Figure 26 depicts the crystal structure of Cu (111) in water as the studied solvent in this reaction.

The DFT investigation was accomplished using the GGA method and the PBE exchange–correlation functional. Also, ultrasoft pseudopotential was applied to study the nuclei and core electrons, and the Kohn–Sam equations are solved by a plane-wave basis set. Two overall mechanisms including (a) CH_2O and (b) CHOH pathways in the presence of the simulated low overpotential have been considered for the reaction. Common intermediates for the first and second mechanisms are CHO and CH_2, respectively (Fig. 27). The results showed that both considered mechanisms may happen in a parallel way in the presence of the simulated low overpotential. Also, the formation of CO is the potential-limiting step, which is according to the observed

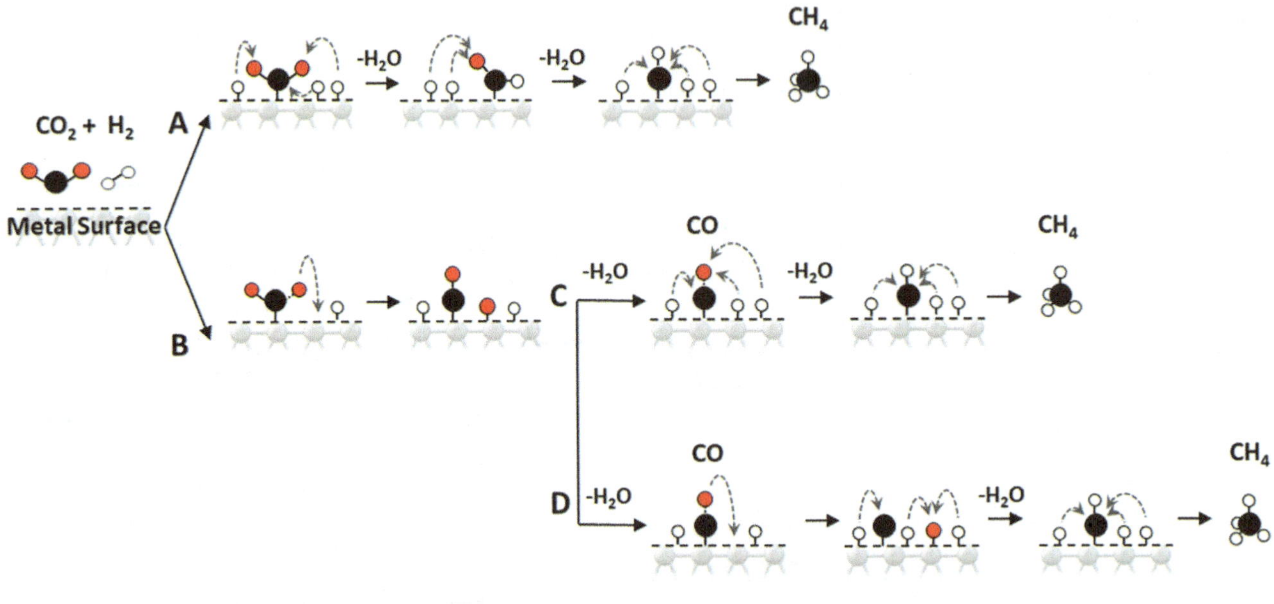

Fig. 17 Two general mechanisms of CO_2 methanation on the solid surfaces (Podrojkova et al. 2020)

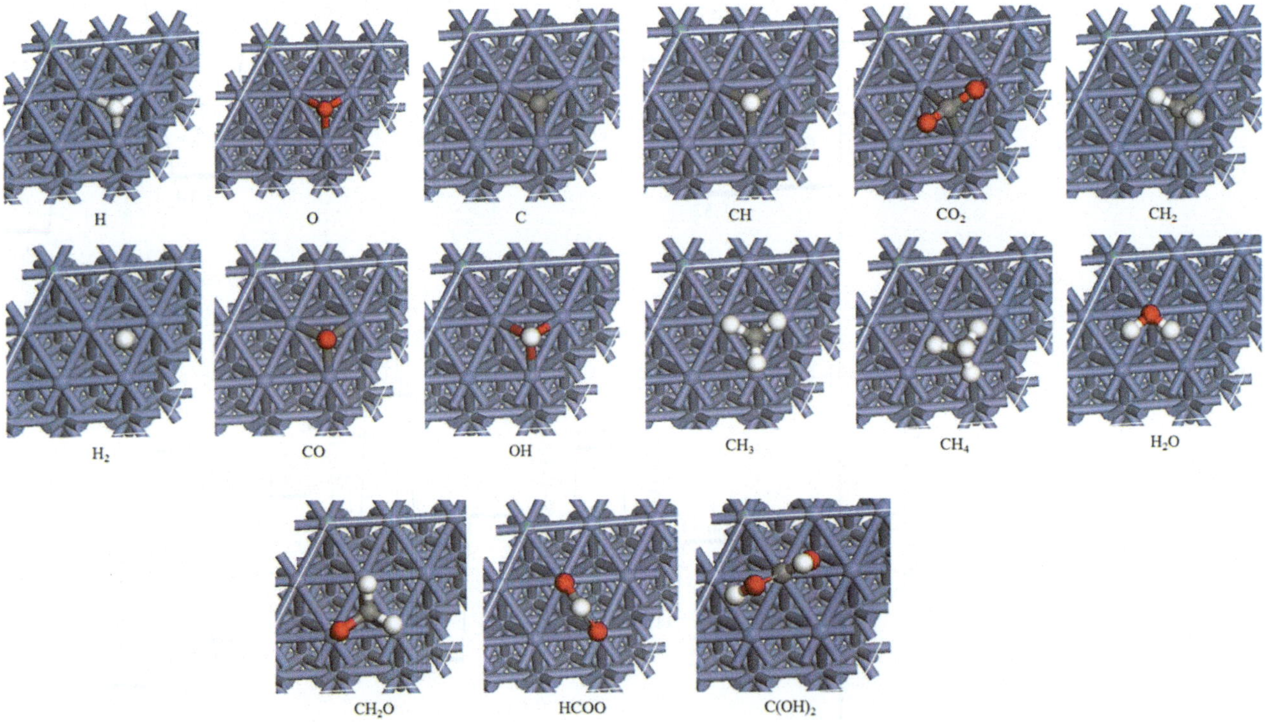

Fig. 18 Involved species in the mechanism of CO_2 methanation on Ni(111) surfaces, in which C, O, H, and Ni atoms are depicted by as the gray, red, white, and blue colors, respectively (Ren et al. 2015)

experimental results. Based on the studied mechanisms, methanol formation is a side reaction, which can interrupt the methane formation; however, in the simulated low overpotential, methanol formation on Cu(111) is prohibited, kinetically. Figure 28 depicts the obtained PED of CO_2 reduction into CH_4 and CH_3OH on Cu(111) in the presence of the considered low overpotential: (a) CH_2O pathway; (b) CHOH pathway.

4.1.3 Methanol

The vision of liquid sunshine was applied by Shih and coworkers that mentioned that solar energy is an enormous

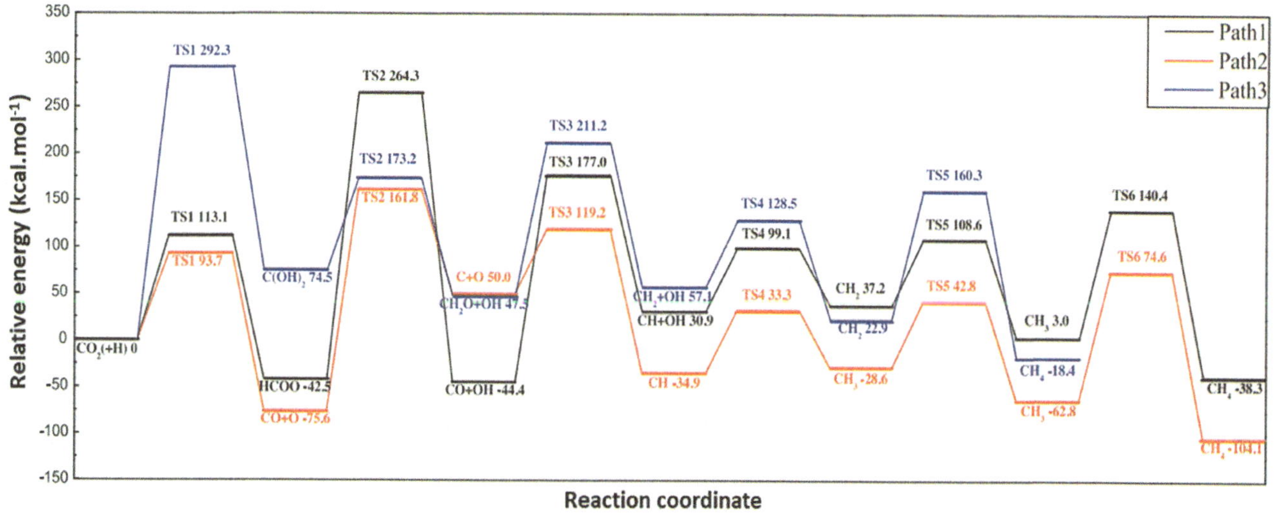

Fig. 19 PED diagrams for the methanation of CO_2 on the Ni(111) surfaces (Ren et al. 2015)

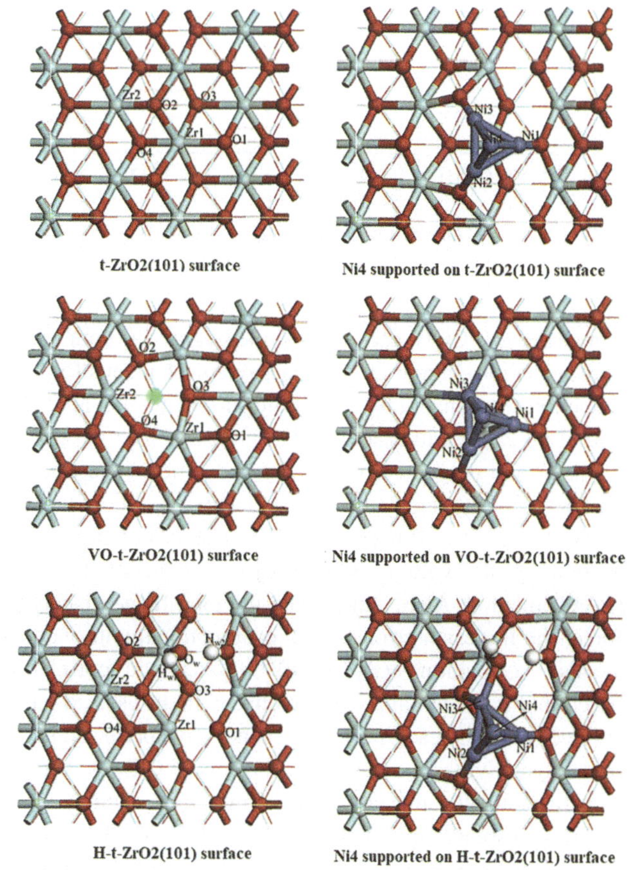

Fig. 20 Optimized crystal structures of the Ni_4/t-ZrO_2(101), Ni_4/VO-t-ZrO_2(101), and Ni_4/H-t-ZrO_2(101) surfaces (Han et al. 2017)

resource to make renewable alcohol fuels, such as methanol (Shih et al. 2018). Also, regarding obtaining a sustainable energy source, the "Methanol Economy" suggested by George A. Olah has been widely accepted among the scientific community (Olah et al. 2018).

Nowadays, the main starting material for the industrial production of methanol is mostly from syngas (a mixture of CO and H_2 molecules). Syngas is mainly produced by coal and natural gas, as fossil resources, due to the gasification of coal and the reforming of natural gas by steam (Zhong et al. 2020). To improving the kinetic aspect and to achieve a desired stoichiometry for the reaction, a few amounts of CO_2 (about 2–8%) are added to the CO/H_2 mixture. Equation 7 shows the stoichiometric ratio for methanol synthesis from CO and CO_2. It can be concluded that in comparison with syngas (CO), methanol synthesis from CO_2 needs an excessive value of hydrogen. Because an excess amount of H_2 is required to eliminate one oxygen atom from CO_2 via the production of water as a by-product. Also, the thermodynamic aspect of methanol formation from CO_2 is not favorable, similar to CO. Thus, methanol production from CO_2 leads to a lower yield than that of the syngas reaction (Mikkelsen et al. 2010; Macquarrie 2005).

$$\begin{aligned} CO + 2H_2 &\rightleftharpoons CH_3OH \\ \Delta H_{298} &= -90.6\ \text{kJ mol}^{-1} \\ CO_2 + 3H_2 &\rightleftharpoons CH_3OH + H_2O \\ \Delta H_{298} &= -49.5\ \text{kJ mol}^{-1} \end{aligned} \quad (7)$$

The lower temperature and higher pressure for the reaction favor the methanol production from CO_2, thermodynamically. Accordingly, an increase in reaction temperature is helpful for CO_2 activation and methanol formation, subsequently. Also, other by-products are formed. However, the formation of undesired by-products, such as CO, hydrocarbons, and higher alcohols, causes the use of a highly

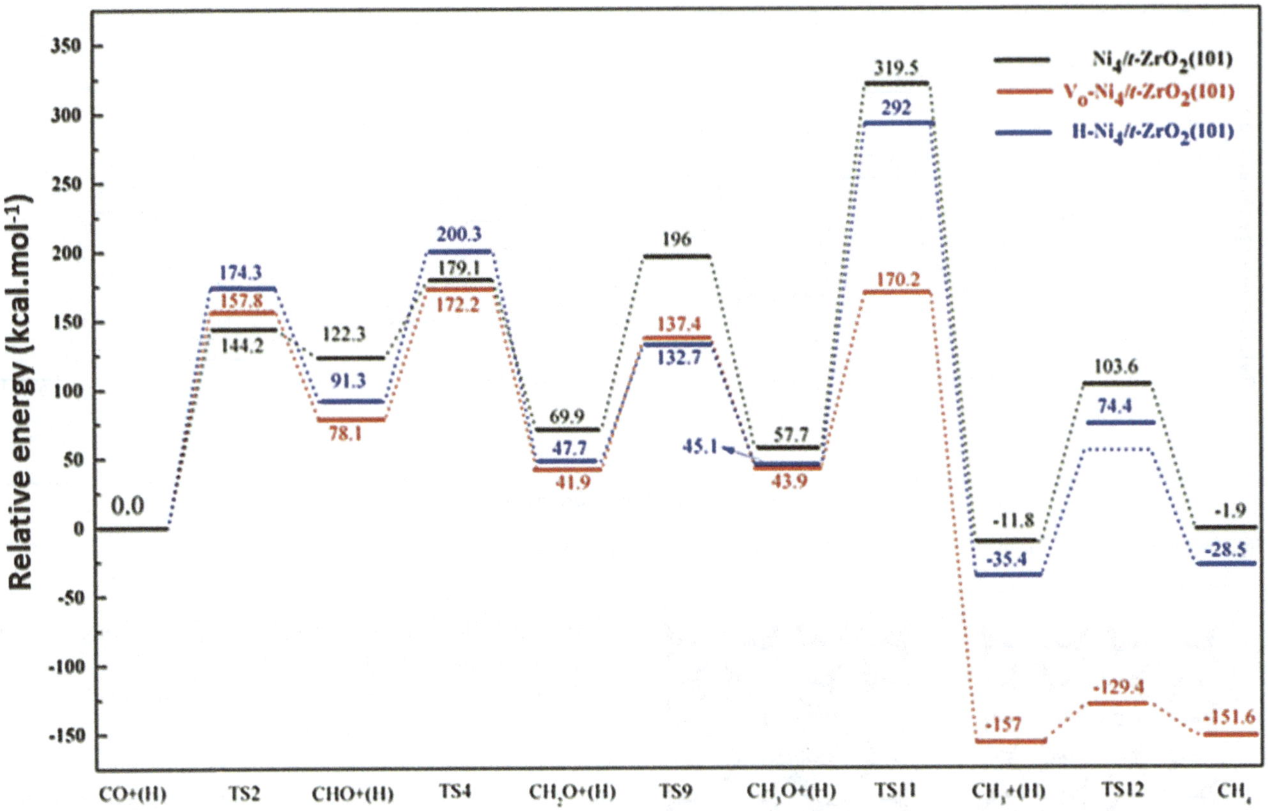

Fig. 21 PED of the reaction for hydrogenation of CO to CH_4 on $Ni_4/t\text{-}ZrO_2(101)$, $Ni_4/VO\text{-}t\text{-}ZrO_2(101)$, and $Ni_4/H\text{-}t\text{-}ZrO_2(101)$ surfaces (Han et al. 2017)

selective catalyst to become an undeniable necessity. Regarding the various investigation, methanol synthesis over an active surface catalyst such as Cu-based can proceed through three different reversible reaction paths including (1) formate production as intermediate via CO_2 reaction with surface atomic H, (2) the formation of carboxyl intermediate through RWGS mechanism, (3) the formation of $*C(OH)_2$ as the produced intermediate by CO_2 hydrogenation (Dang et al. 2019; Saeidi et al. 2014; Wu et al. 2017). The progress of the hydrogenation for the mentioned three mechanisms leads to the formation of formyl (H_2CO*), methoxy (H_3CO*), and methanol (CH_3OH), respectively. Mechanism 1 passes through the chemisorbed formate, which can be produced from the CO_2 reaction with dissociated surface hydrogen. Afterward, the hydrogenation of surface-bound formate causes dioxomethylene formation that H_2CO* is formed by the elimination of hydroxyl as H_2O. The hydrogenation of H_2CO* intermediate can be continued to methoxy and methanol formation. Mechanism 2 proceeds through the CO* formation by the elimination of hydroxyl from hydrocarboxyl. The sequential hydrogenation of the formed HCO intermediate leads to formyl and methanol. However, in mechanism 3, the hydrogenation of hydrocarboxyl leads to COOH* intermediate formation, which is converted to *COH and hydroxymethylene by continuous hydrogenation. Figure 29 depicts three mechanisms for CO_2 reduction to methanol by a catalytic surface (Wu et al. 2017; Qiu et al. 2016; Liu et al. 2019).

Yang and coworkers reported a theoretical investigation on the CO_2 hydrogenation to methanol on the PdIn(310) surface (Fig. 30) (Wu et al. 2019). The calculations were accomplished within the projector-augmented wave (PAW) method. Also, considering the long-range dispersion force, the Bayesian error estimation functional with van der Waals correlation (BEEF-vdW) exchange−correlation functional has been applied, which is widely used to study surface catalysis reactions by the plane-wave basis set. In the reported study, three overall mechanisms of CO_2 conversion were investigated that are divided into four distinguishing paths corresponding to I, II, III, IV in Fig. 31. The results showed that the reaction mechanism of methanol formation on the clean PdIn(310) proceeds through the COOH intermediate and CO stepwise hydrogenation. Also, based on the applied microkinetic analysis, it was illustrated that Pd atom is the active catalytic site in comparison with In atom.

However, on this basis of DFT calculations and microkinetic analysis, it was illustrated that the differential adsorption energies of formate at the Pd or the In site are

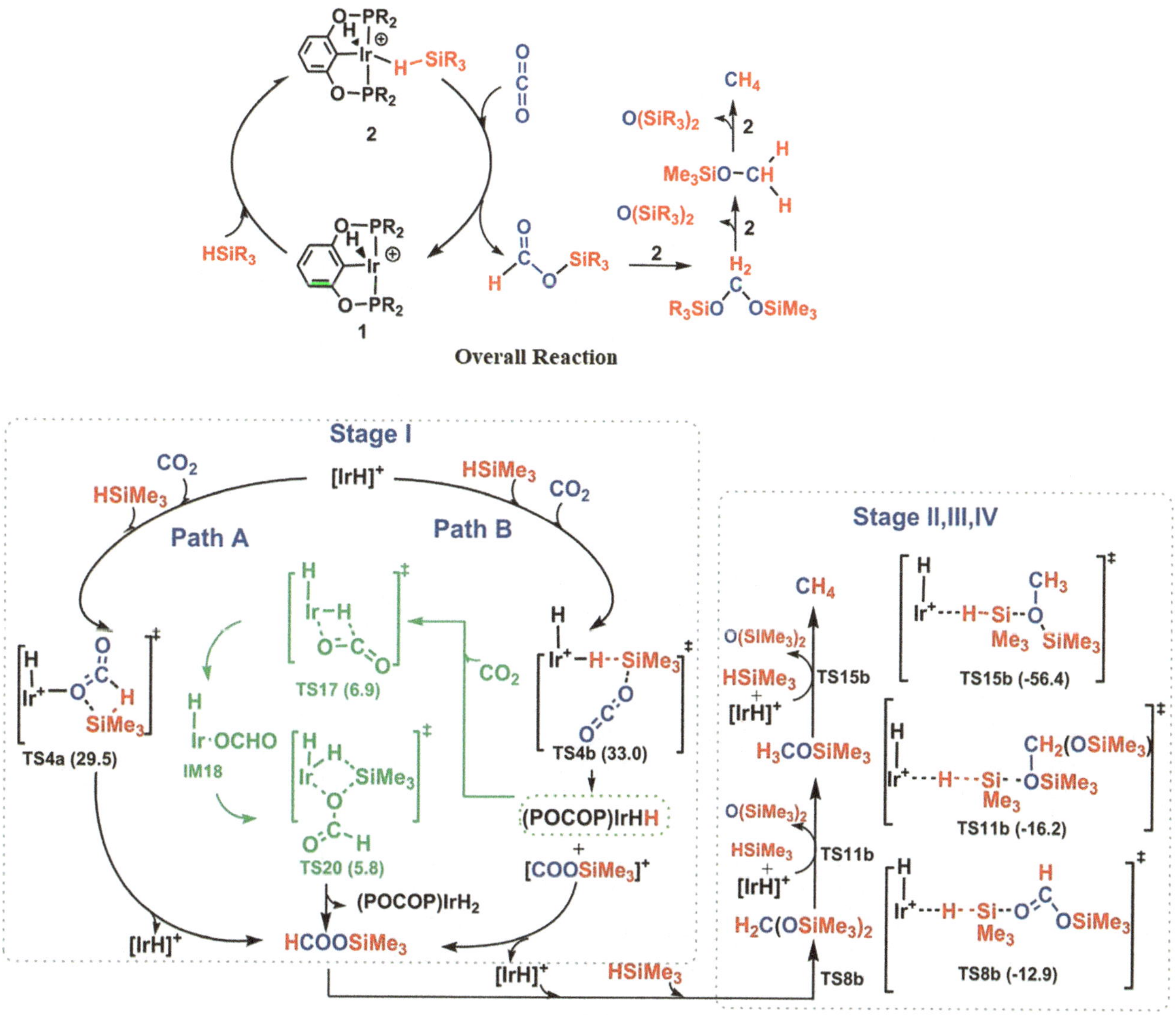

Fig. 22 Overall CO_2 methanation reaction by the Ir-pincer complex and studied four-step mechanism for the reaction (Fang et al. 2018)

close to each other when two formates are preadsorbed at the In step-bridge site. Moreover, in the presence of formate on the surface, the favorable path of methanol formation changes from the COOH/CO path to HCOO/HCOOH. Figure 32 depicts the obtained PED diagram for the studied paths (Wu et al. 2019).

In another DFT calculation, the consequence of the size of the Cu cluster on the reduction of CO_2 to methanol was studied by using PAW potentials and the PBE functional. Figure 33 depicts the optimized structures of the adsorbed species on the crystal structure of Cu clusters having different sizes and corresponding adsorption energies. The studied mechanism reveals direct CO_2 decomposition to CO and O rather than hydrogenation of CO_2 to COOH because it has been reported that the former mechanism has lower barrier energy than the latter one on the Cu (111) and Cu (211) (Zhang et al. 2018a). Figure 34 illustrates a schematic presentation of the mechanism of CO_2 reduction to methanol on the Cu clusters.

The DFT results showed that all barriers of the reaction, involved in the CO_2 transformation to methanol, represent a linear correlation with the adsorption energies of CO and O. Also, based on the results of the microkinetics simulations, it can be concluded that Cu_{19} clusters are appropriate candidates for CO_2 reduction. Moreover, particle measure of the Cu alters the adsorption energies of the involved intermediates, which can be associated with the position of the d-band of the Cu clusters. Thus, the upshift of the d-band center has remarkable effects on the strengths of the bonding interaction between the metal and intermediates, which has a substantial influence on the CO_2 reduction activity. Figure 35 depicts the PED of the CO_2 reduction to methanol in the presence of different size of Cu clusters and involved species in each step.

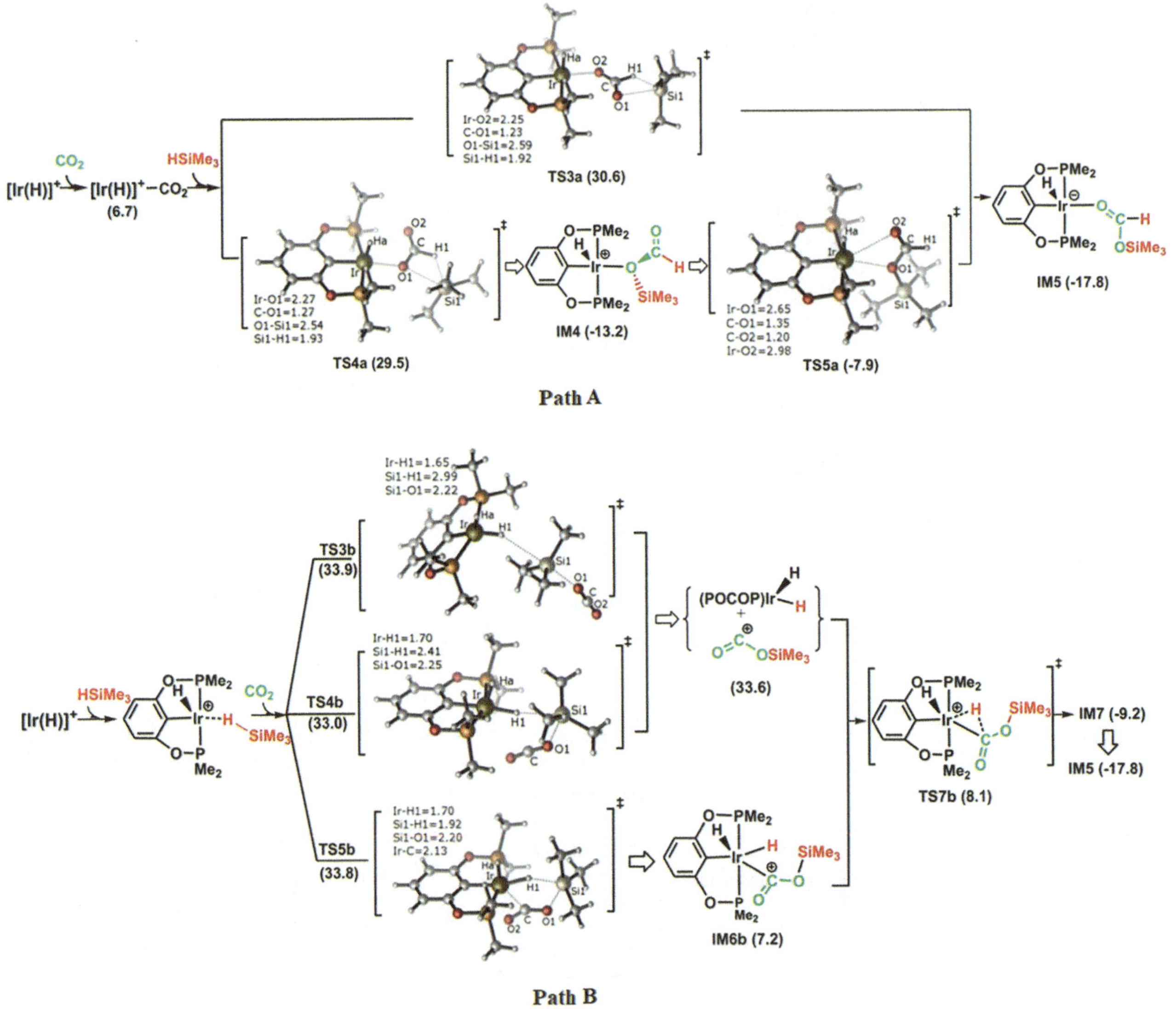

Fig. 23 Schematic representation of the involved species and corresponding relative energies in stage 1 silylformate ($HCOOSiMe_3$) formation via pathways A and B (Fang et al. 2018)

The electrochemical approach based on the Cu surface is another method in CO_2 hydrogenation to methanol. The next example is DFT calculation based on the PAW method to understand and comparison of the catalytic activity and mechanism investigation of CO_2 reduction on the Cu_{85} nanocluster and Cu (111) surface. Also, the exchange–correlation interaction is treated by the generalized gradient approximation of Perdew–Burke–Ernzerhof (GGA-PBE) (Rawat et al. 2017).

Equation 8 shows electrochemical CO_2 reduction in which two-electron reduction leads to H_2, CO, and HCOOH formations. However, four- ($4e^-$) and six-electron ($6e^-$) reduction leads to CH_2O and CH_3OH formation, respectively. The authors describe two overall electrochemically C–O bond dissociation, as direct and indirect. The direct C–O bond dissociation ($*CO_2 \rightarrow *CO + *O$) improves the four- and six-electron reduction processes, kinetically. On the other hand, indirect C–O bond dissociation by hydrogenation ($*CO_2 + *H \rightarrow *COOH$) and then COOH dissociation ($*COOH \rightarrow *CO + *OH$) are more favorable than direct C–O bond dissociation. Also, the direct C–O bond dissociation has a remarkable effect on the product selectivity.

$$2H^+ + 2e^- \rightarrow H_2 \quad E_0 = -0.000V_{RHE} \text{ Step 1} \tag{8}$$

$$CO_2(g) + 2H^+ + 2e^- \rightarrow HCOOH(l)$$
$$E_0 = -0.250V_{RHE} \text{ Step 2}$$

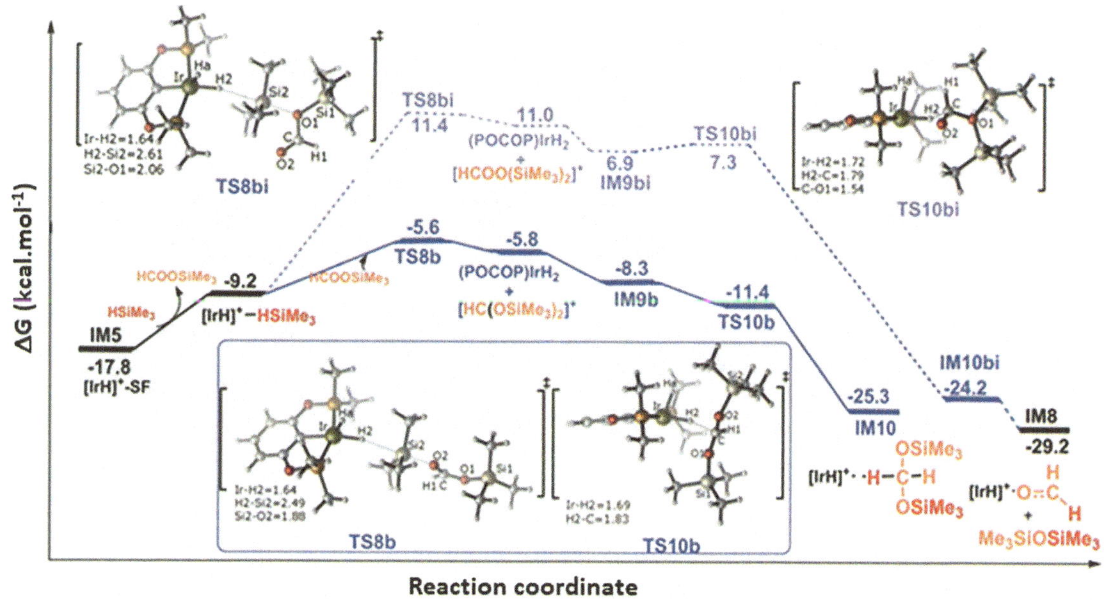

Fig. 24 PED of the silylformate conversion to bis(silyl)acetal or formaldehyde in step II via path B (Fang et al. 2018)

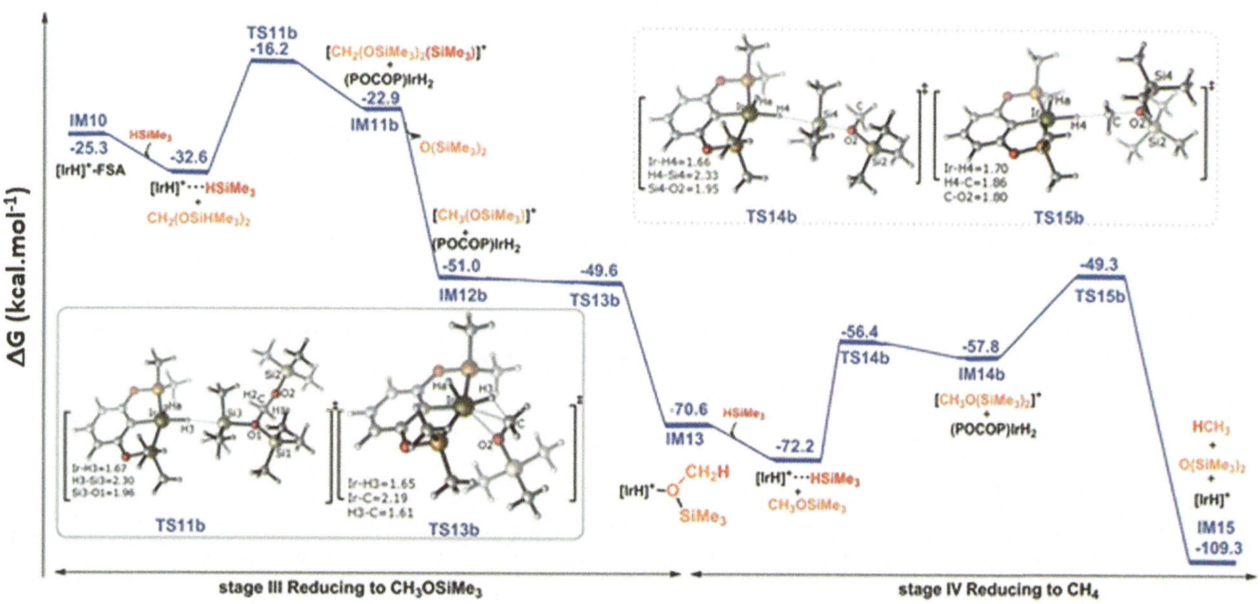

Fig. 25 PED of the bis(silyl)acetal conversion to methoxysilane and methane within steps 3 and 4 through the ionic outer-sphere mechanistic routes (Fang et al. 2018)

$$CO_2(g) + 2H^+ + 2e^- \rightarrow CO(g) + H_2O(l)$$
$$E_0 = -0.106V_{RHE} \text{ Step 3}$$

$$CO_2(g) + 4H^+ + 4e^- \rightarrow CH_2O(l) + H_2O(l)$$
$$E_0 = -0.070V_{RHE} \text{ Step 4}$$

$$CO_2(g) + 6H^+ + 6e^- \rightarrow CH_3OH(l) + H_2O(l)$$
$$E_0 = 0.016V_{RHE} \text{ Step 5}$$

The obtained results show that Cu_{85} nanocluster is more active than the Cu(111) surface in the reduction reaction. Also, the reduction of *CO to *CHO/*COH is the RDS of the hydrogenation reaction, whose barrier energy for Cu_{85} nanocluster is lower than that of Cu(111) surface. However, the formation of *CHO is more favorable than *COH, followed by CH_3OH formation. The next obtained result is a lower required overpotential of Cu_{85} nanocluster for progressing the potential-limiting step of the reaction. Indeed,

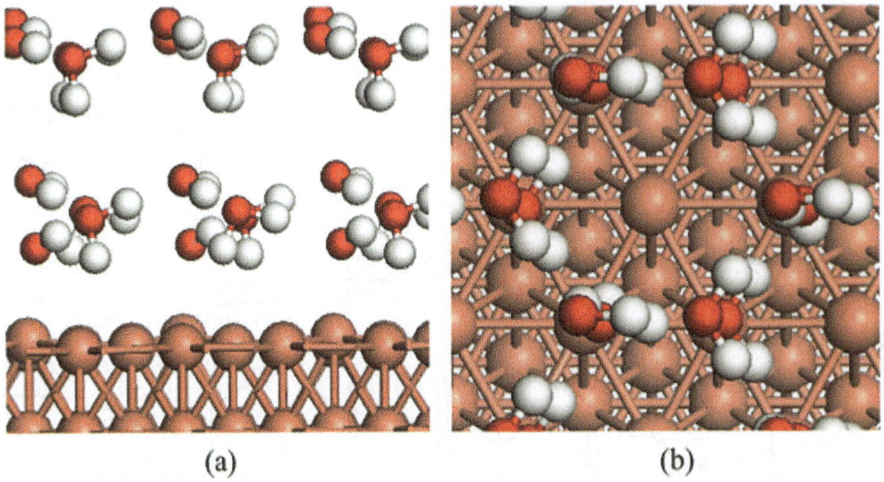

Fig. 26 Arrangement of solvent molecule on Cu(111): **a** side view; **b** top view (Ou et al. 2019)

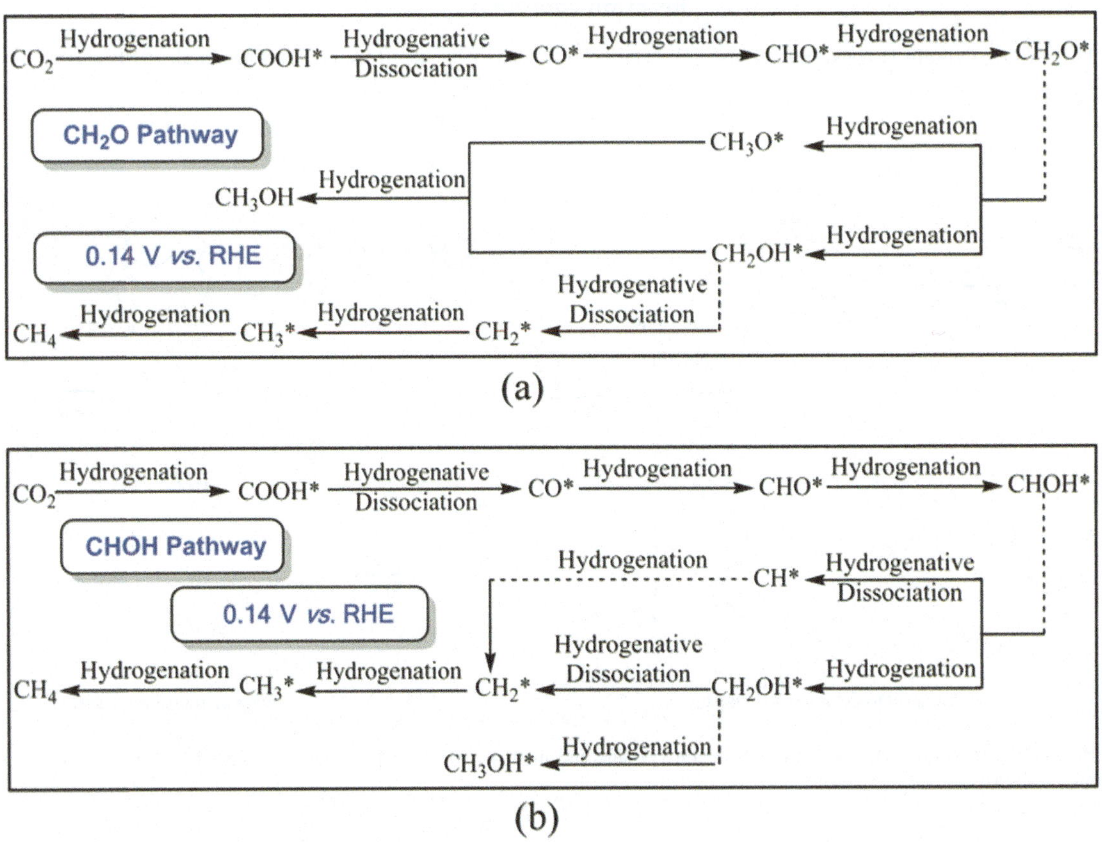

Fig. 27 Considered CO_2 reduction mechanisms on Cu(111): **a** CH_2O pathway; **b** CHOH pathway (RHE = reversible hydrogen electrode) (Ou et al. 2019)

the reduction of *CO to *CHO is a potential-limiting step on both Cu(111) and Cu_{85} surfaces. The overpotential of this step for Cu(111) is 0.71 eV, which is the reaction Gibbs energy for the development of *CO to *CHO. While the reaction Gibbs energy for this reaction on Cu_{85} nanocluster is 0.53 eV. Figure 36 depicts the PED of the reduction reaction on the Cu_{85} nanocluster and their dependence on the applied electrode potentials (Rawat et al. 2017).

Nowadays, the photocatalytic CO_2 conversion to methanol has been attracted specific attention among the scientific

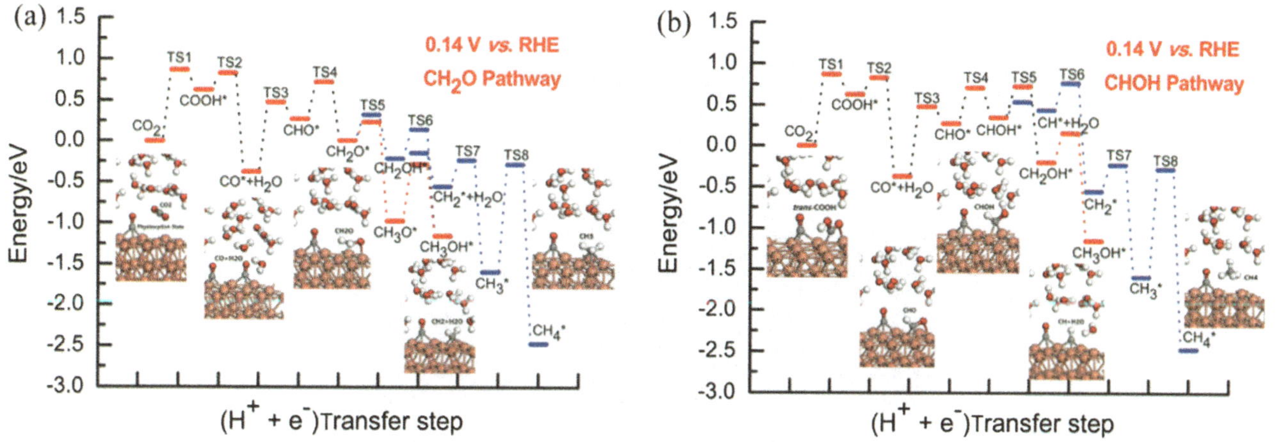

Fig. 28 Obtained PEDs of CO_2 reduction into CH_4 and CH_3OH by Cu(111) in the presence of the considered low overpotential: **a** CH_2O pathway; **b** CHOH pathway (Ou et al. 2019)

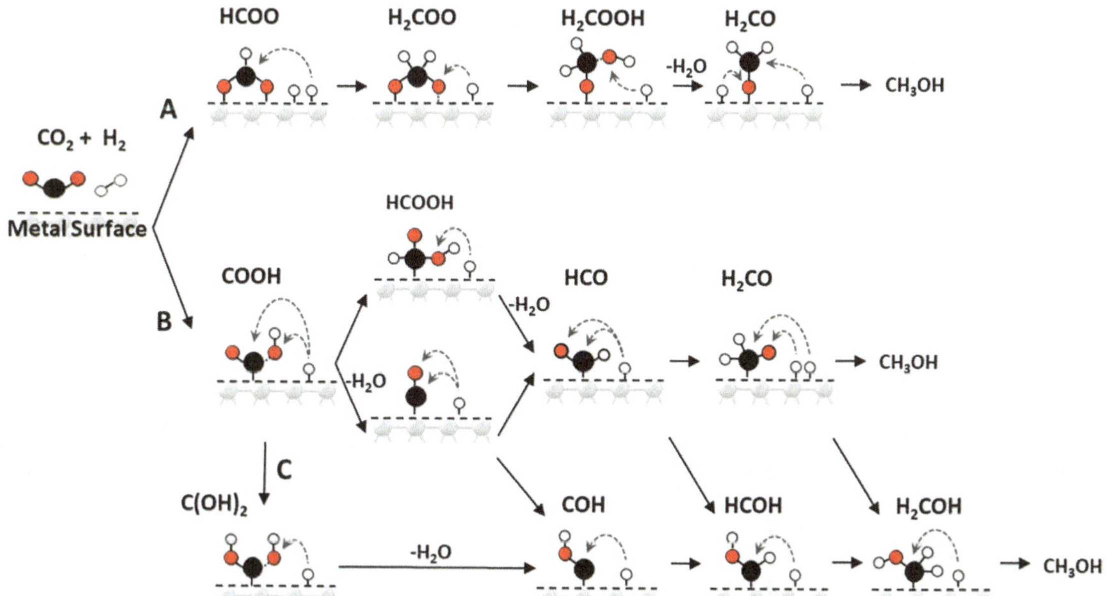

Fig. 29 Three possible mechanisms in CO_2 reduction to methanol on the catalytic surface: **a** formate, **b** RWGS, and **c** hydrocarboxyl mechanisms (Wu et al. 2019)

community (Wang et al. 2011; Bensaid et al. 2012; Centi et al. 2013) because, by using sustainable solar energy, methanol can be formed without any emission of greenhouse gases. Recently, a defect-laden indium oxide, $In_2O_{3-x}(OH)_y$, with a rod-like nanocrystal superstructure was reported, which is a photocatalyst in the reduction of CO_2 to methanol with 50% selectivity under the atmospheric pressure. This report is a suitable investigation for the formation of a low-pressure solar methanol process using CO_2 and renewable H_2 sources (Wang et al. 2018b).

To investigate the mechanism based on the DFT calculation, all involved species were optimized by the PBE exchange–correlation functional, together with the Rappe–Rabe–Kaxiras–Joannopoulos (RRKJ) ultrasoft pseudopotentials. Structural analysis shows that the surface of $In_2O_{3-x}(OH)_y$ possesses hydroxide groups that coordinate to unsaturated indium positions. These positions behave as a Lewis base and a Lewis acid, respectively, which form a surface-frustrated Lewis pairs (SFLP) that plays a key role in catalytic hydrogenation of CO_2 to CO and CH_3OH. Also, compared with the ground state, Lewis acidity and Lewis basicity of the SFLPs increase the excited state leading to a remarkable facility of the photochemical CO_2 transformation to methanol. Figure 37 shows the investigated mechanism of CO_2 reduction on the $In_2O_{3-x}(OH)_y$. As shown in this figure, CO_2 reduction to CO is a side reaction developed by CO_2

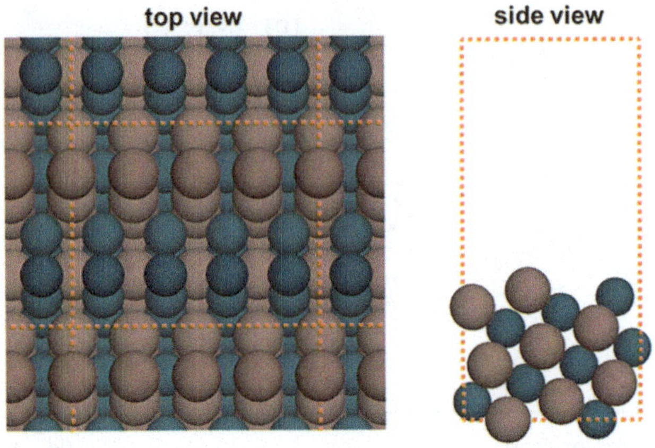

Fig. 30 Optimized crystal structure of PdIn(310), in which Pd and In atoms are depicted in blue and brown colors, respectively. (Left) top and (right) side views (Wu et al. 2019)

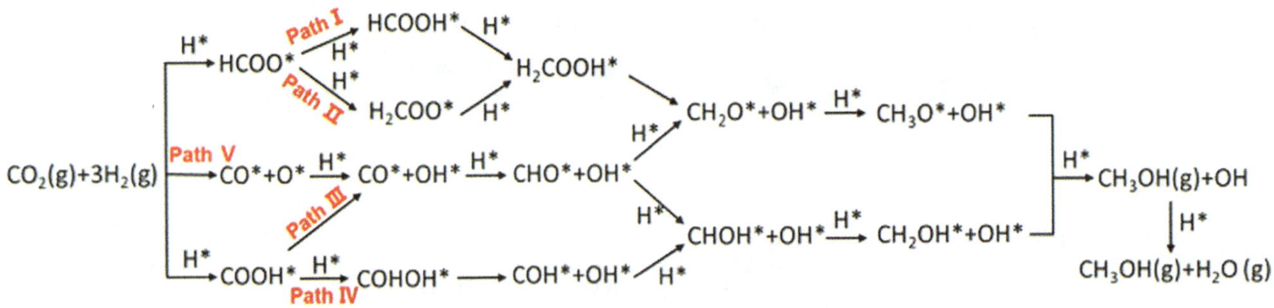

Fig. 31 Three overall mechanisms of CO_2 conversion on the PdIn(310) surface (Wu et al. 2019)

Fig. 32 Obtained PEDs of the studied paths for methanol synthesis in the Pd and In sites over PdIn(310) with two preadsorbed formates at the In step-bridge site (Wu et al. 2019)

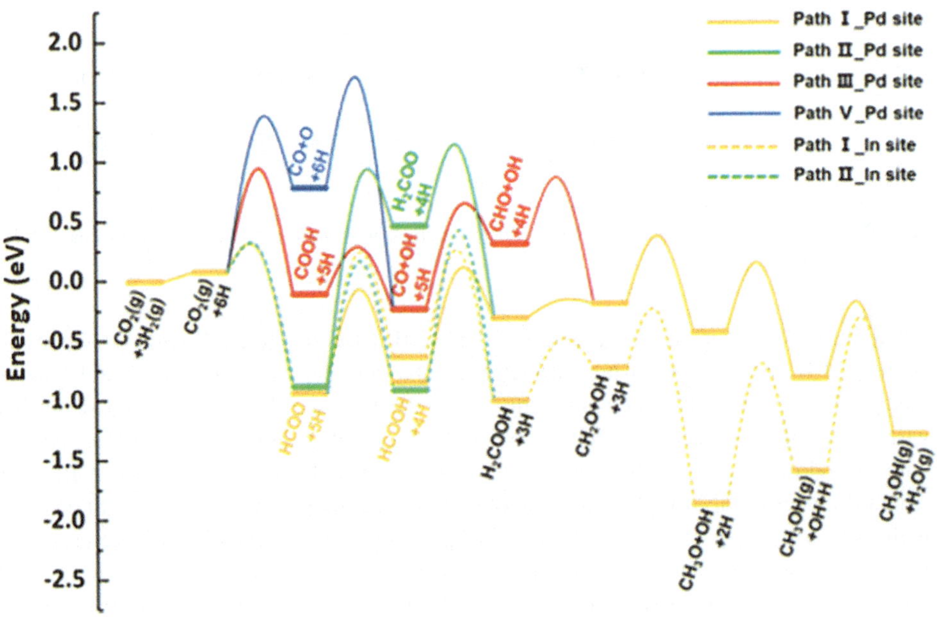

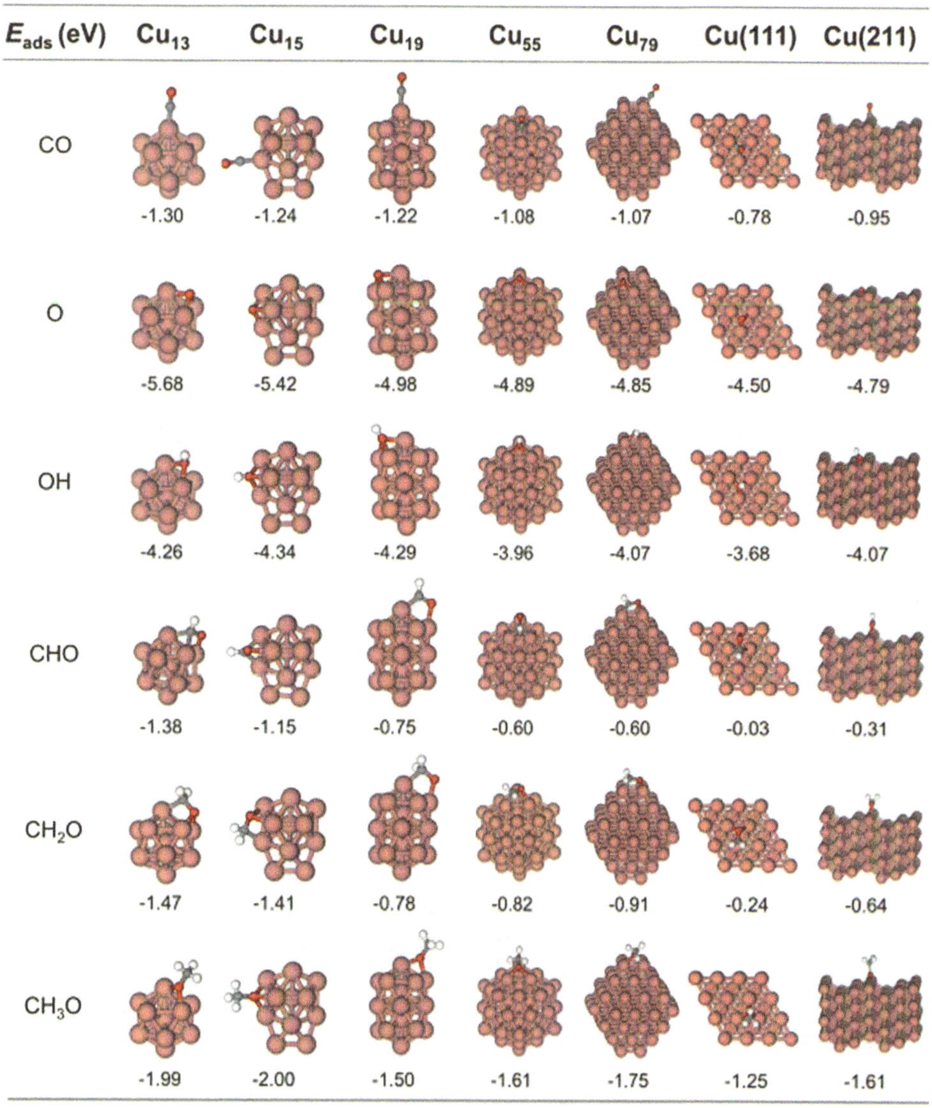

Fig. 33 Adsorbed key intermediates on the Cu_n (n = 13, 15, 19, 55, and 79) clusters and corresponding adsorption energies (Zhang et al. 2018a)

conversion to methanol. However, the results illustrate that methanol selectivity increases from 40% for nanocrystals to >50% for nanocrystal superstructure of $In_2O_{3-x}(OH)_y$, which shows that the nanocrystal superstructures enhance both yield and selectivity toward the methanol production. Figure 38 depicts the PED diagram of the CO_2 reduction to methanol over $In_2O_{3-x}(OH)_y$ (Wang et al. 2018b).

The application of organic molecules, as a catalyst in CO_2 transformation, has remarkable properties, such as cost-effectiveness, metal-free status, and sustainability (Fiorani et al. 2015). The investigations show that, in comparison with the organo-based catalysts, the activation of the substrates by metal-based catalysts, via the coordination of the functional group to metal, is more effective. But, organocatalysts have some privileges, such as non-toxicity, stability along with the reaction, and resistance toward moisture and air. Thus, they can be considered safe catalysts (Fiorani et al. 2015). Moreover, in some cases, they can be readily provided from renewable chemicals. Based on DFT studies, our group investigated the performances of the proton sponges in CO_2 reduction to methanol through the B3LYP/6–311 + + G (d,p) level of theory. Thermodynamic and kinetic aspects of the CO_2 reduction to methanol were investigated in the presence of six proton sponges, as the organocatalysts, and borane molecule as the reducing reagent. Figure 39 illustrates the investigated proton sponges and the overall reduction of CO_2 to methoxyborane $(MeOBO)_3$, as the final product of the reaction, which is decomposed to methanol and boric acid by the reaction with water (Sabet-Sarvestani et al. 2018).

Two different cycles can be considered for the catalytic reaction, and the performance of considered proton sponges was studied in gas and solvent phases. BH_3OCOH is the output of cycle 1, which can be considered as the starting material of cycle 2. Based on the kinetic studies, the development of boronium–borohydride ion pair (step 2) is the key

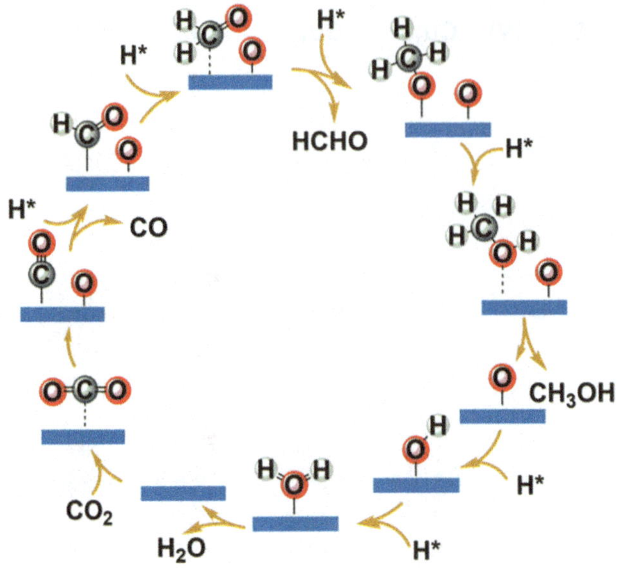

Fig. 34 Schematic representation of the reduction mechanism of CO_2 transformation to methanol on the Cu clusters (Zhang et al. 2018a)

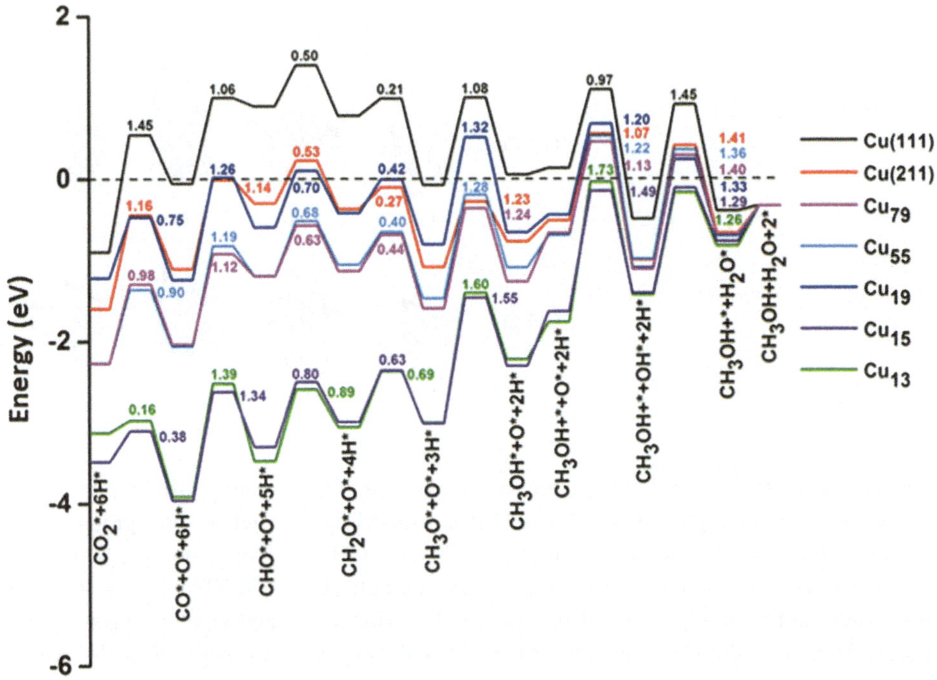

Fig. 35 PED of the CO_2 reduction to methanol for different sizes of the Cu clusters (Zhang et al. 2018a)

step of the reaction in cycle 1. Also, the activation energy of this step is affected by the steric effects of the linked methyl groups to the proton sponges. Structural analysis of the proton sponges reveals that dihedral angles (θ) between two aromatic rings of the proton sponges have a key factor in the $\Delta G^{\neq}$ value of step 2. Figure 40 illustrates the involved species of cycle 1 (Sabet-Sarvestani et al. 2018).

In the next part of the study, the probable pathways of the $(MeOBO)_3$ formation from BH_3OCOH were investigated. Three paths that are nominated as green, blue, and red routes were considered for this conversion (Fig. 41). BH_3OCOH is the starting material for all these paths. Figure 41 depicts the involved species in red (above), green and blue paths (down) in cycle 2.

Based on the PED of cycle 2 (Fig. 42), the red route is not an appropriate path for the $(MeOBO)_3$ formation, thermodynamically and kinetically. Also, in comparison with the green path, the blue route is not a probable path for the reaction,

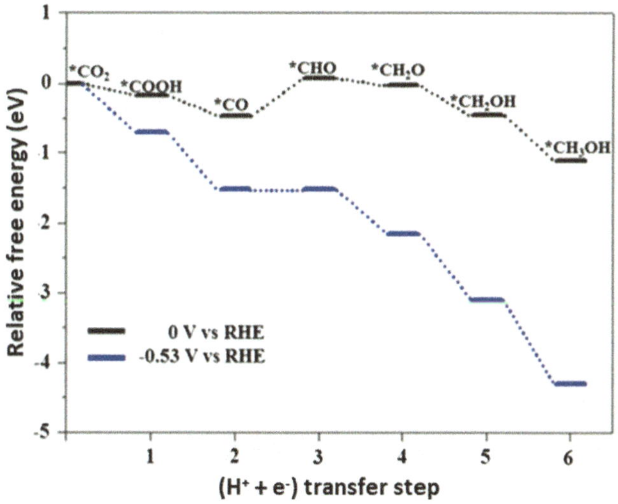

Fig. 36 PED of the reduction reaction on the Cu_{85} nanocluster and the difference between the applied electrode potentials (Rawat et al. 2017)

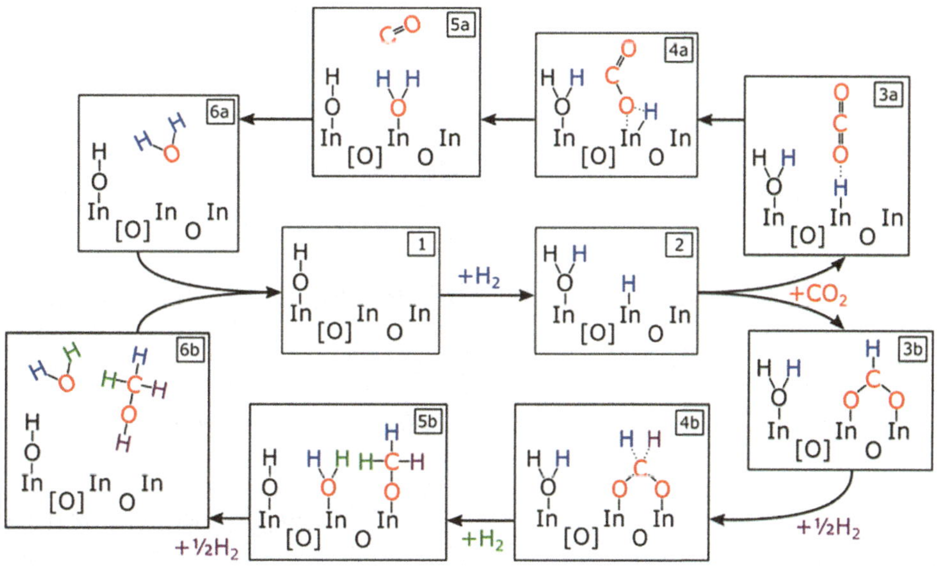

Fig. 37 Investigated mechanism of CO_2 reduction on the $In_2O_{3-x}(OH)_y$ surface (Wang et al. 2018b)

kinetically. Therefore, the green route is the most probable pathway for cycle 2, thermodynamically and kinetically.

4.1.4 Formic Acid

Due to the wide application of formic acid as a preservative, insecticide, and industrial material, the development of the efficient approaches of synthesis has been considered. The formic acid fuel cell is the next usage of this compound to provide electricity (Wang et al. 2018c). Moreover, properties such as non-toxicity, biodegradability, being liquid at the ambient conditions, easiness of storage and transportation, relatively high hydrogen capacity (4.4 wt%), and sustainability cause it to become one of the most proper materials for hydrogen storage. In the gas phase, direct CO_2 hydrogenation to formic acid is endergonic; however, in the aqueous phase or the presence of an inorganic base, such as ammonia, the reaction becomes exergonic and feasible (Eq. 9) (Wang et al. 2018c). The use of an inorganic base, as a catalyst for the reaction, leads to formate production, which by adding a strong acid is converted to formic acid. Thus, the inorganic base changes the reaction equilibrium toward higher selectivity to formic acid. Other approaches, such as the application of buffers, basic ionic liquids, and solvents addition (like DMSO as coordinating agent), are alternative methods for the production of pure formic acid, without the use of amine or other strong bases.

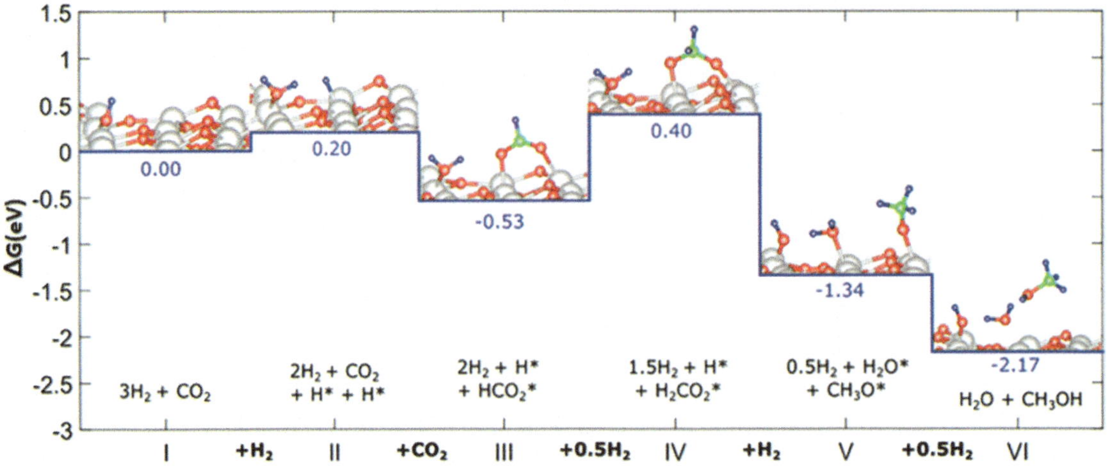

Fig. 38 PED of the CO_2 reduction to methanol over the $In_2O_{3-x}(OH)_y$, in which gray, red, green blue atoms corresponds to In, O, C, and H, respectively (Wang et al. 2018b)

Fig. 39 Overall reduction of CO_2 within the studied proton sponges (Sabet-Sarvestani et al. 2018)

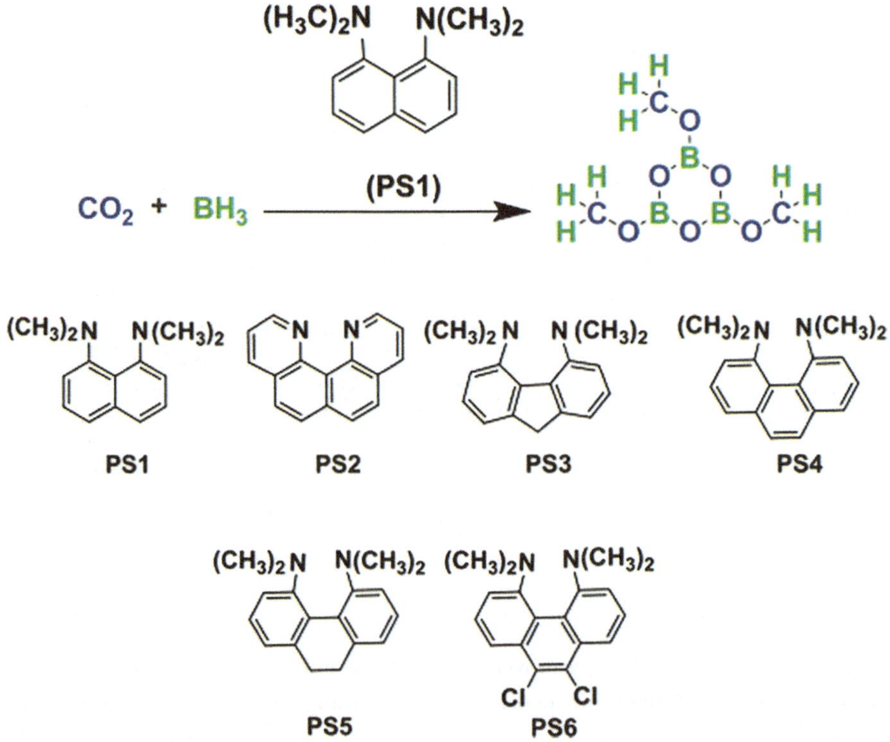

$$CO + H_2 \rightleftharpoons HCO_2H$$
$$G_{298} = +33\,\text{kJ.mol}^{-1}$$
$$CO_2 + H_2 + NH_3 \rightleftharpoons HCO_2^- + NH_4^+ \quad (9)$$
$$G_{298} = -9.5\ \text{kJ mol}^{-1}$$

In 1976, the first study of the CO_2 hydrogenation to formic acid was reported by Inoue and et al that applied the complexes of Ru, Rh, and Ir and triphenylphosphine (PPh_3). This report did not attract significant regard until the 1990s, in which the attention to the CO_2 transformation to formate was revived (Inoue et al. 1976). Based on the recent investigations, two different paths can be considered in CO_2 hydrogenation to formic acid on a catalytic surface. These pathways depend on the different adsorption mode of the CO_2 and generated formate. As shown in Fig. 43, in paths A and B, monodentate formate and bidentate formate are generated, respectively, in which formic acid is produced after hydrogenation of these intermediates (Peng et al. 2012; Chiang et al. 2018; Zhang et al. 2018b; Filonenko et al. 2016).

A DFT investigation based on the M06-L density functional is reported to study the efficacy of the copper atoms

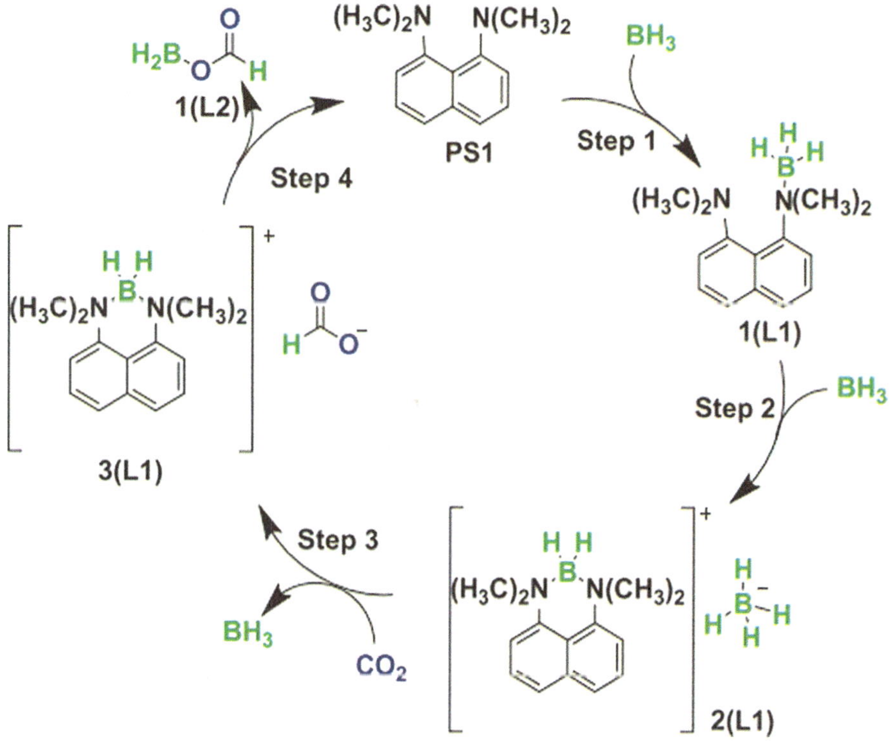

Fig. 40 Involved species of cycle 1 in CO_2 reduction in the presence of PS1 as a typical proton sponge (Sabet-Sarvestani et al. 2018)

implanted in the plane of graphene (Cu/dG) in catalytic hydrogenation of CO_2 to formic acid. For this study, the Stuttgart–Dresden effective core potential and 6-31G(d,p) basis set are considered for copper and normal atom, respectively. Cu atom on the Cu/dG surface is an active site for the adsorption and heterolytic cleavage of the hydrogen molecule. Figure 44 depicts the optimized structure of (Cu/dG), atomic natural bond orbital (NBO) charges, and bond lengthen of Cu bond with the nearby atoms(Sirijaraensre and Limtrakul 2016).

Two different paths have been considered for CO_2 hydrogenation. In the initial path, CO_2 is reduced via the bimolecular adsorption, without H_2 activation. As depicted in Fig. 45, this path passes through the barrier energy of 34.6 kcal mol^{-1} for the first hydrogenation of CO_2, which produces an unsteady H-Cu-COOH intermediate. However, in the first step of the second path, the H_2 molecule splits into coordinated hydride and proton on the Cu atom. The calculated barrier energy for this step is 19.7 kcal mol^{-1}. The H_2-activated Cu/dG can make easy the CO_2 reduction to the formate kinds.

In the second step of path 2, CO_2 is hydrogenated by H_2-activated Cu/dG, yielding a bidentate complex of formate on the Cu, which is more stable than the monodentate HCOO– complex. After the formation of formate intermediate (**INT2b**), two routes are possible for path 2. The first route includes the protonation of **INT2b** through the hydrogenated site of graphene to an oxygen atom of the HCOO–moiety, and the second route is the further reduction by the second hydrogen molecule. The obtained results show that hydrogenation of **INT2b** through the adsorbed hydrogen on the graphene has lower barrier energy than another hydrogen molecule. Also, as depicted in the PED (Fig. 46), the obtained energy of the RDS of path 2 (19.7 kcal mol^{-1}) is lower than the first one (Sirijaraensre and Limtrakul 2016).

In another report based on the DFT calculation, CO_2 activation and reduction to formic acid on the hybrid metal –organic frameworks (MOFs), namely Mo-Cu-BTC and W-Cu-BTC, has been investigated in which the Cu-BTC is formed from copper carboxylate dimers, $[Cu_2(COO)_4 - (H_2O)_2]$, and organic linker, benzene-1,3,5-tricarboxylate (BTC). The metallic property of the bond of M-Cu dimer provides the unsaturated metal site which is distinguished from the metallic site of the pure M-M-BTC (Fig. 47) (Dong et al. 2018), in which a unit cell of Cu-BTC consisting of 48 coppers, 192 oxygens, 288 carbons, and 96 hydrogens is chosen for the calculation by exchange–correlation functional, the GGA-PW91 method, and double numeric plus polarization (DNP) basis set.

Two potential reaction paths were considered for the CO_2 reduction to formic acid, called formate (HCOO*) and carboxyl (COOH*) mechanisms. Each path can be accomplished by two hydrogenation modes, and CO_2 is hydrogenated either by H_2 molecule (mode 1) or by activated atomic hydrogen (H*) linked to the metallic site (mode 2). Regarding the CO_2 hydrogenation through the H_2 molecule, the calculated energy barrier on the W-Cu-BTC via carboxyl path (1.27 eV) is lower than that of formate

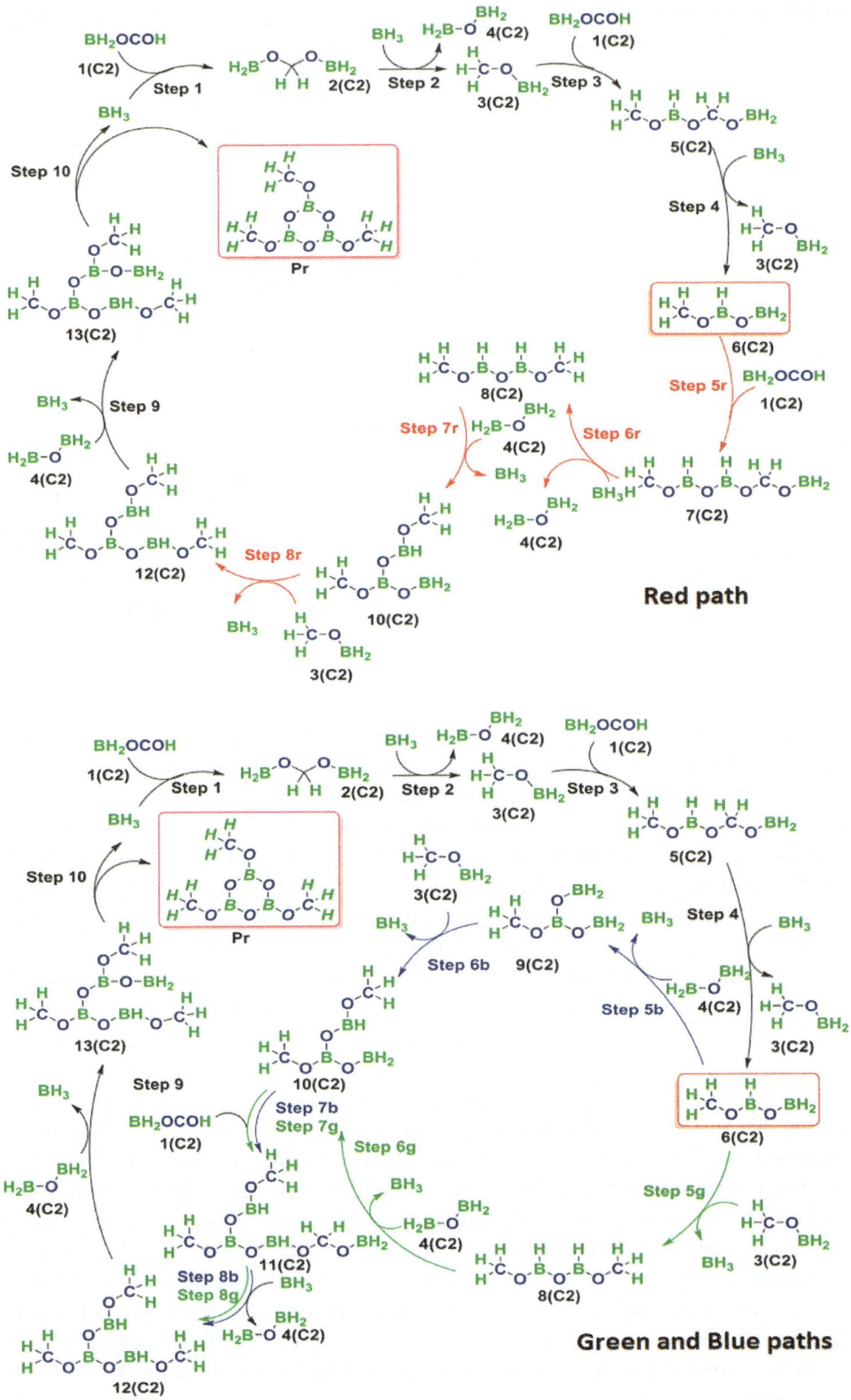

Fig. 41 Involved species in red (above), green, and blue pathways (down) of cycle 2 (Sabet-Sarvestani et al. 2018)

(1.48 eV), while on the Mo-Cu-BTC, the energy barriers in both paths are nearly equal. Thus, Mo-Cu acts as a relatively ideal catalytic system in hydrogenation via H_2 molecule mode. However, in the case of hydrogenation through H*, there is a remarkable reduction of the activation energy between CO_2 and H* in the formate (HCOO*) path for two bimetallic MOFs. The energy barrier of the carboxyl (COOH*) path stays almost similar. Also, the calculated energy barrier for CO_2 reduction along the HCOO* path is 0.30 eV. Figure 48 shows the PED of Mo-Cu-BTC and

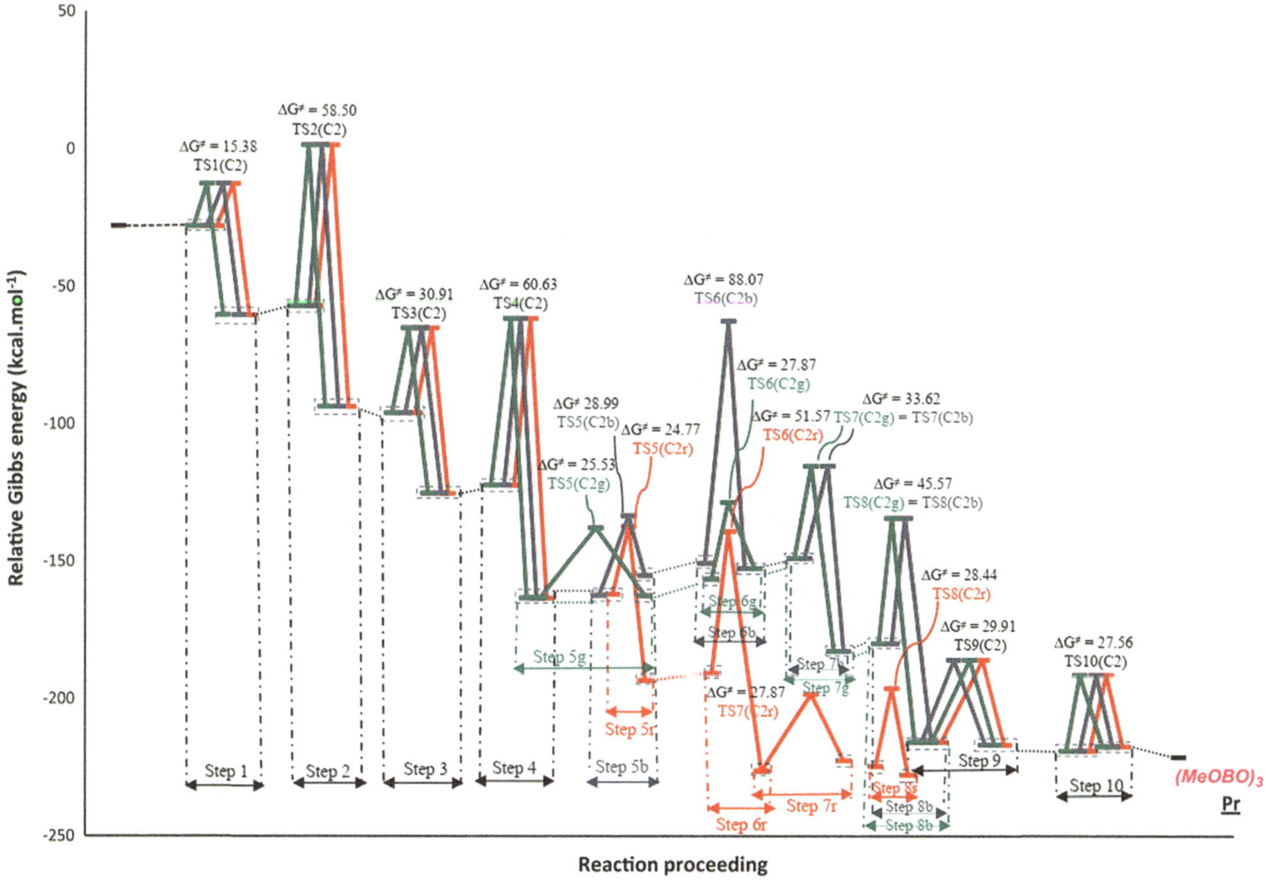

Fig. 42 PED of cycle 2 corresponding to the green, blue, and red paths (Sabet-Sarvestani et al. 2018)

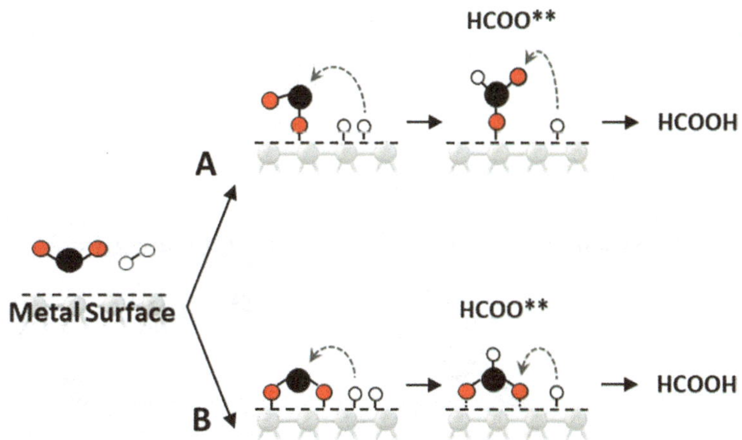

Fig. 43 Two different paths for CO_2 reduction to formic acid on a catalytic surface (Podrojkova et al. 2020)

W-Cu-BTC for formate (HCOO*) and carboxyl (COOH*) paths (Dong et al. 2018).

In an electrochemical study, In–Zn bimetallic nanocrystals were evaluated as the catalysts for CO_2 conversion to formic acid. Electro-reduction of CO_2 to formic acid formation by using several transition metals is an easy and highly efficient method ($CO_2 + 2H^+ + 2e \rightarrow HCOOH$; E = −0.608 V versus normal hydrogen electrode (NHE) at pH = 7). However, the hydrogen evolution reaction (HER, $2H^+ + 2e \rightarrow H_2$) is a competitive reaction that distributes the high efficiency of the CO_2 to HCOOH conversion. A composition ratio of In: Zn = 0.05, like $Zn_{0.95}In_{0.05}$, shows the highest catalysts selectivity toward HCOOH. In this research, a DFT calculation based on the PAW method with the PBE basis set was employed. Also, the effect of attractive van der Waals (vdW) interaction was considered

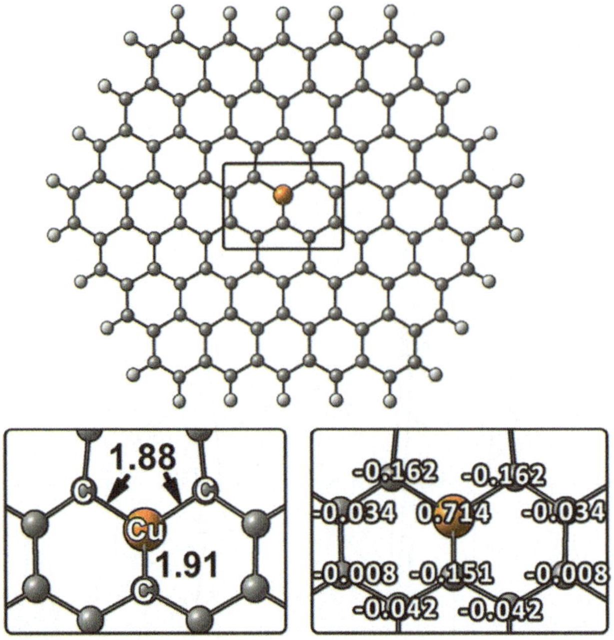

Fig. 44 Optimized structures and atomic NBO charges of applied Cu/dG surface in CO_2 hydrogenation to formic acid (Sirijaraensre and Limtrakul 2016)

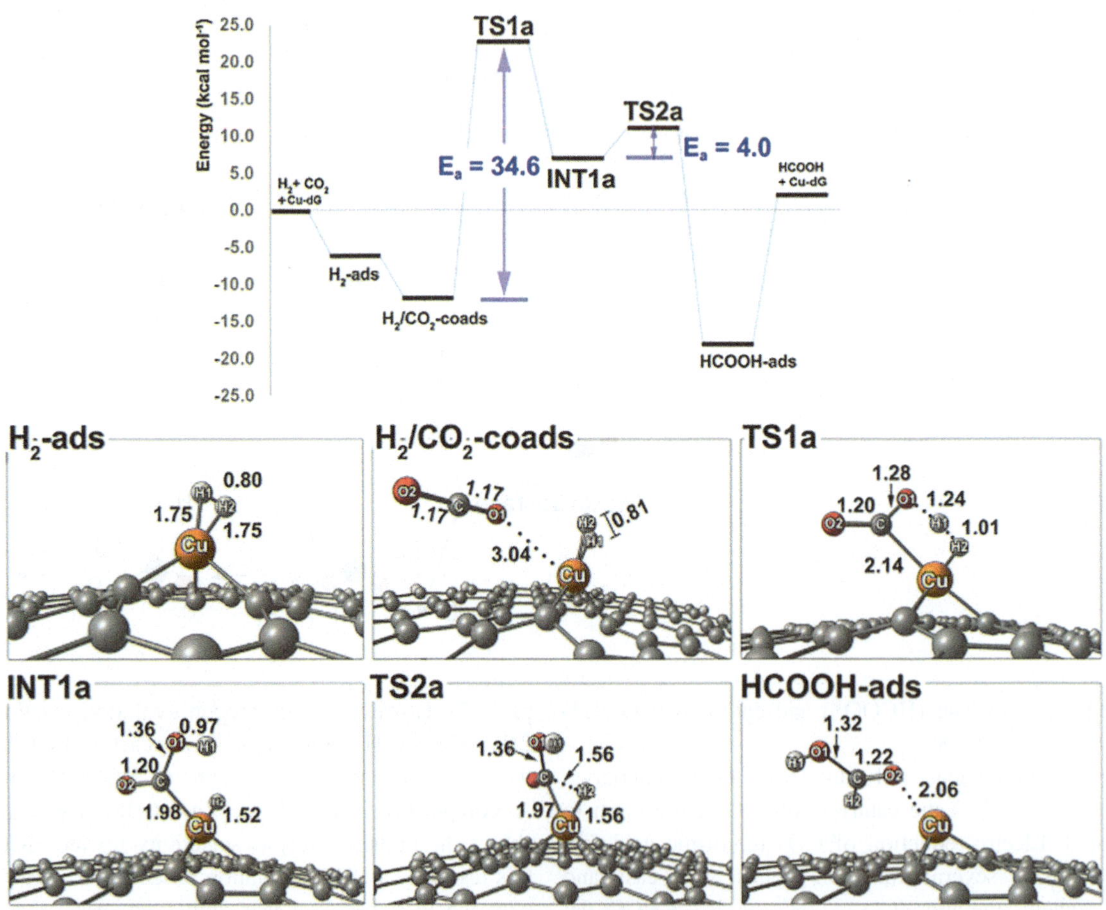

Fig. 45 PED and optimized involved species for path 1 of CO_2 hydrogenation to formic acid (Sirijaraensre and Limtrakul 2016)

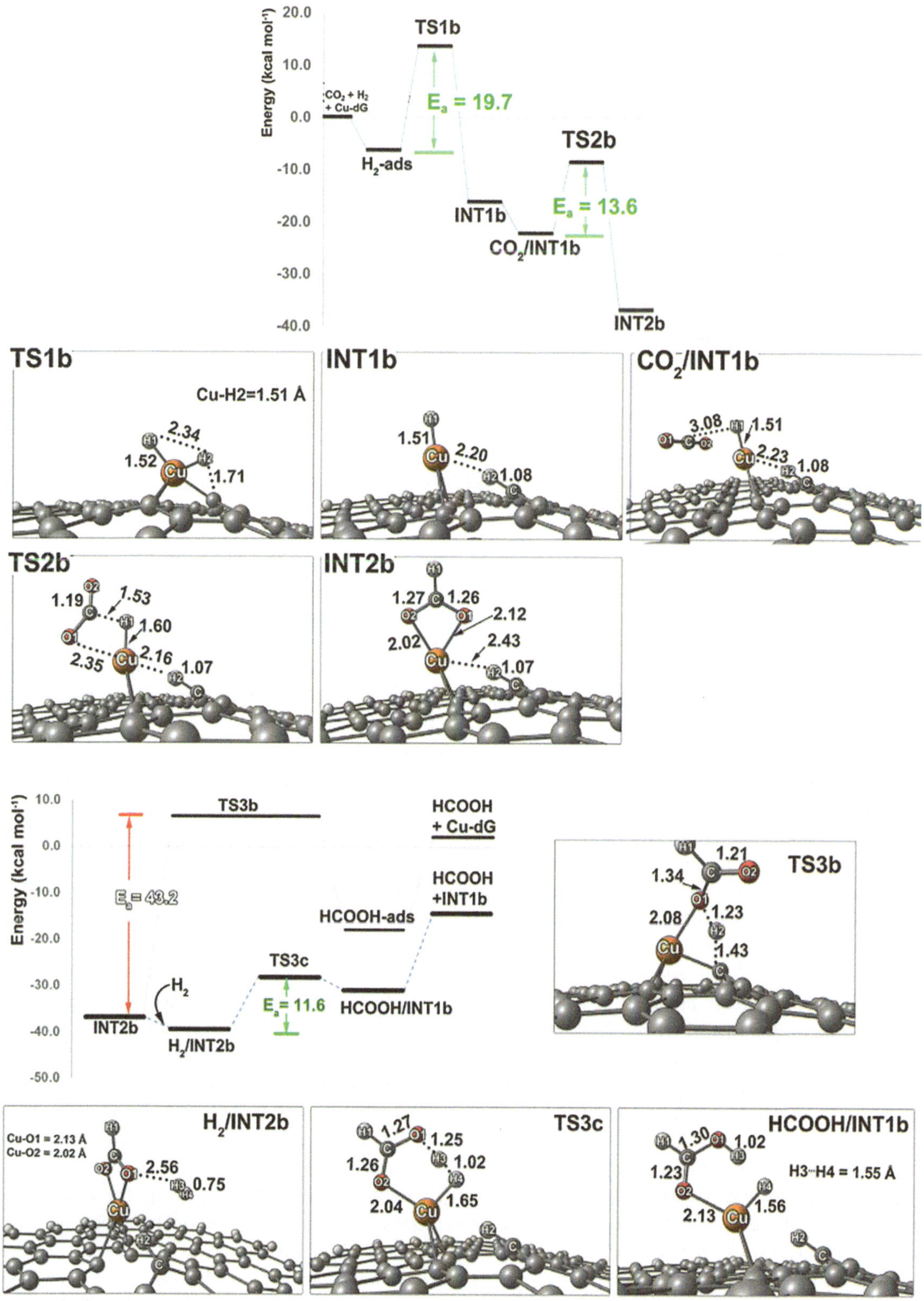

Fig. 46 PED and optimized species for path 2 of the CO_2 hydrogenation to formic acid (Sirijaraensre and Limtrakul 2016)

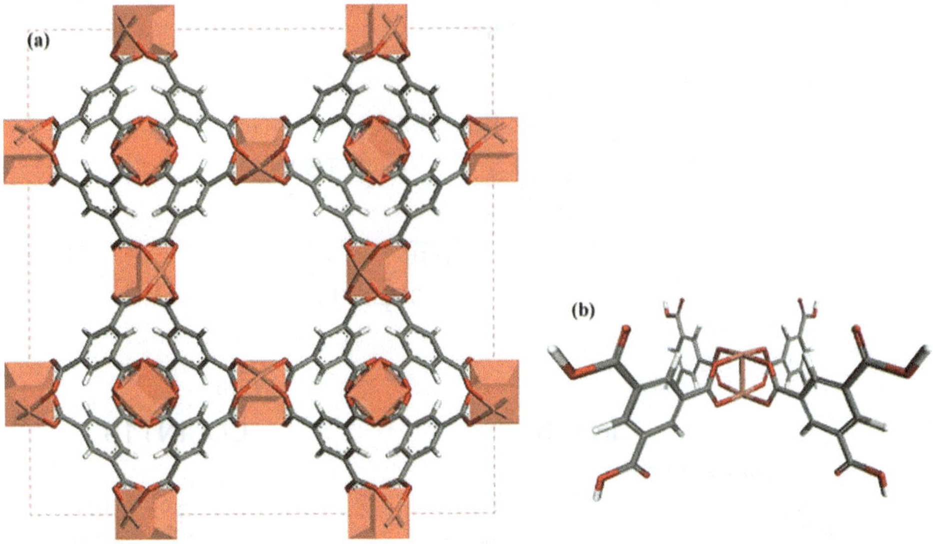

Fig. 47 Structure of the periodic model (**a**) and cluster model (**b**) of Cu-BTC. Orange, red, gray, and white colors indicate copper, oxygen, carbon, and hydrogen atoms, respectively (Dong et al. 2018)

by applying Grimme's D3 correction (PBE-D3). Based on the obtained results, the energy of formate intermediate (*OCHO), (produced by $CO_2 + H^+ + e \rightarrow$ *OCHO) is a key factor in $Zn_{0.95}In_{0.05}$ selectivity. The relative Gibbs energy diagram of the CO_2 hydrogenation to formic acid over Zn (002), Zn (101), In (110), In (101), (112) facets (Fig. 49), and $Zn_{0.95}In_{0.05}$ was calculated, in which deposited four-atom In monolayer on a three-layer (5 × 5) Zn (002) surface is used to model the $Zn_{0.95}In_{0.05}$ nanocrystals (Fig. 50) (Kwon et al. 2019).

Based on ΔG values, the strong adsorption strength of *OCHO acts as a limiting factor in HCOOH formation. Based on these diagrams, the energy of *OCHO intermediate on the $Zn_{0.95}In_{0.05}$ surface is higher than those of other studied surfaces, and therefore, the adsorbed *OCHO easily releases. These results reveal that the Zn–In interface makes the HCOOH production energy-favorable, which causes a higher reaction rate of $Zn_{0.95}In_{0.05}$ (Kwon et al. 2019).

CO_2 reduction by photocatalysts is the next approach toward formic acid production. Zhao and coworkers reported a theoretical study, in which the facet-dependent photocatalytic activity of TiO_2 (101) and (001) was investigated in photocatalytic CO_2 reduction to formic acid. All calculations were carried out using the GGA-PBE exchange–correlation potential, and the PAW method was applied to consider the effect of core electrons. Two TiO_2 facets including (101) and (001) in this investigation are depicted in Fig. 51 (Ma et al. 2016).

By analogy between the obtained results for the surface and the bulk of TiO_2 (101) and (001), it can be deduced that the excited electron tends to stay on the surface than the bulk. Also, the barriers for transferring the excited electron through the bulk and among the surface are similar for both facets. Moreover, the energy of Ti^{3+} that appeared on the (001) facets is lower than that on the (101) facets, and thus, (101) facets possess a higher conduction band minimum (CBM). Figure 52 shows the PED of CO_2 reduction to formic acid over both facets of TiO_2 (101) and (001). The energy barrier of CO_2 reduction to HCOOH over (101) facets is lower than that of the (001). A higher CBM generates stronger reducing electrons. Thus, the lower energy barrier of the (101) facets and a faster rate of formic acid production on this surface can be related to a higher level of CBM (Ma et al. 2016).

4.1.5 Heterocycles

Heterocyclic compounds are an important class of chemicals that possess various biological and pharmacological properties (Joule and Mills 2010). Value-added heterocycles such as carbonates, carbamates, carboxylic acids, and numerous sophisticated heterocyclic rings having "CO_2", "CO", "CH_2" or "CH" ingredients are the products of CO_2 reaction with different substrates such as amino groups, hydroxyl groups, and carbon nucleophiles. CO_2 incorporation by nucleophiles, as an efficient and green approach in heterocycle synthesis, has been investigated in recent years, frequently (Wang and Xi 2019). Based on the investigations, CO_2 incorporation proceeds through three modes including (I) nucleophilic addition to CO_2 followed by intramolecular cyclization to carboxylated cycle production; (II) concerted two nucleophilic site attacks on CO_2 affording cyclic carbonyl compound; (III) cyclization due to nucleophilic attack on the reduced CO_2 (Wang and Xi 2019). Figure 53 represents different approaches of CO_2 incorporation cyclization in the formation of the heterocyclic compounds. The carbon atom of CO_2 is an electrophile, and therefore, CO_2 reaction

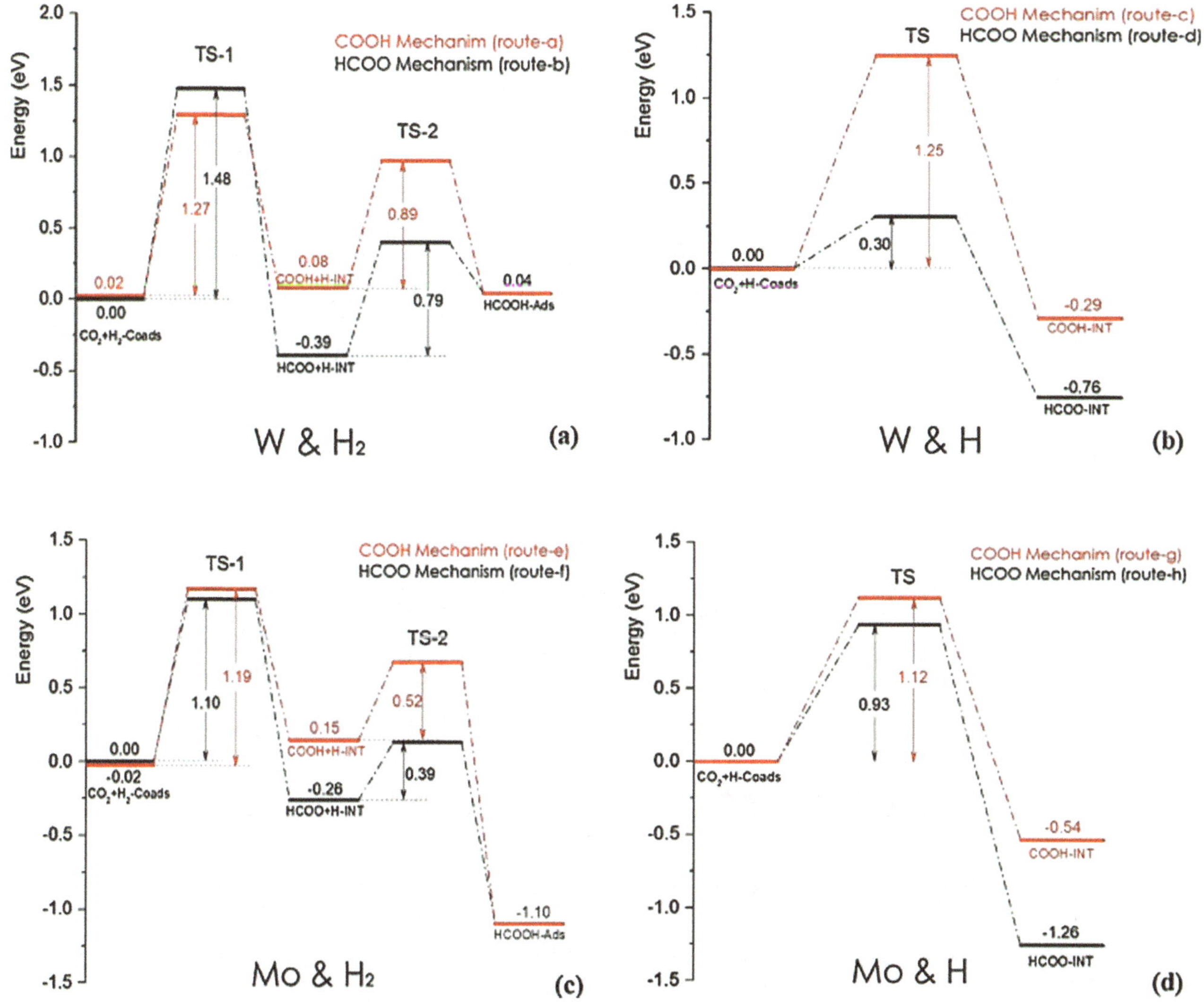

Fig. 48 PED of the CO_2 hydrogenation, in which **a**, **b**, **c**, and **d** are used for hydrogenation over W site by mode 1; over W site by mode 2; **c** over Mo site by mode 1 and **d** over Mo site by mode 2, respectively. (Black and red lines indicate HCOO* and COOH*mechanisms, respectively (Dong et al. 2018)

with strong nucleophiles leads to the development of the C–N, C–O or C–C bond and the negative charge over the O atom of CO_2. However, due to CO_2 stability, an efficient catalyst is required for CO_2 transformation to heterocycles.

Cyclic Carbonates

Low vapor pressure, high boiling point, low toxicity, and environment-friendly are appealing characters of cyclic carbonates, which cause extensive applications of this compound. The use of the cyclic carbonates as the high-boiling polar solvents, additives for fuel, and precursor of plastics are the well-known applications of cyclic carbonates. Moreover, these compounds are intermediates for the formation of other valuable compounds like dialkyl carbonates, glycols, carbamates, and pyrimidines (Calabrese et al. 2019). Traditionally, cyclic carbonates are prepared by phosgene as highly toxic, corrosive, and difficult to handle chemicals. Thus, the formation of cyclic carbonates via carbon dioxide and epoxide, as a low-toxicity and sustainable choice, has been attracted very much attention among scientists. Also, this approach is known as a high-yielding catalytic procedure which shows an atom economy of 100%. However, the use of efficient catalysts is an important necessity for this conversion (Calabrese et al. 2019).

Xia and coworkers report an amine-functionalized ionic liquid (AFIL) to produce 3-chloro-1,2-propylene via CO_2 cycloaddition with epichlorohydrin (Chen et al. 2019). All the structures of reactants, intermediates, transition states, and products were optimized in the solution phase by employing ωB97X-D/6–31 + G(d) level of calculation. Also, the solvent effects were described by the solvation model based on density (SMD) to improve the accuracy of

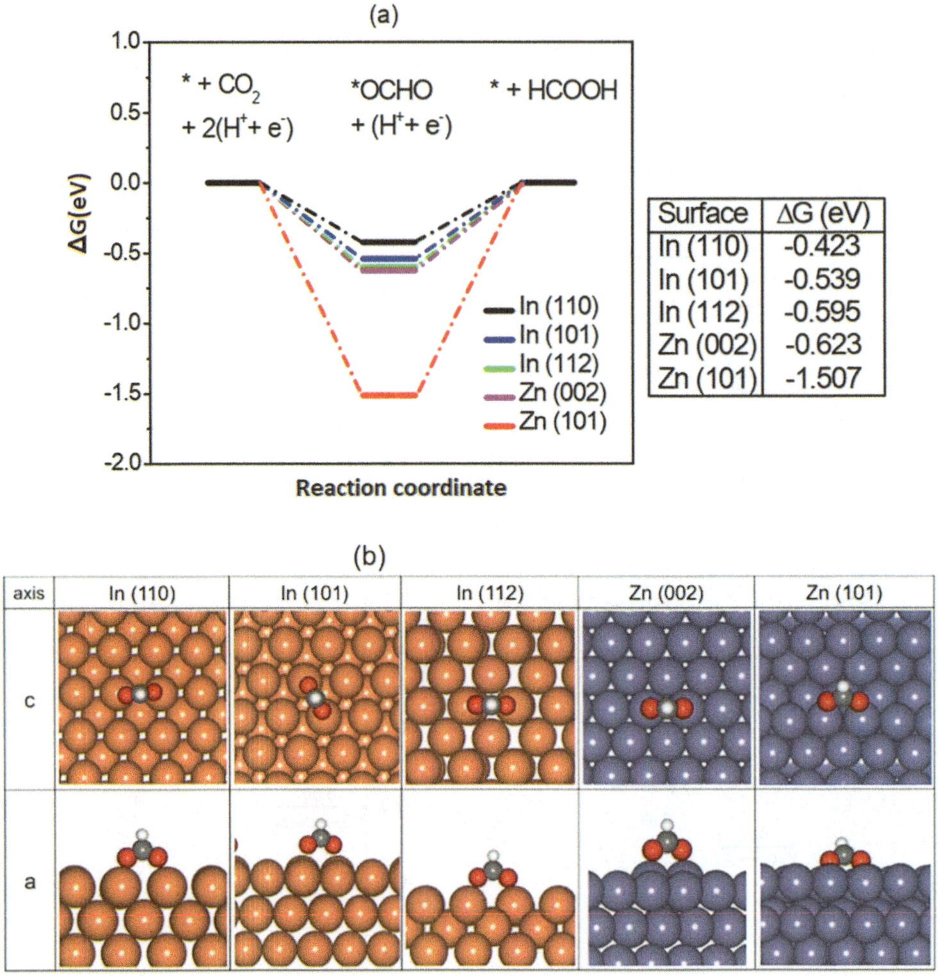

Fig. 49 **a** ΔG diagrams for CO_2 hydrogenation to HCOOH over In (110), In (101), In (112), Zn (002), and Zn (101) surfaces. **b** Optimized structures of the reaction intermediates (*OCHO) on each surface in the top (c-axis) and side (a-axis) views. Zn: bluish-purple, In: brown, C: gray, O: red, and H: white (Kwon et al. 2019)

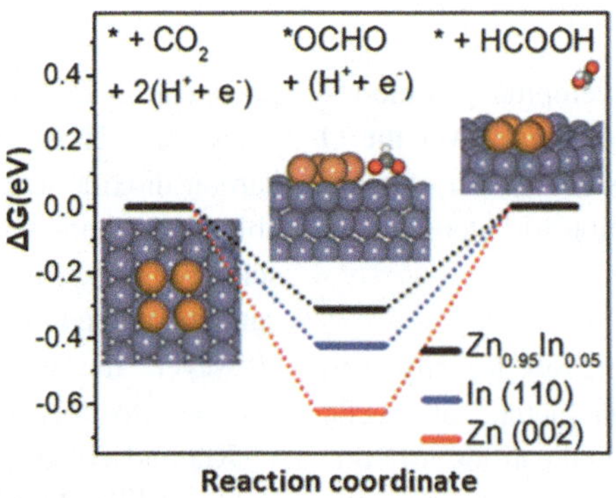

Fig. 50 Relative Gibbs energy diagrams for the CO_2 hydrogenation to HCOOH over Zn (002), In (110), and $Zn_{0.95}In_{0.05}$ surfaces in which Zn, In, C, O, and H atoms are indicated by bluish-purple, brown, gray, red, and white colors (Kwon et al. 2019)

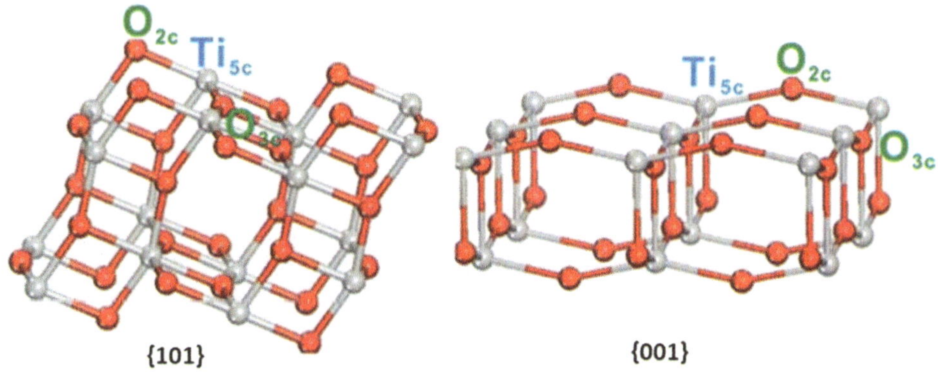

Fig. 51 Optimized structure of TiO_2 (101) and (001) (Ma et al. 2016)

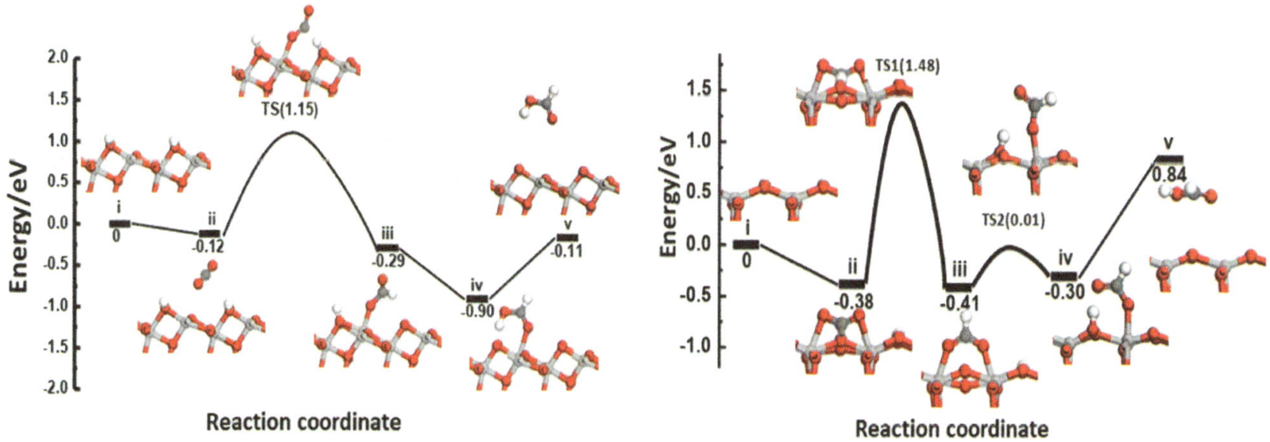

Fig. 52 PED and the optimized structure of CO_2 reduction to formic acid over both facets of TiO_2 (101) (right) and (001) (left) (Ma et al. 2016)

the calculated results. Based on the obtained results, the Br^- of the AFIL acts as a nucleophile. However, hydrogen bond formation between the oxygen atom of the epoxide and a hydrogen atom of the catalyst, as a Lewis acid, is a key interaction to epoxide activation against a nucleophilic addition. As depicted in Fig. 54, the authors described a reaction between AFIL and carbonic acid (H_2CO_3) to protonation of tertiary amine groups of the ionic liquids (Chen et al. 2019).

Figure 55 shows the PED of the ring-opening step of the reaction of epichlorohydrin facilitated by the hydrogen bond interaction of AFIL. The protonated tertiary amine group of AFIL stabilizes Br^- along with the ring-opening reaction of epichlorohydrin through the Br^- nucleophilic attack to the carbon atom of epichlorohydrin. 2-buthylimidazole is another moiety of the ionic liquid, which has an important effect in the ring-opening procedure of epoxides via hydrogen bonding interaction. After hydrogen bond formation, due to the nucleophilic attack of Br^-, the O–C bond of epichlorohydrin is broken, and yielding **Int-b3** consists of the protonated AFIL and ring-opened epichlorohydrin. After coming back the protonated AFIL to the catalytic cycle, **Int-b6** is produced due to the nucleophilic reaction between the negatively charged oxygen atom of the **Int-b4** and CO_2. **Int-b6** is an intermediate, which can be considered as the starting material for 3-chloro-1,2-propylene formation in the final step. The obtained results show that the ring-opening process is the RDS of the whole reaction. Figure 56 illustrates the PED of the final step and the optimized structure of the involved species.

The MOFs are the next efficient catalysts in CO_2 transformation to cyclic carbonate. Park and coworkers reported a joint of experimental and theoretical studies on novel adenine-based Zn-(II)/Cd(II), namely PNU-21 and PNU-22 assisted by tetrabutylammonium bromide (TBAB) as the cocatalyst. Based on structural analysis, both MOFs possess unsaturated Lewis acidic metal centers [Zn(II) and Cd(II)], free basic N atoms of adenine molecules, and auxiliary dicarboxylate ligand (Fig. 57). Experimental results show that both catalysts have similar efficiency in the cycloaddition reaction of CO_2 and epichlorohydrin (ECH) as the heterogeneous catalysts. In the theoretical section, a

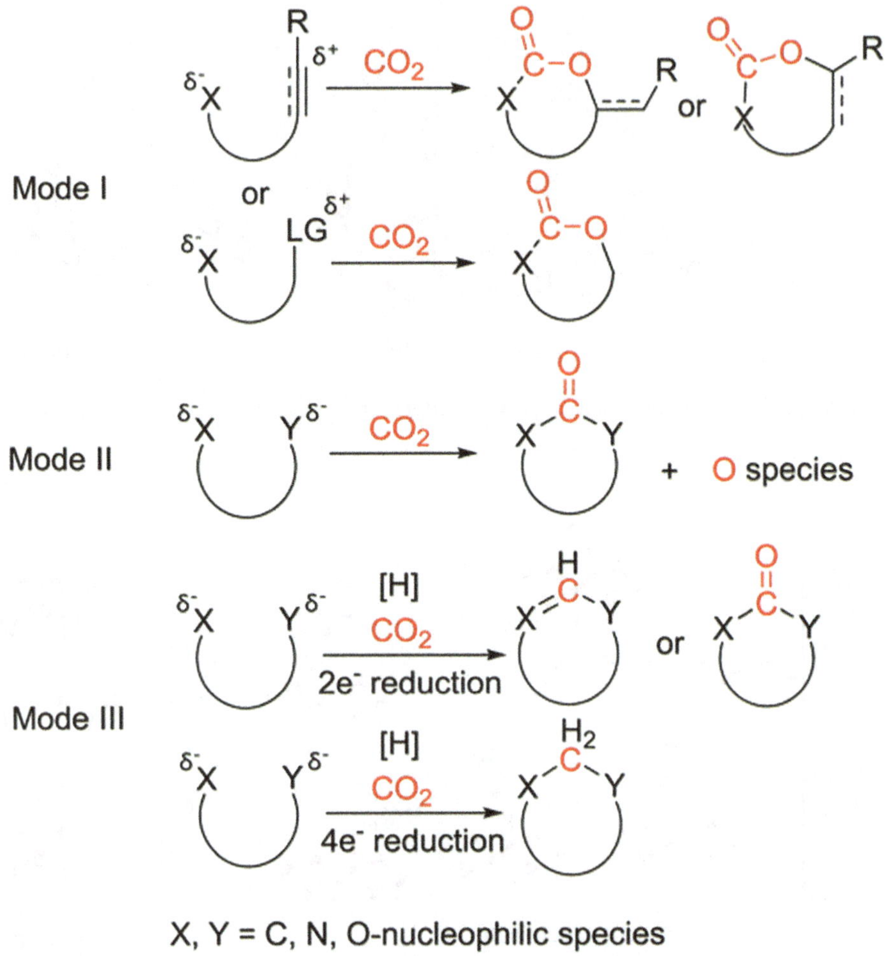

Fig. 53 Different approaches of CO_2 incorporation cyclization in the formation of a heterocyclic compound (Wang and Xi 2019)

comparative DFT-based investigation was performed on the mechanistic aspects of the cycloaddition reaction in three paths, namely noncatalyzed, TBAB-catalyzed, and PNU-21/TBAB cocatalyzed. For this purpose, meta-hybrid GGA functional M06 was employed, in which 6-31G(d) basis set was applied for H, C, N, O, and Cl atoms, while the heavier atoms (Zn and Br) were studied by LanL2DZ as a "double-ξ" quality basis set. Figure 58 depicts the investigated mechanism for the cycloaddition reaction by PNU-21/TBAB catalytic system (Rachuri et al. 2019).

In the reaction mechanism, firstly, the oxygen atom of the epoxide is coordinated to the Zn site via the oxygen atom of the epoxide. Afterward, due to the nucleophilic attack of the activated Br^- anion of TBAB, three-membered epoxide rings are opened. Then, alkyl carbonate anion is formed through another nucleophilic attack of oxide anion of the opened epoxy to the polarized CO_2 molecule by one of the $-NH_2$ groups of adenine molecules. Finally, the corresponding cyclic carbonate is produced by an ultimate ring-closure step. The obtained results show that the activation energy of the ring epoxide opening for the noncatalyzed pathway, as the RDS of the cycloaddition reaction, is 61.96 kcal mol^{-1}. This value for TBAB-catalyzed pathways is reduced to 39.60 kcal mol^{-1}. However, as depicted in the PED of PNU-21/TBAB cocatalyzed cycloaddition pathway (Fig. 59), the energy barrier of the ring-opening of epoxide by nucleophilic attack of Br^- is 7.58 kcal mol^{-1}.

Organocatalysts are the other kind of the applied catalysts in CO_2 transformation to cyclic carbonates by ethylene oxide (Galvan et al. 2014; Aoyagi et al. 2013; Wang et al. 2012; Wong et al. 2008; Girard et al. 2014; Yu et al. 2010; Tsutsumi et al. 2010; Chatelet et al. 2013). We reported a DFT study on the kinetics and mechanism of the cyclic carbonate formation by carbonyl-stabilized phosphonium ylides as a helpful organocatalyst (Fig. 60). M06-2X/6-31G(d,p) level of the theory was employed for geometry optimization of the starting materials, intermediates, products, and transition states. Also, single-point energies were obtained to reinforce the accuracy of the theoretical results, at the MPW1PW91/6–311 + + G(d,p) (Sabet-Sarvestani et al. 2020).

This transformation can be considered by two mechanisms (Fig. 61). The oxygen atom of the organocatalyst has a nucleophilic character, and the first mechanism is begun

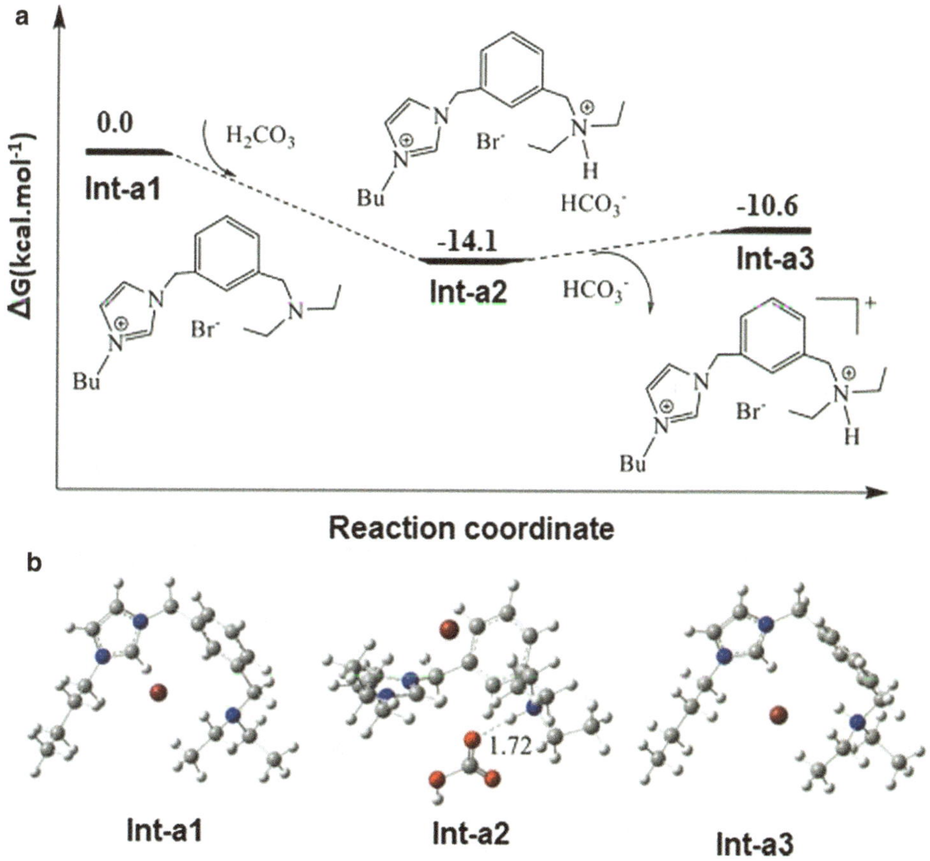

Fig. 54 PED and optimized structures of protonated AFIL via reaction with H_2CO_3 (Chen et al. 2019)

via the nucleophilic attack of this atom to the carbon atom of CO_2 (mechanism A). This pathway is followed by another nucleophilic attack of the oxygen atom of **In1A**, as the product of step 1, to the epoxide ring. In the second mechanism (mechanism B), the reaction is initiated by the oxygen attack of organocatalysts to the three-membered ring of ethylene oxide, followed by the next nucleophilic attack to the carbon atom of CO_2.

The energetic span model (ESM) is a quantum chemistry descriptor to evaluate turnover frequency (TOF) based on the obtained energy profile from the electronic structure calculations, which is used for investigation of the kinetic aspects of the proposed mechanisms (Kozuch and Shaik 2011). In this model, the degree of the TOF control (X_{TOF}) is applied to determine the effects of each involved species in reaction on the TOF values. Then, TOF-determining intermediate (TDI) and the TOF-determining transition state (TDTS) can be identified by the X_{TOF} values (Kheirabadi et al. 2018; Falivene et al. 2018).

Figure 62 illustrates the obtained TDI and TDTS of mechanisms A and B on the PEDs of four typical organocatalysts. The results show that the transition state of **In2A** formation through **In1A** in step 2 is TDTS for mechanism A. In the case of mechanism B, the energy barrier of step 1 is the TDTS. Also, **In2B** is considered as the TDI of the mechanism. Finally, based on the calculations, it can be concluded that linked electron-donating groups to carbonyl-stabilized phosphonium ylides have an efficient effect on the barrier energy of the cyclic carbonate formation.

Cyclic Carbamate

This class of heterocyclic rings is found in numerous valuable chemicals. For example, a five-membered carbamate ring (2-oxazolidinone) is a heterocyclic core of antibiotics such as Linezolid, Posizolid, and Tedizolid and also antidepressants like Toloxatone (Brickner 1996; Schaadt et al. 2009; Chen et al. 2017). Traditionally, chemicals such as isocyanates, phosgene, urea, or organic carbonates are used to synthesize cyclic carbamates. However, due to high toxic and expensive starting materials, CO_2-based synthetic paths, as sustainable, cost-effective, and highly efficient methods, can be an alternative for the cyclic carbamates synthesis. Thus, the investigation of novel catalysts to overcome the kinetic and thermodynamic stability of the CO_2 conversion to cyclic carbamates is of interest to scientists (Niemi 2019).

Lan and coworkers reported a DFT study on the mechanism of CO_2 (1 atm) transformation to cyclic carbamates via the copper complex and Togni's reagent in the presence

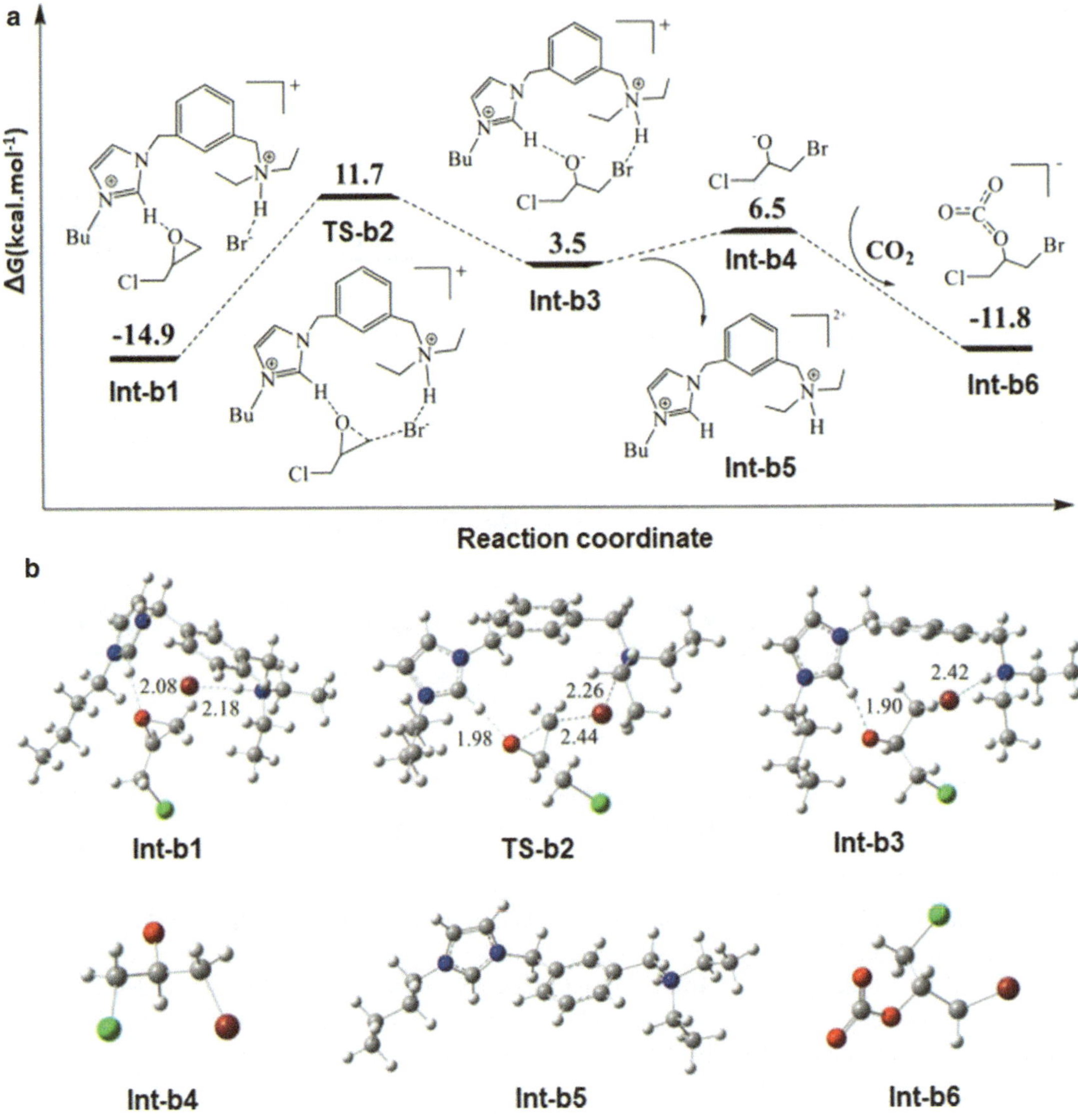

Fig. 55 PED and the involved compounds in the ring-opening process of epichlorohydrin through the hydrogen bond interaction of AFIL (Chen et al. 2019)

of DBU (1,8-diazabicyclo[5.4.0]undec-7-ene) (Fig. 63). The calculations are based on B3LYP functional with the standard 6-31G(d) basis set, for normal atoms, and SDD basis set for Cu and I. Also, single-point calculations by the integral equation formalism (IEF) polarizable continuum (PC) model (IEF-PCM) were performed to evaluate the solvent effects (Zhu et al. 2018).

Based on obtained results, the reaction is progressed by a mutual conversion of Cu(I)/Cu(II) in the catalytic cycle. As shown in Fig. 63, coordination of the nitrogen atom of **1a** to the Cu(I) center of catalyst gives an amine-coordinated intermediate. This step is followed by the deprotonation of amine by the DBU base, as the RDS of the reaction. The next step is an insertion reaction, in which CO_2 is inserted between the Cu(I)-N bond of intermediate **II**, yielding copper(I) carboxylates (**III**). Then, the Cu (I) atom of the later intermediate is oxidized by Togni reagent (2) to copper (II) dicarboxylate intermediate (**IV**) and free trifluoromethyl radical. Subsequently, radical intermediate (**V**) is produced through the trifluoromethyl radical attack to the alkene moiety of **IV**. Finally, the final oxytrifluoromethylation product and the released active Cu(I) intermediate are produced via a radical–radical cross-coupling reaction. Moreover, regarding the global electrophilicity, ω°, and global nucleophilicity, N°, indices, proposed by Domingo and coworkers, ω° and N° indices of the CF_3 radical are 2.36 eV and 0.74 eV, respectively, which suggest that this radical possesses a strong electrophilicity character. Also, based on the local electrophilicity analysis of intermediate **IV**, the ω° index of the Cu(II) atom is 2.16 eV. Therefore, the

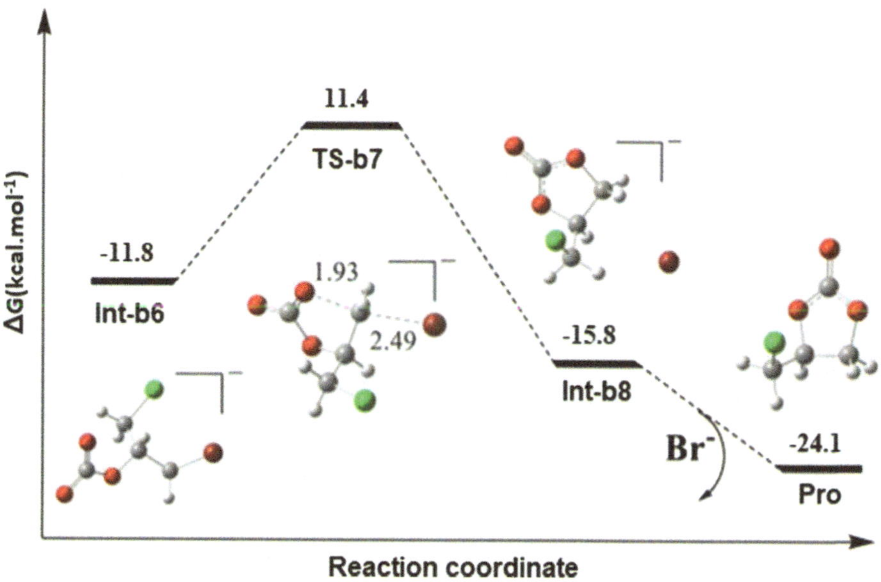

Fig. 56 PED and the involved species of 3-chloro-1,2-propylene formation through intramolecular nucleophilic reaction of Int-b6 (Chen et al. 2019)

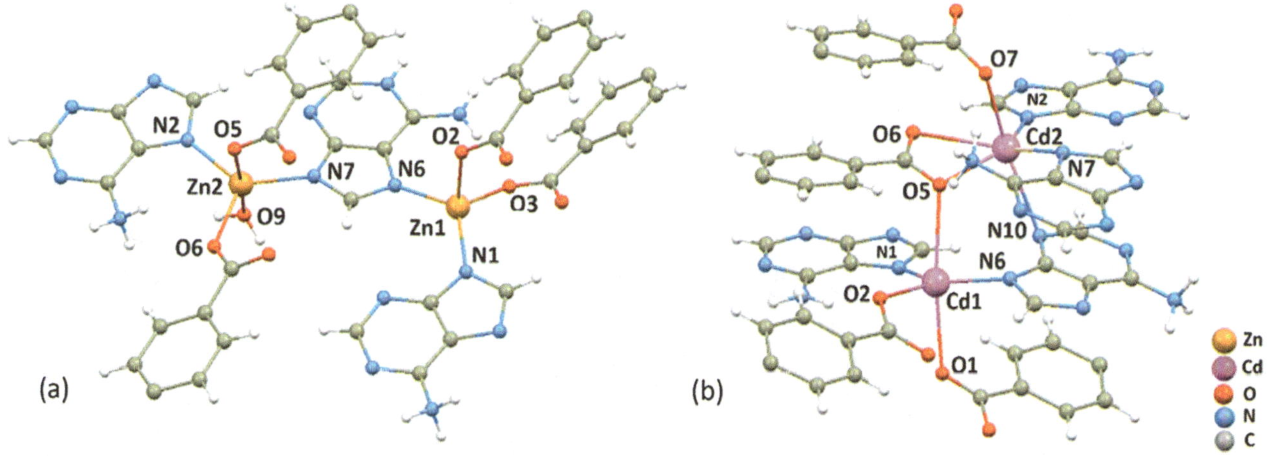

Fig. 57 Bonded ligands to Zn and Cd atoms in PNU-21 (**a**) and PNU-22 (**b**) structures (Rachuri et al. 2019)

combination of the strong electrophile ($CF_3\cdot$) and the electron-deficient Cu center is unfavorable. Thus, electrophilic addition to the C–C double bond of intermediate **IV** simply occurs to produce the diradical intermediate **V**.

In another study, the formation of cyclic carbamates from carbon dioxide and amino alcohols, in the presence of p-Toluenesulfonyl Chloride (TsCl) as a hydroxyl protecting agent, was reported (Fig. 64). Some properties, such as mild situations, appropriate yields, high enantiomeric excess, and stereoselectivity, were considered as the advantages of this method. The results show that the tosylation of the alcohol group of the amino alcohols improves the leaving group character, and also, an external base is required for the reaction progress toward the corresponding cyclic carbonates. However, tosylation of the amine group is a side reaction that can reduce the yield of the cyclic carbonate formation (Niemi et al. 2018).

DFT calculations were carried out based on the dispersion corrected meta-hybrid functional M06-2X by the standard 6–31 + G(d) basis set for all atoms. According to the PED of the reaction (Fig. 65), the energy barrier of the nucleophilic attack of a nitrogen atom to MsCl (30.3 kcal mol^{-1}) is higher than that of nucleophilic attack to CO_2 molecule (17.5 kcal mol^{-1}). The linked proton atom to **INT1** is abstracted by the carbonate part of Cs_2CO_3 as an external base. Finally, cyclic carbonate is formed due to the

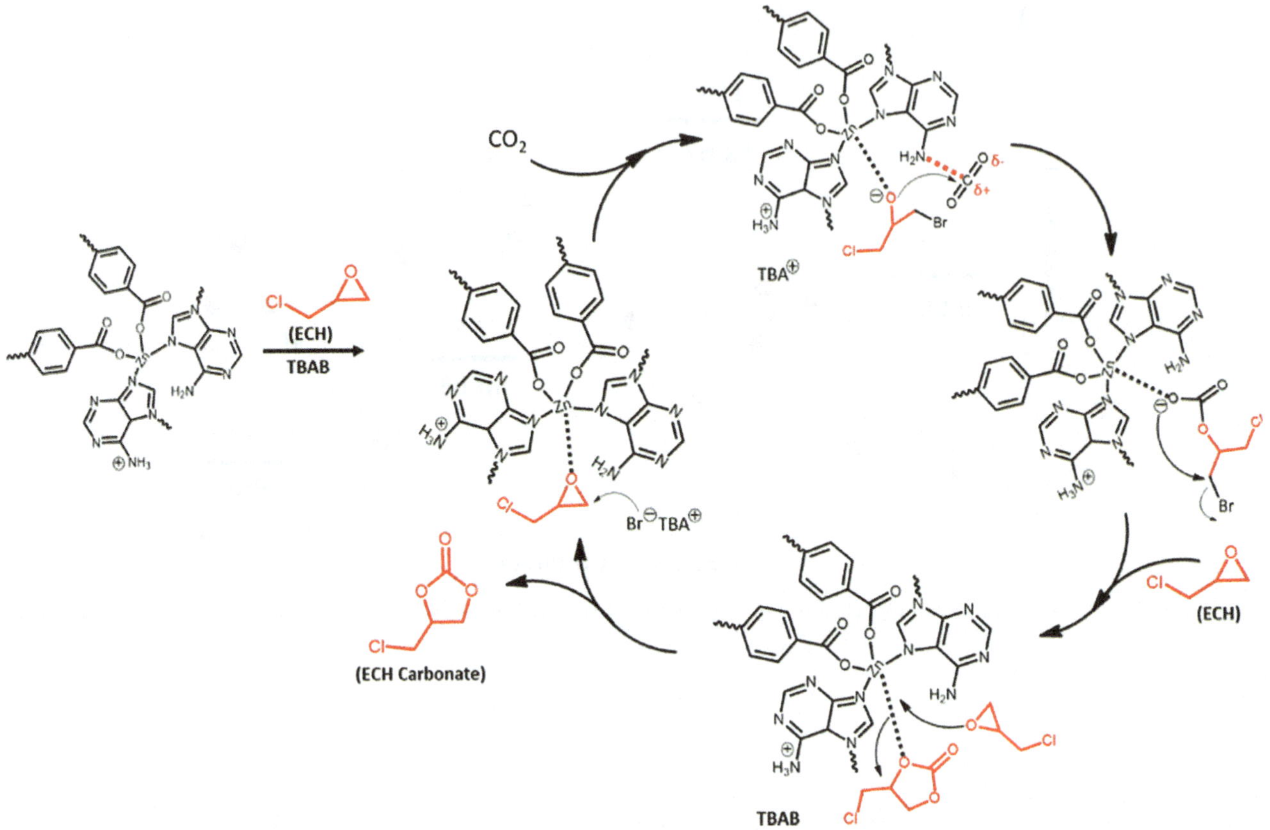

Fig. 58 Proposed mechanism for cycloaddition of CO_2 and epichlorohydrin in the presence of PNU-21/TBAB catalytic system (Rachuri et al. 2019)

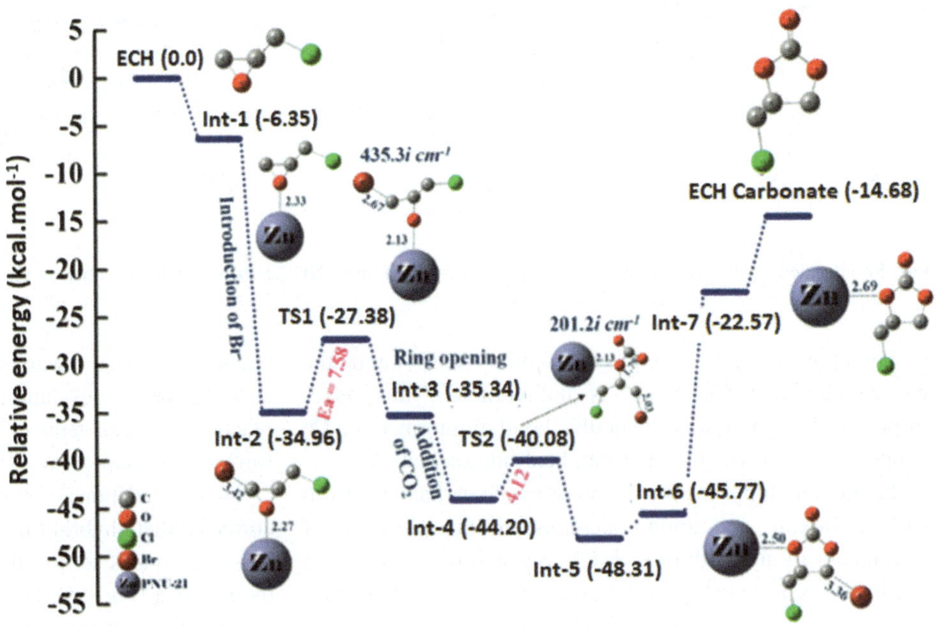

Fig. 59 PED and optimized involved structures of the PNU-21/TBAB-catalyzed cycloaddition (Rachuri et al. 2019)

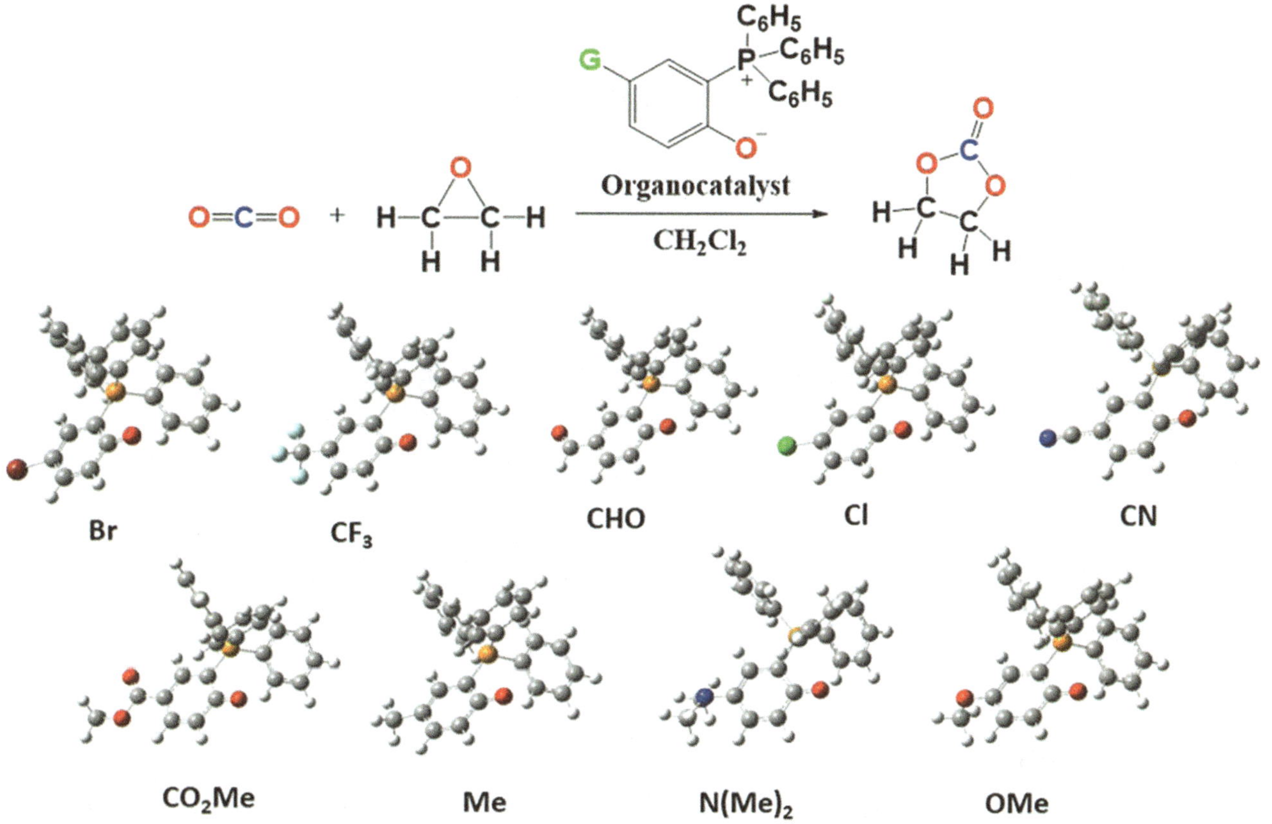

Fig. 60 Applied carbonyl-stabilized phosphonium ylides as the efficient catalysts in the cyclic carbonate synthesis (Sabet-Sarvestani et al. 2020)

tosylation of the alcohol group and an S_N2-type reaction mechanism during the ring-closing stage.

Kukhtin–Ramirez adducts, as organophosphorus compounds, are other intermediates in carbamate synthesis (Wang and Radosevich 2013). Between the various organophosphorus having distinct valence states, trivalent phosphorus derivatives are known as compounds that undergo an addition reaction with 1,2-dicarbonyl structures, yielding 1:1 adduct, formulated as either dioxaphospholene (A) or oxyphosphonium enolate (B) (Fig. 66) (Zhao et al. 2013). The reaction is named as Kukhtin–Ramirez addition. The reaction of **B** with the electrophilic agents produces alkoxyphosphonium **C**, which has an electrophile character for nucleophilic displacement. We reported the thermodynamic and kinetic parameters of the carbamate formation through various phosphorous reagents (PRs), as a Kukhtin–Ramirez addition (Sabet-Sarvestani et al. 2017a). The produced carbamate is a valuable starting material for the formation of oxazolidine-2,4-dione as a bioactive compound. DFT calculations are carried out at the B3LYP/6–31 + G(d, p) level of theory. Also, MPWB95, as a meta-GGA functional, is employed for single-point energy calculations. Figures 67 and 68 show the studied PRs in overall reaction and considered mechanism, respectively (Sabet-Sarvestani et al. 2017a).

In 1 is formed through the nucleophilic addition of the phosphorous atom of PR to the carbonyl group of α-keto ester (1), with two isomeric structures of **In 1(A)** and **In 1 (B)**. The corresponding absolute Gibbs energies of these isomeric forms are −1318.207 and −1318.211 (a.u), respectively. Thus, **In 1(B)** is the most stable isomer of **In 1**. In step 2, **In 2** is produced through the proton abstraction of phenyl carbamic acid (2) by **In 1**. However, **In 1(B)** possesses two prochiral centers, *re* and *si* faces, which make step 2 able to proceed via *re* face or *si* face approaches. Finally, in step 3, the final product **3** is developed via the nucleophilic attack of phenyl carbamate part of **In 2** and PRs abstraction as oxide forms (PORs). Kinetic results show that step 1 is RDS of the reaction. Based on quantum chemistry descriptors such as local nucleophilicity indices (N_k), Mulliken atomic spin density (P_k^-)(Domingo et al. 2013), and donor–acceptor orbital interactions of the phosphorous atom of the PRs, it has been concluded that the PRs having nitrogen atom such as $P(NC_4H_8)_3$ and $P(NEt)_3$ are appropriate agents in the reaction, due to larger P_k^- values. Finally, it has been concluded that $P(NEt)_3$ is a more

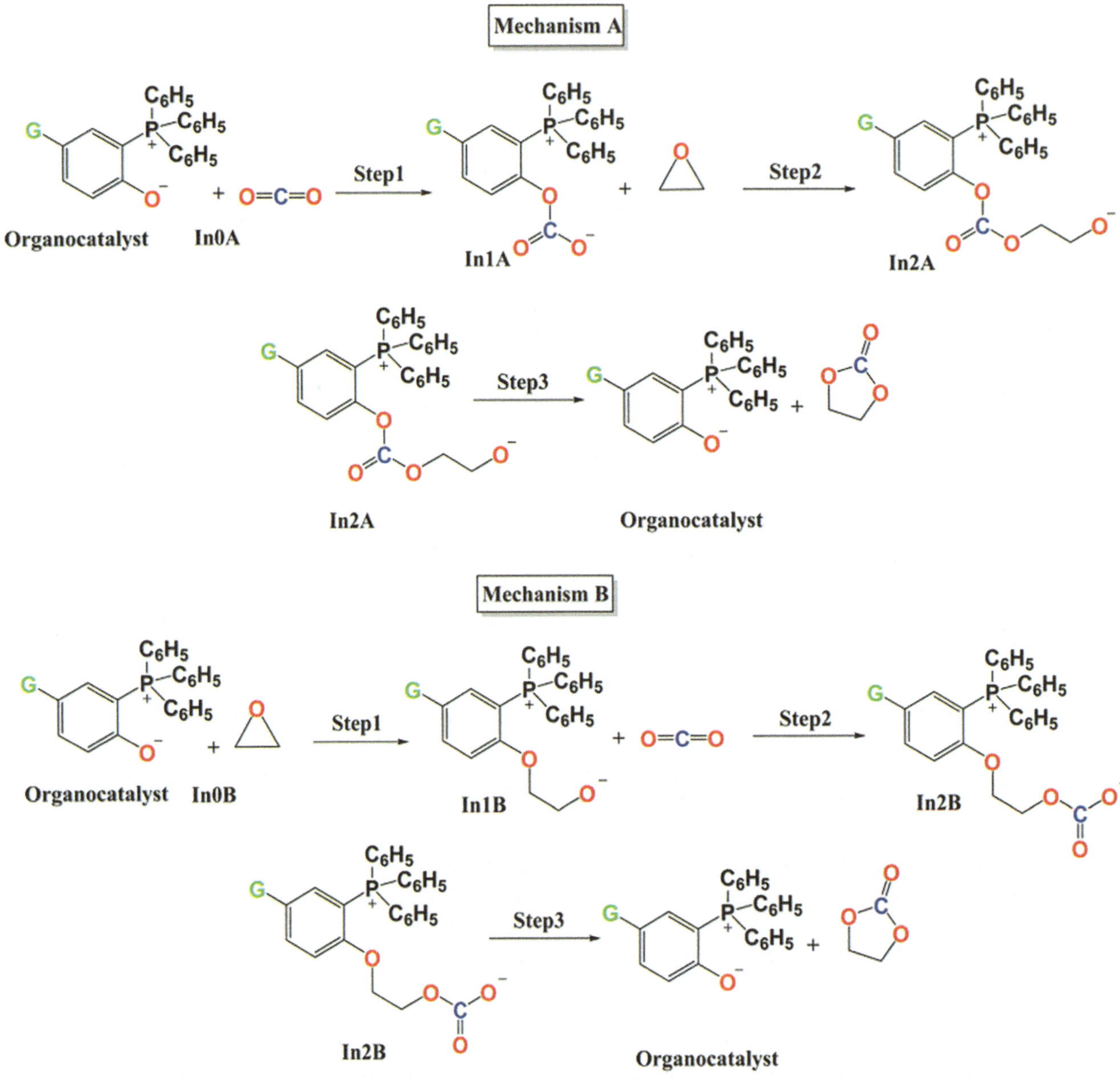

Fig. 61 Two investigated mechanisms for cyclic carbonate formation in the presence of phosphonium ylides as the organocatalyst (Sabet-Sarvestani et al. 2020)

efficient phosphorus agent for the reaction, kinetically and thermodynamically.

Quiznazoline-2,4(1H,3H)-Dione

Because of various biological activities and being a basic moiety of numerous available drugs, such as Ketanserin, Cloperidone, and Pelanserin this heterocyclic system has attracted the attention of chemists and medicinal chemists (Biswas et al. 2020). The reactions between anthranilamide and phosgene, anthranilic acid and potassium cyanate or chlorosulfonyl isocyanate, aromatic amino nitriles and diethylformamide, methyl anthranilate and various iso (thio)cyanates as well as 2-nitrobenzamide and CO are traditional approaches for the synthesis of quinazoline-2,4 (1H,3H)-dione (Nale et al. 2014). Thus, a synthetic methodology based on carbon dioxide as a feedstock is much more admitted than other methods. The majority of the traditional approaches have drawbacks, such as the use of highly toxic and specialized reagents like phosgene. Therefore, researchers focus on designing efficient catalysts for CO_2 utilization to various derivatives of this heterocyclic system.

Islam and coworkers reported a heterogeneous catalyst based on incorporated palladium metal to aminically

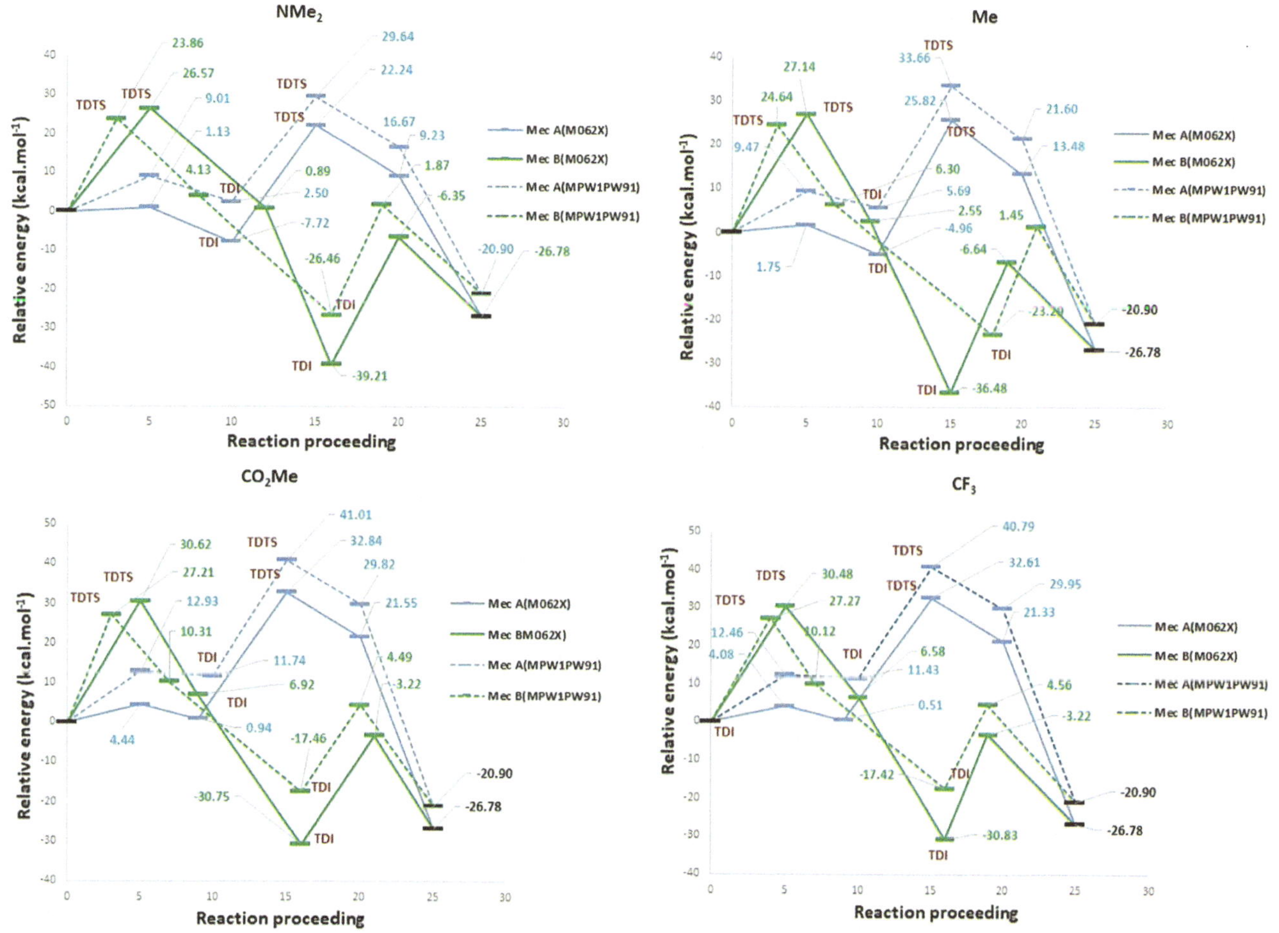

Fig. 62 PED of the substituted organocatalysts (G = CF_3, CO_2Me, Me and NMe_2) and corresponding TDI and TDTS of mechanisms A and B (Sabet-Sarvestani et al. 2020)

improved graphene oxide (GO). They used Pd(II)EN@GO for the formation of quinazoline-2,4(1H,3H)-dione derivatives through a one-pot reaction of atmospheric carbon dioxide, isocyanides, and 2-iodoaniline (Fig. 69) (Biswas et al. 2020). In the mechanism study, a DFT-based investigation was carried out by the B3LYP function and LanL2DZ basis set. The structural analysis of the Pd(II)EN@GO shows that Pd(II) atoms are stabilized by nitrogen atoms of ethylene diamine moieties on the EN@GO composite. At the first step of reaction, Pd(II)EN@GO is reduced to Pd(0) EN@GO through the elimination of two chlorine atoms as a Cl_2 molecule. The next step is an oxidation addition in which intermediate-**1** is produced via o-iodoaniline coordinates with Pd(0)EN@GO. Subsequently, due to the tertbutyl isocyanide attack to intermediate-**1**, the next intermediate is formed. The coordination of the oxygen atom of CO_2 with Pd increases the p-accepting capacity of the carbon atom of carbon dioxide. Thus, the carbon atom of CO_2 can be readily attacked by the nitrogen atom of the amine group of intermediate-**2**, yielding intermediate-**3**, which is converted to intermediate-**4** due to the intramolecular rearrangement. Finally, after proton attraction by the DBU as the applied based in the reaction and a ring-closing/ring-opening rearrangement, intermediate-**4** is converted to the final product (Fig. 70). Figure 71 depicts the PED of the reaction, in which the attack of tertbutyl isocyanide to intermediate-**1** is the RDS of the reaction (Biswas et al. 2020).

In the next report, the water molecule has been considered as a catalyst that activates CO_2 as carbonic acid (H_2CO_3) in the formation of various substituents of quinazoline-2,4 (1H,3H)-diones using CO_2 and 2-aminobenzonitriles (Ma et al. 2013). Based on the DFT calculations at the M06-2X/D95(d,p) and M06-2X/aug-cc-pVDZ levels of theory, the mechanism of the reaction was investigated. Moreover, the single-point energies were evaluated at the M06-2X/aug-cc-pVTZ level. The CPCM model was applied to investigate the effects of water as the solvent. Figure 72 depicts the considered mechanism for this reaction. H_2CO_3 is produced via the nucleophilic attack of H_2O to CO_2. Then, the hydrogen bond formation between the OH moiety of H_2CO_3 and CN segment of the amino nitrile activates the CN group against the next nucleophilic attack by H_2CO_3 in step 2. Step 3 consists of

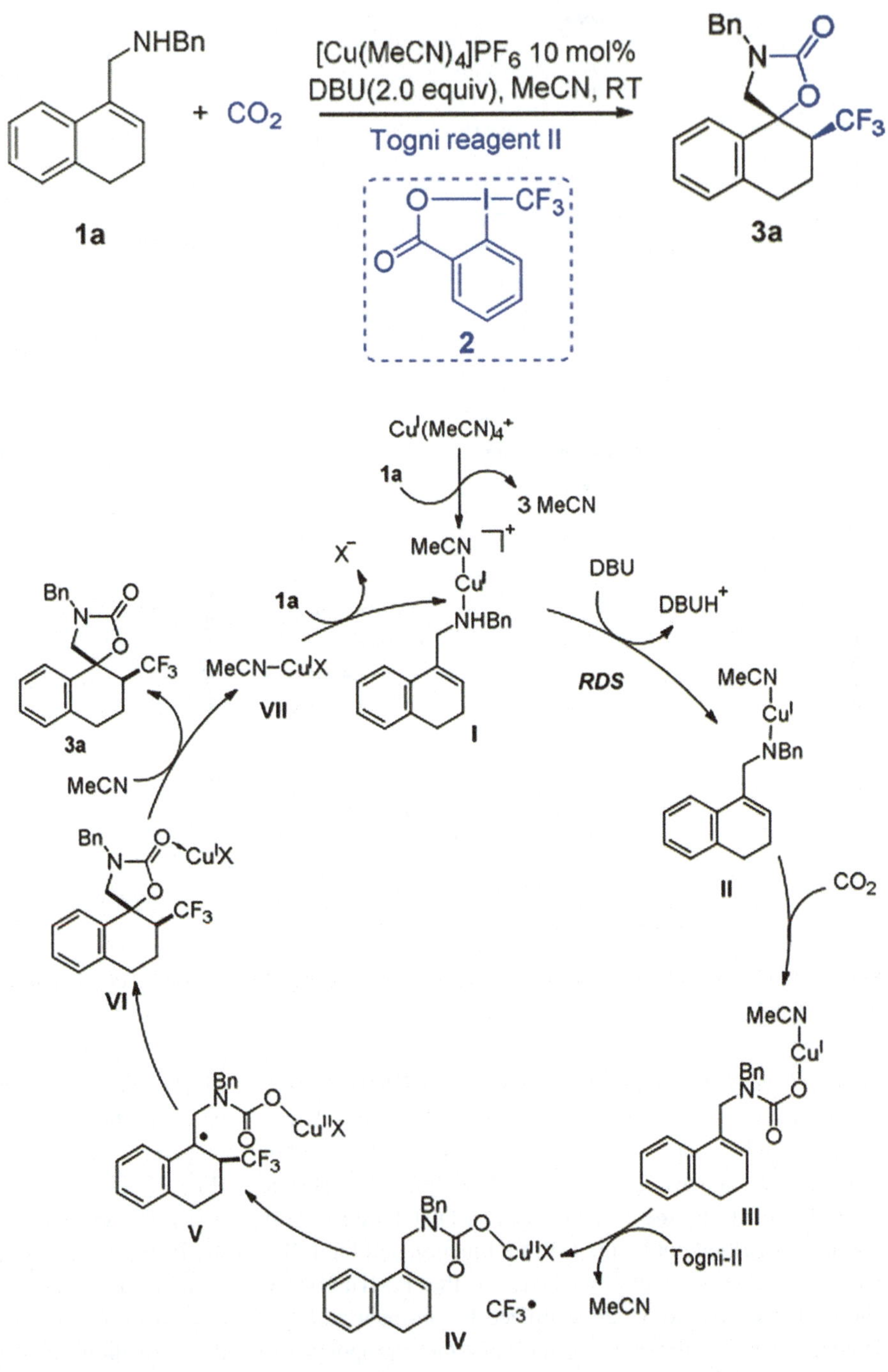

Fig. 63 Mechanism of the CO_2 conversion to cyclic carbamates via copper complex (Zhu et al. 2018)

continuous rearrangements including oxime group conversion to amide and reorientation of the –COOH group by the C–N bond rotation. Step 4 is an intramolecular nucleophilic addition, which is started by hydrogen bonding formation between the intermediate **6** and new formic acid molecules. Thus, along with the reaction, H_2CO_3 has two distinct behaviors. First, instead of CO_2, H_2CO_3 interacts with 2-aminobenzonitriles, and second, H_2CO_3 plays as a proton bridge that improves the nucleophilic addition, efficiently. As depicted in the PED of the reaction (Fig. 73), oxime group rearrangement to amide in step 3 is the RDS of the reaction (Ma et al. 2013).

Guanidines-based organic superbases (SBs) are the other efficient organocatalysts in the formation of quinazoline-2,4 (1H,3H)-diones derivatives (Mizuno et al. 2000a,b; Gao et al. 2010). We investigated the performances of four SBs, including DBU, 1,5- diazabicyclo[4.3.0]non-5-ene (DBN), 1,1,3,3-tetramethylguanidine (TMG), and 1,5,7-triazabicyclo [4.4.0]dec-5-ene (TBD) in the conversion of CO_2 and 2-aminobenzonitrile to quinazoline-2,4(1H, 3H)-diones.

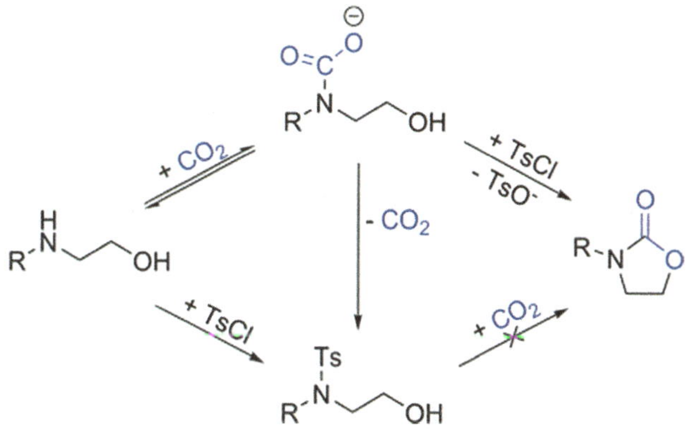

Fig. 64 Cyclization of carbon dioxide and amino alcohols in the presence of an external base and N-tosylation reaction, as a disrupting reaction (Niemi et al. 2018)

Fig. 65 PED of the reaction of amino alcohol 1a and CO_2 in the presence of MsCl and other compounds (Niemi et al. 2018)

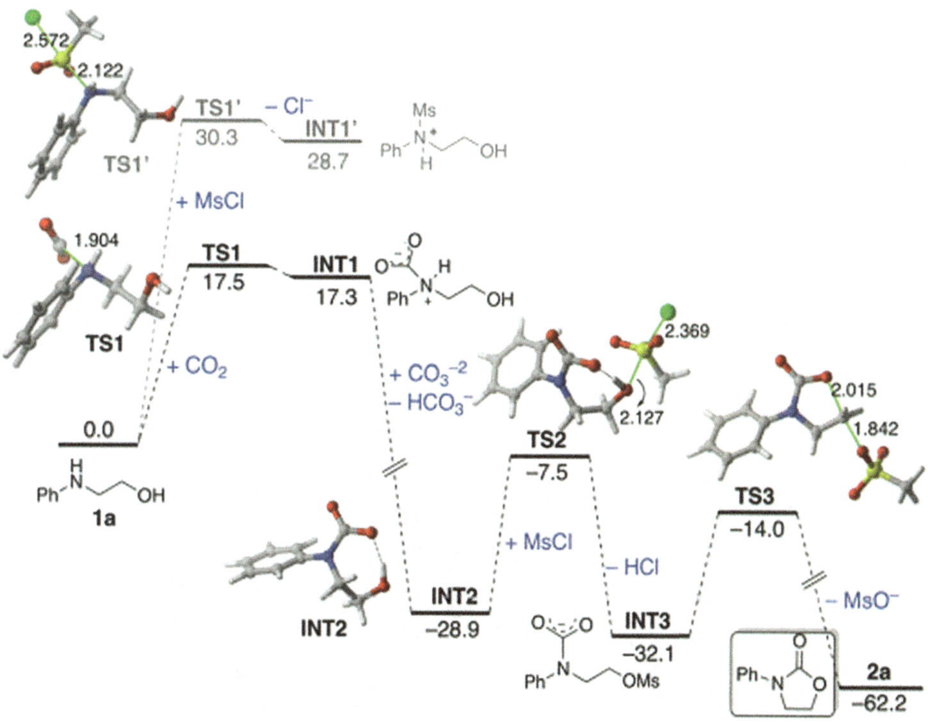

The DFT study was carried out at the M06-2X/6-31G(d) level, in which some quantum chemistry descriptors were analyzed to justify the obtained kinetic parameters. Mechanisms A and B were investigated for the reaction. **In 1A** in the mechanism A is produced by the nucleophilic attack of 2-aminobenzonitrile to CO_2 and proton adsorption by the SB. Step 2 in the mechanism A is an intramolecular nucleophilic attack to the CN group, yielding **In 2A**. The formation of isocyanate and the iminol groups in **In 3A** via intramolecular rearrangement is the output of step 3. Finally, due to the tautomerization of the iminol group to amide and another intramolecular nucleophilic attack, the final product is produced. However, mechanism B is started by the nucleophilic attack of SBs to CO_2, and the obtained SB^+–COO^- intermediates act as the base through the reaction. In step 2 of this mechanism, via another nucleophilic addition reaction, SB molecule is liberated, and **In 1B** is formed. The other stages of this mechanism are the same as mechanism A; however, the SB^+–COO^- plays as a base. Figure 74 illustrates the mechanisms A and B in the presence of the SBs as the organocatalysts (Sabet-Sarvestani et al. 2017b).

The theoretical results show that mechanism B passes through higher barrier energy than mechanism A. Thus, along with the reaction, the basic character of the SBs is more dominant than the nucleophilic character. The quantum theory of atoms in molecules (QTAIM) (Bader 1990; Henkelman et al. 2006) and NBO analysis (Sabet-Sarvestani et al. 2014) are two powerful quantum chemistry descriptors

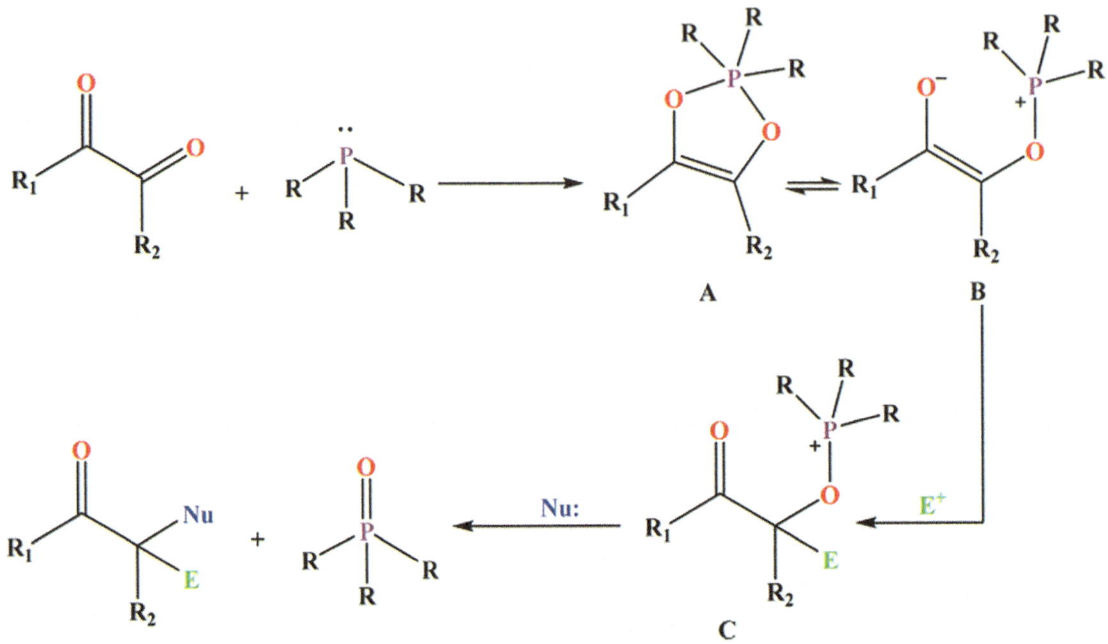

Fig. 66 Kukhtin–Ramirez addition via trivalent phosphorus derivatives (Sabet-Sarvestani et al. 2017a)

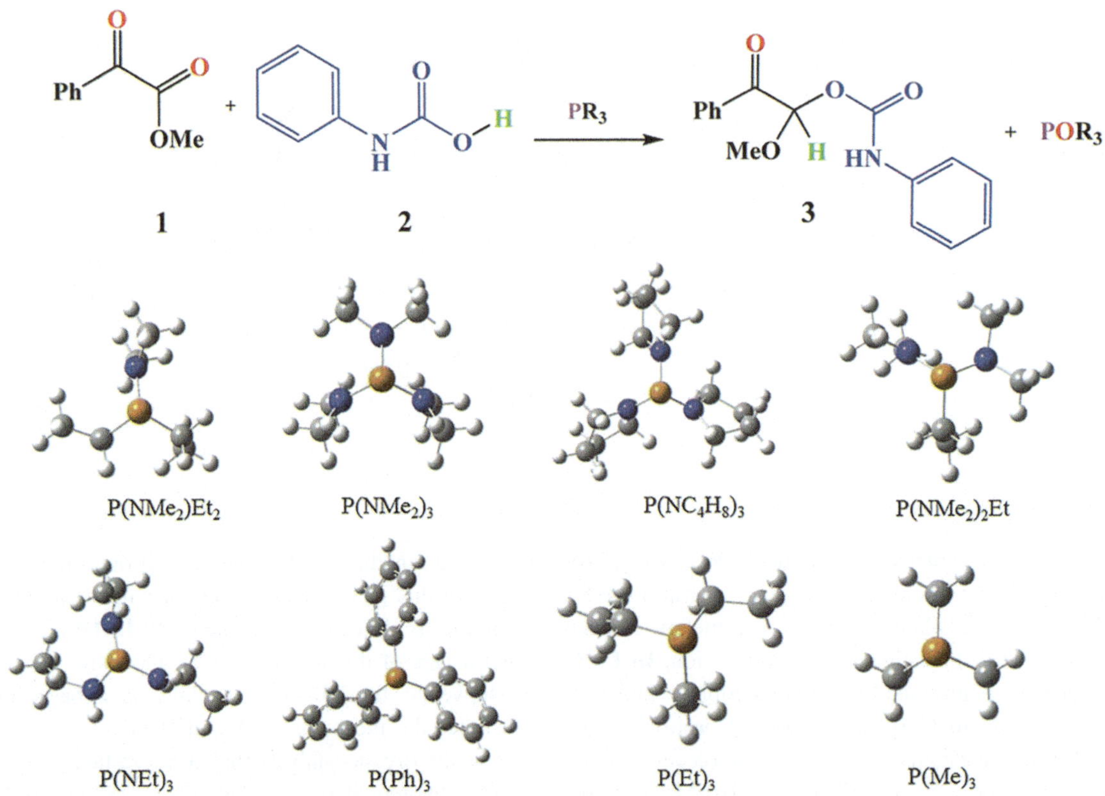

Fig. 67 Overall reaction and studied trivalent phosphorus in this reaction (Sabet-Sarvestani et al. 2017a)

to describe the behavior of the SBs in mechanism A. In this mechanism, the transition state of step 1 includes the N–H bond diminishing and a new bond developing between the dissociated proton of the N–H bond. Figure 75 depicts the obtained QTAIM graph, the bond critical points (BCPs), and the natural atomic charge of nitrogen atom corresponding to studied SBs at TS1. According to the results, when DBU is the involved base, the H–N developing bond and N–H

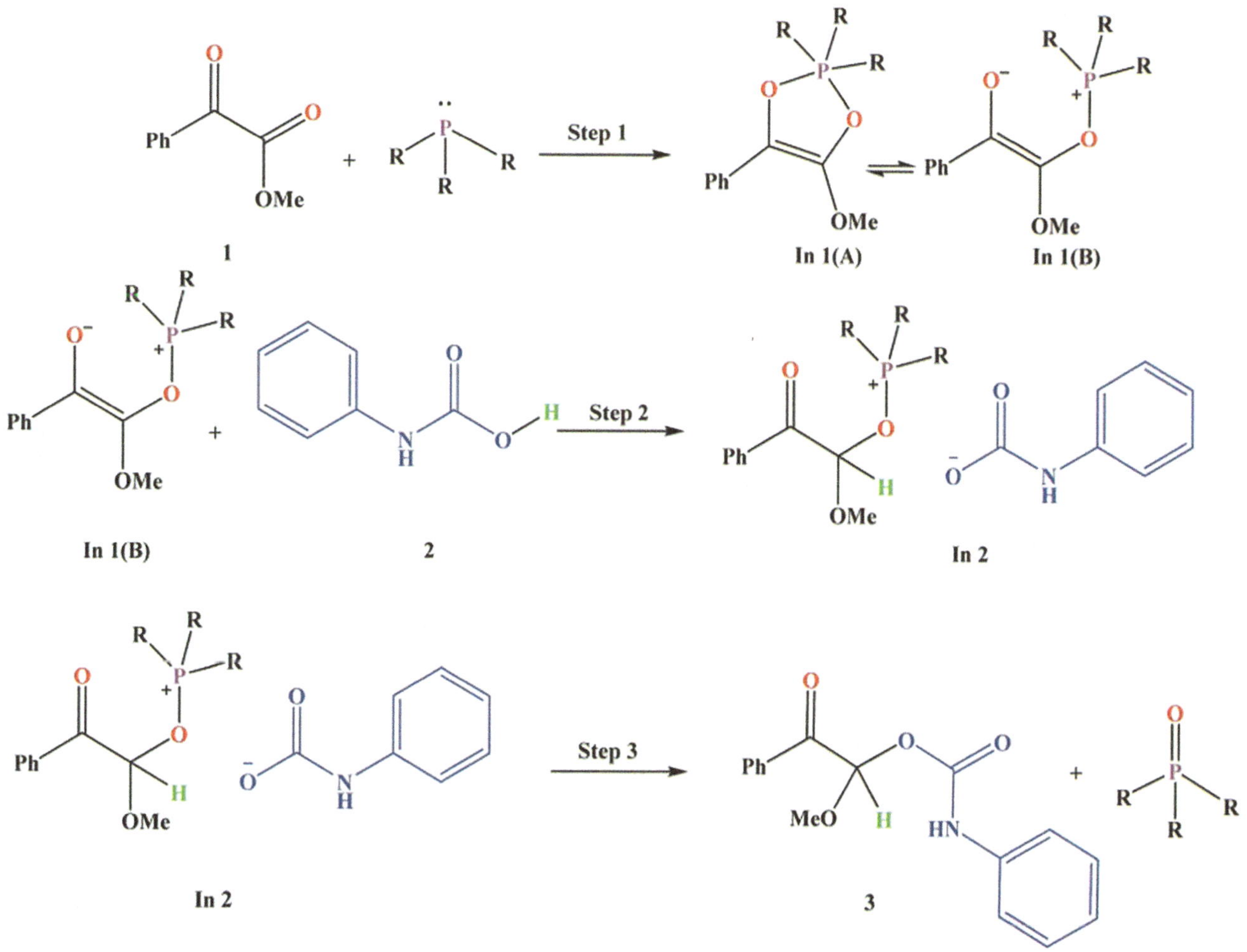

Fig. 68 Proposed mechanism of Kukhtin–Ramirez addition (Sabet-Sarvestani et al. 2017a)

Fig. 69 Formation of quinazoline-2,4(1H,3H)-dione derivatives in the presence of Pd (II)EN@GO (Biswas et al. 2020)

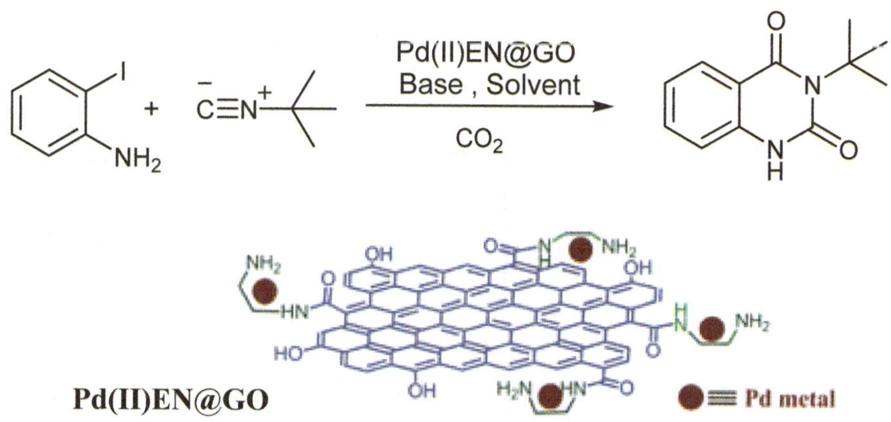

diminishing bond possess the highest and lowest electron density values, respectively. Also, the natural atomic charge of the nitrogen atom of 2-aminobenzonitrile, when DBU has been used as the base, is the maximum value than the next SBs. Thus, DBU is a more effective base than the other SBs at the TS1 (Sabet-Sarvestani et al. 2017b).

4.1.6 Summary and Outlook

It is undeniable that DFT calculations can be applied in exploring new intermediates, designing the novel catalysts, and determining the reaction pathways. Obviously, identifying the reaction intermediate, reaction steps, RDS of a catalytic cycle and most favorable reaction pathway based

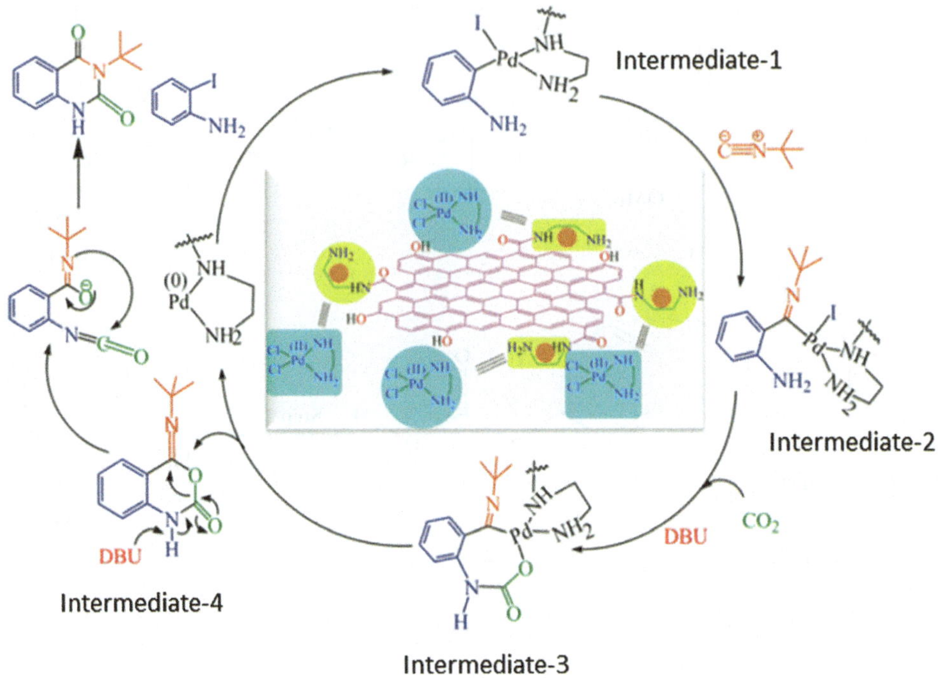

Fig. 70 Mechanism of quinazoline-2,4(1H,3H)-dione formation through a cycloaddition reaction using Pd(II)EN@GO and DBU as the catalyst and base, respectively (Biswas et al. 2020)

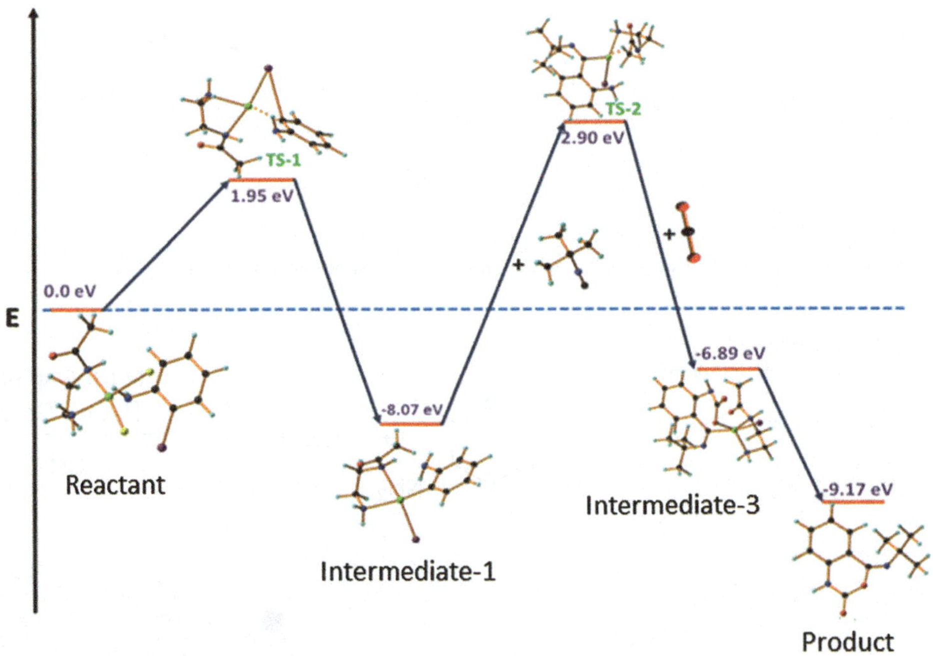

Fig. 71 PED of quinazoline-2,4 (1H,3H)-dione formation (Biswas et al. 2020)

on energy barrier for each step of the reaction has remarkable advantages of the DFT methods, which significantly improve our insights about the most favorable conditions for the reaction progress. However, it is noticeable that notwithstanding various theoretical reports, an obvious agreement on a unit reaction mechanism is not achieved by the researchers. Because there are not integrative opinions and principles, which can lead the theoretical and experimental researchers to the same outcomes. In the catalytic CO_2 transformation, the high chemical durability of the CO_2 molecule is another problem in the mechanism investigation. Indeed CO_2 activation is an essential step that shows a high chemical energy barrier in CO_2 conversions. Moreover, some issues such as active possible path in a given experimental condition and active site of the catalytic surfaces are the other problems in mechanism investigation of catalytic CO_2 transformations. Notably, computational costs are also serious problems in the mechanism studies. For a given

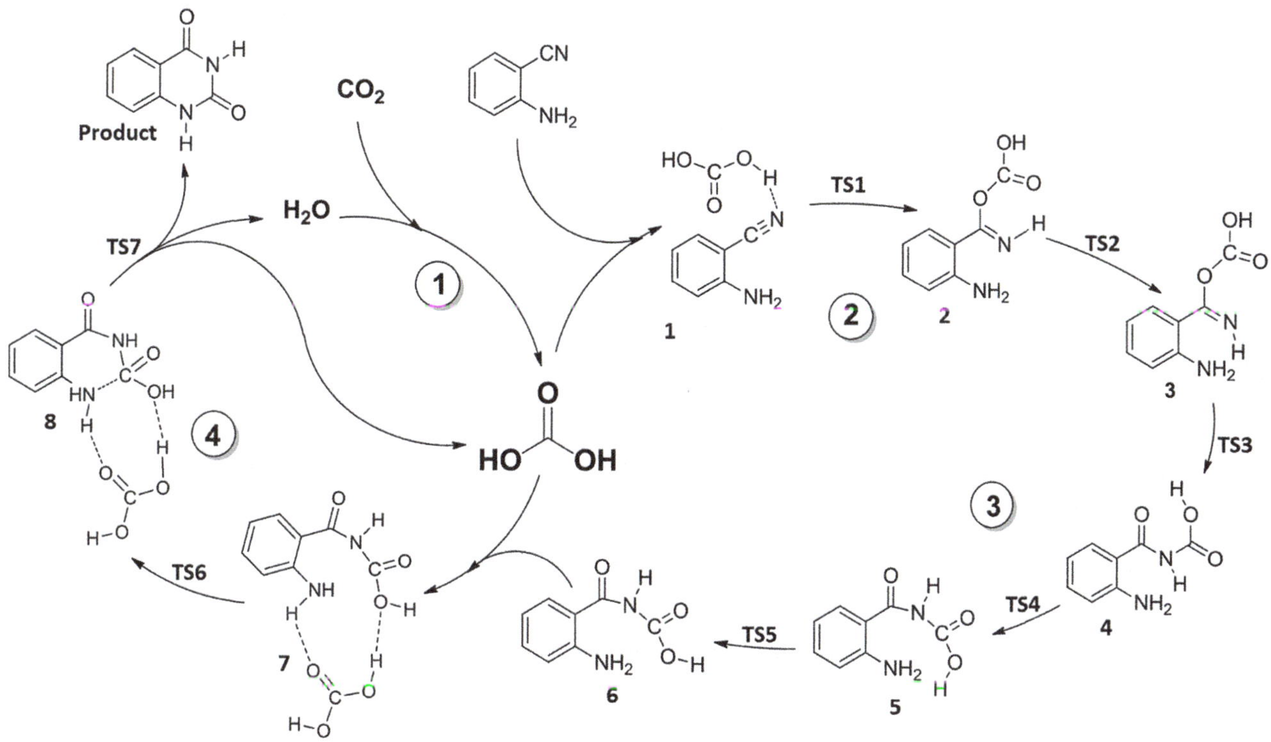

Fig. 72 Studied mechanism for the reaction of activated CO_2 molecule as H_2CO_3 in the synthesis of quinazoline-2,4(1H,3H)-diones (Ma et al. 2013)

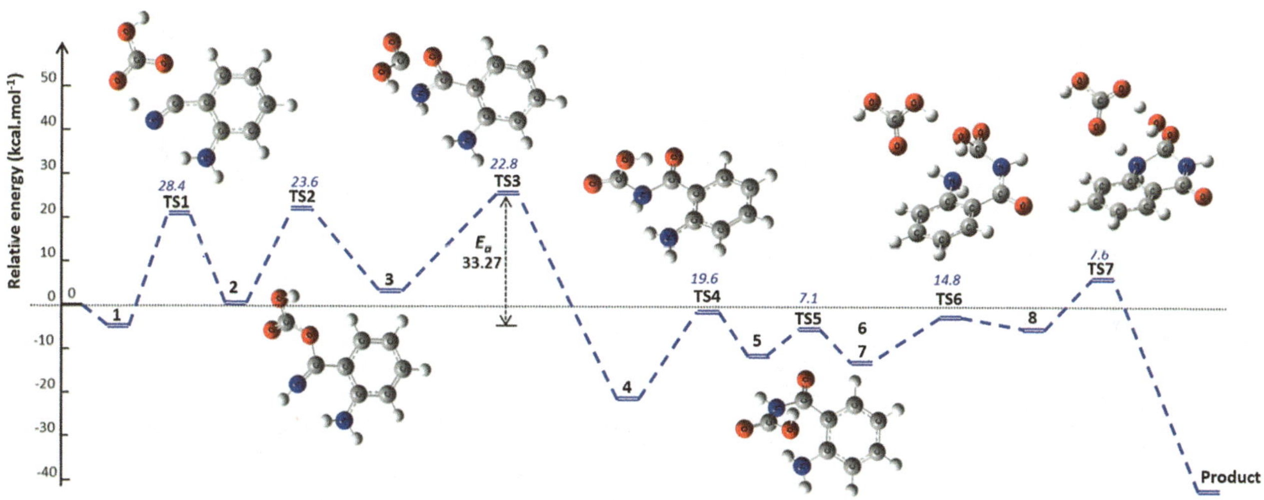

Fig. 73 PED of the studied mechanism of CO_2 and 2-aminobenzonitrile reaction in water (Ma et al. 2013)

accomplished reaction, the mechanism must be determined either by the comparison of the previously reported reactions, which is based on guesses or by systematical investigations based on computational approaches. Due to difficulties and being time-consuming, the computational expense of the calculation of activation energy and the actual pathway of the reaction are limiting issues to employing these approaches. Thus, it can be anticipated that new unexplored mechanism may exist, and more attempts should be carried out along with this aim in the future. However, some critical ignored subjects such as synergetic factors of multiple CO_2 adsorptions on a catalytic surface, ligand–

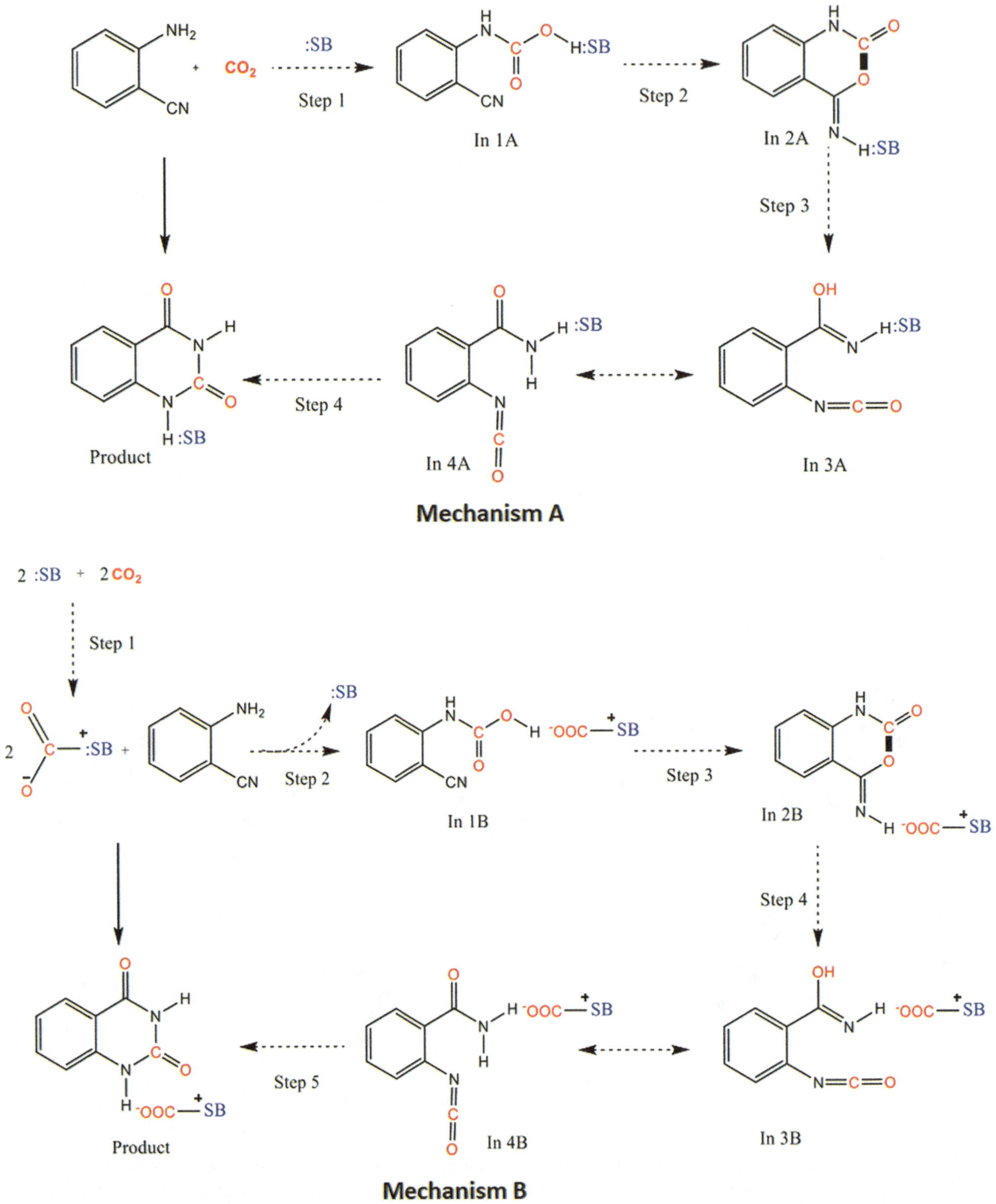

Fig. 74 Studied mechanisms A and B in the presence of SBs as organocatalysts in the formation of quinazoline-2,4(1H,3H)-diones (Sabet-Sarvestani et al. 2017b)

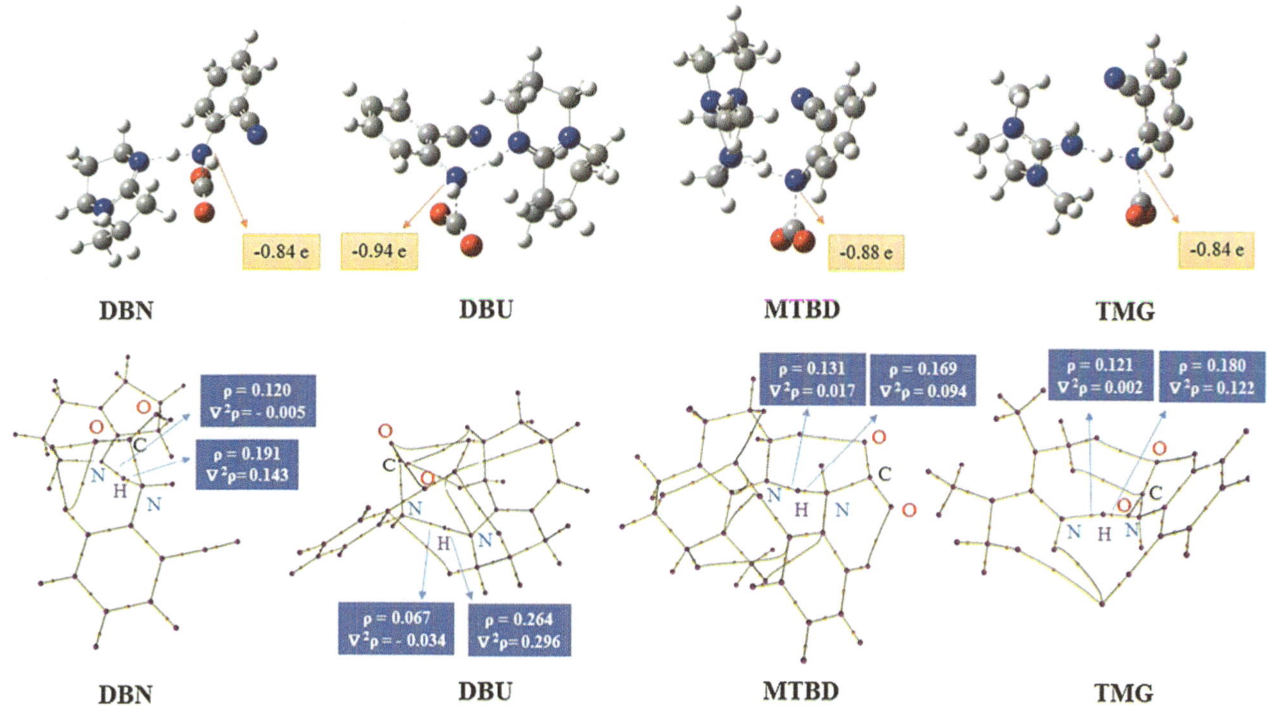

Fig. 75 QTAIM graph, the bond critical points (BCPs), and natural atomic charge of nitrogen atom corresponding to studied SBs at TS1 (Sabet-Sarvestani et al. 2017b)

ligand interactions on a typical catalyst, and unrealistic assumptions in the reaction conditions are effective factors on the accuracy of computational results.

5 Theoretical Designing of Novel Catalysts Based on DFT Studies

Despite defects such as being expensive and time-consuming, the main outcome of the previous section is that DFT calculations are efficient tools to investigate the mechanism of CO_2 transformation to value-added materials. Numerous clear potential advantages of DFT optimization of new catalysts and virtually designing novel procedures in the computer before the experimental evaluations can be considered (Streitwieser 2009). Against the experimental efforts, whenever a mechanism is explored based on DFT calculations, it is readily possible to make changes in the catalytic structures and reactant structures to evaluate the effect of these variations. Virtual exploring attempts with the aim of discovering novel catalysts in CO_2 conversion have been rapidly achieved popularity. Because of the importance of CO_2 transformation, mechanistic studies often require specific regards (Cheng et al. 2013). Besides the energies and optimized structures of the intermediates and transition states that cannot be achievable experimentally, approaches based on DFT provide a qualitative/quantitative investigation of the new optimized structures that are responsible for the observed kinetic and thermodynamic behaviors (Ahn et al. 2019). Although there are numerous reported catalysts in CO_2 transformation, the greater selectivity and reactivity of the catalysts are also considered as the main issues among the researchers (Sakakura et al. 2007; Wang et al. 2011; Centi and Perathoner 2009). Hence, catalyst designing based on the DFT calculations can be regarded as a helpful tool beyond tedious trial-and-error experimental approaches. The opportunities and challenges of this approach have been discussed in the following section.

Nørskov and coworkers are the pioneers in catalyst designing based on the DFT approach (Nørskov et al. 2009, 2011). Considering the explained principles, an intuition into the kinetic aspects of a catalytic reaction, including optimized structures, reaction energies, the energy barrier of each elementary step of the reaction are necessities in designing a catalyst based on the DFT approach. However, to obtain a convenient agreement with the experimental results, the kinetic results must be calculated by an appropriate computational technique (Andersson et al. 2006; Jones et al. 2008). In the kinetic investigation, energies of intermediates, reactants, products, energy barriers of the elementary steps, and the energy of RDS are criteria to evaluate the catalytic activity and selectivity (Nørskov et al. 2009; Sehested et al. 2007). Moreover, quantum chemical

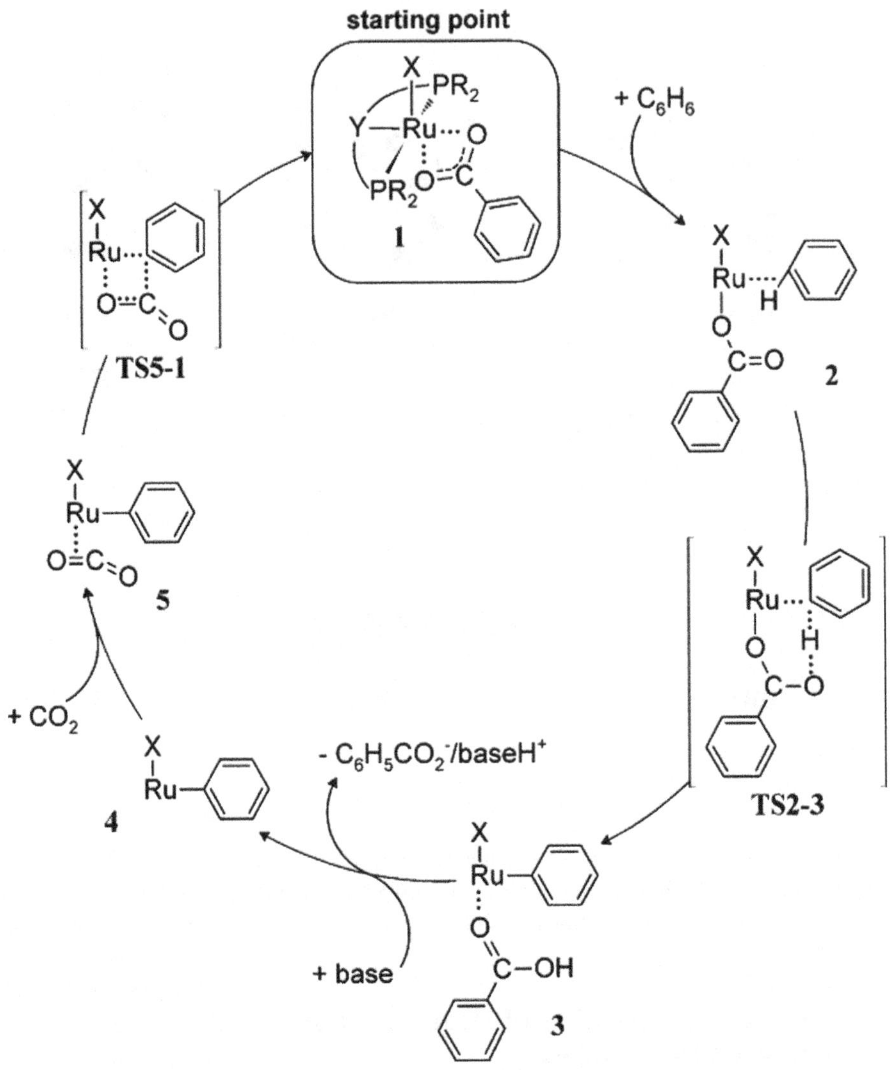

Fig. 76 Studied mechanism for the direct benzene carboxylation with CO_2, in which X^- corresponds to a monodentate anionic ligand (Uhe et al. 2012)

descriptors are good criteria in the description of the designed catalysts. In the following, some DFT-based designing of catalysts and prediction of their performances in CO_2 transformation are described.

Leitner and coworkers reported a DFT investigation on a catalytic cycle, progressed by ruthenium (II) pincer complexes as catalysts to the direct carboxylation of the C–H bonds of an arene with CO_2 (Uhe et al. 2012). They applied the ESM to evaluate the TOF of the designed catalysts and their efficiency in the studied mechanism, in which TOF values in the range of 10^{-5}–10^{-7} h^{-1} were predicted for the best systems. Moreover, for modification of the obtained results, the effects of the substrate, the solvent, and a base on the thermodynamic and kinetic aspects of the reaction were investigated, computationally. The DFT calculations carried out based on the Grimme's B97-D functional that considers the empirical dispersion correction, explicitly. Also, the def2-TZVP basis set (Weigend and Ahlrichs 2005) was employed for normal atoms and the associated ECPs for ruthenium and iodine atoms. Figure 76 depicts the studied mechanism for the direct benzene carboxylation by CO_2 (Uhe et al. 2012).

In the first step of the catalytic cycle, the benzene molecule is replaced by the oxygen atom of a coordinated benzoate ligand, yielding a C–H bond σ-complex (complex **2**). Then, through a formed six-membered transition state via σ-bond-metathesis type CH-activation step, Ru–C bond forms (complex **3**). Complex **4**, as a phenyl hydride complex, is the outcome of leaving the coordinated benzoic acid from complex **3**, which is accomplished by deprotonation via a base in the reaction mixture. Subsequently, the CO_2 molecule can link in a ^{2} mode, which results in compound **5**. Finally, the C–C bond and complex **1** are formed, due to passing the final step through TS5–1.

Two different complexes A and B having 2,6-bis (di-tert-butylphosphino) methylpyridine ligand, PNP(tBu), and 2,6-bis(diphenylphosphino)methylpyridine, PNP(Ph), respectively, were considered as the catalysts of the reaction

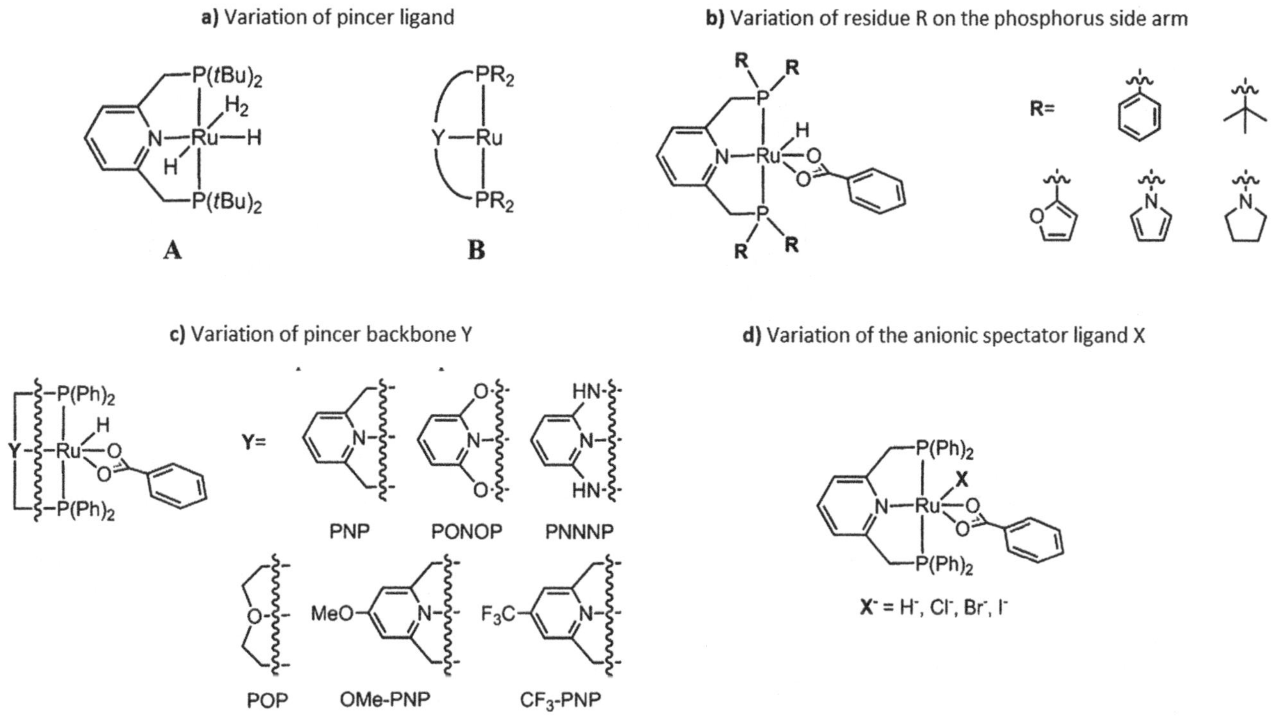

Fig. 77 Studied variations in the analysis of the TOF values of the direct carboxylation of the C–H bonds with CO_2 (Uhe et al. 2012)

Fig. 78 PED of the reaction in the presence of PNP (tBu) and PNP(Ph) ligands (Ge et al. 2016)

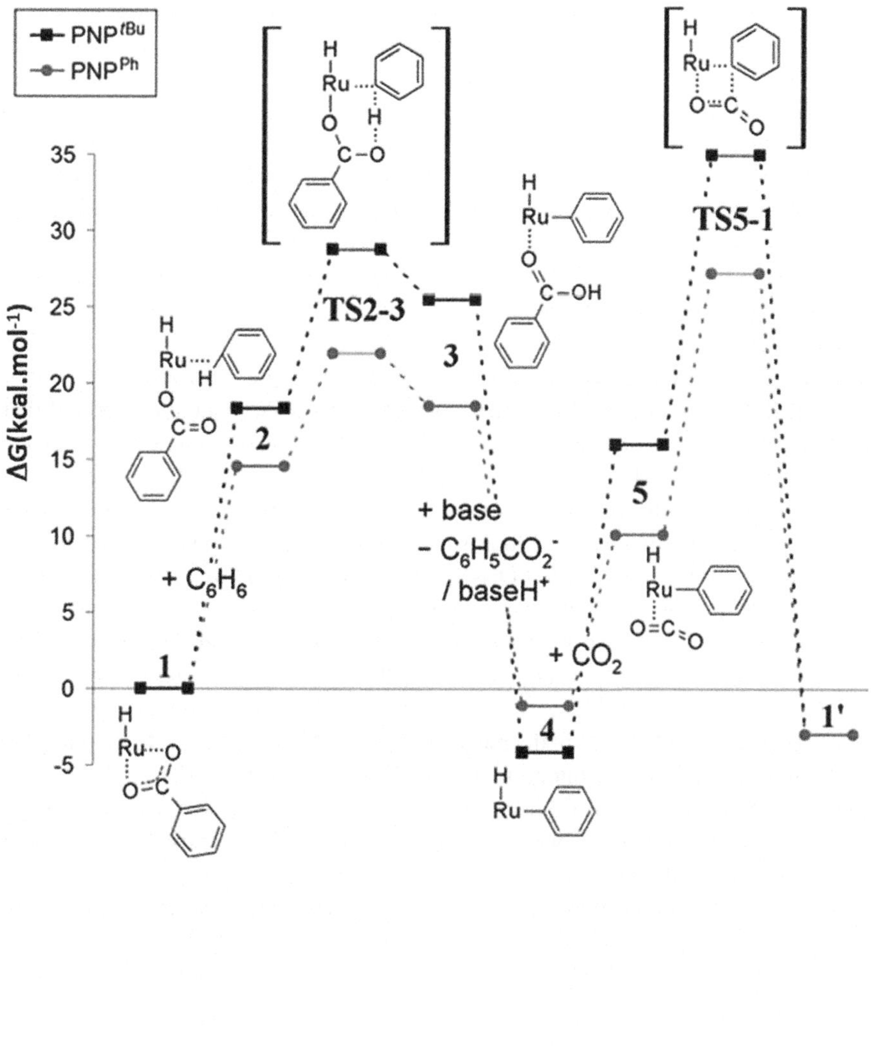

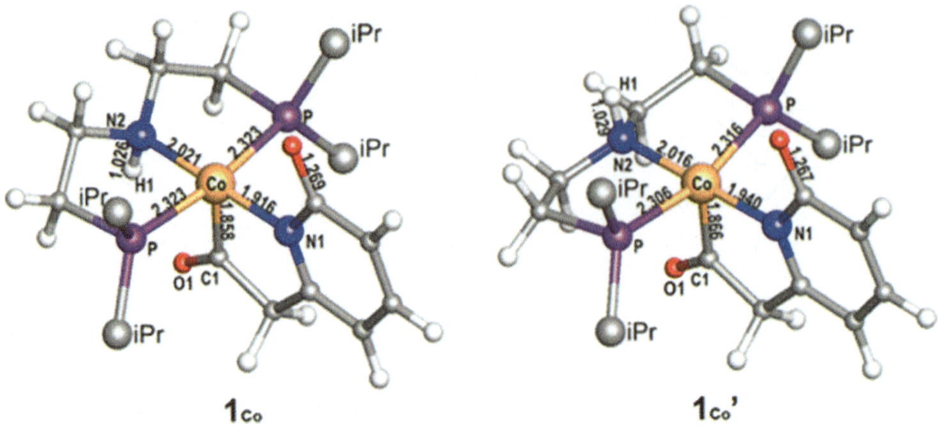

Fig. 79 Optimized structures of PNP cobalt pincer complexes as the designed catalysts in CO_2 conversion to formic acid (Ge et al. 2016)

(Fig. 77a). Figure 78 depicts the PED of the reaction in the presence of these catalysts. Considering the ESM, the calculated TOF values of PNP(tBu) and PNP(Ph) are 1.6×10^{-9} h^{-1} and 7.9×10^{-3} h^{-1}, respectively. This result shows that a small change in catalyst structure alters the TOF by six orders of magnitude. The next studied change is the outcome of the bonded substituents (R) to the phosphorous atom (Fig. 77b), in which the results show that the pyrrolidinyl group has the maximum increase in TOF = 8.6 h^{-1} relative to the phenyl group. The calculated results of TOF values corresponding to the variation of pincer backbone Y (Fig. 77c) show that the PONOP structure with an electron-releasing residue at the 4-position of the pyridine possesses the maximum TOF values (3.7×10^{-1} h^{-1}). The variation of the anionic ligand X^- (Fig. 77d) illustrates that the TOF quantity of the halide systems strongly related to the pK_B value of the applied base, and thus, the studies are carried out in trimethylamine and 2,7-substituted 1,8-bis (diethylamino) naphthalenes (BDN). TOF values of Cl^- and Br^- in the presence of DBN are larger than trimethylamine. Therefore, these anions as the X ligand would be the favorite selection for experimental investigations. Moreover, the calculated TOF values in the gas phase, water, acetonitrile, acetone, and anisole illustrate that acetone strongly affects the TOF (1.6×10^{-2} h^{-1}).

As the final result, the most favorable TOF can be provided by the conditions that include substitute R = 1-pyrrolidinyl, backbone Y = PONOP, having an electron-releasing group at the 4-position, X = bromide as an anionic ligand in composition with a powerful base, and acetone as the solvent of reaction. Based on these choices, synthetic conditions could be recognized that leading to efficient catalysts with TOFs in the range of 10^{-2} to 10^{-5} h^{-1}.

In another catalyst designing, a class of the PNP cobalt pincer complexes, inspired from the structure of the active site of [Fe]-hydrogenase, was reported for reversible base-free hydrogenation of CO_2 to formic acid (Ge et al. 2016). All calculations were carried out by applying the M06-2X/6–31 + + G(d,p) level of theory for all atoms, in which the solvent effects of water on the optimized structures were considered by the integral equation formalism polarizable continuum model (IEFPCM) with SMD atomic radii solvent effect corrections. Figure 79 depicts the optimized structures of 1_{Co} and its isomer $1_{Co'}$. Also, Fig. 80 illustrates the studied mechanism of the CO_2 hydrogenation (Ge et al. 2016).

The first step has been considered as a σ-complex formation between hydrogen molecule and cobalt complex, yielding intermediate **2**. Then, similar to the FLPs, the ortho oxygen atom of acylmethylpyridinol ligand in intermediate **2** assists the H_2 splitting by establishing a $Co{-}H^{\delta-}{\cdots}H^{\delta+}{-}O$ dihydrogen bond. Subsequently, intermediate **4** is produced through the CO_2 activation as the formate anion by releasing the hydride from the Co atom. This step is considered as the CO_2 insertion process, which is the RDS of the overall reaction. A formic acid molecule in intermediate **5** is produced due to the transformation of hydroxyl proton in acylmethylpyridinol to one of the oxygen atoms. Finally, the catalyst is regenerated by the release of HCOOH from **5**. Regarding the principle of microscopic reversibility, the authors reported that the designed catalyst could be applied as an effective catalyst for the hydrogen abstraction of formic acid in organic solvents with a similar energy barrier. Also, this DFT-based design and mechanism investigation produce favorable catalysts for the affordable and high-efficiency reduction of CO_2 and hydrogen abstraction of formic acid.

Mizuta and coworkers reported the CO_2 conversion to methoxyborane in the presence of the borane molecules and sodium borohydride as a reducing agent, experimentally (Fujiwara et al. 2014). We reported a DFT study, in which the role of other borohydride salts of $M[RBH_3]$ (M = Li^+, Na^+, and R = Ph, CH_3, CN, and H), as the reductive reagent, was evaluated (Fig. 81). The optimization of structures was carried out by the B3LYP/6–311 + + G(d,p) level of the theory, and the effect of tetrahydrofuran (THF) as the solvent of the reaction was considered by the CPCM model.

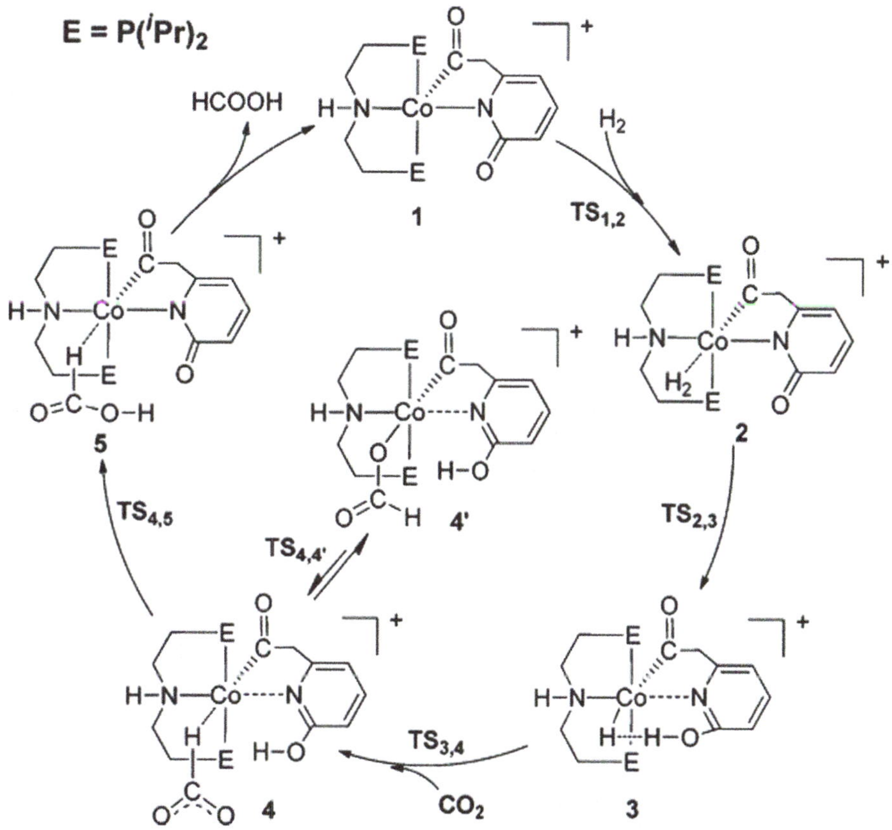

Fig. 80 Studied mechanism of the CO_2 hydrogenation by PNP cobalt pincer complex (Ge et al. 2016)

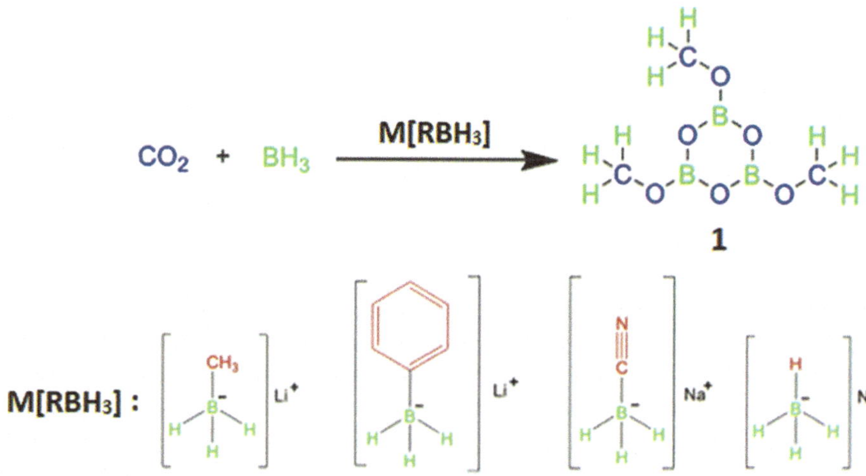

Fig. 81 Studied borohydride salts in the CO_2 reduction reaction (Sabet-Sarvestani et al. 2016)

Figure 82 depicts the studied mechanism for the CO_2 reduction reaction (Sabet-Sarvestani et al. 2016).

In 3 is the product of the sequential CO_2 reduction from steps 1 to 3 by hydride ions of the studied salts. In this intermediate, three hydride ions of the salt are replaced by three units of the formate ion. In step 4, **In 5** (formate salt) and **In 4** are made by the cleavage of the B–O bond of **In 3**. Then, **In 5** is reduced by a borane molecule in step 5. Regarding the role of **In 6** as the reducing agent, in step 6, the obtained formaldehyde molecule (**In 7**) is reduced to **In 8**, having a linked methoxy group to the boron atom. Steps 7 and 8 proceed through the oxidant role of **In 4**, in which other methoxy groups are formed. Finally, in step 9, due to hydride ion absorption by **In 4** and a ring-closing reaction, $(MeOBO)_3$ is formed (Sabet-Sarvestani et al. 2016).

Based on the roles of the linked substituents to the borohydride salts, the steps of reaction can be classified into

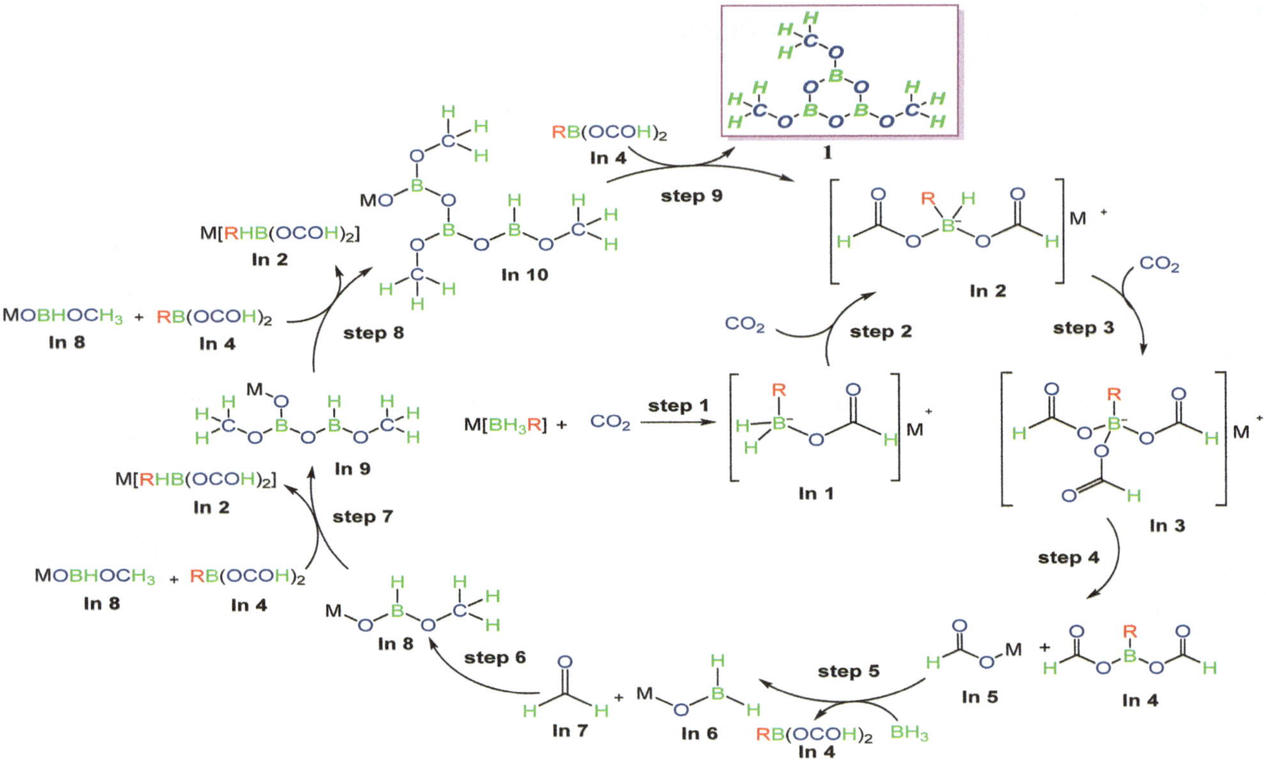

Fig. 82 Proposed mechanism for the CO_2 reduction (Sabet-Sarvestani et al. 2016)

three categories. Category 1 consists of steps 1–3, in which $M[RB(OCHO)_3]$ is produced due to the replication of hydride ions by formate ion. The results show that steps 5 and 6, classified as category 2, are independent of the substituent groups. These steps are similar for all considered borohydride salts. The steps of category 3 include the sequential oxidation of **In 4**, which leads to the ultimate product with three methoxy groups, and **In 2** category includes steps 7–9. According to the calculated data, the reactions of categories 1 and 3 are influenced by the R substituents, only. Based on the theoretical results, the trend of $\Delta G^{\neq}$ values in category 1 for the substituted moieties of the borohydride salt is CN > H > Ph > Me. Cyano and methyl substituents as the electron-withdrawing and electron-donating groups, respectively, have the highest and the lowest value of $\Delta G^{\neq}$ in category 1. Also, it can be concluded that in this category, $Li[MeBH_3]$ is the best reductive agent for these kinds of reactions, kinetically. However, in category 3, electron-withdrawing substituent increases the tendency of hydride absorption via the boron atom of **In 4**. When $Na[CNBH_3]$ is used, the obtained $\Delta G^{\neq}$ values are lower than the other reagents. Finally, from the kinetic and thermodynamic viewpoints of all steps for the general reaction, it is specified that $Li[MeBH_3]$ is the best borohydride salts for carbon dioxide transformation to methanol.

The role of the phosphorus ylides (P-ylides) chemistry in the novel synthetic protocols is undeniable. P-ylides are used in numerous reactions, mainly in the formation of natural products, biological and pharmacological compounds (Kolodiazhnyi 2008). In another DFT-based investigation, a new class of the P-ylides was designed for CO_2 activation in which B3LYP/6–31 + G(d,p) and B3LYP/6–311 + G (2d,2p) levels were employed for the optimization of the involved compounds. Figure 83 depicts the CO_2 activation reaction and the considered P-ylides (Sabet-Sarvestani et al. 2017c).

Based on the obtained results, P-ylides having the aliphatic chains linked to the phosphorous atom have lower ΔG values than the others. The minimum value of the molecular electrostatic potential (MESP) (Ajitha and Suresh 2012) was employed for justification of the calculated ΔG values. MESP, as a quantifiable physical character, can be measured experimentally by the X-ray diffraction tool or theoretically determined by the electron density (Ajitha and Suresh 2012). Figure 84 depicts the obtained MESP isosurfaces and the values of minimum MESP (V_{min}) at the bonded carbon atom to the phosphorous atom of P-ylides (C1). The results reveal that P-ylides having aliphatic chains, such as methyl, ethyl, and propyl, show the lower V_{min} values. Generally, in the CO_2 activation process, P-ylides having a more negative character of V_{min} possess the lower ΔG. Thus, $P(Me)_3CH_2$,

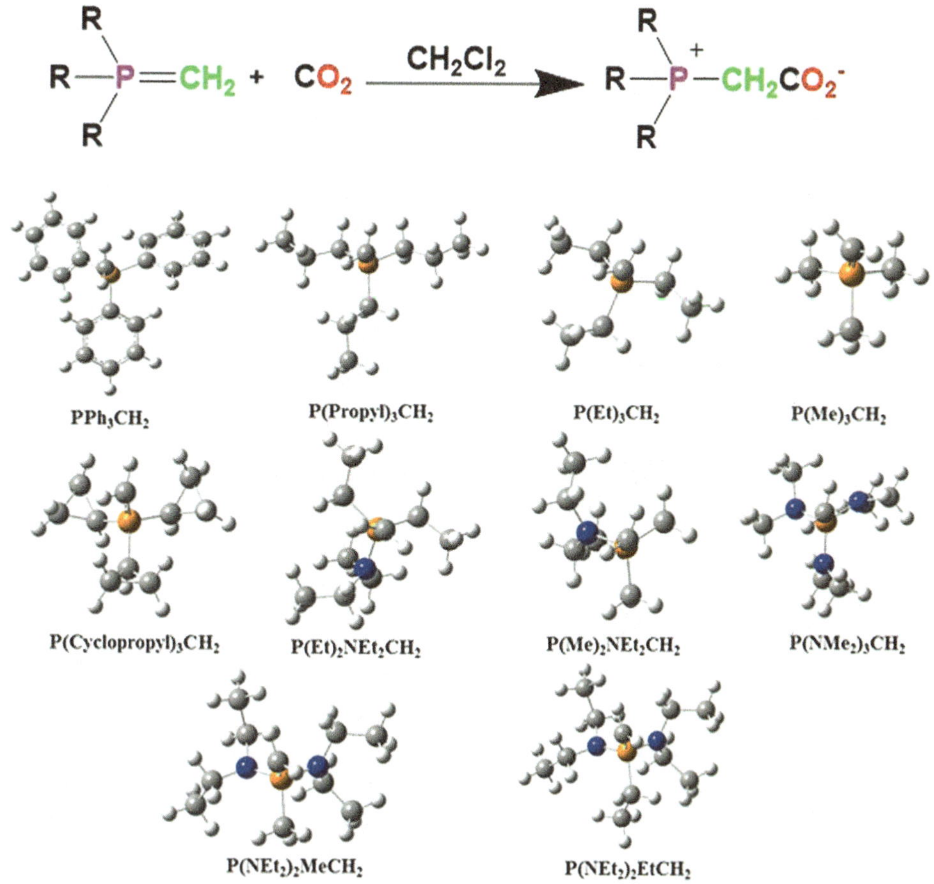

Fig. 83 CO_2 activation reaction and the considered P-ylides (Sabet-Sarvestani et al. 2017c)

$P(Et)_3CH_2$, and $P(Propyl)_3CH_2$ with the minimum values of V_{min} have the lowest ΔG. As a result, among the studied P-ylides, this class is the most efficient in the CO_2 activation.

Experimental investigations show that $PPh_3CH_2CO_2$ adduct, as the product of CO_2 activation by P-ylides, can be used as an organocatalyst in CO_2 conversion to cyclic α–alkylidene carbonates, oxazolidinone, and N-methylated and N-formylated amines (Zhou et al. 2015). These investigations reveal that in CO_2 conversion, $PPh_3CH_2CO_2$ adduct is a more prosperous catalyst than PPh_3CH_2 ylides. It is clear that the charge densities of the oxygen atoms of the P-ylide-CO_2 adducts are more localized and have a lower interaction with the near orbitals than the carbon atom of P-ylides (C1). Figure 85 depicts the MESP isosurfaces and V_{min} values (a.u) at the nucleophilic oxygen atom corresponding to the $P(Me)_3CH_2CO_2$, $P(Et)_3CH_2CO_2$, $P(Propyl)_3CH_2CO_2$, and $PPh_3CH_2CO_2$. As a result, the oxygen atom of P-ylide-CO_2 adducts possesses a greater negative character of the V_{min}, which makes them stronger nucleophilic catalysts than the P-ylides. Finally, thermodynamic and kinetic data show that P-ylides having the aliphatic chains, including $P(Me)_3CH_2$ and $P(Et)_3CH_2$, are the best choices for CO_2 transformation.

5.1 Theoretical Designing: Problems and Opportunities

Several studies involving mechanism investigations and DFT-based designing have been presented in this chapter. Nowadays, computational molecular modeling and catalyst designing have matured and become an inseparable part of investigations. The developments in creating a realistic model for exploring the complex mechanisms and catalytic cycles have been significantly increased, and the field has been grown considerably. Regarding many reported and successful instances, it can be concluded that computational techniques can be applied in an anticipating sense to the logical design and optimization of novel categories of efficient catalysts (Ahn et al. 2019). However, the investigation of numerous possibilities in a DFT-based design is a challenge ahead.

In a catalytic CO_2 transformation, depending on the given experimental conditions, various reaction paths might be involved and compete with each other. Thus, kinetic investigation for all possible paths, linking between the composition factors of the reactant, size, and the shape of the designed catalyst to predict the optimal catalytic conditions,

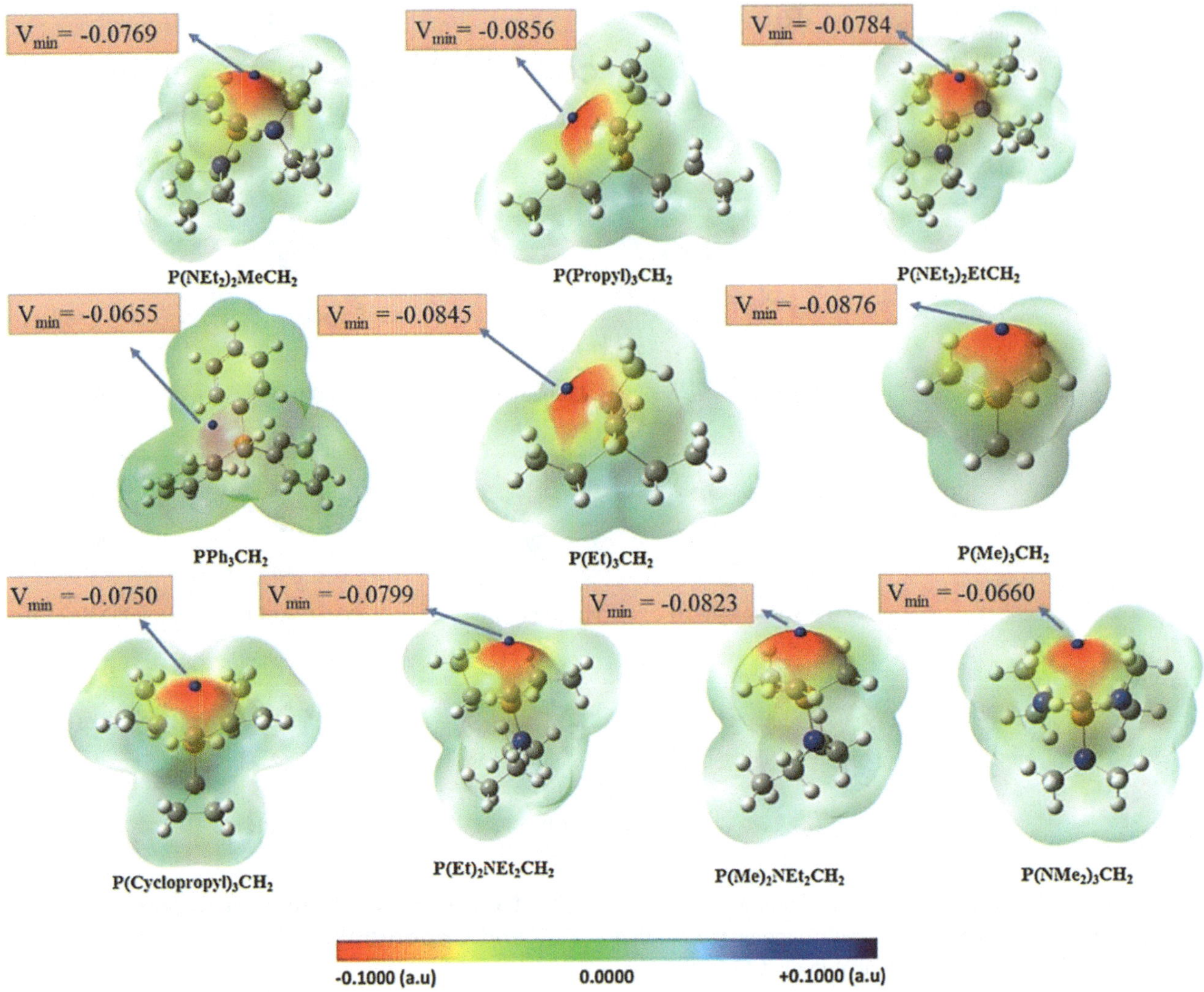

Fig. 84 Calculated MESP isosurfaces and corresponding $V_{\min}$ values (a.u) at the C1 atom of P-ylides (Sabet-Sarvestani et al. 2017c)

needs a perfect evaluation. Hence, the obtained DFT predictions need to be validated by a complete experimental procedure (Bell et al. 2008). Unfortunately, the validation of the DFT-based design has been investigated very partially in the experimental reports, and more comprehensive investigations are remarkably required.

In an experimental study, the catalyst structures are critically significant for the design of a new catalytic cycle. Also, in a typical theoretical design, the prediction of the properties of the structure, size, composition, and shape of the catalysts has important in successful designing (Bell et al. 2008; Guo and Wang 2011; Cuenya 2010). Thus, the reported experimental catalytic structures based on characterization techniques, such as vibrational spectroscopy, X-ray crystallography, nuclear magnetic resonance (NMR), X-ray photoelectron spectroscopy (XPS), scanning electron microscopy (SEM) and transmission electron microscopy (TEM) are good inspirational criteria for DFT-based catalytic design. The combination of the developed theoretical and experimental characterization techniques will finally produce new visions into the design of more efficient catalysts for CO_2 transformation (Somorjai and Li 2010).

The efficiency and the cost of the catalyst are also important issues in catalyst efficiency. In a DFT design, higher activity and lower cost must be considered in exploring more economical and efficient pathways. Especially, regarding industrial applications, it would be crucial to design non-expensive-metal catalysts instead of expensive-metal ones. Considering the global issue that the catalysts are used to solve, cost-effectiveness and higher activity are critical purposes in exploring and designing more economical and efficient paths. Moreover, the selectivity of the catalyst is the next important issue. This means that an increase in the yield of the reaction, energy

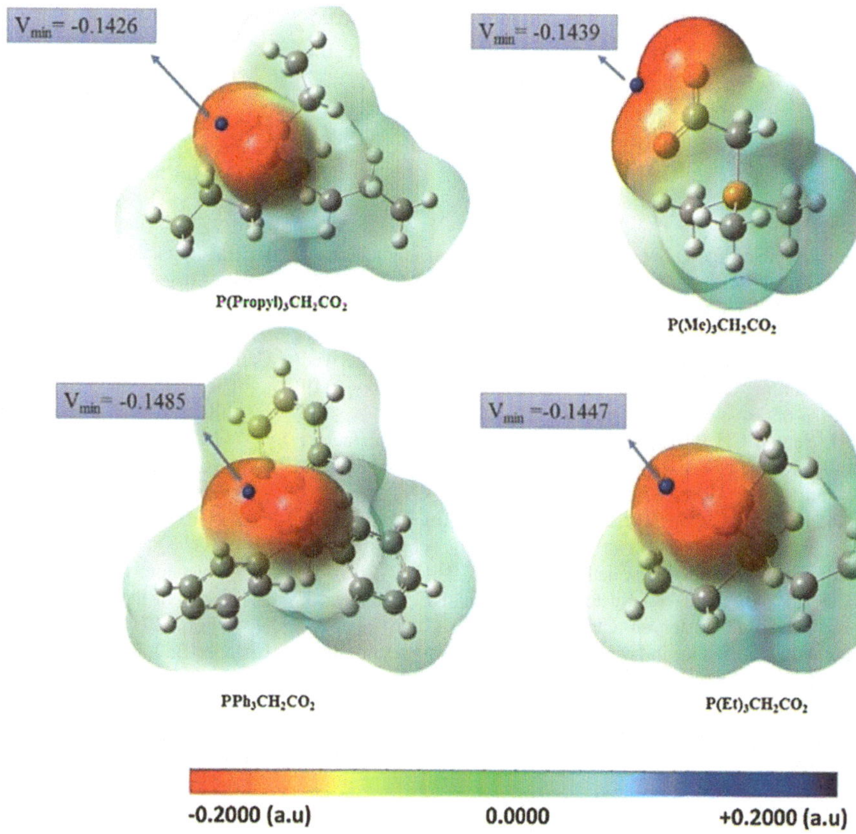

Fig. 85 MESP isosurfaces and the corresponding V_{min} values (a. u) at the nucleophilic oxygen atom of P-ylide-CO_2 adducts (Sabet-Sarvestani et al. 2017c)

efficiency, and higher selectivity decreases the operating costs of the catalytic cycle and produced by-products. Thus, the designing of the catalysts having efficient active sites is a method to increase the selectivity.

6 Conclusion

Global warming is a serious problem for humanity, and the CO_2 transformation into fuels and value-added chemicals is a valuable approach for the limitation of atmospheric CO_2 as a greenhouse gas. Furthermore, CO_2 transformation can be considered as a cost-effective method for the manufacture of hydrocarbons and thus to attain sustainable CO_2 recycling. Unfortunately, experimental catalytic cycles have drawbacks, such as low activity and selectivity of the applied catalysts. These issues are the main challenges in the CO_2 utilization process. Accordingly, deep insights into conversion processes and novel catalyst evaluations are an undeniable requirement in this field. It can be concluded that understanding the reaction mechanisms of the catalytic conversion and the logical design of efficient catalysts are two essential concepts in CO_2 conversions. Experimental approaches, based on trial-and-error methodologies, are very tedious, time-consuming, and frustrating. DFT-based techniques are promising strategies to overcome the disadvantages of the experimental approaches which could significantly improve the experimental sampling.

In this chapter, recent developed computational studies on the catalytic CO_2 transformation to value-added materials, such as CO, CH_3OH, CH_4, HCOOH, and heterocyclic compounds, have been considered. The investigated catalysts include heterogeneous and homogeneous forms, organocatalysts, and also, the photo- and electro- catalysts. DFT calculations are a successful tool in specifying intermediates, exploring the efficient catalysts, and finding the probable reaction paths. Generally, the goals of a DFT study can be considered as an investigation of the kinetic aspects of a catalytic cycle, providing the optimized structures of the catalyst and helpful descriptors to explore the activity and selectivity of a catalyst. The validity of DFT results can be evaluated through the characterization of the catalytic structures and performing a controlled synthesis.

It can be anticipated that the future of using computational techniques in CO_2 conversion reactions is brighter than ever. Improved software, availability of the required hardware, and development in the efficient electronic structure programs based on modern quantum chemistry are the effective factors in this field.

References

Abeydeera UW, Hewage L, Wadu Mesthrige J, Samarasinghalage TI (2019) Global Research on Carbon Emissions: A Scientometric Review. Sustainability 11(14):3972

Ahn S, Hong M, Sundararajan M, Ess DH, Baik M-H (2019) Design and optimization of catalysts based on mechanistic insights derived from quantum chemical reaction modeling. Chem Rev 119 (11):6509–6560

Ajitha MJ, Suresh CH (2012) Assessment of stereoelectronic factors that influence the CO_2 fixation ability of N-heterocyclic carbenes: A DFT study. J Org Chem 77(2):1087–1094

Álvarez A, Borges M, Corral-Pérez JJ, Olcina JG, Hu L, Cornu D, Huang R, Stoian D, Urakawa A (2017) CO_2 activation over catalytic surfaces. ChemPhysChem 18(22):3135–3141

Andersson MP, Bligaard T, Kustov A, Larsen KE, Greeley J, Johannessen T, Christensen CH, Nørskov JK (2006) Toward computational screening in heterogeneous catalysis: pareto-optimal methanation catalysts. J Catal 239(2):501–506

Aoyagi N, Furusho Y, Endo T (2013) Effective synthesis of cyclic carbonates from carbon dioxide and epoxides by phosphonium iodides as catalysts in alcoholic solvents. Tetrahedron Lett 54 (51):7031–7034

Aresta M (2010) Carbon dioxide as chemical feedstock, Wiley Online Library

Aresta M, Dibenedetto A (2007) Utilisation of CO_2 as a chemical feedstock: opportunities and challenges. Dalton Trans 28:2975–2992

Aresta M, Nobile CF, Albano VG, Forni E, Manassero M (1975) New nickel–carbon dioxide complex: synthesis, properties, and crystallographic characterization of (carbon dioxide)-bis (tricyclohexylphosphine) nickel. J Chem Soc. Chem Commun 15:636–637

Aresta M, Dibenedetto A, Quaranta E (2016) Reaction mechanisms in carbon dioxide conversion. Springer, Heidelberg

Artz J, Müller TE, Thenert K, Kleinekorte J, Meys R, Sternberg A, Bardow A, Leitner W (2018) Sustainable conversion of carbon dioxide: an integrated review of catalysis and life cycle assessment. Chem Rev 118(2):434–504

Aziz M, Jalil A, Triwahyono S, Ahmad A (2015) CO_2 methanation over heterogeneous catalysts: recent progress and future prospects. Green Chem 17(5):2647–2663

Richard F, Bader R (1990) Atoms in molecules: a quantum theory, Oxford University Press

Bartlett RJ, Musiał M (2007) Coupled-cluster theory in quantum chemistry. Rev Mod Phys 79(1):291

Bell AT, Gates BC, Ray D, Thompson MR (2008) Basic research needs: catalysis for energy. The U.S. Department of Energy's Office of Scientific and Technical Information

Ben-Nun M, Martínez TJ (2002) Ab initio quantum molecular dynamics. Adv Chem Phys 121:439–512

Bensaid S, Centi G, Garrone E, Perathoner S, Saracco G (2012) Towards artificial leaves for solar hydrogen and fuels from carbon dioxide. Chemsuschem 5(3):500–521

Bertau M, Offermanns H, Menges G, Keim W, Effenberger F (2010) Methanol needs more attention as a fuel and raw material for the future. Chem Ing Tech (weinh) 82(12):2055–2058

Biswas S, Khatun R, Dolai M, Biswas IH, Haque N, Sengupta M, Islam MS, Islam SM (2020) Catalytic formation of N3-substituted quinazoline-2, 4 (1H, 3H)-diones by Pd (ii) EN@ GO composite and its mechanistic investigations through DFT calculations. New J Chem 44(1):141–151

Bontemps S (2016) Boron-mediated activation of carbon dioxide. Coord Chem Rev 308:117–130

Brickner SJ (1996) Oxazolidinone antibacterial agents. Curr Pharm Des 2(2):175–194

RJ BS, Loganathan M, Shantha MS (2010) A review of the water gas shift reaction kinetics. Int J Chem React Eng 8(1):1–32

Calabrese C, Giacalone F, Aprile C (2019) Hybrid catalysts for CO_2 conversion into cyclic carbonates. Catalysts 9(4):325

Cao Z, Guo L, Liu N, Zheng X, Li W, Shi Y, Guo J, Xi Y (2016) Theoretical study on the reaction mechanism of reverse water–gas shift reaction using a Rh–Mo 6 S 8 cluster. RSC Adv 6 (110):108270–108279

Centi G, Perathoner S (2009) Opportunities and prospects in the chemical recycling of carbon dioxide to fuels. Catal Today 148(3–4):191–205

Centi G, Quadrelli EA, Perathoner S (2013) Catalysis for CO_2 conversion: a key technology for rapid introduction of renewable energy in the value chain of chemical industries. Energy Environ Sci 6(6):1711–1731

Chatelet B, Joucla L, Dutasta J-P, Martinez A, Szeto KC, Dufaud V (2013) Azaphosphatranes as structurally tunable organocatalysts for carbonate synthesis from CO_2 and epoxides. J Am Chem Soc 135 (14):5348–5351

Chen C-S, Cheng W-H, Lin S-S (2000) Mechanism of CO formation in reverse water–gas shift reaction over Cu/Al_2O_3 catalyst. Catal Lett 68(1–2):45–48

Chen Y, Hone CA, Gutmann B, Kappe CO (2017) Continuous flow synthesis of carbonylated heterocycles via Pd-catalyzed oxidative carbonylation using CO and O_2 at elevated temperatures and pressures. Org Process Res Dev 21(7):1080–1087

Chen J, Gao H, Ding T, Ji L, Zhang JZ, Gao G, Xia F (2019) Mechanistic studies of CO_2 cycloaddition reaction catalyzed by amine-functionalized ionic liquids. Front Chem 7:615

Cheng D, Negreiros FR, Apra E, Fortunelli A (2013) Computational approaches to the chemical conversion of carbon dioxide. Chemsuschem 6(6):944–965

Chiang C-L, Lin K-S, Chuang H-W (2018) Direct synthesis of formic acid via CO_2 hydrogenation over $Cu/ZnO/Al_2O_3$ catalyst. J Clean Prod 172:1957–1977

Choi S, Sang B-I, Hong J, Yoon KJ, Son J-W, Lee J-H, Kim B-K, Kim H (2017) Catalytic behavior of metal catalysts in high-temperature RWGS reaction: In-situ FT-IR experiments and first-principles calculations. Sci Rep 7:41207

Chorkendorff I, Niemantsverdriet JW (2017) Concepts of modern catalysis and kinetics. John Wiley & Sons

Cokoja M, Bruckmeier C, Rieger B, Herrmann WA, Kühn FE (2011) Transformation of carbon dioxide with homogeneous transition-metal catalysts: a molecular solution to a global challenge? Angew Chem Int Ed 50(37):8510–8537

Costentin C, Robert M, Savéant J-M (2013) Catalysis of the electrochemical reduction of carbon dioxide. Chem Soc Rev 42 (6):2423–2436

Cuenya BR (2010) Synthesis and catalytic properties of metal nanoparticles: Size, shape, support, composition, and oxidation state effects. Thin Solid Films 518(12):3127–3150

Dai Z, Stauffer PH, Carey JW, Middleton RS, Lu Z, Jacobs JF, Hnottavange-Telleen K, Spangler LH (2014) Pre-site

characterization risk analysis for commercial-scale carbon sequestration. Environ Sci Technol 48(7):3908–3915
Dang S, Yang H, Gao P, Wang H, Li X, Wei W, Sun Y (2019) A review of research progress on heterogeneous catalysts for methanol synthesis from carbon dioxide hydrogenation. Catal Today 330:61–75
Darensbourg DJ (2010) Chemistry of carbon dioxide relevant to its utilization: a personal perspective. Inorg Chem 49(23):10765–10780
Domingo LR, Pérez P, Sáez JA (2013) Understanding the local reactivity in polar organic reactions through electrophilic and nucleophilic Parr functions. RSC Adv 3(5):1486–1494
Dong X, Liu X, Chen Y, Zhang M (2018) Screening of bimetallic M-Cu-BTC MOFs for CO_2 activation and mechanistic study of CO_2 hydrogenation to formic acid: A DFT study. J CO_2 Util 24:64–72
Falivene L, Kozlov SM, Cavallo L (2018) Constructing bridges between computational tools in heterogeneous and homogeneous catalysis. ACS Catal 8(6):5637–5656
Fang S, Chen H, Wei H (2018) Insight into catalytic reduction of CO_2 to methane with silanes using Brookhart's cationic Ir (iii) pincer complex. RSC Adv 8(17):9232–9242
Filonenko GA, Vrijburg WL, Hensen EJ, Pidko EA (2016) On the activity of supported Au catalysts in the liquid phase hydrogenation of CO_2 to formates. J Catal 343:97–105
Fiorani G, Guo W, Kleij AW (2015) Sustainable conversion of carbon dioxide: the advent of organocatalysis. Green Chem 17(3):1375–1389
Frontera P, Macario A, Ferraro M, Antonucci P (2017) Supported catalysts for CO_2 methanation: a review. Catalysts 7(2):59
Fujiwara K, Yasuda S, Mizuta T (2014) Reduction of CO_2 to Trimethoxyboroxine with BH_3 in THF. Organometallics 33 (22):6692–6695
Galvan M, Selva M, Perosa A, Noè M (2014) Toward the design of halide-and metal-free ionic-liquid catalysts for the cycloaddition of CO_2 to epoxides. Asian J Org Chem 3(4):504–513
Gao J, He L-N, Miao C-X, Chanfreau S (2010) Chemical fixation of CO_2: efficient synthesis of quinazoline-2, 4 (1H, 3H)-diones catalyzed by guanidines under solvent-free conditions. Tetrahedron 66(23):4063–4067
Gattrell M, Gupta N, Co A (2007) Electrochemical reduction of CO_2 to hydrocarbons to store renewable electrical energy and upgrade biogas. Energy Convers Manag 48(4):1255–1265
Ge H, Jing Y, Yang X (2016) Computational design of cobalt catalysts for hydrogenation of carbon dioxide and dehydrogenation of formic acid. Inorg Chem 55(23):12179–12184
Gershikov A, Spiridonov V (1983) Anharmonic force field of CO_2 as determined by a gas-phase electron diffraction study. J Mol Struct 96(1–2):141–149
Girard A-L, Simon N, Zanatta M, Marmitt S, Gonçalves P, Dupont J (2014) Insights on recyclable catalytic system composed of task-specific ionic liquids for the chemical fixation of carbon dioxide. Green Chem 16(5):2815–2825
Guharoy U, Ramirez Reina T, Gu S, Cai Q (2019) Mechanistic Insights into Selective CO_2 Conversion via RWGS on Transition Metal Phosphides: A DFT Study. J Phys Chem C 123(37):22918–22931
Guo S, Wang E (2011) Noble metal nanomaterials: controllable synthesis and application in fuel cells and analytical sensors. Nano Today 6(3):240–264
Han X, Yang J, Han B, Sun W, Zhao C, Lu Y, Li Z, Ren J (2017) Density functional theory study of the mechanism of CO methanation on Ni4/t-ZrO_2 catalysts: roles of surface oxygen vacancies and hydroxyl groups. Int J Hydrog Energy 42(1):177–192
Henkelman G, Arnaldsson A, Jónsson H (2006) A fast and robust algorithm for Bader decomposition of charge density. Comput Mater Sci 36(3):354–360
Hofmann DJ, Butler JH, Tans PP (2009) A new look at atmospheric carbon dioxide. Atmospheric Environ 43(12):2084–2086
Hu B, Guild C, Suib SL (2013) Thermal, electrochemical, and photochemical conversion of CO_2 to fuels and value-added products. J CO_2 Util 1:18–27
Huang C-H, Tan C-S (2014) A review: CO_2 utilization. Aerosol Air Qual Res 14(2):480–499
Hunt AJ, Sin EH, Marriott R, Clark JH (2010) Generation, capture, and utilization of industrial carbon dioxide. Chemsuschem 3(3):306–322
Inoue Y, Izumida H, Sasaki Y, Hashimoto H (1976) Catalytic fixation of carbon dioxide to formic acid by transition-metal complexes under mild conditions. Chem Lett 5(8):863–864
Jing H, Li Q, Wang J, Liu D, Wu K (2018) Theoretical study of the reverse water gas shift reaction on copper modified β-Mo_2C (001) surfaces. J Phys Chem C 123(2):1235–1251
Jones G, Jakobsen JG, Shim SS, Kleis J, Andersson MP, Rossmeisl J, Abild-Pedersen F, Bligaard T, Helveg S, Hinnemann B (2008) First principles calculations and experimental insight into methane steam reforming over transition metal catalysts. J Catal 259(1):147–160
Joule JA, Mills K (2010) Heterocyclic chemistry, 5th (Ed.), John Wiley & Sons, Publication Ltd., Blackwell Publishing Ltd, West Sussex, UK
Kayaki Y, Yamamoto M, Ikariya T (2009) N-heterocyclic carbenes as efficient organocatalysts for CO_2 fixation reactions. Angew Chem Int Ed 48(23):4194–4197
Keim W, Offermanns H (2010) Beyong oil and gas: early visions of the natives of Aachen. Nachr Chem 58(4):434–435
Kheirabadi R, Izadyar M, Housaindokht MR (2018) Computational kinetic modeling of the catalytic cycle of glutathione peroxidase nanomimic: effect of nucleophilicity of thiols on the catalytic activity. J Phys Chem A 122(1):364–374
Kohn W (1999) Nobel Lecture: Electronic structure of matter—wave functions and density functionals. Rev Mod Phys 71(5):1253
Kohn W, Sham LJ (1965) Self-consistent equations including exchange and correlation effects. Phys Rev 140(4A):A1133
Kolodiazhnyi OI (2008) Phosphorus ylides: chemistry and applications in organic synthesis. John Wiley & Sons
Kozuch S, Shaik S (2011) How to conceptualize catalytic cycles? The energetic span model. Acc Chem Res 44(2):101–110
Kwon IS, Debela TT, Kwak IH, Seo HW, Park K, Kim D, Yoo SJ, Kim J-G, Park J, Kang HS (2019) Selective electrochemical reduction of carbon dioxide to formic acid using indium–zinc bimetallic nanocrystals. J Mater Chem A 7(40):22879–22883
Lapidus A, Gaidai N, Nekrasov N, Tishkova L, Agafonov YA, Myshenkova T (2007) The mechanism of carbon dioxide hydrogenation on copper and nickel catalysts. Pet Chem 47(2):75–82
Lee S-Y, Park S-J (2015) A review on solid adsorbents for carbon dioxide capture. J Ind Eng Chem 23:1–11
Liu Q, Wu L, Jackstell R, Beller M (2015) Using carbon dioxide as a building block in organic synthesis. Nat Commun 6:5933
Liu M, Yi Y, Wang L, Guo H, Bogaerts A (2019) Hydrogenation of carbon dioxide to value-added chemicals by heterogeneous catalysis and plasma catalysis. Catalysts 9(3):275
Luther Iii GW (2004) Kinetics of the reactions of water, hydroxide ion and sulfide species with CO_2, OCS and CS_2: frontier molecular orbital considerations. Aquat Geochem 10(1–2):81–97
Ma J, Sun N, Zhang X, Zhao N, Xiao F, Wei W, Sun Y (2009) A short review of catalysis for CO_2 conversion. Catal Today 148(3–4):221–231

Ma J, Hu J, Lu W, Zhang Z, Han B (2013) Theoretical study on the reaction of CO_2 and 2-aminobenzonitrile to form quinazoline-2, 4 (1 H, 3 H)-dione in water without any catalyst. Phys Chem Chem Phys 15(40):17333–17341

Ma S, Song W, Liu B, Zhong W, Deng J, Zheng H, Liu J, Gong X-Q, Zhao Z (2016) Facet-dependent photocatalytic performance of TiO_2: A DFT study. Appl Catal B 198:1–8

Maslin M (2008) Global warming: a very short introduction. Oxford University Press

Mikkelsen M, Jørgensen M, Krebs FC (2010) The teraton challenge. A review of fixation and transformation of carbon dioxide. Energy Environ Sci 3(1):43–81

Mizuno T, Okamoto N, Ito T, Miyata T (2000a) Synthesis of 2, 4-dihydroxyquinazolines using carbon dioxide in the presence of DBU under mild conditions. Tetrahedron Lett 41(7):1051–1053

Mizuno T, Okamoto N, Ito T, Miyata T (2000b) Synthesis of quinazolines using carbon dioxide (or carbon monoxide with sulfur) under mild conditions. Heteroat Chem 11(6):428–433

Mukherjee A, Okolie JA, Abdelrasoul A, Niu C, Dalai AK (2019) Review of post-combustion carbon dioxide capture technologies using activated carbon. J Environ Sci 83:46–63

Muradov N (2014) Industrial utilization of CO_2: a win–win solution. In: liberating energy from carbon: introduction to decarbonization. Springer, pp 325–383

Murphy LJ, Robertson KN, Kemp RA, Tuononen HM, Clyburne JA (2015) Structurally simple complexes of CO_2. Chem Commun 51 (19):3942–3956

Nakamura S, Hatakeyama M, Wang Y, Ogata K, Fujii K (2015) A basic quantum chemical review on the activation of CO_2. In: advances in CO_2 capture, sequestration, and conversion. ACS Publications, Chapter 5, pp 123–134

Nale DB, Saigaonkar SD, Bhanage BM (2014) An efficient synthesis of quinazoline-2, 4 (1H, 3H)-dione from CO_2 and 2-aminobenzonitrile using [Hmim] OH/SiO_2 as a base functionalized supported ionic liquid phase catalyst. J CO_2 Util 8:67–73

Niemi T, Repo T (2019) Antibiotics from carbon dioxide: sustainable pathways to pharmaceutically relevant cyclic carbamates. Eur J Org Chem 2019(6):1180–1188

Niemi T, Fernández I, Steadman B, Mannisto JK, Repo T (2018) Carbon dioxide-based facile synthesis of cyclic carbamates from amino alcohols. Chem Commun 54(25):3166–3169

Nørskov JK, Bligaard T, Rossmeisl J, Christensen CH (2009) Towards the computational design of solid catalysts. Nat Chem 1(1):37

Nørskov JK, Abild-Pedersen F, Studt F, Bligaard T (2011) Density functional theory in surface chemistry and catalysis. Proc Natl Acad Sci USA (PNAS) 108(3):937–943

IEA OECD (2016) Energy and air pollution: world energy outlook (Special Report)

Ola O, Maroto-Valer MM, Mackintosh S (2013) Turning CO_2 into valuable chemicals. Energy Procedia 37:6704–6709

Olah GA, Goeppert A, Prakash GS (2009) Chemical recycling of carbon dioxide to methanol and dimethyl ether: from greenhouse gas to renewable, environmentally carbon neutral fuels and synthetic hydrocarbons. J Org Chem 74(2):487–498

Olah GA, Goeppert A, Prakash GS (2018) Beyond oil and gas: the methanol economy. Third updated and enlarged edition, John Wiley & Sons (US)

Ou L, Chen J, Chen Y, Jin J (2019) Mechanistic study on Cu-catalyzed CO_2 electroreduction into CH_4 at simulated low overpotentials based on an improved electrochemical model. Phys Chem Chem Phys 21(28):15531–15540

Paparo A, Okuda J (2017) Carbon dioxide complexes: bonding modes and synthetic methods. Coord Chem Rev 334:136–149

Peng G, Sibener S, Schatz GC, Mavrikakis M (2012) CO_2 hydrogenation to formic acid on Ni (110). Surf Sci 606(13–14):1050–1055

Podrojkova N, Sans V, Orinak A, Orinakova R (2020) Recent developments in heterogeneous catalysts modelling for CO_2 conversion to chemicals. ChemCatChem 12(7):1802–1825

Pople JA (1999) Nobel lecture: Quantum chemical models. Rev Mod Phys 71(5):1267

Qiu M, Tao H, Li R, Li Y, Huang X, Chen W, Su W, Zhang Y (2016) Insight into the mechanism for the methanol synthesis via the hydrogenation of CO_2 over a Co-modified Cu (100) surface: A DFT study. J Chem Phys 145(13):134701

Rachuri Y, Kurisingal JF, Chitumalla RK, Vuppala S, Gu Y, Jang J, Choe Y, Suresh E, Park D-W (2019) Adenine-based Zn (II)/Cd (II) metal-organic frameworks as efficient heterogeneous catalysts for facile CO_2 fixation into cyclic carbonates: a DFT-supported study of the reaction mechanism. Inorg Chem 58(17):11389–11403

Rafiee A, Khalilpour KR, Milani D, Panahi M (2018) Trends in CO_2 conversion and utilization: a review from process systems perspective. J Environ Chem Eng 6(5):5771–5794

Rashidi NA, Yusup S (2016) An overview of activated carbons utilization for the post-combustion carbon dioxide capture. J CO_2 Util 13:1–16

Rawat KS, Mahata A, Pathak B (2017) Thermochemical and electrochemical CO_2 reduction on octahedral Cu nanocluster: role of solvent towards product selectivity. J Catal 349:118–127

Ren J, Guo H, Yang J, Qin Z, Lin J, Li Z (2015) Insights into the mechanisms of CO_2 methanation on Ni (111) surfaces by density functional theory. Appl Surf Sci 351:504–516

Riduan SN, Zhang Y (2010) Recent developments in carbon dioxide utilization under mild conditions. Dalton Trans 39(14):3347–3357

Sabet-Sarvestani H, Eshghi H, Bakavoli M, Izadyar M, Rahimizadeh M (2014) Theoretical investigation of the chemoselectivity and synchronously pyrazole ring formation mechanism from ethoxymethylenemalononitrile and hydrazine hydrate in the gas and solvent phases: DFT, meta-GGA studies and NBO analysis. RSC Adv 4(82):43485–43495

Sabet-Sarvestani H, Eshghi H, Izadyar M, Noroozi-Shad N, Bakavoli M, Ziaee F (2016) Borohydride salts as high efficiency reducing reagents for carbon dioxide transformation to methanol: Theoretical approach. Int J Hydrog Energy 41(26):11131–11140

Sabet-Sarvestani H, Eshghi H, Izadyar M (2017a) Understanding the mechanism, thermodynamic and kinetic features of the Kukhtin-Ramirez reaction in carbamate synthesis from carbon dioxide. RSC Adv 7(3):1701–1710

Sabet-Sarvestani H, Eshghi H, Izadyar M (2017b) A theoretical study on the efficiency and role of guanidines-based organic superbases on carbon dioxide utilization in quinazoline-2, 4 (1H, 3H)-diones synthesis. Struct Chem 28(3):675–686

Sabet-Sarvestani H, Izadyar M, Eshghi H (2017) Phosphorus ylides as a new class of compounds in CO_2 activation: Thermodynamic and kinetic studies. J CO_2 Util 21:459–466

Sabet-Sarvestani H, Izadyar M, Eshghi H, Noroozi-Shad N, Bakavoli M (2018) Proton sponge as a new efficient catalyst for carbon dioxide transformation to methanol: theoretical approach. Fuel 221:491–500

Sabet-Sarvestani H, Izadyar M, Eshghi H, Norozi-Shad N (2020) Evaluation and understanding the performances of various derivatives of carbonyl-stabilized phosphonium ylides in CO_2 transformation to cyclic carbonates. Phys Chem Chem Phys 22(1):223–237

Saeidi S, Amin NAS, Rahimpour MR (2014) Hydrogenation of CO_2 to value-added products—a review and potential future developments. J CO_2 Util 5:66–81

Saeidi S, Najari S, Fazlollahi F, Nikoo MK, Sefidkon F, Klemeš JJ, Baxter LL (2017) Mechanisms and kinetics of CO_2 hydrogenation to value-added products: a detailed review on current status and future trends. Renew Sust Energ Rev 80:1292–1311

Sakakura T, Choi J-C, Yasuda H (2007) Transformation of carbon dioxide. Chem Rev 107(6):2365–2387

Saxena R, Singh VK, Kumar EA (2014) Carbon dioxide capture and sequestration by adsorption on activated carbon. Energy Procedia 54:320–329

Schaadt R, Sweeney D, Shinabarger D, Zurenko G (2009) In vitro activity of TR-700, the active ingredient of the antibacterial prodrug TR-701, a novel oxazolidinone antibacterial agent. Antimicrob Agents Chemother 53(8):3236–3239

Schilling W, Das S (2018) CO_2-catalyzed/promoted transformation of organic functional groups. Tetrahedron Lett 59(43):3821–3828

Scibioh M, Viswanathan B (2018) Chapter 5—heterogeneous hydrogenation of CO_2. Carbon Dioxide Chem Fuels 191–253

Sehested J, Larsen KE, Kustov AL, Frey AM, Johannessen T, Bligaard T, Andersson MP, Nørskov JK, Christensen CH (2007) Discovery of technical methanation catalysts based on computational screening. Top Catal 45(1–4):9–13

Shih CF, Zhang T, Li J, Bai C (2018) Powering the future with liquid sunshine. Joule 2(10):1925–1949

Siqueira RM, Freitas GR, Peixoto HR, do Nascimento JF, Musse APS, Torres AE, Azevedo DC, Bastos-Neto M (2017) Carbon dioxide capture by pressure swing adsorption. Energy Procedia 114:2182–2192

Sirijaraensre J, Limtrakul J (2016) Hydrogenation of CO_2 to formic acid over a Cu-embedded graphene: A DFT study. Appl Surf Sci 364:241–248

Somorjai GA, Li Y (2010) Major successes of theory-and-experiment-combined studies in surface chemistry and heterogeneous catalysis. Top Catal 53(5–6):311–325

Song Q-W, Zhou Z-H, He L-N (2017) Efficient, selective and sustainable catalysis of carbon dioxide. Green Chem 19 (16):3707–3728

Spigarelli BP, Kawatra SK (2013) Opportunities and challenges in carbon dioxide capture. J CO_2 Util 1:69–87

Spinner NS, Vega JA, Mustain WE (2012) Recent progress in the electrochemical conversion and utilization of CO_2. Catal Sci Technol 2(1):19–28

Streitwieser A (2009) Perspectives on computational organic chemistry. J Org Chem 74(12):4433–4446

Su X, Yang X, Zhao B, Huang Y (2017) Designing of highly selective and high-temperature endurable RWGS heterogeneous catalysts: recent advances and the future directions. J Energy Chem 26 (5):854–867

Topham S, Bazzanella A, Schiebahn S, Luhr S, Zhao L, Otto A, Stolten D (2000) Carbon dioxide. Ullmann's encyclopedia of industrial chemistry, wiley-VCH Verlag GmbH & Co. KGaA, Weinheim, Germany

Tsutsumi Y, Yamakawa K, Yoshida M, Ema T, Sakai T (2010) Bifunctional organocatalyst for activation of carbon dioxide and epoxide to produce cyclic carbonate: betaine as a new catalytic motif. Org Lett 12(24):5728–5731

Uhe A, Hoelscher M, Leitner W (2012) Carboxylation of arene C. H Bonds with CO_2: aDFT-based approach to catalyst design. Chem–A Eur J 18(1):170–177

Wang SR, Radosevich AT (2013) Reductive homocondensation of benzylidene-and alkylidenepyruvate esters by a P (NMe2) 3-mediated tandem reaction. Org Lett 15(8):1926–1929

Wang S, Xi C (2019) Recent advances in nucleophile-triggered CO_2-incorporated cyclization leading to heterocycles. Chem Soc Rev 48 (1):382–404

Wang W, Wang S, Ma X, Gong J (2011) Recent advances in catalytic hydrogenation of carbon dioxide. Chem Soc Rev 40(7):3703–3727

Wang J-Q, Dong K, Cheng W-G, Sun J, Zhang S-J (2012) Insights into quaternary ammonium salts-catalyzed fixation carbon dioxide with epoxides. Catal Sci Technol 2(7):1480–1484

Wang Y-B, Wang Y-M, Zhang W-Z, Lu X-B (2013) Fast CO_2 sequestration, activation, and catalytic transformation using N-heterocyclic olefins. J Am Chem Soc 135(32):11996–12003

Wang L, Sun W, Liu C (2018a) Recent advances in homogeneous carbonylation using CO_2 as CO surrogate. Chinese J Chem 36 (4):353–362

Wang L, Ghoussoub M, Wang H, Shao Y, Sun W, Tountas AA, Wood TE, Li H, Loh JYY, Dong Y (2018b) Photocatalytic hydrogenation of carbon dioxide with high selectivity to methanol at atmospheric pressure. Joule 2(7):1369–1381

Wang W-H, Feng X, Bao M (2018) Transformation of CO_2 to formic acid or formate with homogeneous catalysts. In: transformation of carbon dioxide to formic acid and methanol. Springer, pp 7–42

Wei W, Jinlong G (2011) Methanation of carbon dioxide: an overview. Front Chem Sci Eng 5(1):2–10

Weigend F, Ahlrichs R (2005) Balanced basis sets of split valence, triple zeta valence and quadruple zeta valence quality for H to Rn: Design and assessment of accuracy. Phys Chem Chem Phys 7 (18):3297–3305

Wong WL, Chan PH, Zhou ZY, Lee KH, Cheung KC, Wong KY (2008) A robust ionic liquid as reaction medium and efficient organocatalyst for carbon dioxide fixation. Chemsuschem 1(1–2):67–70

Wu Z, Cole J, Fang HL, Qin M, He Z (2017) Revisiting catalyst structure and mechanism in methanol synthesis. J Adv Nanomater (JAN) 2(1):1–10

Wu P, Zaffran J, Yang B (2019) Role of surface species interactions in identifying the reaction mechanism of methanol synthesis from CO_2 hydrogenation over intermetallic PdIn (310) steps. J Phys Chem C 123(22):13615–13623

Yaashikaa P, Kumar PS, Varjani SJ, Saravanan A (2019) A review on photochemical, biochemical and electrochemical transformation of CO_2 into value-added products. J CO_2 Util 33:131–147

Yan X, Guo H, Yang D, Qiu S, Yao X (2014) Catalytic hydrogenation of carbon dioxide to fuels. Curr Org Chem 18(10):1335–1345

Yang Z-Z, He L-N, Zhao Y-N, Li B, Yu B (2011) CO_2 capture and activation by superbase/polyethylene glycol and its subsequent conversion. Energy Environ Sci 4(10):3971–3975

Yang L, Pastor-Pérez L, Gu S, Sepúlveda-Escribano A, Reina T (2018) Highly efficient Ni/CeO_2-Al_2O_3 catalysts for CO_2 upgrading via reverse water-gas shift: effect of selected transition metal promoters. Appl Catal B 232:464–471

Yaumi A, Bakar MA, Hameed B (2017) Recent advances in functionalized composite solid materials for carbon dioxide capture. Energy 124:461–480

Yu KK, Curcic I, Gabriel J, Morganstewart H, Tsang SC (2010) Catalytic coupling of CO_2 with epoxide over supported and unsupported amines. J Phys Chem A 114(11):3863–3872

Zhang Y-J, Da Y-B (2015) The decomposition of energy-related carbon emission and its decoupling with economic growth in China. Renew. Sust. Energ. Rev. 41:1255–1266

Zhang Q, Guo L (2018) Mechanism of the reverse water-gas shift reaction catalyzed by Cu_{12}TM bimetallic nanocluster: a density functional theory study. J Clust Sci 29(5):867–877

Zhang X, Liu J-X, Zijlstra B, Filot IA, Zhou Z, Sun S, Hensen EJ (2018a) Optimum Cu nanoparticle catalysts for CO_2 hydrogenation towards methanol. Nano Energy 43:200–209

Zhang W, Wang S, Zhao Y, Ma X (2018b) Hydrogenation of CO_2 to formic acid catalyzed by heterogeneous Ru-PPh3/Al2O3 catalysts. Fuel Process Technol 178:98–103

Zhao W, Fink DM, Labutta CA, Radosevich AT (2013) A Csp^3–Csp^3 bond forming reductive condensation of α-Keto esters and enolizable carbon pronucleophiles. Org Lett 15(12):3090–3093

Zhong J, Yang X, Wu Z, Liang B, Huang Y, Zhang T (2020) State of the art and perspectives in heterogeneous catalysis of CO_2 hydrogenation to methanol. Chem Soc Rev 49(5):1385–1413

Zhou H, Wang G-X, Zhang W-Z, Lu X-B (2015) CO_2 adducts of phosphorus ylides: highly active organocatalysts for carbon dioxide transformation. ACS Catal 5(11):6773–6779

Zhu L, Ye J-H, Duan M, Qi X, Yu D-G, Bai R, Lan Y (2018) The mechanism of copper-catalyzed oxytrifluoromethylation of allylamines with CO_2: a computational study. Inorg Chem Front 5 (4):633–639

Carbon Dioxide Conversion Methods

Sidra Saqib, Ahmad Mukhtar, Sami Ullah, Muhammad Sagir, M. Bilal Tahir, Rabia Amen, Muhammad Babar, Abdullah G. Al-Sehemi, Muhammad Ali Assiri, and Muhammad Ibrahim

Abstract

The carbon capture and sequestration (CCS) community has searched with the economic viability of CO_2 sequestration over the past few decades. In reality, however, it was not technically feasible until a recovered CO_2 sector, like in the beverage or improved recovery of oil and gas, was created. The scientific community and business are slowly converging to the view that the demand offer is significantly constrained by the sequestration of CO_2. Besides, the development of new consumer markets and income sources will prevail in the economic viability of this climate reduction strategy. The recovery of almost pure CO_2 is a potential alternative. In the energy and chemical fields work is shown to shift both explicitly and indirectly towards CO_2 consumption in the field of carbon capture and control. Within this segment, we have examined and measured advancement in research and development in this respect, the literature related to carbon capture, recycling, and consumption routes. Pathways for the usage of both physical and chemical CO_2 are researched and main pathways are defined. The research is often reviewed to look at conditions for method development and success assessment.

Keywords

CO_2 • Conversion • Green synthesis • Climate change • Greenhouse gases

1 Introduction

Environmental change is also due to pollution of other GHG's including CO_2, phosphorus, nitrous oxides, HFC, perfluorocarbons, and SF6, the so-called "anthropogenic" greenhouse gas. The normal influence of greenhouse by added gasses is broken down by ozone (3–7%), water vapor (36–70%), methane (4–9%) and carbon dioxide (9–26%) (Zheng et al. 2017a). From the beginning of the industrial era and fossil fuels discovery, the anthropogenic GHG emissions have gradually amplified. As indicated by the Intergovernmental Panel on Climate Change, the 5th valuation study "The human impact on the global climate is the current anthropogenic greenhouse gas releases in the past" (Liu et al. 2016). The most significant in terms of emission rates is carbon dioxide (CO_2) among GHGs.

CO_2 is one of the noteworthy elements present in the Earth's composition in the heart, soil, and climate. This may be absorbed in water, reacted to, solidified in carbonated rocks or it can be contained in the natural atmosphere. This pattern has continued for millions of years before the Industrial Revolution. At that point, CO_2 releases have first risen by < 1 part per million (ppm/y) each year before the 1960s and after that by 2 ppm/y (Banerjee et al. 2016). In the past, over-emission of CO_2 has little effect. However, the first human in the 1880s estimated that CO_2 emissions from

S. Saqib
Department of Chemical Engineering, COMSATS University Islamabad, Lahore Campus, Lahore, 54000, Pakistan

A. Mukhtar · M. Babar
Department of Chemical Engineering, Universiti Teknologi PETRONAS, Bandar, 32610 Seri Iskandar, Perak, Malaysia

S. Ullah · A. G. Al-Sehemi · M. A. Assiri
Department of Chemistry, College of Science, King Khalid University, 9004 Abha, 61413, Saudi Arabia

M. Sagir (✉)
Department of Chemical Engineering, Khwaja Fareed University of Engineering and Information Technology, Ryk, Pakistan

M. B. Tahir
Department of Physics, Khwaja Fareed University of Engineering and Information Technology, RyK, 50700, Pakistan

R. Amen
Institute of Soil and Environmental Sciences, University of Agriculture Faisalabad, Faisalabad, 38040, Pakistan

M. Ibrahim
School of Environmental Sciences and Engineering, Government College University, Faisalabad, 38000, Pakistan

Inamuddin et al. (eds.), *Carbon Dioxide Utilization to Sustainable Energy and Fuels*,
Advances in Science, Technology & Innovation, https://doi.org/10.1007/978-3-030-72877-9_9

burnt fossil fuels were sufficiently high to induce global warming, was Svante Arrhenius (Aresta et al. 2016a). This hypothesis, however, was later confirmed in 1938 by Guy Stewart Callendar, who in his Callendar hypothesis checked the projections of Arrhenius by witnessing the climatic increase (Aresta et al. 2014).

The key choices to replace fossil fuels and mitigate carbon pollution are (1) lowering electricity usage, e.g. increasing performance, (2) substituting alternative energy including renewables for fossil fuels as well as (3) CO_2 capture and storage (Ishida et al. 2003). While renewable energy sources are the best alternative round in solving the issue of GHG pollution, most of these renewables have less economic viability and reliability compared to fossil fuels (especially natural gas). Several reports indicated that the goal of 100% renewability in short to medium terms is not realistic in the overall transformation to zero net future electricity and that fossil fuels remain a dominant source (Schuchmann and Müller 2013).

This chapter is a comprehended analysis concentrating on the valuable fuels/chemicals technologies of CO_2 conversion. The main topics discussed in this chapter include CO_2, thermodynamics, and kinetic oxidation in chemical formats. Enzyme, mineralization, photocatalysis, and thermo-catalysis are outlined with terms to CO_2 conversion technologies. However, an economic study is made of the transition of CO_2 into usable energies/chemicals. The technological and implementation issues are evaluated and outlined, and design recommendations are introduced to promote potential study and development for CO_2 transformation techniques to address the issues.

2 Molecular Structure of CO_2

The CO_2 is a spherical 44 Da triatomic particle. The nucleus is characterized by strong electrical connections that are accompanied by electric interaction between carbon and oxygen atoms. It has an axial symmetry on the triangle, a base of symmetry, and a horizontal axis (Armenise and Kustova 2013). O atoms will capture the pair of electrons compared to atoms of C and the electrons are partly separated from the C atom, which causes fairly low energy levels for the C atom (Rothman et al. 1996). Of notice, in molecular CO_2 the bond duration among C and O is 116 p.m. The bond is weaker compared to usual dual bonds (e.g. acetaldehyde C = O 128 p.m. and similar to triple bonds (e.g. CO, 112.8 p.m.). This connection is less than average. CO_2 bonding energy is 803 kJ/mole (Stanbury et al. 2017) which is far greater than the O–H bonding energy of H_2O (463 kJ/mole) molecules. The formation of two π-bonds (π_3^4) and two σ bonds in CO_2 tends to be due to these effects (Ishida et al. 2003; Hamilton 1957).

3 Thermo-Kinetics of CO_2 Conversion

CO_2 in every combustion reaction is one of the most common end products, be it chemical or biological. The handling of CO_2 is complicated from a thermodynamic point of view. As for thermodynamics, when an O/C ratio of one component is <2, the transmission process of CO_2 is typically endergonic (Aresta et al. 2014). In contrast, if CO_2 is stored in the +4-oxidizing environment, the reactions appear to be exergonic, with different O/C ratios in the substance (for example, carbonate or urea). The thermodynamics (such as free energy, enthalpy, and entropy change, etc.) include the heat and temperature function. Conversion to the CO_2 can take place at 298.15 K or a certain temperature in terms of thermodynamic equations and carbon dioxide and other reactants or components (Barin 1989). Therefore, even at higher temperatures, negative thermodynamics will make conversion difficult.

The molecular C atom is usually initially targeted at CO_2 reactions. The nucleophile often strikes the atom C at a more positive charge in response to a reaction. An entity with a C-metal relation (e.g. Grignard), a single pair of electrons (e.g. an amine) and or a phenolate coupling (e.g.) may be a nucleophile. Throughout this case, the structure of CO_2 is defined by the polarization of O- and C-bonds which regulates both the CO_2 and nucleophile kinetics. In terms of CO_2 chemistry, its interaction with metals is the other significant factor. Through binding of CO_2 to the surface, it may change the distribution of electron and molecular form in a CO_2 molecule. The electrons are used as electron donors in the bonding orbits of CO_2, which break up the links between C and O and help increase chemical reactivity. As discussed above, thermodynamically a reaction is random, where the free transition of energy (soft energy) is beneficial. But these responses may also be spontaneous due to intense activating forces. Such carbon reactions also require an efficient catalyst. Catalysts can reduce energy activation and quicken the reaction at appropriate temperatures. A catalyst for $\Delta G -$ 144 kJ/mol and (−56 kJ/mol) appropriate for the reaction is essential for the development of ethylene carbonate of ethylene oxides and CO_2, for instance (North 2015).

4 CO_2 Conversion Methods and Products

As described above, carbon dioxide is thermodynamically and kinetically inert, rendering it difficult to transfer to other substances. However, the electron deficit in carbonyl carbons may be due to CO_2 attacks by a nucleophile, and electron-donated reagents (Sakakura et al. 2007). Consequently, facing technical obstacles, the synthesis of CO_2 is in principle achievable (Aresta et al. 2016a). Typical methods are discussed in the following, in particular for the development of high-energy fuels.

4.1 Fischer–Tropsch Gas-to-Liquid (GTL)

To save the global reliance on liquid fuels for transport of Fischer and Tropsch in the initial twentieth century, Syngas conversion to intensively chained carbons was built up over almost a hundred years. The main component of the FT reactor is syncrode that is converted by the separation methods into other hydrocarbons, like wax, LPG, naphtha, and diesel. The Fischer–Tropsch GTL process is a technology that enables the conversion of natural gas (NG) into syngas first by one of the many reform procedures (Wilhelm et al. 2001). In the so-called FT reactors and in the presence of iron or cobalt catalysts the syngas is transformed into a liquid blend of hydrocarbons (syncrudes) (Laan and Beenackers 1999). Generation by several research groups like Giesen et al. of liquid CO_2 fuels utilizing solar energy. CO_2 is generated in a water gas change reactor from CO_2 in this that routine and H_2 is created by electrical water electrolysis. The LCA measured the output potential of 140,000 barrels of oil liquid fuel each day for five method configurations Giesen and its co-authors. Such cycle paths are: (1) CO_2 and electricity and from the combustion of natural gas and the processing of fuel for hydrocarbon fluids, (2) solar photovoltaic power for additional energy, (3) energy and CO_2 from biomass combustion, (4) photovoltaics for supplementary electricity despite combustion, (5) CO_2 from the procurement of additional electricity For any process path, the authors measured the sum of emissions of CO_2 and accumulated energy demand. The design in the above phase was less energy usage. The CO_2 emissions from routes 4 and 5 as a combined device are 0.03 kg per 1 MJ of fluid air (Giesen et al. 2014).

4.2 Mineralization

Xie et al. (2014) suggested the theory and implementation of CO_2 mineralization for power production. The mineralization of CO_2 or carbonation reaction is a CO_2 transformation into CO_2 and the release of electricity. Specific CO_2 (mixed N_2/CO_2) amounts have been studied. The findings revealed that a large concentration of CO_2 could reach a higher power density. New research directed by Ebrahimi et al. (2017) has suggested the production of "natural" construction materials utilizing fly ash (FA) carbonation at an energy plant in Australia. Jang et al. (2016) analyzed recent developments in CO_2 confiscation and usage of cement-based equipment focusing on manufacturing problems, capacity-building and performance, economic viability, environmental effects, and a few other primary concerns. Bertos et al. (2004) examined the advancement of rapid CO_2 sequestration carbonation technologies. They defined CO_2 as the key control parameters of the carbonation process as diffusivity and reactivity. Park et al. (2017) examined the production of porosity during calcium aluminum cement hydration at 10% CO_2 concentration. They stated that it was possible to turn the metastable stages into polymorphs of stable calcium carbonate phases without major changes in the pore structure. The CO_2enhanced concrete preserving in a fluid reactor and the records of carbonation output roughly were examined by the Kashef-Haghighi and the Ghoshal (2010). 18 and 20% carbon dioxide. Galan et al. (2010) stated that carbon depth in concrete cannot be treated as the right sequestered CO_2 measure. The effect on the physical–chemical characteristics of cement morter was studied by Jange and Lee (2016). The investigators investigated the microstructure, CO_2, reaction, and mechanical power.

4.3 Chemical Looping Dry Reforming

Three steps are involved in chemical-loop dry-forming: (1) the elimination of methane; (2) carbon dioxide-reforming; and (3) oxidation (Sudiro and Bertucco 2009). This procedure is known as an effective revolutionary CO_2 recovery technology. In a chemical loop dry-reforming method, Huang et al. (2016) researched the feasibility of using the spinal ferry ($NiFe_2O_4$). At 900 °C, air oxidation was conducted. We decided that $NiFe_2O_4$ was a successful candidate. With the usage of CO_2 by chemical looping Galvita et al. (2013) used CeO_2, adjusted Fe_2O_3. Throughout the quartz tube microreactor, having an inner diameter of 10 mm, the operation tests were carried out at ambient pressure. In a furnace, the microreactor was installed. A 20-wt% $CeO_2{-}Fe_2O_3$ maximum CO production was recorded. $CeO_2{-}Fe_2O_3$ are 50 and 70 wt.% the most stable varieties. On the development of CO_2 from CO_2, the impact of CeO_2 on Fe_2O_3 was examined. In contrast to single compounds Fe_2O_3 and CeO_2, the mixed $CeO_2{-}Fe_2O_3$ catalysts developed enhanced performance. Thermodynamic test studies have also explored the possibility of chemical looping dry reforming. The findings indicated the probability of strong CO_2 conversions. We determined that iron-based transporters are the most auspicious for dry chemicals change.

4.4 Enzymatic Conversion

The synthesis of carbon dioxide is a key method of integrating CO_2 into the form of organic matter. In contrast to CO_2 processing processes, CO_2 enzymes make available an effective and environment—friendly approach because of their outstanding selectivity and specificity (Shi et al. 2015; Demars et al. 2016; Zhou et al. 2016). There are several strategies for the enzyme transfer of CO_2 in cells, such as the Calvin process and the 3-hydroxypropionate process. For plants, cyanobacteria, bacteria, and so on, the process Calvin, the most effective known biosynthetic method, has been commonly used in photosynthesis (Zheng et al. 2017b). In addition to CO_2 conversion in cells, alternate methods to enzyme conversion with various enzymes occur in nature for chemicals/fuels synthesis. Reda et al. (2008) has a method for format processing concerning CO_2 incorporation into single enzyme systems. This book provides divided approaches for the enzyme conversion of CO_2, but scientific and technological advances in the enzyme conversion process and increased output rates, in particular in the field of industrialization, must be necessary because these are environmental friendly and efficient methods for turning CO_2 into chemical substances/fuel.

4.5 Photocatalytic and Photo-Electrochemical Conversion

A great deal of work has been performed on photocatalytic and photo-electrochemical methods for the incorporation of CO_2 into carbon-based fuels since the last centuries. In 1979 Honda et al. accomplished CO_2 photo-catalytical reduction to hydrocarbons while particles of oxides and non-oxide semiconductors likeCdS, TiO_2, ZnO, and others were still to grow. Since then, ongoing studies have been conducted and published (Tu et al. 2012; Fujita 1999; Li et al. 2017; Yu et al. 2017) on photocatalytic carbon reduction. The mechanism causes electrons to excite from their balancing band (VB) to the conducting band (CB) as catalysts (for example, the semiconducting material) are subjected to photon irradiation, creating concurrently a positively charged electron holes in the CB. Such electron–hole pairs are isolated and pass aggressively to the active sites on the catalyst sheet. CO_2 is reduced to many usable fuels with or without H_2O at the catalytic surface with the aid of certain photogenerated electrons (e.g. CO, CH_4, and CH_3OH). The goods vary relying on the frequency of the electrode in the chemical reaction. Such as, CO or HCOOH are produced by $2e^-$, whereas CH_3OH is created by $6e^-$ and CH_4 requires $8e^-$. In regulating its kinetics in the stage of photocatalysis, the behavior of the catalyst plays an important part. Here are discussed three major groups: semiconductor-based materials, nano-catalysts based on graphs, and organometal. Half-manufacturing products involve various forms, such as TiO_2, ZnO, CdS, SYS, and other inorganic binary compounds. On the surface of a semiconductor, a connection between the energy rates of the semiconductor of the redox agents of the resolution was suggested (Inoue et al. 1979). It is the kinetic transfer charge between the photo-produced carrier (e^- and h^+). (1) the width of the semi-conductor band distance compared to the absorption spectrum of photons in the generation of electron–hole pairs, (2) the efficient isolation and migration of electron–hole paired band; and (3) the volume and number of active sites that can pair with each other. (4) The performance of the semiconductor photocatalyst is acknowledged. In this portion, several methods may be used to achieve a photo-electric catalytic reduction of CO_2. The key research aim in this area is to develop the Photoelectric Catalyst rationally by integrating photocatalytical and electrocatalytic techniques to achieve strong CO_2 reduction in performance and strong selectivity. Of this reason, more study and mechanisms analysis of reducing photo-electrocatalytic CO_2, the different mechanisms for joint effects, or the structures of the reaction equipment are required. In this area, it is particularly important to understand and develop various reaction routes in various reaction media for photocatalytic CO_2 reductions and guidance to high efficiency and high-selectiveness CO_2 conversions.

4.6 Thermo-Chemical Conversion

With a long history of study, the thermochemical method (especially the CO_2 hydrogenation technique) is an important solution among many approaches to turn carbon dioxide into hot fuels. According to the intrinsic properties of CO_2 molecules, the absorption of electrons at catalyst sites by CO_2 is more likely than the donation (Sakakura et al. 2007; Valverde and Medina 2017; Rao and Dey 2016; Zhang et al. 2016; Takacs et al. 2015). To achieve successful CO_2 thermochemical transitions, electrons must then be introduced into the CO_2 antibonding orbital. $H_2(g)$ must be isolated from the atomic H* absorb in active areas for the thermochemical processing of CO_2 as a reducer and as a high-energy material (Li et al. 2015).

Sabatier and Senderens first reported hydrogenation of CO_2 in the early twentieth century (Gnanamani et al. 2014) as a typical mechanism on the thermochemical route for carbon transformations. Numerous works on this topic have subsequently been conducted (Li et al. 2015; Fidalgo and Menendez 2011; Grabow and Mavrikakis 2011; Ashcroft et al. 1991). In specific, the catalytic conversion mechanism occurring on a solid–gas interface is investigated with VIIIB metals on numerous oxide cares. In this section, three primary fuels/chemicals are produced: carbon monoxide (CO), methanol (CH_3OH) and formic acid (HCOOH) and.

4.7 Hydrogenation

CH_3OH from carbon combustion has been widely researched (Zheng et al. 2017b; An et al. 2017; Larmier et al. 2017; Li et al. 2016; Martin et al. 2016; Rungtaweevoranit et al. 2016; Lei et al. 2015) in certain manufacturing processes as an alternative fuel and raw material. Wang et al. (2011) examined many CH_3OH synthesis catalytic systems increasing hydrogenation from CO_2, stressing catalytic sensitivity, and mechanisms of reactors. Compared to other metals, Cu is chiefly studied in laboratories for its superior catalytic efficiency and used in industrial applications. This is also used in conjunction with numerous editors including Si, Ti, Al, Zn, and so on (Kobl et al. 2016; Gao et al. 2013; Arena et al. 2007). The Catalytic Process $Cu/ZnO/Al_2O_3$ for industrial CH_3OH, CO_2, and H_2 synthesis in mild terms (200–300 °C and 5–10 MPa) is a common illustration of this. While various experiments are performed on CO_2 hydrogenation reaction processes, the principle of CO_2 transitions is still complex and challenging to grasp. More studies are therefore required, especially at the atomic level.

4.8 Reforming

In addition to the hydrogenation of CO_2 described above, the reformation of CO_2 is a further essential approach to the thermochemical transfer of different fuels. The syngas produced by Fischer–Tropsch (Pakhare and Spivey 2014) can be used in this method to produce several products such as higher alkanes. Fischer and Tropsch are (Failed 1928) believed to have first researched CO_2 methane reform (DRM) in 1928 using Ni and Co Catalysts. In this analysis, carbon accumulation was detected, and the catalysts were seriously damaged. Several experiments have also been undertaken to deal with this question to enhance the effectiveness and quality of this reform mechanism. For catalyst systems, there might be other variables that affect the overall functionality and quality of the catalyst used in the DRM process, including the product's intrinsic properties, product particle size, and support, morphological frameworks, including the size of particle and surface area, and the support and interaction between the support and the metal (Jafarbegloo et al. 2015; Kathiraser et al. 2015; Barroso Quiroga and Castro Luna 2007). The most widely employed catalysts for methane reforming CO_2 are non-based. However, the main constraint for realistic use is extreme decommissioning induced by carbon deposition (Avetisov et al. 2010; Yuan et al. 2016). Catalysts are also commonly produced based on such metals as Pd, Ce, Pt, Ru, and so on. The development of Ni catalysts using other processes, for example, Pd, Pt, and Ru is a viable option to achieve greater output and lower costs. In conjunction with the CO_2 reforming methane, alternate approaches to CO_2 alteration to generate organic chemicals should be added.

5 Economic Assessment of CO_2Alteration to Valuable Products

In addition to the hydrogenation of CO_2 described above, the reformation of CO_2 is a further essential approach to the thermochemical transfer of different fuels. The syngas produced by Fischer–Tropsch can be used in this method to produce several products such as higher alkanes. Fischer and Tropsch are believed to have first researched CO_2 methane reform (DRM) in 1928 using Ni and Co Catalysts. In this analysis, carbon accumulation was detected, and the catalysts were seriously damaged. Several experiments have also been undertaken to deal with this question to enhance the effectiveness and quality of this reform mechanism. New technology must also be introduced to satisfy anticipated consumer requirements, in particular concerning gasoline. Such technologies must be commercially sustainable, and they need to be commercially evaluated to facilitate more advances in CO_2 conversions. The fixed costs for investments, variable/fixed manufacturing costs, commodity rates, etc., depend heavily on the form of product, plant productivity, processing methods and also on the disparity among different provinces and countries regarding the economic examination of the refining of carbon dioxide into usable energies and chemicals in particular manufacturing techniques or chemical plant.

The price of transforming CO_2 into different usable chemicals/fuels are also hard to measure systematically. Consequently, in specific situations, this segment should analyze the economic study of some common chemicals/fuels. All economic analyzes in the literature should be remembered that they are focused on standard energy prices. The economic effects of the plant will vary as surplus power is used from renewable energy or nuclear supplies so that the method becomes more competitive. However, in the subsequent study, the environmental effects of CO_2 elimination are not usually addressed. Aside from all factors, the advantages of the ecosystem are much higher than the potential gains from the decline of carbon dioxide.

5.1 Syngas

In comparison to low-temperature electro-chemicals CO_2 production, the chemical selectivity, for example, CO (electrolyzes) and CO/H_2(co-electrolyzes CO_2/H_2O) at elevated SOC temperatures is beneficial in the Manufacture of the advanced high-temperature electrolysis (steam electrolysis or H_2O/CO_2 SOC co-Electrolysis). The impacts, cell degradation levels, CO_2 feedstuff costs, and electrolytic stack values of the following essential parameters are further analyzed. The most significant element here is undoubtedly the energy. The cost of electricity, maintenance, and operation for the co-electrolysis is higher. The energy demand for co-electrolysis is estimated to be \$67/MWh and \$8/MWh at the average and state-of-the-art prices. The combination of Fischer–Tropsch and the co-electrolysis of high temperatures H_2O/CO_2 with SOC processes may also generate syngas for electricity storage effectively. The economic analyses by Pérez-Fortes et al. 2016a indicate that the chosen power sources are surplus nuclear capacity and wind and that fuel-making prices are competitive with the expense of the 'biomass to liquid' methods.

This section addresses the economic study of carbon processing based on some common industrial situations, comprising syngas, methanol, dimethyl carbonate, urea, and formic acid, Besides the relative additional advantage of the different goods. Although more research is still needed for the economic viability of chemical/fuel processing, costs are expected to decline more, with the further advancement of the relevant technologies.

5.2 Methanol

Because of its durability, methanol (CH_3OH) is a popular feedstock and a building block of chemical production, since they are compatible with a range of major chemicals like acetic acid, chloro-methanol, alkyl halides, formaldehyde and even higher hydrocarbons (Armenise and Kustova 2013; Zheng et al. 2017b; Rungtaweevoranit et al. 2016; Aresta et al. 2016b; Kiss et al. 2016). In 2014 about 60% of the methanol generated was used as a solvent and the remaining 40% was used to manufacture fuels including biodiesel, dimethyl ether (DME), marine fuels, etc. (Kourkoumpas et al. 2016). This amount is projected to be about 60%. Methanol is usually generated via the hydrogenation technique ($CO_2 + H_2$) and bi-reforming cycle (CO_2 + NG). Kourkoumpas etc. have recently conducted a technological and cost-effective study into the usage of CO_2 from Lignite Power Plants to introduce "Power to Methanol (PtM)." Their principle involves the generation of energy, capture, and isolation of CO_2 (CCS), development of H_2, and main steps for processing and application of methanol (Wiesberg et al. 2016; Pérez-Fortes et al. 2016b).

5.3 Formic Acid

A material contained commonly in clothes, food, and pharmaceutical items is formic acid (or HCOOH). This can also be used as a hydrogen and oxygen carrier and might be utilized by fuel cells in particular. Formic acid has a wide, proven, small substitution market and is a high-value drug (Pérez-Fortes et al. 2016a). The main contributors to operating costs are products and consumables for example steam and electricity. Sales of formic acid and other by-products compensate for almost 50% of production costs. There is also a need for more initiatives to minimize costs of processing from CO_2 to formic acid, comprising R&D for electrolyzer, coupled with renewable energy and the advancement of other associated equipment.

5.4 Urea

Urea is generally sold as a solid, normally granules or pills (NH_2CONH_2). The development of urea requires four steps, namely CO_2 reaction and ammonia, recirculation, purification, and granulation of low pressure (Pérez-Fortes et al. 2014). The coal-based poly-generation suggested by Bose et al. (2015) first captured CO_2 to generate urea, which is a traditional fertilizer, then used captured CO_2.

5.5 Dimethyl Carbonate (DMC)

Dimethyl carbonate ($(CH_3O)_2CO$) can be synthesized by 4 separate processes based on a CO_2 conversion: (1) from urea; (2) by ethylene carbonate; (3) directly from CO_2 to methanol and (4) from propylene to carbonate. The techno-economic evaluation of the cycle utilizing modern catalytic membrane reactors was reviewed, in particular concerning investment costs and operational costs as a standard to convert $CO_2$for the formation of dimethyl carbonates (Kuenen et al. 2016). It can be argued that the manufacturing cycle, as a consequence of the expense of depreciation, repair, and services, is unrentable under the conditions obtained.

6 Conclusions and Future Perspective

CO_2 reduction presents the chemical and oil industry with a threat and a potential. Despite three decades of CCS development, this technique poses challenges as in the majority of cases the cost of funding and service remains large and inefficient. Work in the area of carbon control is now moving to the use of CO_2 in chemical and fuel syntheses to boost CO_2 capture efficiency. Within this chapter, we objectively analyzed and evaluated improvements within our work and growth on carbon recovery, recycling, and consumption paths. The literature indicates that the practical usage of CO_2 is an established market with a weak requirement for creative core scientific work. Instead, these routes focus primarily on market conditions linked to global and local oil markets and policies on mitigation of climate change such as carbon taxation frameworks. Nevertheless, the industrial usage of CO_2 is a growing area with significant competition for research and production, except for certain specific goods. Work is still underway in every field, from catalysis to unit architecture, optimization of systems, lifecycle review, environmental benchmarking, and policy study.

Acknowledgements The authors gratefully acknowledge the Departments at their respective universities for providing state of the art research facilities.

References

An B, Zhang J, Cheng K, Ji P, Wang C, Lin W (2017) Confinement of ultrasmall Cu/ZnO_x nanoparticles in metal-organic frameworks for selective methanol synthesis from catalytic hydrogenation of CO_2. J Am Chem Soc 139:3834–3840

Arena F, Barbera K, Italiano G, Bonura G, Spadaro L, Frusteri F (2007) Synthesis, characterization and activity pattern of Cu–ZnO/ZrO_2 catalysts in the hydrogenation of carbon dioxide to methanol. J Catal 249:185–194

Aresta M, Dibenedetto A, Angelini A (2014) Catalysis for the valorization of exhaust carbon: from CO_2 to chemicals, materials, and fuels. Technological use of CO_2. Chem Rev 114:1709–1742

Aresta M, Dibenedetto A, Quaranta E (2016) One-and multi-electron pathways for the reduction of CO_2 into C1 and $C1^+$ energy-richer molecules: some thermodynamic and kinetic facts. In: Reaction mechanisms in carbon dioxide conversion. Springer, pp 311–345

Aresta M, Dibenedetto A, Quaranta E (2016b) State of the art and perspectives in catalytic processes for CO_2 conversion into chemicals and fuels: the distinctive contribution of chemical catalysis and biotechnology. J Catal 343:2–45

Armenise I, Kustova E (2013) State-to-state models for CO_2 molecules: from the theory to an application to hypersonic boundary layers. Chem Phys 415:269–281

Ashcroft A, Cheetham A, Green M, Vernon P (1991) Partial oxidation of methane to synthesis gas using carbon dioxide. Nature 352:225–226

Avetisov A, Rostrup-Nielsen J, Kuchaev V, Hansen J-HB, Zyskin A, Shapatina E (2010) Steady-state kinetics and mechanism of methane reforming with steam and carbon dioxide over Ni catalyst. J Mol Catal A Chem 315:155–162

Banerjee A, Dick GR, Yoshino T, Kanan MW (2016) Carbon dioxide utilization via carbonate-promoted C-H carboxylation. Nature 531:215–219

Barin I (1989) Thermochemical data of pure substances. VCH

Barroso Quiroga MM, Castro Luna AE (2007) Kinetic analysis of rate data for dry reforming of methane. Ind Eng Chem Res 46:5265–5270

Bertos MF, Simons S, Hills C, Carey P (2004) A review of accelerated carbonation technology in the treatment of cement-based materials and sequestration of CO_2. J Hazard Mater 112:193–205

Bose A, Jana K, Mitra D, De S (2015) Co-production of power and urea from coal with CO_2 capture: performance assessment. Clean Technol Environ Pol 17:1271–1280

Demars BO, Gíslason GM, Ólafsson JS, Manson JR, Friberg N, Hood JM et al (2016) Impact of warming on CO_2 emissions from streams countered by aquatic photosynthesis. Nat Geosci 9:758–761

Ebrahimi A, Saffari M, Milani D, Montoya A, Valix M, Abbas A (2017) Sustainable transformation of fly ash industrial waste into a construction cement blend via CO_2 carbonation. J Clean Prod 156:660–669

Fisher F, Tropsch H (1928) Conversion of methane into hydrogen and carbon monoxide. Brennst Chem 9

Fidalgo B, Menendez JÁ (2011) Carbon materials as catalysts for decomposition and CO_2 reforming of methane: a review. Chin J Catal 32:207–216

Fujita E (1999) Photochemical carbon dioxide reduction with metal complexes. Coord Chem Rev 185:373–384

Galan I, Andrade C, Mora P, Sanjuan MA (2010) Sequestration of CO_2 by concrete carbonation. Environ Sci Technol 44:3181–3186

Galvita VV, Poelman H, Bliznuk V, Detavernier C, Marin GB (2013) CeO_2-modified Fe_2O_3 for CO_2 utilization via chemical looping. Ind Eng Chem Res 52:8416–8426

Gao P, Li F, Zhao N, Xiao F, Wei W, Zhong L et al (2013) Influence of modifier (Mn, La, Ce, Zr and Y) on the performance of Cu/Zn/Al catalysts via hydrotalcite-like precursors for CO_2 hydrogenation to methanol. Appl Catal A 468:442–452

van der Giesen C, Kleijn R, Kramer GJ (2014) Energy and climate impacts of producing synthetic hydrocarbon fuels from CO_2. Environ Sci Technol 48:7111–7121

Gnanamani MK, Jacobs G, Pendyala VRR, Ma W, Davis BH (2014) Hydrogenation of carbon dioxide to liquid fuels. Green Carbon Dioxide 2:99–118

Grabow L, Mavrikakis M (2011) Mechanism of methanol synthesis on Cu through CO_2 and CO hydrogenation. ACS Catal 1:365–384

Hamilton WC (1957) Equivalent orbital bond moments: applications to carbon dioxide and water. J Chem Phys 26:345–351

Huang Z, Jiang H, He F, Chen D, Wei G, Zhao K et al (2016) Evaluation of multi-cycle performance of chemical looping dry reforming using CO_2 as an oxidant with Fe–Ni bimetallic oxides. J Energ Chem 25:62–70

Inoue T, Fujishima A, Konishi S, Honda K (1979) Photoelectrocatalytic reduction of carbon dioxide in aqueous suspensions of semiconductor powders. Nature 277:637–638

Ishida S, Iwamoto T, Kabuto C, Kira M (2003) A stable silicon-based allene analogue with a formally sp-hybridized silicon atom. Nature 421:725–727

Jafarbegloo M, Tarlani A, Mesbah AW, Sahebdelfar S (2015) Thermodynamic analysis of carbon dioxide reforming of methane and its practical relevance. Int J Hydrogen Energ 40:2445–2451

Jang JG, Lee H-K (2016) Microstructural densification and CO_2 uptake promoted by the carbonation curing of belite-rich Portland cement. Cem Concr Res 82:50–57

Jang JG, Kim G, Kim H, Lee H-K (2016) Review on recent advances in CO_2 utilization and sequestration technologies in cement-based materials. Constr Build Mater 127:762–773

Kashef-Haghighi S, Ghoshal S (2010) CO_2 sequestration in concrete through accelerated carbonation curing in a flow-through reactor. Ind Eng Chem Res 49:1143–1149

Kathiraser Y, Oemar U, Saw ET, Li Z, Kawi S (2015) Kinetic and mechanistic aspects for CO_2 reforming of methane over Ni based catalysts. Chem Eng J 278:62–78

Kiss AA, Pragt J, Vos H, Bargeman G, De Groot M (2016) Novel efficient process for methanol synthesis by CO_2 hydrogenation. Chem Eng J 284:260–269

Kobl K, Thomas S, Zimmermann Y, Parkhomenko K, Roger A-C (2016) Power-law kinetics of methanol synthesis from carbon dioxide and hydrogen on copper–zinc oxide catalysts with alumina or zirconia supports. Catal Today 270:31–42

Kourkoumpas D, Papadimou E, Atsonios K, Karellas S, Grammelis P, Kakaras E (2016) Implementation of the power to methanol concept by using CO_2 from lignite power plants: techno-economic investigation. Int J Hydrogen Energ 41:16674–16687

Kuenen H, Mengers H, Nijmeijer DC, van der Ham AG, Kiss AA (2016) Techno-economic evaluation of the direct conversion of CO_2 to dimethyl carbonate using catalytic membrane reactors. Comput Chem Eng 86:136–147

Van Der Laan GP, Beenackers A (1999) Kinetics and selectivity of the Fischer-Tropsch synthesis: a literature review. Catal Rev 41:255–318

Larmier K, Liao WC, Tada S, Lam E, Verel R, Bansode A et al (2017) CO_2-to-methanol hydrogenation on zirconia-supported copper nanoparticles: reaction intermediates and the role of the metal-support interface. Angew Chem Int Ed 56:2318–2323

Lei H, Nie R, Wu G, Hou Z (2015) Hydrogenation of CO_2 to CH_3OH over Cu/ZnO catalysts with different ZnO morphology. Fuel 154:161–166

Li Y, Chan SH, Sun Q (2015) Heterogeneous catalytic conversion of CO_2: a comprehensive theoretical review. Nanoscale 7:8663–8683

Li MM-J, Zeng Z, Liao F, Hong X, Tsang SCE (2016) Enhanced CO_2 hydrogenation to methanol over CuZn nanoalloy in Ga modified Cu/ZnO catalysts. J Catal 343:157–167

Li M, Yang Y, Ling Y, Qiu W, Wang F, Liu T et al (2017) Morphology and doping engineering of Sn-doped hematite nanowire photoanodes. Nano Lett 17:2490–2495

Liu C, Colón BC, Ziesack M, Silver PA, Nocera DG (2016) Water splitting–biosynthetic system with CO_2 reduction efficiencies exceeding photosynthesis. Science 352:1210–1213

Martin O, Martín AJ, Mondelli C, Mitchell S, Segawa TF, Hauert R et al (2016) Indium oxide as a superior catalyst for methanol synthesis by CO_2 hydrogenation. Angew Chem Int Ed 55:6261–6265

North M (2015) What is CO_2? Thermodynamics, basic reactions and physical chemistry. In: Carbon dioxide utilisation. Elsevier, pp 3–17

Pakhare D, Spivey J (2014) A review of dry (CO_2) reforming of methane over noble metal catalysts. Chem Soc Rev 43:7813–7837

Park S, Jang J, Son H, Lee H-K (2017) Stable conversion of metastable hydrates in calcium aluminate cement by early carbonation curing. J CO_2 Utilization 21:224–226

Pérez-Fortes M, Bocin-Dumitriu A, Tzimas E (2014) CO_2 utilization pathways: techno-economic assessment and market opportunities. Energ Procedia 63:7968–7975

Pérez-Fortes M, Schöneberger JC, Boulamanti A, Harrison G, Tzimas E (2016) Formic acid synthesis using CO_2 as raw material: techno-economic and environmental evaluation and market potential. Int J Hydrogen Energ 41:16444–16462

Pérez-Fortes M, Schöneberger JC, Boulamanti A, Tzimas E (2016b) Methanol synthesis using captured CO_2 as raw material: techno-economic and environmental assessment. Appl Energ 161:718–732

Rao C, Dey S (2016) Generation of H_2 and CO by solar thermochemical splitting of H_2O and CO_2 by employing metal oxides. J Solid State Chem 242:107–115

Reda T, Plugge CM, Abram NJ, Hirst J (2008) Reversible interconversion of carbon dioxide and formate by an electroactive enzyme. Proc Natl Acad Sci 105:10654–10658

Rothman LS, Rinsland C, Goldman A, Massie S, Edwards D, Flaud J et al (1998) The HITRAN molecular spectroscopic database and HAWKS (HITRAN atmospheric workstation): 1996 edition. J Quant Spectr Radiat Transf 60:665–710

Rungtaweevoranit B, Baek J, Araujo JR, Archanjo BS, Choi KM, Yaghi OM et al (2016) Copper nanocrystals encapsulated in Zr-based metal–organic frameworks for highly selective CO_2 hydrogenation to methanol. Nano Lett 16:7645–7649

Sakakura T, Choi J-C, Yasuda H (2007) Transformation of carbon dioxide. Chem Rev 107:2365–2387

Schuchmann K, Müller V (2013) Direct and reversible hydrogenation of CO_2 to formate by a bacterial carbon dioxide reductase. Science 342:1382–1385

Shi J, Jiang Y, Jiang Z, Wang X, Wang X, Zhang S et al (2015) Enzymatic conversion of carbon dioxide. Chem Soc Rev 44:5981–6000

Stanbury M, Compain J-D, Trejo M, Smith P, Gouré E, Chardon-Noblat S (2017) Mn-carbonyl molecular catalysts containing a redox-active phenanthroline-5, 6-dione for selective electro-and photoreduction of CO_2 to CO or HCOOH. Electrochim Acta 240:288–299

Sudiro M, Bertucco A (2009) Production of synthetic gasoline and diesel fuel by alternative processes using natural gas and coal: process simulation and optimization. Energy 34:2206–2214

Takacs M, Scheffe JR, Steinfeld A (2015) Oxygen nonstoichiometry and thermodynamic characterization of Zr doped ceria in the 1573–1773 K temperature range. Phys Chem Chem Phys 17:7813–7822

Tu W, Zhou Y, Liu Q, Tian Z, Gao J, Chen X et al (2012) Robust hollow spheres consisting of alternating titania nanosheets and graphene nanosheets with high photocatalytic activity for CO_2 conversion into renewable fuels. Adv Func Mater 22:1215–1221

Valverde JM, Medina S (2017) Limestone calcination under calcium-looping conditions for CO_2 capture and thermochemical energy storage in the presence of H_2O: an in situ XRD analysis. Phys Chem Chem Phys 19:7587–7596

Wang W, Wang S, Ma X, Gong J (2011) Recent advances in catalytic hydrogenation of carbon dioxide. Chem Soc Rev 40:3703–3727

Wiesberg IL, de Medeiros JL, Alves RM, Coutinho PL, Araújo OQ (2016) Carbon dioxide management by chemical conversion to methanol: hydrogenation and bi-reforming. Energ Convers Manag 125:320–335

Wilhelm D, Simbeck D, Karp A, Dickenson R (2001) Syngas production for gas-to-liquids applications: technologies, issues and outlook. Fuel Process Technol 71:139–148

Xie H, Wang Y, He Y, Gou M, Liu T, Wang J et al (2014) Generation of electricity from CO_2 mineralization: principle and realization. Sci CHINA Technol Sci 57:2335–2343

Yu Y, Zhang Z, Yin X, Kvit A, Liao Q, Kang Z et al (2017) Enhanced photoelectrochemical efficiency and stability using a conformal TiO_2 film on a black silicon photoanode. Nat Energ 2:1–7

Yuan K, Zhong J-Q, Zhou X, Xu L, Bergman SL, Wu K et al (2016) Dynamic oxygen on surface: catalytic intermediate and coking barrier in the modeled CO_2 reforming of CH_4 on Ni (111). ACS Catal 6:4330–4339

Zhang Y, Nie T, Wang Z, Liu J, Zhou J, Cen K (2016) Splitting of CO_2 via the heterogeneous oxidation of zinc powder in thermochemical cycles. Ind Eng Chem Res 55:534–542

Zheng Y, Wang J, Yu B, Zhang W, Chen J, Qiao J et al (2017a) A review of high temperature co-electrolysis of H_2O and CO_2 to produce sustainable fuels using solid oxide electrolysis cells (SOECs): advanced materials and technology. Chem Soc Rev 46:1427–1463

Zheng Y, Zhang W, Li Y, Chen J, Yu B, Wang J et al (2017b) Energy related CO_2 conversion and utilization: advanced materials/nanomaterials, reaction mechanisms and technologies. Nano Energ 40:512–539

Zhou H, Li P, Liu J, Chen Z, Liu L, Dontsova D et al (2016) Biomimetic polymeric semiconductor based hybrid nanosystems for artificial photosynthesis towards solar fuels generation via CO_2 reduction. Nano Energ 25:128–135

Closing the Carbon Cycle

Anoop Singh, Prerna Mahajan, and Sandeep Arya

Abstract

The energy demand of the world can be fulfilled by burning fossil fuels such as natural gas, petroleum, and coal on a large scale. However, the burning of these energy resources produces a large amount of carbon dioxide in the atmosphere, which is outpacing the natural carbon cycle and changes the natural environment. Recently, the atmospheric concentration of carbon dioxide has surpassed the four hundred parts per million marks. Despite producing much energy, these fossil fuel reserves are limited. We must capture carbon dioxide to get our environment neat and clean. For capturing carbon dioxide, various technologies such as adsorption on solids, separation of the membrane, and absorption into liquids could be utilized after knowing some parameters like the existence of impurities, temperature, the concentration of carbon dioxide, etc. The extracted carbon dioxide can be utilized for various applications such as the conversion of carbon dioxide into dimethyl ethanol, methanol, and hydrocarbons. There are various types of energy resources that we can utilize instead of carbon-emitting energy resources. By using such type of resources, we can make our environment neat and clean. While doing so, we can capture carbon more easily.

Keywords

Methanol • Carbon cycle • Dimethyl ether • Renewable energy

1 Introduction

The population of the world is increasing day by day, due to which energy demand is also increasing. To satisfy their energy demand, fossil fuels such as natural gas, petroleum, and coal are being burnt on a large scale. Presently, ten billion cubic meters of natural gas, twelve million tonnes of oil, and twenty-two million tonnes of coal are used in everyday life to fulfill near about 82% demand for energy (Abas et al. 2015). Simultaneously, they release thirty billion tonnes of carbon dioxide annually into the environment (Pachauri et al. 2014). Some part of this carbon dioxide is absorbed by the oceans for acidification of seawater and few plants in the world are also utilizing carbon dioxide. Nearly, half of the carbon dioxide released in the world remains as it is in the atmosphere. The concentration of carbon dioxide in the atmosphere has just gone beyond the four hundred parts per million marks. Carbon dioxide is one of the greenhouse gases and it plays an important role to trap the heat radiations radiated by the Sun. The change in the climatic phenomena is observed due to the presence of carbon dioxide in the atmosphere. Because of the large amount of anthropogenic emission of carbon dioxide in the atmosphere, the natural carbon cycle has been greatly disturbed i.e. exceeded. Numerous solutions are given by scientists to overcome this disturbance. Conservation of energy is helpful up to some extent but, it will not fully solve this issue relating to the carbon cycle. One of the important solutions is to reduce the usage of fossil fuel-based energy resources, and instead of using these resources, we can use nuclear energy resources as well as renewable energy resources.

These fossil fuel reserves, regardless of producing so much energy, are inadequate. The fossil fuels such as shale gas and shale oil are exploiting day by day; also their impact on the environment raises numerous challenges. Almost half of the countries of the world including Germany and France banned the usage of a new technology commonly called fracking. These newly developed technologies have not been

A. Singh · P. Mahajan · S. Arya (✉)
Department of Physics, University of Jammu,
Jammu and Kashmir, 180006, India

Inamuddin et al. (eds.), *Carbon Dioxide Utilization to Sustainable Energy and Fuels*,
Advances in Science, Technology & Innovation, https://doi.org/10.1007/978-3-030-72877-9_10

proved safe for the environment and have also exploited shale resources. The greenhouse gases are emitted by utilizing fossil fuel resources and hence, it can become the center of various problems. If these problems are overcome, then humankind is permitted to utilize fossil fuels, otherwise, the usage of fossil fuel resources would be banned. The most general solution to control this problem is carbon capture and sequestration (CCS). In this case, the carbon dioxide is captured from numerous sources such as industrial facilities and power plants followed by dumping this captured carbon dioxide in the earth. The enhanced oil recovery can be done by utilizing the captured carbon dioxide (Khoo and Tan 2006). Different methods are available to capture carbon dioxide that includes adsorption on solids, separation of the membrane, and absorption into liquids. For this, several parameters such as the existence of impurities, temperature, the concentration of carbon dioxide, etc. are required to be known. Nowadays, various technologies are developed to capture carbon dioxide in the air (Goeppert et al. 2012; Meth et al. 2012). The direct air capture (DAC) technique is very helpful to capture the carbon dioxide that arises from the burning of fossil fuels in the trains, airplanes, cars, domestic heating, and stoves (Lackner 2009,2010; Keith et al. 2006; Stolaroff et al. 2008; Mahmoudkhani et al. 2009; Mahmoudkhani and Keith 2006). It is not economically feasible and also impractical to capture carbon dioxide at the source.

After extracting the carbon dioxide, it can be efficiently transformed to dimethyl ethanol, methanol, and hydrocarbons and hence can be utilized in different applications. All these applications are also playing an important role in determining the cost associated with DAC. This will help to decrease the cost of DAC. The specific term used for the various types of heterogeneous materials is biomass; hence it is convenient, advantageous, and easy to convert biomass into products like methanol and dimethyl ether (DME). Nowadays, more efficient conversion of biomass into bio-methanol can take place (Hamelinck and Faaij 2002). There are various technologies such as combustion, gasification, liquefaction, and pyrolysis that are used for the conversion of biomass. In diesel engines, one of the prominent candidates which are efficiently used is methanol. It is also used in direct methanol fuel cells (DMFC) which use methanol and directly generate the electrical power at ambient temperature (Bromberg and Cohn 2010; McGrath et al. 2004). The dehydration of methanol gives DME. It exists in gaseous form and hence moderate pressure is required to liquefy this gas easily. The properties of DME are similar to liquefied petroleum gas (LPG). DME is a better substitute for diesel fuel and possesses high octane rating that's why it attracts much attention (Semelsberger et al. 2006; Arcoumanis et al. 2008). In most of the applications like cooking and heating, it efficiently replaces LPG.

A variety of energy resources are there that would facilitate a neat and clean environment and can be accessed as a competent replacement for carbon-emitting energy resources. It is economically feasible to replace fossil fuels with renewable sources of energy. Life becomes easier with the help of such type of energy. It helps us to regulate the consumption of fossil fuels and make them available for future generations. In this chapter, we discuss the capture of carbon dioxide and its conversion into methanol and DME and also a replacement of fossil fuels by renewable energy resources that help in closing the carbon cycle.

2 Methods to Capture CO_2

For the handling of carbon dioxide that releases through heat or power generation, two methods are available. These are:

(i) Storage in geological strata
(ii) Chemical conversion of CO_2 to products that are easier to store and can be utilized effortlessly.

An immense amount of energy is needed in both the methods and in several cases; the energy required is even more than the energy that releases during the combustion process. Therefore, the likelihood of CO_2 capture i.e. the storage or conversion of carbon dioxide significantly depends on the utilization of extremely economical 'waste' energy. This 'waste' energy could be either superfluous electrical energy generated from wind power or from technical systems that are completely dependent on direct solar energy (MacElroy 2016). The carbon capture technology must be established without any holdup because of the existing as well as expected monetary pressures for the continuous utilization of carbon-based fuels. Out of the two elements of CCS, the carbon capture component is more expensive. Further, it is believed that if we depend on carbon fuels as a primary energy source for the long term, then we must find some ways to re-use or recycle the released CO_2.

3 CO_2 Capture Technologies

Generally, three technologies can be practiced for the capture of CO_2 from process streams or the atmosphere. These are:

(i) Post-combustion
(ii) Pre-combustion
(iii) Oxy-fuel combustion.

All these three technologies add a considerable cost to carbon fuel utilization during power generation, heating, and transportation. Thus, the basic physical chemistry behind the combustion processes relies on cost and technology.

The existing capture technologies are typically operating at an efficiency of 30–40%. As a consequence, around three times extra power would essentially be needed than that specified in the ideal estimation based on thermodynamics. In all the three capture processes mentioned above, the foremost operation includes separating the gas mixtures for the production of pure carbon dioxide and after the separation process, the gas is finally compressed. The minimum amount of power required for the separation process can be predicted based on reversible work i.e. free energy change needed during demixing (Goeppert et al. 2012).

These technologies possess some individual technical advantages as well as disadvantages; however, there is a considerable opportunity for developing a reduction in the cost and enhancement in the efficiencies of the process. For instance, while managing and matching the peak and off-peak power loads and optimizing the processes for internal heat recovery via the suitable implementation of process integration, breakthroughs would be requisite (Harkin et al. 2010; Kemp 2011). These advancements (Aro 2016) would undoubtedly play significant functions. The solvent or the solution could be effectively replaced with, for example, ionic liquids in the post-combustion technology. The membrane applications like complex membranes used for the transportation of hydrogen in the pre-combustion method and oxygen in the oxyfuel combustion process can also be replaced with new applications such as adsorption processes that can be utilized during small-to-medium-sized distributed emissions. For instance, these novel applications can be employed while capturing (location independent) CO_2 from the atmosphere (Goeppert et al. 2012). New carbon capture technologies are also under progress that would certainly play an important part in driving successful carbon capture power demands in the vicinity of the ideal limits.

4 CO_2 Capture from the Air

CO_2 from both the natural and industrial sources can be separated economically when a high concentration of carbon dioxide is available in these sources. Anthropogenic emissions of CO_2 can be broadly divided into two portions. The first portion consists of large manufacturing sources likely cement factories and large power plants while the second portion is distributed into various sectors such as the transportation sector and cooling or heating in homes and offices. It's very difficult for us to collect CO_2 from numerous small scale units of fossil fuel burning. It is also a tedious task to capture CO_2 from a flying airplane. Moreover, the transportation of captured CO_2 from the collection site to the conversion site requires expensive as well as enormous infrastructure. To avoid the requirement of enormous infrastructure for the collection of CO_2, we must set up an apparatus to capture CO_2 straight out of the atmosphere. The collection of CO_2 directly from the atmosphere helps to save transportation costs from the collection center to the conversion site. DAC technique as mentioned earlier also was reported in the past to collect CO_2 independently from the source of CO_2 (Specht and Bandi 1999). This indicates that the CO_2 can be captured easily from various sources such as dynamic or stable, large or small, etc. CO_2 extraction plants can be constructed anywhere in the world because of the similar concentration of CO_2 throughout the globe. It is also mentioned in this chapter that DAC is free from the sources of carbon dioxide. This leads to comparably more extraction of CO_2 than the production of it from human activities. This technique is not only to control the production of CO_2 but also helps in the reduction of the concentration of atmospheric CO_2. Even though the atmospheric concentration of CO_2 is very low near about 400 parts per million, the theoretical requirements for energy for DAC to go from a minimum amount of 400 parts per million to pure gas are near about 5 kilocalories per mole (Zeman 2007). Theoretically, DAC would need near about 3.4 times the energy compared with the CO_2 captured from the power plant (fossil fuel) (Ranjan and Herzog 2011). To capture the CO_2 from the atmosphere various chemical substances are used but practically utilized chemical substances are only sorbents. These sorbents can accumulate the molecules of another species by the phenomena of sorption. Sorbent also soaks up molecules of oil production from water. This can happen through both the phenomena of absorption and adsorption. The oil from the water is penetrating through pores into the sorbent by the process of absorption while adsorption only helps to attract oil on the surface but does not permit it to enter into the sorbent.

Broadly speaking, monoethanolamine dependent sorbents are used. These types of sorbents have some limitations such as less stability, unfavorable evaporation. While some bases with high pH values such as hydroxide of sodium, calcium, and potassium are generally used. Instead of these hydroxides, we also use an aqueous solution of some strong bases. These bases can strongly bind CO_2 and help us to make our environment neat and clean. Some other sorbents developed through the utilization of amine and polyamine can either be adsorbed physically or chemically with the help of some supporting agents like mesoporous solids, silica, carbon fibers as well as polymers, have a potential for DAC (Didas et al. 2012; Sculley and Zhou 2012; Chen et al. 2013; Goeppert et al. 2011,2014). Hyperbranched aminosilicas that can be synthesized with the help of polymerization of organic compound (aziridine) are also a potential candidate for DAC (Choi et al. 2011). Some adsorbents which are based on networks of polymer, ammonium dependent resin, and metal–organic frameworks are also having the possibility for DAC (wang et al. 2013; He et al. 2013). However, new techniques were also developed for the removal of CO_2 from both the spacecraft and submarines (Goeppert et al. 2012).

There should not arise any question against the possibility of capturing carbon dioxide from the atmosphere, but still many researches are required to make this technology economically beneficial. By using the latest technologies, the estimated expenditure of removing one ton of carbon dioxide from small power plants is more than 30 dollars. But there is a lot of variation in the estimated cost of removing one ton of carbon dioxide in the case of DAC (Simon et al. 2011; Pielke 2009). Various important factors play an effective role to manage the cost of a commercial power plant. Such factors are labor cost, the method used, energy used, and the requirement of material. A very few companies use carbon engineering to tackle carbon emissions (Keith et al. 2006) for example Climeworks, Global Thermostat, and Kilimanjaro Energy have been established with such efforts. We can utilize the extracted carbon dioxide by converting it into dimethyl ethanol, methanol, and hydrocarbons and therefore can be exploited in a variety of applications which also play a significant role in the determination of the cost associated with DAC. Consequently, this would facilitate a reduction in the cost of DAC. Rely on the process, not only the carbon dioxide is separated from the air but also at the same time, water is separated. This could generate purified water with added value.

5 Biomass and Waste-Based Chemicals

Rely on the recycling of carbon dioxide; biomass is defined as a material obtained from animals or plants. This comprises crops, forests with their byproducts, animal waste, algae, aquatic plants, and municipal solid waste. Originally, methanol was popular with the name of wood alcohol, it is generally because methanol is obtained from the wood through thermal distillation. This thermal distillation method was abandoned at the starting of the twentieth century with the development of the latest synthetic process for the development of syngas. Biomass is a definite term generally utilized for a variety of different heterogeneous materials. The conversion of biomass into products like methanol and DME is therefore beneficial, suitable, and effortless. Moreover, biomass is being more efficiently converted into bio-methanol these days. Different technologies are available to convert the biomass. These are combustion, gasification, liquefaction, and pyrolysis and are shown in Fig. 1 (Bridgwater 1994; Chmielniak and Sciazko 2003; Bulushev and Ross 2011; Larson 1993). The cellulosic material and wood are the solid feedstock that requires technology similar to the used one for the conversion of coal product to hydrocarbon (methanol). Another type of feedstock is animal manure and converts it into biogas which is a mixture of carbon dioxide and methane, and in this process finally, methanol is synthesized. An overview of methanol produced from biomass is shown in Fig. 2 (Reschetilowski 2013).

6 Advantages of Biomass-Based Chemicals

Various energy sources such as geothermal, hydro, solar, and wind are bounded to fulfill the requirements of energy in the world by generating electricity. The electrical energy can be transferred only in a short distance but its storage in a huge quantity is still a big challenge. Batteries, flywheels, compressed air storage, and hydro pumped are all possible but these are available with limited capacity. The two main renewable sources of energy such as wind and solar are highly fluctuating and irregular, hence it is problematic to use these sources in a huge amount to meet the requirements of energy in the world. Solar energy produced under the condition of a cloudy atmosphere is very less and also produces a negligible amount of energy at night. The same is the case with wind energy; the energy is not produced consistently or constantly because the wind is not blown at a constant speed. The electrical energy can be stored by converting it into chemical energy in some compounds like methane, methanol, hydrogen, and long-chained hydrocarbons. These chemicals are easier to store or transport in comparison to electrical energy and hence this chemical energy is further used for various applications such as cooking, heating, cooling, etc.

By splitting water, it is simplest for us to get hydrogen from electrical energy. Electrolysis is the sole practical method these days to get hydrogen by splitting the water molecules, but there are other methods under investigation such as high temperature, thermal splitting, and photochemical. The electrolysis method is practiced since 1900, but the efficiency of conversion of this method is very high i.e. more than 80%. The energy stored in the form of hydrogen gas is environment friendly. Based on the conversion of water into hydrogen gas, the simplest theory of hydrogen economy is originated that would be helpful to understand this conversion technique (Rifkin 2002). However, because of the physical properties and chemical properties of hydrogen, it has few limitations (Bossel 2006). Its storage is very problematic because of its very low volumetric density. It is also difficult to handle such gas because of its explosive and highly flammable nature. The requirement of infrastructure for storing, transporting, and converting it safely is very expensive. Hence for storing energy in the form of chemical bonds, it requires less manageable

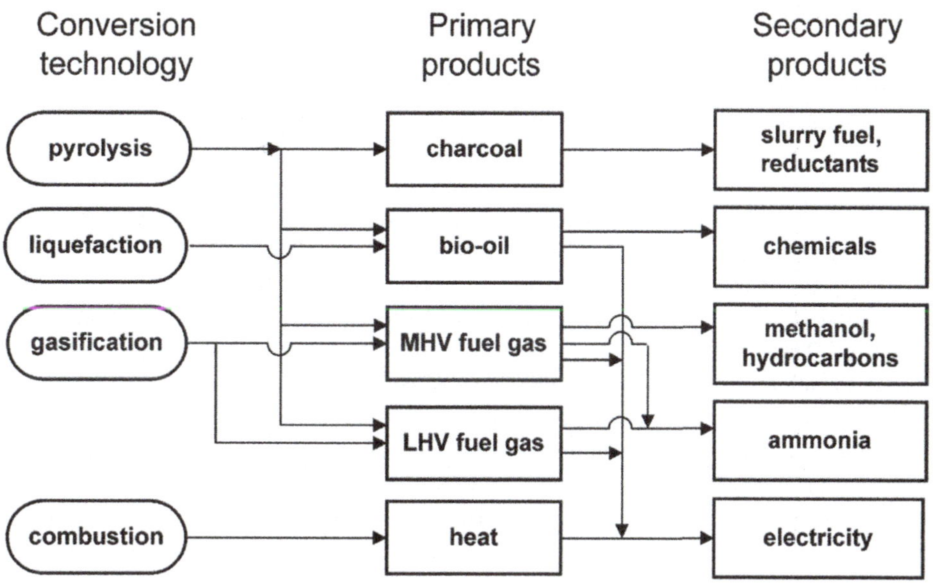

Fig. 1 Various technologies for the conversion of biomass into chemicals. Reproduced with permission from Ref. (Chmielniak and Sciazko 2003). Copyright @ Elsevier B.V. (2003)

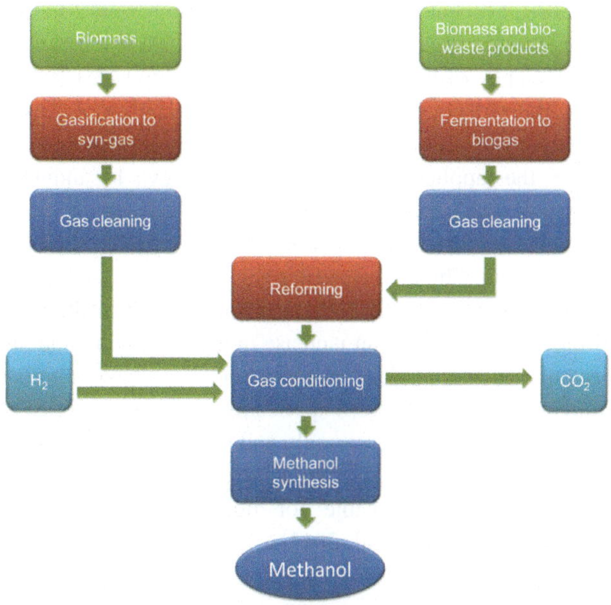

Fig. 2 Simplified schematic illustrating the methanol production from the biomass-based feedstock. Reproduced with permission from Ref. (Reschetilowski 2013). Copyright @ IOP Publishing (2013)

ways. In most of the applications, the chemicals in liquid form are preferred for storing energy over the chemicals in the gaseous form. However, in the transportation sector, it is highly required to use renewable liquid fuel instead of using kerosene, diesel fuel, and gasoline. Among the various chemicals, methanol is the prominent candidate to fulfill the requirements (Bromberg and Cheng 2010). Only one carbon atom is present in the methanol and hence it is simplest among other chemical compounds. It is also produced very easily by the chemical process (hydrogenation) of carbon dioxide whereas another technique such as electrochemical reduction is also being developed (Olah et al. 2011). There are various advantages of methanol and also promoted in the past for its use as an energy carrier (Reed and Lerner 1973; Moffat 1991; Gray and Alson 1985; Kohl 1990). The octane rating of methanol is very high; hence it is an important substitute or additive for gasoline internal combustion engines (ICE).

Methanol is a well-known candidate that is proficiently used in diesel engines. Further, methanol is also utilized in DMFC where it is employed to directly generate the electrical power at some ambient temperature conditions (McGrath et al. 2004). Methanol exists in gaseous form and therefore, this gas can be liquefied easily under moderate pressure. Its dehydration results in the formation of DME whose properties are similar to the LPG. Moreover, in most home applications like cooking and heating, it competently replaces LPG. (Arcoumanis et al. 2008). Furthermore, both DME as well as methanol fuel are superior to the generation of electric power in gas turbines (Temchin 2003). Besides its applications as a fuel, methanol is also used in various chemicals like methyl tert-butyl ether, acetic acid, and formaldehyde. Due to these entire advantageous features, methanol is therefore considered a better substitute for diesel fuel and is attracting much consideration.

Gasification is an easy method to make methanol from biomass. It is a process (thermochemical) that involves the conversion of biomass into a mixture of gases such as hydrogen, syngas, carbon monoxide, and carbon dioxide. In this process, firstly solid biomass is dried and after that crushed to get particles with uniform shape and size. To get optimal results, the moisture content in the sample is not more than 20%. The pretreated biomass is passed into the gasifier where it is treated with oxygen and water at a particular pressure. The biomass starts burning when it interacts with oxygen due to which heat evolves and hence this heat is used for gasification. Some combustion gases are generated such as steam and carbon dioxide which after reacting with biomass generate other secondary gases such as carbon monoxide and hydrogen. In this process, no external heat sources are required; biomass itself acts as a fuel source. The single partial oxidation process is a very good step for the generation of syngas but this process has some technical problems. Therefore, the two-step process is conducted for the gasification of biomass. The first step is commonly known as destructive distillation or pyrolysis, 400–600 °C temperature is required for heating the dried biomass and hence for avoiding complete combustion, the oxygen-deficient atmosphere is generated. The pyrolysis gas is composed of steam, carbon dioxide, methane, hydrogen, carbon monoxide, and volatile tars. A total of 10–25% of the biomass is left after the complete process and this leftover residue is known as charcoal. Char conversion is the second step of the process, in this step the charcoal residue is treated with oxygen at 1300–1500 °C to get carbon monoxide. The obtained syngas has to be adjusted in the right amount of hydrogen, carbon monoxide, and carbon dioxide. These gases can be purified before passing through the methanol production center. The biomass has an advantage over other commonly used coals. The biomass contains a very low amount of sulfur and also a very less amount of metallic impurities (arsenic, mercury, etc.). Despite that, biomass gasification has some problems related to it. The formation of tar faces various challenges such as restriction in flow and clogging to commercialize the biomass-based gasification process. The nature as well as the amount of tars depends on the composition of biomass and technology of gasification (Milne et al. 1998; Williams 1995; Azar et al. 2003).

7 Replacement of Carbon-Based Energy Resources

Different kinds of energy resources are accessible that can be exploited as an alternative to carbon-emitting energy resources. These resources assist in closing the carbon cycle more effortlessly and consequently result in making our environment neat and clean. These energy resources are biomass energy, wind energy, solar energy, hydrothermal energy, geothermal energy, and ocean energy. Further, there is an extensive range of technologies that do either exist or under development and offer economical, consistent, and sustainable energy services using the above-mentioned renewable energy sources. However, these technologies differ significantly in their stage of progress, performance, and competitiveness which depend on the local conditions, physical and socioeconomic as well as the local availability of fossil fuels. Table 1 summarizes the categories of renewable energy conversion technologies (Goldemberg 2001).

8 Biomass Energy

To fulfill energy requirements in the world with sustainable development, biomass energy has a greater potential to achieve energy needs. Presently, in the primary energy of the world, biomass energy contributes to about 14%, but it is expected that in the year 2050, this energy will contribute up to 40%. Various benefits are obtained from biomass energy such as environmental, economic, and social. Solid biomass has vital traditional applications such as lighting, cooking, and small scale industries (cottage industries) having very less energy efficiency that results in a bad impact on the environment by excreting a large amount of waste product. Traditional bioenergy has efficiency varying from 2 to 20% in developing countries. But, by using modern technology, the efficiency of bioenergy can be reached up to 80–90% in industrialized countries like Sweden, Finland, and Austria (Rosillo-Calle 2006). That's why it's our priority to modernize the application of biomass energy. In industrial countries, the traditional application of biomass is greatly changed by modern applications. For instance, in Sweden, Finland, and China, the traditional application of biomass energy is rapidly declined. Continuous growth in the usage of bioenergy is due to an increase in the population and also because of the lack of cheaper and accessible alternate sources of energy. Modern applications of these energies are confined in the urban areas only and the lack of some factors such as skills, capital, market structure, technology, etc. in the village is responsible for utilizing only traditional applications (Chum et al. 2011).

9 Wind Energy

Wind energy is the most important and vital resource of energy, because of its high neatness and negligible production of pollutants (Wang et al. 2011). The amount of this energy is generally relying on the environmental conditions as well as on the speed of wind (Dongmei et al. 2011; Sideratos and Hatziargyriou 2007). As we know, a nuclear fusion reaction continuously takes place in the sun which causes the electromagnetic radiations that reach the earth and it becomes warmer. As we are familiar that the earth is not equally distributed; it includes high mountains as well as plains so the environmental conditions vary from one place to another. Low-pressure regions are created in hilly areas. This is the responsible factor for the blowing of

Table 1 Categories of renewable energy conversion technologies (Goldemberg 2001)

Technology	Energy product	Application
Biomass energy Combustion (domestic scale) Combustion (industrial scale) Gasification/power production Gasification/fuel production Hydrolysis and fermentation Pyrolysis/production of liquid fuels Pyrolysis/production of solid fuels Extraction Digestion	Heat (cooking, space heating) Process heat, steam, electricity Electricity, heat (CHP) Hydrocarbons, methanol, H_2 Ethanol Bio-oils Charcoal Biodiesel Biogas	Widely applied; improved technologies available Widely applied; potential for the improvement Demonstration phase Development phase Commercially applied for sugar/ starch crops; production from wood under development Pilot phase; some technical barriers Widely applied; wide range of efficiencies Applied; relatively expensive Commercially applied
Wind energy Water pumping, battery charging Onshore wind turbines Offshore wind turbines	Movement, power Electricity Electricity	Small wind machines, widely applied Widely applied commercially Development and demonstration phase
Solar energy Photovoltaic solar energy conversion Solar thermal electricity Low-temperature solar energy use Passive solar energy use Artificial photosynthesis	Electricity Heat, steam, electricity Heat (water and space heating, cooking, drying) and cold Heat, cold, light, ventilation H_2 or hydrogen-rich fuels	Widely applied; rather expensive; further development needed Demonstrated; further development needed Solar collectors commercially applied; solar cookers widely applied in some regions; solar drying demonstrated and applied Demonstrations and applications; no active parts Fundamental and applied research
Hydropower	Power, electricity	Commercially applied; small and large scale applications
Geothermal energy	Heat, steam, electricity	Commercially applied
Marine energy Tidal energy Wave energy Current energy Ocean thermal energy conversion Salinity gradient/osmotic energy Marine biomass production	Electricity Electricity Electricity Heat, electricity Electricity Fuels	Applied; relatively expensive Research, development, and demonstration phase Research and development phase Research, development, and demonstration phase Theoretical option Research and development phase

the wind. So we can establish windmills in such areas where the speed of the wind is generally high. Hence, it is a better replacement for the carbon dioxide producing resources for the elimination of pollution. There are various types of windmills but presently the most efficient turbine unit has three cutting edges that rotate on the top of the steel tower which is as shown in Fig. 3 (Kumar et al. 2018). Its blades are made up of fiberglass. The rotor blades can be turned to optimize power and the entire tower is rotated to face the wind.

10 Solar Energy

On our planet earth, solar energy is the inexhaustible source of energy procured from the sun. Presently, various modern technologies are being used to get electric energy from solar energy. These techniques have been utilized in almost all countries in the world as an alternative to non-hydro technologies. Hypothetically, solar energy has the potential to cover the energy demands of this planet if technologies for

Fig. 3 NEG Micon 1.5 MW Wind Turbine, 68 m diameter Rotor, Tubular Steel Tower. Reproduced from Ref. (Kumar et al. 2018) under Creative Commons license. Copyright @ Gyancity Research Lab (2018)

supplying, storing, distributing were enthusiastically available (Blaschke et al. 2013). Solar radiations take around 8 min to reach the surface of the earth. Near about four million exajoules of energy from the sun reaches the blue planet yearly, out of which fifty thousand exajoules are harvested easily. Even we get such a huge amount of energy from the sun but the contribution of this energy to the total energy used in the world is very low. One of the major issues regarding research is connected with the movement of reducing the emission of carbon at a global level, which has been a social, environmental, economic, and political issue in the past decade. For instance, through the installation of household solar panels (113,533) in California in the USA, about 696,544 metric tons of carbon dioxide have been reduced. Hence, by using such types of technologies based on solar energy, various issues related to unemployment, climate change, and energy security, etc. have been greatly reduced. It is also predicted that solar energy will play a crucial role in the tertiary sector of the economy (transportation sector) in the future. Funding, investment, policies, and planning from organizations either governmental or private for such technologies strengthen the foundation for using such type of renewable energy system. Rebate and incentives also play an important role in the development of such markets. Additionally, for the generation of electric energy from solar energy, subsidies must be provided to the residents. Solar energy is the best replacement for the pollution causing sources of energy because it never harms our environment. Life becomes easier with the help of such type of energy. It helps us to regulate the consumption of fossil fuels and make them available for future generations. Figure 4 shows the conversion of solar energy into solar fuel (Steinfeld and Palumbo 2001). Concentrated solar radiation is used as the energy source for high-temperature process heat to drive chemical reactions towards the production of storable and transportable fuels.

11 Ocean Energy

On earth, two types of crust are found viz. continental crust and oceanic crust. A large part of the earth is covered with an ocean. In the ocean, a huge amount of renewable energy sources are present. Rely on characteristics and origin, there are six types of ocean energy: tidal current, ocean wave, salinity gradient, tidal range, ocean thermal energy, and ocean currents (Huckerby et al. 2011). Nowadays, technologies related to ocean energy other than tidal range are supposed to be an early stage of development. Both energies such as tidal current and ocean wave are very important and advanced energy sources that are also expected to significantly utilize in the future. In this chapter, along with other renewable resources, we also focused on these two types of energy, because these two sources generate green power. Recently, some industries which are based on oceanic energy sources have acquired much attention and achieved significant growth but it has not reached its zenith (Magagna et al. 2014). Some challenges were faced for the development of this technology such as the reduction in the cost, reduction of risk, and creation of an environment-friendly atmosphere.

Oceanic energy has the potential to generate a huge amount of electricity if we properly utilize all the sources present in the ocean. Tidal energy plays an effective role and hence its contribution is much greater than the other sources of oceanic energy. Tidal energy can be directly generated through the tides in the ocean. Tide arises in the ocean due to the position of the earth, moon, and sun. If all of these bodies i.e. the earth, moon, and sun lie in a straight line then, in that case, higher tides are generated. The gravitational force of both sun and the moon are added up in that case and hence, the net gravitational force is exerted on earth. But, when the position of the sun, moon, and earth are not linear, then lower tides are generated. The higher tides not only help to generate electricity in the ocean but also help the fishermen in catching the fishes, navigation, etc. This type of energy is also a good source of energy and helps us to minimize the use of fossil fuels (non-renewable sources of energy).

12 Geothermal Energy

The shortage of energy, increase in the population of the planet and rapid increase in environmental pollution are the major problems that the world is facing for the attainment of

Fig. 4 Schematic of solar energy conversion into solar fuels. Reproduced with permission from Ref. (Steinfeld and Palumbo 2001). Copyright @ Elsevier Inc. (2004)

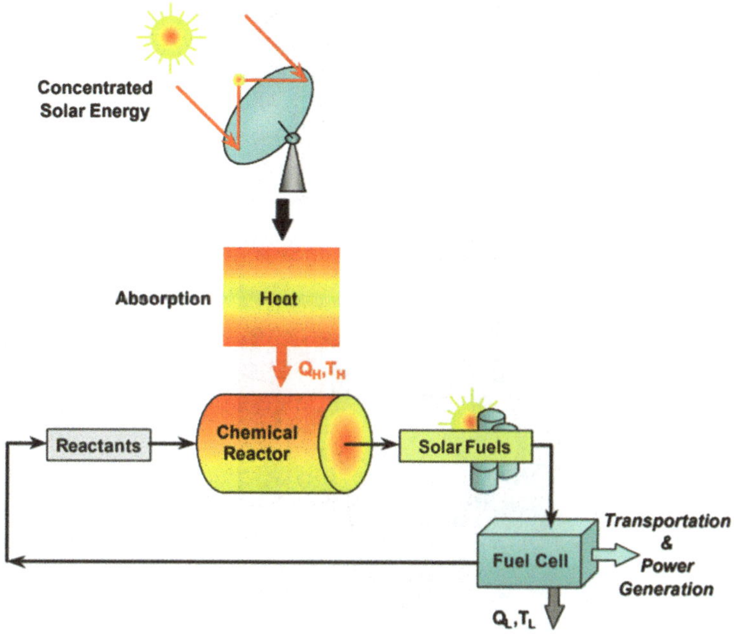

sustainable development. Presently, almost all the countries in the world are indulged in research on renewable energy such as tidal energy, wave energy, solar energy, wind energy, geothermal energy, and hydrothermal energy for the reduction of usage of traditional fossil fuels (Hepbasli 2004; Bahadori et al. 2013; Aikins and Choi 2012). It is also published in the research papers that the generation of energy from renewable resources is anticipated to double or triple before 2035 and reach near about 31% of the total production of electricity. Geothermal energy is produced 0.3% of the power generated in the world and it is 1.5% of the total power produced from renewable resources. Geothermal energy attracts more attention because of its stability, abundance, and CO_2 emission-free as compared to wind, solar, and hydraulic energy (Li 2015). The core of the earth has a temperature of more than 5800 °C and because of that very high temperature; the earth's interior possesses a huge amount of energy which is inexhaustible in comparison to other energy resources (Barbier 2002). More than 80 countries in the world directly use geothermal energy and about 30 countries use geothermal energy for the generation of power (Unverdi and Cerci 2013; Kuo et al. 2011; Li et al. 2012; Lund et al. 2010; Yang 2013; Lund and Boyd 2016). The United States is the largest country in the world for the production of power through geothermal power plants and contributes about 28.8% to the total production capacity of the world. For direct utilization of geothermal energy, it stands at rank 2nd in the world and contributes near about 24.6% of the total. China is the no. 1 country in the world for direct geothermal utilization and contributing near about 25% of the total. In the past few years, it is observed that there is an increase in the growth of geothermal power plants. One example of such a power plant is shown in Fig. 5 (Zhu et al. 2015). It is confirmed that the geothermal energy resources are pollution-free and hence, the geothermal energy can be effectively used for sustainable development.

13 Hydrothermal Energy

Non-renewable sources like petroleum, natural gas, and coal are mainly used as the main sources for the generation of power in the world (Lazkano et al. 2017). The total dependence on these fossil fuels from the past few decades results in global warming and scarcity of fossil fuels and thus it reduces the usage of these non-renewable resources (Chala et al. 2019). In the united state of America, the generation of power requires near about 600 billion US dollar and in South Africa, the generation of power require near about 150 billion US dollar for the fulfillment of power density which are based on one dollar per watt capital cost in a power plant (conventional gas-based power plant). Nowadays, it becomes a trend to switch towards renewable sources from fossil fuel technologies, and hence the development of other alternate technologies and energy sources for sustainable development starts growing day by day (Guangul et al. 2019). Shifting towards pollution-free energy, like solar, wave, geothermal and water energy attracts more attention these days in the world. The replacement of fossil fuels over the past decades is facing up and down in the way of progress. These alternative energy sources are adopted by the various nations of the world for sustainable development. Indonesian National Energy Policy provides the data which

Fig. 5 300 kW geothermal power station in Fengshun, Guangdong. Reproduced with permission from Ref. (Zhu et al. 2015). Copyright @ Elsevier Inc. (2015)

shows that in 2050; the renewable resources of energy would contribute 31% of total energy, from which contribution of biomass energy would be 23%, biodiesel 21%, hydropower 10%, coal gas methane 6%, geothermal 20%, nuclear 7%, bioethanol 4% and the remaining from wind, solar, sea, biogas, etc. (Erinofiardi et al. 2017). Environmental protection and economic growth are the major challenges that are faced by these sources of energy (Ma'arof et al. 2018). The two major barriers that come in the way of promoting the usage of these energy sources are the difficulties to control and irregular distribution of these energy sources (Kocaman and Modi 2017). For the development of the hydropower project, various technological facilities are required such as stream technology, running rivers, dammed reservoirs, new gravitational vortex, and pumped storage. In the European continent, the research on this topic is primarily focused on the component of electromechanical apparatus like pumps, generators, and turbines. Nowadays, reaction turbines and impulse turbines are mainly used for the generation of energy from hydropower plants. The impulse turbines are further classified as crossflow, turbo, and Pelton turbines. Impulse turbines are less efficient in high flow and low head sites in comparison with reaction turbines (Yaakob et al. 2014). Through the channel, the continuously flowing water strikes with the specific parts of the wheel due to which the turbine starts to rotate. The rotation of the turbine converts the potential energy of water into kinetic energy which is further converted into electrical energy with the help of a generator that is connected with a rotating shaft. The benefits of hydroelectric power such as environmental, economical, and technical make it a crucial contributor to sustainable development mainly in underdeveloped countries. The popularity of the hydropower plant technique is confined only in the hilly areas that possess numerous rivers, for instance, India, Nigeria, Brazil, Ethiopia, and Malaysia. In 2013, 16.3% of total electricity is generated through hydropower plants which are nearly 75.1% of natural energy resources (Hidrovo et al. 2017). In the past decade, the generation of electricity from Nepal Electricity Authority (NEA) is near about 539 MW power out of which 485 MW are produced through hydropower plants and the remaining 54 MW from liquid fuel. Similar research papers are also published which report that Independent Power Producers (IPPs) generate 362 megawatts of power by utilizing hydro sources (Alam et al. 2017). However, the development of a large hydropower plant needs much amount of money and it also causes environmental instability which leads to less demand on the earth. In this situation, various countries adopt hydropower plant technique for the generation of electricity only at a small scale. For the generation of electricity from water, the hydropower plant at the mini/micro level has attracted various countries because of its usage in a rural area and environment-friendly nature.

The increase in the amount of carbon dioxide in our environment is the major factor for global warming and this arises a major problem in the world. Every country in the world either developed or developing is facing this dangerous problem. Various organizations are running in the world either governmental or nongovernmental to tackle this disaster. The efforts of these organizations have been proved successful up to some extent. Various awareness programs such as Van Mahotsav, save tree save the environment and many more are awaking almost all the people in the world for the betterment of the environment. In the environment, carbon dioxide is present at nearly about 0.03%. This amount of carbon dioxide is good for the nourishment of the earth. It makes our planet warm and makes life possible only in this planet but if the amount of carbon dioxide increase slightly more than 0.03%, it would cause global warming and would be harmful to living being on the earth. The electromagnetic radiations come from the sun and only a few amounts of its light radiation reach the earth. These light radiations are taking near about eight minutes to reach the surface of the earth. The surface of the earth starts absorbing these radiations in a day time and this process is known as isolation. During the night time, the earth starts emitting these absorbed radiations. Due to the presence of carbon dioxide in the environment, these radiations are not successful to escape from the earth. The carbon dioxide helps to keep this radiation in the atmosphere and hence our environment remains warm. This is the reason for the existence of life on the earth among the other planets. So it is up to us, to keep this amount of carbon dioxide constant. To achieve this goal, we have to use natural resources in a good manner. For this reason, we must shift from using fossil fuels to green power resources.

These days, green power is marked as the essential power source in developing countries like Bangladesh, northern India, and central-southern China. The prosperity of Nepal mainly depends on hydropower resources because the potentiality of hydropower is huge in Nepal. Thus, this strength of Nepal makes them the main producer and exporter of green power and for better trades of power, Nepal has purchased secured power agreements with neighboring countries like Bangladesh and India. The revenue generated from exporting power supplies is very helpful for developing countries to generate funds for housing, agriculture, education, infrastructures, and healthcare (Manzano-Agugliaro et al. 2017). There is no consumption of water for the production of power through the installation of hydropower plant that helps the developing countries in views of economic, social, and environmental perspective (Faria et al. 2017). Inadvertent of the huge capital cost of a hydropower plant, it has low maintenance and renewing cost and also the favorable rainfall that makes it productive in comparison to other green energy production sources. Hydropower plays an effective role in controlling floods and is also used for irrigation purposes.

Micro-hydro technology is flourishing nowadays because of its low maintenance and environmental friendly. Due to the negligible cost of fuel, efficient technology, and very low operative cost, the hydropower plant attracts much attention and hence, it is a better alternative energy source. Both William and Simpson observed the new scheme (Pico-Hydro) and told them that this new scheme is cheap and hence it is a good option for some areas. The small hydropower plant is very helpful to improve the life quality of the villagers. This power plant creates a job for the local people due to which they run their life easily. These small power plants also gain the support of the government to run smoothly in the village over large hydropower plants. The stored water in the dams is not only used to generate electricity but also used for other purposes such as irrigation, fish production, and recreational tourism. In comparison with other sources of energy, the generation of electricity through hydropower plant has very low emission of methane and carbon dioxide and no emission of NO_2, NO, and SO_2, that leads to its better performance. It is also noted that the sustainable usage of H_2O for irrigation, water supply, and hydropower plant have high investment returns. The growth in the economy and the consumption of hydropower energy are complementary and hence, with the rise in the consumption of hydropower energy, there is also an increase in economic growth.

14 Conclusions

The burning of fossil fuels harvested the energy that allowed the unprecedented and rapid development of human society. But, the utilization of fossil fuels globally changes the environment. Numerous solutions are given by scientists to overcome this disturbance. Conservation of energy is helpful up to some extent but, it will not fully solve this issue relating to the carbon cycle. One of the important solutions is to reduce the usage of fossil fuel-based energy resources, and instead of using these resources, we can use nuclear energy resources as well as renewable energy resources. The most general solution to control this problem is CCS. In this case, the carbon dioxide is captured from numerous sources such as industrial facilities and power plants facilities followed by dumping this captured carbon dioxide in the earth. The enhanced oil recovery can be done by utilizing the captured carbon dioxide. DAC is very helpful to capture the carbon dioxide that arises from the burning of fossil fuels in the trains, airplanes, cars, domestic heating, and stoves. We can utilize the extracted carbon dioxide by converting it into dimethyl ethanol, methanol, and hydrocarbons and therefore can be used in a variety of applications which also play a significant role in the determination of the cost associated with DAC. Different kinds of energy resources are accessible that can be employed as an alternative to carbon-emitting energy resources. These resources assist in closing the carbon cycle more effortlessly and consequently result in making our environment neat and clean.

References

Abas N, Kalair A, Khan N (2015) Review of fossil fuels and future energy technologies. Futures 69:31–49

Aikins KA, Choi JM, (2012) Current status of the performance of GSHP (ground source heat pump) units in the Republic of Korea. Energy 47:77–82

Alam F, Alam Q, Reza S, Khurshid-ul-Alam SM, Saleque K, Chowdhury H (2017) A review of hydropower projects in Nepal. Energ Procedia 110:581–585

Arcoumanis C, Bae C, Crookes R, Kinoshita E (2008) Fuel 87:1014–1030

Aro EM (2016) From first generation biofuels to advanced solar biofuels. Ambio 45(1):24–31

Azar C, Lindgren K, Andersson BA (2003) Energ Pol 31:961–976

Bahadori A, Zendehboudi S, Zahedi G (2013) A review of geothermal energy resources in Australia: current status and prospects. Renew Sustain Energ Rev 21:29–34

Barbier E (2002) Geothermal energy technology and current status: an overview. Renew Sustain Energ Rev 6:3–65

Blaschke T, Biberacher M, Gadocha S, Schardinger I (2013) Energy landscapes: meeting energy demands and human aspirations. Biomass Bioenerg 55:3–16

Bossel U (2006) Proc IEEE 94:1826–1837

Bridgwater AV (1994) Appl Catal 116:5–47

Bromberg L, Cheng WK (2010) Methanol as an alternative transportation fuel in the US: Options for sustainable and/or energy-secure transportation. Cambridge, MA: Sloan Automotive Laboratory, Massachusetts Institute of Technology

Bromberg L, Cohn DR (2010) Heavy duty vehicles using clean, high efficiency alcohol engines, PSFC/JA-10-43. MIT, Cambridge

Bulushev DA, Ross JRH (2011) Catal Today 171:1–13

Chala GT, Guangul FM, Sharma R (2019) Biomass energy in malaysia-A SWOT analysis. In 2019 IEEE Jordan international joint conference on electrical engineering and information technology (JEEIT) (pp 401–406). IEEE

Chen Z, Deng S, Wei H, Wang B, Huang J, Yu G (2013) Polyethylenimine-impregnated resin for high CO2 adsorption: an efficient adsorbent for CO2 capture from simulated flue gas and ambient air. ACS applied materials & interfaces 5(15): 6937–6945

Chmielniak T, Sciazko M (2003) Appl Energ 74:393–403

Chum H, Faaij A, Moreira J, Berndes G, Dhamija P, Dong H, Pingoud K (2011) Bioenergy. In: Renewable energy sources and climate change mitigation: special report of the intergovernmental panel on climate change (pp 209-332). Cambridge University Press.

Choi S, Drese JH, Eisenberger PM, Jones CW (2011) Environ Sci Technol 45:2420–2427

Didas SA, Kulkarni A, Sholl DS, Jones CW (2012) Chemsuschem 5:2058–2064

Dongmei Z, Yuchen Z, Xu Z (2011) Research on wind power forecasting in wind farms. In 2011 IEEE Power Engineering and Automation Conference (Vol. 1, pp 175–178). IEEE

Edenhofer O, Pichs-Madruga R, Sokona Y, Seyboth K, Kadner S, Zwickel T, Matschoss P (Eds.) (2011) Renewable energy sources and climate change mitigation: Special report of the intergovernmental panel on climate change. Cambridge University Press.

Erinofiardi PG, Date A, Akbarzadeh A, Bismantolo P, Suryono AF, Mainil AK, Nuramal A (2017) A review on micro hydropower in Indonesia. Energ Procedia 110:316–321

de Faria FA, Alex Davis M, Severnini E, Jaramillo P (2017) The local socio-economic impacts of large hydropower plant development in a developing country. Energ Econ 67:533–544

Goeppert A, Czaun M, May RB, Prakash GKS, Olah GA, Narayanan SR (2011) J Am Chem Soc 133:20164–20167

Goeppert A, Czaun M, Prakash GS, Olah GA (2012) Air as the renewable carbon source of the future: an overview of CO_2 capture from the atmosphere. Energ Environ Sci 5(7):7833–7853

Goeppert A, Zhang H, Czaun M, May RB, Prakash GKS, Olah GA, Narayanan SR (2014) Chem SusChem 7:1386–1397

Goldemberg J (2001) World energy assessment. Energ Challenge sustain

Gray CL Jr, Alson JA (1985) Moving America to Methanol. The University of Michigan Press, Ann Arbor

Guangul FM, Chala GT (2019) Solar energy as renewable energy source: SWOT analysis. In 2019 4th MEC international conference on big data and smart city (ICBDSC) (pp 1–5). IEEE

Hamelinck CN, Faaij APC (2002) J Power Sour 111:1–22

Harkin T, Hoadley A, Hooper B (2010) Reducing the energy penalty of CO_2 capture and compression using pinch analysis. J Clean Prod 18 (9):857–866

He H, Li W, Zhong M, Konkolewicz D, Wu D, Yaccato K, Rappold T, Sugar G, David NE, Matyjaszewski K (2013) Energ Environ Sci 6:488–493

Hepbasli A, Ozgener L (2004) Development of geothermal energy utilization in Turkey: a review. Renew Sustain Energ Rev 8(5):433–460

Hidrovo B, Andrei JU, Amaya M-G (2017) Accounting for GHG net reservoir emissions of hydropower in Ecuador. Renew Energ 112:209–221

Huckerby J, Jeffrey H, Jay B (2011) An international vision for ocean energy (Phase 1)

Keith DW, Ha-Duong M, Stolaroff JK (2006) Clim Change 74:17–45

Kemp IC (2011) Pinch analysis and process integration: a user guide on process integration for the efficient use of energy. Elsevier

Khoo HH, Tan RBH (2006) Energ Fuels 20:1914–1924

Kocaman AS, Modi V (2017) Value of pumped hydro storage in a hybrid energy generation and allocation system. Appl Energ 205:1202–1215

Kohl WL (1990) Methanol as an alternative fuel choice: an assessment

Kumar A, Khan MZU, Pandey B, Mekhilef S (2018) Wind energy: a review paper. Gyancity J Eng Technol 4(2):29–37

Kuo CR, Hsu SW, Chang KH, Wang CC (2011) Analysis of a 50 kW organic Rankine cycle system. Energ 6(10):5877–5885

Lackner KS (2009) Eur Phys J Spec Top 176:93–106

Lackner KS (2010) Washing carbon out of the air. Sci Am 302(6):66–71

Larson ED (1993) Technology for electricity and fuels from biomass. Annu Rev Energ Environ 18(1):56–7630

Lazkano I, Nøstbakken L, Pelli M (2017) From fossil fuels to renewables: the role of electricity storage. Eur Econ Rev 99:113–129

Li K, Bian H, Liu C, Zhang D, Yang Y (2015) Comparison of geothermal with solar and wind power generation systems. Renew Sustain Energ Rev 42:1464–1474

Li T, Zhu J, Zhang W (2012) Cascade utilization of low temperature geothermal water in oilfield combined power generation, gathering heat tracing and oil recovery. Appl Therm Eng 40:27–35

Lund JW, Boyd TL (2016) Direct utilization of geothermal energy 2015 worldwide review. Geothermics 60:66–93

Lund JW, Freeston DH, Boyd TL(2010) Direct utilization of geothermal energy 2010 worldwide review. Proceedings World Geothermal Congress, Bali, Indonesia, 25–29 April 2010, 1–23

Ma'arof MIN, Chala GT, Ravichanthiran S (2018) A study on microbial fuel cell (MFC) with graphite electrode to power underwater monitoring devices. Int J Mech Technol 9(9):98–105

MacElroy JD (2016) Closing the carbon cycle through rational use of carbon-based fuels. Ambio 45(1):5–14

Magagna D, MacGillivray A, Jeffrey H, Hanmer, C, Raventos A, Badcock-Broe A, Tzimas E (2014) Wave and Tidal Energy Strategic Technology Agenda, published by SI Ocean

Mahmoudkhani M, Keith DW (2009) Int J Greenhouse Gas Control 3:376–384

Mahmoudkhani M, Heidel KR, Ferreira JC, Keith DW, Cherry RS (2009) Energ Procedia 1:1535–1542

Manzano-Agugliaro F, Taher M, Zapata-Sierra A, Juaidi A, Montoya FG (2017) An overview of research and energy evolution for small hydropower in Europe. Renew Sustain Energ Rev 75:476–489

McGrath KM, Prakash GKS, Olah GA (2004) J Ind Eng Chem 10:1063–1080

Meth S, Goeppert A, Prakash GKS, Olah GA (2012) Energ Fuels 26:3082–3090

Milne TA, Evans RJ, Abatzaglou N (1998) Biomass gasifier"Tars": their nature, formation, and conversion

Moffat AS (1991) Science 251:514–515

Olah GA, Prakash GKS, Goeppert A (2011) J Am Chem Soc 133:12881–12898

Ortiz FG, Serrera A, Galera S, Ollero P (2013) Methanol synthesis from syngas obtained by supercritical water reforming of glycerol. Fuel 105:739–751

Pachauri RK, Allen MR, Barros VR, Broome J, Cramer W, Christ R, van Ypserle JP (2014) Climate change 2014: synthesis report. Contribution of Working Groups I, II and III to the fifth assessment report of the Intergovernmental Panel on Climate Change (p 151). Ipcc

Pielke RA Jr (2009) Environ Sci Pol 12:216–225

Ranjan M, Herzog HJ (2011) Energ Procedia 4:2869–2876

Reed TB, Lerner RM (1973) Science 182:1299

Reschetilowski W (2013) Russ Chem Rev 82:624–634

Rifkin J (2002) The hydrogen economy. Tarcher/Putnam, New York

Rosillo-Calle F (2006) Energy technologies. In: Biomass energy, in Landolf-Bornstein handbook, vol 3. Springer, Germany. https://doi.org/10.1007/10858992_13

Sculley JP, Zhou HC (2012) Enhancing amine-supported materials for ambient air capture. Angew Chem Int Ed 51(51):12660–12661

Semelsberger TA, Borup RL, Greene HL (2006) J Power Sour 156:497–511

Sideratos G, Hatziargyriou ND (2007) An advanced statistical method for wind power forecasting. IEEE Trans power syst 22(1):258–265

Simon AJ, Kaahaaina NB, Friedmann SJ, Aines RD (2011) Energ Procedia 4:2893–2900

Specht M, Bandi A (1999) The'methanol cycle'-sustainable supply of liquid fuels; Der"Methanol-Kreislauf"-nachhaltige Bereitstellung fluessiger Kraftstoffe

Steinfeld A, Palumbo R (2001) Solar thermochemical process technology. Encycl Phy Sci Technol 15(1):23–756

Stolaroff JK, Keith DW, Lowry GV (2008) Environ Sci Technol 42:2728–2735

Temchin J (2003) Analysis of market characteristics for conversion of liquid fueled turbines to methanol prepared for the methanol foundation and methanex by electrotek concepts

Unverdi M, Cerci Y (2013) Performance analysis of Germencik geothermal power plant. Energ 52:192–200

Wang X, Guo P, Huang X (2011) A review of wind power forecasting models. Energ procedia 12:770–778

Wang T, Lackner KS, Wright AB (2013) Phys Chem Chem Phys 15:504–514

Williams RH (1995) Energ Sustain Dev 1.18–34

Yaakob OB, Ahmed YM, Elbatran AH, Shabara HM (2014) A review on micro hydro gravitational vortex power and turbine systems. Jurnal Teknologi 69(7):1–7

Yang W (2013) Experimental performance analysis of a direct-expansion ground source heat pump in Xiangtan, China. Energ 59:334–339

Zeman F (2007) Environ Sci Technol 41:7558–7563

Zhu J, Kaiyong Hu, Xinli Lu, Huang X, Liu K, Xiujie Wu (2015) A review of geothermal energy resources, development, and applications in China: current status and prospects. Energy 93:466–483

Carbon Dioxide Utilization to Energy and Fuel: Hydrothermal CO_2 Conversion

Demet Ozer

Abstract

CO_2 reduction has gained great attention due to its effects on climate changes, ecological environment, and human health. In addition to these, finding an alternative energy source to the non-renewable fossil fuels has been a crucial topic. For greenhouse gas control and supplanting fossil fuel, one promising technique is to transform CO_2 into fine chemicals, building materials, energy, and fuels using different reductants. Metals, catalysts, organic wastes, inorganic wastes, and biomass were used as reductant and converted CO_2 to alkanes, formic acid, methanol, phenol, syngas, and ethylene glycol. This chapter briefly surveys the reduction, conversion, and utilization of the CO_2 with these reductants through the hydrothermal method. The obtained high-temperature water treats as both a reaction ambient and a cheap hydrogen resource and gained economical and practical advantages in hydrothermal CO_2 conversion.

Keywords

CO_2 • Hydrothermal • Conversion • Utilization • Chemical • Energy • Fuel

1 Introduction

The increasing carbon dioxide levels in the Earth has been an important problem and finding new solutions to reduce the CO_2 amount as a greenhouse gas is crucial to control global warming (Philander and Philander 2008). From the viewpoint of the environmental safeguard, it is valuable to reduce the exhaust amount of CO_2 and reuse as carbon resources of low cost and vast quantities (Onoki et al. 2006). Due to its abundance, the CO_2 is an attractive raw material that is colorless, odorless, harmless, nontoxic, non-flammable, eco-friendly, inexhaustible, and renewable carbon source (Song 2002). Carbon dioxide has great thermodynamic stability and −390 kJ/mol free formation energy (Jagadeesan et al. 2015) that lets it transform high external energy substances like hydrogen, organometallic, and small-membered ring compounds, etc. (Andérez-Fernández et al. 2018). CO_2 is generally used in different industrial processes such as food and beverage industries as antimicrobial (Ravindra et al. 2014) and cryogenic freezing gas (Mascheroni 2012). CO_2 has also been applied as a surface cleaning agent (Sherman 2007), CO_2-enhanced oil recovery (North and Styring 2015), urea production (Hao et al. 1996), water treatment (Eskandarloo and Badiei 2015), the production of fire retardants and coolants (Zhu 2019), reactant and soft oxidant (Ansari and Park 2012). To remove and utilize CO_2 from the atmosphere, the most effective technique is to transform CO_2 into fine chemicals, building materials, energy, and fuels. For CO_2 conversion, several types of methods were successfully applied (Fig. 1). Fischer-Trospch reactions (Perez-Alonso et al. 2008), hydrogenations (Jessop et al. 2004; Gao et al. 2018), photochemical conversions (Denning et al. 2015), electrochemical reductions (Sanchez-Sanchez et al. 2001), photoelectrochemical reduction (Zhang et al. 2004), biological process supercritical CO_2 usage (Jessop et al. 1996), catalytic hydrogenation reduction (Inui et al. 1999) and various catalytic processes (Matsuo and Kawaguchi 2006) were used to obtain valuable products like methane (Tambach et al. 2009), methanol (Kobayashi and Takahashi 2004), ethanol (Inui et al. 1999) and phenol (Constantinou et al. 2012). However, these technologies have some limitations, selectivity, and productivity problems. New technically and economically feasible researches concerning process enrichment and energy integration are required (Navarrete et al. 2017). The hydrothermal reactions have

D. Ozer (✉)
Department of Chemistry, Hacettepe University, Ankara, 06590, Turkey
e-mail: demetbaykan@hacettepe.edu.tr

Inamuddin et al. (eds.), *Carbon Dioxide Utilization to Sustainable Energy and Fuels*,
Advances in Science, Technology & Innovation, https://doi.org/10.1007/978-3-030-72877-9_11

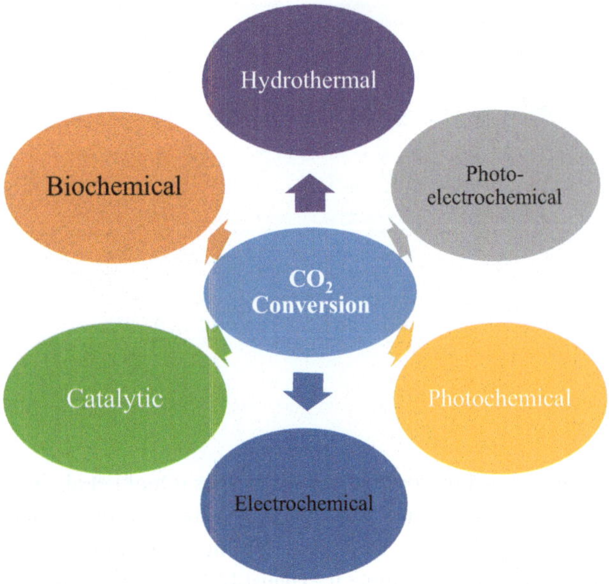

Fig. 1 The CO_2 conversion methods

gained great attention for the reduction of CO_2 to chemicals like alkanes, phenol, formic acid, methane, and methanol. High-temperature water that has lower dielectric constant, higher isothermal compressibility, and fewer hydrogen bonds than ambient water (Akiya and Savage 2002) are used as an environmentally benign solvent and enables to occur various reactions without any catalyst. In the hydrothermal method, gaseous CO_2, or basic CO_2 solution ($NaHCO_3$) is generally reduced using a different catalyst, organic reductants, organic wastes, and biomasses to eliminate the need for hydrogen gas that has economic and safety problems. Thus, CO_2 hydrothermal reduction can be simple, rapid, and efficient, which are important attributes for industry applications (Luo et al. 2017).

This chapter aims to explicate the hydrothermal conversion of CO_2 into valuable products as chemicals, fuels, and energy using various biomasses, organic wastes, inorganic wastes, and catalysts as reductants. Using these reductants, the formation of alkanes, methanol, formic acid, phenol, syngas, and ethylene glycol were detailed explained.

2 Hydrothermal CO_2 Conversion

To transform CO_2 into valuable chemicals, energy, and fuels via the hydrothermal method is an important process for the deduction of the CO_2 amount in the atmosphere to decrease global carbon and hence global warming. Besides, the global energy need is an important problem, and to find new methods have gained accelerating research attention. In recent times, the hydrothermal method has been used as an effective process in various industries. Hydrothermal CO_2 conversion has been impressed by the formation of fossil fuel in the global cycle. In the global cycle, the combination of CO_2, water, and solar energy form biomass that turns to waste with the effect of the human, waste get into the soil and turn to fossil fuels (petroleum, natural gas, coal, etc.) with the effect of geothermal energy. These fuels are consumed and turn to a large amount of CO_2 and waste and carbon cycle start again (Jin et al. 2010). To convert the carbon and obtain fuel and energy, various electrochemical, photochemical, and catalytic reactions have been widely used. The hydrothermal conversion is an alternative method with low cost and high efficiency. The major problem is to find an economic hydrogen source. The usage of hydrogen as an energy carrier suggests a considerable decrement in CO_2 levels (Arakawa et al. 2001). In the hydrothermal method, the water was used as an efficient hydrogen source and as a solvent, and hydrogen acts as a highly efficient reductant and easily reduces CO_2 into valuable products. In addition to water, organic wastes, inorganic wastes, and biomasses were used as a hydrogen source and reductant for CO_2 conversion. Metals and catalysts also act as a reductant to increase conversion of CO_2 amount. Some examples of hydrothermal conversion of CO_2, obtained products, using reductants and synthesis procedures were given in Table 1.

2.1 Metals and Catalysts as Reductant

Hydrothermal carbon reduction processes were successfully performed using various catalysts and metals. The catalysts and metals enable to obtain the appropriate reduction conditions. The chemical, energy, and fuel conversion of CO_2 has great potential for industrial processes. Taking into account the highly oxidizability and thermodynamically stability of CO_2, to find or synthesis highly reactive metals and catalysts is needed for the synthesis of different chemicals (Omae 2006).

Hydrocarbons are the major products acquired through hydrothermal CO_2 reduction. With changing the synthesis condition, different **alkanes** were synthesized. In methane synthesis, the dissolved CO_2 has been reduced to CH_4 with nickel–iron alloy (Ni_3Fe, awaruite) at 400 °C and 50 MPa inside a closed gold-cell hydrothermal reactor (Horita and Berndt 1999). The CO_2 and H_2 bearing aqueous fluids were applied to synthesis more abundant hydrocarbons (C_1-C_3) under hydrothermal conditions (400 °C and 500 bars) using a magnetite mineral catalyst that is a ubiquitous component of ultramafic-hosted hydrothermal systems (Fu et al. 2007). These types of geochemical and geomicrobiological hydrothermal processes were widely used and important for submarine (Karl et al. 1980) and subareal vent systems (Aharon 1994). At 300 °C and 30 MPa, C_4, and C_5 were

Table 1 Hydrothermal CO_2 conversion of products, using reductants and reaction conditions

Products	Reductants	Reaction conditions	% Yield	References
Formic acid	Iron metal	Hydrothermal CO_2 reduction (300 °C, 2 h)	92	(Duo et al. 2017)
Formic acid	Iron nanoparticles	Hydrothermal CO_2 reduction (200 °C, 72 h, 1.4 MPa)	8.5	(He et al. 2010)
Formic acid	Zinc metal	Hydrothermal CO_2 reduction (300 °C, 2 h, 1.4 MPa)	57.5	(Yang et al. 2017)
Formic acid	Zinc	Hydrothermal conditions (275 °C, 21 MPa, 15 min)	75	(Roman-Gonzalez et al. 2018)
Formic acid	Manganese	Autocatalytic hydrothermal CO_2 reduction (300 °C, 2 h)	43	(Lyu et al. 2017)
Formic acid	Aluminum	Autocatalytic hydrothermal CO_2 Reduction (300 °C, 2 h)	64	(Jin et al. 2017)
Formic acid	Glycerin	Hydrothermal CO_2 reduction (300 °C, 60 min, 9 MPa)	90	(Shen et al. 2017)
Formic acid	Microalgae	Hydrothermal conditions (300 °C, 2 h)	15.6	(Yang et al. 2019)
Formic acid	Hexanehexol	Hydrothermal $NaHCO_3$ reduction (300 °C, 1 h)	80	(Yang et al. 2018)
Formic acid	Isopropanol	Hydrothermal $NaHCO_3$ reduction (300 °C, 2.5 h)	70	(Shen et al. 2011)
Formic acid	Waste plastic	Hydrothermal $NaHCO_3$ reduction (300 °C, 8 h)	16.0	(Lu et al. 2020)
Methanol	Copper catalyst + zinc metal	Hydrothermal CO_2 reduction (350 °C, 2 h)	11.4	(Huo et al. 2017a)
Methanol	Cu supported $La_2O_2CO_3$	Hydrogenation of CO_2, 240 °C, in 5% H_2 95% N_2, 3.0 MPa	5.6	(Chen et al. 2018)
Methane	Iron metal	Hydrothermal CO_2 reduction (200 °C, 72 h, 0.14–1.4 MPa)	1.96	(Tian et al. 2017)
Methane	Nickel	Hydrothermal conditions (205 °C, 2 MPa, 60 min)	6.2	(Takahashi et al. 2008)
Phenol	Iron metal	Hydrothermal CO_2 reduction (200 °C, 120 h, 1.41 MPa)	1.21	(Tian et al. 2017)

prepared from CO_2 and H_2 using cobalt-bearing magnetite catalyst (Ji et al. 2008). The other transition metals (iron and chromium) bearing catalysts also assisted the abiotic formation of organic compounds in hydrothermal conditions (390 °C and 400 bars) (Foustoukos and Seyfried 2004). FeO- and Cr_2O_3-bearing systems affected hydrocarbon chain growth mechanisms (Novak and Madon 1984) and used as an influential catalyst for Fischer–Tropsch reaction of long-chain hydrocarbons (Foustoukos and Seyfried 2004).

The other hydrothermal converted chemical is **methanol** that is an important raw material and energy carrier for chemical industries like flavor, dye, medicine, and gunpowder, etc. (Liu et al. 2003). Methanol is widely used to produce solvents such as acetic acid that presents 10% of the global demand (Al-Saydeh and Zaidi 2018), methyl tert-butyl ether, and chloromethane (Olah 2005). It is also used as a fuel which is converted to high-octane gasoline, ethylene to olefins, and other petrochemicals (Goeppert et al. 2018) Compared with gasoline, methanol is more biodegradable and environmentally benign. The carbon dioxide converted the methanol using the hydrothermal method in the presence of Cu/Zn metals. Sodium bicarbonate is used as a carbon dioxide source, copper and zinc act as catalyst and reductant, respectively. Water takes both solvent and hydrogen source. This provides to avoid using gaseous hydrogen that is flammable, explosive, and not easy to operate during transportation (Huo et al. 2017b). The methanol yield was found 11.4% at 350 °C for 3 h using Cu catalyst and Zn reductant (Huo et al. 2012). Liu and coworkers showed that the ZnO has a synergistic effect with Cu to synthesize methanol (Liu et al. 2003). Gao and coworkers used a hydrotalcite-derived catalyst to synthesis methanol from CO_2 hydrogenation. Cu^{2+} didn't join layers, CuO was formed and showed good catalytic activity because of the maximum content of active species (Gao et al. 2012).

The most obtained chemical from the reduction of CO_2 is **formic acid** and it is used as a strong electrolyte, hydrogen storage material (Fellay et al. 2008), food additive (Food), excellent fuel for fuel cells (Rice et al. 2002), and an important commercial product for various industries (Shen et al. 2012). Formic acid has not been only an important chemical but also a useful reducing agent and carbon source in synthetic chemical industries (Duo et al. 2017). Farlow and Adkins synthesized formic acid using CO_2 and H_2 with RANEY® nickel catalyst via the direct synthesis in 1935 (Farlow and Adkins 1935). To produce formic acid from CO_2, the zero-valent metal/metal oxide redox cycles were used with high yield under hydrothermal conditions (Fig. 2).

The cycle depends on the oxidation and reduction efficiencies. Firstly, the water was used as a hydrogen source, and maximum hydrogen formation was achieved as 99.4%. Secondly, organic chemicals like glycerin were used as a reductant. In this cycle, CO_2 was turned to formic acid through oxidation of the metal to metal oxide that was formed its zero-valent state through oxidation of glycerin to lactic acid. Zinc, aluminum, iron, and manganese were used as metal source. The relative efficiency arrangement was as Al>Zn≫Mn>Fe (Jin et al. 2011). The iron nanoparticles used both catalyst and a reducing agent to reduce CO_2 to formic acid and acetic acid with 8.5 and 3.5 $mmol.L^{-1}$ yields, respectively. With increasing reaction temperature, the CO_2 conversion also increased as a result of increasing H_2 formation that ensures a stronger reducing environment. With increasing pressure, the yield of formic acid and conversion of CO_2 was increased because the number of active centers increased at the iron surfaces (He et al. 2010). In another study, nickel used with iron and formic acid produced. After the hydrothermal reaction, Fe-powder was converted to Fe_3O_4 and $FeCO_3$, while Ni showed no apparent change. With increasing nickel amounts, formic acid is reduced to form methane at the basic conditions up to 300 °C (Takahashi et al. 2006). Wu and coworkers used sodium bicarbonate as a carbon dioxide source and successfully converted carbon dioxide to formic acid with more than 98% selectivity and 15.6% yield under mild conditions (Fe/Ni ratio = 1:1, $NaHCO_3$/Fe ratio = 1:6, 300 °C, and 120 min) using Fe (reductant) and Ni (catalyst) (Wu et al. 2009). Zhong and coworkers were used Fe as a reductant, Cu as a catalyst, and water which acted not only the reaction media but also the source of hydrogen to convert CO_2 to formic acid through the hydrothermal method. The highest yield was found as 46% for formic acid under the following condition, Fe/Cu ratio of 1:1, CO_2/Fe mole ratio of 1:6, 35% filling rate, 300 °C, and 120 min (Zhong et al. 2010). Wang and coworkers examined the zinc effect on the enhancement of the CO_2 conversion to formic acid with different catalysts (Cu, Fe, and Sn). As a result, the reaction temperature was decreased to a low temperature (250 °C). Fe or Sn additions didn't remarkability affect on yield. However, the formic acid yield has increased to 60.47% using Cu that acts as an effective catalyst with higher catalytic activity in the conversion of CO_2 into formic acid (Wang et al. 2015). Manganese, as a first-row transition metal, has a reactive redox capacity and it mediates the splitting of water to provide electrons for photosynthesis. Lyu and coworkers used manganese to reduce CO_2 and produce formic acid. CO_2 has been used as a carbon source. It also enhances the production of hydrogen from water. The manganese oxidized to manganese oxide through the formation of $Mn(OH)_2$. The manganese oxide provides the catalytic activity to reduce CO_2 to formic acid. In experimental produce, an appropriate amount of manganese, sodium bicarbonate, and deionized water were put into the Teflon-lined reactor at 300 °C for 2 h. The formic acid was obtained with high selectivity (98%) and yield (43%) (Lyu et al. 2017). The Mn powder sizes (ranging from 50 to 1400 mesh) has no significant effect on formic acid conversion. With increasing Mn amount, the yield of formic acid increased evidently from 13 to 43%. After that, the yield remained nearly constant due to a stronger reduction condition. Manganese has high efficiency to reduce CO_2 to formic acid than iron (16%) (Wu et al. 2009). The reduction of CO_2 with aluminum (200 mesh) is an exothermic process under the hydrothermal method, the Al oxide converted to Al through solar energy. CO_2 dissolved in water and produced HCO_3^-. Sodium bicarbonate was used as a CO_2 source and aluminum powder was used as the reductant. The effects of reaction parameters were also investigated. The formic acid yield increased considerably with increasing the Al loading amount, reaction time, and temperature. The yield decreased with an increasing $NaHCO_3$ amount (Jin et al. 2017). In the hydrothermal CO_2 reduction with Zn, zinc hydride (Zn–H) is a key intermediate and the formation of formic acid is an SN2 mechanism (Zeng et al. 2014). Roman-Gonzalez and coworkers examined the reaction parameters to produce formic acid and the reaction conditions were optimized as 300 °C with rapid heating using 10 μm particle size. The conversion is up to 60% with 100% selectivity (Roman-Gonzalez et al. 2018). Zinc metal is also used as a reductant for direct and efficient conversion of ethylene carbonate to **ethylene glycol** that is an important intermediate and solvent that has been widely used for different purposes like antifreeze, detergents, pharmaceuticals, explosives, and plasticizers, etc. (Xu et al. 1995). A high yield of 99% was obtained at 250 °C for 2 h (Fu et al. 2016).

The **phenol** was synthesized by the hydrothermal reaction of CO_2 and H_2O using iron powder as a catalyst. The reaction was heated at 200 °C for 120 h with 1.4 MPa. The phenol yield was found as 1.21 mol % as to CO_2. The carbon dioxide chemisorbs onto the iron surface, as a result, iron atoms oxides and the carbon dioxide reduces. Additionally, other metals (Al, Zn, Co, Ni), metal-doped ZSM-5, MCM-41, and zeolites were applied under the same reaction conditions and no significant amount of phenol was obtained, although trace amounts of phenol were observed using Co and Ni catalysts (Tian et al. 2010). Tian and coworkers explained the reaction mechanism and showed in Fig. 3. Sodium bicarbonate was converted to CO_2 through the hydrothermal method and CO_2 adsorbed on the catalyst (iron powder) surface. Metal iron reacted with water, H_2 formed and attacked CO_2 to form formaldehyde. In an acidic medium, the phenol was obtained and yield increased (Tian et al. 2007).

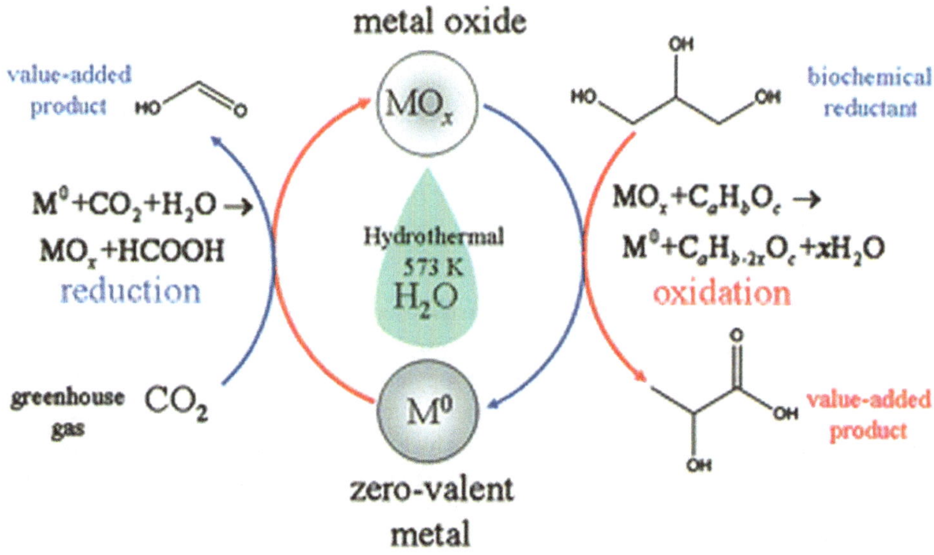

Fig. 2 Carbon and metal/metal oxide redox cycles. Reproduced from Jin et al. (2011) with permission from RSC

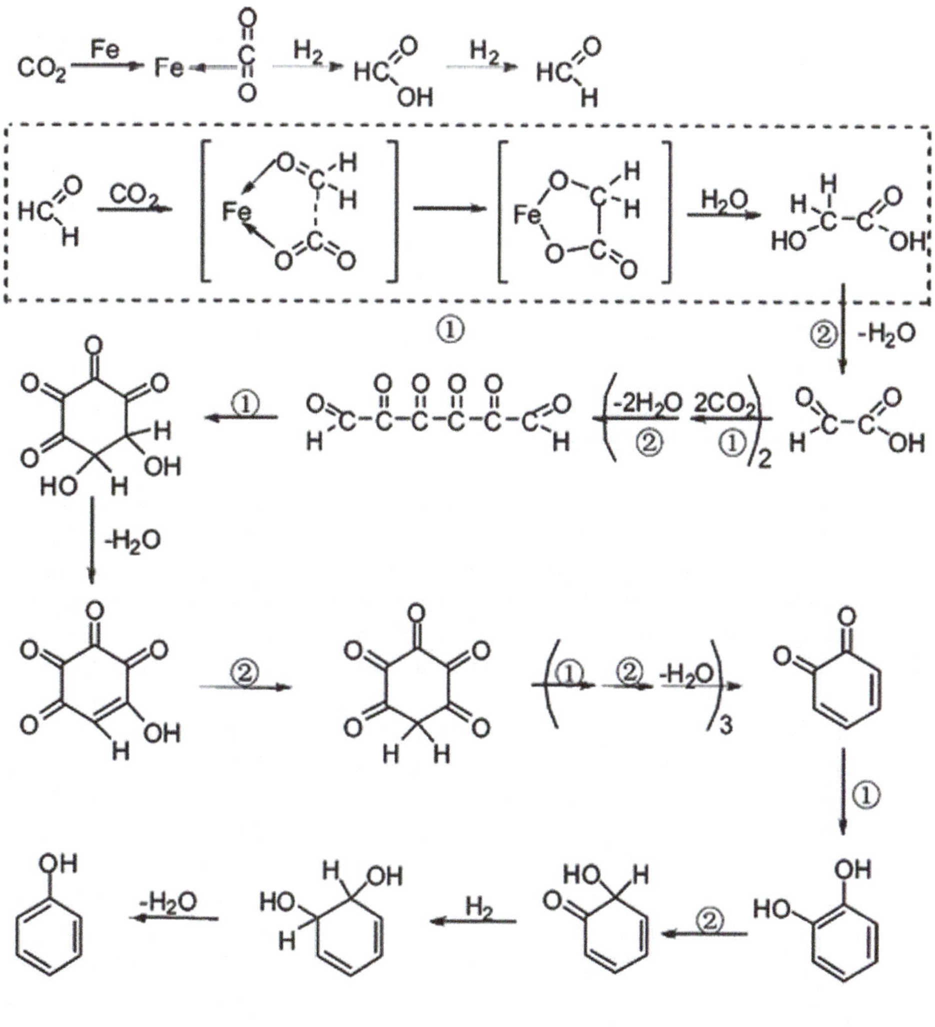

Fig. 3 Formation of phenol by hydrothermal method (1. The oxidative coupling reaction, 2. Rearrangement reaction) Copyright (Tian et al. 2007) from ACS

The conversion of carbon dioxide and methane that are the main greenhouse gases to **syngas** (H_2 and CO) is one of the most environmentally benign routes for methane reforming. Syngas is a source of clean energy and synthetic chemicals. CO_2 reforming of methane reactions was performed with Ni-doped silicalite zeolite as a catalyst that has good activity as well as the high surface area under atmospheric pressure in a fixed bed reactor. The pore of zeolites (0.56 nm) fastly diffused the H_2, CO, CO_2, and CH_4 and nickel loading was increased the H_2/CO ratio and hydrogen-rich syngas was formed (Bawah et al. 2018).

2.2 Organic Wastes as Reductant

The hydrothermal cracking of **polymer wastes** used as an efficient reductant to reduce CO_2. Zeng and coworkers used polyethylene (PE) as plastic and ethylene propylene diene monomer (EPDM) as sulfur-containing rubber waste for the reduction of CO_2 (Zeng et al. 2011). When the hydrothermal cracking of PE was rarely reduced CO_2, EPDM ensured a significant reduction in the CO_2 level. During the hydrothermal cracking of EPDM, hydrogen sulfide was released and reduced H_2O to hydrogen. Hydrogen reduced CO_2 to organic products and the hydrothermal reduction of EPDM into fuel oils.

Shen and coworkers used **organic hydrogen sources** as a reducing agent to produce chemicals from CO_2. $NaHCO_3$ was used as a CO_2 source and converted into formate using isopropanol as a reductant and also a catalyst in high-temperature water with 9 MPa reaction pressure and 300 °C reaction time. An equal amount of formate and acetone were formed via hydrogen-transfer reduction of $NaHCO_3$ by hydrothermal method and high-temperature water act like a catalyst (Shen et al. 2011).

2.3 Inorganic Wastes as Reductant

Hydrazine hydrate is an inorganic waste that has an amine group as a safe and simple hydrogen source for the reduction of CO_2. It was also used as hydrogen storage material (Lan et al. 2012). Yao and coworkers were used hydrazine hydrate for the reduction of CO_2 to chemicals. In a typical procedure, $N_2H_4.H_2O$, $NaHCO_3$, Ni, ZnO, and deionized water were put into the batch reactor at 300 °C within 20 s. They obtained no formic acid without $N_2H_4.H_2O$ and an 18% yield of formic acid with $N_2H_4.H_2O$. These results show that $NaHCO_3$ can be selectively reduced into formic acid by $N_2H_4.H_2O$. Nickel and zinc oxides were used as a catalyst. The 55.5% yield of formic acid was achieved with Ni and ZnO catalysts, whereas a 49% formate yield was obtained only with Ni (Yao et al. 2017).

2.4 Biomass as Reductant

Biomass has gained considerable attention as a sustainable and inexpensive feedstock. Up to now, the gasification and reforming process is widely used to produce hydrogen using biomass. The hydrogen directly transfers from biomass and CO_2 reduces to different chemicals at mild conditions. This solves the potential problems of CO_2 hydrogenation like thermodynamic limitation and a hydrogen source. Hydrogen abundant biomasses and their derived materials are successfully used in the biofuels industry as a byproduct and to turn biomass wastes into valuable chemicals is very precious for the biofuel economy. The hydrothermal method has been widely used to convert biomass to intermediates or valuable products. The hot compressed water has a lower dielectric constant than ambient water (Kruse and Dinjus 2007) and is used as a catalyst and also solvent. Using the hydrothermal method, biomass conversion, and CO_2 reduction in the same reaction is a remarkable topic for lignocellulosic residues valorization and the CO_2 reduction.

Glycerol is a potentially important biorefinery feedstock, available as a by-product in the production of biodiesel by transesterification of vegetable oils or animal fats (Behr et al. 2008). Su and coworkers used **glycerol** biomass as a hydrogen donor and produced valuable carboxylic acids through a "one-pot" aqueous-phase hydrogen transfer (APHT). The lactate (55.0%), pyruvate (5.0%), propylene glycol (6.3%), and formate (29.4%) are formed by combining glycerol conversion and bicarbonate reduction using the activated carbon-supported palladium catalyst (5% Pd/AC) through hydrothermal method for 12 h at 240 °C (Su et al. 2015). Shen and coworkers employed **glycerin** as the feedstock and reductant to produce formate and lactate through the alkaline hydrothermal method with the 9 MPa reaction pressure and 300 °C reaction temperature. The reaction mechanism (Fig. 4) depends on the hydrogen transfer from glycerin to carbon dioxide and produce high-value chemicals (pyruvaldehyde, formate, and lactate) (Shen et al. 2012; Kishida et al. 2005; Costine et al. 2013; Ramírez-López et al. 2011). In the first step, glycerin converted to acetol through dehydration and keto-enol tautomerization. Secondly, two H-bonds formed between molecules (acetol, H_2O, and CO_2). The carbonyl carbon on CO_2 and the hydride ion on the acetol can turn to more positive and a cyclic transition state may be formed. At last, pyruvaldehyde and formate are obtained, and pyruvaldehyde undergoes a benzilic acid rearrangement to generate the lactate salt (Shen et al. 2014; Wang et al. 2016).

Glucose is used as a carbohydrate-based biomass and formed hydrogen that can reduce CO_2 to fine chemicals. In an example, $NaHCO_3$, glucose and deionized water were put in a reactor at 573 K for 20 h. Without CO_2, glucose turned to lactic acid, acetic acid, and formic acid. With CO_2, the main product was formic acid. With increasing alkalinity of the reaction (1 M NaOH), formic acid concentration also increases to 7423 ppm that was higher than the obtained formic acid without CO_2 (1600 ppm) (Jin et al. 2010).

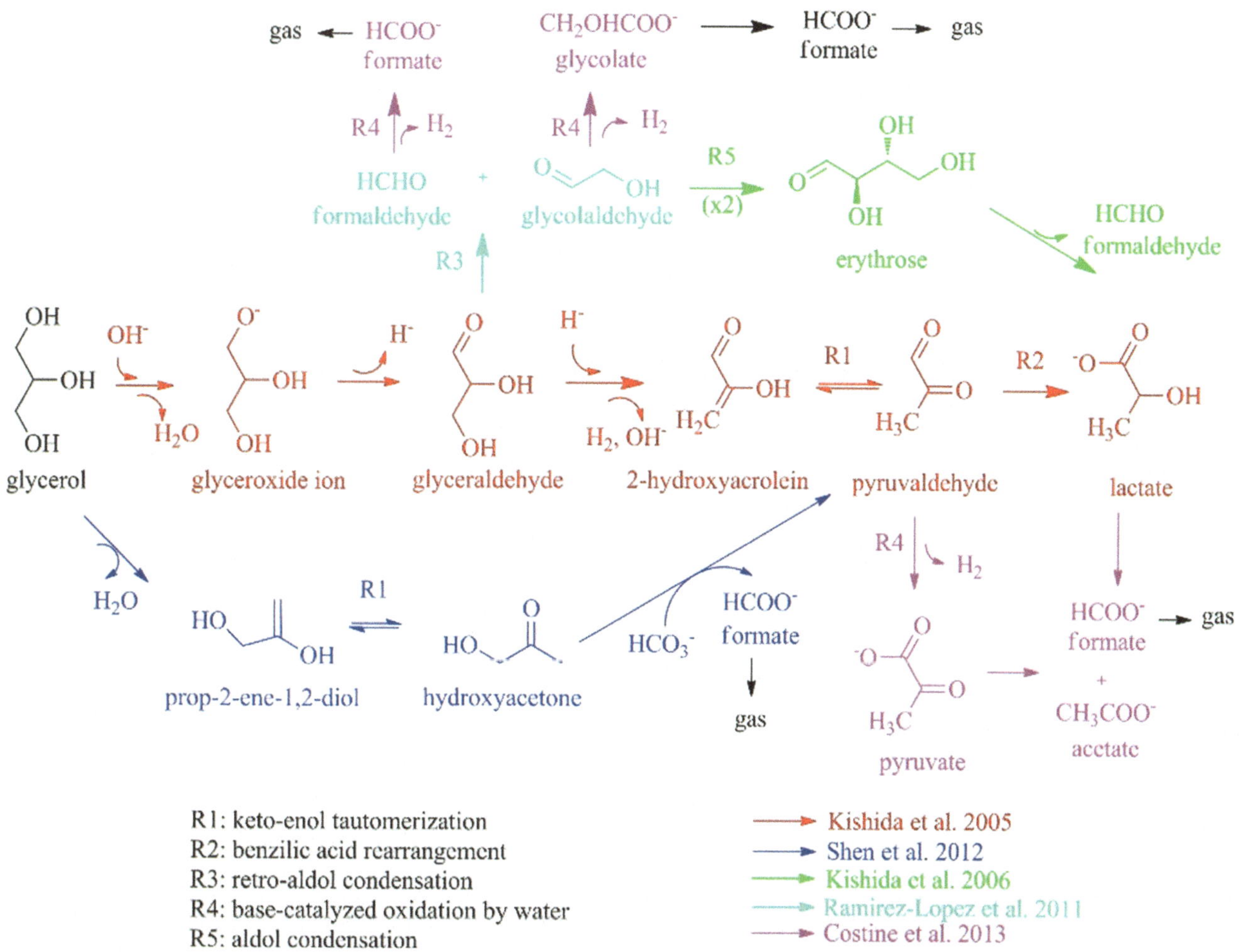

Fig. 4 The reduction reaction of CO_2 with glycerin in alkaline medium. Reprinted from Wang et al. (2016) Copyright (2016), with permission from Elsevier

Fernandez and coworkers were applied to different **lignocellulosic** biomass to reduce CO_2. They used three different organic substances categories as saccharides, phenolic derivatives, and simple molecules (C_1-C_3 alcohols, aldehydes, and ketones) derived from lignocellulosic biomass and were obtained formic acid. Using **simple molecules**, $NaHCO_3$ was reduced with n-propanol (reductant) under hydrothermal conditions at 300 °C 180 min. The formic acid yield was found as 36% that is higher than the yield of iPrOH (secondary alcohol) because of the steric effect. When C_2 and C_3 alcohols were used, similar yields were obtained because of alcohol moiety. When derivatives or intermediates (glycolaldehyde, glyceraldehyde, lactic acid, and pyruvaldehyde) were used as reductant, obtaining yields were found as 26%, 32%, 26%, and 15%, respectively. The primary differences between the formic acid yields were related to their molecular structures. With using **monosaccharides** (glucose and fructose) and **disaccharides** (sucrose and cellobiose), high CO_2 reduction results were obtained. If starting from glucose or fructose, the yields were found as 65% and 35%. The monosaccharides divide into smaller compounds and form 5-HMF and furfural. When disaccharides sucrose (glucose-fructose) and cellobiose (glucose-glucose) were used as reductants, the formic acid yield was found as 60% and 35% respectively. The yield differences came from their thermal stability differences and the main by-products were acetic and lactic acids. When using **phenolic lignin compounds**, the lowest yield (2%) was achieved from phenol. Higher yields were getting from resorcinol and catechol solutions (19% and 9%). For formic acid formation, the highest yield was found as 51% when using vanillin solution, followed by guaiacol (24%) (Andérez-Fernández et al. 2018).

The conversion of CO_2 with **microalgae** (Spirulina) is an alternative method for CO_2 reduction using indirect solar energy (Yang et al. 2019). Through the hydrothermal method, the microalgae were turned to organic acids, N-substituted lactam, and CO_2 reduced into formate. Yang and coworkers also tried bovine serum albumin, glucose, and palmitic acid as biomass. When bovine serum albumin was applied, N-containing compounds and organic acids were formed. When glucose used, formate and acetate were the major products. No water-soluble product was observed with palmitic acid that was unconverted and found in the liquid phase (Yang et al. 2019).

3 Conclusion

The hydrothermal CO_2 conversion is a highly effective technique using metals, catalysts, organic wastes, inorganic wastes, and biomass as a reductant. The high-temperature water produced in the hydrothermal method is a cheap hydrogen source and has gained economical and practical advantages to reduce CO_2. The alkanes, methanol, formic acid, phenol, syngas, and ethylene glycol were successfully synthesized using the hydrothermal method with high efficiency and selectivity according to selected reductant. The hydrothermal method can be used the convert CO_2 to different chemicals, fuels, and energies with new reductants in the future studies to control greenhouse effects.

References

Aharon P (1994) Geology and biology of modern and ancient submarine hydrocarbon seeps and vents: an introduction. Geo-Mar Lett 14(2–3):69–73

Akiya N, Savage PE (2002) Roles of water for chemical reactions in high-temperature water. Chem Rev 102(8):2725–2750

Al-Saydeh SA, Zaidi SJ (2018) Carbon dioxide conversion to methanol: opportunities and fundamental challenges. In: Carbon dioxide chemistry, capture and oil recovery, pp 41

Andérez-Fernández M, Pérez E, Martín A, Bermejo MD (2018) Hydrothermal CO_2 reduction using biomass derivatives as reductants. J Supercritical Fluids 133:658–664

Ansari MB, Park S-E (2012) Carbon dioxide utilization as a soft oxidant and promoter in catalysis. Energy Environ Sci 5(11):9419–9437

Arakawa H, Aresta M, Armor JN, Barteau MA, Beckman EJ, Bell AT, Bercaw JE, Creutz C, Dinjus E, Dixon DA (2001) Catalysis research of relevance to carbon management: progress, challenges, and opportunities. Chem Rev 101(4):953–996

Bawah A-R, Malaibari ZO, Muraza O (2018) Syngas production from CO2 reforming of methane over Ni supported on hierarchical silicalite-1 fabricated by microwave-assisted hydrothermal synthesis. Int J Hydrogen Energy 43(29):13177–13189

Behr A, Eilting J, Irawadi K, Leschinski J, Lindner F (2008) Improved utilisation of renewable resources: new important derivatives of glycerol. Green Chem 10(1):13–30

Chen K, Duan X, Fang H, Liang X, Yuan Y (2018) Selective hydrogenation of CO_2 to methanol catalyzed by Cu supported on rod-like $La_2O_2CO_3$. Catal Sci Technol 8(4):1062–1069

Constantinou DA, Álvarez-Galván MC, Fierro JLG, Efstathiou AM (2012) Low-temperature conversion of phenol into CO, CO_2 and H_2 by steam reforming over La-containing supported Rh catalysts. Appl Catal B 117:81–95

Costine A, Loh JS, Busetti F, Joll CA, Heitz A (2013) Understanding hydrogen in Bayer process emissions. 3. Hydrogen production during the degradation of polyols in sodium hydroxide solutions. Ind En Chem Res 52(16):5572–5581

Denning DM, Thum MD, Falvey DE (2015) Photochemical reduction of CO_2 using 1, 3-dimethylimidazolylidene. Org Lett 17(17):4152–4155

Duo J, Yao G, Jin F, Zhong H (2017) Hydrothermal CO_2 reduction with iron to produce formic acid. In: Hydrothermal reduction of carbon dioxide to low-carbon fuels. CRC Press, pp 61–78

Eskandarloo H, Badiei A (2015) Fabrication of an inexpensive and high efficiency microphotoreactor using CO_2 laser technique for photocatalytic water treatment applications. Environ Technol 36(8):1063–1073

Farlow MW, Adkins H (1935) The hydrogenation of carbon dioxide and a correction of the reported synthesis of urethans. J Am Chem Soc 57(11):2222–2223

Fellay C, Dyson PJ, Laurenczy G (2008) A viable hydrogen-storage system based on selective formic acid decomposition with a ruthenium catalyst. Angew Chem Int Ed 47(21):3966–3968

Food, Administration D Code of Federal Regulations, 21 CFR 172.515. Chapter I–Food and Drug Administration Title

Foustoukos DI, Seyfried WE (2004) Hydrocarbons in hydrothermal vent fluids: the role of chromium-bearing catalysts. Science 304 (5673):1002–1005

Fu Q, Lollar BS, Horita J, Lacrampe-Couloume G, Seyfried WE Jr (2007) Abiotic formation of hydrocarbons under hydrothermal conditions: Constraints from chemical and isotope data. Geochim Cosmochim Acta 71(8):1982–1998

Fu J, Jiang N, Ren D, Song Z, Li L, Huo Z (2016) Hydrothermal conversion of ethylene carbonate to ethylene glycol. Int J Hydrogen Energy 41(21):9118–9122

Gao P, Li F, Xiao F, Zhao N, Wei W, Zhong L, Sun Y (2012) Effect of hydrotalcite-containing precursors on the performance of Cu/Zn/Al/Zr catalysts for CO_2 hydrogenation: introduction of Cu^{2+} at different formation stages of precursors. Catal Today 194(1):9–15

Gao P, Dang S, Li S, Bu X, Liu Z, Qiu M, Yang C, Wang H, Zhong L, Han Y (2018) Direct production of lower olefins from CO_2 conversion via bifunctional catalysis. ACS Catal 8(1):571–578

Goeppert A, Olah GA, Prakash GS (2018) Toward a sustainable carbon cycle: the methanol economy. In: Green Chemistry. Elsevier, pp 919–962

Hao Z, An L, Zhou J, Wang H (1996) A supported gold catalyst for the elimination of hydrogen from CO_2 feed gas in the production of urea. React Kinet Catal Lett 59(2):295–300

He C, Tian G, Liu Z, Feng S (2010) A mild hydrothermal route to fix carbon dioxide to simple carboxylic acids. Org Lett 12(4):649–651

Horita J, Berndt ME (1999) Abiogenic methane formation and isotopic fractionation under hydrothermal conditions. Science 285 (5430):1055–1057

Huo Z, Hu M, Zeng X, Yun J, Jin F (2012) Catalytic reduction of carbon dioxide into methanol over copper under hydrothermal conditions. Catal Today 194(1):25–29

Huo Z, Ren D, Yao G, Jin F, Hu M (2017) 10 Cu-Catalyzed hydrothermal CO_2. In: Hydrothermal reduction of carbon dioxide to low-carbon fuels, p 141

Huo Z, Ren D, Yao G, Jin F, Hu M (2017) Cu-Catalyzed hydrothermal CO_2 reduction with zinc to produce methanol. In: Hydrothermal reduction of carbon dioxide to low-carbon fuels. CRC Press, pp 141–152

Inui T, Yamamoto T, Inoue M, Hara H, Takeguchi T, Kim J-B (1999) Highly effective synthesis of ethanol by CO_2-hydrogenation on well balanced multi-functional FT-type composite catalysts. Appl Catal A 186(1–2):395–406

Jagadeesan D, Joshi B, Parameswaran P (2015) Chemical utilisation of CO_2. Resonance 20(2):165–176

Jessop PG, Hsiao Y, Ikariya T, Noyori R (1996) Homogeneous catalysis in supercritical fluids: hydrogenation of supercritical carbon dioxide to formic acid, alkyl formates, and formamides. J Am Chem Soc 118(2):344–355

Jessop PG, Joó F, Tai C-C (2004) Recent advances in the homogeneous hydrogenation of carbon dioxide. Coord Chem Rev 248(21–24):2425–2442

Ji F, Zhou H, Yang Q (2008) The abiotic formation of hydrocarbons from dissolved CO_2 under hydrothermal conditions with cobalt-bearing magnetite. Orig Life Evol Biosph 38(2):117–125

Jin F, Gao Y, Jin Y, Zhang Y, Cao J, Wei Z, Smith RL Jr (2011) High-yield reduction of carbon dioxide into formic acid by zero-valent metal/metal oxide redox cycles. Energy Environ Sci 4 (3):881–884

Jin F, Huo Z, Zeng X, Enomoto H (2010) Hydrothermal conversion of CO_2 into value-added products: a potential technology for improving global carbon cycle. In: Advances in CO_2 conversion and utilization. ACS Publications, pp 31–53

Jin B, Yao G, Jin F, Zhong H (2017) Autocatalytic hydrothermal CO_2 reduction with aluminum to produce formic acid. In: Hydrothermal reduction of carbon dioxide to low-carbon fuels. CRC Press, pp 127–140

Karl D, Wirsen C, Jannasch H (1980) Deep-sea primary production at the Galapagos hydrothermal vents. Science; (United States) 207

Kishida H, Jin F, Zhou Z, Moriya T, Enomoto H (2005) Conversion of glycerin into lactic acid by alkaline hydrothermal reaction. Chem Lett 34(11):1560–1561

Kobayashi T, Takahashi H (2004) Novel CO2 Electrochemical reduction to methanol for H2 storage. Energy Fuels 18(1):285–286

Kruse A, Dinjus E (2007) Hot compressed water as reaction medium and reactant: properties and synthesis reactions. J Supercrit Fluids 39(3):362–380

Lan R, Irvine JT, Tao S (2012) Ammonia and related chemicals as potential indirect hydrogen storage materials. Int J Hydrogen Energy 37(2):1482–1494

Liu X-M, Lu G, Yan Z-F, Beltramini J (2003) Recent advances in catalysts for methanol synthesis via hydrogenation of CO and CO_2. Ind Eng Chem Res 42(25):6518–6530

Lu L, Zhong H, Wang T, Wu J, Jin F, Yoshioka T (2020) A new strategy for CO_2 utilization with waste plastics: conversion of hydrogen carbonate into formate using polyvinyl chloride in water. Green Chem 22(2):352–358

Luo L, Jin F, Zhong H (2017) Perspectives and challenges of CO_2 hydrothermal reduction. In: Hydrothermal reduction of carbon dioxide to low-carbon fuels. CRC Press, pp 199–204

Lyu L, Jin F, Yao G (2017) Autocatalytic hydrothermal CO_2 reduction with manganese to produce formic acid. In: Hydrothermal reduction of carbon dioxide to low-carbon fuels. CRC Press, pp 109–126

Mascheroni RH (2012) Operations in food refrigeration. CRC Press

Matsuo T, Kawaguchi H (2006) From carbon dioxide to methane: homogeneous reduction of carbon dioxide with hydrosilanes catalyzed by zirconium−borane complexes. J Am Chem Soc 128 (38):12362–12363

Navarrete A, Centi G, Bogaerts A, Martin A, York A, Stefanidis GD (2017) Harvesting renewable energy for carbon dioxide catalysis. Energ Technol 5(6):796–811

North M, Styring P (2015) Perspectives and visions on CO_2 capture and utilisation. Faraday Discuss 183:489–502

Novak S, Madon RJ (1984) Models of hydrocarbon product distributions in Fischer-Tropsch synthesis. 2. Model for hydrocarbon chain growth and cracking. Ind Eng Chem Fundam 23(3):274–280

Olah GA (2005) Beyond oil and gas: the methanol economy. Angew Chem Int Ed 44(18):2636–2639

Omae I (2006) Aspects of carbon dioxide utilization. Catal Today 115 (1–4):33–52

Onoki T, Takahashi H, Kori T, Yamasaki N, Hashida T (2006) Effects of NaOH concentration on CO_2 reduction via hydrothermal water. In: AIP conference proceedings, vol 1. American Institute of Physics, pp 61–64

Perez-Alonso F, Ojeda M, Herranz T, Rojas S, González-Carballo J, Terreros P, Fierro J (2008) Carbon dioxide hydrogenation over Fe–Ce catalysts. Catal Commun 9(9):1945–1948

Philander G, Philander SG (2008) Encyclopedia of global warming and climate change: AE, vol 1. Sage

Ramírez-López CA, Ochoa-Gómez JR, Gil-Río S, Gómez-Jiménez-Aberasturi O, Torrecilla-Soria J (2011) Chemicals from biomass: synthesis of lactic acid by alkaline hydrothermal conversion of sorbitol. J Chem Technol Biotechnol 86(6):867–874

Ravindra MR, Rao KJ, Nath BS, Ram C (2014) Carbonated fermented dairy drink–effect on quality and shelf life. J Food Sci Technol 51 (11):3397–3403

Rice C, Ha S, Masel R, Waszczuk P, Wieckowski A, Barnard T (2002) Direct formic acid fuel cells. J Power Sources 111(1):83–89

Roman-Gonzalez D, Moro A, Burgoa F, Pérez E, Nieto-Márquez A, Martín Á, Bermejo MD (2018) 2Hydrothermal CO2 conversion using zinc as reductant: batch reaction, modeling and parametric analysisof the process. J Supercrit Fluids 140:320–328

Sanchez-Sanchez C, Montiel V, Tryk D, Aldaz A, Fujishima A (2001) Electrochemical approaches to alleviation of the problem of carbon dioxide accumulation. Pure Appl Chem 73(12):1917–1927

Shen Z, Zhang Y, Jin F (2011) From $NaHCO_3$ into formate and from isopropanol into acetone: hydrogen-transfer reduction of $NaHCO_3$ with isopropanol in high-temperature water. Green Chem 13 (4):820–823

Shen Z, Zhang Y, Jin F (2012) The alcohol-mediated reduction of CO_2 and $NaHCO_3$ into formate: a hydrogen transfer reduction of $NaHCO_3$ with glycerine under alkaline hydrothermal conditions. RSC Adv 2(3):797–801

Shen Z, Gu M, Zhang M, Sang W, Zhou X, Zhang Y, Jin F (2014) The mechanism for production of abiogenic formate from CO_2 and lactate from glycerine: uncatalyzed transfer hydrogenation of CO_2 with glycerine under alkaline hydrothermal conditions. RSC Adv 4 (29):15256–15263

Shen Z, Gu M, Xia M, Zhang W, Zhang Y, Jin F (2017) Hydrothermal reduction of CO_2 with glycerine. In: Hydrothermal reduction of carbon dioxide to low-carbon fuels. CRC Press, pp 153–184

Sherman R (2007) Carbon dioxide snow cleaning. Part Sci Technol 25 (1):37–57

Song C (2002) CO_2 conversion and utilization: an overview. In: ACS Publications

Su J, Yang L, Yang X, Lu M, Luo B, Lin H (2015) Simultaneously converting carbonate/bicarbonate and biomass to value-added carboxylic acid salts by aqueous-phase hydrogen transfer. ACS Sustain Chem Eng 3(1):195–203

Takahashi H, Liu L, Yashiro Y, Ioku K, Bignall G, Yamasaki N, Kori T (2006) CO_2 reduction using hydrothermal method for the selective formation of organic compounds. J Mater Sci 41(5):1585–1589

Takahashi H, Kori T, Onoki T, Tohji K, Yamasaki N (2008) Hydrothermal processing of metal based compounds and carbon dioxide for the synthesis of organic compounds. J Mater Sci 43 (7):2487–2491

Tambach TJ, Mathews JP, Van Bergen F (2009) Molecular exchange of CH_4 and CO_2 in coal: enhanced coalbed methane on a nanoscale. Energy Fuels 23(10):4845–4847

Tian G, Yuan H, Mu Y, He C, Feng S (2007) Hydrothermal reactions from sodium hydrogen carbonate to phenol. Org Lett 9(10):2019–2021

Tian G, He C, Chen Y, Yuan HM, Liu ZW, Shi Z, Feng SH (2010) Hydrothermal reactions from carbon dioxide to phenol. ChemSusChem Chem Sustain Energy Mater 3(3):323–324

Tian G, He C, Liu Z, Feng S (2017) Hydrothermal reduction of CO_2 to low-carbon compounds. In: Hydrothermal reduction of carbon dioxide to low-carbon fuels. CRC Press, pp 79–90

Wang LY, Yao GD, Jing ZZ (2015) Jin FM Hydrothermal conversion of CO_2 into formic acid with zinc and copper powders under low temperature. In: Advanced materials research. Trans Tech Publ, pp 39-42

Wang Y, Wang F, Li C, Jin F (2016) Kinetics and mechanism of reduction of CO_2 by glycerol under alkaline hydrothermal conditions. Int J Hydrogen Energy 41(21):9128–9134

Wu B, Gao Y, Jin F, Cao J, Du Y, Zhang Y (2009) Catalytic conversion of $NaHCO_3$ into formic acid in mild hydrothermal conditions for CO_2 utilization. Catal Today 148(3–4):405–410

Xu G-h, Li Y-c, Li Z-h, Wang H-J (1995) Kinetics of the hydrogenation of diethyl oxalate to ethylene glycol. Ind Eng Chem Res 34 (7):2371–2378

Yang Y, Zhong H, Yao G, He R, Jin B, Jin F (2018) Hydrothermal reduction of $NaHCO_3$ into formate with hexanehexol. Catal Today 318:10–14

Yang Y, Zhong H, He R, Wang X, Cheng J, Yao G, Jin F (2019) Synergetic conversion of microalgae and CO_2 into value-added chemicals under hydrothermal conditions. Green Chem 21(6):1247–1252

Yang Y, Yao G, Jin B, He R, Jin F, Zhong H (2017) Hydrothermal CO_2 reduction with zinc to produce formic acid. In: Hydrothermal reduction of carbon dioxide to low-carbon fuels. CRC Press, pp 91–108

Yao G, Chen F, Duo J, Jin F, Zhong H (2017) 12 Hydrothermal reduction of CO_2 with Compounds containing nitrogen. In: Hydrothermal reduction of carbon dioxide to low-carbon fuels, p 185

Zeng X, Jin F, Huo Z, Mogi T, Kishita A, Enomoto H (2011) Reduction of carbon dioxide in hydrothermal cracking of polymer wastes. Energy Fuels 25(6):2749–2752

Zeng X, Hatakeyama M, Ogata K, Liu J, Wang Y, Gao Q, Fujii K, Fujihira M, Jin F, Nakamura S (2014) New insights into highly efficient reduction of CO_2 to formic acid by using zinc under mild hydrothermal conditions: a joint experimental and theoretical study. Phys Chem Chem Phys 16(37):19836–19840

Zhang XV, Martin ST, Friend CM, Schoonen MA, Holland HD (2004) Mineral-assisted pathways in prebiotic synthesis: Photoelectrochemical reduction of carbon (+IV) by manganese sulfide. J Am Chem Soc 126(36):11247–11253

Zhong H, Jin F, Wu B, Chen H, Yao G (2010) Hydrothermal conversion of CO_2 into formic acid on the catalysis of Cu. In: AIP conference proceedings, vol 1. American Institute of Physics, pp 213–216

Zhu Q (2019) Developments on CO_2-utilization technologies. Clean Energy 3(2):85–100

Ethylenediamine–Carbonic Anhydrase Complex for CO_2 Sequestration

Egwim Evans Chidi, G. K. Ezikanyi, Onyeaku Ugoona Sandra, and Joseph Peter Shaba

1 Introduction

The earth's temperature has been continuously increasing with rising humanitarian activities. This has led to the increase in climate change and resulting in global warming (Rachael et al. 2015). This effect has been a major concern globally. Environmental changes affect the earth's temperature, thereby resulting in the rise of sea level, greenhouse gases, disease patterns, and global warming. In our world today, the rise of greenhouse gasses through environmental changes contributes directly to the rise in global warming. There are several types of greenhouse gases that have been identified and reported to be a contributing factor in creating an extra burden to the environment and increasing global warming. These gases includes carbon dioxide (CO_2), sulfur dioxide (SO_4), hydrocarbons, etc. Among these emitted gases, carbon dioxide (CO_2) has contributed massively to the raise in the earth's temperature which has resulted in what we called "global warming." Gases can be sourced from industries, power plants, and the fossil fuel combustion process (Figueroa et al. 2008; Rayalu et al. 2012; Shekh et al. 2012). Carbon dioxide (CO_2) gases emitted in the environment through industrial activities have created a serious debate on either as a source of energy or its impact on the environment. With this regard, carbon dioxide (CO_2) may pose a serious environmental and health crisis (Lackner et al. 2010). With the emerging problem of carbon dioxide (CO_2) emission, several approaches have been proposed by switching to non-fossil fuel sources such as wind energy as an alternative to fossil fuel sources (Sims et al. 2003). Recent development has been employed to salvage this problem by involving enzymatic and chemical processes to capture and sequestrate carbon dioxide (CO_2). In this regard, enzymes such as carbonic anhydrase and ethylenediamine (amine-based solvent) can be used to capture and sequestrate carbon dioxide (CO_2). The end product of carbon dioxide (CO_2) capturing and sequestration has been regarded to be less toxic and eco-friendly which makes it to be more suitable for industrial purposes and environmentally safe for long-term storage (Shekh et al. 2012; Ramanan et al. 2010). Capturing and sequestration of carbon dioxide (CO_2) can be achieved by immobilization of carbonic anhydrase as a catalyst on a bioreactor and CO_2 as the substrate, thereby leading to the formation of solid carbonate. (Favre et al. 2009; Mirjafari et al. 2007). With this technological approach of capturing and sequestration of carbon dioxide (CO_2), research groups and chemical industries across the globe are working on the modalities for large scale carbon dioxide (CO_2) capture (Kim et al. 2012; Bryce et al. 2015). In the carbon dioxide (CO_2) capturing process, one of the key technologies used is the solvent system with many advantages compared to other systems of separation. The most considered amine-based system is the monoethanolamine. Recently, there are other amine-based solvent systems such as ethylenediamine which have been reported to have good physical and chemical characteristics needed for carbon dioxide (CO_2) capture (Gary 2009). This present work is aimed at addressing the importance of using ethylenediamine–carbonic anhydrase complex system under good industrial operation. The importance of adopting this system is the unique characteristics of the enzyme used which are stability and reusability. The regeneration of amine solvent at low-temperature desorption is another importance of using ethylenediamine as an

E. E. Chidi (✉)
Center for Genetic Engineering and Biotechnology (CGEB), Federal University of Technology, PMB 65, Minna, Niger State, Nigeria
e-mail: c.egwim@futminna.edu.ng

G. K. Ezikanyi
Biochemistry Department, Federal University of Technology, PMB 65, Minna, Niger State, Nigeria

O. U. Sandra
Chemistry Department, Federal University of Technology, PMB 65, Minna, Niger State, Nigeria

J. P. Shaba
Biochemistry Department, Ibrahim Badamasi Babangida University Lapai, Lapai, Nigeria

Inamuddin et al. (eds.), *Carbon Dioxide Utilization to Sustainable Energy and Fuels*, Advances in Science, Technology & Innovation, https://doi.org/10.1007/978-3-030-72877-9_12

amine-based solvent for carbon dioxide (CO_2) capture (Prakash et al. 2017).

2 An Overview of Carbonic Anhydrase (CA)

Carbonic anhydrase also known as carbonate dehydratase (E.C 4.2.1.1) is responsible in catalyzing the reversible conversion of carbon dioxide (CO_2) and water, forming carbonic acid (H_2CO_3) and eventually dissociate into hydrogen ions and bicarbonate. It is also a metalloenzyme with a zinc metal ion (Zn^{2+}) at its active site (Sawaya et al. 2006). The most important function of this enzyme in animals is to maintain a balance between acid–base concentrations in the blood and transport carbon dioxide and bicarbonate in various metabolizing tissues (Hilvo et al. 2005). Carbonic anhydrase is known to possess lots of physiological roles in animals which includes electrolyte secretion, carbon dioxide (CO_2) fixation, pH regulation, secretion of gastric, cerebrospinal fluid, and pancreatic juices within the body system (Chegwidden et al. 2002; Kyllonen et al. 2003) (Fig. 1).

3 Mechanism of Action for Biocarbonate Formation

Carbonic anhydrase catalyzes a reversible reaction between carbon dioxide (CO_2) and water to form bicarbonate and hydrogen ions. This enzyme contains a central metal ion

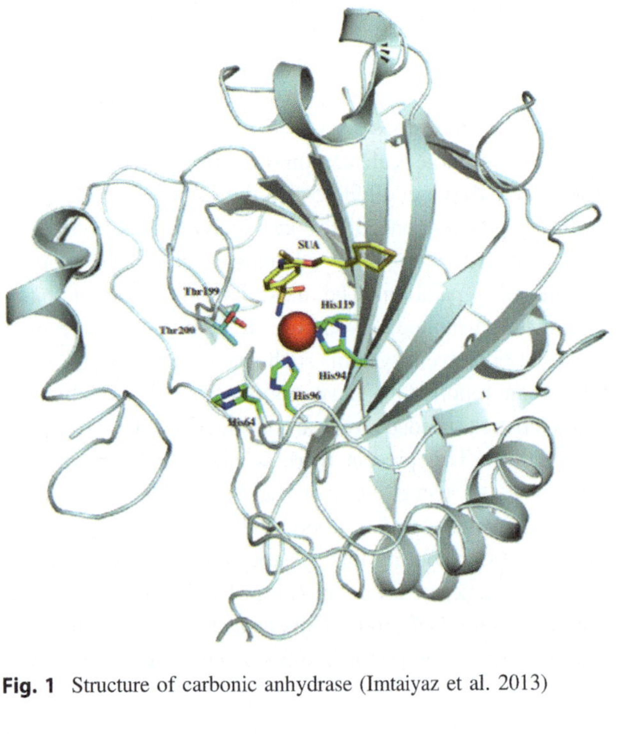

Fig. 1 Structure of carbonic anhydrase (Imtaiyaz et al. 2013)

(zinc ion) which is a prosthetic group responsible for the positioning of histidine side chains. During enzyme reaction, the enzyme active site "E" usually contains a pocket which is specific for carbon dioxide (CO_2) and at the process moves to the hydroxide group attached to the central metal ion which is zinc ion. Therefore, this makes the electron-rich hydroxide ion to attack the carbon dioxide (CO_2) molecules and alongside create a bicarbonate molecule. The general equation for carbonic anhydrase hydration and dehydration is as follow (Fig. 2):

$$CO_2 + H_2O \rightleftharpoons HCO_3^{--} + H^+ \quad (1)$$

$$HCO_3^{--} + H^+ CO_2 + H_2O \quad (2)$$

4 Historical Background of Carbonic Anhydrase

Carbonic anhydrase has been known to be an ancient enzyme that can be found in plants, prokaryotes, and eukaryotes. It is known to catalyze the reversible reaction involving hydration and dehydration of carbon dioxide and bicarbonate (Smith and Ferry 2000; Tripp et al. 2001). This enzyme was discovered by renowned scientists Meldrum and Roughton in 1933 while in the process of studying the transportation mechanism of carbon dioxide (CO_2) in the blood and across the lungs surface. At the active site of the enzyme, central metal (Zn^{2+}) is found which plays an important role during catalytic reactions. Through the process of carbonic anhydrase discovery, Keilinand and Mann in 1939 were the first scientists to report on carbonic anhydrase in eukaryotes, and prokaryotes was reported by Blankenship in 1963 (Smith and Ferry 2000). Carbonic anhydrase can be found in plants, microorganisms, and humans. The existence of carbonic anhydrase in the eukaryotic cells and plant cells as well was known to be detected during the early periods of the twentieth century. Carbonic anhydrase is an enzyme to have different distinctive properties regardless were located. In plant cells, it was discovered to possess different properties when compared to the animal cells. During the early sixties, carbonic anhydrase was not extensively characterized due to its complex nature with i different properties existing in plants and animals. Since carbonic anhydrase was found to be important in the field of science, this led researchers to undergo serious laboratory research in the isolation of the enzyme from different sources such as bacteria, plants etc. purified the enzyme using different steps, and finally characterized it. During the nineties, carbonic anhydrase in *N. sicca* and *E. coli*, were successfully purified and characterized (Adler et al. 1972; Guilloton et al. 1992). Based on the protein sequences, it was characterized into five

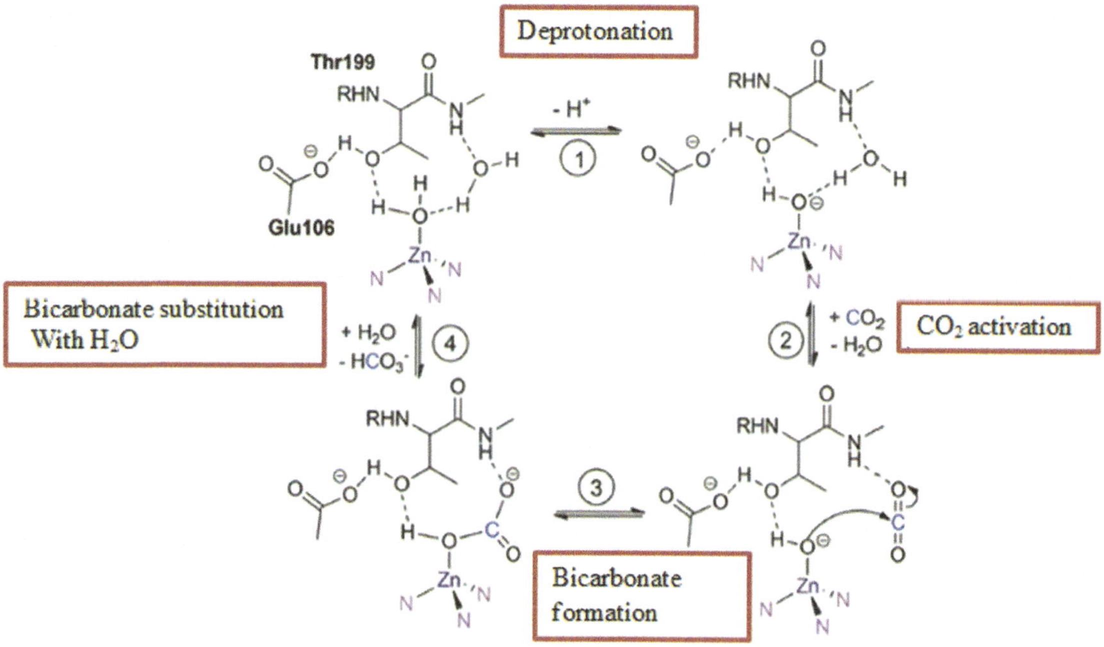

Fig. 2 Mechanism of action of bicarbonate formation (Rachael et al. 2015)

(5) distinct classes, namely (α, β, γ, δ, and ζ). Within the classes, α, β, and γ are widely distributed both in plant and in animal cells, whereas the remaining two classes δ and ζ are found in diatoms and marine phytoplankton (Syrjänen et al. 2010; Roberts et al. 1997), purified and sequenced carbonic anhydrase partially from microorganism *T. weissflogii* which was later identified and reported to be the new class of carbonic anhydrase as δ-class. The research insight of carbonic anhydrase for carbon dioxide (CO_2) sequestration was successful in the year 2001 and opened a new field of research. With the new development on the area of research, it created an avenue for lots of research groups on this novel work on carbon dioxide (CO_2) capturing, sequestration as well the reduction of atmospheric flue gases. On the other hand, bovine carbonic anhydrase (BCA) was quite expensive commercially (Raju et al. 2014). Therefore, several research groups developed a means of large production of carbonic anhydrase from bacteria origin due to its fast growth rate, easy cultivation, and high over expression rate (Bond et al. 2001).

5 Sources of Carbonic Anhydrase

Carbonic anhydrase can be sourced from microorganisms, plants, and humans. In microogarnisms, carbonic anhydrase can be found in different species of bacteria. There are multiple isoforms of carbonic anhydrase located in different human organs of the body system. Generally, plants have three types of carbonic anhydrase which are alpha, beta, and gama.

6 Carbonic Anhydrase in Microorganism

Carbonic anhydrase can be found in both prokaryotic and eukaryotic cells. Carbonic anhydrase in bacteria *N. sicca* was first discovered by Veitch and Blankenship in the year 1963 (Smith and Ferry 2000). Subsequently, further research was carried out in bacteria with carbonic anhydrase which are listed below:

6.1 Micrococcus Lylae, Micrococcus Luteus, and Pseudomonas Fragi

In *Micrococcus lylae, and Micrococcus luteus,* carbonic anhydrase was isolated, purified, and immobilized on the different supporting matrix for carbon dioxide (CO_2) sequestration (Sharma et al. 2009). Reported on the isolation and purification of carbonic anhydrase from *Pseudomonas fragi* which was characterized and purified with the molecular weight to be 31 kDa with the use of an electrophoretic technique (SDS-PAGE). Its enzyme activity was active with the temperature ranging from 7–8.5 and 35 to 45 °C, respectively. Metal ion such as Pb^{2+} and Hg^{2+} inhibited its activity while Zn^{2+}, Fe^{3+} acts as activators of the enzyme.

6.2 Bacillus Subtilis and Citrobacter Freundii

Carbonic anhydrase from *Bacillus subtilis* was isolated, purified, and characterized as reported by Ramanan et al. (2009). The molecular weight was found to be 37 kDa with Hg^{2+}, Cl^{-} HCO_3^{-} acting as inhibitors and Zn^{2+} as activators. The kinetic parameters were determined using p-nitro phenyl acetate, and V_{max} and K_m values were found to be 714.28 mmol/mg/min and 9.09 mM, respectively. The optimum pH and temperature was found to be 8.3 and 37 °C, respectively. Carbonic anhydrase from *Citrobacter freundii* was also isolated and purified for bicarbonate formation which yielded 225mg carbonate/mg. The molecular weight of the enzyme was found to be 37 kDa by using the electrophoretic technique. Carbonic anhydrase from *Bacillus subtilis* was sequenced and showed similarity with plant carbonic anhydrase (Ramanan et al. 2009).

6.3 Neisseria Gonorrhoeae

Chirica et al. 1997), reported the activity of carbonic anhydrase from *Neisseria* gonorrhoeae. The protein was isolated and purified for molecular studies. It was observed that the protein sequence of the enzyme was found to be similar to the human carbonic anhydrase. The molecular weight of the enzyme was shown to be 28 kDa.

6.4 Helicobacter Pylori

Helicobacter pylorus is a gram-negative bacterium that is found mainly in the mucus layer of the stomach region. It is spiral in shape and associated with metabolic disorders such as peptic ulcer and gastric cancer. It aids in maintaining the acidic environment of the stomach in basic condition. It maintains the stomach pH to 6.4 and helps in the metabolism of urea and bicarbonate (Marcus et al. 2005).

7 Plant Carbonic Anhydrase

Carbonic anhydrase is also referred to as a metalloenzyme containing a zinc metal ion that catalyzes the reversible conversion of carbon dioxide (CO_2) and bicarbonate (HCO_3^{-}). In nature, carbonic anhydrase are found to be ever presented with concurrent evolution, multiple isoforms, and structurally different but catalyze the same reaction. Generally, there are three major forms of carbonic anhydrase, namely alpha (α), beta (β), and gamma (γ) (Moroney et al. 2001). α-carbonic anhydrase was the first to be discovered as its family in erythrocytes (Brinkman et al. 1932; Meldrum and Roughton 1932). α-carbonic anhydrase is mainly composed of ten (10) β strands which created β sheet at the central region with seven (7) α helices in a large portion of the protein shown in Fig. 3. The active site contains mainly three histidine residues and water molecules which are responsible for its coordination with zinc present (Liljas et al. 1972). β-carbonic anhydrase was first discovered in plants showing differences in its protein sequence and structure when compared with α-carbonic anhydrase (Burnell et al. 1990). It contains two cysteine residues, one histidine residue, and a molecule of water which is responsible for coordinating the metal ion found at the active site of the enzyme (Fig. 3) (Kimber and Pai 2000). β-carbonic anhydrase is structured in four parallel β strands consisting of an α helix surrounded by β sheet. The monomer structure of β-carbonic anhydrase is composed of α-helices surrounded by β-sheet with four parallel β strands. The dimer is formed through the interaction by α helices N-terminal with a monomer surrounding the next monomer unit via hydrogen bond formation existing between each of the monomers. The tetramers are formed from the C-terminal of β strands with interaction at the fifth monomer. Octamers are formed via slightly different interactions with these fifth β-strand extensions (Kimber and Pai 2000; Rowlett 2010). γ-carbonic anhydrase is found mainly in photosynthetic bacteria and in plants (Parisi et al. 2004) but was first discovered in archaea (Alber and Ferry 1994) (Fig. 3). The active site contains a zinc ion (Zn^{2+}), fused together by three histidine molecule and a molecule of water (Kisker et al. 1996).

The protein structures of the different carbonic anhydrases in plants have distinct differences but catalyze the same reaction. In tobacco, the activity of carbonic anhydrase enzyme follows the pattern: leaves > stem > pods, as they are found in different compartments of plant species (Diamantopoulos et al. 2013). Chloroplastic carbonic anhydrase contents in higher plants are highly saturated adding with the process of photosynthetic reaction. Carbonic anhydrase helps in the stomatal closure of plant leaves and in converting carbon dioxide to bicarbonate during photosynthesis. Carbonic anhydrase activity in the plant helps in the stomata opening and closure through which the exchange of gases takes place. They are mostly found on the under surface of plant leaves. During the dark phase of photosynthesis, the amount of fixation gets depleted while the process of respiration continues which leads to an increase in carbon dioxide (CO_2) level due to the shift in the reaction mechanism centrally toward the right, resulting in a decrease of pH. The decreasing pH causes the inactivation of various enzymes such as amylase, which stops the mechanism of

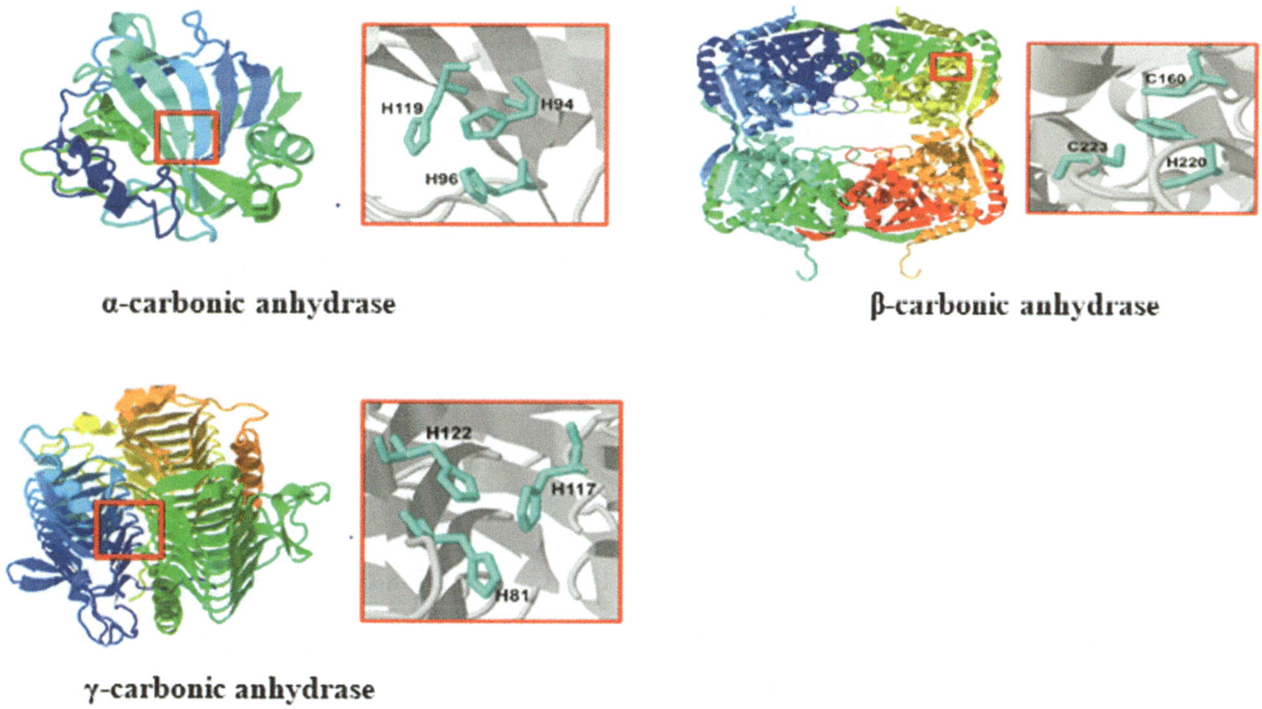

Fig. 3 **Structures of α-, β-, and γ-carbonic anhydrase** (Robert et al. 2017)

hydrolysis of starch and glucose, and in turn decreases the osmotic gradient in occlusive cells resulting in loss of water and closure of ostiole (Casson and Gray 2008). The rate of photosynthesis and carbon dioxide (CO_2) fixation (under limited CO_2 condition) gets directly affected due to changes in the activity of carbonic anhydrase. It is the only carbon metabolism enzyme which shows fluctuations in the activity among different species with varying carbon dioxide (CO_2) concentration in the environment (Fig. 4).

8 Overview of CO_2

Carbon dioxide is a colorless, odorless inert gas which has a density of 1.977 kg/m^3. Carbon dioxide (CO_2) is 60% higher than that of air. It is a major heat-trapping greenhouse gas at room temperature consisting of two molecules of oxygen covalently bounded to a carbon atom. It is soluble in water and occurs naturally as a trace element in the atmosphere (Bryngelsson and Westermark 2009). In nature, carbon dioxide (CO_2) is the most used supercritical fluid which is converted solid ice form and to bicarbonates for various industries purposes. Generally, carbon dioxide (CO_2) does not support combustions which are neither explosive nor flammable (Smit et al. 2014). Through the carbon circle, carbon dioxide stays balanced with other chemical reactions in the air and bodies of water.

9 Sources of Carbon Dioxide (CO_2)

Carbon dioxide can be sourced from both natural and anthropogenic emissions.

Natural sources: Carbon dioxide can be produced from the respiration process in plants and animals. In animals, carbon dioxide (CO_2) is produced aerobically from carbohydrate, lipid, and protein in a process to produce an energy form of ATP (adenosine triphosphate). Aquatic animals such as fish and frogs produce CO_2 through their gills and lungs in the process of evaporation. It can also be released through the process of organic matter decomposition into the atmosphere at a favorable concentration as emission from forest fires and volcanic eruptions.

Anthropogenic sources: Carbon dioxide can be produced from humanitarian activities such as electricity generation, burning of woods, agricultural processing, food processing, ethanol production, deforestation, fossil fuel combustion, petroleum, and natural gases used in transportation, electrical generation, heating, and cooling. The release of by-products from industrial activities involving a large scale of oxidation processes rises the concentration of carbon dioxide (CO_2) in the atmosphere resulting to change in environmental temperature. Industries involving in the large-scale production of acrylic acid, food, and beverages through fermentation release carbon dioxide (CO_2) in large quantities in the environment.

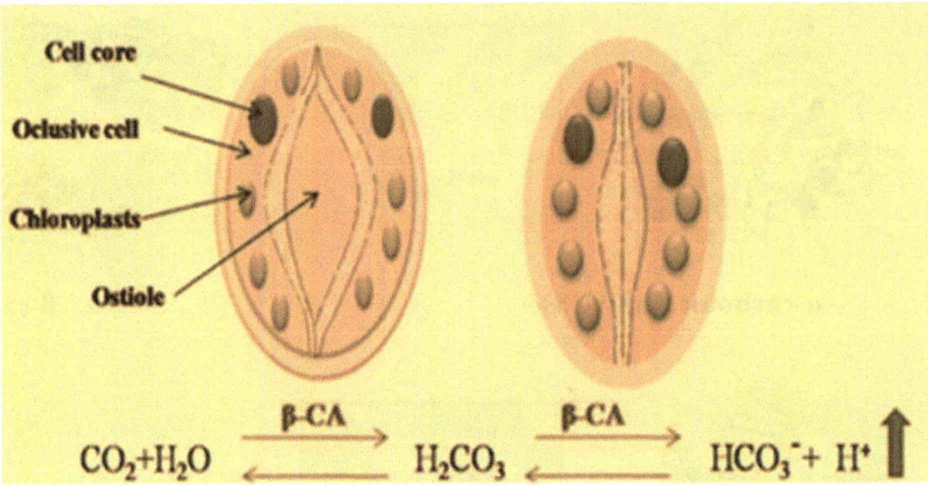

Fig. 4 Carbonic anhydrase activity on stomatal opening and closing

10 Effect of Carbon Dioxide (CO_2)

Carbon dioxide (CO_2) has its negative and positive effects on the environment and to humans too. In human, when exposed to a high concentration of carbon dioxide (CO_2) may result to deleterious effect in body system by increasing venous PO_2 through shifting of hemoglobin oxygen saturation curve and affecting the delivery of oxygen to the tissues (Willie et al. 1999). In the environment, an increase in atmospheric carbon dioxide (CO_2) concentration causes global climatic warming which is known as the greenhouse effect. This global warming has an adverse effect on both plants and human causing increased natural disasters such as extreme weather events, sea-level rise, altered crop growth, and disrupted water systems. Carbon dioxide in the other hand can be seen to be relevant in the reproduction, growth, and survival of living things on earth. It is a major component of living things and also a finite resource in different forms transferring from living things to nonliving things. In plants, carbon dioxide (CO_2) is a major reactant that is needed during the process of photosynthesis to provide glucose as food for human consumption. In humans, carbon dioxide (CO_2) is essential for metabolic pathways such as Kreb's cycle for maintaining the normal metabolic activities in the body system. In the environment, carbon dioxide (CO_2) captures heat radiated from earth's surface and keeping the planet warm enough to support life on earth (Millward and Yaghi 2005). Its also can be used in different industrial applications such as wine production, as a coolant for refrigerators, carbonated drinks, frozen foods, as compressed gas for pneumatic systems, fire extinguishers, and as gaseous blanket to prevent substances from decaying, etc.

11 Carbon Dioxide Capturing

The major contributor of climate change is the release of atmospheric carbon dioxide (CO_2) through the greenhouse effect. Carbon dioxide (CO_2) capturing is a technology that deals with the reduction in atmospheric carbon dioxide (CO_2) from industrial plants such as coal, fossil fuel combustion, and gas-fired power plants (Bryngelsson and Westermark 2009; Kulkarni and Sholl 2012) and converting the captured carbon dioxide (CO_2) into industrial products. This use of carbon-capturing has been employed in the oil and gas industries as a way of recovering its products. The technology of carbon capture is basically adopted to isolate the generated carbon dioxide (CO_2) from different sources to a form that is suitable for transportation and subsequent storage (a supercritical form of CO_2) for further applications. This involves various separation strategies that involve separating carbon dioxide (CO_2) from different gas mixtures. The process of separation involves three main steps which include trapping and separating carbon dioxide (CO_2) from flue gases in a power plant, i.e., membrane-based separation, transporting the captured and compressed carbon dioxide (CO_2) to a location where it is stored away from the atmosphere (geologic sequestration). There are several methods that have been adopted for CO_2 capture. Carbon dioxide (CO_2) can be captured and sequestrated before, during, or after fuel is combusted. Before the combustion of fossil fuels is taken place which is also known as pre-combustion carbon, carbon dioxide (CO_2) can be trapped from fossil fuels by chemical absorption (solvent system) with chemicals such as potassium carbonate and ethylenediamine (Anusha 2010). For the standpoint of energy consumption, the use of a solvent system for carbon dioxide (CO_2) capturing has appeared to be an attractive alternative method used in industries. Carbon dioxide (CO_2) can also be captured during the fossil fuel burning by power plants compression of different gas mixtures such as steam and carbon dioxide (CO_2). Both gases can be separated by a cooling and compressing process, alongside with carbon dioxide (CO_2) seized from the flue gases which are released after the fossil fuel is burned (Plasynski et al. 2007). There is another method that has been adopted for carbon dioxide (CO_2) capture industrially which is the enzyme system such as carbonic anhydrase. The use of an enzyme system for CO_2 capture has created a promising view and has been integrated into the field of science using enzyme for large scale of carbon dioxide (CO_2) capturing and sequestration. This can be carried out with the use of a bioreactor with immobilization of enzyme and gas mixtures passing through it. The immobilized enzyme converts carbon dioxide (CO_2) as its substrate to carbonates which can be exploited for industrial purposes.

12 Carbon Dioxide (CO_2) Sequestration

Carbon dioxide sequestration is a process that deals with conversion of trapped carbon dioxide (CO_2) from the atmosphere into industrial purposes (i.e., the net removal of CO_2) (Raju et al. 2014). It can further be described as the conversion of carbon dioxide (CO_2) emissions into useful compounds using biological, physical, or chemical methods (Skjånes et al. 2007). Carbon dioxide (CO_2) is sourced from different industrial plants such as power plants, cement plants, and refineries. (Rao and Rubin 2002). There are different methods adopted to trap and sequester carbon dioxide (CO_2); this includes absorption, adsorption, cryogenic, and membrane. However, these methods at some point are not viable for large carbon dioxide (CO_2) sequestration scales. Therefore, this tends to pull the minds of researchers in developing economical and sustainable carbon dioxide (CO_2) sequestration technologies. Elevated carbon dioxide (CO_2) levels in the atmosphere can lead to serious adverse effects on man even at the point of causing death. Proper control of elevated carbon dioxide (CO_2) in confined space is critically important where there is low buffering ability in absorbing gas such as spacecraft. Using carbonic anhydrase to trap carbon dioxide (CO_2) into carbonates has developed an effective and successful emergence for further technological advancement (Domsic et al. 2008). The essence for utilizing carbonic anhydrase for carbon dioxide (CO_2) sequestration is due to fact that the hydrophobic region of the enzyme active site is capable to sequestrate carbon dioxide (CO_2) and orients the carbon atoms suitably for nucleophilic attack by the zinc-bound hydroxide (Loferer et al. 2003). It has been shown that the crystal structure of the ordered water molecule in the hydrophobic pocket of the amino acid residue Thr^{199} is stabilized by central metal (Zn^{2+}). Interestingly, the water molecule present is displaced by carbon dioxide (CO_2) into the binding pocket (Fierke et al. 1991). This displacement reaction in the active site of the enzyme makes capturing of carbon dioxide (CO_2) possible.

13 Carbon Dioxide (CO_2) Sequestration by Carbonic Anhydrase

Carbon dioxide (CO_2) sequestration is a process that involves capturing carbon dioxide (CO_2) by using a reactor system either chemical or bioreactor for industrial purposes. In recent times, industrial and humanitarian activities have led to the release of carbon dioxide (CO_2) which is considered as a major anthropogenic gas. With these activities, research has shown that CO_2 concentration in the earth plant has increased by 40% ranging from 380 to 360 ppm from the preindustrial level and has contributed to some environmental changes (Mirjafari et al. 2007). This has drawn an environmental concern toward its effect on humans and to the environment. The by-product carbon dioxide (CO_2) is the reaction of the combustion process in the air without any fuel value, and this has been known to be the major cause of global warming in our world today. Therefore, reducing the release of carbon dioxide (CO_2) into the atmosphere by capturing and sequestration process from industrial base sources will go a long way in salvaging its effect. There are several methods that have been exploited for the reduction of carbon dioxide (CO_2) in the atmosphere which involves capturing and sequestration technologies (Kothandarama et al. 2010). This process proved to be costly and not eco-friendly due to the processes involved. One of the methods used is the solvent system which seems to be more tedious and not eco-friendly. This led to the development of a carbonic anhydrase system which is considered to be an eco-friendly and low cost-effective approach for carbon dioxide (CO_2) capturing and sequestration technology (Mirjafari et al. 2007; Bond et al. 2001; Ramanan et al. 2009). Carbonic anhydrase has catalyzing rates as high as $1.4 \times 106\ M/s^{-1}$ thus plays a vital role during carbon dioxide (CO_2) capturing and sequestration. The importance of using carbonic anhydrase for industrial carbon dioxide (CO_2) sequestration is the fact that it can be regenerated and still maintain its activity and stability after being used. With this, the use enzyme system reduces the cost-effectiveness

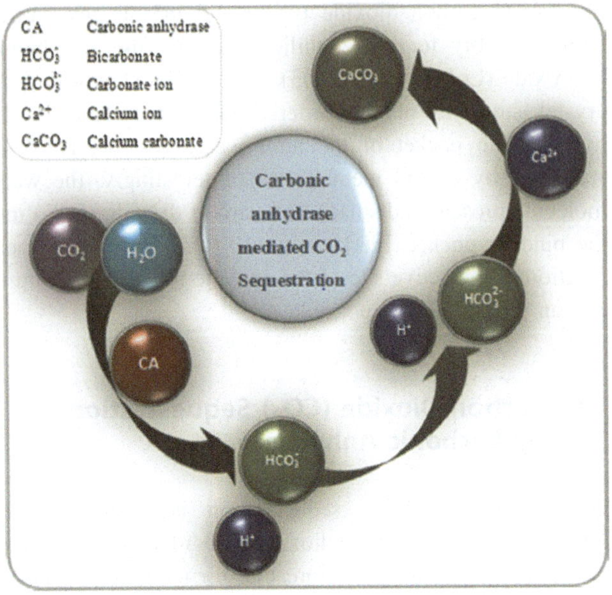

Fig. 5 Mechanism of carbon dioxide (CO_2) sequestration by carbonic anhydrase (Raju et al. 2014)

for capturing and sequestration carbon dioxide (CO_2) in a large-scale industry (Fig. 5).

The general equation for CO_2 sequestration and formation of limestone by carbonic anhydrase **is** given below:

$$CO_{2(aq.)} + H_2O \leftarrow k \rightarrow 1k^{-1}H^{+} {}^{+}HCO^{-3} \quad (3)$$

$$Ca^{2+} + HCO_2^{-3} \rightarrow CaCO_3 \downarrow \quad (4)$$

In the year 2001, CO_2 sequestration was demonstrated using bovine carbonic anhydrase by Bond et al. (2001) and also evaluated the availability of bovine carbonic anhydrase in the presence of different ions. Subsequently, carbonic anhydrase was successfully extracted, purified, and characterized from biological systems and non-biological systems for its utilization in carbon dioxide (CO_2) sequestration. The bicarbonate generated serves as a raw material for solid mineral formation. Several studies reported the use of microorganisms as a source of carbonic anhydrase for CO_2 sequestration. (Ramanan et al. 2009), demonstrated bio-sequestration of carbon dioxide (CO_2) using carbonic anhydrase from a microorganism such as *C. freundii*. Carbonic anhydrase was purified and characterized from *P. fragi* showing promising potentials for carbon dioxide (CO_2) sequestration. Other studies reported bacterial such as *M. lylae and M. Luteus* as a source of carbonic anhydrase for sequestration of CO_2 in lime stone (Sharma et al. 2011). The use of either bovine carbonic anhydrase or microbial carbonic anhydrase has stated to be both faster in terms of precipitation.

14 Separation System for CO_2 Sequestration

There are different methods of carbon dioxide (CO_2) capturing that have been adopted. This includes cryogenic, membrane, absorption, and adsorption separation.

15 Cryogenic Separation

The principle of condensation and cooling is associated with the cryogenic separation of carbon dioxide (CO_2) from flue gases containing high concentration of carbon dioxide (CO_2) (Tan et al. 2016; Chu et al. 2016). Carbon dioxide (CO_2) capture cannot be done using cryogenic separation from power plants due to the presence of dilute carbon dioxide (CO_2) stream which may affect the condensation and cooling process (Kwak et al. 2006; Hu et al. 2016). This technology has been seen to be quite expensive requiring a high amount of energy needed for carbon dioxide (CO_2) to be captured. Coal-fired power plants with low temperature is also not suitable using this type of technique. However, cryogenic separation has been known to be effective when carried out at low temperatures (Rochelle 2009). There are various steps that are used for carbon dioxide (CO_2) capturing with cryogenic separation. This includes compression, refrigeration, and separation steps (Mangindaan et al. 2015). The presence of impurities affects the separation by lowering the phase transition temperature of carbon dioxide (CO_2) to as low as -80^0C. At this point, the energy is increased during the process of refrigeration leading to the formation of carbon dioxide (CO_2) which might affect the safety of the equipment been used (Cai and Yao 2010; Mondal et al. 2015).

16 Membrane Separation

The principle of membrane separation is based on the physical and chemical changes in the interaction between the membrane material and carbon dioxide (CO_2) gas. The membrane material is formed in such a way that the gas passing through the column moves indirectly proportional to each other; i.e., one gas moves faster than the others. Previous studies reported by Corti and Lombardi (2004) show the use of membrane technology for carbon dioxide (CO_2) capture. This application might be competitive if the flue gas percentage is less than 10 of CO_2. This type of technique is poor in selectivity due to its high requirements during separation and as such poses a major setback in the membrane technology for CO_2 capturing and sequestration (Ang et al. 2015; Cheah et al. 2015; Cheng et al. 2013). More so, this type of separation uses a ceramic membrane which can be organic or inorganic in nature depending on the type of separation required (Cheah et al. 2015; Chiang et al. 2011). Another limitation regarding the use of this type of technology is that they are quite expensive with difficulties in attaining higher degrees and purity during capturing processes. This technology for post-combustion carbon dioxide (CO_2) capture cannot be successful, especially in developing countries such as Nigeria, due to its expensive nature. Therefore, there are limitations with membrane technology application for post-combustion carbon dioxide (CO_2) capture in power plants which is similar to other developing countries. Membrane technology for carbon dioxide (CO_2) capture with low-cost and high capture potential is needed to be studied as the best alternative (Fulke et al. 2015).

17 Absorption

This type of separation technology deals with the use of chemical-based solvent for carbon dioxide (CO_2) capturing. It is mainly grouped into two categories, namely chemical and physical absorptions. Absorption method is pressure and acid–base neutralization dependent which occurs at high pressure and low temperature as well (Corsten et al. 2013; Nykvist 2013). Amines, chilled methanol, and ammonia solution are commonly used as an absorbent for carbon dioxide (CO_2) capture (Kelvin and Patrick 2016). The corrosiveness and high energy demand for absorption make it necessary during solvent regeneration. The major challenge for using absorption technology is the use of liquid absorbent in the power plant. Absorption technology remains used for carbon dioxide (CO_2) capturing in most countries with power plants. This is because it is non-corrosive with minimum energy required for the regeneration of solid adsorbent making it less expensive (Little and Jackson 2010). Monoethanolamine is one of the frequently used absorbents for carbon dioxide (CO_2) capture. It requires high heat energy during the process of capturing which makes it not viable for power plants considering its cost implication. In order to reduce the high heat energy consumption for power plants during post-combustion carbon dioxide carbon dioxide (CO_2) capture, it is suggested that a high concentration of monoethanolamine be used and inhibitors to remove the corrosive effect of the absorbent.

18 Adsorption

Adsorption technology is a type of separation that is used mainly in chemical and environmental processes which involves in the use of mainly adsorbents such as zeolites, activated carbon, and polyaspartamide for carbon dioxide (CO_2) capture (Mondal et al. 2012). This technique uses activated carbon fiber in power plants (Rubin et al. 2007). This use adsorption technique for carbon dioxide (CO_2) capture has proven to be the best method adopted due to its characters exhibited during capturing processes. This characteristic includes minimum energy requirement, easy handling, simple operation, and flexibility (Cau et al. 2014). Studies have also shown that temperature swing adsorption consumes less thermal energy and is less expensive which makes it suitable for carbon dioxide (CO_2) sequestration and capturing but makes the heating and cooling process take a longer time(Scheffknecht et al. 2011). Therefore, when incorporated into coal-fired plants will reduce the cost of operation (Wang et al. 2011). Another type of adsorbent includes vacuum swing adsorption which is less expensive than pressure swing adsorption for carbon dioxide (CO_2) capture (Kundu et al. 2014). Nevertheless, vacuum swing adsorption has its disadvantages as it is sensitive in feeding gas temperature, thereby requiring more heat before injecting the vacuum swing adsorption plant thereby affecting the efficiency of separation. On the other hand, pressure swing adsorption has a wide range of temperatures and pressure requiring less energy for operation and less investment cost. This has created a promising approach for carbon dioxide (CO_2) capture and sequestration (Rehfeldt et al. 2011). This technique has its drawbacks such as poor heat transfer and slow kinetics. But despite this drawback, its advantages far outweigh its disadvantages (Cormos et al. 2013) (Fig. 6).

19 Bioreactors for CO_2 Sequestration

The use of reactors as a separation system has contributed positively to creating a friendly environment for carbon dioxide (CO_2) sequestration. Studies have revealed that both free and immobilized carbonic anhydrase on bioreactors has been known to possess a good foundation for carbon dioxide (CO_2) sequestration technology. Low-cost implication, low maintenance, and environmental safety are the importance of using a biochemical system for sequestration which was found to be of great industrial advancement (Cowan et al. 2003). Carbonic anhydrase was tested on a membrane reactor which constitutes a phosphate buffer with polypropylene and carbonic anhydrase in between (Cowan et al. 2003). An annular spacer was used to vary the thickness of the aqueous phase in the range of 70–670 μm. The reactor contains an air-mixed serving as the feed gas in the range 0.04–1.0% with argon used as a sweep gas. Through the process, a controller was used to monitor the flow of gas in the reactor system. At the end of the sequestration process, gases such as N_2, O_2, and CO_2 were analyzed by mass spectrophotometer to evaluate the amount of carbon dioxide (CO_2) been sequestrated alongside with another parameters

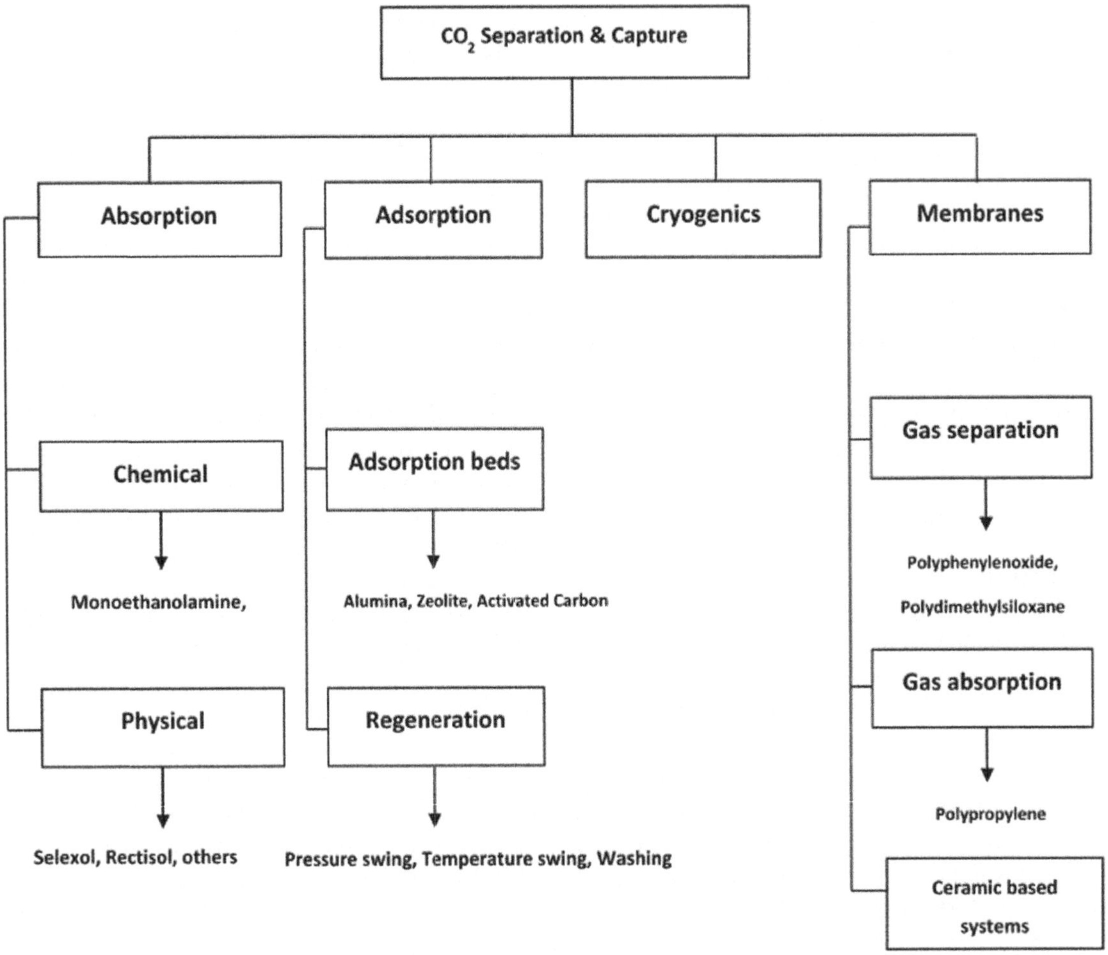

Fig. 6 Separation system for carbon dioxide (CO_2) sequestration (Kelvin and Patrick 2016)

such as carbonic anhydrase concentration, buffer concentration, pH of the medium, temperature, and humidity. Furthermore, a spray reactor was implemented with an immobilized carbonic anhydrase used in capturing and sequestering carbon dioxide (CO_2). A silica-coated steel matrix was used as a carrier and immobilized carbonic anhydrase on it using the coupling method (Sharma et al. 2009). Different conditions were employed during operation by subjecting the bioreactor to different gas outlets such as exhaust/emission gas stream and water flow with carbon dioxide (CO_2) reduction being obtained in the reactor design. The performance of the bioreactor was evaluated by variation through the flow rate and maintaining the concentration of carbon dioxide (CO_2) emission at a constant in the range of 33–40%. Pyrogallol spray column system is a type of reactor system that is used in the presence of oxygen and sulfur been extremely high (Sharma et al. 2011). This new approach of carbon dioxide (CO_2) sequestration was used over time until in early 2010 the development of a hollow-fiber membrane reactor was launched for low removal of carbon dioxide (CO_2) concentration from a mixture of flue gases. (Sharma et al. 2011). However, capturing and sequestration of carbon dioxide (CO_2) has been studied differently in the rector system comparatively which can be found to be efficient for the removal of carbon dioxide (CO_2) from gas streams.

20 Carbonic Anhydrase Immobilization

Over the years, carbonic anhydrase has been seen as an ancient enzyme and well-known for various industrial applications. It is used mainly in its free nature which seems to be very expensive and tedious to isolate, purify, and characterize. In the end, this enzyme is not recovered from the reaction medium after being used. In order to salvage this particular problem, the idea of immobilization was considered. Immobilization involves the attachment of an enzyme into a supporting matrix for delivery in a reaction medium. This technique of whole-cell and enzyme has been reported. Generally, the immobilization technique is divided into two major types, namely chemical and physical immobilizations.

Chemical immobilization technique involves the application of cross-linkages and covalent bonds between the matrix and the enzyme, while physical immobilization technique on the other hand involves the use of a supporting aid or matrix with the enzyme alongside (Krajewska 2004). There are several reports on different immobilization methods giving rise to the use of different supporting matrix and linkers. Bovine carbonic anhydrase was immobilized on silica beads and graphite rods which was reported by Crumbliss et al. (1988), taking cognizant of the benefit as well the non-toxic nature of these materials. The material used for immobilization shows to be suitable and coffined as a packaging aid due to its high surface area serving as an important property for enzyme immobilization in a flow bioreactor. The bovine carbonic anhydrase immobilization process, an amide linkage was formed between the primary amines lysyl residues of the enzyme and the activated N-hydroxysuccinimide ester surface of the graphite rod (Ajam et al. 2012). The immobilized enzyme was stored for about 50 days of storage at pH 8.0, 4°C, and assayed. Prior to the enzyme activity, silica support was found to be more active than that of graphite. (Jovica and Nenad 1999) reported the immobilization of carbonic anhydrase within the silica monoliths sol–gel method as the supporting matrix. The above report suggests that carbonic anhydrase after been encapsulated retained its conformation and remain stable up to 64 °C. (Bond et al. 2001) immobilized bovine carbonic anhydrase in chitosan alginate beads system and shown in retaining its activity after been kept for a long period of time. Through the process of immobilization, a crosslinking solution was under continuous stirring, and samples were taken for enzyme quantification. About forty (40) beads were synthesized per milliliter of alginic acid solution. After been prepared, the synthesized beads were washed and taken for enzyme quantification. Protein loss was evaluated in the sample and stored for (37oC, 110 rpm). Nitrogen in beads was quantified as well which was found to increase as chitosan molecular weight increased. An absorptive immobilization of carbonic anhydrase was done on hydrophobic absorbents reported by Hosseinkhani and Gorgani (2003). This type of immobilization was partially unfolded due to the fact that carbonic anhydrase in its native form does not interact with the hydrophobic absorptive surface. To overcome this, heat was applied to the native carbonic anhydrase forming a heat-denatured carbonic anhydrase which was immobilized on the sepharose 4B (Azari and Gorgani 1999). The presence of long alkyl chains such as the pamityl group in the sepharose 4B was responsible for sufficient interaction leading to an irreversible adsorption process. (Cheng et al. 2008), reported the use of carbonic anhydrase entrapment on acrylic acidoacryamide. Bovine carbonic anhydrase was immobilized on polyurethane foam at room temperature reported by Ozdemir (2009). This type of immobilization possess good characteristics for immobilization by creating support due to its hydrophilic and porous polymeric properties. Enzyme immobilization with polyurethane have been proven to maintain its activity perfectively at room temperature for 45 days tested period. Biopolymer materials for immobilization were developed by Prabhu et al. (2009) for effective carbonic anhydrase immobilization and for industrial application in mineral carbonates formation. Different tools for the characterization of material were used. However, the materials used showed better properties during immobilization. Supporting beads such as chitosan-ammonium hydroxide, multilayered, and alginate presented a better result with high affinity for carbonic anhydrase. (Zhang et al. 2011) reported the immobilization of carbonic anhydrase of a nano-composite (acrylic acid-coacrylamide/hydrotalcite) and its N-hydroxysuccinamide and N, N-dicyclohexylcarbodiamide used for activation. (Vinoba et al. 2012), reported carbonic anhydrase immobilization on SBA-15 and utilized silver and gold nanoparticles as well. (Wanjari et al. 2011), immobilized carbonic anhydrase on chitosan beads for carbonation enhancement. (Yadav et al. 2012) reported the use of nanoparticle for biomimetic CO_2 sequestration and immobilization on a chitosan stabilized iron nanoparticles for biomimetic carbonation reaction. Immobilizing carbonic anhydrase on a controlled pore glass and carbon surfaces which was activated by silane and aldehyde shows a stability of the activity of the enzyme (Zhang et al. 2011; Vinoba et al. 2013), studied the immobilization of bovine carbonic anhydrase on a magnetic nanoparticle matrix resulting in a good reusable catalyst. Immobilized bovine carbonic anhydrase showed 82% activity up to 30 days on this matrix which shows its potentials for CO_2 capturing, sequestration, and transformation of carbon dioxide to mineral carbonates (Yadav et al. 2012). Chitosan–alginate polyelectrolyte complex was used as a supporting matrix for enzyme immobilization with 93.9% potential rate (Oviya et al. 2012) (Table 1).

21 Ethylenediamine for Carbon Dioxide (CO_2) Capturing

Ethylenediamine is a synthetic hygroscopic chemical solvent, yellowish in color with an ammonia-like odor. It has chemical formula ($C_2H_8N_2$) and a molecular weight of 60.12g/mol. Its melting point is about 8.5 °C, and a boiling point of 116 °C. Ethylenediamine belongs to the "amine" family and is grouped as a primary diamine. It has been shown to be reactive with carbon dioxide (CO_2) detailing a large experimental mechanism and rates of reactions. As an amine-based solvent, ethylenediamine has been studied over the years and its utilization has increased for CO_2 capture. The primary interest of reaction between post-combustion carbon dioxide (CO_2) gas and ethylenediamine occurs in a step-wise reactions. These reactions are summarized in a three-step process. At first, carbamate is formed, following the formation and deprotonation of zwitterion (Niall et al. 2010). The reaction mechanism is shown below:

$$CO_2 + 2R_1R_2NH \rightarrow R_1R_2NCO_2^{--} + R_1R_2NH_2^+ \quad (5)$$

From Eq. (5), carbon dioxide reacts with ethylenediamine in a reversible reaction leading the formation of carbamate. The other reaction involves the hydrolysis reaction of carbamate leading to the formation of bicarbonate (Eq. (6)).

$$R_1R_2NCO_2^{---} + H_2O\ R_1R_2NH + HCO_3^{--} \quad (6)$$

In the formation of bicarbonate, the reaction is favored with a methyl group by reducing the stability of bond within the carbamate which at the end result in the formation of bicarbonate at a high loading capacity of the solvent system. The above reaction mechanism does not act with carbon dioxide (CO_2) directly but rather acts as a base medium in the hydration reaction in an aqueous medium of ethylenediamine (Niall et al. 2010). Carbon dioxide (CO_2) capturing technology through the use of absorbent such as ethylenediamine has been reported by researchers since the year 1955. This technology also contributed to the aspect of carbon dioxide (CO_2) capturing mainly from power plants with coal-fired practice (Shan et al. 2010). There are advanced amines that have been reported to reduce the energy of amine scrubbing such as monoethanolamine (30wt %, 7M) and concentrated piperazine (40w%, 8M). Piperazine in a concentration of 8M is considered a good solvent for carbon dioxide (CO_2) capture. The rate of the reaction is twice that of monoethanolamine and comparable to its amine volatility. Similarly, ethylenediamine and piperazine have two amino groups that react with carbon dioxide to form carbonate (Bishnoi and Rochelle 2002). Prior to this, ethylenediamine have also been reported to have smaller weight, greater carbon dioxide (CO_2) capacity and highly concentrated in alkali at a given amine weight concentration than piperazine making it more preferred for carbon dioxide (CO_2) capturing technology (Shan et al. 2010). One of the importance of using ethylenediamine as an amine-based solvent system for carbon dioxide (CO_2) capture is the ability of ethylenediamine been regenerated after hydration to the formation of carbonate molecule. This can be considered a promising prospect for further research. Using this solvent technology, the carbamate formed can be integrated into the industrial applications for the production of insecticides. One of the drawbacks of using this type of solvent technology is due to the fact that ethylenediamine is prepared from monoethanolamine which might seem to be quite expensive as a solvent used for carbon dioxide (CO_2) capture. To solve this problem, ethylenediamine can be synthesized from other sources such as 1,2-dichloroethane with ammonia in an aqueous medium which might require less time and low capital cost for production.

22 CO_2 Capturing and Sequestration with Ethylenediamine–Carbonic Anhydrase Complex

This release of anthropogenic gases such as carbon dioxide (CO_2) by industrial power plants has created a lot of challenges which has led to the change in environmental temperature. This has led to much intervention in salvaging this problem by utilizing flue gases from industrial sources for the purpose of limiting the effect of these gases on humans and the environment. There are several systems that have been adopted to capture and sequestrate carbon dioxide (CO_2) ranging from the use of a solvent systems such as amides, ammonia, etc., and enzyme systems such as carbonic anhydrase. Amine-based chemical absorbents such as ethylenediamine has shown to be employed typically for commercial technology of carbon dioxide (CO_2) capture due to its distinct properties as regards its good carbon dioxide (CO_2) capturing capacity compared to other amine-based system such as monoethanolamine (MEA) and piperazine. Ethylenediamine solvent system requires a lower energy regeneration and is highly effective toward forming carbamates which are stable when reacting with carbon dioxide and forming carbonates through hydrolysis and regenerating ethylenediamine for subsequent usage. Commercially, amine-based solvent such as monoethanolamine can generate a hundred tons of carbon dioxide, where a coal combustion plant typical with 550 MW emits up to 12,800 metric tons of carbon dioxide per day (Klara 2007). In this contempt, mitigating the effect of gases from power plants, carbon dioxide (CO_2) capture solvent systems such as ethylenediamine should be adopted in other to generate thousand tons of carbon dioxide (CO_2) from flue gases for industrial purposes. The use of enzyme systems such as carbonic anhydrase in its free and immobilized has been adopted and seems to be fast growing when compared to other carbon dioxide (CO_2) capturing and sequestration technologies (Fradette et al. 2017; Yuan et al. 2017; Yeh et al. 2005). As earlier defined, carbonic anhydrase catalyzes the reversible hydration of CO_2 to yield bicarbonate (Tripp et al. 2001). The principle is based on binding carbon dioxide (CO_2) molecules to the hydrolyzed water molecule to release a proton. The hydroxide is hydrated from carbon dioxide (CO_2) to yield HCO_3. Since the development of an enzymatic system of carbon dioxide (CO_2) post-combustion capture dates back in the mid-1960s, there are lots of scientific approaches that have been put in place to promote the usability of carbonic anhydrase by creating bioreactor system and immobilizing the enzyme on a supporting matrix. This technology was adopted, and lots of advancement was made to better carbon dioxide (CO_2) capturing processes. The use of immobilizing enzymes is considered to be inexpensive and can be explored in diverse applications. With the effect of greenhouse gas, a novel system has been prompted in designing a liquid system that is enhanced by carbonic anhydrase (Prakash et al. 2017). These designs are proposed by immobilizing carbonic anhydrase on the reactor column in other to promote the effectiveness of the carbonic anhydrase in the solvent system. Previous studies have shown the co-immobilization of enzyme and amine-based solvent system on a bioreactor for carbon dioxide (CO_2) post-combustion capture process. Despite the development, lots of consideration were made to optimize the solvent system that will be best in order to reduce the regeneration energy and improve solvent system efficiency. This was based on the physical and chemical properties of the solvent in question. In addition to this, there are advancements that offer a better result in providing a value added technology. This type of technology is through co-immobilization with different systems in order to create a better result for capturing and sequestration of carbon dioxide (CO_2) technology. Research has been carried out on co-immobilization. Example of this is the co-immobilization of zinc complexes carried out on the same support in order to improve carbon dioxide (CO_2) absorption and desorption ability. Studies have also reported the kinetics of carbon dioxide (CO_2) hydration with the use of carbonic anhydrase with high-molarity chemical sorbents such as aqueous ammonia and temperature unit operators. This advancement has been tested in the USA and China using (Niall et al. 2010). Co-immobilization of enzyme complex on hetero-functionalized support was carried in a hybrid solvent to accelerate in carbon dioxide (CO_2) capture (Parissa et al. 2007). The equilibrium of carbon dioxide (CO_2) capture in the hybrid solvent was measured by pressure and gravimetric drop test. The incorporation of carbonic anhydrase-Zn-Im in the supporting matrix showed good stability and absorption of CO_2 capturing from 13.75 wt% to 18.65 wt% of carbonic anhydrase. With the incorporation of zinc complex, the carbon dioxide (CO_2) uptake was improved further to 21.6 wt% (Parissa et al. 2007). The co-immobilization of an amine-based solvent and carbonic anhydrase will exhibit a remarkable increase in carbon dioxide (CO_2) absorption giving rise to high carbon dioxide (CO_2) capture rate. Therefore, carbonic anhydrase has been suggested to increase the CO_2 absorption rate of the amine-based solvent such as diethanol amine, N-methyl-2, 2′- iminodiethanol, and 2-amino-2-methyl-1-propanol (Vinoba et al. 2013; Heupen and Kenig 2010). Despite the effective proof, the enzyme carbonic anhydrase has some intrinsic limitations like stability and reusability with an amine solution. Therefore, in a diverse approach, the co-immobilization of carbonic anhydrase with amine-based solvent ethylenediamine will facilitate and promote good carbon dioxide (CO_2) capture application industrially due to its chemical and physical properties when compared with other amine solvent systems. In a typical co-immobilization system of ethylenediamine–carbonic anhydrase for carbon dioxide (CO_2) in Fig. 7, flue gases are passed through a reactor containing an immobilized ethylenediamine as the amine solvent with carbon dioxide-rich amine through a stripper where low-pressure steam is applied. Hot lean amine is released containing (95% methane and ethane) from the steaming chamber through the desorber system and carbon dioxide (CO_2) is captured and sent to the immobilized carbonic anhydrase chamber where enzymatic processes take place to release carbonate for industrial application.

23 CO_2 Capturing and Sequestration Design and Optimization: Challenges and Future Prospects

The design of the carbon dioxide (CO_2) post-combustion capture and sequestration process has been described in the above section. The use of complex modeling for carbon dioxide (CO_2) capture and sequestration has been exploited for various industrial applications driving the carbon dioxide (CO_2) capture to a greater height. This view has prompted the use of solvent systems such as amine-based solvent with carbonic anhydrase as complex systems for carbon dioxide (CO_2) capture and sequestration. Basically, in most amine-based solvent systems, it is necessary to consider in understanding the rate-limiting steps in the transfer of carbon dioxide (CO_2) between phases and making sure the equipment is sized fittingly. Several models for mass transfer of carbon dioxide (CO_2) and operational analysis have been using amine-based solvent system and carbonic anhydrase designed by research groups (Kvamsdal et al. 2009). Other researchers have also studied the transient process behavior (Bardow and K., & Gross, J., 2010). In this review, the advantage of the co-immobilization of ethylenediamine–carbonic anhydrase as a complex system for carbon dioxide (CO_2) capture and sequestration is established. This complex model can be used to estimate the cost of energy requirement and capital need to capture carbon dioxide (CO_2) and sequestrate it in flue gas carbon dioxide (CO_2) concentration. Considering the cost capital and energy requirement as performance parameters are paramount because the optimal production of the carbon

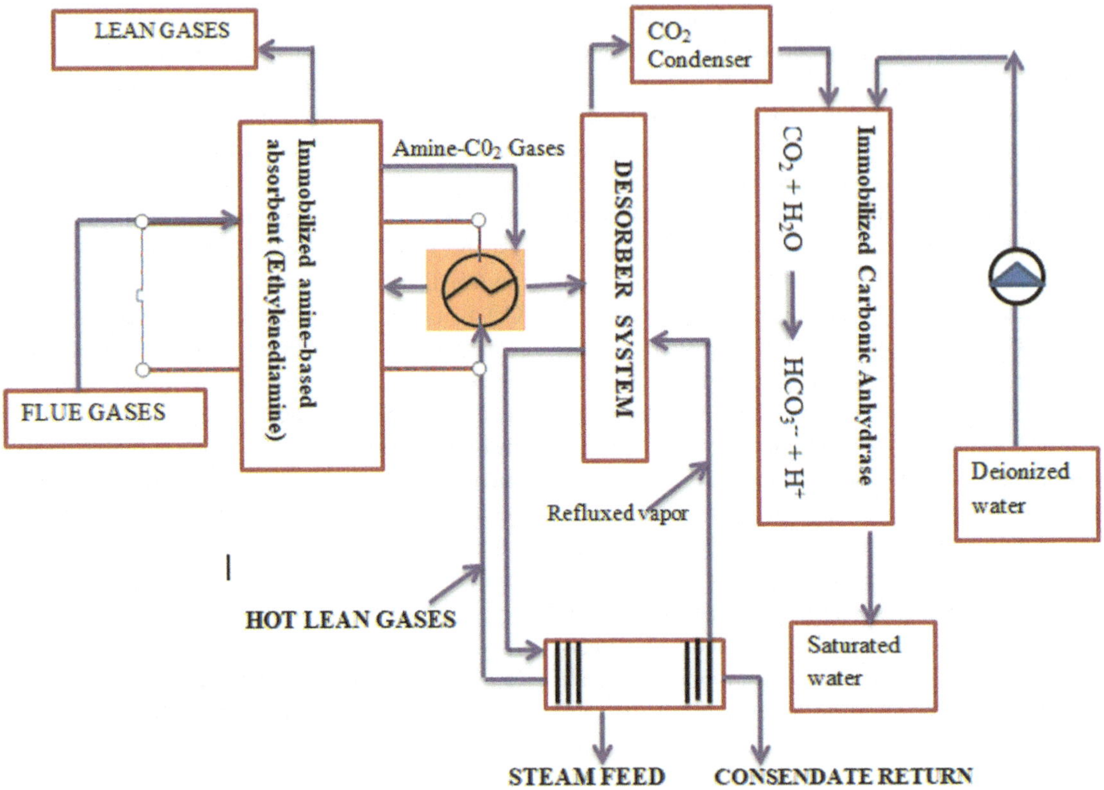

Fig. 7 Schematic diagram for ethylenediamine–carbonic anhydrase complex for CO_2 capturing and sequestration

dioxide (CO_2) will not be ascertained until its infrastructure is examined. However, the selection on the type of solvent system is key in evaluating the efficiency of the overall equipment performance (Niall et al. 2010). A complex model system of ethylenediamine and carbonic anhydrase will reduce the cost capital and energy requirement needed for carbon dioxide (CO_2) capture and sequestration due to their physical and chemical properties. The immobilized solvent system and the enzyme are regenerated and can be reused during the carbon dioxide (CO_2) capture and sequestration. Research has shown the importance of using an integrated design of processes and solvents system serving as a means of achieving truly standardized solvent-based processes (MacDowell et al. 2010). One of the main principle drawbacks for using a solvent system for carbon dioxide (CO_2) capture is predicting the effect of solvent used for the separation process which also has a significant amount of experimental data required. Furthermore, the use of the enzyme system has also developed some drawbacks during the sequestration process. The major problem is the buildup of HCO_3^{--} in the bioreactor serving as an inhibitor for carbonic anhydrase, poor mass transfer, and instant backward dehydration of carbon dioxide (CO_2). For better immobilization, the use of ion exchangers severing as anti-inhibitors can be impregnated into the bioreactor to prevent loss of carbonic anhydrase activity and to enhance the mass production of carbon dioxide (CO_2). The use of a highly alkaline base range of pH in carbonation formation is favorable. Therefore, alkaliphilic bacterial carbonic anhydrase will be effective for carbon dioxide (CO_2) sequestration (Raju et al. 2014). This will provide produce the abundant enzyme for the bioreactor process. However, [121] used the physically based approach for the absorption of carbon dioxide (CO_2) in aqueous solutions of monoethanolamine which is a primary amine-based solvent. Similarly, using the complex modeling method between solid ethylenediamine and carbonic anhydrase, the rate processes of carbon dioxide (CO_2) transfer is known to be more complex with gas phase mass transfer, pore, and ash layer diffusion (Niall et al. 2010). Nevertheless, a multi-design system may be put in place to access the efficiency of the separation process in other to determine the optimal design parameters such as temperature and equipment geometrics. Complex design for carbon dioxide (CO_2) capture and sequestration tend to focus on the efficiency of the system and should be integrated into process-level models in a multi-scale system in order to create and support the optimized design structure.

Table 1 Different materials for immobilization of carbonic anhydrase

Carbonic anhydrase source	Technique of immobilization/support	References
BCA	Silica beads/graphite	Crumbliss et al. 1988)
M. thermoautotrophicum	Steel matrix	Bhattacharya et al. 2003)
BCA	Chitosan/alginate	Bond et al. 2001)
BCA	Polymeric supports	Drevon et al. 2003)
BCA	Sepharose 4B	Hosseinkhani and Gorgani 2003)
Bacillus pumilus	Chitosan-NH_4OH beads	Vinoba et al. 2012)
Bacillus pumilus	Multilayered beads	Prabhu et al. 2009)
Sulfurihydrogenibium sp.	Polyurethane foam	Capassoa et al. 2012)
BCA	Polyurethane	Ozdemir 2009)
Bacillus pumilus	Chitosan stabilized iron nanoparticles (CSIN)	Yadav et al. 2011)
Bacillus pumilus	Single enzyme nanoparticle particle (SEN)	Yadav et al. 2012)
Bacillus pumilus	Chitosan beads	Wanjari et al. 2011)
HCA	SBA-15	Vinoba et al. 2011)
HCA	Magnetic nanoparticle	Vinoba et al. 2012)
HCA	SBA-15 silver and gold nanoparticles	Vinoba et al. 2011)

KEY: *BCA*—bovine carbonic anhydrase; *CA*—carbonic anhydrase

24 Conclusion

In conclusion, the use of carbonic anhydrase and solvent system (ethylenediamine) has been shown in most literature to possess good qualities of carbon dioxide (CO_2) capturing potentials that are integrated into the industrial application for carbon dioxide (CO_2) capture. In this review carbon dioxide (CO_2) post-combustion capturing and sequestration from industrial carbon dioxide (CO_2) emission sources using ethylenediamine and carbonic anhydrase complex through immobilization system have been elaborated. This complex system for carbon dioxide (CO_2) capture and sequestration possess some distinct advantages compared to other alkanolamine solvents. The use of immobilization in a rector system promotes the mass transfer of carbon dioxide (CO_2) and reduces the energy requirement needed for carbon dioxide (CO_2) capture. Furthermore, immobilization also promotes the reusability of carbonic anhydrase and ethylenediamine and for effective modeling design of the carbon dioxide (CO_2) capturing and sequestration processes.

References

Adler L, Brundell J, Falkbring SO, Nyman PO (1972) Carbonic anhydrase from Neisseria sicca, strain 6021. I. Bacterial growth and purification of the enzyme. Biochem Biophys Acta 284:298–310

Ajam YS, Kannan K, Sandeep NM, Raju RY, Abhay BF, Sivanesan SD, Tapan C (2012) Recent advancements in carbonic anhydrase–driven processes for CO_2 sequestration: minireview. Crit Rev Environ Sci Technol 42(14):1419–1440

Alber BE, Ferry JG (1994) A carbonic anhydrase from the archaeon Methanosarcina thermophila. Proc Nat Acad Sci 91:6909–6913

Ang RR, Sin LT, Bee ST, Tee TT, Kadhum A, Rahmat A, Wasmi BA (2015) A review of copolymerization of greenhouse gas carbon dioxide and oxiranes to produce polycarbonate. J Clean Prod 102:1–17

Anusha K (2010) Carbon dioxide capture by chemical absorption: a solvent comparison study. Massachusetts Inst Technolo 3:1–263

Azari F, Gorgani MN (1999) Reversible denaturation of carbonic anhydrase provides a method for its adsorptive immobilization. Biotechnol Bioeng 62:193–199

Bardow S, K., & Gross, J., (2010) Continous-Molecular targeting for integrated Solvent and process Design. Ind Eng Chem Res 49:2834–2840

Bhattacharya S, Nayak A, Schiavone M, Bhattacharya SK (2003) Solubilization and concentration of carbon dioxide: novel spray reactors with immobilized carbonic anhydrase. Biotechnol Bioeng 86:37–46

Bishnoi S, Rochelle GT (2002) Absorption of carbon dioxide aqueous piperazine/methyldiethanolamine. AIChE J 48:2788–2799

Bond GM, Stringer J, Brandvold DK, Simsek FA, Medina M, Egeland G (2001) Development of integrated system for biomimetic CO_2 sequestration using the enzyme carbonic anhydrase. Energy Fuel 15:309–316

Brinkman R, Margaria R, Meldrum N, Roughton F (1932) The CO_2 catalyst present in blood. J Physiol 75:3–4

Bryce D, Maohong F, Armistead GR (2015) Amine-based CO_2 capture technology development from the beginning of 2013: a review. ACS Appl Mater Interfaces 7(4):2137–2148

Bryngelsson M, Westermark M (2009) CO_2 capture pilot test at a pressurized coal fired CHP plant. Energy Procedia. 1:1403–1410

Burnell JN, Gibbs MJ, Mason JG (1990) Spinach chloroplastic carbonic anhydrase nucleotide sequence analysis of cDNA. Plant Physiol 92:37–40

Cai Y, Yao J (2010) Effect of proteins on the synthesis and assembly of calcium phosphate nanomaterials. Nanoscale 20:1842–1848

Capassoa C, Lucaa VD, Carginalea V, Caramuscioc P et al (2012) Characterization and properties of a new thermoactive and thermostable carbonic anhydrase. Chem Eng Trans 27:271–276

Casson S, Gray JE (2008) Influence of environmental factors on stomatal development. New Phytol 178:9–23

Cau G, Tola V, Deiana P (2014) Comparative performance assessment of USC and IGCC power plants integrated with CO_2 capture systems. Fuel 116:820–833

Cheah WY, Show PL, Chang JS, Ling TC, Juan JC (2015) Bio-sequestration of atmospheric CO_2 and flue gas-containing CO_2 by microalgae. Bio-Resources Technol 184:190–201

Chegwidden WR, Carter ND, Edwards YH, Birkhäuser V (2002) The enzyme carbonic anhydrase mediates the taste sensation of carbonated drinks Boston. MA, USA 2000:221–240

Cheng LH, Zhang L, Chen HL, Gao C (2008) Hollow fiber contained hydrogel-CA membrane contactor for carbon dioxide removal from the enclosed spaces. J Membr Sci 324:33–43

Cheng J, Huang Y, Feng J, Sun J, Zhou J, Cen K (2013) Mutate *Chlorella sp.* by nuclear irradiation to fix high concentrations of CO2. Bio-Resources Technol 136:496–501

Chiang CL, Lee CM, Chen PC (2011) Utilization of the cyanobacteria *Anabaena sp.* CH1 in biological carbon dioxide mitigation processes. Bio-Resources Technol 102:5400–5405

Chirica LC, Elleby B, Jonsson BJ, Lindskog S (1997) The complete sequence, expression in Escherichia coli, purification and some properties of carbonic anhydrase from *Neisseria gonorrhoeae.* Eurology J Biochem 244:755–760

Chu F, Jon C, Yang L, Du X, Yang Y (2016) CO_2 absorption characteristics in ammonia solution inside the structured packed column. Ind Eng Chem Res 55:3696–3709

Cormos CC, Vatopoulos K, Tzimas E (2013) Assessment of the consumption of water and construction materials in state-of-the-art fossil fuel power generation technologies involving CO_2 capture. Energy 51:37–49

Corsten M, Ramírez A, Shen L, Koornneef J, Faaij A (2013) Environmental impact assessment of CCS chains-Lessons learned and limitations from LCA literature. Int J Greenhouse Gas Control 13:59–71

Corti A, Lombardi L (2004) Reduction of carbon dioxide emissions from a SCGT/CC by ammonia solution absorption-Preliminary results. Int J Thermodyn 4:173–181

Cowan RM, Ge JJ, Qin YJ, McGregor ML, Trachtenberg MC (2003) CO_2 Capture by means of an enzyme-based reactor. Annal New York Acad Sci 984:453–469

Crumbliss AL, McLachlan KL, O'Daly JP, Henkens RW (1988) Preparation and activity of carbonic anhydrase immobilized on porous silica beads and graphite rods. Biotechnol Bioeng 31:796–801

Diamantopoulos PD, Aivalakis G, Flemetakis E, Katinakis P (2013) Expression of three β-type carbonic anhydrase in tomato fruits. Mol Biol 40:4189–4196

Domsic JF, Avvaru BS, Kim CU, Gruner SM, Agbandje-McKenna M, Silverman DN, McKenna RJ (2008) The hydration of CO_2 or dehydration of HCO_3^- is efficiently done without significant structural changes to the protein. Biol Chem 283,30766

Drevon G, Fnbarke C, Russell A (2003) Enzyme containing michael-adduct-based coatings. J Biomacromol. 4:675–682

Favre N, Christ ML, Pierre AC (2009) Biocatalytic capture of CO_2 with carbonic anhydrase and its transformation to solid carbonate. J Molecular Catalyst β-Enzyme 60:163–170

Fierke CA, Calderone TL, Krebs JF (1991) Inhibition and catalysis of carbonic anhydrase. Biochemistry 30:11054–11063

Figueroa JG, Fout T, Plasynski S, McIlvried H, Srivastava RD (2008) Advances in CO_2 capture technology—the U.S. department of energy's carbon sequestration program. Int J Greenhouse Gas Control 2:9–20

Fradette S, Gingras J, Carley J, Kelly GR, Ceperkovic O (2017) Process for capturing CO_2 from a gas using carbonic anhydrase and potassium carbonate. Patent 9(533):258

Fulke AB, Krishnamurthi K, Giripunje MD, Devi SS, Chakrabarti T (2015) Biosequestration of carbon dioxide, biomass, calorific value and biodiesel precursors production using a novel flask culture photobioreactor. Biomass Bioenergy 72:136–142

Gary TR (2009) Amine scrubbing for CO_2 capture. Science 325:1652–1654

Guilloton MB, Korte JJ, Lamblin AF, Fuchs JA, Anderson PM (1992) Carbonic anhydrase in Escherichia coli, a product of the cyn operon. J Biol Chem 267:3731–3734

Heupen B, Kenig EY (2010) An overview of CO_2 capture technologies. Ind Eng Chem Res 49:772–779

Hilvo M, Tolvanen M, Clark A, Shen B, Shah GN, Waheed A, Halmi P, Hänninen M, Hämäläinen JM, Vihinen M, Sly WS, Parkkila S (2005) Characterization of CA XV, a new GPI-anchored form of carbonic anhydrase. Biochem J 392:83–92

Hosseinkhani S, Gorgani M (2003) Partial unfolding of carbonic anhydrase provides a method for its immobilization on hydrophobic adsorbents and protects it against irreversible thermo-inactivation. Enzyme Microbiol Technol 33:179–184

Hu G, Nicholas NJ, Smith KH, Mumford KA, Kentish SE, Stevens GW (2016) Carbon dioxide absorption into promoted potassium carbonate solutions: A review. Int J Greenhouse Gas Control 53:28–40

Imtaiyaz HM, Bushra S, Abdul W, Faizan A, William SS (2013) Structure, function and applications of carbonic anhydrase isozymes. Bioorg Med Chem 21:1570–1582

Jovica DB, Nenad MK (1999) Effects of encapsulation in sol-gel silica glass on esterase activity, conformational stability and unfolding of bovine carbonic anhydrase II. Chem Matter 11:3671–3679

Kelvin OY, Patrick TS (2016) The potential of CO_2 capture and storage technology in south Africa's coal-fired thermal power plants. Environments 3:24

Kim IG, Jo BH, Kang DG, Kim CS, Chio YS, Cha HJ (2012) Bio-mineralization-based conversion of carbon dioxide to calcium carbonate using recombinant carbonic anhydrase. Chemosphere 87:1091–1096

Kimber MS, Pai EF (2000) The active site architecture of Pisum sativum beta-carbonic anhydrase is a mirror image of that of alphacarbonic anhydrases. EMBO J 19:1407–1418

Kisker C, Schindelin H, Alber BE, Ferry JG, Rees DC (1996) A left-handed beta-helix revealed by the crystal structure of a carbonic anhydrase from the archaeon Methanosarcina thermophile. EMBO J 15:2323–2330

Klara JM (2007) Volume 1: Bituminous Coal and natural gas to electricity, in cost and performance baseline for fossil energy plants. Nation Energy Technol Lab. DOE

Kothandarama A, Nord L, ollad, O., Herzog, H.j., & McRae, G. J. (2010) Comparism of solvent for post-combustion capture of CO_2 by chemical absorption. Energy Procedia 4:1373–1380

Krajewska B (2004) Application of chitin- and chitosan-based materials for enzyme immobilizations: A review. Enzyme Microbiol Technol 35:126–139

Kulkarni A, Sholl D (2012) Analysis of equilibrium-based TSA processes for direct capture of CO_2 from Air. Ind Eng Chem Res 51 (25):8631–8645

Kundu PK, Chakma A, Feng X (2014) Effectiveness of membranes and hybrid membrane processes in comparison with absorption using amines for post combustion CO2 capture. Int J Greenhouse Gas Control 28:248–256

Kvamsdal HM, Jakobsen JP, Hoff KA (2009) Dynamic modeling and simulation of a CO_2 absorber column for post-combustion CO_2 capture. Chem Eng Process 48:135–144

Kwak KO, Jung SJ, Chung SY, Kang CM, Huh YI, Bae SO (2006) Optimization of culture conditions for CO_2 fixation by a chemoautotrophic microorganism, strain YN-1 using factorial design. Biochem Eng J 31:1–26

Kyllonen MS, Parkkila S, Rajaniemi H, Waheed A, Grubb JH, Shah GN, Sly WS, Kaunisto KJ (2003) Localization of carbonic anhydrase XII to the basolateral membrane of H^+ secreting cells of mouse and rat kidney. J Histochem Cytochem 51:1217

Lackner KS (2010) Comparative impacts of fossil fuels and alternative energy sources. Issues in environmental science and technology, carbon capture: sequestration and storage. In: Hester RE, Harrison RM (eds) Carbon Capture and Storage. Royal Society of Chemistry, Great Britain, pp 1–40

Liljas A, Kannan KK, Bergsten PC, Waara I, Fridborg K, Strandberg B, Carlbom U, Jrup L, Lvgren S, Petef M (1972) Crystal-structure of human carbonic anhydrase-C. Nat New Biol 235:131–137

Little MG, Jackson RB (2010) Potential impacts of leakage from deep CO2 gas sequestration on overlying freshwater aquifers. Environ Sci Technol 44:9225–9232

Loferer MJ, Tautermann CS, Loeffler HH, Liedl KR (2003) Infulence of backbone conformation of human carbonic anhydrase II on carbonic carbon dioxide : hydration pathways and binding of biocarbonate. Journal American Chemical Society 125:8921–8927

MacDowell N, Llovell F, Adjiman CS, Jackson G, Galindo A (2010) 11th International sysmposium on process systems Engineering. Ind Eng Chem Res 49:1883–1899

Mangindaan DW, Woon NM, Shi GM, Chung TS (2015) P84 polyimide membranes modified by a tripodal amine for enhanced pre-evaporation dehydration of acetone. Chem Eng Sci 122:14–23

Marcus EA, Moshfegh AP, Sachs G, Scott DR (2005) The periplasmic a-carbonic anhydrase activity of Helicobacter pylori is essential for acid acclimation. J Bacteriol 187:729–738

Meldrum N, Roughton F (1932) Some properties of carbonic anhydrase, the CO_2 enzyme present in blood. J Physiol 75:15

Millward A, Yaghi O (2005) Metal−Organic Frameworks with Exceptionally High Capacity for Storage of Carbon Dioxide at Room Temperature. J Am Chem Soc 51:17998–17999

Mirjafari P, Asghari K, Mahinpey N (2007) Investigating the application of enzyme carbonic anhydrase for CO_2 sequestration purposes. Ind Eng Chem Res 46:921–926

Mondal MK, Balsora HK, Varshney P (2012) Progress and trends in CO_2 capture/ separation technologies: a review. Energy 46:431–441

Mondal B, Song J, Neese F, Ye S (2015) Bio-inspired mechanistic insights into CO_2 reduction. Current Opinion Chem Biology 25:103–109

Moroney JV, Bartlett SG, Samuelsson G (2001) Carbonic anhydrases in plants and algae. Plant Cell Environment 24:141–153

Niall M, Nick F, Antoine B, Jason H, Amparo G, George J, Claire SA, Charlotte KW, Nilay S, Paul F (2010) An overview of CO_2 capture technologies Energy Environ Sci 3:1645–1669

Nykvist B (2013) Ten times more difficult: Quantifying the carbon capture and storage challenge. Energy Policy 55:683–689

Oviya M, Sib SG, Sukumaran V, Natarajan P (2012) Immobilization of Carbonic Anhydrase Enzyme Purified From *Bacillus Subtilis* Vsg-4 And Its Application As Co2 Sequesterer. Prep Biochem Biotechnol 42(5):462–475

Ozdemir E (2009) Biomimetic CO_2 Sequestration: Immobilization of carbonic anhydrase within polyurethane foam. Energy Fuels 23:5725–5730

Parisi G, Perales M, Fornasari MS, Colaneri A, Gonzalez-Shain N, Gomez-Casati D, Zimmermann S, Brennicke A, Araya A, Ferry JG, Zabalata E (2004) Gamma carbonic anhydrases in plant mitochondria. Plant Mol Biol 55:193–207

Parissa M, Koorosh A, Nader M (2007) Investigating the application of enzyme carbonic anhydrase for CO_2 sequestration purposes. Ind Eng Chem Res 46(3):921–926

Plasynski S, Lang W, Richard A (2007) Carbon Dioxide Separation with Novel Microporous Metal Organic Frameworks. Nat Energy Technol Lab 2:724–781

Prabhu C, Wanjari S, Gawande S, Das S, Labhsetwar N, Kotwal S, Puri AK, Satyanarayana T, Rayalu S (2009) Immobilization of carbonic anhydrase enriched microorganism on biopolymer based materials. J Molecular Catalyts β-Enzyme 60:13–21

Prakash CS, Manoj K, Amardeep S, Singh MP, Puri SK, Ramakumar SSV (2017) Accelerated CO_2 capture in hybrid solvent using co-immobilized enzyme/complex on a hetero-functionalized support. J CO_2 Utilization 21:77–81

Qi G, Liu K, Frimpong RA, House A, Salmon S, Liu K (2016) Integrated bench-scale parametric study on CO_2 capture using a carbonic anhydrase promoted K_2CO_3 solvent with low temperature vacuum stripping. Ind Eng Chem Res 55:12452–12459

Rachael AK, David AM, Sean RP, Kun L, Joe ER, Yue Y, Felice CL, Kunlei L, Cameron AL, Susan AO (2015) Carbonic anhydrase mimics for enhanced CO_2 absorption in an amine-based capture solvent. Royal Soc Chem 94550–9234

Raju RY, Kannan K, Sandeep NM, Saravana DS, Pravin KN, Amit B, Tapan C (2014) Carbonic anhydrase mediated carbon dioxide sequestration: promises, challenges and future prospects. J Basic Microbiol 54:472–481

Ramanan R, Kannan K, Devi SS, Mudliar S, Kaur S, Tripathi AK, Chakrabarti T (2009) Bio-sequestration of carbon dioxide using carbonic anhydrase enzyme purified from *Citrobacter freundii*. World Journal Microbiology Biotechnology. 25:981–987

Ramanan R, Kannan K, Deshkar A, Yadav R, Chakrabarti T (2010) Enhanced algal CO_2 sequestration through calcite deposition by Chlorella sp. and Spirulina platensis in a mini-raceway pond. Bio-Resource Technol 101:2616–2622

Rao AB, Rubin ES (2002) A technical, economic, and environmental assessment of amine-based CO_2 capture technology for power plant greenhouse gas control. Environ Sci Technol 36:4467–4475

Rayalu S, Yadav R, Wanjari S, Prabhu C, Mushnoori SC, Labhsetwar N, Satyanaratanan T, Kotwai S, Wate SR, Hong S, Kim J (2012) Nanobiocatalysts for carbon capture, sequestration and valorisation. Top Catal 55:1217–1230

Rehfeldt S, Kuhr C, Schiffer FP, Weckes P, Bergins C (2011) First test results of oxy-fuel combustion with Hitachi's DST-burner at Vattenfall's 30 MWth pilot plant at schwarze pumpe. Energy Procedia 4:1002–1009

Robert JD, Harmony C, Ananya M, Martha L, James VM (2017) Plant carbonic anhydrases: structures, locations, evolution, and physiological roles *2017. Molecular Plant 10:30–46

Roberts SB, Lane TW, Morel FMM (1997) Carbonic anhydrase in the marine diatom Thalassiosira weissflogii (*Bacillariophyceae)*. J Phycol 33:845–850

Rochelle GT (2009) Amine scrubbing for CO2 capture. Science 325:1652–1654

Rowlett RS (2010) Structure and catalytic mechanism of the b-carbonic anhydrases. Biochemical Biophysics, Acta 1804:362–373

Rubin ES, Chen C, Rao AB (2007) Cost and performance of fossil fuel power plants with CO_2 capture and storage. Energy Policy 35:4444–4454

Sawaya MR, Cannon GC, Heinhorst S, Tanaka S, Williams EB, Yeates TO, Kerfeld CA (2006) The structure of beta-carbonic anhydrase from the carboxysomal shell reveals a distinct subclass with one active site for the price of two. J Biochem 281(11):7546–7555

Scheffknecht G, Al-Makhadmeh L, Schnell U, Maier J (2011) Oxy-fuel coal combustion—A review of the current state-of-the-art. Int J Greenhouse Gas Control 5:16–35

Shan Z, Xi C, Thu N, Alexander KV, Gary TR (2010) Aqueous Ethylenediamine for CO_2 Capture. Sustain Chem 3:913–918

Sharma A, Bhattacharya A, Singh S (2009) Purification and characterization of an extracellular carbonic anhydrase from *Pseudomonas fragi*. Process Biochem 44:1293–1297

Sharma A, Bhattacharya A, Shrivastava A (2011) Biomimetic CO_2 sequestration using purified carbonic anhydrase from indigenous bacterial strains immobilized on biopolymeric materials. Enzyme Microbiology Technol 48:416–426

Shekh AY, Krishnamurthi K, Mudliar SN, Yadav RR, Fulke AB, Devi SS, Chakrabarti T (2012) Recent advancements in carbonic anhydrase driven processes for CO_2 sequestration: Minireview. Crit Rev nvironm Sci Technol 42:1419–1440

Sims REH, Rogner HH, Gregory K (2003) Carbon emission and mitigation cost comparisons between fossil fuel, nuclear and renewable energy resources for electricity

Skjånes K, Lindblad P, Muller J (2007) Bio CO_2 a multidisciplinary, biological approach using solar energy to capture CO_2 while producing H_2 and high value products. Bio Molecular Engineering 24:405–413

Smit B, Reimer J, Oldenburg C, Bourg I (2014) Introduction to Carbon Capture and Sequestration. Imperial College Press

Smith KS, Ferry JG (2000) Prokaryotic carbonic anhydrases. Proc National Acad Sci USA 96:15184–15189

Syrjänen L, Tolvanen M, Hilvo M, Olatubosun A, Innocent A, Scozzafava A, Lappiniemi J, Niederhauser B, Hytonen VP, Gorr TA, Parkkila S, Supuran CT (2010) Characterization of the first beta-class carbonic anhydrase from an arthropod (*Drosophila melanogaster*) and phylogenetic analysis of beta-class carbonic anhydrases in invertebrates. BMC Biochem 11:28

Tan Y, Nookuea W, Li H, Thorin E, Yan J (2016) Property impacts on carbon capture and storage (CCS) processes: a review. Energy Conversation Manag 118:204–222

Tripp BC, Smith KS, Ferry JG (2001) Carbonic anhydrase: new insights for an ancient enzyme, a minireview. J Biol Chem 276:48615–48618

Vinoba M, Kim DH, Lim KS, Jeong SK et al (2011) Biomimetic sequestration of CO_2 and reformation to CaCO3 using bovine carbonic anhydrase immobilized on SBA-15. Energy Fuel 25:438–445

Vinoba M, Bhagiyalakshmi M, Jeong SK, Yoon YI, Nam SC (2012) Carbonic anhydrase conjugated to nanosilver immobilized onto mesoporous SBA-15 for sequestration of CO_2. J Molecular Catalyst β- Enzyme 75:60–67

Vinoba M, Bhagiyalakshmi M, Grace AN, Kim DH, Yoon Y, Nam SC, Jeong SK (2013) Carbonic anhydrase promotes the absorption rate of CO_2 in post-combustion processes. J Phys Chem 117(18):5683–5690

Wang M, Lawal A, Stephenson P, Sidders J, Ramshaw C (2011) Post-combustion CO2 capture with chemical absorption: a state-of-the-art review. Chemical Engineering Resources 89:1609–1624

Wanjari S, Prabhu C, Yadav R, Satyanarayana T, Labhsetwar N, Rayalu S (2011) Immobilization of carbonic anhydrase on chitosan beads for enhanced carbonation reaction. Process Biochem 46:1010–1018

Willie S, Sallie L, Arthur B, Zachary W (1999) Environmental Effects of Increased Atmospheric Carbon dioxide. Climate Res 13:149–164

Yadav R, Satyanarayanan T, Kotwal S, Rayalu S (2011) Enhanced carbonation reaction using chitosan based carbonic anhydrase nanoparticles. Curr Sci 100:520–524

Yadav R, Joshi M, Wanjari S, Prabhu C, Kotwai S, Satyanarayanan T, Rayalu S (2012) Immobilization of carbonic anhydrase on chitosan stabilized iron nanoparticles for the carbonation reaction. Water Air Soil Pollution 223:5345–5356

Yeh JT, Renisk KP, Rygle K, Pennline HW (2005) Semi-batch absorption and regeneration studies for CO_2 capture by aqueous ammonia. Fuel Process Technol 86(14–15):1533–1546

Yuan H, Xing C, Fan Y, Ruimin RC, Zhan Y, Peng F, Qie F (2017) Carbon dioxide controlled assembly of water-soluble conjugated polymers catalyzed by carbonic anhydrase. Macromolecular Rapid Commun 38

Zhang S, Zhang Z, Lu Y, Rostam-Abadi M, Jones A (2011) Activity and stability of immobilized carbonic anhydrase for promoting CO_2 absorption into a carbonate solution for post-combustion CO2 capture. Bio-Resources Technol 102:10194–10201

Green Pathway of CO_2 Capture

Amita Chaudhary

Abstract

Global warming reaches an average of up to 425 ppm in May 2020 with an increasing rate of CO_2 emissions. The outcome is the melting of icebergs and seriously affected biodiversity. Therefore, demand was raised for the creation of new mitigation approaches to effectively catch it. Developing cost-effective, environmentally sustainable, and productive CO_2 capture technologies are required. In this chapter, we illustrated alternate green CO_2 sequestration routes such as adsorption via, MOFs, green solvent absorption, and biological CO_2 conversion method. Unlike biotic sequestration, abiotic sequestration is a method of engineering. It is possible to develop advanced technologies for deep injection into the ocean, geological structures, coal mines, and oil wells, etc. The detailed process and results are discussed in the given chapter.

Keywords

CO_2 sequestration • MOFs • Zeolite • Microalgae • Bio-refineries

1 Introduction

With the continually rising rate of global Carbon dioxide emissions in the air, the whole universe is endangered. The primary source of energy is fossil fuels, which produce a lot of CO_2 gas on combustion. The CO_2 emission rate is increased by 90% since 1900, out of which approximately 70% was increased from the starting of the industrial era to 2010 (Boden et al. 2016). Atmospheric CO_2 accumulation is continuously increasing from 280 to 427 ppm globally as predicted from Fig. 1. As a result of which the melting of icebergs causes, and biodiversity is highly affected.

Global CO_2 emissions in 2008 were approximately 29.4 gigatons (Gt), an enhancement of approximately 40% compared to the 20.9 Gt emissions in 1990 (Walker et al. 2017). The power industry currently accounts for 41% of all the energy-related CO_2 emissions shown in Fig. 2, followed by the transportation sector (23%), the manufacturing sector (20%), the construction sector (10%), and others (US EPA 2015). Hence, the development of strategies to prevent this problem is essential. Techno-economic analysis has shown that the advanced techniques for capturing CO_2 are energy-intensive. Carbon Capture and Storage (CCS) is a potential tool for reducing CO_2 emissions. In petrochemical and chemical fertilizer industries, the processes of scrubbing acidic gases are must as per government norms. There are four main CO_2 capture techniques: absorption, adsorption, membranes, and cryogenic distillation (Leung et al. 2014). The choice of technology opts to capture CO_2 from post-combustion, pre-combustion, or combustion of Oxy-fuel varies based on the process processing conditions. In industries, acid gas removal is typically accomplished by absorption in alkanol-amine (Astarita et al. 1986). Using absorption, the benchmark solvent for CO_2 capture is aqueous amine scrubbing is MEA. The low partial pressure of CO_2 in the flue gas eliminates the use of physical adsorbents. MEA has high CO_2 selectivity, absorption capacity and is relatively low-cost, thus widely used. But on the other side, it has several disadvantages like high volatility, which results in the solvent loss at absorption temperature, and its corrosive nature increases the maintenance cost of the plant. Moreover, due to the exothermic reaction with CO_2 leads to high energy about 25–45% for regeneration (D'Alessandro et al. 2010). Thus, this is high time to develop such absorbents which can overcome all the

A. Chaudhary (✉)
Chemical Engineering Department, Nirma University, Ahmedabad, 382481, India
e-mail: amita.chaudhary@nirmauni.ac.in

Inamuddin et al. (eds.), *Carbon Dioxide Utilization to Sustainable Energy and Fuels*, Advances in Science, Technology & Innovation, https://doi.org/10.1007/978-3-030-72877-9_13

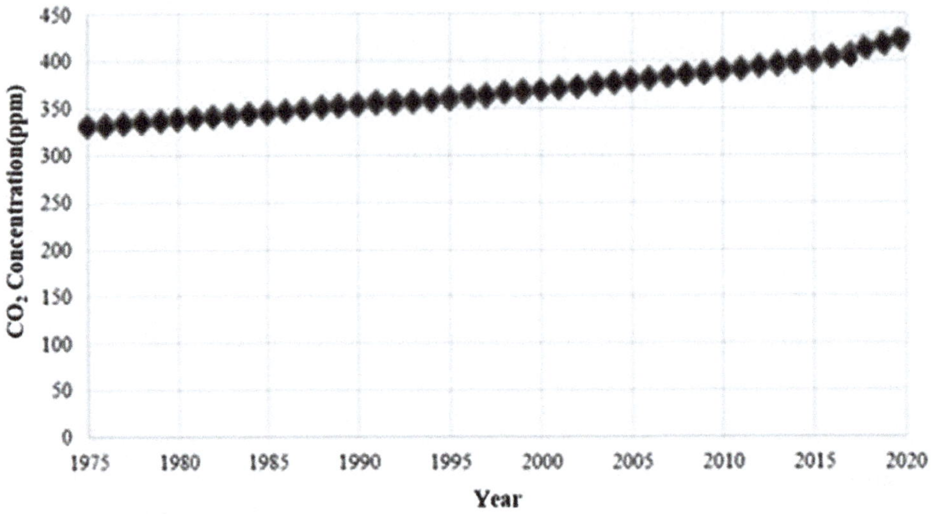

Fig. 1 The plot of atmospheric CO_2 concentration (ppm) verses year as measured at Mauna Loa, Hawaii (In-situ CO_2 data from 1980-present)

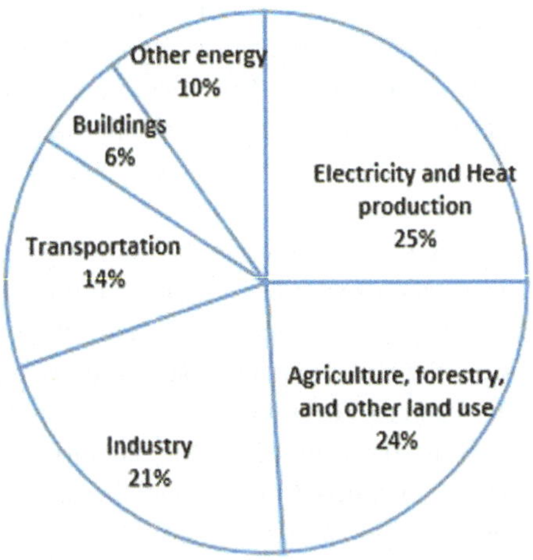

Fig. 2 Global CO_2 emission (%) by different sectors (IPPC 2014)

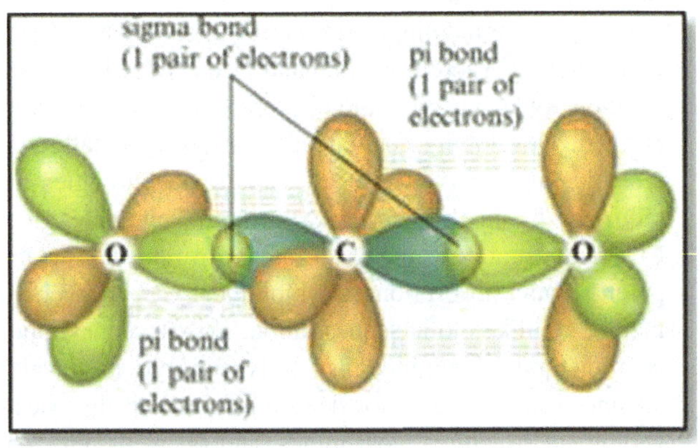

Fig. 3 Bond formation in CO_2 molecule

limitations associated with conventional absorbents and have high selectivity for carbon-dioxide. In absorption, the chemistry of CO_2 molecules plays a great role.

2 Molecular Structure of Carbon Dioxide

The chemistry of CO_2 molecules helps in the creation of a new component for its capturing. The CO_2 molecule has two sigma bonds and two pi bonds formed by the head-on overlapping and side by side overlapping of sp-hybrid carbon atom through the two oxygen atoms in the CO_2 molecule as shown in Fig. 3. As a result, the C = O bond length is observed to be 1.16 Å, in the bond length of normal C-O (1.22 Å) and C≡O (1.10 Å). The C atom is in its highest valence state with I. E_1 (first ionization energy) corresponding to 13.97 eV. The CO_2 molecule possesses an unoccupied orbital bearing with a low energy degree and high electronic affinity (38 eV) (Momeni and Riahi 2015). So, a CO_2 molecule can be triggered by a nucleophile easily. Structurally, $CO_2$is having a plane of symmetry, with asymmetric stretching at about 2270 cm^{-1}. Besides, at interaction with a metal complex, the bond angle of CO_2 may also be changed. Therefore, together with the estimation of the O = C = O bond angle value, the FT-IR study elucidates the bonding of CO_2 and metal, which would be useful in developing efficient catalysts. CO_2 has zero dipole moment; even though, it possesses a large quadrupole moment due to the C = O bonds. The CO_2 molecule is highly polar at short bond length and occurs in two canonical forms which help in the simulation of its reactions. CO_2gas molecule can easily be captured inside the host adsorbent sites as in zeolites, the guest molecules form a strong electrostatic bond with the metal atoms or and other legends present (Wlazło et al. 2017).

3 CO_2 Capture System

3.1 Post-Combustion System

In the post-combustion systems, initially, the burning of coal takes place in the combustion chamber in the presence of pure O_2 stream as shown in Fig. 4. It creates great amounts of flue gas at high temperatures up to 383.15–393.15 K (110–120 °C) and pressure near to 101.325 kPa. The partial-pressures of CO_2 in the flue gas are very low i.e., 10 to 15 kPa (0.1 to 0.15 bars). So, the flue gas is cooled up to 313–333 K (40–60 °C), is inserted from the bottom of an absorption tower and an absorbent is sprayed from the top in the counter-flow mode. The CO_2 present in the exhausted gas is chemically removed by an absorbent. Afterward, the CO_2-rich absorbent mixture is renewed thermally and recycled back to the absorber column (Bailey and Feron 2005).

3.2 Pre-Combustion System

In the pre-combustion system, the integrated gasification combined cycle (IGCC) is in-built into it. Here, the fuel is vaporized in the presence of pure oxygen steam to produce syngas. The syngas is fed into the water–gas-shift (WGS) reactor. In this reactor, the syngas stream is converted to H_2 and CO_2. Afterward, the concentrated stream of $CO_2$and H_2 is used for the production of energy as shown by Fig. 5 (EPRI 2012).

3.3 Oxy-Fuel Combustion System

This process is still under development. In this, a pure oxygen gas flow is used for combustion of a fuel, which produced high-pressure flue- gas stream. The main components of the exhausted gas are H_2O and CO_2. The flue-gas is cooled down and the water vapor gets condensed which results in the production of the concentrated CO_2 stream as shown in Fig. 6. The Oxy-fuel combustion system requires a low operating cost in the CO_2 capture step; merely the procedure needs a costly air-separation unit (Anheden 2005).

The US Department of Energy provides a comprehensive economic assessment of the three systems. These system analysis studies indicate those pre-combustion structures are a comparatively cost-powerful choice for scrubbing the CO_2, on the other hand, the capital expenditure in this method is greater than within the post-combustion and oxyfuel combustion system as proven in Table 1 (Zhou 2005).

However, the post-combustion systems are considered due to its capacity to capture 80–90% diluted CO_2 present in low concentration in the exhausted gas. The common processes applicable to CO_2 mitigation are tabulated in Table 2. The technologies (Leung et al. 2014) applied to capture carbon-dioxide in the post-combustion system are:

- Physisorption and Chemisorption
- Adsorption
- Membrane Permeation and Separation
- Cryogenic Distillation
- Biological Sequestration.

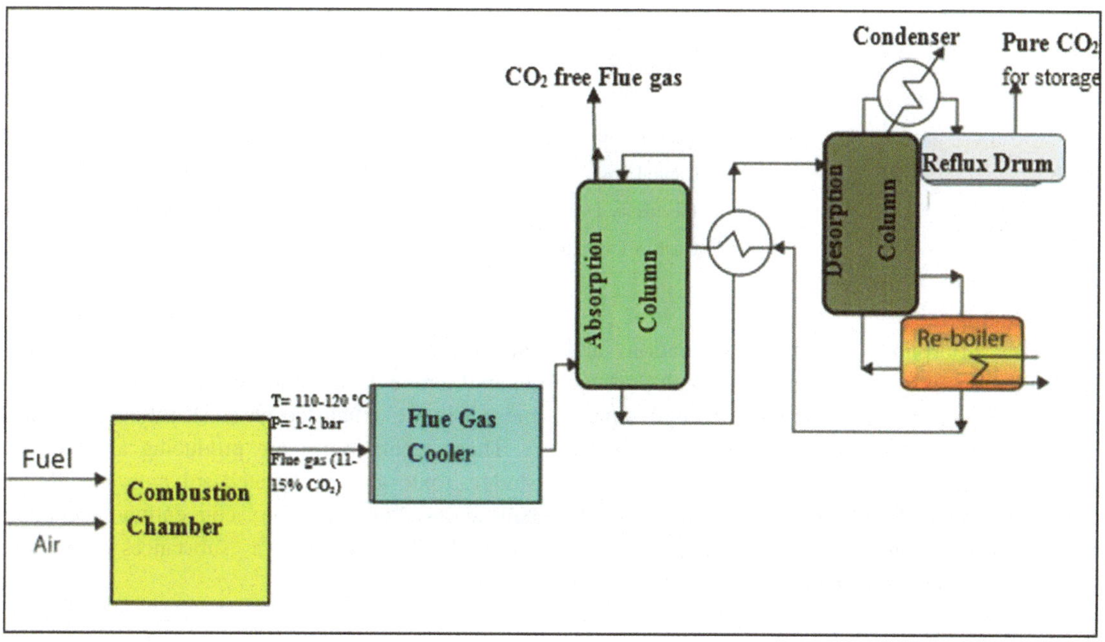

Fig. 4 Schematic representation of post-combustion system

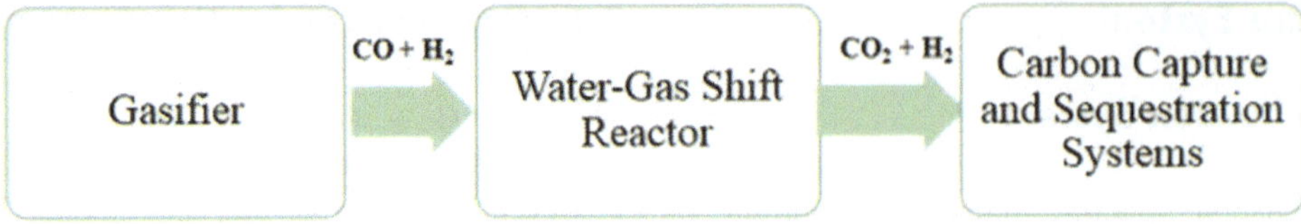

Fig. 5 Schematic representation of pre-combustion system

Fig. 6 Schematic representation of the oxy-fuel combustion system

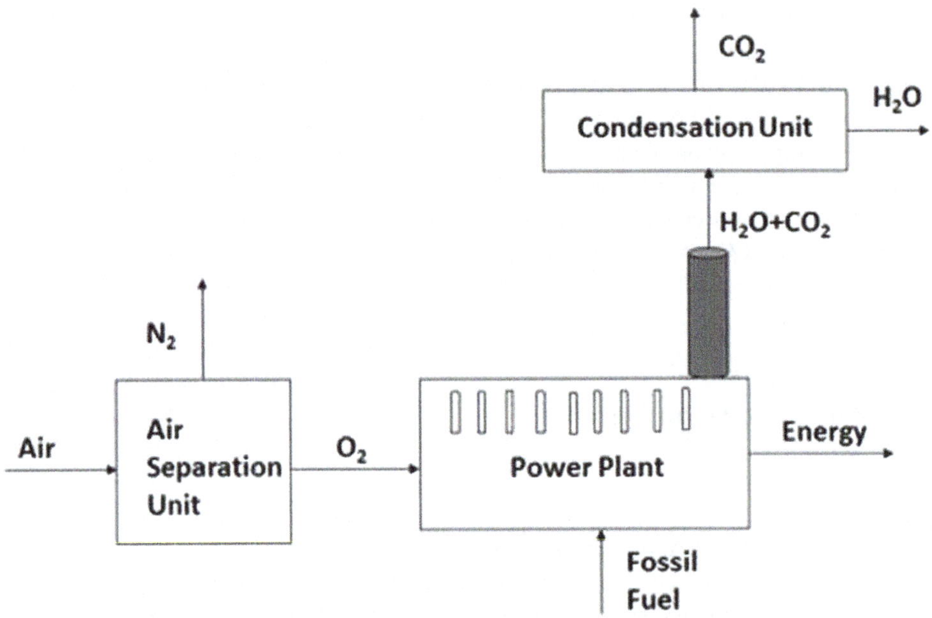

Table 1 Comparative cost-estimation studies of the three carbon capture systems

Source	Factor	CO_2 capturing points		
		Post-combustion	Pre-combustion	Oxy-fuel
Coal-fired	Thermal efficiency	34.8%	31.5%	35.4%
	Capital cost	1980$/kW	1820$/kW	2210$/kW
	Electricity cost	7.5$/kWh	6.9$/kWh	7.8$/kWh
	Cost of CO_2 capture	34$/t CO_2	23$/t CO_2	36$/t CO_2
Gas-fired	Thermal efficiency	47.4%	41.5%	44.7%
	Capital cost	870$/kW	1180$/kW	1530$/kW
	Electricity cost	8.0$/kWh	9.7$/kWh	10.0$/kWh
	Cost of CO_2 capture	58$/t CO_2	112$/t CO_2	102$/t CO_2

Out of these, adsorption, membrane separation, and cryogenic distillation processes require a high-pressure stream of CO_2 for performance whereas microbial separation technique is under evaluation (Cents et al. 2005). Absorption is the most desirable technology to capture low-pressure CO_2 streams from flue gas at a large scale. In absorption, the solvents can be regenerated back to reuse again by removing the CO_2 from the rich solvent by increasing the temperature. An aqueous solution of alkanol-amine is commonly used for scrubbing the CO_2 present in the exhausted gases (Kohl and Nielsen 1997).

These technologies are producing a lot of wastes like sludge, toxic gas, carbonated solvents, etc. So, alternative substances for all the convention technologies are required. The alternative substances are tabulated in Table 3.

Table 2 Conventional methods for carbon capture

S. No.	Carbon capture technology	Material required	Limitation
1	Absorption	Chemically selective absorbents/basic solvents like monoethanolamine (MEA), Methyldiethanolamine (MDEA) and Chilled ammonia, etc	• High solvent loss due to volatile nature • Have an affinity for other acidic gases • High energy consumption in rich-solvent regeneration • Thermally unstable • Corrosive nature
2	Adsorption	Physically or chemically adsorbed on solid adsorbent such as activated carbon, zeolite, polymers, etc	• Energy inefficient • Prior flue gas treatment required to remove moisture and dust particles
3	Membrane technology	Membrane with CO_2 selective permeability. Usually, polymeric membranes used	• Ambient temperature should be maintained before treatment. So, energy-intensive process • Moisture content and dust particles in the flue gas can clog the membrane pores • High cost, chances of frequent fouling of the membrane • Readily exchange
4	Cryogenic separation	Alternative freezing and condensation of the gas mixture is required	• Energy-intensive process for refrigeration • Moisture removal is required before cooling to avoid plugging by ice formation • Solidified CO_2 is continuously deposited on the heat-exchanger surfaces and needs to be removed
5	Geologicalsequestration	Dosing of CO_2 into subterranean geological reservoirs	• High operational cost • Risk of CO_2 leakage problem • Specific geomorphic structure requirement
6	Oceanic injection	Injection of CO_2 into the deep ocean	• Cost intensive • The potential threat to marine life

Table 3 Alternative green and eco-friendly methods to capture CO_2

S. No.	Carbon capture technology	Material required	Advantages	Limitation
1	Absorption	Ionic liquid	Green and eco-friendly absorbents	• Cost intensive • High viscosity • Difficult to scale-up
2	Adsorption	Metal–organic frameworks (MOFs)	High porosity crystallinityand high surface area	Powdered MOFs have low mechanical strength and difficult handling
3	Bioprocess	Microalgae are used	Bioconversion CO_2 into biofuelsand other valuable products	• Culturing and harvesting systems of biomass is a difficult process • NO_2 and SO_2present in the gas stream can contaminate the growth rate of microalgae • Growth also Effected by pH, temperature, salt concentration, etc
4	Electrochemical Reduction	Organic redox nucleophile–electrophile interaction in an electrochemical cell	Selectively CO_2capturefrom flue gas	Energy-intensive process
5	Hybrid capture processes	The mixture of absorbent and adsorbent	• Membrane-distillation • Membrane–pressure swing adsorption • Combination of two or more capturing technologies	Under development

4 Absorption Technology

4.1 Green Absorption with Ionic Liquids

The unique characteristics of ionic liquids (ILs) make them an important and promising solvent for CO_2 capture. Ionic liquids are molten salts composed of ions i.e. anions and cations. Their m.pt. remains less than 100 °C, so they exist in a molten state at room temperature. A significant feature associated with Ionic liquids is that their properties can be modified by adjusting the component ion structure to obtain the desired solvent properties (Zhanga et al. 2006). Due to this reason, Ionic liquids possess a wide range of industrial applications. Several research studies have found CO_2 is highly soluble in many ionic liquids due to their ionic interaction.

4.1.1 Properties and Uses of Ionic Liquids

i. Ionic liquids are non-volatile, non-flammable, high thermal stability, a wide range of solubility and miscibility, wide electrochemical windows, and wide liquid ranges in comparison of organic molecular solvents.
ii. It can be reused and recycled for several rounds of operations.
iii. Ionic liquids can be tailored/tuned for the specific application.
iv. It is widely used as a solvent and/or catalyst for proceeding chemical, physical, or biological reactions.
v. Ionic liquids play a very important role in acidic gas scrubbing.
vi. It is also used for extraction and separation of chemical species from the solutions of chemicals (Mohanty et al. 2010).

Ionic liquids could be safer solvents for reducing and preventing wastage and contaminations.

4.1.2 CO_2 Solubility in PILs

Ionic liquids have very good solubility and miscibility for polar and non-polar solutes. Their properties can be tailored to dissolve organic and/or inorganic liquids, solids, and gases. The power to dissolve any solute depends on its polarity and the coordination ability of their ions.

In this research study, we will mainly explore the absorption of carbon-dioxide in PILs. In 2001, Brennecke et al., checked the solubility of nine gases viz., CO_2, CH = CH, C_2H_6, CH_4, Ar, O_2, CO, H_2, and N_2 in [C_4mim] [PF_6] and the studies reveal that CO_2 has the highest solubility (Yu et al. 2012; Lei et al. 2014). The ionic liquids have the greatest affinity for CO_2 gas in [C_4mim] [BF_4] at ambient conditions (Brennecke and Maginn 2004). A new ionic liquid 1-propylamine-3-butyl-imidazolium tetrafluoroborate, which could chemically absorb CO_2 at ambient pressure (Bates et al. 2002). Ionic liquids are proved as a promising solvent to extract CO_2 from gas composition selectively and efficiently (Costa Gomes 2007). The carbon—dioxide can be taken up physically (Shiflett et al. 2010) or chemically in Ionic liquids.

The nature of the cations and anions in Ionic liquids are very important for the solubility and selectivity of a gas.

The CO_2 selectivity in ionic liquids can be governed by the following parameters:

- Effect of fluorination of the cation: The investigation established that the solubility of CO_2 is high in an ionic liquid having higher fluorinated-cation in comparison to others (Anderson et al. 2007). In the comparative studies of CO_2loading, was found to be more in [C_8F_{13}mim] [Tf_2N] than [$C_8H_4F_9$mim] [Tf_2N], and lowest in [C_8mim] [Tf_2N].
- Alkyl chain length on precursor ions: The studies reveal that CO_2-absorption in an imidazolium-based Ionic liquid is increased as the number of carbon atoms increased. For example, the case of CO_2-absorption in [C_8mim] [Tf_2N] and [C_3mim] [Tf_2N]. It is observed that CO_2 loading in [C_8mim] [Tf_2N] is more than [C_3mim] [Tf_2N] (Baltus 2011).
- Effect of anions: Cadena et al. proved in their research work that the nature of anion plays an important role in CO_2loading. The CO_2loading on the sites of different anions in the increasing order are: $[NO_3]^- < [DCA]^- < [BF_4]^- < [PF_6]^- < [CF_3SO_3]^- < [Tf_2N]^- < [methide]^-$ with same cation (Cadena et al. 2004).
- Combinations of different solvents: Camper et al. 2008 explored a few composite solvents to capture CO_2 most of which are composed of imidazolium-based Ionic liquids and alkanolamines. They worked on the combinations of 1-hexyl-3-methyl-imidazolium-bis (trifluoromethylsulfonyl) imide and ([Hmim] [Tf_2N]) with monoethanolamine (MEA). The viscosity has been one of the main hurdles to working with ionic liquids. In this study, this was observed that the addition of MEA reduces the viscous nature of Ionic liquids and thus increases carbon dioxide absorption (Cadena et al. 2004). Ahmady et al. 2011, did the same kind of studies with an aqueous solution of 1-butyl-3-methyl-imidazolium tetrafluoroborate ([Bmim] [BF_4]) and methyldiethanolamine (MDEA) (Galan Sanchez et al. 2011). Zhang et al. used the amino acids for CO_2-absorption which shows better absorption capability in comparison of MEA (Ma 2011). The effect of dilution with water in the solution of IL and MEA was studied by Yang et al. in 2014. In such mixtures, less energy is consumed for the regeneration of riches solvents

in comparison to conventional molecular solvents. (Xue et al. 2011)

Mohanty et al. (2010), explored different basic sites of amino acids to check the influence of anionic site in the loading of CO_2. The FT-IR results have shown that more the amino sites in amino acids, higher CO_2-absorption capacity.

4.1.3 CO_2 Absorption in PILs with Carboxylate Anion

The carboxylate anion containing ionic liquids shows very high CO_2-absorption capacity as carboxylate anion can be co-ordinated chemically with CO_2 more efficiently than other anions (Maginn 2005). Various other research groups also made with an acetate based Ionic liquids and found that the reaction of such Ionic liquids and CO_2 is reversible and so absorbent can be easily recycled with relatively low-cost (Stevanovic et al. 2013). Chaudhary et al., used aliphatic polyammonium-based cations have low toxicity with moderate viscosity (Chaudhary and Bhaskarwar 2015).

4.2 Reaction Mechanism Involved in CO_2-Absorption

All Alkanol amine have two reacting sites, $-NH_2$ and –OH, which are likely to react with carbon dioxide, according to the following reactions:

$$-NH_2 + CO_2 \rightarrow -NHCOOH, \text{substituted carbamic acid}$$

$$-OH + CO_2 \rightarrow -OCOOH, \text{substituted carbonic acid}$$

Both products are acidic, and can be neutralized by the –NH_2 group when it hydrolyzes to $-NH_3^+$ (Astarita 1967). The chemical absorption reactions of CO_2 with 1 and 2° amine sites are identified well in literature initially by Caplow (1968) followed by Danckwerts (1979). This mechanism involves two steps, beginning with the formation of the CO_2-amine complex i.e. zwitterion formation followed by de-protonation of protonated amines with the base present as a solvent in reacting solution (Caplow 1968). The reactions are shown below:

(i) Formation of zwitterion intermediate complex:

$$CO_2 + R_1R_2NH \underset{k_{-1}}{\overset{k_2}{\rightleftharpoons}} R_1R_2NH^+COO^- \quad (1)$$

(ii) De-protonation of the zwitterion (Danckwerts 1979):

$$R_1R_2NH^+COO^- + : B \rightarrow^{k_B} R_1R_2NCOO^- + BH \quad (2)$$

By steady-state approximation, the overall forward reaction rate (r) can be expressed as

$$r = \frac{k_2[CO_2][R_1R_2NH]}{1 + \frac{k_{-1}}{\sum k_B[B]}} \quad (3)$$

The second, single-step mechanism was given by Crooks and Donnellan (1989) which was further supported by Alper (Crooks and Donnellan 1989).

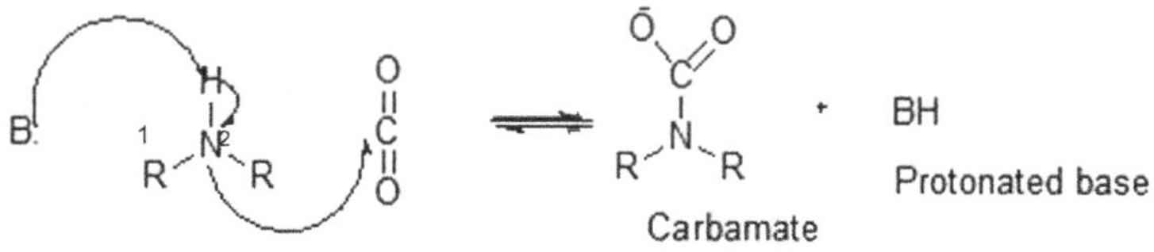

(4)

Here,

$$2R_1R_2NH + CO_2 \underset{k_{-1}}{\overset{k_2}{\rightleftharpoons}} R_1R_2NH_2^+ + R_1R_2NCOO^- \quad (5)$$

$$R_1R_2NH + CO_2 + : B \underset{k_{-B}}{\overset{k_B}{\rightleftharpoons}} R_1R_2NCOO^- + BH \quad (6)$$

Accordingly, the rate of forwarding reaction will be

$$r = (k_1[R_1R_2NH] + k_B[B])[R_1R_2NH][CO_2] \quad (7)$$

Of these, mechanisms, da Silva and Svendsen (2004) advised that the most likely mechanism of the CO_2 response with aqueous primary and secondary amines is much like the single-step termolecular mechanism from their ab-initio calculations and the solvation model studies (Silva and Svendsen 2004).

5 Adsorption Technology

Adsorption technologies are better than other CO_2 scrubbing technologies in the case of pre-combustion and oxy-fuel systems. Solid adsorbents like ZSM-5, activated charcoal, CNTs, MOFs, and silicates. These adsorbents are chosen based on some peculiar properties like nucleophilic surface sites, extend of porosity, total adsorption area, bonding metals and ligands, electrostatic interactions which helps in determining the potential of CO_2 adsorption. At low temperatures, Modified Organic Frameworks (MOFs) are the best candidate for physical adsorption of CO_2. MOFs composed of metal ions forms a coordination bond with nucleophilic ligands. MOFs are solid porous organic adsorbent consist of large voids that help in high capacity and selectively for binding CO_2 molecules.

Porous solid adsorbents comprise of nano-scale morphological features and are usually organized into different systems based on pore size: pores diameter less than 2 nm called microporous, pores diameter ranges from 2 to 50 nm are mesoporous materials and greater than 50 nm for macroporous materials. Sorbents having large adsorbent area has a greater capacity for CO_2 loading (Zdravkov et al. 2007). The interaction between the adsorbent and adsorbate can be explain using Lennard–Jones term and an electrostatic term in the given equation:

$$u_{ij}(r) = \sum_{i,j} \left\{ 4\varepsilon_{ij} \left(\frac{\sigma_{ij}}{r_{ij}} \right)^{12} - \left(\frac{\sigma_{ij}}{r_{ij}} \right)^{6} + \frac{q_i q_j}{4\pi\varepsilon_0 r_{ij}} \right\}$$

$\varepsilon_0 = 8.8542 \times 10^{-12} C^2 N^{-1} m$, the permittivity of vacuum.

Gas adsorption on the surface of adsorbents can either by chemisorption or physisorption. During physical adsorption, weak Van Der Waals forces are responsible for interaction between adsorbent and the adsorbate molecules. Thus the adsorbents' surface area and pore size play a significant role in the physisorption process. Whereas, in the chemisorption, the adhered functional group's interaction with CO_2 plays a significant role. The interaction between both is established by a chemical bond. Due to the presence of weak bonds in physisorption, the adsorbents can be readily regenerated with low energy. However, in the chemisorption process, large energy is required for breaking the chemical bond between adsorbate and adsorbent. Activated charcoal or ZSM-5 can adsorb the CO_2 either physically or chemically depending on the groups present in their site, whereas MOFs binds the CO_2gas via., weak physical forces.

5.1 Organic Adsorbents

Several organic adsorbents having micro or mesoporous range are promising adsorbents for CO_2. These adsorbents have unique properties like high surface area, specific selectivity, and easy regeneration.

5.1.1 Activated Charcoal

Activated charcoal (AC) is generally used in several applications as an adsorbent, such as the treatment of wastewater, paper industry, etc. It is vital that the adsorbent has a high surface to volume ratio and particularly evolved porosity, especially micro and mesoporosity, to maximize the adsorption of AC (Pellerano et al. 2009). The charcoal is activated either physically using hot steam or chemically by some chemicals. During activation the porosity of activated charcoal is increases and chemical functional groups are adhere to its site. The AC is used for CO_2 capture on the laboratory scale but so far the studies indicate the successful results for CO_2 sequestration.

5.1.2 Biochar

Biochar is a suitable and reliable material, well known for its agricultural application. Highly porous nature makes it a suitable candidate for carbon capture. It is prepared by the pyrolysis of biomass available from agricultural wastes. It is the best way of storing CO_2 in biological systems. In the investigations, it was concluded that biochar can store 30.6 kg carbon per GJ created (Mulabagal et al. 2015).

Unlike activated carbon, biochar has a highly porous structure that has gained publicity for its ability to eliminate toxins, sequestrate carbon in soils, and at the same time improving the productivity of soils (Jeffery et al. 2011). The pyrolysis process under anaerobic conditions is used to produce it. Biochar is less prone to degradation. Biochar can trap carbon for a long time and mixed with soil to increase its fertility. Predicated upon European carbon-dioxide emissions report, about 1.1 gigatons of carbon per year. It is projected that the implementation of biochar land could mitigate about 9% of the emissions from Europe. (Barrow 2012).

Biochar has excellent potential for CO_2 capture because of its non-polar and high porosity (Guo et al. 2020). Biochar is more biocompatible and cheaper in comparison to AC. Experimental data showed that biochar adsorbs CO_2 by physical adsorption. The presence of the nitrogenous group in biochar enhances the adsorption ability for carbon-dioxide.

5.1.3 Metal–Organic Frameworks (MOFs)

Recently a new class of nanoporous materials has emerged in the field of adsorbents called MOFs. It is a promising material for carbon capture at a large scale due to its high surface to volume ratio. It can hold remarkably high volumes of gases in comparison to its surface. The MOFs have a very high surface area in comparison to its volume. About 10–15% of power plant exhaust gases contain CO_2. Snurr and his team have studied several MOFs that can trap CO_2 before entering the atmosphere (Hu et al. 2019). The exhaust gases also contain hydrogen and water vapors along with carbon dioxide. After the MOFs traps the carbon dioxide, the hydrogen gas is burned, and the only by-product is water which condensed and collect for further used.

To improve carbon dioxide adsorption efficiency and selectivity, different nucleophilic groups with a high CO_2 affinity can be implemented at MOF sites (Zhang et al. 2013). A fine-tuned, or quickly stimulus-responsive MOFs are designed by molecular modeling which has shown complementary benefits for selective adsorption and separation of CO_2. They are synthesized and tested for their ability to sustain at high temperature and pressure. Due to high thermal stability, the MOFs can be used to capture directly the CO_2 gas from flue gases which are vented out at high temperatures nearly 110–120 °C (Burns 2020). MOFs can be better modified to absorb CO_2 at different pressures compared to zeolites because the pore structure and chemical compositions can easily tailor. MOFs can be synthesized with low cost in comparison to ZSMs. Amines-based MOFs and their derivatives are used to capture CO_2. The heat of adsorption is 90 kJ/mol which is less in comparison to the heat of absorption when using amine-modified MOFs (Megías-Sayago et al. 2019).

5.2 Other CO_2 Adsorbents

5.2.1 Metal Oxide-Based Absorbents

In metal oxides, CaO and MgO are mainly used to scrub CO_2 to form carbonates (Kumar and Saxena 2014). The uptake of CO_2 in CaO-based sorbents can be analyzed by its morphological properties, such as pore structures, pore-volume, surface area, and particle size. With increasing absorption, the carbonate forms a powder layer, and thus decreases the pore size and surface area of absorption. Due to it, the CO_2 uptake decreases significantly. The metal oxides tend to form composites due to their electrostatic interaction and types of packing which further enhances the absorption of CO_2. For example, the composite metal oxide $Ca_{12}Al_{14}O_{33}$ captures more CO_2 in comparison to CaO. The mechanism of composite with CO_2 is composed of two steps. In the first step, it captures the CO_2 and form carbonates followed by calcination reaction, which releases the CO_2 under N_2 flow at high temperatures (1173 K). In the regeneration of CO_2 rich CaO and $Ca_{12}Al_{14}O_{33}$ metal-oxide composites is very expensive because the bond formed between CO_2 and composites is very strong and it requires lots of energy to break and release CO_2 (Jensen et al. 2005).

5.2.2 Zeolites

Zeolites are the micro-porous alumino-silicate adsorbents, which have been used extensively at lab scale for the CO_2 absorption. Adsorption required a concentrated stream of gas for adsorption, but zeolites can capture high CO_2 even when it is present in low concentration. To increase the capturing capacity of zeolites, different functional groups and structural frameworks with different metal oxides, pore size, pore volume can be tuned. The addition of amino sites in zeolite like zeolitic imidazolate can hold up to 83 L of CO_2 with only 1 L of material, due to their high porosity (Siriwardane et al. 2005). The Zeolite also shows the specific selectivity towards CO_2. Therefore, it also can be used in the separation of CO_2 from a CO_2/CO mixture at room temperature which shows complete holding of CO_2 and spurges CO through the pores. Due to the flexibility in their structures, it can be used as a catalyst for the conversion of CO_2 into alternative fuels (Wang et al. 2018).

5.3 Biological Processes of CO_2Sequestration

In geological formations, there are mainly three pathways through which carbon dioxide can be trapped, namely physical trapping, chemical trapping, and hydrodynamic trapping based on the properties of CO_2 capturing source (IPCC 2013). Physical trapping corresponds to immobilizing carbon dioxide in geological deposits in the gaseous or lid states, which can be either static trapping in rocks or residual gas trapping in porous structures. Chemical trapping can be done by dissolution or ionic interaction of carbon dioxide gas with reservoir fluids results in formations of carbonated compounds which are soluble in water. In hydrodynamic trapping, low-velocity upward movement of carbon dioxide can result in intermediate-layer trapping.

Other than physical and chemical processes of carbon capture and sequestration, eco-friendly biological processes can be used. Some of the processes are as follows:

(i) Afforestation and agricultural practices can enhance the capturing of CO_2 via the photosynthesis process to produce abundant biomass which can be further used in the generation of fine chemicals (Lin and Lin 2013).
(ii) Ocean Fertilization: With some important nutrients like iron added to the oceans can help in the growth of phytoplankton which can increase the uptake of CO_2 (Zeebe and Archer 2005).
(iii) Microalgae cultivation (Pavlik 2017).

Biological processes comprise the main activities of forest and sequestration of agricultural materials. This can also be done by using unique CO_2 scavenger microorganisms which are available in numerous ecosystems. Sometimes also, natural enzymes can turn carbon dioxide into practical uses. One way of reducing the GHG effect is by carbon cycle reforms. Ocean, atmosphere, plants, and soil are the major elements of the carbon cycle, which actively exchange carbon between them. Massive CO_2 storage potential in oceans, geological, and terrestrial components (Michalak et al. 2011). While the biomass of the ocean accounts for about 0.05% of the total, it converts about 50 gigatons of inorganic carbon into organic carbon annually. For instance, green photosynthesizing microscopic organisms ingest up to 60 gigatons of carbon for each year into natural carbon about a similar sum created via land plants and almost multiple times the sum delivered by human action. Because much of the dissolved organic carbon is stored for just a brief period, marine species must transform it into hard to digest types known as refractory dissolved organic carbon; this conversion-driven entity has been related to as the 'jelly pump' and the microbial carbon pump (MCP) (Jiao and Zheng 2011). When converted into "indigestible" compounds, these DOCs settled deep in the ocean regions and reside there for a long time and used by the aquatic plants and animals and forms the ocean food chain (Ma'mum et al. 2005). Coastal sediments and sedimentary rocks have a huge capacity for CO_2 accumulation. It is the typical organic siphon that moves the carbon to the base of the sea. Carbon sequestration techniques manage to change or improve this regular organic CO_2 deposition by taking various systems. Some methods are available for immediate fixation of CO_2 at different depths, such as (i) moderate-depth (1000–2000 m), (ii) greater-depth (>3000 m), (iii) in the ocean bottom, and (iv) in deep-water surface layer (Adams and Caldeira 2008). Ametistova et al. (2002) investigated the process of transferring emitted carbon dioxide from different sources to the section underneath 800 m above ocean level straightforwardly (Ametistova et al. 2002). The best injection approach would be to dissolve liquid carbon dioxide using a fixed solid form at depths between 1000–1500 m. The deposited CO_2 is then suspended as small spherical molecules. The benefits of this strategy are that carbon dioxide can be delivered at quite slow release rates and under the carbonate dissolution cap with limited environmental impacts. It can be stored there for a prolonged period in the hydrated form at the depth of 3000 m or deeper than it (Qanbari et al. 2011). CO_2 deposited at this depth on the seabed is heavier than the surrounding seawater and is supposed to fill topographical depressions, settling as a CO_2 lake on which a thin layer of hydrate can be formed. This layer of hydrate would slow dissolution, but would not separate the lake from the surrounding water, and therefore dissolve into the surrounding water. Nevertheless, by the creation of new crystals, the hydrate layer will be constantly refreshed (Bose and Satyanarayana 2017). Laboratory studies and low, deep ocean studies indicate that CO_2 deep-sea storage would result in the formation of CO_2 hydrate and eventual dissolution. More than 30 years of a long time is taken by the CO_2 to saturate completely the water available at the depth of 50 m ranges from 30 to 400 years. The dissolution period is also subjected to the various other factors like the dissolution process, the carbon dioxide concentration, the pressure and temperature, and the seabed surface morphology (Lee et al. 2003).

5.3.1 Carbon Utilization by Forest and Agricultural Management

Overall, the agriculture land area accounts for approximately 35% of the global land area. The various types of soil are recognized as a probable sink for removing atmospheric CO_2 present in the terrestrial biosphere. Soil can capture atmospheric carbon dioxide during the carbon cycle, by converting it into a soil consuming form, perhaps by humidifying photosynthetic residues or by forming supplementary bicarbonate in a fairly stable manner. In plants the production of CO_2 increases due to the oxidation of soil organic content, fertilizers and pesticides used, and equipment used for crop cultivation (Deviram et al. 2020). In most regions, the subdividing of agricultural land results in the decline of carbon storage in soil. Among the many methods that exist to improve the storage of carbon in soils, the most feasible would be done by rising the intake among carbon or reducing the decay. Effective and improved management practices will convert soil into potential carbon sinks to increase the quality of the upper layer of the soil and to reduce the decomposition levels. Implementing quality improvement such as the reduction of the leftover land fallows, the use of substantial drilling, the inclusion of legumes and grasses in crop rotations, the cultivation of perennial grasses over the land helps the soil for prolonged carbon recovery. The introduction of cropland cover crops helps to increase the carbon concentration and also increase soil quality. Growing production of winter crops also enhances the carbon content of the soil especially in comparison to monoculture and thus enhances the sequestration of CO_2 in the soil. Changing the cultivation practices will achieve complete carbon sequestration in cultivated fields. Plowing disintegrates the soil aggregates and causes decomposition of the organo-mineral compounds. It helps to expose the organic matter to microbial decomposition and to minimize soil erosion. Conversion of crops into grasslands is another most successful alternative for carbon mitigation, which can be done by keeping a large land for it (Hernandez-Soriano et al. 2018). The storage of CO_2 in soils can be achieved by improvements in farming methods, such as the efficient use of pesticides, fertilizers, field equipment, and tillage management. The reclaiming of polluted and poorly maintained land for the CO_2 storage through the use of fossil fuels, biomass, and agricultural products from wastewater treatment helps increase the condition of the soil. But it also enhances soil microbial activity and leads to the production of global house gas. The carbon deposited in the leaves that contribute to the atmosphere during the production of cellulose, and when they fall and decompose, they add up the fertility to the soil. The duration for which carbon stays trapped in the trees is based on the tree's growth rate, atmospheric conditions, and methods of cultivation. At the end of the carbon cycle, biomass becomes a part of the food web and ultimately mixes with soil as a nutrient. The composting of biomass recycled back the CO_2 into the environment, and thus became the part ofthe carbon cycle again. Some of the potential ways of reducing the stored CO_2 in the environment is the accumulation and preservation of carbon in vegetation or by the renovation of carbon by organic matter. Through reforestation, afforestation, and more environmentally sustainable logging activities, carbon sinks can be preserved or increased. Restoration of forest loss or deforestation not only increases the CO_2 loading but also improves the quality of wood. The pattern of land usages, reforestation, and soil conservation are widely recognized as techniques for mitigating CO_2 emission. The dense adequate soil layer that provides for a primarily anaerobic condition that prevents the decomposition of buried wood. Since a significant flux of carbon dioxide is continuously assimilated into world forests by photosynthesis, cutting its return route to the atmosphere forms an important CO_2 storage strategy. The agroforestry program has the great potential to minimize the CO_2 accumulation rate in the atmosphere.

5.3.2 Ocean Fertilization

Some of the ways of promoting carbonate neutralization are through gradual oxidation of calcium carbonate in sediments and lands on the seafloor. This will neutralize ocean acidity as a result of the injection of CO_2 from a prolonged period (Archer et al. 2009). Due to the ready conversion of CO_2 into bicarbonates in the presence of water makes this sequestration process more demanding (Sheps et al. 2009). Studies were conducted to artificially facilitate the dissolution of mineral carbonate at an exponential rate for carbon dioxide sequestration. Enhanced mineral weathering reactions have been observed in the intake of high CO_2. Consequently, all these ingredients viz., water, and waste gases and discharge help in the accumulation of bicarbonates on the seabed. This helps in the production of carbonic acid at greater quantities in comparison to the atmosphere, results in the decrease in pH and deposition of bicarbonate increases. Rau and Calderia (1999) suggested a carbon dioxide reaction isolated from flue gas with crushed calcareous and seawater. As a result of this reaction, the carbonic acid formed will decompose the minerals present in the complex form in oceans (Rau and Caldeira 1999).

5.3.3 CO_2 Capture by Microalgae

Microalgae is a part of the microorganisms present in water. They carry out photosynthesis, as do green plants, to prepare their food (Stamenković and Hanelt 2013). This group includes cyanobacteria, diatoms, and other protists (Bahadar and Bilal Khan 2013). Micro-algal cells are photosensitized and are using carbon dioxide (CO_2) and sunlight to grow. The microalgae derived biomass could be a very valuable source of biofuels, e.g. bioethanol and biodiesel, etc. (Razzak et al. 2013). The capacity of these microbial cells to absorb CO_2, in particular, suggests this as an effective option for CO_2 storage in the form of biological sequestration. Fixing and storing CO_2 via microalgae is a fundamental approach to transform CO_2 into organic compounds. Especially in comparison to other CO_2 capture techniques, microalgae CO_2 fixation has several advantages, such as a high photosynthesis rate (6.9 per 104 cells per mL per hour), a fast-growth (0.7–3.2 per day), better environmental ability to adapt and low operating costs (Khan et al. 2018). The high demand for CO_2 and microalgae nutrient needs can be met by releases of flue gases and the discharge of effluent from coal power plants. The utilization of CO_2 in microalgae growth can be well depicted in Fig. 7.

The cultivation of microalgae in the presence of sunlight by imbibing nutrients suspended in water or wastewater and CO_2 from the surrounding atmosphere or the post-combustion exhaust gases. The biomass is eventually converted into energy and fuel. Microalgae cultivation furthermore provides three significant benefits, including (1) extracting of CO_2 from flue gases, (2) treatment of wastewater, and (3) renewable energy sources generation (Razzak et al. 2013). Microalgal biomass contains about 50% dry mass carbon (Govindan et al. 2020). Usually, this energy is derived from CO_2; about a hundred tons of it is equivalent to 183 tons CO_2 (Klinthong et al. 2015). Microalgae convert inorganic carbon to organic matter via. the following mechanism: (1) direct assimilation of CO_2 using the plasma membrane; (2) the assimilated CO_2 converts into bicarbonates (HCO_3^-) by selective enzymes; and (3) formed bicarbonates are transferred to the plasma membrane of plant cells.

5.4 Electrochemical Ways for CO_2 Capture

Alternative approaches should be based upon sources of renewable energy. Electrochemistry has received great attention in this field in recent years. This research dates back to the nineteen, whenever a molten carbonate fuel cell was designed to extract CO_2 from flue gas and the electrochemical reduction followed. The nucleophilic sites in the oxidant react with reducing site in CO_2 during the redox reaction, which tends to result in the specific capture of CO_2 from flue gas. Electrochemical CO_2 reduction is a multi-stage reaction cycle that typically involves an elementary reaction mechanism, frequently occurring at the electrode–electrolyte interface, in which the electrode is a solid electro-catalyst, while the electrolyte is normally a water-saturated solution of CO_2.

This catalytic process usually consists of three main steps:

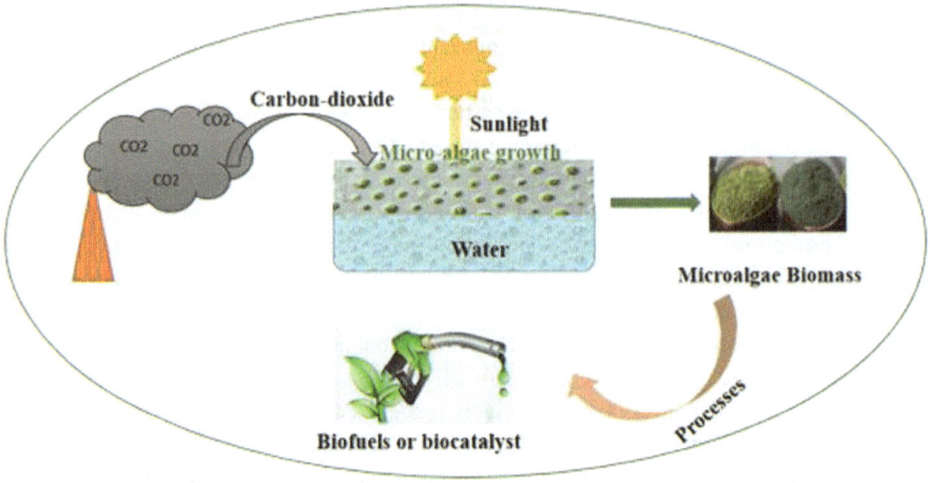

Fig. 7 Conversion of captured CO_2 into the useful product by Microalgae

I Chemisorption of carbon-dioxide on a solid perforated electrode coated by a catalyst.
II Breaking of C-O bonds and formation of C-H bonds followed by electron transfer.
III Transfer the product from electrode to electrolyte present in the electrochemical cell.

In the end, the captured CO_2 is stripped out using either electrochemically mediated amine regeneration (EMAR) or membrane capacitive deionization (MCDI) techniques. Within the EMAR cycle, an electrochemical cell is inserted into a CO_2-rich amine stream in which cu ions produced from a copper anode dissociate the carbamate bond formed between amine and CO_2. The EMAR technique is better than the thermal treatment method for the regeneration of the CO_2 rich amine stream. The amines used for CO_2 absorption has high vapor pressure and degrade at high temperature. MCDI technique can capture $CO_2$from the salts of bicarbonates and carbonates. In this system, the ions of bicarbonates and carbonates are adsorbed on the electrical double layer membrane presented at a perforated graphite electrode surface. Then the membrane is regenerated to remove CO_2 via energy-efficient methods. Although MCDI processes are used at the pilot-stage it shows a great opportunity for carbon capture in the future (Nitopi et al. 2019).

In this study, the process is completed in three electrochemical reactions which take place in an absorber, an electrode, and electrolyte. The chemical absorber used KOH solution to CO_2 capture results in the formation of K_2CO_3. In this method, CO_2 is changed to bicarbonate (HCO_3^-) and carbonate (CO_3^{2-}) after reacting with the KOH solution. The CO_2–rich gas stream is then sent to the anode of an electrochemical cell where the proton is released into the solution resulting in the decrease of pH of the solution. The electrolyte used in this process is potassium chloride (KCl), by which the Cl^- ion permeates through an anion-selective membrane. It helps in maintaining the neutrality of the charges in the solution. To renew the sorbent, the lean stream is recycled back to the cathode where the reduction reaction takes place, and the pH increases. The regenerated high pH solvent is being sent back further for the absorption of more CO_2 to the absorber column. Therefore CO_2 capture and release could be accomplished by chemisorption, accompanied by an electrochemical renewal of the adsorbent material by increasing the protons concentration in the solution. The polarity of the electrode is important to minimize the cathode from being fully saturated with proton and anode reduction. Researchers around the world have suggested and continue to develop new reaction mechanisms for ECR based on their knowledge of metallic electrode CO_2 conversion (Amos et al. 2018).

6 Conclusion

Overall, when the chapter is concluded, CCS technologies are proved to be essential for large-scale applications. There is a lot of difference between what we want to do and what we are doing technically. High costs, limited capital, remaining difficulties in regulatory and legal frameworks for the implementation of CCS, and uncertainties about widespread support are barriers to worldwide large-scale applications of CCS. Overcoming the technological, legislative, political, and social obstacles is crucial. By comparison, biotic techniques are processes that are natural and cost-effective, have various indirect benefits, are immediately available, but have limited sink power. Especially in comparison with other sequestration approaches, biological CCS is the largest ecologically viable process. Significant research has now been undertaken to develop biological technologies and processes that improve the effectiveness of capture systems while keeping costs down. Applications like ocean fertilization have an unpredictable impact as there is a lack of proper or minimal studies. Use all these biological tools successfully around the globe will turn the destiny of our world into a healthy state. Biological microalgae CCS have several benefits over traditional carbon sequestration approaches as CO_2 is used to generate high-value biomass with a variety of applications in the development of energy and fine chemicals.

References

Adams E, Caldeira K (2008) Ocean storage of CO2. Elements 4:319–324

Afshin AM, Ali HMK, Aroua (2011) Absorption of carbon dioxide in the aqueous mixtures of methyldiethanolamine with three types of imidazolium-based ionic liquids. Fluid Phase Equilibria 309 (1):76-82. https://doi.org/10.1016/j.fluid.2011.06.029

Ametistova L, Twidell J, Briden J (2002) The sequestration switch: removing industrial CO2 by direct ocean absorption. Sci Total Environ 289(1–3):213–223

Amos P, Louis H, Adesina Adegoke K, Eno EA, Udochukwu AO, Odey Magub T (2018) Understanding the mechanism of electrochemical reduction of CO2 Using Cu/Cu-based electrodes: a review. Asian J Nanosci Mater 1(4):183–224

Anderson JL, Dixon JK, Brennecke JF (2007) Solubility of CO2, CH4, C2H6, C2H4, O2, and N2 in 1-Hexyl-3-methylpyridinium Bis (trifluoromethylsulfonyl)imide: comparison to other ionic liquids. Acc Chem Res 40(11):1208–1216

Anheden M et al (2005) CO2 quality requirement for a system with CO2 capture, transport and storage BT—greenhouse gas control technologies 7. Elsevier Science Ltd., Oxford, pp 2559–2564

Archer D et al (2009) Atmospheric lifetime of fossil fuel carbon dioxide. Annu Rev Earth Planet Sci 37:117–134

Astarita G (1967) Mass transfer with chemical reaction. Elsevier, Amsterdam, London

Bahadar A, Bilal Khan M (2013) Progress in energy from microalgae: a review. Renew Sustain Energy Rev 27:28–148

Bailey D, Feron P (2005) Post-combustion decarbonisation processes. Oil Gas Sci Technol IFP 60(3):461–479

Baltus R et al (2011) Examination of the potential of ionic liquids for gas separations. Sep Sci Technol 40(1–3):525–541

Barrow C (2012) Biochar: potential for countering land degradation and for improving agriculture. Appl Geo 34(2012):21–28. https://doi.org/10.1016/j.apgeog.2011.09.008

Bates ED, Mayton RD, Ntai I, Davis JH (2002) CO2 capture by a task-specific ionic liquid. J Am Chem Soc 124(6):926–927

Boden TA, Marland G, Andres RJ (2016) Global, regional, and national fossil-fuel CO2 emissions. carbon dioxide information analysis center

Bose H, Satyanarayana T (2017) Microbial carbonic anhydrases in biomimetic carbon sequestration for mitigating global warming: prospects and perspectives. Front Microbiol 8:1615

Brennecke JF, Maginn EJ (2004) Ionic liquids: innovative fluids for chemical processing. AIChE J 47(11):2384–2389

Burns TD et al (2020) Prediction of MOF performance in vacuum swing adsorption systems for postcombustion CO2 capture based on integrated molecular simulations, process optimizations, and machine learning models. Environ Sci Technol 54(7):4536–4544

Cadena C, Anthony JL, Shah JK, Morrow TI, Brennecke JF, Maginn EJ (2004) Why is CO2 so soluble in imidazolium-based ionic liquids? J Am Chem Soc 126(16):5300–5308

Caplow M (1968) Kinetics of carbamate formation and breakdown. J Am Chem Soc 90(24):6795–6803

Cents A, Brilman W, Versteeg G (2005) CO2 absorption in carbonate/bicarbonate solutions: the Danckwerts-criterion revisited. Chem Eng Sci 60(21):5830–5835

Chaudhary A, Bhaskarwar AN (2015) A novel ionic liquid for carbon capture. Athens J Sci 2(3):187–202

Chakraborty AK, Astarita Bischoff GB (1986) CO_2 absorption in aqueous solutions of hindered amines. Chem Eng Sci 41(4):997–1003. https://doi.org/10.1016/0009-2509(86)87185-8

Costa Gomes MF (2007) Low-pressure solubility and thermodynamics of solvation of carbon dioxide, ethane, and hydrogen in 1-hexyl-3-methylimidazolium Bis(trifluoromethylsulfonyl)amide between temperatures of 283 and 343 K. J Chem Eng Data 52 (2):472–475

Crooks JE, Donnellan JP (1989) Kinetics and mechanism of the reaction between carbon dioxide and amines in aqueous solution. J Chem Soc Perkin Trans 2(4):331–333

D'Alessandro JRLD, Smit B (2010) Carbon dioxide capture: prospects for new materials. Angew Chemie Int Ed 49(35):6058–6082

Danckwerts PV (1979) The reaction of CO2 with ethanolamines. Chem Eng Sci 34(4):443–446

Deviram G, Mathimani T, Anto S, Ahamed TS, Ananth DA, Pugazhendhi A (2020) Applications of microalgal and cyanobacterial biomass on a way to safe, cleaner and a sustainable environment. J Clean Prod 253.119770

Dean, Camper JE, Bara DL, Gin RD, Noble (2008) Ind Eng Chem Res 47(21):8496–8498. https://doi.org/10.1021/ie801002m

EPRI (2012) CO2 capture technologies-pre combustion capture—report sponsored by the Global Carbon Capture and Storage Institute, Canberra, Australia

Galan Sanchez LM, Meindersma GW, de Haan AB (2011) Kinetics of absorption of CO2 in amino-functionalized ionic liquids. Chem Eng J 166(3):1104–1115

Govindan N et al (2020) Statistical optimization of lipid production by the diatom Gyrosigma sp. grown in industrial wastewater. J Appl Phycol 32(1):375–387

Guo X, Liu H, Zhang J (2020) The role of biochar in organic waste composting and soil improvement: a review. Waste Manag 102:884–899

Hernandez-Soriano MC et al (2018) Soil organic carbon stabilization: mapping carbon speciation from intact microaggregates. Environ Sci Technol 52(21):12275–12284

Hu Z, Wang Y, Shah BB, Zhao D (2019) CO2 capture in metal-organic framework adsorbents: an engineering perspective. Adv Sustain Syst 3(1):1800080

IPCC (2013) Climate change 2013: the physical science basis. contribution of working group I to the fifth assessment report of the intergovernmental panel on climate change. Cambridge University Press, Cambridge, UK and New York, NY, USA

IPCC (2014) Climate Change 2014: mitigation of climate change. Contribution of working group III to the fifth assessmentreport of the intergovernmental panel on climate change [Edenhofer O, Pichs-Madruga R, Sokona Y, Farahani E, Kadner S, Seyboth K, Adler A, Baum, Brunner S, Eickemeier P, Kriemann B, Savolainen J, Schlömer S, von Stechow C, Zwickel T, Minx JC (eds)]. Cambridge University Press, Cambridge, United Kingdom and New York, NY, USA

Jeffery S, Verheijen FGA, van der Velde M, Bastos AC (2011) A quantitative review of the effects of biochar application to soils on crop productivity using meta-analysis. Agric Ecosyst Environ 144 (1):175–187

Jensen MB, Pettersson LGM, Swang O, Olsbye U (2005) CO2 sorption on MgO and CaO surfaces: a comparative quantum chemical cluster study. J Phys Chem B 109(35):16774–16781

Jiao N, Zheng Q (2011) The microbial carbon pump: from genes to ecosystems. Appl Environ Microbiol 77(21):7439–7444

Khan MI, Shin JH, Kim JD (2018) The promising future of microalgae: current status, challenges, and optimization of a sustainable and renewable industry for biofuels, feed, and other products. Microb Cell Fact 17(1):36

Klinthong W, Yang Y-H, Huang C-H, Tan C-S (2015) A review: microalgae and their applications in CO2 capture and renewable energy. Aerosol Air Qual Res 15(2):712–742

Kohl AL, Nielsen RB (1997) Chapter 2—Alkanolamines for hydrogen sulfide and carbon dioxide removal BT—gas purification, 5th edn. Gulf Professional Publishing, Houston, pp 40–186

Kumar S, Saxena SK (2014) A comparative study of CO2 sorption properties for different oxides. Mater Renew Sustain Energy 3(3):30

Lee S, Liang L, Riestenberg D, West OR, Tsouris C, Adams E (2003) CO2 hydrate composite for ocean carbon sequestration. Environ Sci Technol 37(16):3701–3708

Lei Z, Dai C, Chen B (2014) Gas solubility in ionic liquids. Chem Rev 114(2):1289–1326

Leung DYC, Caramanna G, Maroto-Valer MM (2014) An overview of current status of carbon dioxide capture and storage technologies. Renew Sustain Energy Rev 39(Supplement C):426–443

Lin C, Lin C-H (2013) Comparison of carbon sequestration potential in agricultural and afforestation farming systems. Sci Agricola 70:93–101

Ma J et al (2011) Ditetraalkylammonium amino acid ionic liquids as CO2 absorbents of high capacity. Environ Sci Technol 45 (24):10627–10633

Maginn E (2005) Design and evaluation of ionic liquids as novel CO2 absorbents.University of Notre Dame

Ma'mum S, Svendsen HF, Hoff KA, Juliussen O (2005) Selection of new absorbents for carbon dioxide capture BT—greenhouse gas control technologies 7. Elsevier Science Ltd, Oxford, pp 45–53

Megías-Sayago C, Bingre R, Huang L, Lutzweiler G, Wang Q, Louis B (2019) CO2 adsorption capacities in Zeolites and layered double hydroxide materials. Front Chem 7:551

Michalak AM, Jackson R, Marland G, Sabine C, Group CCSW (2011) A US carbon cycle science plan. University Corporation for Atmospheric Research, Boulder, Colorado

Mohanty S, Banerjee T, Mohanty K (2010) Quantum chemical based screening of ionic liquids for the extraction of phenol from aqueous solution. Ind Eng Chem Res 49(6):2916–2925

Momeni M, Riahi S (2015) An investigation into the relationship between molecular structure and rich/lean loading of linear amine-based CO2 absorbents. Int J Greenh Gas Control 42:157–164

Mulabagal V, Baah D, Egiebor N, Chen WY (2015) Biochar from biomass-a strategy for carbon dioxide sequestration, soil amendment, power generation and CO2 utilization. Handb Clim Mitig Adapt

Nitopi S et al (2019) Progress and Perspectives of Electrochemical CO2 Reduction on Copper in Aqueous Electrolyte. Chem Rev 119 (12):7610–7672

Pavlik D et al (2017) Microalgae cultivation for carbon dioxide sequestration and protein production using a high-efficiency photobioreactor system. Algal Res 25:413–420

Pellerano M, Pré P, Kacem M, Delebarre A (2009) CO2 capture by adsorption on activated carbons using pressure modulation. Energy Procedia 1(1):647–653

Qanbari F, Pooladi-Darvish M, Hamed Tabatabaie S, Gerami S (2011) Storage of CO2 as hydrate beneath the ocean floor. Energy Procedia 4:3997–4004

Rau GH, Caldeira K (1999) Enhanced carbonate dissolution: a means of sequestering waste CO2 as ocean bicarbonate. Energy Convers Manag 40(17):1803–1813

Razzak SA, Hossain MM, Lucky RA, Bassi AS, de Lasa H (2013) Integrated CO2 capture, wastewater treatment and biofuel production by microalgae culturing—a review. Renew Sustain Energy Rev 27:622–653

Sheps KM, Max MD, Osegovic JP, Tatro SR, Brazel LA (2009) A case for deep-ocean CO2 sequestration. Energy Procedia 1(1):4961–4968

Shiflett M, Niehaus A, Yokozeki A (2010) Separation of CO2 and H2S using room-temperature ionic liquid [bmim] [MeSO4]. J Chem Eng Data 55(11):4785–4793

da Silva EF, Svendsen HF (2004) Ab Initio Study of the Reaction of Carbamate Formation from CO2 and Alkanolamines. Ind Eng Chem Res 43(13):3413–3418

Siriwardane RV, Shen M-S, Fisher EP, Losch J (2005) Adsorption of CO2 on Zeolites at moderate temperatures. Energy Fuels 19 (3):1153–1159

Stamenković M, Hanelt D (2013) Adaptation of growth and photosynthesis to certain temperature regimes is an indicator for the geographical distribution of Cosmarium strains (Zygnematophyceae, Streptophyta). Eur J Phycol 48(1):116–127

Stevanovic S, Podgorsek A, Moura L, Santini CC, Padua AAH, Costa Gomes MF (2013) Absorption of carbon dioxide by ionic liquids with carboxylate anions. Int J Greenh Gas Control 17(Supplement C):78–88

US EPA (US Environmental Protection Agency) (2015) Inventory of US greenhouse-gas emissions and sinks. US EPA, Washington, DC

Walker SJ, Keeling RF, Piper SC (2017) Reconstruction of the Mauna Loa carbon dioxide record using High Frequency APC data

Wang Y, Du T, Qiu Z, Song Y, Che S, Fang X (2018) CO2 adsorption on polyethylenimine-modified ZSM-5 zeolite synthesized from rice husk ash. Mater Chem Phys 207:105–113

Wlazło M, Siklitskaya A, Majewski JA (2017) Ab initio studies of carbon dioxide affinity to carbon compounds and minerals. Energy Procedia 125:450–456

Xue Z, Zhang Z, Han J, Chen Y, Mu T (2011) Carbon dioxide capture by a dual amino ionic liquid with amino-functionalized imidazolium cation and taurine anion. Int J Greenh Gas Control 5(4):628–633

Yu CH, Huang CH, Tan CS (2012) A review of CO2 capture by absorption and adsorption. Aerosol Air Qual Res

Jie YX, Jinyue Y, Shan-Tung Y, Tu (2014) Ind Eng Chem Res 53 (7):2790–2799. https://doi.org/10.1021/ie4040658

Zdravkov B, Čermák J, Šefara M, Janků J (2007) Pore classification in the characterization of porous materials: a perspective. Cent Eur J Chem 5:385–395

Zeebe RE, Archer D (2005) Feasibility of ocean fertilization and its impact on future atmospheric CO2 levels. Geophys Res Lett 32(9)

Zhang Z, Zhao Y, Gong Q, Li Z, Li J (2013) MOFs for CO2 capture and separation from flue gas mixtures: the effect of multifunctional sites on their adsorption capacity and selectivity. Chem Commun 49 (7):653–661

Zhanga S, Sun N, He X, Lu X, Zhang X (2006) Physical properties of ionic liquids: database and evaluation. J Phys Chem Ref Data 35:1475

Zhou W et al (2005) The IEA Weyburn CO2 monitoring and storage project—modeling of the long-term migration of CO2 from Weyburn BT—greenhouse gas control technologies 7. Elsevier Science Ltd., Oxford, pp 721–729

Carbon Derivatives from CO_2

Abbas Ghareghashi and Ali Mohebbi

Abstract

Today, the most demanding energy sources are hydrocarbon fuels because of their high energy density, stability, and natural abundance. Increased carbon dioxide emissions from them into the atmosphere has become one of the concerns in global warming. To solve this problem, various strategies need to be developed to CO_2 capture, storage, and utilization. One of the methods for reducing this gas is CO_2 capture and storage; however, the storage has not attracted the attention of artisans because of its high cost. The conversion of CO_2 into valuable products is a promising choice.

We reviewed various processes for converting CO_2 into carbon derivatives. In addition, products of each of processes and their advantages and disadvantages are evaluated. The effects of several catalysts on selectivity and efficiency of products are discussed. Overview of the challenges and problems of CO_2 reduction is another objective of this chapter.

Keywords

Carbone dioxide • Carbon derivatives • Product selectivity • Global warming • Hydrocarbon fuels • Storage

1 Introduction

Our conventional energy sources are mainly non-renewable natural gas, coal, and petroleum, including a maximum amount of energy consumed (Hu and Suib 2014). Given the depletion of fossil energy resources, we will face many challenges in the future of energy resources (Hu and Suib 2014). The alternative energy sources are wind power, biofuel, solar, water power and nuclear power that reduce CO_2 emissions (Hu and Suib 2014). Carbon dioxide is an important component of the Earth, presents in the atmosphere, crust, and core, significantly (Rafiee et al. 2018). Reaction with other substances, dissolution in water, and freely exist in the atmosphere are specifications of carbon dioxide (Rafiee et al. 2018). However, the emission of human-made CO_2 has disadvantages, including global warming, ocean acidification and climate changes (Hu and Suib 2014). Indeed, CO_2 is the main contributor to global warming compared to other greenhouse gases (Concepcion et al. 2012).

To reducing CO_2 emission, the mainly accepted approaches are CO_2 capture and storage [CCS], CO_2 conversion to feedstock for fuels and chemicals (Muthuraj and Mekonnen 2018). Because of the large scale of carbon produced from CO_2 emission reductions, a lot of space is required to CCS in geological formations; as a result, the evaluation of the methods of captured CO_2 conversion into valuable products is essential (Wang et al. 2013a). In recent years, significant efforts have been made to convert CO_2 as an attractive, abundant, safe, non-oxidant, and inexpensive feedstock for making fuels, solvents, commodity chemicals, and so on (Finn et al. 2012; Inglis et al. 2012; Schneider et al. 2012). The production of industrial chemicals by converting of carbon dioxide has attracted the attention of many researchers who are researching modern organic synthetic chemistry (Vessally et al. 2017). Currently, the use of CO_2 feedstock for useful products is insignificant; however, the most significant consumption of this gas is urea production in organic synthesis (Muthuraj and Mekonnen 2018).

A. Ghareghashi
Department of Chemical Engineering, Faculty of Engineering, Velayat University, Iranshahr, Iran

A. Mohebbi (✉)
Department of Chemical Engineering, Faculty of Engineering, Shahid Bahonar University of Kerman, Kerman, Iran
e-mail: amohebbi@uk.ac.ir

Inamuddin et al. (eds.), *Carbon Dioxide Utilization to Sustainable Energy and Fuels*,
Advances in Science, Technology & Innovation, https://doi.org/10.1007/978-3-030-72877-9_14

Besides, CO_2 gas is used to produce carbonates and salicylic acid in large scale (Muthuraj and Mekonnen 2018). One of the difficulties of CO_2 reduction reactions is a high energy consumption; as a result using catalysts to overcome this limitation is suggested (Muthuraj and Mekonnen 2018). In recent years, the volume of researches done on CO_2 reduction, including several methods, has become more and more (Fu et al. 2019).

This chapter is a comprehensive review of the CO_2 reduction processes. Indeed, the conversion of CO_2 as undesirable gas to useful products was evaluated. Different methods were categorized, and each of them was described in general terms. The selectivity and efficiency of output products from each process were generally discussed. Finally, the advantages and disadvantages of the methods and the best option were introduced according to the environmental conditions and available facilities.

2 Artificial Photoreduction

To decrease the rate of CO_2 concentration in the atmosphere, the reduction of CO_2 has attracted the attention of many active academic and non-academic researchers. It has been studied widely in the last decade (Indrakanti et al. 2009a). Converting CO_2 to fuel for energy production, which is currently fed by fossil fuels, is an exciting idea (Varghese et al. 2009a). Nevertheless, because CO_2 is a relatively stable and inert gas, its conversion processes require a large amount of energy, which is not economical and cause many environmental problems (Chen and Cheng 2002). Photocatalytic CO_2 reduction is one of the methods of carbon conversion that requires less energy than other processes and is a sustainable and environment-friendly process (Wu and Lin 2005). Besides, using solar energy as a cheap, abundant, ecologically safe, and clean source to activate the reaction in this process makes it an available process and a continuous power supply. Implementation steps for a typical system of semiconductor-based photocatalyst are (a) light energy absorption for catalyst, (b) pairs of electron–hole generation, (c) separation of them spatially, and (d) transfer of them across the interface to redox-active species (Fujishima et al. 2008). The photocatalytic pathways are very complex despite its environmental and economic benefits. Besides, several limiting factors cause low efficiency, such as complicated backward reactions, fast recombination rates of electron–hole, low CO_2 dependency of the photocatalysts, and narrow range of light absorption wavelength. The basics of CO_2 photoreduction pathways are listed as (1) organic photocatalysts, (2) inorganic photocatalysts, (3) inorganic and organic/biological hybrid (biomimetic systems), and (4) biological systems (Indrakanti et al. 2009a; Sutin et al. 1997; Song 2006; Usubharatana et al. 2006; Kitano et al. 2007). CO_2 photoreduction is carried out in multiple steps. The first step is in the following reaction, the e^--h^+ pairs generation occurs (Eq. 1).

$$Photocatalyst + hv \rightarrow e^- + h^+ \quad (1)$$

In the following, spatial e^--h^+ pairs are separated and transferred to the interface in the form of reduced active species; as a result, the minimization of electron–hole recombination is done.

$$e^- + h^+ \rightarrow heat(recombination) \quad (2)$$

This step is considered as an essential limiting step in this process because the time scale of e^--h^+ recombination is several times faster than that of other electron transfer processes (Indrakanti et al. 2009b). In the next step, oxygen and protons are formed by oxidation of water by holes (H^+).

$$H_2O + 2h^+ \rightarrow 1/2O_2 + 2H^+ \quad (3)$$

Besides, there are several reverse reactions that consume H^+ and e^- and produce H_2 and H_2O_2, respectively.

$$O_2 + 2H^+ + e^- \rightarrow H_2O_2 \quad (4)$$

$$2h^+ + e^- \rightarrow H_2 \quad (5)$$

A broad research was performed on reducing undesired reactions such as replacing water with methanol and ethanol, which leads to increase yield and selectivity (Liu et al. 1998). Replacing water with carbon dioxide in photoreduction reaction pathways was suggested by several researchers (Yahaya et al. 2004).

$$CO_2 + e^- \rightarrow CO_2^- \quad (6)$$

As shown in Eqs. 7–12, the multielectron transfer [MET] photoreduction reactions for CO_2 form various products.

$$CO_2 + 2H^+ + 2e^- \rightarrow HCO_2H \quad (7)$$

$$CO_2 + 2H^+ + 2e^- \rightarrow CO + H_2O \quad (8)$$

$$CO_2 + 4H^+ + 4e^- \rightarrow HCHO + H_2O \quad (9)$$

$$CO_2 + 6H^+ + 6e^- \rightarrow CH_3OH + H_2O \quad (10)$$

$$CO_2 + 8H^+ + 8e^- \rightarrow CH_4 + 2H_2O \quad (11)$$

$$CO_2 + 12H^+ + 12e^- \rightarrow C_2H_5OH + 3H_2O \quad (12)$$

CO and HCOOH are usual products, because these reactions need only two protons and two electrons that CO formation requires less reduction potential than HCOOH. The other products need more electrons and protons; as a result, they are not easy to form. In addition to reaction (11), methane can also be produced from the following multiple reactions (Varghese et al. 2009b; Roy et al. 2010).

$$CO + 6H^+ + 6e^- \rightarrow CH_4 + H_2O \quad (13)$$

$$CH_3OH + 2H^+ + 2e^- \rightarrow CH_4 + H_2O \quad (14)$$

Factors influencing reaction (11) are CO_2 activation, product desorption, reactant adsorption, reaction temperature, incident light fraction adsorbed by the photocatalyst, incident light intensity, the photocatalyst specific area adsorbing the light, and photocatalysts crystalline properties. For CO_2 photoreduction, the e^--h^+ recombination and the transfer of an electron from the activated photocatalyst to CO_2 are significant limiting steps; moreover, adsorption of CO_2 on the photocatalyst surface is a critical step.

The conversion efficiency of CO_2 in artificial photosynthesis remains low overall. The main reason for this problem could be the limited ability of traditional semiconductors to activate CO_2 molecules, which are thermodynamically stable. A wide range of studies have been conducted on developing a stronger photocatalyst or constructing strategies such as band-gap engineering, combined semiconductors and co-catalyst loading to overcome the weakness of traditional photocatalysts. Recently a lot of attention is focused on the enhancement of CO_2 activation through strategies such as nanostructuring, surface modification, or dispersing the semiconductor to increase CO_2 adsorption (Yuan et al. 2016).

3 Electrochemical Reduction

The direct heterogeneous electrochemical reduction of CO_2 is an attractive method for CO_2 conversion. Several products, including carbon monoxide (CO), methane (CH_4), formic acid (HCOOH), methanol (CH_3OH), ethanol (C_2H_5OH), acetate (CH_3COOH), and ethylene (C_2H_4) can be produced by this method (Nielsen and Leung 2010; Leung et al. 2010; Azuma et al. 1990; Furuya et al. 1997; Hori et al. 1994; Ikeda et al. 1987; Mahmood et al. 1987; Udupa et al. 1971). Electrochemical reduction of CO_2 requires a platform of high activity catalysts include homogeneous or heterogeneous. As the name implies, this process requires electricity, and in the production of electricity, a lot of carbon dioxide is released. As a result, given that our goal in this process is to reduce carbon dioxide levels, the electricity used in this process should be from nuclear, that is the carbon–neutral source, or be renewable. The advantages of this method are: (1) using wind, solar, geothermal, hydroelectric, tidal, and heat processes, the required electricity of this process can be supplied. As a result, carbon dioxide cannot be produced; (2) no heat source is needed for this process, as this process takes place at room temperature; and (3) these reaction systems have advantages such as modular, easy for scale-up application, on-demand and compact, and so on (Yamamoto et al. 2002; Akahori et al. 2004; Subramanian et al. 2007).

Significant improvements in technology are necessary for this process which is economically justified. Recent researches on this process have focused mainly on the various catalysts used and manufactured products. The design of the proposed reactors for the reduction process has also been another recent study in which the products of these reactors have been formic acid or gas synthesis. Besides, various studies were performed on the effects of parameters such as temperature and electrolyte on product selectivity (Kaneco et al. 2006; Hori et al. 1989; Li and Oloman 2006). Product generation rate or high reactivity has a direct impact on the reactor size. By increasing the electrode active area, the electrode reactivity can be significantly improved; as a result, constant selectivity and electrochemical polarization behavior are achieved. Several studies were performed on the effects of catalysts on selectivity and the mechanism of CO_2 reduction (Sridhar et al. 2011; Gattrell et al. 2006; Chaplin and Wragg 2003). Ikeda et al. (Ikeda et al. 1987) reviewed several metallic electrodes for the electrochemical reduction of CO_2. The selectivity of formic acid and carbon monoxide in the aqueous and non-aqueous electrolyte was examined (Figs. 1 and 2). Sen et al. (Sen et al. 2014) used copper foams for the electrochemical reduction of CO_2. The hierarchical porosity of copper foams caused a change in the CO_2 electroreduction mechanism. Decreasing of faradaic efficiency for CH_4, C_2H_4 and CO, C_3 product generation, production of saturated hydrocarbons, and increasing of formate faradaic efficiency were factors that changed this mechanism. They found that the current density of copper nanofoam increases more than that of a smooth copper electrode. This comparison is accurate at concentrations higher than 0.5 M $KHCO_3$ (Sen et al. 2014).

4 Hydrogenation

One of the methods of carbon dioxide recycling that has attracted the attention of many researchers is hydrogenation reactions. Hydrogenation methods, compared to other methods of carbon dioxide stabilization, have the advantage of being connected to existing technologies for converting synthetic gas (CO and H_2) into methanol, methane, alcohol, DME, and higher hydrocarbons. This connection is through the water–gas-shift reaction.

$$CO_2 + H_2 \leftrightharpoons CO + H_2O \quad (15)$$

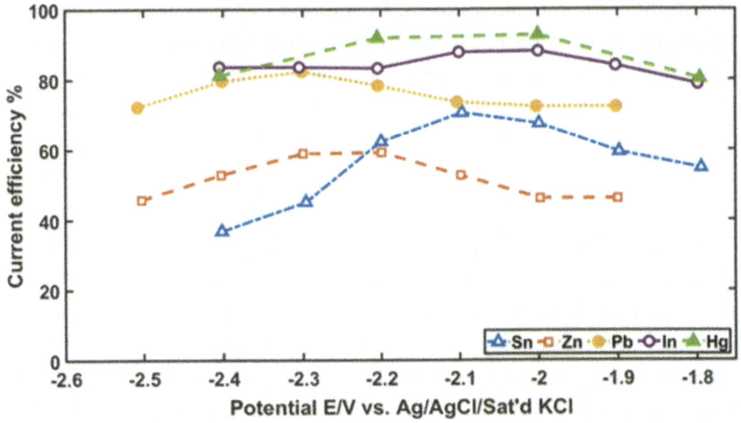

Fig. 1 Effect of various metallic electrodes in 0.1 M TEAP/H_2O on formic acid current efficiency. Quantity of electricity passed: 100 C, TEAP: Tetraethylammonium perchlorate. Modified after (Ikeda et al., 1987) (Ikeda et al. 1987).

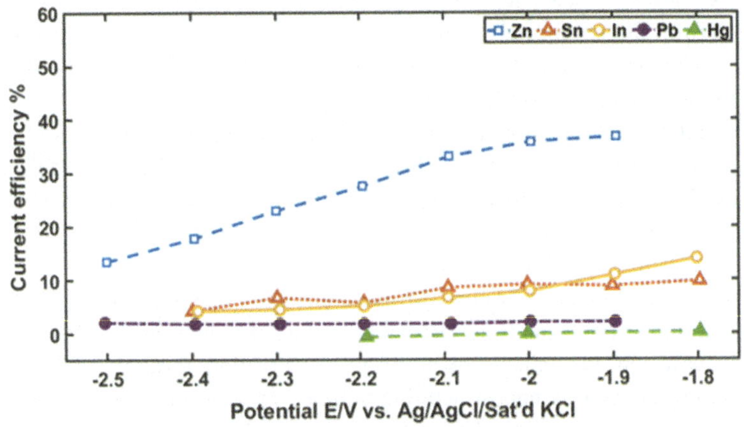

Fig. 2 Effect of various metallic electrodes in 0.1 M TEAP/H_2O on carbon monoxide current efficiency. Quantity of electricity passed: 100 C, TEAP: Tetraethylammonium perchlorate, Modified after (Ikeda et al., 1987) (Hori et al. 1994)

Carbon dioxide first reacts with hydrogen to become carbon monoxide. The synthesis gas (the combination of CO and H_2) is converted to valuable hydrocarbons in a Fischer–Tropsch process. In this process, syngas in the presence of catalysts such as iron-based catalysts is converted to light and heavy hydrocarbons as follow (Montazer-Rahmati and Bargah-Soleimani 2001):

$$\text{Step 1}: CO + 3H_2 \xrightarrow{R_1} CH_4 + H_2O \quad (16)$$

$$\text{Step 2}: 2CO + 4H_2 \xrightarrow{R_2} C_2H_4 + 2H_2O \quad (17)$$

$$\text{Step 3}: 2CO + 5H_2 \xrightarrow{R_3} C_2H_6 + 2H_2O \quad (18)$$

$$\text{Step 4}: 3CO + 7H_2 \xrightarrow{R_4} C_3H_8 + 3H_2O \quad (19)$$

$$\text{Step 5}: 4CO + 9H_2 \xrightarrow{R_5} n - C_4H_{10} + 4H_2O \quad (20)$$

$$\text{Step 6}: 4CO + 9H_2 \xrightarrow{R_6} i - C_4H_{10} + 4H_2O \quad (21)$$

$$\text{Step 7}: 6.05CO + 12.23H_2 \xrightarrow{R_7} C_{6.05}H_{12.36}\left(C_5^+\right) + 6.05H_2O \quad (22)$$

$$\text{Step 8}: CO + H_2O \overset{R_8}{\leftrightarrow} CO_2 + H_2O \quad (23)$$

For each reaction step, the reaction rate equations are as below:

$$R_i = 0.278K_i \exp\left(\frac{-E_i}{RT}\right) P_{co}^m P_{H_2}^n \quad (24)$$

Selectivity and the type of hydrocarbon produced depend on several factors like the type of reactor, catalytic bed, catalyst, flow (co-current or counter-current), and so on. Numerous studies have been done in this field (Forghani et al. 2009; Marvast et al. 2005; Bhatia et al. 2009; Stansch et al. 1997; Itoh 1987; Ghareghashi et al. 2013, 2017). Forghani et al. (Forghani et al. 2009) suggested a new structure for the Fischer–Tropsch reactor. This proposed reactor improved the selectivity and yield of valuable hydrocarbons (Fig. 3).

One of the important intermediates for petrochemicals production or useful chemicals is dimethyl ether [DME]. Producing of DME from CO_2 is performed using either direct one-step or two-step technologies. In the two-step technology (the conventional route), catalytic converting of syngas to methanol and catalytic dehydration of methanol to

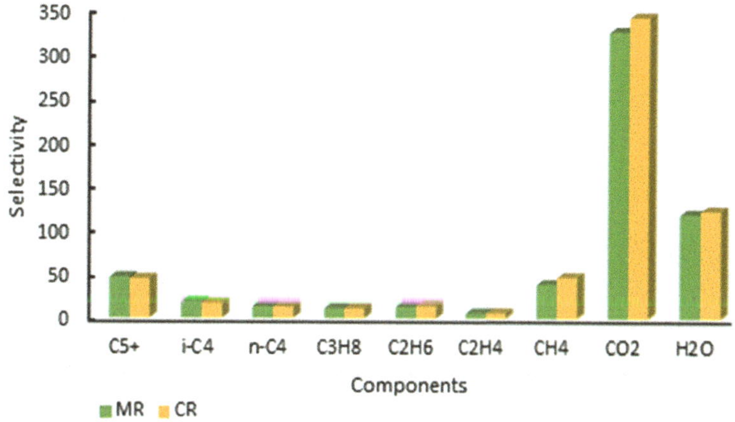

Fig. 3 Selectivity (%) comparison of the components for the membrane and conventional reactor systems, MR = membrane reactor, CR = conventional reactor, modified after (Forghani et al., 2009) (Chaplin and Wragg 2003)

DME are the first and second steps, respectively (Ma et al. 2009; Bonura et al. 2014).

One-step technology:

$$2CO_2 + 6H_2O \rightleftarrows CH_3OCH_3 + 3H_2O \quad (25)$$

Two-step technology:

$$CO_2 + 3H_2 \rightleftarrows CH_3OH + H_2O \quad (26)$$

$$CO_2 + H_2 \rightleftarrows CO + H_2O \quad (27)$$

$$3CH_3OH \rightleftarrows CH_3OCH_3 + H_2O \quad (28)$$

CO_2 methanation as a method of storing renewable energy was evaluated in synthetic natural gas form. A variety of supported catalysts based on Ru, Ni, Rh, Fe, Pd, or Co for CO_2 methanation has been investigated (Oyola-Rivera et al. 2015; Li et al. 2014; Silva et al. 2015; Herwijnen et al. 1973; Lunde and Kester 1974; Karelovic and Ruiz 2013, 2012; Deleitenburg and Trovarelli 1995; Schild et al. 1991; Park and McFarland 2009). Conversion reaction of carbon dioxide to methane is as follows (Hwang et al. 2013; Erdöhelyi et al. 1986):

$$CO_2 + 4H_2 \rightleftarrows CH_4 + 2H_2O \quad (29)$$

CO_2 hydrogenation to higher alcohols synthesis from the catalytic reaction is as follow (Zhou et al. 2013):

$$CO_2 + H_2 \rightleftarrows CO + H_2O \quad (30)$$

$$nCO + 2nH_2 \rightleftarrows C_nH_{2n+2}OH + (n-1)H_2O \quad (31)$$

5 Synthesis of Organic Carbonates

Organic carbonates that have a large market constitute an important compound class that they can be used as selective reagents, solvents, fuel additives, monomers for polymers, and intermediates (Aresta et al. 2002).

Organic carbonates are the main target of synthesis in terms of market size, thermodynamics, and environmental demand. In terms of market size, organic carbonates are industrially useful due to their low molecular weight. Moreover, aromatic polycarbonates are considered as important and practical plastics. In thermodynamics terms, the carbonyl carbon chemical bonds in organic carbonates turn these molecules into a heavily oxygenated molecule; as a result, it is considered as an attractive synthetic target. Observing the energy difference among the bonds of two molecules shows that the conversion of carbon dioxide to carbonate is possible because molecular bonds are stronger in organic carbonates. In environmental demand terms given that phosgene used as a reactive reagent to produce organic carbonate is a toxic and lethal substance. Using carbon dioxide as an alternative, which is an economical, safe, and abundant substance, is highly desirable. Organic carbonates are used in various cases, including fuel additives, electrolyte solvents for lithium-ion batteries, green reagents, organic solvents, and the production of engineering plastics. Carbon dioxide as a new feedstock for chemicals synthesis can be used to produce organic carbonates. There are several benefits to using this method, such as prevention of waste production and energy saving (Aresta et al. 2002). The carbonate synthesis via carbon dioxide fixation into epoxides is an attractive methodology. The catalysts used for this process include $Bu_2Sn(OMe)_2$, ZrO_2, $Ni(OAc)_2$, $Bu_2Sn(OMe)_2$, $ZrO_2{}^+H_3PO_4$, $Bu_2Sn(OMe)_2$, $ZrO_2{}^+CeO_2$ and so on (Sakai et al. 1975; Yamazaki et al. 1978; Tomishige et al. 1999, 2000; Zhao et al. 2000; Tomishige and Kunimori 2002; Ballivet-Tkatchenko et al. 2000; Choi et al. 2002). An attractive methodology for the carbonate synthesis is direct olefines oxidative carboxylation that combines carbonation and epoxidation processes. Various studies were performed on the improvement of different catalysts in this process. Yoshida et al. (Yoshida et al. 2006) investigated CeO_2 catalysts for the direct synthesis of dimethyl carbonate [DMC] from CO_2. They found that increasing of CO_2 amount in

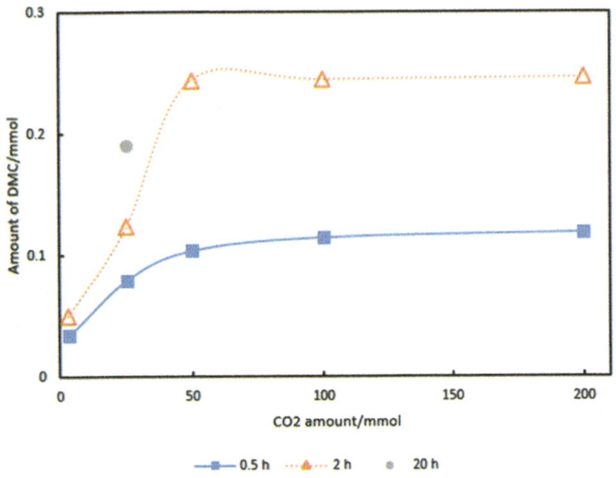

Fig. 4 Variation of DMC formation as a function of CO_2 amount in $CH_3OH^+CO_2$reactions on CeO_2-HS calcined at 873 K. (Catalyst weight: 10 mg, reaction temperature: 403 K, and CH_3OH: 200 mmol, Reaction time: 0.5 h, 2 h, and 20 h), modified after (Yoshida et al., 2006)(Ballivet-Tkatchenko et al. 2000)

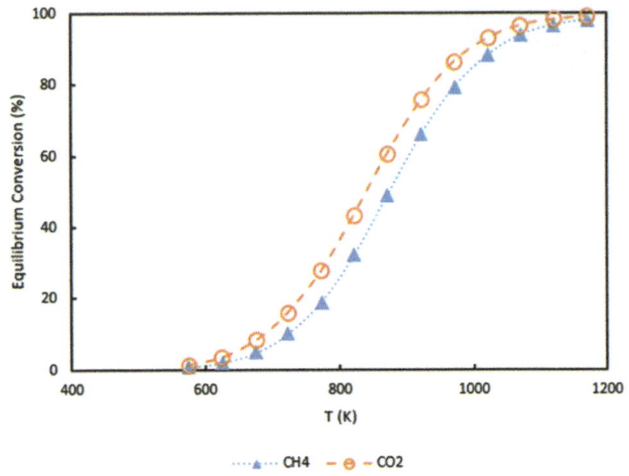

Fig. 5 Equilibrium conversion of CO_2 and CH_4 in reforming reaction, modified after (Bradford et al. 1999; Gadalla and Bower 1988)

quantities less than 50 mmol, increased the formation of DMC. However, the amount of DMC was constant above 50 mmol of CO_2 in both reaction times (Fig. 4).

6 Reforming

Steam reforming of methane is a conventional process for the production of synthesis gas in the industrial scale. One of the main components of this process is water, which can be attractive if the water is not available and using carbon dioxide as an appropriate alternative to water. The use of carbon dioxide instead of water in the reforming process causes syngas yields with lower H_2/CO ratios, which the use of this feed in the Fischer–Tropsch reactor leads to the production of long-chain hydrocarbons (Gadalla and Bower 1988). Although a few industrial reforming processes of CO_2 and CH_4 exist, the use of this process to convert methane and carbon dioxide to other chemicals has a significant effect on reducing these gases in the atmosphere (Teuner 1985; Töpfer 1976). The equilibrium reaction of synthesis gas production from CO_2 and CH_4 is as follows:

$$CH_4 + CO_2 \rightleftarrows 2H_2 + 2CO \tag{32}$$

The equilibrium conversion percentages of CO_2 and CH_4 in this reaction are shown in Fig. 5 As indicated in this figure, the CO_2 conversion rate of carbon is higher than that of methane; as a result, CO_2 is a limiting reactant.

One of the reasons that prevents this process from being commercialized is the endothermic nature of this process. Besides, to increase the conversion rate, the temperature needs to be close to 1073 K, which leads to the deactivation of the catalyst. As a result, several researches have been performed on process catalysts (Stagg-Williams et al. 2000; Stagg and Resasco 1998). Chen et al. (1999) compared the performance of two catalysts: reduced $Ni_xMg_{1-x}O$ solid solution and magnesia-supported Ni metal (Ni/MgO) catalysts. Results showed that Ni/MgO catalyst has lower resistance than reduced $Ni_xMg_{1-x}O$ solid solution catalyst. Besides, the contact time affects the CO/H_2 ratio. The catalyst activity was investigated in both catalysts (Fig. 6). CO_2 reforming of methane reaction was also examined by Osaki and Mori (Osaki and Mori 2001). The catalyst considered in this process was Ni/Al_2O_3 catalysts promoted by potassium. The rates of CH_4-CO_2 reaction were investigated by variation of potassium wt % in the catalyst. The effect of potassium in the catalyst is visible as CO_2 adsorption increasing, however do not have a significant impact on CO_2 dissociation to CO and O_{ads} (adsorbed O).

7 Photocatalytic Reduction of CO_2 with Water

One of the attractive methods of converting CO_2 into the more useful products is the photocatalytic conversion of CO_2. Conversion into valuable products such as formaldehyde, methanol, and methane, and removing this gas from the atmosphere, are advantages of this method. Using solar energy in this process, and thus reducing energy consumption distinguishes this process from other processes. Inoue et al. (1979), were the first researchers that obtained organic compounds from the photocatalytic reduction of CO_2. Subsequently, this process attracted the attention of many researchers, and extensive researches have been conducted in this field because of environmental problem solving and renewable energy utilization.

Fig. 6 Variation of catalytic activity (blue color) and CO/H_2 ratio (red color) with contact time over $Ni_{0.03}Mg_{0.97}O$ solid solution and 3 mol% Ni/MgO catalysts at 773 Kin CO_2 reforming of methane. (Catalyst weight, 0.05 g,0.1 MPa, CH_4/CO_2 = 1/1), modified after (Chen et al., 1999) (Teuner 1985)

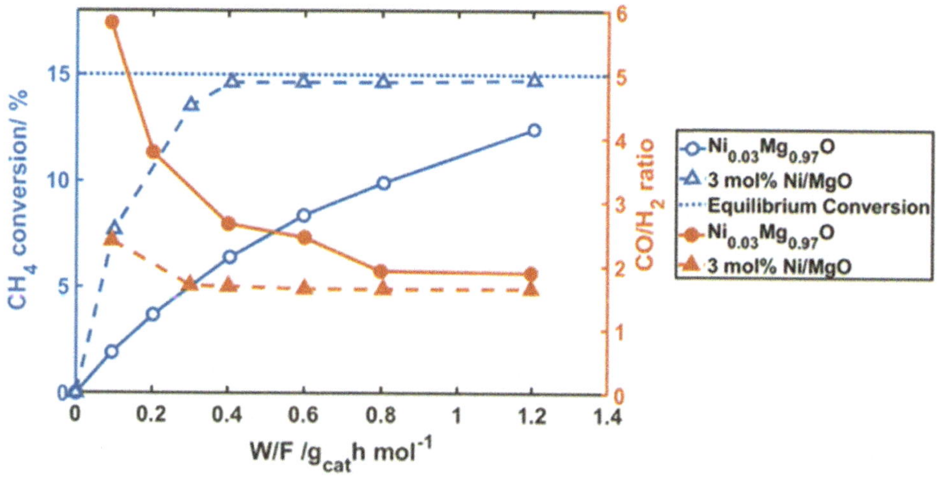

Many types of catalysts, like MgO, TiO_2, ZrO_2, Ti-MCM, CdS, ZnO, Tisilicalite molecular sieve, and $NiO/InTaO_4$, are evaluated for photocatalytic conversion of CO_2 (Chen et al. 1999; Kohno et al. 2001; Yamashita et al. 1998; Sasirekha et al. 2006; Lin et al. 2004; Sayama and Arakawa 1993). Among these catalysts, TiO_2 appears to be the best option for this process because of its chemical stability, electronic qualities, and commercial availability. Various studies have been performed on titanium catalysts, which are listed in Table 1. (Zhai et al. 2013).

Nevertheless, some disadvantages of this catalyst are led we have low efficiency of photocatalytic reduction of CO_2. Several researches have been done to improve the performance of the TiO_2 catalyst, including the metal oxides (Yui et al. 2011; Wang et al. 2013b), loading of metals (Varghese et al. 2009a; Diebold 2003; Wang et al. 2012), and so on. Using basic oxide in the TiO_2 catalyst for CO_2 photoreduction reaction has attracted the attention of many researchers. This additive destabilized the CO_2 molecule and promoted the CO_2 adsorption (Kočí et al. 2008; Fujiwara et al. 1997).

Gui et al. (2015) investigated the effect of Ag loadings in silver (Ag)-doped MWCNT@TiO_2 (multiwalled carbon nanotubes@titanium dioxide) core–shell nanocomposites in the photoreduction of CO_2. It discovered that at optimum loading of Ag (2 wt%), the highest yields of CH_4 (6.34 μmol/g-catalyst) and C_2H_4 (0.68 μmol/g-catalyst) were obtained (Figs. 7 and 8).

8 Biological Fixation

Biological fixation of CO_2 is an environmentally friendly technology to the conversion of CO_2 as undesirable gas to desirable products; as a result, reduction of global warming. In this method, CO_2 in the flue gas and atmosphere is converted into a variety of products. Various products such as biohydrogen, ethanol, and oil can be achieved from higher plants, microalgae, and photosynthetic bacteria. One of the feedstocks of CO_2 biological fixation to bio-oil generation is microalgae (Chisti 2007). Microalgae advantages compared to other plant feedstocks include high production rates of biomass, thriving ability in different ecosystems, high efficiencies in photosynthetic conversion, and the ability to produce biofuel feedstock in a wide variety of ways (Jansson and Northen 2010). The oil productivity and CO_2 capturing capacity of various kinds of plants are provided in Table 2 (Zeng et al. 2011).

The process steps of the microalgae biodiesel production system involve dewatering, harvesting, cultivation, transesterification, and extraction. The cultivation conditions, the cultivation system, and microalgae strain choice are the critical process considerations to achieve CO_2 fixation capacity and high oil yields during cultivation. Despite the various technologies for microalgae biotechnology, the optimum targets for industrial application are low CO_2 emission, high efficiency, and energy-saving technologies.

9 Conclusions

Carbon dioxide, one of the causes of global warming, has attracted the attention of many researchers in recent years. The conversion of this gas into valuable products such as light and heavy hydrocarbons has been the focus of scientists using various methods. The topics discussed in this chapter indicate efforts made to converting CO_2 into carbon derivatives by different green processes. These processes have two main advantages: (1) eliminating undesirable carbon dioxide gas and (2) production of valuable products sustainably. Due to the enormous impacts of CO_2 on earth life, many efforts have been performed to optimize and design new processes. In this chapter, we overviewed the

Table 1 Researches done on photocatalytic reduction of CO_2 over TiO_2 catalyst (Zhai et al. 2013)

Photocatalyst	Reductant	Light source	Primary products
$TiO_2/Pd/Al_2O_3$, TiO2/Pd/SiO_2, Li_2O-TiO_2/MgO, Li_2O-TiO_2/Al_2O_3	Water	250 mW Hg arc lamp	Acetone, ethanol, methanol, formaldehyde, formic acid, methane, ethane
TiO_2/zeolite, TiO_2/molecular sieve	Water vapor	High-pressure Hg lamp, $\lambda > 280$ nm	Methanol, methane
TiO_2nanocrystals in SiO_2matrices	Lithium nitrate and propan-2-ol	500 W high-pressure Hg arc lamp, $\lambda = 280$ nm	Formate, carbon monoxide, ammonia, urea,
TiO_2	1 M propan-2-ol solution	4.2 kW Xe lamp	Methane
TiO_2nanocrystals in SiO_2matrices	1 M propan-2-ol solution	500 W high-pressure Hg arc lamp	Formate, carbon monoxide
TiO_2	Liquid CO_2, water	990 W Xe lamp	Formic acid
TiO_2/zeolite	Water	75 W high-pressure Hg lamp, $\lambda > 280$ nm	Methane, methanol
TiO_2	Water, 0.2 M NaOH solution	4.5 kW Xe lamp	Formic acid, methanol, ethanol, methane, ethane, ethylene
TiO_2	Water	75 W high-pressure Hg lamp, $\lambda > 280$ nm	Methane, methanol, carbon monoxide
TiO_2	Water	75 W high-pressure Hg lamp, $\lambda > 280$ nm	Methane, methanol, carbon monoxide
Cu–TiO_2	Water	450 W Xe lamp	Methane, ethylene
TiO_2	Water	15 W UVC lamp, $\lambda =$ 365 nm	Methane, formic acid, ethanol
TiO_2	Water vapor	UVC lamp $\lambda = 253.7$ nm	Methane, carbon monoxide, hydrogen
TiO_2	Water vapor	3 W UVA lamp $\lambda =$ 365 nm, 4.8 W UVC lamp $\lambda = 254$ nm	Methane
TiO_2/SiO_2, Ru-TiO_2/SiO_2	Water	1000 W high-pressure Hg lamp, $\lambda = 365$ nm	Formic acid, formaldehyde, hydrogen, methane, methanol
TiO_2, Cu/TiO_2	5 M NaOH solution	Hg lamp, $\lambda = 365$ nm	Methanol
Sol–gel prepared Cu/TiO_2	0.2 N NaOH solution	Hg lamp, $\lambda = 254$ nm, 365 nm	Methanol
TiO_2, Cu/TiO_2	0.2 N NaOH solution	8 W Hg lamp, $\lambda = 254$ nm	Methanol, oxygen

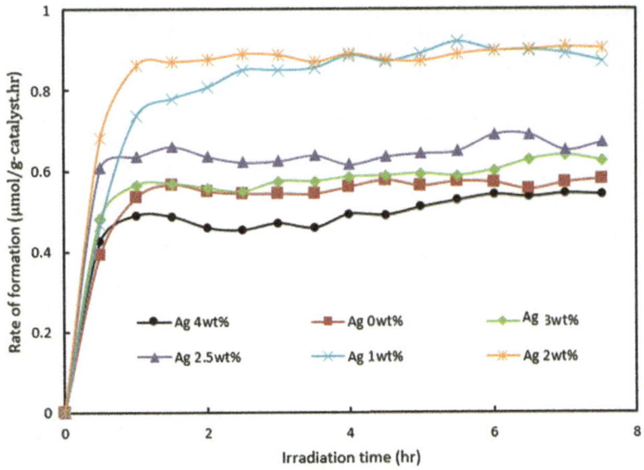

Fig. 7 Variation of CH_4 formation rate with irradiation time (various Ag loadings in MWCNT@TiO_2 modified after Direct) (Gui et al., 2015) (Yamashita et al. 1998)

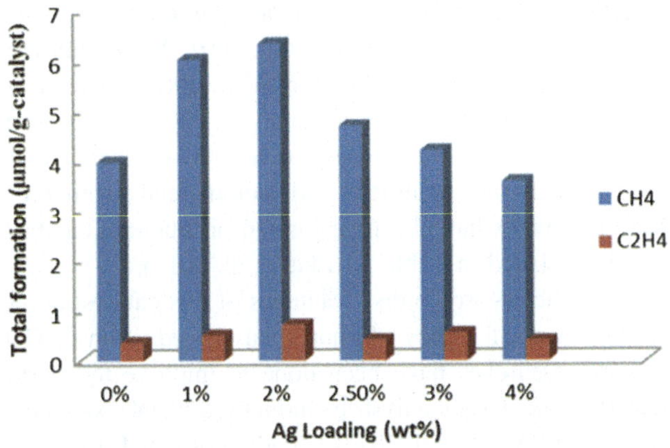

Fig. 8 Variation of CH_4 and C_2H_4 formation in photocatalytic reaction of CO_2 (various Ag loadings in Ag-MWCNT@TiO_2) (reaction duration = 7.5 h),modified after (Gui et al., 2015) (Yamashita et al. 1998)

challenges and problems of CO_2 reduction and converting to useful products; moreover, we introduced some solutions to overcome these problems. In summary, according to researches done, the green processes used in the CO_2 conversion should be optimized in terms of product selectivity and efficiency, carbon dioxide conversion rates, economic considerations, and so on. Summary of the processes examined in this chapter includes:

- The artificial photoreduction process of CO_2, which produces synthesis gas (CO/H_2), formic acid, and carbon monoxide (CO) depends on the metal photoelectrodes.
- Electrochemical reduction of CO_2 is another process for converting CO_2 to numerous products, including HCOOH, CO, light hydrocarbons, and alcohols that need improvement in thermodynamic efficiency.

Table 2 Amounts of CO_2 capturing capacity and biodiesel productivity of microalgae and higher plants (Zeng et al. 2011)

Plants	CO_2 capture capacity (%)[a]	Productivity (g/m^2 d)	Productivity of the theoretical max (%)	Lipid content of biomass (%)	Biodiesel productivity (g/m^2 d)[b]
Theoretical	100	77	100	-	N/A
Jatropha	N/A	2.2 (of seed)	2.8 (of seed)	30 (of seed)	0.66
Soya	N/A	0.65 (of seed)	0.8 (of seed)	20 (of seed)	0.2
Oilseed rape	1–2	0.8–1.6 (of seed)	1.0–1.2 (of seed)	40–44 (of seed)	0.24–0.48
Miscanthusgiganteus	10–22	8–17	11–22	N/A	2.4–5.4
Chlorella	22–26	17–20	22–27	40–60	5.4–6
Green microalgae	26–52	20–40	27–54	30–45	6–12

[a]In the theoretical view, to produce 77 g biomass, 3.6 mol (158.4 g) of CO_2 is captured by autophyte (Wang et al. 2013b)
[b] It is assumed that 30% of biomass is lipid (Yui et al. 2011)

- Hydrogenation of CO_2, a heterogeneous catalytic process produces various products such as dimethyl ether, methanol, methane, higher hydrocarbons, and alcohols. Hydrogen cost in this process is a challenge.
- Synthesis of organic carbonates from CO_2 is including dialkyl carbonates and dimethyl carbonate. One of the problems with this process is that metal oxides catalyst, which is active for the direct synthesis of carbonate from CO_2 and alcohol are very few.
- CO_2 reforming of hydrocarbons includes the reaction of hydrocarbons (C_nH_m) with CO_2 to produce synthesis gas. One of the limitations of this process is the catalyst of this process, which is under development.
- Photocatalytic reduction of CO_2 with water is another process that produces CO and CH_4 when electron donor of the process is water vapor and renewable and carbon–neutral fuels. A slight reduction of CO_2 by typical TiO_2 catalyst is one of the limitations of this process.
- Biological fixation process of CO_2 produces chemicals and biofuels. In this process, CO_2 is used as a C_1 source of microalgae for biofuels. Limitations of this process are: C_1 feedstock is not energy-dense and gas fermentation technology is under development.

Each of these methods has advantages and disadvantages that, according to environmental conditions and facilities, the best approach can be applied to reduce carbon dioxide and produce valuable green products.

References

Akahori Y, Iwanaga N, Kato Y, Hamamoto O, Ishii M (2004) New electrochemical process for CO_2 reduction to from formic acid from combustion flue gases. Electrochemistry 72:266–270. https://doi.org/10.5796/electrochemistry.72.266

Aresta M, Dibenedetto A (2002) Carbon dioxide as building block for the synthesis of organic carbonates: behavior of homogeneous and heterogeneous catalysts in the oxidative carboxylation of olefins. J Mol Catal a: Chem 182:399–409. https://doi.org/10.1016/S1381-1169(01)00514-3

Azuma M, Hashimoto K, Hiramoto M, Watanabe M, Sakata T (1990) Electrochemical reduction of carbon dioxide on various metal electrodes in low-temperature aqueous $KHCO_3$ media. J Electrochem Soc 137:1772–1778. https://doi.org/10.1149/1.2086796

Ballivet-Tkatchenko D, Douteau O, Stutzmann S (2000) Reactivity of carbon dioxide with n-butyl (phenoxy)-,(alkoxy)-, and (oxo) stannanes: insight into dimethyl carbonate synthesis. Organometallics 19:4563–4567. https://doi.org/10.1021/om000397f

Bhatia S, Thien CY, Mohamed AR (2009) Oxidative coupling of methane (OCM) in a catalytic membrane reactor and comparison of its performance with other catalytic reactors. Chem Eng J 148:525–532. https://doi.org/10.1016/j.cej.2009.01.008

Bonura G, Cordaro M, Cannilla C, Mezzapica A, Spadaro L, Arena F, Frusteri F (2014) Catalytic behaviour of a bifunctional system for the one step synthesis of DME by CO_2 hydrogenation. Catal Today 228:51–57. https://doi.org/10.1016/j.cattod.2013.11.017

Bradford MCJ, Vannice MA (1999) CO_2 reforming of CH_4. Catal Rev 41:1–42. https://doi.org/10.1081/CR-100101948

Chaplin RPS, Wragg AA (2003) Effects of process conditions and electrode material on reaction pathways for carbon dioxide electroreduction with particular reference to formate formation. J Appl Electrochem 33:1107–1123. https://doi.org/10.1023/B:JACH.0000004018.57792.b8

Chen CS, Cheng WH (2002) Study on the mechanism of CO formation in reverse water gas shift reaction over Cu/SiO_2 Catalyst by pulse reaction, TPD and TPR. Catal Let 83:121–126. https://doi.org/10.1023/A:1021006718974

Chen YG, Tomishige K, Yokoyama K, Fujimoto K (1999) Catalytic performance and catalyst structure of nickel–magnesia catalysts for CO_2 reforming of methane. J Catal 184:479–490. https://doi.org/10.1006/jcat.1999.2469

Chisti Y (2007) Biodiesel from microalgae. Biotechnol Adv 25:294–306. https://doi.org/10.1016/j.biotechadv.2007.02.001

Choi JC, He LN, Yasuda H, Sakakura T (2002) Selective and high yield synthesis of dimethyl carbonate directly from carbon dioxide and methanol. Green Chem 4:230–234. https://doi.org/10.1039/B200623P

Concepcion JJ, House RL, Papanikolas JM, Meyer TJ (2012) Chemical approaches to artificial photosynthesis. Proc Natl Acad Sci USA 109:15560–15564. https://doi.org/10.1073/pnas.1212254109

Da Silva DC, Letichevsky S, Borges LE, Appel LG (2015) Elimination of acetaldehyde from hydrogen rich streams employing Ni/ZrO_2. Int J Hydrogen Energy 40:8706–8712

Deleitenburg C, Trovarelli A (1995) Metal-support interactions in Rh/CeO_2, Rh/TiO_2, and Rh/Nb_2O_5 catalysts as inferred from CO_2 methanation activity. J Catal 156:171–174. https://doi.org/10.1006/jcat.1995.1244

Diebold U (2003) The surface science of titanium dioxide. Surf Sci Rep 48:53–229. https://doi.org/10.1016/S0167-5729(02)00100-0

Erdöhelyi A, Pásztor M, Solymosi F (1986) Catalytic hydrogenation of CO_2 over supported palladium. J Catal 98:166–177. https://doi.org/10.1016/0021-9517(86)90306-4

Finn C, Schnittger S, Yellowlees LJ, Love JB (2012) Planned maintenance close the message box. Chem Commun 48:1392–1399. https://doi.org/10.1039/C1CC15393E

Forghani AA, Elekaei H, Rahimpour MR (2009) Enhancement of gasoline production in a novel hydrogen-permselective membrane reactor in Fischer-Tropsch synthesis of GTL technology. Int J Hydrog Energy 34:3965–3976. https://doi.org/10.1016/j.ijhydene.2009.02.038

Fu Z, Yang Q, Liu Z, Chen F, Yao F, Xie T, Zhong Y, Wang D, Li J, Li X, Zeng G (2019) Photocatalytic conversion of carbon dioxide: From products to design the catalysts. J CO_2 Util 34: 63–73. https://doi.org/10.1016/j.jcou.2019.05.032

Fujishima A, Zhang X, Tryk DA (2008) TiO_2 photocatalysis and related surface phenomena. Surf Sci Rep 63:515–582. https://doi.org/10.1016/j.surfrep.2008.10.001

Fujiwara H, Hosokawa H, Murakoshi K, Wada Y, Yanagida S, Okada T, Kobayashi H (1997) Effect of surface structures on photocatalytic CO_2 reduction using quantized CdS nanocrystallites. J Phys Chem B 101:8270–8278. https://doi.org/10.1021/jp971621q

Furuya N, Yamazaki T, Shibata M (1997) High performance Ru-Pdcatalysts for CO_2reduction at gas-diffusion electrodes. J Electroanal Chem 431:39–41. https://doi.org/10.1016/S0022-0728(97)00159-9

Gadalla AM, Bower B (1988) The role of catalyst support on the activity of nickel for reforming methane with CO_2. Chem Eng Sci 43:3049–3062. https://doi.org/10.1016/0009-2509(88)80058-7

Gattrell M, Gupta N, Co A (2006) A review of the aqueous electrochemical reduction of $CO_2$to hydrocarbons at copper. J Electroanal Chem 594:1–19. https://doi.org/10.1016/j.jelechem.2006.05.013

Ghareghashi A, Ghader S, Hashemipour H (2013) Theoretical analysis of oxidative coupling of methane and Fischer Tropsch synthesis in two consecutive reactors: Comparison of fixed bed and membrane reactor. J Ind Eng Chem 19:1811–1826. https://doi.org/10.1016/j.jiec.2013.02.025

Ghareghashi A, Shahraki F, Razzaghi K, Ghader S, Torangi MA (2017) Enhancement of gasoline selectivity in combined reactor system consisting of steam reforming of methane and Fischer-Tropsch synthesis. Korean J Chem Eng 34:87–99. https://doi.org/10.1007/s11814-016-0242-z

Gui MM, Wong WMP, Chai SP, Mohamed AR (2015) One-pot synthesis of Ag-MWCNT@TiO_2 core–shell nanocomposites for photocatalytic reduction of CO_2 with water under visible light irradiation. ChemEng J 278:272–278. https://doi.org/10.1016/j.cej.2014.09.022

Hori Y, Murata A, Takahashi R (1989) Formation of hydrocarbons in the electrochemical reduction of carbon dioxide at a copper electrode in aqueous solution. J Chem Soc Faraday Trans 85:2309–2326. https://doi.org/10.1039/F19898502309

Hori Y, Wakebe H, Tsukamoto T, Koga O (1994) Electrocatalytic process of CO selectivity in electrochemical reduction of CO_2 at metal-electrodes in aqueous-media. Electrochim Acta 39:1833–1839. https://doi.org/10.1016/0013-4686(94)85172-7

Hu B, Suib SL (2014) Synthesis of useful compounds from CO_2. Green Carbon Dioxide: Adv CO_2 Utilization 51–97. https://doi.org/10.1002/9781118831922

Hwang S, Hong UG, Lee J, Seo JG, Baik JH, Koh DJ, Lim H, Song IK (2013) Methanation of carbon dioxide over mesoporous Ni–Fe–Al_2O_3 catalysts prepared by a coprecipitation method: Effect of precipitation agent. J Ind Eng Chem 19:2016–2021. https://doi.org/10.1016/j.jiec.2013.03.015

Ikeda S, Takagi T, Ito K (1987) Selective formation of formic- acid, oxalic-acid, and carbon-monoxide by electrochemical reduction of carbon-dioxide. Bull Chem Soc J 60:2517–2522. https://doi.org/10.1246/bcsj.60.2517

Indrakanti VP, Kubicki JD, Schobert HH (2009a) Photoinduced activation of CO_2 on Ti-based heterogeneous catalysts: Current state, chemical physics-based insights and outlook. Energy Environ Sci 2:745–758. https://doi.org/10.1039/B822176F

Indrakanti VP, Kubicki JD, Schobert HH (2009b) Photoinduced activation of CO_2 on Ti-based heterogeneous catalysts: current state, chemical physicsbased insights and outlook. Energy Environ Sci 2:745–758. https://doi.org/10.1039/B822176F

Inglis JL, MacLean BJ, Pryce MT, Vos JG (2012) Electrocatalytic pathways towards sustainable fuel production from water and CO_2. Coord Chem Rev 256:2571–2600. https://doi.org/10.1016/j.ccr.2012.05.002

Inoue T, Fujishima A, Konishi S, Honda K (1979) Photoelectrocatalytic reduction of carbon dioxide in aqueous suspensions of semiconductor powders. Nature 277:637–638. https://doi.org/10.1038/277637a0

Itoh N (1987) A membrane reactor using palladium. AIChE J 33:1576–1578. https://doi.org/10.1002/aic.690330921

Jansson C, Northen T (2010) Calcifying cyanobacteria-the potential of biomineralization for carbon capture and storage. Curr Opin Biotechnol 21:365–371. https://doi.org/10.1016/j.copbio.2010.03.017

Kaneco S, Katsumata H, Suzuki T, Ohta K (2006) Electrochemical reduction of CO_2 to methane at the Cu electrode in methanol with sodium supporting salts and its comparison with other alkaline salts. ENERG FUEL 20:409–414. https://doi.org/10.1021/ef050274d

Karelovic A, Ruiz P (2012) CO_2 hydrogenation at low temperature over Rh/γ-Al_2O_3 catalysts: Effect of the metal particle size on catalytic performances and reaction mechanism. Appl Catal B Environ 113:237–249. https://doi.org/10.1016/j.apcatb.2011.11.043

Karelovic A, Ruiz P (2013) Mechanistic study of low temperature CO_2 methanation over Rh/TiO_2 catalysts. J Catal 301:141–153. https://doi.org/10.1016/j.jcat.2013.02.009

Kitano M, Matsuoka M, Ueshima M, Anpo M (2007) Recent developments in titanium Oxide-based photocatalysts. ApplCatal A 325:1–14. https://doi.org/10.1016/j.apcata.2007.03.013

Kočí K, Obalová L, Lacný Z (2008) Photocatalytic reduction of CO_2 over TiO_2 based catalysts. Chem Papers 62:1–9. https://doi.org/10.2478/s11696-007-0072-x

Kohno Y, Ishikawa H, Tanaka T, Funabiki T, Yoshida S (2001) Photoreduction of carbon dioxide by hydrogen over magnesium oxide. PhysChem 3:1108–1113. https://doi.org/10.1039/b008887k

Leung K, Nielsen IM, Sai N, MedforthC SJA (2010) Cobalt–porphyrin catalyzed electrochemical reduction of carbon dioxide in water. 2. Mechanism from first principles. J Phys Chem A 114:10174–10184. https://doi.org/10.1021/jp1012335

Li H, Oloman C (2006) Development of a continuous reactor for the electro-reduction of carbon dioxide to formate–Part 1: Process variables. J Appl Electrochem 36:1105. https://doi.org/10.1007/s10800-006-9194-z

Li YN, Ma R, He LN, Diao ZF (2014) Homogeneous hydrogenation of carbon dioxide to methanol. Catal Sci Tech 4:1498–1512. https://doi.org/10.1039/C3CY00564J

Lin WY, Han HX, Frei H (2004) CO_2 splitting by H_2O to CO and O_2 under UV light in TiMCM-41 silicate sieve. J PhysChem B 108:18269–18273. https://doi.org/10.1021/jp040345u

Liu BJ, Torimoto T, Yoneyama H (1998) Photocatalytic reduction of carbon dioxide in the presence of nitrate using TiO_2 nanocrystal photocatalyst embedded in SiO_2 matrices. J Photochem Photobiol A 115:227–230. https://doi.org/10.1016/S1010-6030(98)00272-X

Lunde PJ, Kester FL (1974) Carbon dioxide methanation on a ruthenium catalyst. Ind Eng Chem Proc Design Dev 13:27–33. https://doi.org/10.1021/i260049a005

Ma J, Sun N, Zhang X, Zhao N, Xiao F, Wei W, Sun Y (2009) A short review of catalysis for CO_2 conversion. Catal Today 48:221–231. https://doi.org/10.1016/j.cattod.2009.08.015

Mahmood MN, Masheder D, Harty CJ (1987) Use of gas- diffusion electrodes for high-rate electrochemical reduction of carbon-dioxide. 1. Reduction at lead, indium- impregnated and tin-impregnated electrodes. J Appl Electrochem 17:1159–1170. https://doi.org/10.1007/BF01023599

Marvast MA, Sohrabi M, Zarrinpashne S, Baghmisheh G (2005) Fischer-Tropsch Synthesis: Modeling and performance study for Fe-HZSM5 bifunctional catalyst. Chem Eng Tech: Ind Chem-Plant Equip-Process Eng-Biotechnol 28:78–86. https://doi.org/10.1002/ceat.200407013

Montazer-Rahmati MM, Bargah-Soleimani M (2001) Rate equations for the Fischer-Tropsch reaction on a promoted iron catalyst. Can J Chem Eng 79:800–804. https://doi.org/10.1002/cjce.5450790515

Muthuraj R, Mekonnen T (2018) Recent progress in carbon dioxide (CO_2) as feedstock for sustainable materials development: Co-polymers and polymer blends. Polymer 145:348–373. https://doi.org/10.1016/j.polymer.2018.04.078

Nielsen IM, Leung K (2010) Cobalt− porphyrin catalyzed electrochemical reduction of carbon dioxide in water. 1. A density functional study of intermediates. J Phys Chem A 114:10166–10173. https://doi.org/10.1021/jp101180m

Osaki T, Mori T (2001) Role of potassium in carbon-free CO2 reforming of methane on K-promoted Ni/Al_2O_3 catalysts. J Catal 204(1):89–97. https://doi.org/10.1006/jcat.2001.3382

Oyola-Rivera O, Baltanás MA, Cardona-Martínez N (2015) CO_2 hydrogenation to methanol and dimethyl ether by $Pd–Pd_2Ga$ catalysts supported over Ga_2O_3 polymorphs. J CO_2 Util 9: 8–15. https://doi.org/10.1016/j.jcou.2014.11.003

Park JN, McFarland EW (2009) A highly dispersed $Pd–Mg/SiO_2$ catalyst active for methanation of CO_2. J Catal 266:92–97. https://doi.org/10.1016/j.jcat.2009.05.018

Rafiee A, Khalilpour KR, Milani D, Panahi M (2018) Trends in CO_2 conversion and utilization: A review from process systems perspective. J Environ Chem Eng 6:5771–5794. https://doi.org/10.1016/j.jece.2018.08.065

Roy SC, Varghese OK, Paulose M et al (2010) Toward solar fuels: photocatalytic conversion of carbon dioxide to hydrocarbons. ACS Nano 4:1259–1278. https://doi.org/10.1021/nn9015423

Sakai S, Fujinami T, Yamada T, Furusawa S (1975) Reaction of organotin alkoxides with carbon-disulfide, carbonyl sulfide of or carbon-dioxide C. Nippon Kagaku Kaishi 10:1789–1794

Sasirekha N, Basha SJS, Shanthi K (2006) Photocatalytic performance of Ru doped anatase mounted on silica for reduction of carbon dioxide. App Catal b: Environ 62:169–180. https://doi.org/10.1016/j.apcatb.2005.07.009

Sayama K, Arakawa H (1993) Photocatalytic decomposition of water and photocatalytic reduction of carbon-dioxide over zirconia catalyst. J PhysChem 97:531–533. https://doi.org/10.1021/j100105a001

Schild C, Wokaun A, Baiker A (1991) Surface species in CO_2 methanation over amorphous palladium/zirconia catalysts. J Mol Catal 69:347–357. https://doi.org/10.1016/0304-5102(91)80115-J

Schneider J, Jia H, Muckerman JT, Fujita E (2012) Thermodynamics and kinetics of CO_2, CO, and H^+ binding to the metal centre of CO_2 reduction catalysts. Chem Soc Rev 41:2036–2051. https://doi.org/10.1039/C1CS15278E

Sen S, Liu D, Palmore GTR (2014) Electrochemical reduction of CO_2 at copper nanofoams. Acs Catal 4:3091–3095. https://doi.org/10.1021/cs500522g

Song CS (2006) Global Challenges and strategies for control, conversion and utilization of CO_2 for sustainable development involving energy, catalysis, adsorption and chemical processing. Catal Today 115:2–32. https://doi.org/10.1016/j.cattod.2006.02.029

Sridhar N, Hill D, Agarwal A, Zhai Y, Hektor, E (2011) Carbon dioxide utilization. Electrochemical conversion of CO_2–opportunities and challenges. Research and Innovation, Position Paper, 7

Stagg SM, Resasco DE (1998) Effect of promoters on supported Pt catalysts for CO_2 reforming of CH_4. Studies in surface science and catalysis, 813–818.

Stagg-Williams SM, Noronha FB, Fendley G, Resasco DE (2000) CO_2 reforming of CH_4 over Pt/ZrO_2 catalysts promoted with La and Ce oxides. J Catal 194:240–249. https://doi.org/10.1006/jcat.2000.2939

Stansch Z, Mleczko L, Baerns M (1997) Comprehensive kinetics of oxidative coupling of methane over the La_2O_3/CaO catalyst. Ind Eng Chem Res 36:2568–2579. https://doi.org/10.1021/ie960562k

Subramanian K, Asokan K, Jeevarathinam D, Chandrasekaran M (2007) Electrochemical membrane reactor for the reduction of carbondioxide to formate. J Appl Electrochem 37:255–260. https://doi.org/10.1007/s10800-006-9252-6

Sutin N, Creutz C, Fujita E (1997) Photo-induced generation of dihydrogen and reduction of carbon dioxide using transition metal complexes. Comments Inorg Chem 19:67–92. https://doi.org/10.1080/02603599708032729

Teuner S (1985) Make CO from CO_2. Hydrocarbon Processing (international Ed.) 6:106–107

Tomishige K, Kunimori K (2002) Catalytic and direct synthesis of dimethyl carbonate starting from carbon dioxide using CeO_2-ZrO_2 solid solution heterogeneous catalyst: effect of H_2O removal from the reaction system. Appl Catal a: Gen 237:103–109. https://doi.org/10.1016/S0926-860X(02)00322-8

Tomishige K, Sakaihori T, Ikeda Y, Fujimoto K (1999) A novel method of direct synthesis of dimethyl carbonate from methanol and carbon dioxide catalyzed by zirconia. Catal Letters 58:225–229. https://doi.org/10.1023/A:1019098405444

Tomishige K, Ikeda Y, Sakaihori T, Fujimoto K (2000) Catalytic properties and structure of zirconia catalysts for direct synthesis of dimethyl carbonate from methanol and carbon dioxide. J Catal 192:355–362. https://doi.org/10.1006/jcat.2000.2854

Töpfer HJ (1976) Gas Erdgas. Gas Wasserfach Wasser Abwasser 117:412. https://doi.org/10.1007/s11356-016-6310-4

Udupa KS, Subramanian GS, Udupa HVK (1971) The electrolytic reduction of carbon dioxide to formic acid. Electrochim Acta 16:1593–1598. https://doi.org/10.1016/0013-4686(71)80028-2

Usubharatana P, McMartin D, Veawab A, Tontiwachwuthikul P (2006) Photocatalytic process for CO_2 emission reduction from industrial flue gas streams. Ind Eng Chem Res 45:2558–2568. https://doi.org/10.1021/ie0505763

Van Herwijnen T, Van Doesburg H, De Jong WA (1973) Kinetics of the methanation of CO and CO_2 on a nickel catalyst. J Catal 28:391–402. https://doi.org/10.1016/0021-9517(73)90132-2

Varghese OK, Paulose M, LaTempa TJ, Grimes CA (2009a) High-rate solar photocatalytic conversion of CO_2 and water vapor to hydrocarbon fuels. Nano Let 9:731–737. https://doi.org/10.1021/nl803258p

Varghese OK, Paulose M, LaTempa TJ, Grimes CA (2009b) High-Rate solar photocatalytic conversion of CO_2 and water vapor to hydrocarbon fuels. Nano Lett 9:731–737. https://doi.org/10.1021/nl803258p

Vessally E, Soleimani-Amiri S, Hosseinian A, Edjlali L, Babazadeh M (2017) Chemical fixation of CO_2 to 2-aminobenzonitriles: A straightforward route to quinazoline-2, 4 (1H, 3H)-diones with green and sustainable chemistry perspectives. J CO_2 Util 21:342–352. https://doi.org/10.1016/j.jcou.2017.08.006

Wang WN, An WJ, Ramalingam B, Mukherjee S, Niedzwiedzki DM, Gangopadhyay S, Biswas P (2012) Size and structure matter: enhanced CO_2 photoreduction efficiency by size-resolved ultrafine Pt nanoparticles on TiO_2 single crystals. J Am Chem Soc 134:11276–11281. https://doi.org/10.1021/ja304075b

Wang WN, Soulis J, Yang YJ, Biswas P (2013a) Comparison of CO_2 photoreduction systems: a review. Aerosol Air Qual Res 14:533–549. https://doi.org/10.4209/aaqr.2013.09.0283

Wang Y, Li B, Zhang C, Cui L, Kang S, Li X, Zhou L (2013b) Ordered mesoporous CeO_2-TiO_2 composites: Highly efficient photocatalysts for the reduction of CO_2 with H_2O under simulated solar irradiation. App Catal b: Environ 130:277–284. https://doi.org/10.1016/j.apcatb.2012.11.019

Wu J, Lin HM (2005) Photo reduction of CO_2 to methanol via TiO_2 photocatalyst. INT J Photoenergy 7:115–119. https://doi.org/10.1155/S1110662X05000176

Yahaya AH, Gondal MA, Hameed A (2004) Selective laser enhanced photocatalytic conversion of CO_2 into methanol. ChemPhys Lett 400:206–212. https://doi.org/10.1016/j.cplett.2004.10.109

Yamamoto T, Tryk DA, Fujishima A, Ohata H (2002) Production of syngas plus oxygen from CO_2 in a gas-diffusion electrode-based electrolytic cell. Electrochim Acta 47:3327–3334. https://doi.org/10.1016/S0013-4686(02)00253-0

Yamashita H, Fujii Y, Ichinashi Y, Zhang SG, Ikeue K, Park DR, Koyano K, Tatsumi T, Anpo M (1998) Selective formation of CH3OH in the photocatalytic reduction of CO_2 with H_2O on titanium oxides highly dispersed within zeolites and mesoporous molecular sieves. Catal Today 45:221–227. https://doi.org/10.1016/S0920-5861(98)00219-3

Yamazaki N, Nakahama S, Higashi F (1978) Rep. Asahi Glass Found. Ind Technol 33:31

Yoshida Y, Arai Y, Kado S, Kunimori K, Tomishige K (2006) Direct synthesis of organic carbonates from the reaction of CO_2 with methanol and ethanol over CeO_2 catalysts. Catal Today 115:95–101. https://doi.org/10.1016/j.cattod.2006.02.027

Yuan L, Han C, Pagliaro M, Xu YJ (2016) Origin of enhancing the photocatalytic performance of TiO_2 for artificial photoreduction of CO_2 through a SiO_2 coating strategy. J Phys Chem C 120:265–273. https://doi.org/10.1021/acs.jpcc.5b08893

Yui T, Kan A, Saitoh C, Koike K, Ibusuki T, Ishitani O (2011) Photochemical reduction of CO_2 using TiO_2: effects of organic adsorbates on TiO_2 and deposition of Pd onto TiO_2. ACS Appl Mater Inter 3:2594–2600. https://doi.org/10.1021/am200425y

Zeng X, Danquah MK, Chen XD, Lu Y (2011) Microalgae bioengineering: from CO_2 fixation to biofuel production. Renew Sustain Energy Rev 15:3252–3260. https://doi.org/10.1016/j.rser.2011.04.014

Zhai Q, Xie S, Fan W, Zhang Q, Wang Y, Deng W, Wang Y (2013) Photocatalytic conversion of carbon dioxide with water into methane: platinum and copper (I) oxide co-catalysts with a core–shell structure. Angew Chem 125:5888–5891. https://doi.org/10.1002/ange.201301473

Zhao T, Han Y, Sun Y (2000) Novel reaction route for dimethyl carbonate synthesis from CO_2 and methanol. Fuel Proc Tech 62:187–194. https://doi.org/10.1016/S0378-3820(99)00118-6

Zhou G, Wu T, Xie H, Zheng X (2013) Effects of structure on the carbon dioxide methanation performance of Co-based catalysts. Int J Hydrogen Energy 38:10012–10018. https://doi.org/10.1016/j.ijhydene.2013.05.130

Catalysis for CO_2 Conversion; Perovskite Based Catalysts

Osarieme Uyi Osazuwa and Sumaiya Zainal Abidin

Abstract

CO_2 conversion processes require the use of active catalysts due to the endothermicity of these reaction processes. The use of heterogeneous catalysts such as perovskites is gaining more attention due to their ease of separation, stability in handling, catalyst reuse and reproductively. Perovskite possesses wide industrial, scientific and commercial importance because of their cost effectiveness and stability in high temperature conditions while offering flexibility of its structures. CO_2 is effective as a reagent for the conversion of hydrocarbons into more useful products like H_2, syngas and liquid fuels. Researchers have focused extensively on CO_2 reforming of methane (CRM) as the process mitigates two readily available greenhouse gases. However, the overall conversion in this process is dependent on factors such as amount/ratio of reactants, reaction conditions and type of catalysts used. Majorly, perovskites have been applied as catalysts for CO_2 reforming because they possess highly mobile oxygen molecules. This property is essential to suppress the formation of carbon during the reaction. Moreover, its stable nature in a reducing environment gives it an edge over other types of catalysts used for CO_2 reforming. Hence, the focus of this chapter is on the role played by perovskites as catalysts in the CO_2 reforming of methane reaction process.

O. U. Osazuwa · S. Z. Abidin (✉)
Department of Chemical Engineering, College of Engineering, Universiti Malaysia Pahang, Lebuhraya Tun Razak, Gambang, 26300 Kuantan, Pahang, Malaysia
e-mail: sumaiya@ump.edu.my

O. U. Osazuwa · S. Z. Abidin
Centre for Research in Advanced Fluid & Processes (FLUID CENTRE), Universiti Malaysia Pahang, Lebuhraya Tun Razak, Gambang, 26300 Kuantan, Pahang, Malaysia

O. U. Osazuwa
Department of Chemical Engineering, Faculty of Engineering, University of Benin, PMB, Benin City, 1154, Edo State, Nigeria

Keywords

CO_2 conversion • Catalysis • Methane • Perovskite • Reforming

1 Introduction

Reactions utilizing CO_2 as a reagent are effective in mitigating and managing the carbon cycle. CO_2 is a thermodynamically stable compound, hence, its partial and complete reduction to its unstable state (CO) or totally reduced state (C and O) requires high energy usage (Ma et al. 2009). The inert and low reactive nature of CO_2 in different reactions is one of the major reasons for its endothermicity, hence the need for catalysis in CO_2 conversion processes. This type of catalyst applied in CO_2 conversion processes can be homogeneous or heterogeneous in nature. For homogeneous catalysts in CO_2 conversion processes, several studies have been reported over time (Omae 2006; Zhang et al. 2006). The production of carbamates, carbonates, lactones, urethanes, pyrones, formic acid and its derivatives are some of the processes in which homogeneous catalysts have been employed in CO_2 conversion processes. Whereas, heterogeneous catalyst offers several merits such as: ease of separation, stability in handling, catalyst reuse, reactor design and enhanced reaction yield. These kinds of catalysts (heterogeneous catalysts) are basically the main solutions for effective cost management and sustained industrial application of conversion processes. Therefore, the journey to discover a stable and efficient heterogeneous catalyst for reaction processes involving CO_2 as a reactant remains an emerging research area.

The use of metals as catalysts offers a likely solution to heterogeneous catalysis. Also, the combination of the oxides of various metal types to form active metallic catalysts called perovskite has also been extensively reported in literature as solutions to catalysis (Pakhare et al. 2012; Osazuwa et al.

Inamuddin et al. (eds.), *Carbon Dioxide Utilization to Sustainable Energy and Fuels*,
Advances in Science, Technology & Innovation, https://doi.org/10.1007/978-3-030-72877-9_15

2017). Perovskites have the general formula ABO_3 or A_2BO_4. "A" stands for rare earth metals (Ce and La) or alkaline earth metals (Ca, Ba and Mg), while "B" represents transition metals (Co, Ni and Fe) (Mishra and Prasad 2014; Zhu et al. 2014). According to Screen (2007), naturally or synthetically occurring perovskite have an expanded range of industrial, scientific and commercial importance. This importance is attributable to its stability in extreme conditions, cost effectiveness and flexibility of its stoichiometric parameters (Labhasetwar et al. 2015). In addition, multi-functional properties of the perovskite resulting in high catalytic activity are drawn from the perovskite's ability to exist with O_2 in a non-stoichiometric form.

The structure of the perovskite is cubic in nature forming oxygen atoms with an octahedral configuration around the "B" atom. The "A" and "B" site ions are varied to allow for alterations to the perovskite's structure in line with the properties of "A" and "B" (Screen 2007; Moradi et al. 2014). The stability and activity of perovskites' are enhanced by the A and B cationic sites, respectively (Mishra and Prasad 2014; Lombardo and a, Ulla M a. 1998). According to Zhu et al. (2014), about 90% of metals in the periodic table can form perovskites because they easily occupy the sites in the perovskite's matrix thereby forming a range of crystallite shapes with varied stoichiometry. The physicochemical properties of the perovskites are strongly linked to its preparation methods. Several preparation procedures have been described to synthesize perovskites. They include; impregnation, co-preparation, citrate sol gel, Pechini's method, solid state reaction and combustion synthesis method, etc. In a typical example, Labhasetwar et al. (2015) used co-precipitation, citrate sol gel and solution combustion methods to prepare perovskite for catalytic applications. In the study, combustion synthesis method was described for the preparation of perovskite using the ignition of combustible constituents which serves as fuel for the process. Also, Shabbir, Qureshi, and Saeed K. (2006) used sol gel synthesis method to synthesize $LaFeO_3$ by thermally decomposing the gel mixture.

Since heterogeneous catalytic reaction occurs at the surface, it is essential to ensure that the catalytic surface is as large as possible in order to promote adequate interaction between the catalytic material and the reactants (Sheng et al. 2017). Therefore, perovskite's that possess high surface area is desirable for catalytic reaction studies and application. Lower calcination temperatures are utilized to enhance the porosity, surface area and particle size of the perovskite structure (Labhasetwar et al. 2015).

Several catalysts have been applied in CO_2 conversion processes ranging from noble metal catalyst (Ghelamallah and Granger 2014), transition metal catalyst (Jiang et al. 2019), Alumina support catalyst (Ryi et al. 2014), Zeolite supported catalyst (Chang et al. 1994), lanthanide series promoted catalyst (Ayodele et al. 2016), alkali earth metals promoted catalyst (Yang et al. 2015) and perovskite based catalyst (Osazuwa and Cheng 2017). The use of various catalysts comes with its merits and demerits. Perovskites have been used as reforming catalysts due to their high oxygen mobility (Mota et al. 2012; Nalbandian et al. 2011; Murugan et al. 2011). Also, the A/B sites of the perovskite can be altered and partially substituted with cations in order to change the electronic configuration and structural buildup. Another property of the perovskite that makes it an ideal catalyst for conversion and reforming processes is that it is thermally and chemically stable in a redox and H_2 rich environment. Perovskites have high oxygen mobility essential for catalytic/reforming processes. This property is very useful in inhibiting/resisting carbon formation during reaction which can lead to catalyst poisoning (Sfeir et al. 2001; Rida et al. 2007; Doggali et al. 2010). However, the ideal environment for utilizing perovskite is still unclear in the production of H_2 from methane. Due to its unique properties, we will focus on perovskite's as catalysts for carbon dioxide (CO_2) and methane (CH_4) conversion.

CO_2 is effective as a reagent for converting hydrocarbons into valuable products like H_2, syngas and liquid fuels. The most common CO_2—hydrocarbon conversion process is the CO_2 reforming of CH_4 (CRM). However, the extent of conversion of this process is determined by several parameters such as the amount of reactants used, reaction operating condition and type of catalyst utilized. In this chapter, we focus on the effects of the catalyst type (perovskite) on CO_2 conversion of CH_4 reaction. The degree to which perovskites affect this reaction is the main focus of this chapter.

2 Perovskite for CO_2 Reforming of Methane (CRM)

2.1 Combined Reforming for Methane

Several studies have been reported using perovskite as catalyst for CRM. For instance, García et al. (2011) used $LaNiO_3$ perovskite oxide for combined (CO_2–O_2) CH_4 reforming at 650–850 °C. Results showed that CO_2 conversion was highest at 750 °C due to the positive effects of the metallic structure of the perovskite. Highest CO_2 conversion of 61% was observed at CH_4/CO_2 ratio of 4 and temperature of 750 °C. Carbon was formed during the reaction but without deactivation of the catalyst because the carbon formed were elongated filaments that did not block the catalyst pores (Rynkowski et al. 2004). The stability test carried out for 52 h showed a steady state value of 60% CO_2 conversion after about 24 h. The stability was ascribed to $La_2O_2CO_3$ formation in-situ reaction (Slagtern et al. 1997). During the reaction, the ability of the perovskite to be

Table 1 Combined reforming for methane

Process	Perovskite catalyst	Conditions (°C)	Results / Observation	References
Combined (CO_2-O_2) CH_4 reforming	$LaNiO_3$	650–850	• CO_2 conversion was highest at 750 °C. • Highest CO_2 conversion of 61% was observed at CH_4/CO_2 ratio of 4 and temperature of 750 °C.	(García et al. 2011)
CO_2 and steam reforming of CH_4	$LaNiO_3$ substituted with various amounts of Ce	600–900	• $La_{0.9}Ce_{0.1}NiO_3$ catalyst was more active. • CH_4 and CO_2 conversion was 49% and 14%, respectively. • Ce substitution enhanced Ni reduction.	(Yang et al. 2015)
Combined reforming of methane (using O_2 and CO_2)	$LaFe_{1-x}Co_xO_3$	750–850	• Substituting the perovskites with Co favoured the reduction of Fe and enhanced catalytic activity. • Reduced amount of carbon formation.	(Goldwasser et al. 2005)
CO_2 Reforming of Methane (CRM)	$La_{1-x}Sm_xNiO_{3-\delta}$, $x = 0$, 0.1, 0.3, 0.5, 0.7, 0.9 and 1	various reaction temperature up to 800	• $La_{1-x}Sm_xNiO_{3-\delta}$ where $x = 0.1$, 0.9 and 1 had better performance.	(Jahangiri et al. 2012)
CRM	$SmCoO_3$	800	• Reactants conversion of over 90% • C–C bonds and O–C = O bonds in samples of the spent catalyst	(Osazuwa et al. 2018)

reduced to uniformly dispersed Ni, La_2O_3 and La (OH) structures were linked to the perovskite's performance in the combined reforming process (Maneerung et al. 2011).

CO_2 and steam reforming of CH_4 at 600–900 °C was studied by Yang et al. (2015) using $LaNiO_3$ substituted with various amounts of Ce. Amongst all the synthesized catalysts, $La_{0.9}Ce_{0.1}NiO_3$ catalyst was more active with CH_4 and CO_2 conversion of 49 and 14%. Ce was present as CeO_2 and not incorporated into the perovskite structure. This was as a result of the differences in ionic radii between La^{3+} and Ce^{4+} (Krishna et al. 2007). When little amount of Ce was used to substitute the perovskite (at the A site), higher amounts of Ni species were reduced resulting in better catalyst performance. Whereas, higher substitution of Ce resulted in CO_2 methanation accompanied by formation of large Ni cluster which negatively affected the catalytic performance. The Ce substituted perovskite resisted carbon better than the non-promoted perovskite.

Combined reforming of methane (using O_2 and CO_2) using $LaFe_{1-x}Co_xO_3$ perovksite as catalyst precursors was carried out at temperatures ranged between 750 and 850 °C by Goldwasser et al. (2005). The study revealed that substituting the perovskites with Co substantially favoured the reduction of Fe. This led to strong Fe-Co interaction and oxygen species variations on the perovskites surface. Hence, oxidation and CRM carried out simultaneously reduced carbon formation drastically and improved syngas production. It has been established that when O_2 is added to the reaction process the reaction temperature is reduced and energy consumption is minimized, thereby promoting the CRM reaction (Tomishige et al. 2002; Tomishige 2004). Furthermore, substituting Fe with Co enhanced catalytic activity. Dispersed Co particles on Fe decreases metal sintering and formation of carbon. Also, CH_4 combustion is decreased while conversion of CO_2 increased.

Perovskite catalyst; $La_{1-x}Sm_xNiO_{3-\delta}$ with $x = 0$, 0.1, 0.3, 0.5, 0.7, 0.9 and 1 were applied as catalyst for CRM at various reaction temperatures up to 800 °C by Jahangiri et al. (2012). The results showed that the catalyst; $La_{1-x}Sm_xNiO_{3-\delta}$ where $x = 0.1$, 0.9 and 1 had better performance than others. Substituting La by Sm (at $x = 0.1$, 0.9 and 1) restructured the catalyst surface and also improved catalyst activity towards reactants conversion. Hence, the reaction could be carried out at lower temperatures with reduced energy consumption. The increased activity can also be related to the uniform dispersion of Ni ions on the catalyst surface which hinders carbon deposition on the active sites (Zhang et al. 2015).

Also, $SmCoO_3$ perovskite was synthesized and applied as catalyst in a 72 h CRM test at 800 °C by Osazuwa et al. (2018). In the study, feed ratios were varied between 0.5 and 2.0 to obtain reactants conversion of over 90% at equimolar feed ratio. Characterization of the spent catalysts revealed the existence of an oxycarbonate phase required to convert deposited carbon. Further analysis showed the presence of C–C bonds and O–C = O bonds in samples of the spent catalyst. The carbon species at 0.5–1.33 reactants ratio were graphitic in nature with C–H and C–C bonds. Also, the contaminant of carbon with O–C = O bonds were found in the spent catalyst with 2.0 reactant ratio. Table 1 summarizes the studies carried out using combined reforming processes.

2.2 "A"-Site Substituted Perovskite Catalyst for CRM

$La_{1-x}Sr_xCoO_3$ perovskite catalysts precursors have been used for CRM by Valderrama et al. (2013). The reaction took place at 800 °C under atmospheric pressure. The reduction process of the catalyst precursors produces metallic Co, SrO and La_2O_3 phases. CO_2 and La_2O_3 interaction yields $La_2O_2CO_3$-SrO matrix responsible for suppressing the formation of carbon. Sr in doping quantity induces catalyst reduction whereas; higher amount of Sr promotes side reactions thereby suppressing the CRM reaction partially. Hence, the quantity of the dopant has to be in traces and strictly monitored.

"A"-site substitution of lanthanum manganite using barium (Ba) in CRM at 800 °C was reported by Bhavani et al. (2013). Results showed that the optimal level of Ba substitution in the perovskite was Ba/Mn = 0.10, 0.15. The enhanced catalysts created a better reducing environment for the Mn ion by providing oxygen vacancies and rapid lattice oxygen migration (Sazonova et al. 2009; Caprariis et al. 2016). The ability to easily supply oxygen (O_2) to the reaction promoted the cracking and dissociation of CH_4 and CO_2, respectively. Hence, there was less carbon deposition which led to increased conversions of the reactants and higher selectivity towards the product (syngas).

The substitution of the A cationic site of Ru perovskites; $AZrRuO_3$ (A = Ba, Sr, Ca) in relation to the influence of synthesis methods have been investigated in CRM by Ruocco et al. (2019). The reactions took place at temperature range 600–750 °C, while various synthesis methods; modified citrate and auto combustion methods were employed to prepare the catalysts. Results revealed that catalysts prepared by the auto combustion method favoured higher surface area and enhanced reducibility resulting in better performance (Pradeep and Chandrasekaran 2006). For the substitution study, $CaZrRuO_3$ had the lowest CH_4 conversion (55%) at 700 °C. This catalyst showed a change of phase from cubic to orthorhombic as a result of its small dimension resulting in a cubic structured perovskite not properly formed. $BaZrRuO_3$ and $SrZrRuO_3$ had similar behaviour with the latter being the best performing catalyst at 750 °C, resulting in 90% of CH_4 converted with much superior stability.

The effects of promoting $LaNi_{0.5}Fe_{0.5}O_3$ with Ce in CRM at 750 °C was investigated by Wang et al. (2018). When some amount of La was substituted with Ce the catalyst's activity was enhanced in CRM. Analyzing the spent catalyst, it was suggested that Ni phase was responsible for the primary catalyst activity with $(LaCe)(NiFe)O_3$ enhancing the catalytic activity. Ce^{3+} ion in $(LaCe)(NiFe)O_3$ supplies additional oxygen vacancies by activating ions in the "B" site. Also, Ce^{3+}/Ce^{4+} present in $(LaCe)(NiFe)O_3$ and CeO_2 respectively, can exist in the redox process which enhances the catalytic activity. Ce solubility differences in the perovksite provide the redox environment required for driving the exchange of Ce ion between CeO_2 and the perovskite.

$Na_{0.5}La_{0.5}Ni_{0.3}Al_{0.7}O_{2.5}$ was applied for CRM by Pérez-Camacho et al. (2014) at 610–880 °C. Carbon formation was highest at 800 °C; however, at 880 °C, the reverse water gas shift reaction was dominant thereby suppressing the graphitic carbon formation rate. Several O_2 carrying promoters were used to promote the catalyst using different preparation techniques. Amongst all, CeO_2 and ZrO_2 decreased the formation of carbon without altering the catalyst activity in CRM. The sample synthesized when the perovskite gel was mixed before drying had the lowest carbon formed at 800 °C. At 880 °C, the non-promoted perovksite resisted carbon more. Other promoted catalysts showed no significant improvement even at higher temperatures of calcination. This behaviour is linked to the reactivity of the La compound with the promoter thereby forming layers of resistance.

Su et al. (2014) synthesized $LaNiO_3$ and $La_{1-x}Ce_xNiO_3$ for CRM at 600 and 800 °C, respectively. CH_4 and CO_2 conversions increased from 66 and 51% at 600 °C to 94% and 92% at 800 °C, respectively for the perovskite catalyst. Also, the need to minimize carbon deposition and metallic particle sintering gave rise to the incorporation of Ce. Ce supplies abundant lattice oxygen vacancies to break C-H bonds (Akri et al. 2016). The adsorption of CO_2 on La_2O_3 led to the formation of the oxycarbonate species; $La_2O_2CO_3$. This species stabilizes the Ni particles thereby avoiding sintering of metal particles (Batiot-Dupeyrat et al. 2005). However, lower amount of Ce in the perovskite translates to better catalyst activity and suppression of carbon. Therefore, $La_{1-x}Ce_xNiO_3$ (X = 0.1) was highly active in the CRM reaction.

Steam and CRM over $La_{1-x}Sr_xNiO_3$ (x = 0–0.5) perovskite have been investigated by Yang et al. (2018). The effect of adding Sr to $LaNiO_3$ perovskite was studied at 650–900 °C and 1 bar. Results revealed that after substitution with Sr, $La_{2-x}Sr_xNiO_{3\pm\delta}$ and $Sr_{0.5}La_{1.5}NiO_4$ phases were formed. The formations of these phases were a result of the differences in the ionic radii of Sr^{2+} and La^{3+}, thereby resulting in the encapsulation of $La_2O_2CO_3$ phase during reaction and the reduction in the metal/support interaction. Hence, an increase in Sr concentration resulted in larger Ni particle size. Furthermore, although the activity of $La_{1-x}Sr_xNiO_3$ (when x > 0) was lower than the activity of $La_{1-x}Sr_xNiO_3$ (when x = 0), their ability to resist carbon was

Table 2 "A"-Site substituted perovskite catalyst for CRM

Process	Perovskite catalyst	Conditions (°C)	Results/Observation	References
CRM	$La_{1-x}Sr_xCoO_3$	800	• The quantity of the dopant must be in traces • CO_2 and La_2O_3 interaction yields the $La_2O_2CO_3$-SrO matrix responsible for suppressing the formation of carbon	(Valderrama et al. 2013)
CRM	Lanthanum manganite using barium (Ba)	800	• Optimal level of Ba substitution in the perovskite was Ba/Mn = 0.10, 0.15. • Less carbon deposition which led to increased conversions of the reactants and higher selectivity towards the product (syngas).	(Bhavani et al. 2013)
CRM (influence of synthesis methods)	Ru perovskites; $AZrRuO_3$ (A = Ba, Sr, Ca)	600–750	• Auto combustion method favoured higher surface area and enhanced reducibility resulting in better performance. • $CaZrRuO_3$ had the lowest CH_4 conversion (55%) at 700 °C. • $SrZrRuO_3$ was the best performing catalyst at 750 °C, resulting in 90% of CH_4 converted with much superior stability	(Ruocco et al. 2019)
CRM	$LaNi_{0.5}Fe_{0.5}$	750	• (LaCe)(NiFe)O_3 enhanced the catalytic activity. • Ce solubility differences in the perovksite provide the redox environment for driving the exchange of Ce ion between CeO_2 and the perovskite.	(Wang et al. 2018)
CRM	$Na_{0.5}La_{0.5}Ni_{0.3}Al_{0.7}O_{2.5}$	610–880	• 880 °C, the reverse water gas shift reaction was dominant thereby suppressing the graphitic carbon formation rate. • CeO_2 and ZrO_2 decreased the formation of carbon without altering the catalyst activity in CRM.	(Pérez-Camacho et al. 2014)
CRM	$LaNiO_3$ and $La_{1-x}Ce_xNiO_3$	600 and 800	• CH_4 and CO_2 conversions increased from 66 and 51% at 600 °C to 94% and 92% at 800 °C, respectively for the perovskite catalyst. • Ce supplied abundant lattice oxygen vacancies to break C–H bonds. • Lower amount of Ce in the perovskite translates to better catalyst activity and suppression of carbon	(Su et al. 2014)
Steam reforming of CH_4 and CRM (Effect of adding Sr to $LaNiO_3$)	$La_{1-x}Sr_xNiO_3$ (x = 0–0.5)	650–900	• $La_{2-x}Sr_xNiO_{3\pm\delta}$ and $Sr_{0.5}La_{1.5}NiO_4$ phases were formed. • An increase in Sr concentration resulted in larger Ni particle size. • The promoted perovskite had better performance with regards to CH_4 and CO_2 activation. • Traces of Sr were required to resist carbon deposition during CRM reaction.	(Yang and hyeok, Noh YS, Hong GH, Moon DJ. 2018)

significantly higher. Hence, the promoted perovskite had better performance with regards to CH_4 and CO_2 activation. It can be deduced that the carbon gasification rate by CO_2 was greater than or equal to the rate of CH_4 decomposition. Therefore, only traces of Sr were required to resist carbon deposition during CRM reaction. Table 2 summarizes studies on "A" site substituted perovskite in CRM processes.

2.3 "B"-Site Substituted Perovskite Catalyst for CRM

According to Screen (2007) and Moradi et al. (2014) substituting the perovskite's active sites can enhance catalyst performance in the CRM reaction. For instance, in the study by Moradi et al. (2014), $LaNi_{1-x}Zn_xO_3$ with values of x from

0.2 to 1.0 were prepared and used for CRM at 750 °C. As Ni was partially substituted up to x = 0.4, conversion of CO_2 and CH_4 increased. In terms of reactants conversion, $LaNi_{0.8}Zn_{0.2}O_3$ was most effective in the reforming process. Substituting Ni by Zn resulted in increased reduction peaks; hence, the perovskite was more stable when compared to $LaNiO_3$. Also, $LaNi_{0.8}Zn_{0.2}O_3$ catalyst was more effective than $LaNiO_3$ towards resisting the formation of carbon species in a 75 h time of stream (TOS) test. The catalyst characterization showed breakdown of the catalyst structure into metallic species (Ni°) and La_2O_3 during the reduction step. These dual phases were responsible for the perovskite resistance to deactivation. Partial substitution of Ni by Zn made reducibility more difficult by altering the reduction temperature, stabilizing the perovskite structure and limiting the migration of active Ni.

Yadav and Das (2019) synthesized $LaNi_xFe_{1-x}O_3$ perovskite using various methods and applied it for CRM at 700 °C. Catalyst analysis showed that the addition of silica support completely changed the perovskite features of the catalyst at high calcination temperature. 40 wt % $laNi_{0.75}Fe_{0.25}O_3/SiO_2$ calcined at 700 °C and 800 °C were more active when equal amount of reactants were used. Results from the study corroborated with the study by Moradi et al. (2014) proved that Ni to Fe ratio, synthesis method and temperature of calcination were important parameters for designing and synthesizing perovskites for CRM.

Valderrama et al. (2018) synthesized $LaNi_{1-x}Mn_xO_3$ (x = 0–1.0) for CRM at 600–750 °C. The addition of Mn to the perovskite increased the temperature at which Ni^{3+} reduced to Ni^0. Hence, Ni particles were dispersed on MnO_x-LaO_3 surface. With $x \leq 0.8$, the perovskite's performed better in the CRM reaction. Therefore, the activity increased and resistance to the formation of carbon was enhanced (Abdulrasheed et al. 2019). Also, the perovskite possessed high O_2 storing capacity enabling in-situ formation of $La_2O_2CO_3$-MnOx with CO_2 which was vital in the elimination of sintering and carbon formation (Bitter et al. 1998).

Touahra et al. (2019) reported the effect of Cu-Co perovskite on CRM at temperatures ranging between 400 and 700 °C. $LaCoO_3$ and $LaCu_{0.55}Co_{0.45}O_3$ were synthesized and tested in CRM reaction. Results from the catalytic test and characterization proved that $LaCu_{0.55}Co_{0.45}O_3$ perovskite showed better catalytic behaviour than $LaCoO_3$. The smaller particle size of Co and the formation of metallic Cu-Co during the catalyst reduction stage were responsible for the outcome. Hence, the formations of CoO and Co sintering were inhibited. Carbon responsible for blocking the Co active sites were eliminated by species of oxygen present in $La_2O_2CO_3$ formed.

The synthesis and catalytic test of Cu/Ni/$La_{0.7}Sr_{0.3}Cr_{0.5}Mn_{0.5}O_{3-\delta}$ (LSCM) for CRM have been carried out by Kang et al. (2019) at 550/650 °C and atmospheric pressure. Results from the study revealed that as the conversion of reactants increased, the resistance against sintering also increased, hence allowing the carbon deposition to drop with the introduction of Cu. Furthermore, the introduction of Cu enhanced the crystal size of Ni, reduced the size of the carbon species and the amount of deposited carbon species (Touahra et al. 2016a). After 13 h of reaction, results also showed that the carbon species in the sample; Cu/Ni-LSCM were smaller in size and quantity than the ones present in sample Ni-LSCM.

The effects of Fe substitution in $LaNi_{1-x}Fe_xO_3$ (x = 0.2–1.0) were studied and used for CRM at 600–800 °C by Jahangiri et al. (2013). Results showed that addition of Fe to the B site of the perovskite did not favour metal reduction. $LaNiO_3$ gave the highest activity when compared to Fe promoted perovskites. The reduction of the perovskite to Ni and carbonate phase during the reaction was responsible for enhanced activity due to increased CH_4 oxidation and removal of carbon species (Valderrama et al. 2005). The synthesized perovskite's activity were in the order; $LaNiO_3 > LaNi_{0.4}Fe_{0.6}O_3 > LaNi_{0.6}Fe_{0.4}O_3 > LaNi_{0.8}Fe_{0.2}O_3 > LaNi_{0.2}Fe_{0.8}O_3 > LaFeO_3$. Table 3 represents a summary of studies on "B" site substituted perovskite catalyst for CRM.

2.4 Supported Perovskite for CRM

Rabelo-Neto et al. (2018) studied the activity of $LaNiO_3$ perovskite on Al_2O_3 and $CeSiO_2$ supports for CRM. In the stability study carried out at 800 °C for 24 h with CH_4/CO_2 = 1.0, initial CO_2 conversion for the $LaNiO_3$ was 35% which increased steadily. CO_2 conversion for the $CeSiO_2$ supported catalyst was about 65% in the first 6 h, while the Al_2O_3 supported catalyst had about 60% conversion in the first 4 h before increasing to about 70%. However, over 24 h TOS, the $CeSiO_2$ supported catalyst was more stable in terms of performance and conversion. Analysis showed that CO_2 oxidized the Ni^0 in the unsupported and Al_2O_3 supported catalysts. The Ce supported catalyst was oxidized preferentially thereby limiting the metallic phase oxidation. The unsupported and Al_2O_3 supported perovskite catalyst had the highest amount of carbon deposited on the surface while the $CeSiO_2$ supported peorvskite inhibited carbon

Table 3 "B"-Site substituted perovskite catalyst for CRM

Process	Perovskite catalyst	Conditions (°C)	Results/Observation	References
CRM	$LaNi_{1-x}Zn_xO_3$, $x = 0.2–1.0$	750	• $LaNi_{0.8}Zn_{0.2}O_3$ was most effective in the CRM. • $LaNi_{0.8}Zn_{0.2}O_3$ catalyst was more effective than $LaNiO_3$ towards resisting the formation of carbon	(Moradi et al. 2014)
CRM	$LaNi_xFe_{1-x}O_3$	700	• Addition of silica support completely changed the perovskite features. • Synthesis method and temperature of calcination were important catalyst parameters for CRM.	(Yadav and Das 2019)
CRM	$LaNi_{1-x}Mn_xO_3$ ($x = 0–1.0$)	600–750	• Addition of Mn to the perovskite increased the temperature at which Ni^{3+} reduced to Ni^0. • Perovskite performance improved when $x \leq 0.8$.	(Valderrama et al. 2018)
CRM	$LaCoO_3$ and $LaCu_{0.55}Co_{0.45}O_3$	400–700	• $LaCu_{0.55}Co_{0.45}O_3$ perovskite showed better catalytic behaviour than $LaCoO_3$. • The formations of CoO and Co sintering were inhibited. • The formation of metallic Cu-Co during the catalyst reduction stage was responsible for the improved catalytic behaviour.	(Touahra et al. 2019)
CRM	$Cu/Ni/La_{0.7}Sr_{0.3}Cr_{0.5}Mn_{0.5}O_{3-\delta}$ (LSCM)	550/650	• Conversion of reactants and the resistance against sintering increased, hence, allowing the carbon deposition to drop with the introduction of Cu. • The carbon species in the sample; Cu/Ni-LSCM were smaller in size and quantity than the ones present in sample Ni-LSCM	(Kang et al. 2019)
CRM	$LaNi_{1-x}Fe_xO_3$ ($x = 0.2–1.0$)	600–800	• The addition of Fe to the perovskite inhibited the metal reduction. • $LaNiO_3$ gave the highest activity when compared to Fe promoted perovskites. • The in-situ reduction of the perovskite to Ni and carbonate phase was linked to the enhanced activity.	(Jahangiri et al. 2013)

deposition due to a continuous change in the oxidation state of Ce in-situ reaction. This meant that Ce was highly mobilized to react and eliminate carbon by inhibiting the formation of nickel carbide and growth of filamentous carbon (Goscianska et al. 2018).

A study to determine the effects of $LaAlO_3$ and commercial α-Al_2O_3 as support on Ni catalysts in CRM was investigated by Figueredo et al. (2018). Catalytic conditions for the stability test were 700 °C, GHSV of 18 and 72 Lg^{-1} h^{-1} for 10 and 20 h, respectively. After 10 h (GHSV of 18 Lg^{-1} h^{-1}), results showed that CH_4 and CO_2 conversions for Ni/$LaAlO_3$ catalyst increased by 7.8 and 11.5%, respectively, when compared with Ni/α-Al_2O_3. The better stability (by the Ni/$LaAlO_3$ catalyst) was linked to the formation of NiO post-catalyst reduction which was absent in the α-Al_2O_3 supported catalyst. The presence of NiO resulted in increased metal phase dispersion and the formation of carbon nanotubes, whereas, Ni/α-Al_2O_3 catalyst deactivated. Hence, the $LaAlO_3$ perovskite possesses advantages such as; structural defects and abundance of O_2 vacancies which make the perovksite a suitable catalyst for applications in catalytic reforming process.

Ruan et al. (2020) synthesized $LaAl_{0.25}Ni_{0.75}O_3$ using SBA-15 for CRM reaction at 800 °C. Upon removal of the templating agent, Si was successfully incorporated into the perovskite structure. This resulted in better catalytic properties such as increased surface area, improved Ni –support interaction and increment in the amount of strong base sites. Results revealed that the catalyst performance improved when compared to using commercial silica as a templating agent. About 75% of reactants were converted after 36 h of CRM at GHSV of 192, 000 mLh^{-1} g^{-1}. Incorporation of Si into the perovskite crystallite lattice ensures the catalyst was strongly alkaline, thereby promoting the CO_2 dissociation and adsorption (Sadykov et al. 2013).

Ni catalyst on Mg-substituted $LaAlO_3$ support for CRM was prepared and used for CRM at 700 °C under atmospheric pressure by Bai et al. (2019a). The Ni/$La_{0.8}Mg_{0.2}AlO_{2.9}$ had better initial activity and stability than the Ni/$LaAlO_3$ catalyst. CH_4 conversion over Ni/$La_{0.8}Mg_{0.2}AlO_{2.9}$ was 75.4% which is seen to be high under the given reaction conditions. Catalyst characterization showed increased surface area upon addition of Mg to the support. Smaller Ni particles sizes were obtained from Mg-substituted perovskite catalyst leading to improved catalytic properties, such as higher resistance to carbon formation and sintering. These improved properties are closely linked to the availability of oxygen vacancies, surface basicity and metal support interactions which facilitated an effective carbon gasification process.

The catalytic performance of Ni on $LaCuO_3$ was tested in CRM in the temperature range 500–700 °C by Touahra et al. (2016b). 5% $NiO/LaCuO_3$, $LaCuO_3$, and $LaCu_{0.53}Ni_{0.47}O_3$ were synthesized and tested in CRM. At 700 °C, CH_4 conversion was least for $LaCuO_3$ while conversion for $LaCu_{0.53}Ni_{0.47}O_3$ was at 60%. The best result was obtained for 5% $NiO/LaCuO_3$ where CH_4 conversion was 16.5% higher when compared to $LaCu_{0.53}Ni_{0.47}O_3$. The 5% $NiO/LaCuO_3$ perovskite was also more resistant to carbon deposits and sintering in-situ reaction. The presence of NiO, high specific area, smaller Ni crystallite and high $NiO–LaCuO_3$ interaction translated to better catalyst performance.

La_xNiO_y and $La_xNiO_y/MgAl_2O_4$ ($x = 1$ or 2 and $y = 3$ or 4) have been synthesized and applied for CRM by Messaoudi et al. (2018). Characterization results showed that the supported perovskite catalyst had higher specific surface area and Ni dispersion than the unsupported catalyst. This translated to enhanced catalyst activity in CRM. Similar activity to the thermodynamic limitations was obtained for the supported catalyst in a 65 h TOS. Furthermore, the kinetic equation used for the 50% $NiLa_2O_3$/50% $MgAl_2O_4$ catalyst revealed that side reaction occurred as CRM and steam CH_4 reforming occurred at 750 °C.

Stabilizing Ni on pyrochlore perovskite structure for CRM was carried out by Saché et al. (2018). Ni (5 and 10 wt%) was used to substitute Zr at varying loading; in the structure $Ka_2Zr_2O_7$. Post synthesis, the structural analysis showed the presence of $La_2Zr_{2-x}Ni_xO_{7-\delta}$ and $LaNiZrO_6$ at high amount of Ni loading. The formulated perovskite drastically reduced the formation of carbon. Hence, the best catalyst showed high activity and stability (350 h) in CRM at reduced temperature of 600 °C. This performance is linked to the reduction of Ni, resulting in small/evenly dispersed Ni particles. Hence, Ni did not sinter or form larger Ni clusters that are detrimental to CRM. The 5 and 10 wt% Ni promoted perovskites were active for CRM. However, the 10 wt% showed outstanding catalytic stability over 360 h TOS. For the 10 wt% Ni doped perovksite catalyst, even though carbon tubes were formed on the catalyst surface, it had little significance on the catalyst activity thereby making this catalyst a better and economically viable option for CRM.

Ni catalysts from $LaNiO_3$ supported on $Ce_{1-x}Zr_{1-x}O_2$ (where $x = 0.5$, 0.75 and 1) for CH_4 bi-reforming have been investigated by Santos et al. (2020). Addition of Zr into Ce lattice effectively enhanced the surface area and support reducibility. The oxidation of Ni ensured all supported catalysts initially had low reactants conversion. This could be ascribed to the surface oxidation of the catalysts resulting from various side reactions. La and Ce suppressed carbon over the catalyst surface. The highest performance was obtained from $LaNiO_3/Ce_{0.75}Zr_{0.25}O_2$ during the bi-reforming reaction carried out at 800 °C.

$La_{0.8}Ca_{0.2}Ni_{0.6}Co_{0.4}O_3$, $LA_{0.8}Ca_{0.2}NiO_3$ and $LaNiO_3$ were prepared and incorporated into SBA-15 silica for CRM at 600–700 °C by Rivas et al. (2010). The incorporation of the perovskite into the support improves catalytic properties such as metal support interaction and increase in Ni reduction temperature. Reactants conversion decreased with the addition of another cation in the A and/or B sites, respectively. SBA-15 silica incorporation into the perovskites allowed for the reaction to take place at a reduced temperature with increased conversion and selectivity towards the products. Table 4 summarizes the studies on supported perovskite for CRM.

2.5 Promoted Perovskite Catalyst for CRM

Bai et al. (2019b) carried out studies on CRM using perovskites doped cerium catalysts prepared by the Pechini sol gel method. Comparison was made with the un-promoted perovskite to ascertain the effects of doping with reaction conditions of $CH_4/CO_2/N_2 = 3:3:2$, at 400–800 °C and 12 h TOS. Results revealed that the initial CO_2 conversion of La_2CoO_4, $La_{1.8}Ce_{0.2}CoO_4$, $La_{1.4}Ce_{0.6}CoO_4$ and $LaCeCoO_4$ were 51, 56, 65 and 85%, respectively. The increased conversion with the addition of Ce was attributed to the oxygen mobility and storage capacity of CeO_2 which inhibits catalyst deactivation due to coking (Abasaeed et al. 2015; Hou et al. 2017). Ce also increased the specific surface area of $La_{2-x}Ce_xCoO_{4\pm y}$ from 0.2 to 8.5 $m^2\ g^{-1}$, thereby resulting in a well dispersed active catalyst.

Noble metal (Ru, Pt, Pd) promoted $LaAlO_3$ was prepared with the combustion synthesis method and applied for CRM by Anil et al. (2020). At 800 °C, Ru promoted $LaAlO_3$ had the highest amount of CO_2 conversion. Ru and Pt substituted $LaAlO_3$ had 100% conversion of CO_2 while Pd had 80% conversion of CO_2. This Ru perovskite catalyst offers adequate thermal stability because of sufficient Ru-Al interaction. During the stability testing carried out for 50 h, the filamentous carbon type was gasified to CO by the active metallic site. For the Pd and Pt perovskite, conversion decreased at 25 h and 35 h, respectively. Sintering and carbon formation was ascribed to the drop in conversion after several analyses were carried out. Carbon formed was 15 mg, 45 mg and 105 mg for Ru, Pt and Pd substituted catalyst, respectively. Reports claim that CO_2 is activated on the support while CH_4 is activated on the metallic site (Kim et al. 2019; Valderrama et al. 2008), thereby leading to various postulated mechanisms like the power law, Langmuir–Hinshelwood mechanism, Eley-Ridel mechanism and micro kinetic analysis (Egawa 2018; Foppa et al. 2016; Niu et al. 2016). Speculations suggest that H_2 from cracked CH_4

Table 4 Supported perovskite for CRM

Process	Perovskite catalyst	Conditions (°C)	Results / Observation	References
CRM	$LaNiO_3$ perovskite on Al_2O_3 and $CeSiO_2$ supports	800	• $CeSiO_2$ supported catalyst was more stable in terms of performance and conversion. • CO_2 oxidized the Ni^0 in the unsupported and Al_2O_3 supported catalysts. • The unsupported and Al_2O_3 supported perovskite had the highest amount of carbon deposited on the surface. • The $CeSiO_2$ supported peorvskite inhibited carbon deposition.	(Rabelo-Neto et al. 2018)
CRM	$LaAlO_3$ and commercial α-Al_2O_3 as support on Ni catalysts	700	• $LaAlO_3$ perovskite was more advantageous in terms of properties such as structural defects and abundance of O_2 vacancies. • CH_4 and CO_2 conversions for Ni/$LaAlO_3$ catalyst increased by 7.8 and 11.5%, respectively, when compared with Ni/α-Al_2O_3.	(Figueredo et al. 2018)
CRM	$LaAl_{0.25}Ni_{0.75}O_3$ using SBA-15	800	• Incorporation of Si into the perovskite crystallite lattice promotes CO_2 dissociation and adsorption.	(Ruan et al. 2020)
CRM	Ni catalyst on Mg-substituted $LaAlO_3$ support	700	• Increased surface area upon addition of Mg to the support. • Ni/$La_{0.8}Mg_{0.2}AlO_{2.9}$ had better initial activity and stability than the Ni/$LaAlO_3$ catalyst. • CH_4 conversion over Ni/$La_{0.8}Mg_{0.2}AlO_{2.9}$ was 75.4%	(Bai et al. 2019a)
CRM	5% NiO/$LaCuO_3$, $LaCuO_3$, and $LaCu_{0.53}Ni_{0.47}O_3$	500–700	• CH_4 conversion was least for $LaCuO_3$ while conversion for $LaCu_{0.53}Ni_{0.47}O_3$ was at 60%. • CH_4 conversion was 16.5% higher for 5% NiO/$LaCuO_3$ when compared to $LaCu_{0.53}Ni_{0.47}O_3$	(Touahra et al. 2016b)
CRM	La_xNiO_y and La_xNiO_y/$MgAl_2O_4$ (x = 1 or 2 and y = 3 or 4)	750	• The supported perovskite catalyst had higher specific surface area and Ni dispersion than the unsupported catalyst.	(Messaoudi et al. 2018)
CRM	Ni on pyrochlore perovskite ($Ka_2Zr_2O_7$)	600	• $La_2Zr_{2-x}Ni_xO_{7-\delta}$ and $LaNiZrO_6$ structures were formed. • Reduction in carbon formation. • 10 wt% Ni promoted perovskites showed outstanding catalytic stability over 360 h.	(Saché et al. 2018)
CH_4 bi-reforming	Ni catalysts from $LaNiO_3$ supported on $Ce_{1-x}Zr_{1-x}O_2$ (where x = 0.5, 0.75 and 1)	800	• Addition of Zr into Ce lattice effectively enhanced the surface area and support reducibility. • La and Ce suppressed carbon over the catalyst surface. • The highest performance was obtained from $LaNiO_3$/$Ce_{0.75}Zr_{0.25}O_2$	(Santos et al. 2020)
CRM	$La_{0.8}Ca_{0.2}Ni_{0.6}Co_{0.4}O_3$, $LA_{0.8}Ca_{0.2}NiO_3$ and $LaNiO_3$ incorporated into silica	600–700	• Reactants conversion decreased with the addition of another cation in the A and/or B site, respectively. • SBA-15 silica incorporation into the perovskites increased conversion and selectivity.	(Rivas et al. 2010)

can form OH species deposited on the support while adsorbed CO_2 reacts with H/OH species forming decompose species that release CO (Dama et al. 2018a).

$LaNiO_3$ catalysts modified with Co and Mn were used for CRM by Kim et al. (2019). $LaNi_{0.34}Co_{0.33}O_3$ precursor was developed and applied for the reaction at 800 °C and 1 bar. Results showed that Mn improved the stability of the catalyst, while Co improved the reaction rates. The catalyst formed a strong metal support interaction controlled by MnO in the metallic catalyst. This contributed to the synergistic effect that sustained the high performance of the catalyst under extreme experimental conditions. Furthermore, CO formation between carbon deposited on the metal sites and oxygen from the support takes place at the metal support interface.

Sr and Ni were used to co-promote $LaCrO_3$ pervoskite for CRM at temperature ranging from 600 to 800 °C under

Table 5 Promoted perovskite catalyst for CRM

Process	Perovskite catalyst	Conditions (°C)	Results / Observation	References
CRM	Perovskites doped cerium catalyst	400–800	• CO_2 conversion of La_2CoO_4, $La_{1.8}Ce_{0.2}CoO_4$, $La_{1.4}Ce_{0.6}CoO_4$ and $LaCeCoO_4$ were 51, 56, 65 and 85%, respectively. • Increased conversion with the addition of Ce. • Ce also increased the specific surface area of $La_{2-x}Ce_xCoO_{4\pm y}$ from 0.2 to 8.5 $m^2\ g^{-1}$.	(Bai et al. 2019b)
CRM	Noble metal (Ru, Pt, Pd) promoted $LaAlO_3$	800	• Ru promoted $LaAlO_3$ had the highest amount of CO_2 conversion. • Ru and Pt substituted $LaAlO_3$ had 100% conversion of CO_2 while Pd substituted $LaAlO_3$ had 80% conversion of CO_2.	(Anil et al. 2020)
CRM	$LaNiO_3$ catalysts modified with Co and Mn	800	• Mn improved the stability of the catalyst, while Co improved the reaction rates.	(Kim et al. 2019)
CRM	$La_{0.8-x}Sr_xCr_{0.8}Ni_{0.15}O_{3-\delta}$ (LSCrN) for x = 0, 0.1, 0.2, 0.3	600–800	• Ni-support interaction favoured the adsorption of CO_2 • Increased activity and stability in the carbon free CO_2 reforming process.	(Wei et al. 2020)
CRM	Ru promoted $SrTiO_3$	940	• CH_4 and CO_2 conversions were 99.5 and approximately 94%, respectively when the catalyst was promoted and tested with 7 wt. % Ru	(Gangurde et al. 2018)

atmospheric conditions by Wei et al. (2020). The resulting synthesized catalysts were $La_{0.8-x}Sr_xCr_{0.8}Ni_{0.15}O_{3-\delta}$ (LSCrN) for x = 0, 0.1, 0.2, 0.3 (P-80, P-71, P-62, P-53), and the reduced catalysts (R-80, R-71, R-62, R-53). The existing Ni-support interaction favoured the adsorption of CO_2. Hence, higher activity and stability in the carbon free CO_2 reforming process. R-53 catalyst gave the highest activity of the reactants at temperature below 650 °C, whereas, R-62 gave the best performance in terms of reactants conversion up to about 90% at 800 °C. This was attributed to the Ni—support interaction coupled with the basic strength and O_2 vacancy of the support (Osman et al. 2017).

Ru promoted $SrTiO_3$ perovskites were applied for CRM by Gangurde et al. (2018) using microwave heating. Results showed that CH_4 and CO_2 conversions were 99.5 and approximately 94%, respectively when the catalyst was promoted and tested with 7 wt. % Ru at 940 °C. The lower CO_2 conversion when compared with CH_4 was because of the availability of excess CO_2 during the stability test (45: 55 vol. %). Table 5 summarizes the studies on promoted perovskite catalysts for CRM.

2.6 Alkaline Earth Metals Perovskite for CRM

Alkaline earth metals substituted perovskite ($MZr_{1-x}Ni_xO_{3-\delta}$, M = Ca, x = 0 and 0.2) have been prepared by Dama et al. (2018b) and compared with other perovskites ($MZr_{1-x}Ni_xO_{3-\delta}$, M = Ba, Sr, x = 0 and 0.2) for use in CRM 800 °C. Results showed that the nature of the alkaline earth metal substituted in the A site of the perovskite was vital to the overall structure, basicity, oxygen deficiency and Ni dispersion of the catalyst. Substitution with calcium ($CaZr_{0.8}Ni_{0.2}O_{3-\delta}$) gave the best activity of 96% for CO_2 conversion with performance remaining constant after 500 h of reaction. The carbon formed on this catalyst was amorphous in nature which was linked to the basicity of the alkaline perovskite oxide. The surface hydroxyl groups of the support are essential for increased removal of carbon from the surface thereby increasing the performance of $CaZr_{0.8}Ni_{0.2}O_{3-\delta}$ catalyst (Ferreira-Aparicio et al. 2000; Bitter et al. 1997).

Ni perovskite doped with M = Sr, Ca (LaN)) were synthesized for CRM by Yang et al. (2015a). A study on the effect of partial substitution of La by Sr/Ca was carried out for CRM reaction to produce syngas at temperatures ranging between 700 and 900 °C. The addition of Sr/Ca was to modify the A site and enhance the performance of the catalyst. Results showed that the doped perovskites had superior resistance to carbon during the reaction. $La_{0.9}ca_{0.1}Ni_{0.5}Fe_{0.5}O_3$ performed best when compared with other synthesized catalysts. Sr and alkaline earth metal based promoters like Ca altered the basicity of the catalysts while preventing metal sintering. This led to better CO_2 adsorption thereby resulting in enhanced carbon removal rate and improved catalytic activity during the CRM reaction. Furthermore, the inclusion of Fe (LaSrNF and LaCaNF) in the perovskite structure also affected the catalysts behaviour particularly in terms of the resistance to carbon formation. LaSrNF perovskite had higher reactants conversion whereas; LaCaNF had lower reactants conversion. This behaviour was ascribed to the different abilities of the catalysts to adsorb CO_2 (Fang et al. 2018).

$LaNiO_3$ as precursor of Ni/La_2O_3 for CRM at 800 °C was synthesized by Pereñiguez et al. (2012). Four different

Table 6 Alkaline earth metals perovskites for CRM

Process	Perovskite catalyst	Conditions (°C)	Results/observation	References
CRM	Alkaline earth metals substituted perovskite ($MZr_{1-x}Ni_xO_{3-\delta}$, M = Ca, $x = 0$ and 0.2)	800	• Substitution with calcium ($CaZr_{0.8}Ni_{0.2}O_{3-\delta}$) gave the best activity of 96% for CO_2 conversion. • The surface hydroxyl groups of the support was essential for increased removal of carbon from the surface	(Dama et al. 2018b)
CRM	Ni perovskite doped with M = Sr, Ca (LaN))	700–900	• The doped perovskites had superior resistance to carbon during the reaction. • $La_{0.9}ca_{0.1}Ni_{0.5}Fe_{0.5}O_3$ had the best performance when compared with other synthesized catalysts. • Sr and alkaline earth metal based promoters like Ca altered the basicity of the catalysts while preventing metal sintering.	(Yang et al. 2015b)
CRM (effect of synthesis method)	$LaNiO_3$ as precursor of Ni/La_2O_3	800	• Sample prepared by hydrothermal and spray pyrolysis methods promoted CRM more effectively. • Sample prepared by combustion method had the lowest activity for CRM. • High amount of NiO in the sample prepared by combustion method allowed for the formation of large Ni particles which resulted in low CH_4 and CO_2 conversion.	(Pereñiguez et al. 2012)

Ni/La_2O_3 samples (prepared by hydrothermal, combustion, spray pyrolysis and spray pyrolysis-combustion methods) were obtained from the reduction of $LaNiO_3$ precursor and tested in the CRM. Results from XAS and TPR showed that in addition to the crystalline $LaNiO_3$ rhombohedrical phase, a significant amount of amorphous NiO phase was present. The sample prepared by hydrothermal and spray pyrolysis methods promoted CRM more effectively, while the sample prepared by combustion method had the lowest activity for CRM. The amount of NiO phase played a vital role in the performance of Ni/La_2O_3 for CRM. The high amount of NiO phase in the sample prepared by combustion method led to the formation of large Ni metallic particles which resulted in reduced reactants (CH_4 and CO_2) conversion. Table 6 summarizes the studies on alkaline earth metals formulated catalyst for CRM.

3 Conclusions

Carbon cycle can effectively be managed by the utilization of CO_2 as a reagent in various reactions. Due to its thermodynamic stability and inert nature, high energy is required for CO_2 conversion to other useful components. Hence, CO_2 conversion processes require the use of catalyst to minimize these challenges. Heterogeneous catalysts provide solutions for CO_2 conversion processes due to their ease of separation, stability in handling, catalyst reuse and reproductively. Perovskite is a type of heterogeneous catalyst that possess a large range of industrial, scientific and commercial importance due to its low cost and stability in high temperature conditions while offering flexibility of its stoichiometric parameters. Also, the multi-functional properties of the perovskite allow it to exist with O_2 in a non-stoichiometric form. These properties can be related to their preparation methods that have been reported to enhance the physico-chemical properties of heterogeneous catalysts. For instance, since the heterogeneous catalytic reaction occurs at the surface, the catalyst should be synthesized with large enough surface to allow for substantial interaction between catalyst and reactants. CO_2 is effective as a reagent for the conversion of hydrocarbons into more useful products like H_2, syngas and liquid fuels. Most common of these conversion processes is CRM. This process is of great interest as the two major greenhouse gases are converted into more useful products. However, the extent of conversion of this process is determined by several parameters such as amount of reactants used, reaction operating condition and type of catalyst utilized. Perovskite's have been used as catalysts for CO_2 reforming due to their high oxygen mobility. This property is essential for suppressing the formation of carbon during CO_2 reforming reactions. Also the flexibility in altering the acidic and basic sites of the perovskite offer several options in changing its electronic configuration and structural buildup. Its thermally and chemically stable nature under redox and H_2 rich environment gives it an edge over other types of catalysts used in CO_2 reforming. However, the inability of perovskite's to form structures with large surface area is its major setback that requires more attention from researchers. Hence, due to the perovskite's unique properties already mentioned, this chapter has focused on the role played by perovskites as catalysts in the conversion of reactants to more valuable products in the CO_2 reforming of CH_4 reaction process.

Acknowledgments The authors extend their gratitude to Universiti Malaysia Pahang, Malaysia for the financial assistance through the international publication research grant (RDU203304) and post-doctoral fellowship for Osarieme Uyi Osazuwa.

References

Abasaeed AE, Al-Fatesh AS, Naeem MA, Ibrahim AA, Fakeeha AH (2015) Catalytic performance of CeO_2 and ZrO_2 supported Co catalysts for hydrogen production via dry reforming of methane. Int J Hydrogen Energy 40:6818–6826. https://doi.org/10.1016/j.ijhydene.2015.03.152

Abdulrasheed AA, Jalil AA, Hamid MYS, Siang TJ, Fatah NAA, Izan SM et al (2019) Dry reforming of methane to hydrogen-rich syngas over robust fibrous KCC-1 stabilized nickel catalyst with high activity and coke resistance. Int J Hydrogen Energy:1–13. https://doi.org/10.1016/j.ijhydene.2019.04.126

Akri M, Chafik T, Granger P, Ayrault P, Batiot-Dupeyrat C (2016) Novel nickel promoted illite clay based catalyst for autothermal dry reforming of methane. Fuel. https://doi.org/10.1016/j.fuel.2016.03.018

Anil C, Modak JM, Madras G (2020) Syngas production via CO_2 reforming of methane over noble metal (Ru, Pt, and Pd) doped $LaAlO_3$ perovskite catalyst. Mol Catal 484:110805. https://doi.org/10.1016/j.mcat.2020.110805

Ayodele BV, Hossain SS, Lam SS, Osazuwa OU, Khan MR, Cheng CK (2016) Syngas production from CO_2 reforming of methane over neodymium sesquioxide supported cobalt catalyst. J Nat Gas Sci Eng 34:873–885. https://doi.org/10.1016/j.jngse.2016.07.059

Bai X, Xie G, Guo Y, Tian L, El-Hosainy HM, Awadallah AE et al (2019a) A highly active Ni catalyst supported on Mg-substituted $LaAlO_3$ for carbon dioxide reforming of methane. Catal Today. https://doi.org/10.1016/j.cattod.2019.12.033

Bai Y, Wang Y, Yuan W, Sun W, Zhang G, Zheng L et al (2019b) Catalytic performance of perovskite—like oxide doped cerium ($La_{2-x}Ce_xCoO_{4\pm y}$) as catalysts for dry reforming of methane. Chinese J Chem Eng 27:379–385. https://doi.org/10.1016/j.cjche.2018.05.016

Batiot-Dupeyrat C, Gallego GAS, Mondragon F, Barrault J, Tatibouët J-M (2005) CO_2 reforming of methane over $LaNiO_3$ as precursor material. Catal Today 107–108:474–480. https://doi.org/10.1016/j.cattod.2005.07.014

Bhavani AG, Kim WY, Lee JS (2013) Barium substituted lanthanum manganite perovskite for CO_2 reforming of methane. ACS Catal 3:1537–1544. https://doi.org/10.1021/cs400245m

Bitter JH, Seshan K, Lercher JA (1997) The state of zirconia supported platinum catalysts for CO_2/CH_4 reforming. J Catal 171:279–286. https://doi.org/10.1006/jcat.1997.1792

Bitter JH, Seshan K, Lercher JA (1998) Mono and bifunctional pathways of CO_2/CH_4 reforming over Pt and Rh based catalysts. J Catal 176:93–101. https://doi.org/10.1006/jcat.1998.2022

Chang JS, Park SE, Lee KW, Choi MJ (1994) Catalytic reforming of methane with carbon dioxide over pentasil zeolite-supported nickel catalyst. Stud Surf Sci Catal 84:1587–1594. https://doi.org/10.1016/S0167-2991(08)63707-6

Dama S, Ghodke SR, Bobade R, Gurav HR, Chilukuri S (2018a) Active and durable alkaline earth metal substituted perovskite catalysts for dry reforming of methane. Appl Catal B Environ 224:146–158. https://doi.org/10.1016/j.apcatb.2017.10.048

Dama S, Ghodke SR, Bobade R, Gurav HR, Chilukuri S (2018) Active and durable alkaline earth metal substituted perovskite catalysts for dry reforming of methane, vol 224. Elsevier BV. https://doi.org/10.1016/j.apcatb.2017.10.048

de Caprariis B, de Filippis P, Palma V, Petrullo A, Ricca A, Ruocco C et al (2016) Rh, Ru and Pt ternary perovskites type oxides $BaZr_{(1-x)}Me_xO_3$ for methane dry reforming. Appl Catal A Gen 517:47–55. https://doi.org/10.1016/j.apcata.2016.02.029

Doggali P, Kusaba S, Teraoka Y, Chankapure P, Rayalu S, Labhsetwar N (2010) $La_{0.9}Ba_{0.1}CoO_3$ perovskite type catalysts for the control of CO and PM emissions. Catal Commun 11:665–669. https://doi.org/10.1016/j.catcom.2010.01.019

Egawa C (2018) Methane dry reforming reaction on Ru(0 0 1) surfaces. J Catal 358:35–42. https://doi.org/10.1016/j.jcat.2017.11.010

Fang X, Zhang J, Liu J, Wang C, Huang Q, Xu X, et al (2018) Methane dry reforming over Ni/Mg-Al-O: On the significant promotional effects of rare earth Ce and Nd metal oxides. J CO_2 Util 25:242–253. https://doi.org/10.1016/J.JCOU.2018.04.011

Ferreira-Aparicio P, Rodríguez-Ramos I, Anderson JA, Guerrero-Ruiz A (2000) Mechanistic aspects of the dry reforming of methane over ruthenium catalysts. Appl Catal A Gen 202:183–196. https://doi.org/10.1016/S0926-860X(00)00525-1

Figueredo GP, Medeiros RLBA, Macedo HP, de Oliveira ÂAS, Braga RM, Mercury JMR et al (2018) A comparative study of dry reforming of methane over nickel catalysts supported on perovskite-type $LaAlO_3$ and commercial $A\text{-}Al_2O_3$. Int J Hydrogen Energy 43:11022–11037. https://doi.org/10.1016/j.ijhydene.2018.04.224

Foppa L, Silaghi MC, Larmier K, Comas-Vives A (2016) Intrinsic reactivity of Ni, Pd and Pt surfaces in dry reforming and competitive reactions: insights from first principles calculations and microkinetic modeling simulations. J Catal 343:196–207. https://doi.org/10.1016/j.jcat.2016.02.030

Gangurde LS, Sturm GSJ, Valero-Romero MJ, Mallada R, Santamaria J, Stankiewicz AI et al (2018) Synthesis, characterization, and application of ruthenium-doped $SrTiO_3$ perovskite catalysts for microwave-assisted methane dry reforming. Chem Eng Process—Process Intensif 127:178–190. https://doi.org/10.1016/j.cep.2018.03.024

García A, Becerra N, García L, Ojeda I, López E, López CM et al (2011) Structured perovskite-based oxides: use in the combined methane reforming. Adv Chem Eng Sci 01:169–175. https://doi.org/10.4236/aces.2011.14025

Ghelamallah M, Granger P (2014) Supported-induced effect on the catalytic properties of Rh and Pt-Rh particles deposited on La_2O_3 and mixed α-Al $2O_3$-La_2O_3 in the dry reforming of methane. Appl Catal A Gen 485:172–180. https://doi.org/10.1016/j.apcata.2014.07.021

Goldwasser MR, Rivas ME, Lugo ML, Pietri E, Pérez-Zurita J, Cubeiro ML et al (2005) Combined methane reforming in presence of CO_2 and O_2 over $LaFe_{1-x}Co_xO_3$ mixed-oxide perovskites as catalysts precursors. Catal Today 107–108:106–113. https://doi.org/10.1016/j.cattod.2005.07.073

Goscianska J, Pietrzak R, Matos J (2018) Catalytic performance of ordered mesoporous carbons modified with lanthanides in dry methane reforming. Catal Today 301:204–216. https://doi.org/10.1016/j.cattod.2017.05.014

Hou W, Wang Y, Bai Y, Sun W, Yuan W, Zheng L et al (2017) Carbon dioxide reforming of methane over $Ni/Mg_{0.4}Al_{0.4}\text{-}La_{0.1}Zr_{0.1}(O)$ catalyst prepared by recombination sol–gel method. Int J Hydrogen Energy 42:16459–16475. https://doi.org/10.1016/j.ijhydene.2017.05.058

Jahangiri A, Pahlavanzadeh H, Aghabozorg H (2012) Synthesis, characterization and catalytic study of Sm doped $LaNiO_3$ nanoparticles in reforming of methane with CO_2 and O_2. Int J Hydrogen Energy 37:9977–9984. https://doi.org/10.1016/j.ijhydene.2012.03.128

Jahangiri A, Aghabozorg H, Pahlavanzadeh H (2013) Effects of Fe substitutions by Ni in La–Ni–O perovskite-type oxides in reforming of methane with CO_2 and O_2. Int J Hydrogen Energy 38:10407–10416. https://doi.org/10.1016/j.ijhydene.2013.05.080

Jiang C, Akkullu MR, Li B, Davila JC, Janik MJ, Dooley KM (2019) Rapid screening of ternary rare-earth–transition metal catalysts for dry reforming of methane and characterization of final structures. J Catal 377:332–342. https://doi.org/10.1016/j.jcat.2019.07.020

Kang D, Yu J, Ma W, Zheng M, He Y, Li P (2019) Synthesis of Cu/Ni-$La_{0.7}Sr_{0.3}Cr_{0.5}Mn_{0.5}O_{3-\delta}$ and its catalytic performance on dry methane reforming. J Rare Earths 37:585–593. https://doi.org/10.1016/j.jre.2018.10.016

Kim WY, Jang JS, Ra EC, Kim KY, Kim EH, Lee JS (2019) Reduced perovskite $LaNiO_3$ catalysts modified with Co and Mn for low coke formation in dry reforming of methane. Appl Catal A Gen 575:198–203. https://doi.org/10.1016/j.apcata.2019.02.029

Krishna K, Bueno-López A, Makkee M, Moulijn JA (2007) Potential rare earth modified CeO_2 catalysts for soot oxidation: I. Characterisation and catalytic activity with O_2. Appl Catal B Environ 75:189–200. https://doi.org/10.1016/J.APCATB.2007.04.010

Labhasetwar N, Saravanan G, Kumar Megarajan S, Manwar N, Khobragade R, Doggali P et al (2015) Perovskite-type catalytic materials for environmental applications. Sci Technol Adv Mater 16:036002. https://doi.org/10.1088/1468-6996/16/3/036002

le Saché E, Pastor-Pérez L, Watson D, Sepúlveda-Escribano A, Reina TR (2018) Ni stabilised on inorganic complex structures: superior catalysts for chemical CO_2 recycling via dry reforming of methane. Appl Catal B Environ 236:458–465. https://doi.org/10.1016/j.apcatb.2018.05.051

Lombardo EA, Ulla MA (1998) Perovskite oxides present and future in catalysis : past, present, and future. Res Chem Intermed 24:581–92

Ma J, Sun N, Zhang X, Zhao N, Xiao F, Wei W et al (2009) A short review of catalysis for CO_2 conversion. Catal Today 148:221–231. https://doi.org/10.1016/j.cattod.2009.08.015

Maneerung T, Hidajat K, Kawi S (2011) $LaNiO_3$ perovskite catalyst precursor for rapid decomposition of methane: influence of temperature and presence of H_2 in feed stream. Catal Today 171:24–35. https://doi.org/10.1016/j.cattod.2011.03.080

Messaoudi H, Thomas S, Djaidja A, Slyemi S, Barama A (2018) Study of La_xNiO_y and $La_xNiO_y/MgAl_2O_4$ catalysts in dry reforming of methane. J CO_2 Util 24:40–49. https://doi.org/10.1016/j.jcou.2017.12.002

Mishra A, Prasad R (2014) Preparation and application of perovskite catalysts for diesel soot emissions control: an overview. Catal Rev 56:57–81. https://doi.org/10.1080/01614940.2014.866438

Moradi GR, Rahmanzadeh M, Khosravian F (2014) The effects of partial substitution of Ni by Zn in $LaNiO_3$ perovskite catalyst for methane dry reforming. J CO_2 Util 6:7–11. https://doi.org/10.1016/j.jcou.2014.02.001

Mota N, Álvarez-Galván MC, Al-Zahrani SM, Navarro RM, Fierro JLG (2012) Diesel fuel reforming over catalysts derived from $LaCo_{1-x}Ru_xO_3$ perovskites with high Ru loading. Int. J. Hydrogen Energy 37:7056–7066. Pergamon. https://doi.org/10.1016/j.ijhydene.2011.12.156

Murugan A, Thursfield A, Metcalfe IS (2011) A chemical looping process for hydrogen production using iron-containing perovskites. Energy Environ Sci 4:4639–4649. https://doi.org/10.1039/c1ee02142g

Nalbandian L, Evdou A, Zaspalis V (2011) $La_{1-x}Sr_xM_yFe_{1-y}O_{3-\delta}$ perovskites as oxygen-carrier materials for chemical-looping reforming. Int J Hydrogen Energy 36:6657–6670. https://doi.org/10.1016/j.ijhydene.2011.02.146

Niu J, Du X, Ran J, Wang R (2016) Dry (CO_2) reforming of methane over Pt catalysts studied by DFT and kinetic modeling. Appl Surf Sci 376:79–90. https://doi.org/10.1016/j.apsusc.2016.01.212

Omae I (2006) Aspects of carbon dioxide utilization. Catal Today 115:33–52. https://doi.org/10.1016/j.cattod.2006.02.024

Osazuwa OU, Cheng CK (2017) Catalytic conversion of methane and carbon dioxide (greenhouse gases) into syngas over samarium-cobalt-trioxides perovskite catalyst. J Clean Prod 148. https://doi.org/10.1016/j.jclepro.2017.01.177

Osazuwa OU, Setiabudi HD, Rasid RA, Cheng CK (2017) Syngas production via methane dry reforming: A novel application of $SmCoO_3$ perovskite catalyst. J Nat Gas Sci Eng 37:435–448. https://doi.org/10.1016/j.jngse.2016.11.060

Osazuwa OU, Khan MR, Lam SS, Assabumrungrat S, Cheng CK (2018) An assessment of the longevity of samarium cobalt trioxide perovskite catalyst during the conversion of greenhouse gases into syngas. J Clean Prod. https://doi.org/10.1016/j.jclepro.2018.03.060

Osman AI, Meudal J, Laffir F, Thompson J, Rooney D (2017) Enhanced catalytic activity of Ni on H-Al_2O_3 and ZSM-5 on addition of ceria zirconia for the partial oxidation of methane. Appl Catal B Environ 212:68–79. https://doi.org/10.1016/j.apcatb.2016.12.058

Pakhare D, Haynes D, Shekhawat D, Spivey J (2012) Role of metal substitution in lanthanum zirconate pyrochlores ($La_2Zr_2O_7$) for dry (CO_2) reforming of methane (DRM). Appl Petrochemical Res 2:27–35. https://doi.org/10.1007/s13203-012-0014-6

Pereñiguez R, Gonzalez-delaCruz VM, Caballero A, Holgado JP (2012) $LaNiO_3$ as a precursor of Ni/La_2O_3 for CO_2 reforming of CH_4: effect of the presence of an amorphous NiO phase. Appl Catal B Environ 123–124:324–332. https://doi.org/10.1016/j.apcatb.2012.04.044

Pérez-Camacho MN, Abu-Dahrieh J, Goguet A, Sun K, Rooney D (2014) Self-cleaning perovskite type catalysts for the dry reforming of methane. Chinese J Catal 35:1337–1346. https://doi.org/10.1016/S1872-2067(14)60187-X

Pradeep A, Chandrasekaran G (2006) FTIR study of Ni, Cu and Zn substituted nano-particles of $MgFe_2O_4$, vol 60. https://doi.org/10.1016/j.matlet.2005.08.053

Rabelo-Neto RC, Sales HBE, Inocêncio CVM, Varga E, Oszko A, Erdohelyi A et al (2018) CO_2 reforming of methane over supported $LaNiO_3$ perovskite-type oxides. Appl Catal B Environ 221:349–361. https://doi.org/10.1016/j.apcatb.2017.09.022

Rida K, Benabbas A, Bouremmad F, Peña MA, Sastre E, Martínez-Arias A (2007) Effect of calcination temperature on the structural characteristics and catalytic activity for propene combustion of sol-gel derived lanthanum chromite perovskite. Appl Catal A Gen 327:173–179. https://doi.org/10.1016/j.apcata.2007.05.015

Rivas I, Alvarez J, Pietri E, Pérez-Zurita MJ, Goldwasser MR (2010) Perovskite-type oxides in methane dry reforming: effect of their incorporation into a mesoporous SBA-15 silica-host. Catal Today 149:388–393. https://doi.org/10.1016/j.cattod.2009.05.028

Ruan Y, Zhao Y, Lu Y, Guo D, Zhao Y, Wang S et al (2020) Mesoporous $LaAl_{0.25}Ni_{0.75}O_3$ perovskite catalyst using SBA-15 as templating agent for methane dry reforming. Microporous Mesoporous Mater 303:110278. https://doi.org/10.1016/j.micromeso.2020.110278

Ruocco C, de Caprariis B, Palma V, Petrullo A, Ricca A, Scarsella M et al (2019) Methane dry reforming on Ru perovskites, $AZrRuO_3$: influence of preparation method and substitution of A cation with alkaline earth metals. J CO_2 Util 30:222–231. https://doi.org/10.1016/J.JCOU.2019.02.009

Ryi S-K, Lee S-W, Park J-W, Oh D-K, Park J-S, Kim SS (2014) Combined steam and CO_2 reforming of methane using catalytic nickel membrane for gas to liquid (GTL) process. Catal Today 236:49–56. https://doi.org/10.1016/j.cattod.2013.11.001

Rynkowski J, Samulkiewicz P, Ladavos A, Pomonis P (2004) Catalytic performance of reduced $La_{2-x}Sr_xNiO_4$ perovskite-like oxides for CO_2 reforming of CH_4. Appl Catal A Gen 263:1–9. https://doi.org/10.1016/j.apcata.2003.11.022

Sadykov V, Rogov V, Ermakova E, Arendarsky D, Mezentseva N, Alikina G et al (2013) Mechanism of CH_4 dry reforming by pulse microcalorimetry: metal nanoparticles on perovskite/fluorite supports with high oxygen mobility. Thermochim Acta 567:27–34. https://doi.org/10.1016/j.tca.2013.01.034

Santos DBL, Noronha FB, Hori CE (2020) Bi-reforming of methane for hydrogen production using $LaNiO_3/Ce_xZr_{1-x}O_2$ as precursor material. Int J Hydrogen Energy 45:13947–13959. https://doi.org/10.1016/j.ijhydene.2020.03.096

Sazonova NN, Sadykov VA, Bobin AS, Pokrovskaya SA, Gubanova EL, Mirodatos C (2009) Dry reforming of methane over fluorite-like mixed oxides promoted by Pt. React Kinet Catal Lett 98:35–41. https://doi.org/10.1007/s11144-009-0064-7

Screen T (2007) Platinum group metal perovskite catalysts. Platin Met Rev 51:87–92. https://doi.org/10.1595/147106707X192645

Sfeir J, Buffat PA, Mockli P, Xanthopoulos N, Vasquez R, Mathieu HJ et al (2001) Lanthanum chromite based catalysts for oxidation of methane directly on SOFC anodes. J Catal 202:229–244. https://doi.org/10.1006/jcat.2001.3286

Shabbir G, Qureshi AH, Saeed K (2006) Nano-crystalline $LaFeO_3$ powders synthesized by the citrate-gel method. Mater Lett 60:3706–3709. https://doi.org/10.1016/j.matlet.2006.03.093

Sheng T, Jiang YX, Tian N, Zhou ZY, Sun SG (2017) Nanocrystal catalysts of high-energy surface and activity, 1st edn, vol 177. Elsevier BV. https://doi.org/10.1016/B978-0-12-805090-3.00012-7

Slagtern A, Schuurman Y, Leclercq C, Verykios X, Mirodatos C (1997) Specific features concerning the mechanism of methane reforming by carbon dioxide over Ni/La_2O_3 catalyst. J Catal 172:118–126. https://doi.org/10.1006/jcat.1997.1823

Su YJ, Pan KL, Chang MB (2014) Modifying perovskite-type oxide catalyst $LaNiO_3$ with Ce for carbon dioxide reforming of methane. Int J Hydrogen Energy 39:4917–4925. https://doi.org/10.1016/j.ijhydene.2014.01.077

Tomishige K (2004) Syngas production from methane reforming with CO_2/H_2O and O_2 over NiO-MgO solid solution catalyst in fluidized bed reactors. Catal Today 89:405–418. https://doi.org/10.1016/j.cattod.2004.01.003

Tomishige K, Kanazawa S, Suzuki K, Asadullah M, Sato M, Ikushima K et al (2002) Effective heat supply from combustion to reforming in methane reforming with CO_2 and O_2: comparison between Ni and Pt catalysts. Appl Catal A Gen 233:35–44. https://doi.org/10.1016/S0926-860X(02)00131-X

Touahra F, Rabahi A, Chebout R, Boudjemaa A, Lerari D, Sehailia M et al (2016a) Enhanced catalytic behaviour of surface dispersed nickel on $LaCuO_3$ perovskite in the production of syngas: an expedient approach to carbon resistance during CO_2 reforming of methane. Int J Hydrogen Energy 41:2477–2486. https://doi.org/10.1016/j.ijhydene.2015.12.062

Touahra F, Rabahi A, Chebout R, Boudjemaa A, Lerari D, Sehailia M et al (2016b) Enhanced catalytic behaviour of surface dispersed nickel on $LaCuO_3$ perovskite in the production of syngas: an expedient approach to carbon resistance during CO_2 reforming of methane. Int J Hydrogen Energy. https://doi.org/10.1016/j.ijhydene.2015.12.062

Touahra F, Chebout R, Lerari D, Halliche D, Bachari K (2019) Role of the nanoparticles of Cu-Co alloy derived from perovskite in dry reforming of methane. Energy 171:465–474. https://doi.org/10.1016/j.energy.2019.01.085

Valderrama G, Goldwasser MR, de Navarro CU, Tatibouët JM, Barrault J, Batiot-Dupeyrat C et al (2005) Dry reforming of methane over Ni perovskite type oxides. Catal Today 107–108:785–791. https://doi.org/10.1016/j.cattod.2005.07.010

Valderrama G, Kiennemann A, Goldwasser MR (2008) Dry reforming of CH_4 over solid solutions of $LaNi_{1-x}Co_xO_3$. Catal Today 133–135:142–148. https://doi.org/10.1016/j.cattod.2007.12.069

Valderrama G, Urbina de Navarro C, Goldwasser MR (2013) CO_2 reforming of CH_4 over Co–La-based perovskite-type catalyst precursors. J Power Sources 234:31–37. https://doi.org/10.1016/j.jpowsour.2013.01.142

Valderrama G, Kiennemann A, de Navarro CU, Goldwasser MR (2018) $LaNi_{1-x}Mn_xO_3$ perovskite-type oxides as catalysts precursors for dry reforming of methane. Appl Catal A Gen 565:26–33. https://doi.org/10.1016/j.apcata.2018.07.039

Wang M, Zhao T, Dong X, Li M, Wang H (2018) Effects of Ce substitution at the A-site of $LaNi_{0.5}Fe_{0.5}O_3$ perovskite on the enhanced catalytic activity for dry reforming of methane. Appl Catal B Environ 224:214–21. https://doi.org/10.1016/j.apcatb.2017.10.022

Wei T, Jia L, Luo JL, Chi B, Pu J, Li J (2020) CO_2 dry reforming of CH_4 with Sr and Ni co-doped $LaCrO_3$ perovskite catalysts. Appl Surf Sci 506:144699. https://doi.org/10.1016/j.apsusc.2019.144699

Yadav PK, Das T (2019) Production of syngas from carbon dioxide reforming of methane by using $LaNi_xFe_{1-x}O_3$ perovskite type catalysts. Int J Hydrogen Energy 44:1659–1670. https://doi.org/10.1016/j.ijhydene.2018.11.108

Yang E, Noh Y, Ramesh S, Lim SS, Moon DJ (2015) The effect of promoters in $La_{0.9}M_{0.1}Ni_{0.5}Fe_{0.5}O_3$ (M=Sr, Ca) perovskite catalysts on dry reforming of methane. Fuel Process Technol 134:404–413. https://doi.org/10.1016/j.fuproc.2015.02.023

Yang E, Kim NY, Noh Y, Lim SS, Jung J-S, Lee JS et al (2015b) Steam CO_2 reforming of methane over $La_{1-x}Ce_xNiO_3$ perovskite catalysts. Int J Hydrogen Energy 40:11831–11839. https://doi.org/10.1016/j.ijhydene.2015.06.021

Yang EH, Noh YS, Hong GH, Moon DJ (2018) Combined steam and CO_2 reforming of methane over $La_{1-x}Sr_xNiO_3$ perovskite oxides. Catal Today 299:242–50. https://doi.org/10.1016/j.cattod.2017.03.050

Zhang S, Chen Y, Li F, Lu X, Dai W, Mori R (2006) Fixation and conversion of CO_2 using ionic liquids. Catal Today 115:61–69. https://doi.org/10.1016/j.cattod.2006.02.021

Zhang R, Xia G, Li M, Wu Y, Nie H, Li D (2015) Effect of support on the performance of Ni-based catalyst in methane dry reforming. J Fuel Chem Technol 43:1359–1365. https://doi.org/10.1016/S1872-5813(15)30040-2

Zhu J, Li H, Zhong L, Xiao P, Xu X, Yang X et al (2014) Perovskite oxides: preparation, characterizations and applications in heterogeneous catalysis. ACS Catal 4:2917–2940. https://doi.org/10.1021/cs500606g

Carbon Dioxide Conversion to Useful Chemicals and its Thermodynamics

Pallavi Jain and A. Geetha Bhavani

Abstract

Carbon dioxide (CO_2) utilization was well-practiced through chemical processes to produce useful chemicals, such as syngas (H_2/CO), hydrogen (H_2), methane (CH_4), methanol (MeOH), ethanol (EtOH), dimethyl ether (DME). These chemicals have wide industrially demanded to use as a raw material for generating various other chemicals. CO_2 utilization through conversion is a substitute and economical foundation of energy source and is reducing CO_2 emissions. Many industries and bulk scale production plants are aimed to produce selective products over the highly active catalyst to trigger the problems. To encounter the industrial reaction parameter challenges, the thermodynamic stimulation is a significant analysis predicting the reaction steps sequences to optimize the parameters and process should be economically viable. The thermodynamic stimulations are mainly, the reaction temperature, reactor pressure, molar ratio of the feed at the inlet, composition/add-on gases to feed ratio, coke formation, and merely depend upon the outlet products stability and selectivity. This chapter comprehensively discussed the thermodynamics of CO_2 hydrogenation, H_2/CO production, MeOH, EtOH, and DME synthesis.

Keywords

Carbon dioxide • Methane • Methanol • Dimethyl ether • Syngas • Thermodynamics

P. Jain (✉)
Department of Chemistry, SRM Institute of Science and Technology Delhi-NCR Campus, Modinagar, Ghaziabad, U.P. 201204, India
e-mail: palli24@gmail.com; pallavij@srmist.edu.in

A. G. Bhavani
Department of Chemistry, School of Sciences, Noida International University, Greater Noida, U.P. 2013086, India

1 Introduction

CO_2 is a major pollutant constituent produced by human activities which augmented the rise in global temperatures and climate change. In the past few years, the government laid many policies and public awareness capacity-building programs to reduce the CO_2 intensities in the atmosphere. The Carbon Capture and Utilization (CCU) process is aimed to capture the atmospheric CO_2 emission and separated it for further utilization. The separated CO_2 is the conversion into energy production, application of various chemical productions and fuels as growing the demand. Pure CO_2 is used as a raw material for many industrial reactions to produce valuable products. The commercial industries already using CO_2 as a raw material to produce H_2/CO, is further used to produce fuel cell plants, CH_4, DME, MeOH, EtOH, and as a chemical feedstock for organic and inorganic products (ammonia, urea) (Alvarez et al. 2017; Kawi and Kathiraser 2015; Dorner et al. 2010). The industrial applications are intended for effectively high conversion with highly selective products. The reaction and product formations are prophesied by thermodynamic analysis. The thermodynamic study provides the information of reaction steps, degree of reactants conversion, percentage of yield, degree of product selectivity, and effects of parameters (temperature, pressure, and reactants ratio) for unit operations (Huang and Tan 2014).

The thermodynamics of CO_2 conversion can be estimated by stoichiometric thermodynamic analysis and non-stoichiometric thermodynamic analysis. The stoichiometric analysis requires information of all the chemical reaction species and reactions (independent) occurring in a system. For complex systems that involve a lot of reactions, the implementation of this method becomes repetitive. It is not possible to apply the equilibrium constant method to systems where almost all reactions are uncertain. Whereas, the non-stoichiometric approach, also known as Gibbs free energy minimization (GFEM), involves only knowledge of

Inamuddin et al. (eds.), *Carbon Dioxide Utilization to Sustainable Energy and Fuels*, Advances in Science, Technology & Innovation, https://doi.org/10.1007/978-3-030-72877-9_16

the chemical species for thermodynamic equilibrium calculations. Therefore, regardless of its complexities, non-stochiometric techniques can be implemented to any system easily without any stoichiometric information of reactants. (Smith et al. 2005). To perform the thermodynamic analysis, Aspen Plus (AP) software was used and further RGIBBS block of AP was used for the total GFEM of the reaction mixture. Soave-Redlich-Kwong model in AP was used for introducing non-ideal performance in Gibbs energy values by the state of equation (Stenger and Askonas 1986). The present chapter is focused on the thermodynamics of four CO_2 conversion processes, which includes CO_2 utilization, CO_2 conversion for the synthesis of DME, MeOH, EtOH, CH_4, and DRM.

2 Thermodynamic of CO_2 Hydrogenation to CH_4

Methanation of CO_2 for the production of synthetic natural gas (SNG) is a significant reaction producing H_2/CO, which is used as a raw material for many industrial applications. The SNG production from CO_2 is identified as the "Power to Gas" (PtG) process, to yield a high concentration of H_2 (Gotz et al. 2016; Lehner et al. 2014; Specht et al. 2016). The production of the pure and high percentage of CH_4 in SNG composition is considered more for years as it has extensive impending applications in the market, which will be achieved only through catalytic CO_2 hydrogenation. The catalytic CO_2 hydrogenation is passed through various ranges of pressure from low to high, for distribution and storage for further applications. The economic way to produce H_2 is the electrolysis of water, manipulating the surplus electricity by nuclear or renewable, once accessible in the power grid. As a low-cost carbon source, the CO_2 is well thought-out for transforming to CH_4 and further effective H_2 storage and contribution in the steadiness of the electric grid. CO_2 methanation has many advantages; it is contemplated as enabling expertise to store the electricity by the chemical process as a long-term process (Geetha et al. 2015). Thermodynamics evaluates the methanation process and predicts the optimal reaction state (Kiewidt and Thoming 2015; Molino and Braccio 2015). The effects of methanation parameters are pressure, temperature, and molar ratio of H_2/CO_x. Additionally, the conversion of CO_x and CH_4 selectivity and residual carbon deposition influenced by adding reactants (CH_4, H_2O, O_2) (Geetha et al. 2013).

The Sabatier reaction, R(1) during the production of CH_4 using CO_2 and H_2 is an exothermic reaction leading to molar reduction of the composition of the reactants (Sabatier 1902). The reaction, R(2) occurs with minor endothermic reverse water gas shift (RWGS) for equimolar CO production throughout the synthesis. CO_2 methanation reaction (1) and (2), thermodynamics indicates exothermic process and also decreasing the number of reactant moles, increases the pressure and lower the temperature, which ultimately improves CH_4 yield with CO as a major and C2+ hydrocarbon as minor by-products. The progress in CO_2 methanation are factually connected to the production of stable, and pure H_2 from H_2/CO. Indeed, production of pure H_2 is produced by the mixed feed of natural gas (CH_4 and CO_2) through one or two reforming steps of DRM, SRM, ATR process. The reaction steps involve CO_2 conversion to CO at high temperatures through the water gas shift (WGS) mechanism. By the adsorption process, CO_2 is captured for the conversion, and the remaining unconverted CO_x is hydrogenated to enhance CH_4 yield. The yield is processed for purification steps under various conditions. During H_2 purification, the inlet CO_x concentration is low while H_2 yield is outsized. In these dilute circumstances, thermodynamics allows high conversion of CO_x nearly 100% in the reforming reactor. Bulk scale production of H_2/CO is in high demand through CO_x methanation. Large and stable production needs a dynamic catalyst that performs in many limitations of the reactor and thermodynamic parameters. However, CO_x methanation reaction gained highly industrial importance from the 1970s while for the oil crisis, H_2/CO was used from gasification of coal (Barbarossa et al. 2013; Kopyscinski et al. 2010). Additionally, increased utilization of biomass opened many industrial plants to process biomass to CH_4 in the last few years. SNG process is considered for large scale industrial applications over CO_2 hydrogenation and is presently practicing for alternative and effective renewable power. Additionally, this process controls immensely atmospheric CO_2 emissions. The emblematic feedstock CO_2 was usually received through flue gases of power plants or biogas processing plants (Mazza et al. 2018; Meylan et al. 2016). However, the feedstock was also obtained from the process in industries CO_2 emissions like steel manufactory, etc. The initial trial plant for CO_2 methanation was constructed during the 1990s, currently, many marketable applications are accessible (Napp et al. 2014).

The CO_2 methanation, R(1) releases high energy while dissociation (exothermic). Methanation reaction must find a suitable and efficient catalyst for stability to produce CH_4 and hasten the rate of the reaction process. Many research reports proved the development of novel stable catalysts to withstand harsh parameters. Ni-based supported catalysts were well-practiced industrially due to feasibility and economic while using in large quantities (Wei and Jinlong 2011). A few limitations were associated with Ni-based catalyst makes industries to find an alternative catalyst. The limitations of catalyst were the fast sintering even at low to high temperatures and deactivation of the catalyst surface, and inaccessible active surface by reactant molecules for the dissociation of CO_2 and decomposition of CH_4 (Agnelli et al. 1994, 1998). The deactivation was very high that pluged the reactor and resulted the pressure in the reactor which was also a safety concern. Many progress were reported to alter the synthesis mechanism of Ni-based catalyst by adding-other metals (Ba, K, Mn, Co, Ca, Pt) and found thermally stable, promising results. In a report, methanation was reported over Ru, Co, and Fe supported catalysts, which demonstrated high performance with stability in various conditions (Upham et al. 2015; Xu et al. 2016). As coke formation and metal sintering, limitations need to analysis through thermodynamic data before any proceedings in methanation plants. Greyson et al. (1955) have assessed the thermodynamic data of the CO methanation reaction based on three reaction equations i.e., R(3), R (4) and the following one:

$$CO + H_2O \rightarrow CO_2 + H_2 \quad R(8)$$

Beuls et al. (2012) applied the thermodynamic evaluation of methanation of CO_2 through Rh/γ–Al_2O_3 catalyst with different parameters. Gao and co-workers (2012a) have analytically performed the thermodynamic data study on CO_x methanation utilizing the total GFEM method. This data was compared with experimental data of two commercialized methanation plants and the report revealed that the methanation reaction preferred optimal conditions like low temperature, high pressure with high H_2/CO_x ratio. CO_2 hydrogenation of thermodynamic stimulations was executed by using Gibb's reactor model obtained by HSC chemistry (entropy, enthalpy, capacity of heat). The inclusive portrayal of this reaction process involves the reactant feed of CO_2 and H_2 with six other products like CH_4, CO_2, CO, H_2, H_2O, C deposition. The methanation reaction proceeds with three independent reactions i.e., R(1), R(2), R(7), which are enumerated in Table 1. Depending on these reactions, the equilibrium composition of dissimilar products was obtained at a stable pressure of 1 bar with various H_2/CO_2 ratios and reaction temperature. The major products were concerned for the methanation study and the trace species amounts were neglected. The data concluded that CO generated from the RWGS reaction regulates the product in the CO_2 hydrogenation mechanism if the reaction temperature >873 K, whereas CH_4 formation is not favored. Meanwhile, it is possible to preclude the formation of carbon deposits to a great degree. Increasing the ratio of H_2/CO_2 has a significant impact on CH_4 yield. All these patterns and measurements are well aligned with the findings stated by Swapnesh et al. (2014) with different reaction conditions.

2.1 Low-Pressure CO_2 Methanation Enhanced by Sorption

The sorption enhanced methanation (SEM) model has been suggested to involve higher CH_4 conversion at lower operative pressure. The in situ removal of steam generated in SEM has been achieved by adding a suitable sorbent material (Borgschulte et al. 2013) or with water-selective membranes (Catarina Faria et al. 2018). In this methanation, the elimination of H_2O shifts the chemical equilibrium for allowing 100% CH_4 conversion at a suitable temperature and moderately low pressure, in place of thermodynamic restrictions. These thermodynamic restrictions would suggestively lower the energy for gas compression work desirable to achieve the pressure level for the natural gas grid (Walspurger et al. 2014). One of the key problems during the methanation reactor operation was the deactivation of the catalyst owing to C deposition (Ronsch et al. 2016; Seemann et al. 2006). Thermodynamic analysis and experimental findings have shown that C deposition during CO methanation could be substantial, although it was typically not an issue during pure CO_2 methanation (Frick et al. 2014). However, SEM model thermodynamic study found that C-formation was even more substantial under SEM conditions applicable to the biogas upgrading process than that in the absence of H_2O removal, due to the massive significance of carbon-generating reactions. This suggested SEM conditions could improve C-formation even for pure CO_2 methanation, which was entirely ignored in earlier experimental data and thermodynamic analysis (Koytsoumpa and Karellas 2018). The thermodynamic studies for conventional

Table 1 Thermodynamic analysis equations of CO_2 hydrogenation for synthesis of methane

Reaction equations during CO_2 methanation	Enthalpy $\Delta H_{298.15\ K}$(KJ/mol)	Reaction number
$CO_2 + 4H_2 \leftrightarrow CH_4 + H_2O$	−164.7	R(1)
$CO_2 + H_2 \leftrightarrow CO + H_2O$	41.2	R(2)
$CO + 3H_2 \leftrightarrow CH_4 + H_2O$	−206.2	R(3)
$2CO \leftrightarrow C + 2CO_2$	−172.4	R(4)
$2CH_4 \leftrightarrow C + 2H_2$	74.9	R(5)
$CO + H_2 \leftrightarrow C + H_2O$	−131	R(6)
$CO_2 + 2H_2 \leftrightarrow C + 2H_2O$	−90	R(7)

Swapnesh et al. (2014)

CO and CO_2 methanation were analyzed by various reports (Vakalis et al. 2018), but with SEM conditions are still unexplored. Massa et al. (2020) calculated the pure CO_2 sorption increased methanation thermodynamics at lower pressure, and correlated the data with the conventional methanation process. The basic aim of the study was to identify the limits of carbon deposition for such a scheme and to verify the feasibility of the composition of the equilibrium product gas for injection into the natural gas grid. The following observations were noticed:

- With reverence to traditional methanation systems, the benefits of the SEM system was the production of a high yield of CH_4 at various temperatures and pressures.
- Steam separation, particularly at high temperatures and low pressure, could lead to significant carbon generation. The molar feed of $H_2/CO_2 = 4$ was found to be in negligible amounts of C formation at partial steam capture.
- The by-product CO concentration was stable and noticeably by operative SEM conditions.
- Entire steam apprehension was potential with less C formation with non-stoichiometric feed in presence high amounts of H_2 ($H_2/CO_2 > 4$) is possible.
- These parameters pointed towards 100% CH_4 conversion without CO_2 balance and high CO in product composition.
- Further both non-stoichiometric and stoichiometric feed, was adequate to achieve high H_2 in product outlet.

2.2 Catalytic CO_2 Methanation

Scientific reports disclosed the conversion of CO_2 to CH_4 over various catalytic supports like ZrO_2, Al_2O_3, MgO, SiO_2 (Shawabkeh et al. 2015; Frey et al. 2015; Toemen and Ali 2014; Toemen et al. 2016; Guo and Lu 2014) at altered operating parameters, but still, many essential improvements of a novel economical, dynamic with the stability of the catalyst with the combination of composite metals catalysts, bimetallic catalysts and/or trimetallic catalysts over the support surface are in need. The optimization of parameters from the catalysts side and optimization of parameters from the reaction side is crucial to evaluate the product stability and yield. Ahmad et al. (2020) synthesized and evaluated diverse catalysts of Cu-K/Al_2O_3 for CO_2 conversion at 503 K, 673 K, 873 K temperature with H_2/CO_2 feed ratio (2:1 and 4:1) to produce CH_4 and CO at 1 and 7 bar pressures. The thermodynamic investigation reported 63% CO_2 conversion with CH_4 selectivity of 39% over $Cu_{1.62}K_{0.5}/Al_2O_3$ at 873 K, $H_2/CO_2 = 4$ with the pressure of 7 bar. By adding another metal K to Cu_1/Al_2O_3 on the Al_2O_3 catalysts surface significantly improved the conversion of CO_2 conversion from 48 to 55% ($Cu_1K_{0.5}/Al_2O_3$), which confirmed the role of the second metal as a promoter during the conversion of reactants as well as product selectivity. Moreover, the stability in CO_2 conversion and CH_4 selectivity was also boosted with synergic amounts of promotor with reaction parameters like temperature, H_2/CO_2 molar ratio, and pressure. Table 2 shows the thermodynamics of some of the catalytic conversion of CO_2 to CH_4.

3 Thermodynamics of CO_2 DRM

CO_2 is utilized to produce CH_4 with mixing of H_2, which is used as feed in the FT process (Lavoie 2014) to produce demanded chemicals (Guczi 2010; Paksoy et al. 2015). Reforming is an important process to convert raw materials, as well as fossil fuels, to valuable and demanded fuels and chemicals. The “reforming” is referred to pyrolysis/gasification and is the conversion of natural gas to H_2/CO as well as reforming of solid feed (biomass, coal, waste, etc.). Synthesis gas (or ‘syngas’) is the result of reformation,

Table 2 Catalytic conversion of CO_2 to CH_4

Catalyst	Temperature (K)	Pressure (bar)	CO_2 conversion (%)	CH_4 selectivity (%)	References
Ni-Ru/CZP	673	1	82.2	99.3	Frey et al. (2015)
Cu/Ce/Al_2O_3	573	1	10.23	2.15	Toemen et al. (2016)
Mn/Ce/Al_2O_3	473	1	56.79	8.12	Toemen et al. (2016)
Ni/MgO	873	1	28.9	9	Guo and Lu (2014)
Fe-Co/K/Al_2O_3	723	1	50.3	63	Satthawong et al. (2013)
Co/K/Al_2O_3	573	11	60.6	96	Satthawong et al. (2013)
Ru/Al_2O_3	623	1	86	100	Garbarino et al. (2015)
Ni_2CeO_2/Al_2O_3	623	1	85	100	Liu et al. (2012)
$Ni_{16}(Ni/Al_2O_3)$	773	1	75	100	Garbarino et al. (2014)
Pt/Co/γ-Al_2O_3	573	0.04	10	10.6	Porosoff and Chen (2013)

with H_2 and CO being the key components, but also having H_2O and CO_2 fractions. The foremost reforming expertise for the production of syngas utilizing natural gas is catalytic steam methane reforming (SRM), dry reforming of methane (DRM), auto-thermal reforming (ATR), and partial oxidation (POX) (Bhavani et al. 2020; Mohamad 2018; Shen et al. 2000). The SRM reaction is a reversible reaction of CO_2 methanation and if, carried out under adiabatic circumstance would lead to an extreme theoretical temperature at thermodynamic equilibrium. SRM process needs high content of both steam and H_2/CO_2 to pass through the reactor, which is on the higher side of FT process necessity. ATR process required pure O_2 with inlet feed and was the expensive reaction for industry and concern of safety aspects. In industrial practice, the coke formation was one of the major challenges (Mette et al. 2014). The coke formation deactivated the catalyst, which ultimately increased the operating cost (Pakhare and Spivey 2014). A lot of reports have come up with dealing to trigger the C formation in the DRM process (Wang et al. 2014). Depending upon the H_2/CO ratio, various applications were used for innumerable purposes. For example, the synthesis of liquid product through the FT process required the H_2/CO molar ratio ranges from 1.0 to 2.5, whereas alkane (C1–C5) required the H_2/CO mole ratio of ≥ 2. The hydrocarbon (HC) and oxygenated compounds required the H_2/CO mole ratio ranges from 1 to 2 (Zennaro et al. 2013). The modeling of the reactor was the most economic method compared to 'Trial and Error' method (Zhao et al. 2016). Table 3 shows the thermodynamic analysis equations of CO_2 for DRM (He and Liu 2017).

The thermodynamic equilibrium of the DRM process was investigated on Fact Sage simulation by reducing the total Gibbs energy of the system. These simulations were used to predict and improve the system parameters for the removal of C-deposition and the modification of the H_2/CO mole ratio. The simulated results predicted that the C-deposition favored at 819–976 K and further reported ideal process parameters i.e., 1 CH_4/CO_2 mole ratio, 1 bar pressure and $\geq$ 1273 K temperature. CH_4 decomposition was encouraged at pressure $\leq$ 1 bar, and 823 K $\leq$ T 1273 K about C-formation participants, while CO dissociation was facilitated at process pressure $\geq$ 1 bar, and $T \leq$ 973 K. These types of reactor settings were frequently used to reduce the coke formation to produce olefin, oxygenated compounds, and heavy HCs. For Fischer-Tropsch (FT) process the H_2/CO ratio was used as a feed to synthesis various applications. The raw materials for the FT process were achieved with the tuning of mole ratio CH_4/CO_2, altering unit operating parameters by < 1173 K, or efficient process. For the applications like olefin synthesis, oxygenated compounds and heavy HCs, the DRM (feed $CH_4/CO_2 \leq 1$) reactor process parameters i.e., $T > 973$ K, H_2/CO ratio $\leq$ 1 and P = 1 bar was practiced for best results. Another application was to produce alkane (from C1 to C5) achieved by altering the H_2/CO feed ratio (Cao et al. 2017).

Thermodynamic simulation test was also applied for POX, combined dry reforming for the functions of various ratios CH_4/CO_2 (0–2), O_2/CH_4 (0–1), pressure (1–25 bar), and temperature (873–1573 K) aimed to reduce coke formation by GEM. These thermodynamic stimulations are related to experimental results over Ru supported catalyst. The correlated reactor pressure of DRM and POX reactions results found that on increasing CH_4 conversion pressure, H_2 yield over time decreased. The thermodynamic stimulation results with the function of O_2/CH_4 resulted in reduction of the coke formation by oxidation for carbon formed in reforming reactor over time on stream. Finally, every reforming parameter played a significant role in the decomposition of CH_4 and the dissociation of CO_2 for stable production of H_2/CO for utilizing many industrial applications (Nematollahi et al. 2012).

Table 3 Thermodynamic analysis equations of CO_2 for DRM (He and Liu 2017)

Reaction equations during DRM	Description of reaction	Enthalpy $\Delta H_{298.15\ K}$ (KJ/mol)	Reaction number
$CO_2 + CH_4 \leftrightarrow 2CO + 2H_2$	CO_2 reforming	260.5	R(8)
$CO_2 + H_2 \leftrightarrow 2CO + H_2O$	RWGS	41.0	R(9)
$CH_4 + H_2O \leftrightarrow CO + 3H_2$	Methane steam reforming	206.50	R(10)
$CH_4 + 2H_2O \leftrightarrow CO_2 + 4H_2$	Methane steam reforming	165	R(11)
$CH_4 + 2O_2 \leftrightarrow CO_2 + 2H_2O$	Methane combustion	–802	R(12)
$CH_4 + \frac{1}{2}O_2 \leftrightarrow CO + 3H_2$	Methane partial oxidation	–36	R(13)
$CH_4 \leftrightarrow C + 2H_2$	Methane decomposition	74.9	R(14)
$2CO \leftrightarrow C + CO_2$	Disproportionation reaction	–172.5	R(15)
$CO + H_2 \leftrightarrow C + H_2O$	CO reduction	–131.5	R(16)

4 Thermodynamics of CO_2 Hydrogenation for MEOH and DME Synthesis

MeOH synthesis is one of the most important industrial process to synthesis many chemicals by utilizing CO_2 as a feed but need to be practiced in a bulk production as commercial grade. The commercial-grade MeOH is synthesized at an adequate temperature with high pressure as a feed of CO, CO_2, and H_2. CO_2 is the source of carbon for MeOH in this method, while CO reacts via WGS reaction with H_2O production in the process to generate CO_2 and H_2. These reaction processes are shown in R(17) and R(18), respectively. Tables 4 and 5 illustrate the thermodynamic analysis equations of CO_2 for MeOH and DME synthesis (He and Liu 2017).

The CO_2 hydrogenation was processed using high-performance catalyst or else electrochemical reaction. On other hand, the thermodynamic stimulation and experimental data confirmed limited synthesis and low yield of MeOH by CO_2 hydrogenation. The MeOH production limitations were removed by processing the dehydration reaction to yield DME and also prevented side reactions. Majorly explored were CO_2/CO hydrogenation catalysts and thermodynamic study of CO hydrogenation (Hus et al. 2017). The thermodynamic stimulation data is good enough to estimate the desire/undesired reaction pathways to produce desired products. Finally, it was a significant strategy to alalyze the thermodynamic stimulation before catalytic CO_2 hydrogenation on an industrial scale. Table 6 demonstrates the various types of reaction processes of catalytic CO_2 hydrogenation conversion to produce MeOH at room pressure. Zohour et al. (2016) used doped Cu, Ga metals over γ-Al_2O_3 for CO_2 hydrogenation, the conversion was found to be very low with 48% of DME and MeOH. Ramírez et al. (2017) screened over Cu/ZnO and Pd/Ga bimetallic catalysts and found 100% MeOH selectivity with considerably very low CO_2 conversion.

Gao et al. (2017) found 13.1% CO_2 conversion over In_2O_3 impregnated over medium pore zeolite HZSM-5 to produce 1% of HCs to gasoline. Iyer et al. (2015) tailored isothermal and adiabatic processes for various compositions of the inlet feed in the reactor environment of vapor–liquid and vapor systems. Miguel et al. (2015) also confirmed the 100% CH_4 selectivity with MeOH formation, which was favored thermodynamically. Jia et al. (2016) examined CO_2 hydrogenation thermodynamics of numerous products including MeOH under various reaction parameters without RWGS influence and reported CH_4 formation as the most feasible reaction. He et al. (2017) generated the thermodynamic data using the combinations of reaction parameters like feed ratio, reactor pressure, reactor temperature, the composition of feed concentrations on CO_2 hydrogenation to form DME. This data proved the DMEs formation at lower reactor temperature with higher reactor pressure favored CO_2 hydrogenation. The most thermodynamic parameters were labeled as $H_2/CO_2 = 3.0–6.0$, $T < 550$ K, $P = 20 \sim 60$ bars. Ahmad et al. (2018) found 373 K, $P = 60$ bar, $H_2/CO_2 = 3:1$ for higher CO_2 conversion but lower product selectivity in DME synthesis. Whereas, the maximum HC production was possible at extreme pressures, lower reactor temperatures and with high ratio H_2/CO_2feed. The addition of other feeds also played a vital role in conversion (CO_2) and selectivity (DMEs), as the addition of CO favored and H_2O addition led to form excess steam in the reactor, which need to be removed with time. For every FT process, stable conversion of CO_2 showed the minimum change in Gibbs free energy without the interference of the reaction pressure over a stable and active catalyst. The various thermodynamic

Table 4 Thermodynamic analysis equations of CO_2 for MeOH synthesis

Reaction equations during MeOH synthesis	Description of reaction	Enthalpy ΔH_{298K} (KJ/mol)	Reaction number
$CO_2 + 3H_2 \leftrightarrow CH_3OH + H_2O$	CO_2 hydrogenation to MeOH	–49.51	R(17)
$CO_2 + H_2 \leftrightarrow CO + H_2O$	RWGS	41.19	R(18)
$CO + 2H_2 \leftrightarrow CH_3OH$	CO hydrogenation to MeOH	–90.7	R(19)

Table 5 Thermodynamic analysis equations of CO_2 for DME synthesis

Reaction equations during DME synthesis	Description of reaction	Enthalpy ΔH_{298K} (KJ/mol)	Reaction number
$2CO_2 + 6H_2 \leftrightarrow CH_3OCH_3 + 3H_2O$	Overall reaction	–122.2	R(20)
$CO_2 + 3H_2 \leftrightarrow CH_3OH + H_2O$	CO_2 hydrogenation to MeOH	–49.1	R(21)
$CO_2 + H_2 \leftrightarrow CO + H_2O$	RWGS	41	R(22)
$2CH_3OH \leftrightarrow CH_3OCH_3 + H_2O$	Dehydration of MeOH	–23.4	R(23)
$CO + 2H_2 \leftrightarrow CH_3OH$	CO hydrogenation to MeOH	–90.6	R(24)

He and Liu (2017)

Table 6 Catalytic hydrogenation of CO_2 to MeOH

Catalyst	Temperature (K)	Pressure (bar)	CO_2 conversion (%)	MeOH selectivity (%)	References
Pd/Zn/ZnO + CdSe	543	20	7	67.3	Liao et al. (2017)
Pd/SiO_2	523	30	Less than 1	52	Collins et al. (2012)
Cu/SiO_2	523	41	2.8	15	Jiang et al. (2015)
$PdCu/SiO_2$	523	41	6.7	30	Jiang et al. (2015)
Cu, Zn, Al, Y-layered double hydroxide derived catalyst	523	50	20.2	69.3	Gao et al. (2015)
Cu/Zn, Al, Zr-layered double hydroxide	523	50	22.2	45.8	Gao et al. (2015)
PdZnGa-CO_3aqueous miscible organic-layered double hydroxide derived CuZn catalyst	543	45	18.8	47.8	Li et al. (2018)
PdZnAl layered double hydroxide derived PdZn catalyst	523	30	0.6	60	Ota et al. (2012)
PdMgAllayered double hydroxide derived Pd catalyst	523	30	3	4	Ota et al. (2012)
PdMgGalayered double hydroxide derived Pd_2Ga catalyst	523	30	1.0	47	Ota et al. (2012)
Pt/SiO_2	473	30	0.4	1.9	Shao et al. (1995)
PdZn/ZnO	523	20	8.7	1	Bahruji et al. (2016)

points to be considered were, (i) the ranges of favorable conditions, (ii) reaction kinetics of the experimental process and (iii) energy balance of the reaction sequences. Table 7 illustrates the thermodynamic analysis of CO_2 utilization reactions for different reaction systems.

5 Thermodynamics of CO_2 Hydrogenation for EtOH Synthesis

The development of high-grade EtOH fuel is used to diminish pollution. Usually, EtOH was prepared through ethylene hydration and crop fermentation. The ethylene hydration was having a setback of formation of excess H_2SO_4 as a side product, which is corrosive for the unit process. Fermentation was processed using precursors like corn and grain crops, which has a setback associated with increases in food prices and security. The trigger of the above challenges the CO_2/CO hydrogenation was developed to produce EtOH with high efficiency. As it has high significance the EtOH product need to study thermodynamically before the commercialization over a suitable catalyst. Conventionally, EtOH was produced over Rh loaded various supported catalysts was utilized through CO_2/CO hydrogenation reaction (Lopez et al. 2015; Izumi 1997; Kurakata et al. 1996; Fan et al. 2009) with thermodynamic interpretations. Jia et al. (2016) compared the experimental data with thermodynamics scrutiny for assessing the CO_2 hydrogenation complications. One of the complications was the product with low carbon alcohols, influenced by similar species in product stream with carbon no of C1, C2, C3 and C4 alcohols. On the table top this reaction process was complex, as it produced mixture of a products at the same time. For EtOH production, the first step is the CO_2 hydrogenation reaction and undergoes intermediate step by the formation of CO($CO_2 + H_2 \leftrightarrow CO + H_2O$) then through CO hydrogenation led to EtOH as a final product. Kusama et al. (1997) observed the CO_2 hydrogenation mechanism over Rh-Fe loaded SiO_2 supported catalyst to EtOH (Table 8).

The CO_2 hydrogenation thermodynamics transformations were simulated by the GFEM principle. This process of CO_2 hydrogenation was analyzed by the rate of CO hydrogenation formation and rate of DME hydrogenation, which ended with a product of CH_4 with no side products. The rate of the reaction equilibrium was influenced by reactor inside pressure, temperature, and composition of precursors (molar ratio of H_2/CO) of the reaction system. These reaction equilibrium conditions were achieved in terms of the rate of conversion and selectivity of product. The main reaction equilibrium analysis was inferences as follows:

(1) During the EtOH production from CO_2 hydrogenation with CH_4, the temperature of the reaction increased and at the same time decrease in the rate of conversion of CO_2, H_2 and increase in CO selectivity was noticed. At these reaction parameters, the CH_4 selectivity was 100% with H_2O formation and side products of

Table 7 Thermodynamic analysis of CO_2utilization reactions for different reaction systems

Reaction system	Method	Temperature	Pressure	Other parameters	References
CO_2 methanation	GFEM	470–1073 K	10,100 bar	2.0–6.0 (H_2/CO_2 ratio)	Borgschulte et al. (2013)
CO_2 methanation	GFEM with C deposition	473–1143 K	1–50 bar	1.0–6.0 (H_2/CO_2 ratio)	Walspurger et al. (2014)
CO_2 reforming of CH_4	GFEM with C deposition	500–1300 K	1–50 bar	0.5–3.0 (CO_2/CH_4 ratio)	Walspurger et al. (2014)
CO_2 reforming of CH_4	GFEM	550–1500 K	0.4–12 bar	1.0–10.0 (CO_2/CH_4 ratio)	Moradi et al. (2010)
CO_2 reforming of CH_4	GFEM with C deposition on catalysts	573–1473 K	–	0.5–3.0 (CO_2/CH_4 ratio)	Li et al. (2008)
CO_2reforming of $CH_4$with SMR	GFEM	700–1200 K	–	0.5–3.0 (CO_2/CH_4 ratio)	Sun et al. (2011)
CO_2 reforming of CH_4 with POX	GFEM	600–1200 K	–	0.5–3.0 (CO_2/CH_4 ratio)	Li et al. (2007)
CO_2 reforming of CH_4	GFEM	823–1473 K	0.5–50 bar	0.5–2.0 (CH_4/CO_2 ratio)	He and Liu (2017)
CO_2 reforming of CH_4 with POX	GFEM	600–1300 K	1–25 bar	0–2.0 (CO_2/CH_4 ratio)	Gao et al. (2017)
CO_2 reforming of CH_4	GFEM	573–1473 K	1–25 bar	0.5–3.0 (CO_2/CH_4 ratio)	Guangxin et al. (2006)
DME synthesis	GFEM with C deposition	473–1143 K	1–50 bar	1.0–6.0 (H_2/CO_2 ratio)	Walspurger et al. (2014)
DME synthesis	Thermodynamic equilibrium model	470–2673 K	20–50 bar	1.0–2.0 (H_2/CO ratio) Catalyst-$CuO/ZnO/Al_2O_3$ (MeOH synthesis) and H-ZSM-5 (MeOH dehydration)	Jia et al (2016)
DME synthesis	GFEM	293–673 K	1–100 bar	1.020.0 0 (H_2/CO_2 ratio)	Moradi et al. (2010)
DME synthesis	Equilibrium constant method	473–573 K	10–90 bar	0.5–5.0 (H_2/CO_2 ratio)	Guangxin et al. (2006)
DME synthesis	–	470–480 K	30–100 bar	0.1–1.0 (CO_2concentration)	Edwards and Maitra (1995)

Table 8 Thermodynamic analysis equations of CO_2 for EtOH synthesis

Reaction equations during DME synthesis	Description of reaction	Enthalpy ΔH_{298K} (KJ/mol)	Reaction number
$2CO_2 + 6H_2 \leftrightarrow C_2H_5OH + 3H_2O$	CO_2 hydrogenation to EtOH	−86.7	R(25)
$2CO + 4H_2 \leftrightarrow C_2H_5OH + H_2O$	CO hydrogenation to EtOH	−253.6	R(26)
$CH_3OH + CO + 2H_2 \leftrightarrow C_2H_5OH + H_2O$	MeOH hydrogenation to EtOH	−165.1	R(27)
$C_2H_5OH \leftrightarrow CH_4 + CO + H_2$	Dehydration of EtOH	−23.4	R(28)

He (2017)

oxygenated HCs in negligible amounts, concluding methanation reaction as thermodynamically favorable.

(2) The EtOH production from the hydrogenation of CO and DME with CH_4 showed similar laws of reactions as CO_2 hydrogenation. The rise in reactor temperature affected the drop in the formation of H_2 and DME. At these reaction parameters, the rate of CO conversion was 100% with 100% CH_4 selectivity, and the side products of oxygenated HCs in negligible amounts.

(3) For EtOH products from CO_2 hydrogenation without CH_4 was derived by the equilibrium transformations. Initially, CO_2 transformations decreased and later

increased with a rise in the reactor temperature. 100% EtOH selectivity was found at low temperature and further drops due to a rise in the reactor temperature. The EtOH selectivity was found to be higher than MeOH. Enhancing the pressure resulted in high selectivity of EtOH at a given temperature range.

(4) The CH_4 was found to be the more common product for CO hydrogenation as compared to CO_2 hydrogenation, which cleared the advantage of the CO hydrogenation process. Apart from that, the CO hydrogenation was conveniently operated at different temperature ranges leading greater rate of conversion and high EtOH selectivity. The optimization thermodynamic parameters for the CO_2 hydrogenation were 1.0–5.0 MPa pressure with >500 K temperature and the precursor's molar ratio of H_2/CO_2 was 3.0–5.0 whereas, for the CO hydrogenation, the pressure was the same i.e., 1.0–5.0 MPa with >650 K temperature and precursor molar ratio of H_2/CO was 0.5–2.0.

(5) The thermodynamic data with experimental data correlation proved that the catalysts played a vital role in hydrogenation reaction, which also increased the rate of conversion as well as desired product selectivity.

(6) The CO_2 hydrogenation process was complex and ended with a lot of products. The large species of product mixture led to the formation of carbon deposition, which favored at low temperature. The formation of carbon deposition completely deactivated the catalyst bed. Finally, the catalyst deactivation needs to trigger by understanding the complexity of the reaction mechanism by kinetic study and thermodynamic assessment to develop the commercial-grade production (He 2017).

6 Conclusion

The CO_2 utilization is an alternative energy source for promising approaches to produce valuable chemicals, and simultaneously reduces greenhouse gas emissions. The CO_2 conversion is very much significant as it has an industrial development and research prospects. CO_2 is transformed as a methanation reaction to produce CH_4 or H_2/CO is produced by mixing the feed with CH_4 to synthesize other industrially demanded chemicals. These types of reactions are depending on catalyst preparation methods, reaction parameters, and thermodynamics data to yield selective and stable product formation. This chapter is intended to highlight the significance of thermodynamics of CO_2 transformation to various other reactions. The thermodynamics data was compared with research reports of experimental data of CH_4 production, H_2/CO formation, MeOH synthesis, DME synthesis through reforming and hydrogenation, methanation reactions. CO_2 hydrogenation is mainly preceded for the production of fuel, olefins, and many other fine chemicals. From reports, few benchmarks are noted that methanation is exothermic and catalytically favorable over lower temperatures and higher pressure with a specific designed reactor to rectify the thermodynamic challenges for MeOH production. All the reactions commonly face low conversion with catalyst deactivation, which affects the production cost. The catalyst deactivation is a major problem of all the reactions by coke formation over the catalyst surface. Finally, the thermodynamic prediction is necessary to identify the actual problem and resolve for economically viable process.

Acknowledgements Dr. Pallavi Jain is grateful to SRM Institute of Science and Technology, Delhi-NCR campus, Modinagar for support and encouragement.

References

Agnelli M, Kolb M, Mirodatos C (1994) Co Hydrogenation on a nickel catalyst: 1. Kinetics and modeling of a low-temperature sintering process. J Catal 148:9–21. https://doi.org/10.1006/jcat.1994.1180

Agnelli M, Swaan HM, Marquez-Alvarez C, Martin GA, Mirodatos C (1998) CO hydrogenation on a nickel catalyst: II. A mechanistic study by transient kinetics and infrared spectroscopy. J Catal 175:117–128. https://doi.org/10.1006/jcat.1998.1978

Ahmad K, Upadhyayula S (2018) Greenhouse gas CO_2 hydrogenation to fuels: a thermodynamic analysis. Environ Prog Sustain Energy 38:98–111. https://doi.org/10.1002/ep.13028

Ahmad W, Al-Matar A, Shawabkeh R, Aslam Z, Malik IA, Irshad HM (2020) Cu-K/Al_2O_3 based catalysts for conversion of carbon dioxide to methane and carbon monoxide. Chem Eng Commun 207:946–960. https://doi.org/10.1080/00986445.2019.1631815

Alvarez A, Bansode A, Urakawa A, Bavykina AV, Wezendonk TA, Makkee M, Gascon J, Kapteijn F (2017) Challenges in the greener production of formates/formic acid, methanol, and DME by heterogeneously catalyzed CO_2 hydrogenation processes. Chem Rev 117:9804–9838. https://doi.org/10.1021/acs.chemrev.6b00816

Bahruji H, Bowker M, Hutchings G, Dimitratos N, Wells P, Gibson E, Jones W, Brookes C, Morgan D, Lalev G (2016) Pd/ZnO catalysts for direct CO_2 hydrogenation to methanol. J Catal 343:133–146. https://doi.org/10.1016/j.jcat.2016.03.017

Barbarossa V, Bassona C, Deiana P, Vanga G (2013) CO_2 conversion to CH_4. In: de Falco M, Iaquaniello G, Centi G (eds) CO_2 a valuable source of carbon. Springer, London, pp 123–145. https://doi.org/10.1007/978-1-4471-5119-7

BeulsA SC, Jacquemin M, Heyen G, Karelovic A, Ruiz P (2012) Methanation of CO_2: further insight into the mechanism over Rh/γ-Al_2O_3catalyst. Appl Catal B 113–114:2–10. https://doi.org/10.1016/j.apcatb.2011.02.033

Bhavani GA, Vats T, Reddy SN (2020) Challenges in CO_2 reforming with methane for production of hydrogen rich, stable syngas. J Nanosci Nanotechnol 20:3943–3950. https://doi.org/10.1166/jnn.2020.17500

Borgschulte A, Gallandat N, Probst B, Suter R, Callini E, Ferri D, Arroyo Y, Erni R, Geerlings H, Zuttel A (2013) Sorption enhanced CO_2 methanation. Phys Chem Chem Phys 15:9620–9625. https://doi.org/10.1039/c3cp51408k

Cao P, Adegbite S, Wu T (2017) Thermodynamic equilibrium analysis of CO_2 reforming of methane: elimination of carbon deposition and adjustment of H_2/CO ratio. Energy Procedia 105:1864–1869. https://doi.org/10.1016/j.egypro.2017.03.546

Faria AC, Miguel CV, Madeira LM (2018) Thermodynamic analysis of the CO_2 methanation reaction with *in situ* water removal for biogas upgrading. J CO_2 Util 26:271–280. https://doi.org/10.1016/j.jcou.2018.05.005

Collins SE, Delgado JJ, Mira C, Calvino JJ, Bernal S, Chiavassa DL, BaltanásMA BAL (2012) Understanding the role of oxygen vacancies in the water gas shift reaction on ceria-supported platinum catalysts. J Catal 292:90–98. https://doi.org/10.1021/cs500323u

Díez-Ramírez J, Dorado F, Osa AR, Valverde JL, Sánchez P (2017) Hydrogenation of CO_2 to methanol at atmospheric pressure over Cu/ZnO catalysts: Influence of the calcination, reduction, and metal loading. Ind Eng Chem Res 56:1979–1987. https://doi.org/10.1021/acs.iecr.6b04662

Dorner RW, Hardy DR, Williams FW, Willauer HD (2010) Heterogeneous catalytic CO_2 conversion to value-added hydrocarbons. Energy Environ Sci 3:884–890. https://doi.org/10.1039/C001514H

Edwards JH, Maitra AM (1995) The chemistry of methane reforming with carbon dioxide and its current and potential applications. Fuel Process Technol 42:269–289. https://doi.org/10.1016/0378-3820(94)00105-3

Fan Z, Chen W, Pan X, Bao X (2009) Catalytic conversion of syngas into C_2 oxygenates over Rh-based catalysts—Effect of carbon supports. Catal Today 147:86–93. https://doi.org/10.1016/j.cattod.2009.03.004

Frey M, Edouard D, Roger AC (2015) Optimization of structured cellular foam-based catalysts for low-temperature carbon dioxide methanation in a platelet milli-reactor. Comptes Rendus Chim 18:283–292. https://doi.org/10.1016/j.crci.2015.01.002

Frick V, Brellochs J, Specht M (2014) Application of ternary diagrams in the design of methanation systems. Fuel Proc Technol 118:156–160. https://doi.org/10.1016/j.fuproc.2013.08.022

Gao J, Wang Y, Ping Y, Hu D, Xu G, Gu F (2012a) A thermodynamic analysis of methanation reactions of carbon oxides for the production of synthetic natural gas. RSC Adv 2:2358–2368. https://doi.org/10.1039/C2RA00632D

Gao P, Zhong L, Zhang L, Wang H, Zhao N, Wei W, Sun Y (2015) Yttrium oxide modified $Cu/ZnO/Al_2O_3$ catalysts via hydrotalcite-like precursors for CO_2 hydrogenation to methanol. Catal Sci Technol 5:4365–4377. https://doi.org/10.1039/C5CY00372E

Gao P, Li S, Bu X, Dang S, Liu Z, Wang H, Zhong L, Qiu M, Yang C, Cai J, Wei W, Sun Y (2017) Direct conversion of CO_2 into liquid fuels with high selectivity over a bifunctional catalyst. Nat Chem 9:1019–1024. https://doi.org/10.1038/nchem.2794

Garbarino G, Riani P, Magistri L, Busca G (2014) A study of the methanation of carbon dioxide on Ni/Al_2O_3 catalysts at atmospheric pressure. Int J Hydrogen Energy 39:11557–11565. https://doi.org/10.1016/j.ijhydene.2014.05.111

Garbarino G, Bellotti D, Riani P, Magistri L, Busca G (2015) Methanation of carbon dioxide on Ru/Al_2O_3 and Ni/Al_2O_3 catalysts at atmospheric pressure: Catalysts activation, behaviour and stability. Int J Hydrogen Energy 40:9171–9182. https://doi.org/10.1016/j.ijhydene.2015.05.059

Bhavani AG, Kim WY, Kim JY, Lee JS (2013) Improved activity and coke resistance by promoters of nanosizedtrimetallic catalysts for autothermal carbon dioxide reforming of methane. Appl Cat A Gen 450:63–72. https://doi.org/10.1016/j.apcata.2012.10.008

Geetha BA, Won YK, Jin WL, Jae SL (2015) Influence of metal particle size on oxidative CO_2 reforming of methane over supported nickel catalysts: effects of second-metal addition. Chem Cat Chem 7:1445–1452. https://doi.org/10.1002/cctc.201500003

Gotz M, Lefebvre J, Mors F, McDaniel Koch A, Graf F, Bajohr S, Reimert R, Kolb T (2016) Renewable power-to-gas: a technological and economic review. Renew Energy 85:1371–1390. https://doi.org/10.1016/j.renene.2015.07.066

Greyson M, Demeter JJ, Schlesinger MD, Johnson GE, Jonakin J, Myers JJ (1955) Synthesis of Methane, D.O. Interior Report 5137, Bureau of Mines, U.S.

Guangxin J, Tan Y, Han Y (2006) A comparative study on the thermodynamics of dimethyl ether synthesis from CO hydrogenation and CO_2 hydrogenation. Ind Eng Chem Res 45:1152–1159. https://doi.org/10.1021/ie050499b

Guczi L et al (2010) Methane dry reforming with CO_2: a study on surface carbon species. Appl Catal A-Gen 375(2):236–246. https://doi.org/10.1016/j.apcata.2009.12.040

Guo M, Lu G (2014) The effect of impregnation strategy on structural characters and CO_2 methanation properties over MgO modified Ni/SiO_2 catalysts. Catal Commun 54:55–60. https://doi.org/10.1016/j.catcom.2014.05.022

He X (2017) CO_2 hydrogenation for ethanol production: a thermodynamic analysis. Int J Oil Gas Coal Eng 5:145–152. https://doi.org/10.11648/j.ogce.20170506.14

He X, Liu L (2017) Thermodynamic analysis on the CO_2 conversion processes of methane dry reforming for hydrogen production and CO_2 hydrogenation to dimethyl ether. IOP Conf Ser: Earth Environ Sci 100:012078. https://doi.org/10.1088/1755-1315/100/1/012078

Huang CH, Tan CS (2014) A review: CO_2utilization. Aerosol Air Qual Res 14:480–499. https://doi.org/10.4209/aaqr.2013.10.0326

Hus M, Dasireddy VDBC, Stefancic NS, Likozar B (2017) Mechanism, kinetics and thermodynamics of carbon dioxide hydrogenation to methanol on $Cu/ZnAl_2O_4$ spinel-type heterogeneous catalysts. Appl Catal B Environ 207:267–278. https://doi.org/10.1016/j.apcatb.2017.01.077

Iyer SS, Renganathan T, Pushpavanam S, Kumar MV, Kaisare N (2015) Generalized thermodynamic analysis of methanol synthesis: Effect of feed composition. J CO_2 Util 10:95–104. https://doi.org/10.1016/j.jcou.2015.01.006

Izumi Y (1997) Selective ethanol synthesis from carbon dioxide: Roles of rhodium catalytic sites. Platin Met Rev 41:166–170

Jia C, Gao J, Dai Y, Zhang J, Yang Y (2016) The thermodynamics analysis and experimental validation for complicated systems in CO_2 hydrogenation process. J Energy Chem 25:1027–1037. https://doi.org/10.1016/j.jechem.2016.10.003

Jiang X, Koizumi N, Guo X, Song C (2015) Bimetallic Pd-Cu catalysts for selective CO_2 hydrogenation to methanol. Appl Catal B-Environ 170–171:173–185. https://doi.org/10.1016/j.apcatb.2015.01.010

Kawi S, Kathiraser Y (2015) CO_2 as an oxidant for high-temperature reactions. Front Energy Res 3:13–30. https://doi.org/10.3389/fenrg.2015.00013

Kiewidt L, Thoming J (2015) Predicting optimal temperature profiles in single-stage fixed-bed reactors for CO_2-methanation. Chem Eng Sci 132:59–71. https://doi.org/10.1016/j.ces.2015.03.068

Kopyscinski J, Schildhauer TJ, Biollaz SMA (2010) Production of synthetic natural gas (SNG) from coal and dry biomass-a technology review from 1950 to 2009. Fuel 89:1763–1783. https://doi.org/10.1016/j.fuel.2010.01.027

Koytsoumpa EI, Karellas S (2018) Equilibrium and kinetic aspects for catalytic methanation focusing on CO_2 derived substitute natural gas (SNG). Renew Sust Energy Rev 94:536–550. https://doi.org/10.1016/j.rser.2018.06.051

Kurakata H, Izumi Y, Aika K (1996) Ethanol synthesis from carbon dioxide on TiO_2-supported $[Rh_{10}Se]$ catalyst. Chem Commun 3:389–390. https://doi.org/10.1039/CC9960000389

Kusama H, Okabe K, Sayama K, Arakawa H (1997) Ethanol synthesis by catalytic hydrogenation of CO_2 over Rh–Fe/SiO_2 catalysts. Energy 22:343–348. https://doi.org/10.1016/S0360-5442(96)00095-3

Lavoie JM (2014) Review on dry reforming of methane, a potentially more environmentally-friendly approach to the increasing natural gas exploitation. Front Chem 2:81. https://doi.org/10.3389/fchem.2014.00081

Lehner M, Tichler R, Steinmuller H, Koppe M (2014) Power-to-gas: Technology and business models, 1st edn. Springer International Publishing, Springer. https://doi.org/10.1007/978-3-319-03995-4

Li Y, Jin B, Xiao R (2007) Carbon dioxide reforming of methane with a free energy minimization approach. Korean J Chem Eng 24:688–692. https://doi.org/10.1007/s11814-007-0027-5

Li Y, Wang Y, Zhang X, Mi Z (2008) Thermodynamic analysis of autothermal steam and CO_2 reforming of methane. Int J Hydrogen Energy 33:2507–2514. https://doi.org/10.1016/j.ijhydene.2008.02.051

Li MM-J, Chen C, Ayvali T, Suo H, Zheng J, Teixeira I, Ye L, Zou H, O'Hare D, Tsang SCE (2018) CO_2 hydrogenation to methanol over catalysts derived from single cationic layer CuZnGa LDH precursors. ACS Catal 8:4390–4401. https://doi.org/10.1021/acscatal.8b00474

Liao F, Wu XP, Zheng J, Li MM-J, Kroner A, Zeng Z, Hong X, Yuan Y, Gong X-Q, Tsang SCE (2017) A promising low pressure methanol synthesis route from CO_2 hydrogenation over Pd/Zn core–shell catalysts. Green Chem 19:270–280. https://doi.org/10.1039/C6GC02366E

Liu H, Zou X, Wang X, Lu X, Ding W (2012) Effect of CeO_2 addition on Ni/Al_2O_3 catalysts for methanation of carbon dioxide with hydrogen. J Nat Gas Chem 21:703–707. https://doi.org/10.1016/S1003-9953(11)60422-2

Lopez L, Velasco J, Montes V, Marinas A, Cabrera S, Boutonnet M, Jaras S (2015) Synthesis of ethanol from syngas over Rh/MCM-41 Catalyst: effect of water on product selectivity. Catalysts 5:1737–1755. https://doi.org/10.3390/catal5041737

Massa F, Coppola A, Scala F (2020) A thermodynamic study of sorption-enhanced CO_2 methanation at low pressure. J CO_2 Util 35:176–184. https://doi.org/10.1016/j.jcou.2019.09.014

Mazza A, Bompard E, Chicco G (2018) Applications of power to gas technologies in emerging electrical systems. Renew Sustain Energy Rev 92:794–806. https://doi.org/10.1016/j.rser.2018.04.072

Mette KK, Dudder S, Kahler H, Tarasov K, MuhlerM BM (2014) Stable performance of Ni catalysts in the dry reforming of methane at high temperatures for the efficient conversion of CO_2 into syngas. Chem Cat Chem 6:100–104. https://doi.org/10.1002/cctc.201300699

Meylan FD, Moreau V, Erkman S (2016) Material constraints related to storage of future European renewable electricity surpluses with CO_2 methanation. Energy Policy 94:366–376. https://doi.org/10.1016/j.enpol.2016.04.012

Miguel CV, Soria MA, Mendes A, Madeira LM (2015) Direct CO_2 hydrogenation to methane or methanol from post-combustion exhaust streams-a thermodynamic study. J Nat Gas Sci Eng 22:1–8. https://doi.org/10.1016/j.jngse.2014.11.010

Mohamad HA (2018) A mini-review on CO_2reforming of methane. Progress Petrochem Sci 2:161–165. https://doi.org/10.31031/PPS.2018.02.000532

Molino A, Braccio G (2015) Synthetic natural gas SNG production from biomass gasification—thermodynamics and processing aspects. Fuel 139:425–429. https://doi.org/10.1016/j.fuel.2014.09.005

Moradi GR, Ahmadpour J, Yaripour F, Wang J (2010) Equilibrium calculations for direct synthesis of dimethyl ether from syngas. Can J Chem Eng 89:108–115. https://doi.org/10.1002/cjce.20373

Napp TA, Gambhir A, Hills TP, Florin N, Fennell PS (2014) A review of the technologies, economics and policy instruments for decarbonising energy-intensive manufacturing industries. Renew Sustain Energy Rev 30:616–640. https://doi.org/10.1016/j.rser.2013.10.036

Nematollahi B, Rezaei M, Lay EN, Khajenoori M (2012) Thermodynamic analysis of combined reforming process using Gibbs energy minimization method: in view of solid carbon formation. Jnat Gas Chem 21:694–702. https://doi.org/10.1016/s1003-9953(11)60421-0

Ota A, Kunkes EL, Kasatkin I, Groppo E, Ferri D, Poceiro B, Navarro Yerga RM, Behrens M (2012) Comparative study of hydrotalcite-derived supported Pd_2Ga and PdZn intermetallic nanoparticles as methanol synthesis and methanol steam reforming catalysts. J Catal 293:27–38. https://doi.org/10.1016/j.jcat.2012.05.020

Pakhare D, Spivey J (2014) A review of dry (CO_2) reforming of methane over noble metal catalysts. Chem Soc Rev 43(22):7813–7837. https://doi.org/10.1039/C3CS60395D

Paksoy AI, Caglayan BS, Aksoylu AE (2015) A study on characterization and methane dry reforming performance of Co–Ce/ZrO_2 catalyst. Appl Catal B Environ 168–169:164–174. https://doi.org/10.1016/j.apcatb.2014.12.038

PorosoffMD CJG (2013) Trends in the catalytic reduction of CO_2 by hydrogen over supported monometallic and bimetallic catalysts. J Catal 301:30–37. https://doi.org/10.1016/j.jcat.2013.01.022

Ronsch S, Schneider J, Matthischke S, Schluter M (2016) Review on methanation-from fundamentals to current projects. Fuel 166:276–296. https://doi.org/10.1016/j.fuel.2015.10.111

Sabatier P (1902) New Synthesis of Methane. Comptesrendus 134:514–516

Satthawong R, Koizumi N, Song C, Prasassarakich P (2013) Bimetallic Fe–Co catalysts for CO_2 hydrogenation to higher hydrocarbons. J CO_2 Util 3–4:102–106. https://doi.org/10.1016/j.jcou.2013.10.002

Seemann M, Schildhauer TJ, Biollaz S, Stucki S, Wokaun A (2006) The regenerative effect of catalyst fluidization under methanation conditions. Appl Catal A Gen 313:14–21. https://doi.org/10.1016/j.apcata.2006.06.048

Shao C, Fan L, Fujimoto K, Iwasawa Y (1995) Selective methanol synthesis from CO_2/H_2 on new SiO_2-supported PtW and PtCr bimetallic catalysts. Appl Catal A-Gen 128:L1–L6. https://doi.org/10.1016/0926-860X(95)00109-3

Shawabkeh RA, Hussein IA, Ahmad W, Rana AG (2015) Synthesis of a new Cu-Aluminosilicate catalyst for CO_2capture and conversion to hydrocarbons. Proceedings of the 4th international gas processing symposium. 49–58. https://doi.org/10.1016/B978-0-444-63461-0.50005-5

Shen WJ, Jun KW, Choi HS, Lee KW (2000) Thermodynamic investigation of methanol and dimethyl ether synthesis from CO_2 hydrogenation. Korean J Chem Eng 17:210–216. https://doi.org/10.1007/BF02707145

Smith JM, Ness HCV, Abbott M (2005) Introduction to chemical engineering thermodynamics. McGraw-Hill chemical engineering series, 7th edn. McGraw-Hill, Boston. https://trove.nla.gov.au/work/6873286

Specht JBM, Frick V, Sturmer B, Zuberbuhler U (2016) Technical realization of power-to-gas technology (P2G): production of substitute natural gas by catalytic methanation of H_2/CO_2. In: Basshuysen VR (ed) Natural gas and renewable methane for powertrains: future strategies for a climate-neutral mobility. Springer, Cham, pp 141–167

Stenger HG, Askonas CF (1986) Thermodynamic product distributions for the Fischer-Tropsch synthesis. Ind Eng Chem Fund 25:410–413. https://doi.org/10.1021/i100023a018

Sun Y, Ritchie T, Hla SS, McEvoy S, Stein W, Edwards JH (2011) Thermodynamic analysis of mixed and dry reforming of methane

for solar thermal applications. J Nat Gas Chem 20:568–576. https://doi.org/10.1016/S1003-9953(10)60235-6
Swapnesh A, Srivastava VC, Mall ID (2014) Comparative study on thermodynamic analysis of CO_2 utilization reactions. Chem Eng Technol 37:1765–1777. https://doi.org/10.1002/ceat.201400157
Toemen S, Bakar WAWB, Ali R (2016) Effect of ceria and strontia over Ru/Mn/Al_2O_3 catalyst: Catalytic methanation, physicochemical and mechanistic studies. J CO_2 Util 13:38–49. https://doi.org/10.1016/j.jcou.2015.11.005
ToemenS BWAWA, Ali R (2014) Investigation of Ru/Mn/Ce/Al_2O_3 catalyst for carbondioxide methanation: catalytic optimization, physicochemical studies and RSM. J Taiwan Inst Chem Eng 45:2370–2378. https://doi.org/10.1016/j.jtice.2014.07.009
Upham DC, Derk AR, Sharma S, Metiu H, McFarland EW (2015) CO_2 methanation by Ru-doped ceria: the role of the oxidation state of the surface. Catal Sci Technol 5:1783–1791. https://doi.org/10.1039/c4cy01106f
Vakalis S, Malamis D, Moustakas K (2018) Thermodynamic modelling of an onsite methanation reactor for upgrading producer gas from commercial small scale biomass gasifiers. J Environ Manag 216:145–152. https://doi.org/10.1016/j.jenvman.2017.06.044
Walspurger S, Elzinga GD, Dijkstra JW, Saric M, Haije WG (2014) Sorption enhanced methanation for substitute natural gas production: experimental results and thermodynamic considerations. Chem Eng J 242:379–386. https://doi.org/10.1016/j.cej.2013.12.045
Wang C, Sun N, Zhao N, Wei W, Zhang J, Zhao T, Sun Y, Sun C, Liu H, Snape CE (2014) The Properties of Individual Carbon Residuals and Their Influence on The deactivation of Ni-CaO/ZrO_2catalysts in CH_4 dry reforming. Chem Cat Chem 6:640–648. https://doi.org/10.1002/cctc.201300754
Wei W, Jinlong G (2011) Methanation of carbon dioxide: an overview. Front Chem Sci Eng 5:2–10. https://doi.org/10.1007/s11705-010-0528-3
Xu J, Su X, Duan H, Hou B, Lin Q, Liu X, Pan X, Pei G, Geng H, Huang Y (2016) Influence of pretreatment temperature on catalytic performance of rutile TiO_2supported ruthenium catalyst in CO_2 methanation. J Catal 333:227–237. https://doi.org/10.1016/j.jcat.2015.10.025
Zennaro R, Ricci M, Bua L, Querci C, Carnelli L, Monforte AA (2013) Syngas: The basis of Fischer–Tropsch. In: Maitlis PM, Klerk A (eds) Greener Fischer-Tropschprocesses for fuels and feedstocks. WILEY-VCH, pp 17–51. https://doi.org/10.1002/9783527656837.ch2
Zhao H, Yang G, Gao X, Pang C, Kingman S, Lester E, Wu T (2016) Hg^0-Temperature-programmed surface reaction and its application on the investigation of metal oxides for Hg^0 capture. Fuel 181:1089–1094. https://doi.org/10.1016/j.fuel.2016.04.095
Zohour B, Yilgor I, Gulgun MA, Birer O, Unal U, Leidholm C, Senkan S (2016) Discovery of superior Cu-GaO_x-HoO_y catalysts for the reduction of carbon dioxide to methanol at atmospheric pressure. Chem Cat Chem 8:1464–1469. https://doi.org/10.1002/cctc.201600020

Carbon Dioxide-Based Green Solvents

Saraswati Soren, Tejaswini Sahoo, Jagannath Panda, Deepak Kumar Senapati, J. R. Sahu, C. K. Rath, and Rojalin Sahu

Abstract

Currently, the 'sustainability' and 'renewability' is most important aspects for the environment and for the living beings. In the chemical industry, the solvent plays a key role in the chemical processes. A large number of chemical processes are carried out in the presence of solvents. In many chemical industries, the volatile organic compounds (VOCs) are used as a solvent which have adverse effect on the environment and human health. As VOCs are hazardous, there is a need of replacement of these traditional, volatile organic solvent. Hence, there is a rising interest in non-volatile solvents. The nature of CO_2 in its liquid and supercritical state as a solvent is explored. CO_2 is abundant in the atmosphere. It can reach its critical states by raising the temperature and pressure greater than its critical value. Supercritical carbon dioxide ($scCO_2$) is readily available. It is cheap, non-flammable, recyclable, and non-toxic. Supercritical fluids (SCFs) have both gaseous and liquid properties, which enable it to penetrate anything and to dissolve materials into their compound, respectively. In addition, supercritical CO_2 and H_2O form a solvent which is organic in nature. These supercritical fluids can be useful in a number of fields like synthesis of various materials, drug delivery, chromatography, processing of polymers, extraction, purification and separation, biomedical applications, etc. This chapter examines the usage of CO_2 as a solvent to generate greener process and to develop different products.

S. Soren · T. Sahoo · J. Panda · D. K. Senapati · J. R. Sahu · C. K. Rath · R. Sahu (✉)
School of Applied Sciences, Kalinga Institute of Industrial Technology, Deemed To Be University, Bhubaneswar, 751024, India
e-mail: rsahufch@kiit.ac.in

Keywords

CO_2 • Green solvent • Supercritical CO_2 ($scCO_2$) • Supercritical fluid (SCF) • Volatile organic compound (VOC) • Sustainability • Renewability • Polymer processing • Chromatography

1 Introduction

Today, the sustainability and renewability are the two watch words. This is the focus in the scientific development on the idea of green chemistry, a perception which encourages the preparation of products which are chemically efficient and these methods also decrease the use or generation of hazardous substances (Anastas and Warner 1998). The attention regarding the same has been drawn on the environment by the help of CO_2 and supercritical fluids which act as a solvent in chemical processing (Eckert et al. 1996). Hence, the different volatile organic compounds (VOCs) are replaced by $scCO_2$ which are currently being used (Sherman et al. 1998). These $scCO_2$ are easily available, of low cost, recyclable, inflammable, and are not restricted by the US Environment Protection Agency (EPA) (Poliakoff et al. 2002; Persichilli et al. 2012; Guan et al. 1993; Saari et al. 2011; DeSimone 2002; Sarbu et al. 2000; Subramaniam et al. 1997). Supercritical CO_2 is also safe as it is non-toxic. Many consumable items like food and pharmaceutical products are manufactured by the help of supercritical CO_2. It is also used in many different processes like extractions (Campbell et al. 2001), nanoparticle production, and modification (Ohde et al. 2002, 2001, 2000) and also in processing different polymers (Yeo and Kiran 2005; Du et al. 2009; Wood et al. 2004; Cao et al. 2006). Many interesting solvents can be prepared from supercritical fluids. This can be attributed toward the opportunities for matching the properties of the solvent by changing the temperature and pressure. The supercritical conditions can be easily calculated for CO_2.

Inamuddin et al. (eds.), *Carbon Dioxide Utilization to Sustainable Energy and Fuels*,
Advances in Science, Technology & Innovation, https://doi.org/10.1007/978-3-030-72877-9_17

The critical temperature (T_c) for CO_2 was found to be 31.1 ° C, and the critical pressure (P_c) was calculated as 72.8 bar (Peach and Eastoe 2014). The development in the design of green chemicals and environmentally improved routes are the two main surfaces of green chemistry to reduce the influence of chemical developments and compounds on the surroundings. A novel enhanced method of synthesis of chemicals encounters social, economic, and ecological standards in order to achieve the goal of sustainable development and for cleaner production. The environmentally benign technology is the design of minimizing wastes in chemical reactions, air emission minimization, and hazardous reactants minimization. The minimization of unused substances and their discharge in the environment is fully dependent on the productivity yield, its selectivity, and the utilization of atoms in the case of an academic chemist. Therefore, in the chemical industry, catalytic processes have significant impact to deliver quick and selective chemical transformation effective recovery for both catalyst and product which is deprived of generating any environmental problems. In most of the cases, the recovery of the catalyst and its recycling process is more important for the economic process owning to its multistage, erudite, and expensive recycling operations. In few cases, alternatively leaving the catalyst is acceptable. For example, some catalysts are so productive like the Ziegler–Natta catalyst and metallocene-based catalysts that their very low concentrations in ppm level are left behind in the polyolefin products (Ballivet-Tkatchenko et al. 2005). Since 1950, CO_2 was investigated as a very good solvent in the industry or the factory. In general, the supercritical fluids show many interesting physical properties (Krukonis 1994; Sako 2002). It displays specific attention for CO_2 which is exaggerated due to its green characteristics. Some of its green features are that CO_2 is not inflammable, relatively non-toxic, and comparatively non-reactive. Along with water, the supercritical regime of carbon dioxide is formed and shows its critical temperature of 304 K. In this very chapter, the main highlight is on the utilization of carbon dioxide which gives the betterment to the greenness and sustainability of the process and product. Carbon dioxide is considered as green solvent all over the material and chemical research world and its utilization is allowed. Utilization of carbon dioxide as solvent in raw material give rise to green or safety benefits, cost effectiveness, process benefits, and property and product benefits (F'Oliakoff et al. 1999). From the time of Mendeleev (Saito 2013) to today, supercritical fluid research has a long history. Supercritical fluids are gases compressed to reach their densities to those of liquids. Such compression can only occur above so called 'critical temperature' of the fluid, at lower temperature; the liquid will liquefy on compression. Michels and Michels (1935), prior to half century, identified liquid carbon dioxide. Similarly, Charles Thilorier in the year 1834 produced solid carbon dioxide, when he opened the container of pressurized liquid carbon dioxide (Michels et al. 1935). Carbon dioxide physical properties are dependent on pressure and temperature. At low temperatures, solid carbon dioxide is favored which undergoes sublimation and gets converted to vapor phase at 5.1 bar lower pressure and heating. In early 1930, carbon dioxide thermo-physical characteristics have been recorded as function of temperature and pressure thermodynamic model (Mohammed 2012; Bansal 2012; Steiner 1993). The extraordinary chemical and physical characteristics of carbon dioxide makes it suitable to be utilized in industrial and laboratories applications in a wide scale. As such, carbon dioxide is utilized as cooling agent for food refrigeration and as cleaning agent in industries (Braithwaite et al. 2014). At constant pressure and temperature, the lower viscosity and surface tension indicates that carbon dioxide is a liquid which is highly compressed in comparison with other refrigerating substance. Carbon dioxide diffusion property has made it suitable for sterilization of juices, extraction of oil and polymer processing (Kazarian and Martirosyan 2002). Apart from chemical and steel industries carbon dioxide can be also applied for extinguishing fire, and it is also utilized for preservation of food (Figs. 1, 2 and 3).

The development of drug releasing system can be assisted in impregnation by supercritical carbon dioxide, which is applied for active pharmaceutical ingredients loading. The application process for the preparation of drug delivery products requires mobile phase. Moreover, it is also essential to swell the matrix of polymer which permits the drug diffusion in polymer bulk which enhances the rate of impregnation (Kazarian 2000). Drug delivery system preparation needs three processes: active pharmaceutical ingredients solubilization in appropriate solvent, active pharmaceutical ingredients diffusion through matrix of polymers, and elimination of remaining solvent (Duarte et al. 2009). The feasibility of supercritical carbon dioxide-assisted impregnation depends on solubility of pharmaceutical ingredients in carbon dioxide, and supercritical fluid should be capable of swelling the polymer. Davies et al. (2008), Champeau et al. (2015) reported two methods of supercritical carbon dioxide-assisted impregnation. Supercritical carbon dioxide is identified as potential candidate for replacing traditional organic solvent for developing the green and sustainable chemical reaction, thereby meeting the required standards

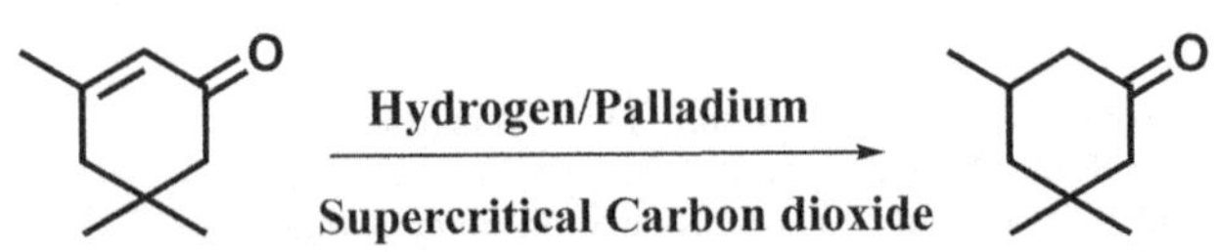

Fig. 1 Isophorone hydrogenation in the presence of supercritical carbon dioxide

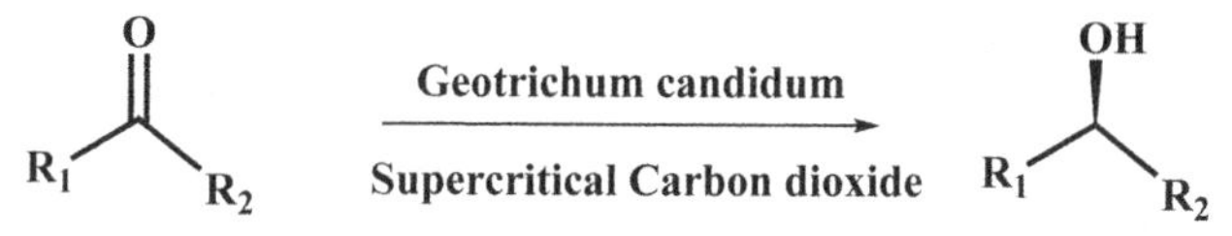

Fig. 2 Prochiral ketones enantioselective reduction by Geotrichum candidum in presence of supercritical carbon dioxide

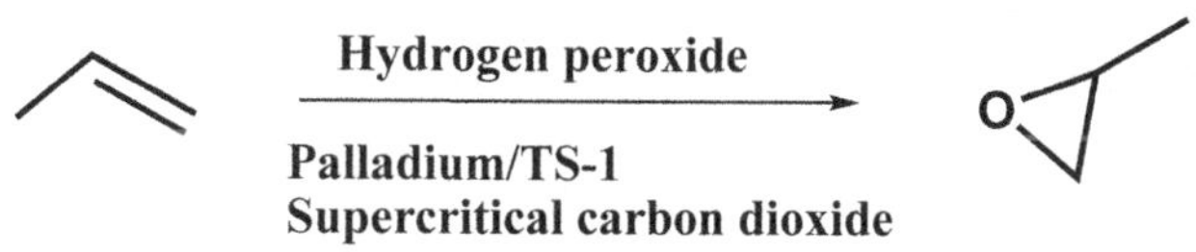

Fig. 3 Propylene epoxidation in presence of hydrogen peroxide and supercritical carbon dioxide

(Ivanovic et al. 2016; Banchero 2013). It is cost effective, highly abundant, and non-flammable solvent. It is easy to separate carbon dioxide from other solvents (Wong and Bioletti 2002). Owing to this benefit, carbon dioxide is used in many fields like textile and food industries (Wong and Bioletti 2002; Garcia Gonzalez et al. 2007; Eren et al. 2020). Supercritical carbon dioxide- assisted impregnation is utilized also in pharmaceutical and biomedical sector primarily for drug delivery and tissue engineering of polymer-based systems.

CO_2 is the waste product of combustion and it is a greenhouse gas. CO_2 is used in many different grounds so far as a supercritical liquid. For example, in case of food industry for the extraction of herbal ingredients, carbon dioxide can be used as a solvent. It is also used in bioscience as an anti-solvent for the purpose of formulation of protein. In case of chemical reactions and synthesis of different materials, it can be used as a solvent or anti-solvent. Recently, CO_2 has been active in the biomass processing via the CO_2 explosion (Anugwom et al. 2014), the $scCO_2$ is extracted from biomass (Reid et al. 1987) and precipitate the dissolved cellulose and lignin. Also, on reaction with bases, CO_2 produces the 'switchable ionic liquids' which is utilized for dissolution of cellulose (Cooper 2000) and biomass fractionation (Tomasko et al. 2003). These processes involve the high temperature and pressure. Environmental Protection Agency (EPA) has created the green chemistry program in order to support by design principle for designing, generating chemical processes (Linthorst 2010). The design for the environment (DFE) program has research and development efforts which are related to advanced technologies, so that they can contribute to the environmental processes by producing different products. The Pollution Prevention Act (PPA) of 1990 emphasizes on the decrease in the amount of pollutants. This was a concept that got ignored because of the main focus of the industry was on the management of waste and controlling pollution. In the year 1992, the EPA's Office of Pollution Prevention and Toxics (OPPT) and National Science Foundation (NSF) had jointly funded for the Green Chemistry research. In 1977, the OPPT was fully convinced that chemicals which were for use and sale does not show any negative effect on the health of human beings and on the environment. Most of the funding was using the exclusive features of supercritical fluids (SCFs) which act as a solvent instead of traditional organic solvents. Supercritical fluid possesses the properties which are transitional between those of liquid and that of gases. This inimitable phase is attained by the exertion of temperature and pressure more than those of critical temperature and critical pressure, respectively. Near the critical point, it can change in temperature and pressure that significantly alter the physicochemical properties such as density, diffusivity, or solubility of the SCFs. Since $scCO_2$ is enormously cheap and owe no danger to the environmental life or human health, $scCO_2$ is an attractive alternative for organic solvents.

From a few decades, the study of polymers has become an inseparable part of daily life. Polymers have various important applications in different fields. The most important attention of polymer is syntheses and processing. The manufactured polymers are helpful in various forms showing different applications. For example, in paint industries or medicine industries, the polymers are used in separation application. Generally, in most of the polymer processing method, they are making use of many volatile organic solvents (VOCs) and chlorofluorocarbons (CFCs) which are hazardous to the environment. Owning to the increase of hazardous wastes emission and rise in generating aqueous waste stream, a new and cleaner method for the polymer processing are invented by the chemist and chemical engineering. For example, today, the supercritical fluids are used as a solvent for polymer processing. The supercritical fluid is used in research laboratory as well as on the commercial scale. Supercritical fluids are obtained when the temperature and pressure are greater than that of critical value. The combination of two properties of supercritical fluids, in case of gas which is viscosity and in case of liquids that is density, makes the supercritical fluid as an outstanding solvent for numerous applications. Due to small change in pressure within the critical region, the density of supercritical fluids can be tuned. These supercritical fluids can be used as anti-solvents or solvents or as plasticizers in processing polymers that is while preparing polymer composites, while doing some modification in polymers, in polymer blending, microcellular foaming and during production of particle and synthesizing polymer (McHugh and Krukonis 1994; Kendall et al. 1999a; Hyatt 1984; Matos et al. 2013; Nalawade et al. 2006). For many polar or non-polar compounds having low molecular weight and for polymers namely amorphous-fluoro-polymers and silicones, the $scCO_2$ is a good solvent. It is not a good solvent for polymers whose molecular weight is very high. Its solubility is reliant on the temperature, pressure, and weak interactions with the chain groups in the polymer. Dissolved carbon dioxide can reduce the viscosity of polymers in molten state because of rise in the free volume. It can change the physical characteristics like diffusive nature, density, and rise in volume of polymers. Hence, these are used as plasticizers at a high temperature in polymer processing. $ScCO_2$ is most commonly used in many fields. This is found to be a clean and a safe solvent than other organic solvents and hazardous chlorofluorocarbons (CFCs). It can be used as a supercritical fluid in the preparation and manufacturing of polymers. It is safe, non-toxic in nature, does not catch fire that is it is non-flammable, chemically non-reactive, and expensive. It is highly rich in atmosphere and also found in the form of a by-product while producing ammonia, hydrogen, and ethanol. Its supercritical behavior is achieved at critical temperature (Tc) of 304 K and critical pressure of 377.38 MPa, and it can be easily taken off from a system by lowering the pressure. The use of $scCO_2$ is not hazardous for environment and human health as it is recovered during processing (DeSimone 2002).

The use of $scCO_2$ as a solvent can also be applied in the manufacturing of several biodegradable or bio-compatible polymeric compounds for pharmaceuticals and medical usage in the form of particle and microcellular form. There is also increasing evidence of $scCO_2$ in the catalytic methodologies that can broaden the scope of application of $scCO_2$. This chapter demonstrates a novel idea for solid phase CO_2 as a 'green' solvent for catalytic oxidation reaction which is homogeneous in nature. Catalytic chemical reactions are substantially replaced by CO_2 at moderate pressure to generate a CO_2-expanded solvent media. The CO_2-expanded solvents deliver ideal properties for maximizing the reaction rates which are characteristically 1–2 orders of magnitude more than those of which are obtained with clean organic solvents or $scCO_2$ as the medium for the reaction (Modell 1982). There was a single green oxidant, i.e., O_2. In general, most of the solvents consisting of $scCO_2$ are solidified to form an organic solvent in order to show advantages which can be prolonged for the many reaction systems. Those mixtures are said to be known as CO_2-expanded organic solvents. The uses of CO_2-expanded solvents are mainly in manufacturing different materials and separating them as well as recently the CO_2-expanded solvents are applied to improve various chemical reactions and different homogeneous catalytic oxidations. The oxidation of phenols is catalyzed by Co-Schiff base, and ferric chloride is used to catalyze cyclohexene. However, in CO_2-expanded CH_3CN, the reaction preceded with turnover frequencies (TOFs), selectivity, and conversions. Since 1990, the organometallic catalysis has broadened the use of CO_2 as a solvent. Production of materials and fuel from feedstock requires designing of biorefinery sustainably. According to International Energy Agency, biorefinery is the sustainable process for biomass conversion into products as foods, materials, chemicals, etc., and into the energy as fuels, power heat, etc. (Diep et al. 2012). In the sustainability point of view, a biorefinery minimizes the waste production and maximizes the outputs that are generally biodegradable and have low toxicity. The biorefinery process provides the application of green chemistry (Aristizábal 2016) for the production of fuels and chemicals.

2 Strategic Organic Solvents Replacement

Solvents like dense phase fluids, supercritical fluid, fluorous fluid, ionic liquids, and water are alternate options to conventional organic solvent as they are environmental friendly (Hobbs and Thomas 2007). All of the solvent has both disadvantage and advantages. Like ionic liquid at room temperature the molten form of organic salt exhibits low vapor pressure. Moreover, their non-volatile property enables reduction of emissions of volatile organic compounds during commercial applications. Supercritical water is also efficient as reaction media for oxidation processes (Lee et al. 1990). Chemical companies during processing of polymers have modified a procedure where carbon dioxide as flowing substance replaces chlorofluorocarbons during the polystyrene foam sheet manufacturing. Intensifying a process requires some essential factors, and they are increment of specificity and selectivity of reaction, no auxiliary fluid net consumption, product quality improvement, and raw materials high conversion (Rothenberg 2017).

3 Physical Properties of CO_2

Since, the year 1930, many PVT properties of CO_2 have been recognized (Hâncu and Beckman 2001). The excessive data sets of carbon dioxide gases are accessible in the procedure while correlating viscosity, calculating density, dielectric constant, etc., as the variation of temperature and pressure (Kendall et al. 1999b). It is found that the vapor pressure, critical pressure, and near-critical or liquid regime are suggestively developed than corresponding values for alkanes, fluoro-alkanes, or hydro-fluoro-alkanes. Quadruple moment is one of its physical features. It is very challenging if the critical pressure is high. During the late 1960s, CO_2 could have signified to be a solvent whose strength should beat alkanes and ketones.

4 CO_2 as a Solvent

Carbon dioxide density at room temperature increases with increase in pressure. Critical pressure and temperature above 72.8 bar and 31.1^{0}C, respectively, makes carbon dioxide a supercritical fluid which had characteristics of gas like diffusion. Owing to these physical characteristics, carbon dioxide can be utilized as solvent which has strength similar or equal to that of ketone and alkane. Presently, carbon dioxide is considered as green solvent which is highly compressible with flexible characteristics. The primary benefit of carbon dioxide as a solvent is its non-flammable property which makes it a better option for industrial processes and extraction. In comparison with the conventional solvent, the surface tension of carbon dioxide is less. Implementation and management of supercritical carbon dioxide in environment is easier owing to its low critical temperature similar to that of room temperature. Carbon dioxide is environmentally friendly and it is not easily oxidized because it is a complete oxidation product of an organic substance. Therefore, it is highly essential solvent for biphasic and oxidation reactions because of the negligible danger of cross contamination.

5 Advantages of CO_2

5.1 Environmental and Safety Advantages of CO_2 in Chemical Processes

Carbon dioxide is largely available in the nature. The threshold limit value of carbon dioxide is 5000 ppm (Tsang and Street 1981), and hence, it is comparatively less hazardous than conventional organic solvents like chloroform (10 ppm), pentane (600 ppm), and acetone (750 ppm). Carbon dioxide remains inactive to reaction compounds and during reaction with carbon dioxide formation of by-products is quite rare. However, it is not inert toward oxygen and hydrogen. Carbon dioxide hydrogenation on platinum catalyst causes carbon monoxide production at 303 K and with palladium catalyst rise to lesser quantity of carbon monoxide at the same temperature. Majority of carbon dioxide comes from natural deposition and effluents of ammonia plants (Beckman 2004).

5.2 CO_2 Cannot be Oxidized

It is already discussed that carbon dioxide is product of organic compound complete oxidation. Therefore, carbon dioxide is utilized as solvent for oxidation processes. Carbon dioxide is recognized as such solvent which is capable for the oxygen and hydrogen reactions to produce hydrogen peroxide (Ali et al. 2015).

5.3 CO_2 is an Aprotic Solvent

In case where labile protons interfere with the reaction, and then, the CO_2 can be applied without any penalty.

5.4 CO_2 is Immune to Free Radical Chemistry

In the free radically initiated polymerization, CO_2 does not support chain transfer to solvent. In such polymerization, CO_2 is an ideal solvent. It is a poor solvent for high molecular weight polymers. A growing chain with a terminal radical abstract is hydrogen from solvent and terminating the first chain in a chain transfer reaction. It was reported that carbon dioxide is inactive toward polymerization reaction which is related to free radical (Saraf et al. 2002), it does not favor chain transfer reaction.

5.5 CO_2 is Miscible with Gases

Oxygen and hydrogen gases are soluble in water and organic solvents. Carbon dioxide in liquid state is capable of absorbing large amount of oxygen and hydrogen in comparison with water and organic solvents (Anderson et al. 2007). It displays that carbon dioxide are capable of mixing with gases above temperature 304 K. The gases are carbon monoxide, oxygen, and hydrogen. Reactant gases like carbon monoxide and oxygen possess lower critical temperature.

5.6 CO_2 Exhibits Solvent Properties that Allow Miscibility with Both Fluorous and Organic Materials

Carbon dioxide mixes with perfluorous or fluorous, organic solvents having lower molecular weight. Carbon dioxide at a lower pressure (20–30 bar) added to fluorous and organic liquid mixture gives rise to formation of single homogeneous phase (Maul et al. 1999).

5.7 CO_2 Exhibits a Liquid Viscosity Only 1/10 that of Water

The viscosity of carbon dioxide is one tenth that of water. In carbon dioxide mixture, heat transfer is very nice (Enick et al. 2012). Carbon dioxide exhibited viscosity like gas and density like liquid. Carbon dioxide surface tension is less in comparison with traditional organic solvents, and solute diffusion is high due to lesser viscosity of carbon dioxide. Carbon dioxide can penetrate easily into geometry of complex than simple liquid in which case carbon dioxide diffusion is faster in pores of catalyst in comparison with analogous system by using traditional liquids.

6 Supercritical Carbon Dioxide

Supercritical fluid refers to those substances whose temperature and pressure both are more than the critical value. For a wide range of applications, excellent integration of liquid like density and gas like viscosity of a supercritical fluid is very suitable. Supercritical fluids density can be modified by minor alterations in pressure in within critical limits. Supercritical fluid are applied as anti-solvents, solvents, or as plasticizers for polymer processes like particle production, microcellular foaming, polymer bending, polymer composites, and polymer modification during polymer synthesis. On selecting supercritical fluid over traditional organic solvent, there is improvement of quality of product. It is a versatile and clean solvent. Supercritical carbon dioxide is a potential substitute for conventional chlorofluorocarbons and organic solvent. Because of its special physical properties, it has gained attention for polymer processing and synthesis areas. Supercritical fluid is cost effective, inert chemically, non-flammable, and non-toxic. Large quantities of this fluid are produced in the form of by-product during ethanol, hydrogen, and ammonia production. It can attain supercritical condition, i.e., 7.38 critical pressure, 304 K critical temperature very easily and can be eliminated out of the system through depressurization simply. Moreover, supercritical carbon dioxide does not create greenhouse issues because it is restored during processing. It is a better solvent for non-polar compounds with lower molecular weighing substances and some polymers like silicones, and amorphous fluoropolymers. For compound with higher molecular weight, it is not a good solvent. Its solubility is substantial in many polymers. The solubility of supercritical carbon dioxide depends on weak interactions with polymer chain groups, pressure, and also temperature. Carbon dioxide after getting dissolved reduces molten polymer viscosity considerably. It modifies physical characteristics of polymers like swollen volume, diffusivity, and density. Hence, it is highly potential as plasticizer for processing of polymers (Sheldon 2005).

6.1 Chemical Reaction in Supercritical Carbon Dioxide

6.1.1 Hydroformylation and Hydrogenation

Supercritical carbon dioxide miscibility with hydrogen leads to high diffusion rate, thereby allowing high rate reaction in comparison with traditional solvents. In case of multifunctional groups, the chemo-selectivity can be tuned by little changes in the parameters of reaction. Supercritical carbon dioxide is also applied for hydroformylation of olefin catalyzed by immobilized rhodium (Sheldon 2005).

6.1.2 Biocatalysis

Bioconversion is also carried out in presence of supercritical carbon dioxide. Biocatalysis in presence of supercritical carbon dioxide was first reported in the year 1985. The stability of enzymes is more in supercritical carbon dioxide in comparison with water. In presence of this solvent, 1-phenylethanol resolution in catalyst Novozyme 435 (Candida antartica lipsase) was performed successfully. Matsuda et al. reported that in presence of supercritical carbon dioxide and Novozyme 435 catalyst, enantio selectivity of alcohol acylation can be controlled by tuning the temperature and pressure. In the same manner, in presence of supercritical carbon dioxide prochiral ketones enantio selective reduction by Geotrichum candidum whole cells in a system of semi-continuous flow was successfully performed. Oxidation reaction with oxygen catalyzed by enzymes was also smoothly carried out in presence of supercritical carbon dioxide by using polyphenol oxidase and cholesterol oxidase. The potentiality of supercritical carbon dioxide solvent for bioconversions reaction is excellent and could provide more application in near future (Sheldon 2005).

6.1.3 Oxidation

Alike water, supercritical carbon dioxide is suitable solvent for carrying out aerobic oxidation catalytically. It mixes with oxygen completely, and also it is inflammable. The catalytic oxidation with H_2O_2, obtained in situ by hydrogen and oxygen reaction catalyzed by palladium in presence of supercritical carbon dioxide and water. This system was efficiently applied for propylene direct epoxidation in presence of palladium/TS-1 catalyst to propylene oxide. Moreover, in these reaction due to reaction between hydrogen peroxide and supercritical carbon dioxide, there is the formation of peroxycarbonic acid intermediate (Sheldon 2005). Comparative study of different solvent applied for propylene oxidation to propylene oxide using carbon dioxide as a solvent is illustrated in Table 1 (Beckman 2003).

Table 1 Comparative study of different solvent applied for propylene oxidation to propylene oxide using CO_2 as the solvent (Beckman 2003)

Sl. No	Solvent	Conversion percentage of Propylene	Selectivity percentage of propylene to propylene oxide	Mass of the catalyst in gram
1	Pure carbon dioxide	7.5	94.3	0.300
2	Pure carbon dioxide	6.5	91.2	0.195
3	Pure carbon dioxide	9.5	77.1	0.15
4	Methanol + water + nitrogen	16.0	3.5	0.199
5	Methanol + water + Carbon dioxide	3.5	17.4	0.197

7 Polymerization and Polymer Processes

7.1 CO_2 as a Solvent for Polymer Systems

Polymers do not dissolve in every solvent. A considerable amount of polymer behaves as supercritical fluid phase in which case minor component is polymer. In some polymer supercritical fluid mixture, at pressure below or above 100 bars, high amount of swelling is observed (Wissinger and Paulaitis 1987). Critical pressure and solubility parameter correlation indicating carbon dioxide solvent strength should be equivalent to pyridine (Giddings 1969). Heller et al. displayed that the strength of carbon dioxide is lower than alkanes. Some polymer tested by Heller resulted solubility in carbon dioxide at intermediate pressure of 200 bar (Marriott and Sin 2012). Johnstone et al. have reported that solubility parameter is not the only factor for determining the carbon dioxide solvent strength (McFann et al. 1994). De Simon et al. have reported that carbon dioxide polymer is polyacrylate which is fluorinated (DeSimone et al. 1992).

7.2 Benefits of Use of Supercritical CO_2 as a Green Reaction Medium

Utilization of carbon dioxide as solvent does not enhance the emission of greenhouse gases because it does not further release carbon dioxide. Plasticizing and swelling impacts of polymer by supercritical carbon dioxide has a vital contribution in polymerization reaction in presence of metal catalyst and compressed carbon dioxide, like rhodium catalyzed polyphenylacetylene formation, ring opening metathesis polymerization catalyzed by ruthenium, etc.

7.3 Improved Reaction Rates

Supercritical carbon dioxide replaces organic solvents as medium of reaction, thereby changing the reaction rate. Based on the physicochemical characteristics of supercritical fluids, rationalization of alterations is possible. In supercritical carbon dioxide, the enhancement of rate is not limited with only gaseous substance. Imines hydrogenation in presence of supercritical carbon dioxide displays increase in the efficiency of catalyst, and this is not dependent on presence of gas (Kainz et al. 1999).

7.4 Carbon Dioxide as a C1-Building Block for Chemicals

Carbon dioxide is readily available C1-building block, non-toxic, cheap, and abundant solvent. It is capable of replacing the majority of waste substance and less suitable reagents like phosgene and carbon monoxide. Utilization of supercritical carbon dioxide as reagents and solvent has increased organic carbonate synthesis and urethane synthesis (Vieville et al. 1998) through non-phosgene process. Styrene carbonates formation from carbon dioxide and styrene in the presence of dimethyl formamide in catalytic amount as a function of carbon dioxide pressure and phase behavior (Yoshida et al. 2000).

8 Application of Carbon Dioxide Solvent in a Biorefinery

8.1 CO_2 and Biorefinery

Carbon dioxide expanded liquids, supercritical carbon dioxide, and sub-critical carbon dioxide are utilized as solvent in biorefinery for reaction, refinement and fractionation, and extraction. Use of carbon dioxide in biorefinery replaces the utilization of toxic solvent and gives potential ways for processing of biomass. Like in extraction of oil volatile, hazardous and highly flammable solvent is substituted by carbon dioxide. Both reaction and extraction need phase behavior understanding which affects the operating conditions and system design. Owing to the solubility of selective low polarity compounds and its capability of mediating conversion process, carbon dioxide is utilized for fuel production (Kawanami and Ikushima 2000; Payne and Kerton 2010).

8.2 Extraction

Use of carbon dioxide for the process of extraction is essential for tenability and increased mass transfer. Soh et al. (2014) provided an overall review of this system, which is important for extraction of biomass by using supercritical carbon dioxide by consideration of mass transfer and internal limitation. At a particular pressure and temperature, solubility parameter is different for every compound because molecular volume divides square root of internal energy of substance (Ramaswamy et al. 2013). Supercritical carbon dioxide is utilized for extraction of biomass compound. At different pressure and temperature, various lipophilic substance like carotenoid (Hansen 1967), tocopherols (Soh and Zimmerman 2011), triglycerides (Gracia 2011), etc., are extracted out of the biomass by utilization of supercritical carbon dioxide. In Table 2, solubility parameter of biological extract of some compound is displayed. By, increasing the system pressure and utilization of co-solvent, the solubility parameter of carbon dioxide can be enhanced and that is permitted for extraction of tocopherols and carotenoids. In Table 3, biomass feedstock which is extracted by using supercritical carbon dioxide has been displayed. Miscible co-solvent like ethyl acetate, ethanol, and methanol enhances the polarity and permits the molecule with intermediate polarity extraction. In Table 4, solubility parameter for common extraction solvents and carbon dioxide co-solvent are displayed (Soh 2014).

8.3 Fractionation and Refinement

Carbon dioxide can be utilized for modifying the materials which are extracted by refinement and fractionation. The process of fractionation could be performed with the substances which are alike in property buy differ in polarity which permits flexible solubility in carbon dioxide. Like, production of ethyl ester out of oil could be subjected to fractionation depending on the alkyl chain length (Mendes et al. 1995). It is reported that specific omega-3 fatty acid and polyunsaturated fatty acid enrichment is provided with suitable optimization of flow rays, pressure, and temperature of carbon dioxide (Nilsson et al. 1989). Brunner et al. displayed the outcome of lipid mixture fractionation and also of tocopherols and free fatty acids (Perretti et al. 2007; Brunner 2000). Carbon dioxide is also utilized for purification of crude biodiesel which contained biodiesel or methyl ester and glycerol and unreacted triglycerides (Soh 2014).

9 Conclusion

With respect to green chemistry, utilization of carbon dioxide has many kinds of milestones like commercialization, determination of carbon dioxide use information, and purely scientific milestones. Purely scientific milestones revealed that the efficiency of carbon dry as solvent is similar to n-alkanes. Later, from 1988 to 1992, some of the scientists reported that silicones and fluorinated materials have better thermodynamic compatibility with carbon dioxide in comparison with alkanes. Carbon dioxide is used in wide scale for both industrial and academic communities. It is

Table 2 Solubility parameters of biological extract classes. Soh 2014 Copyright © 2014, American Chemical Society ()

Compound	δ, MPa½
Hydrocarbons	17.06
Carotenoids	17.84
Tocopherols	18.12
Triglcyerides	18.23
Fatty Acids	18.61
Diglycerides	19.33
Sterols	19.47
Monoglyericdes	20.86
Glycerol	36.16

Table 3 Biological extraction by super critical CO_2. Copyright © 2014, American Chemical Society (Soh 2014)

Feedstock	Examples	Extracts
Algae	C. vulgaris B. braunii S. dimorphus	**Lipids:** Triglycerides, Hydrocarbons, Polyunsaturated Fatty Acids **Pigments:** Lutein, Carotenoids, Chlorophyll
Fish	Mackerel Salmon	Triglycerides Polyunsaturated fatty acids
Oil seeds	Jatropha Seed Palm kernel Soybeans	**Lipids:** Triglycerides, Polyunsaturated Fatty Acids Carotenoids
Leaves	Chamomile Citronella Sage	"Essential oils"
Other plant matter	Chili pepper Turmeric	
Wood	Pine Spruce	Terpenoids Sterols
	Eucalyptus	Phenolics
Waste	Olive	Beta carotene, lycopene
	Tomato	Tocopherols
	Wine	Ethanol

Table 4 Solubility parameters of common extraction solvents and CO_2 co-solvents. Copyright © 2014, American Chemical Society (Soh 2014)

Compounds	δ (in $MPa^{1/2}$)
Hexane	14.90
Ethyl Acetate	18.15
Toluene	18.16
Benzene	18.51
Chloroform	18.95
Acetone	19.94
Dichloromethane	20.20
Ethanol	26.52
Methanol	29.61
Water	47.81

believed that more advancement can be achieved by crossing successfully technical obstacles, thereby enhancing the utilization of carbon dioxide in green chemistry.

References

Ali HM, Ali H, Liaquat MHTB, Nadir MA (2015) Experimental investigation of convective heat transfer augmentation for car radiator using ZnO–water nanofluids. Energy 84:317–324

Anastas PT, Warner JC (1998) Principles of green chemistry. Green Chem: Theory Pract 29–56

Anderson JL, Dixon JK, Brennecke JF (2007) Solubility of CO2, CH4, C2H6, C2H4, O2, and N2 in 1-Hexyl-3-methylpyridinium Bis (trifluoromethylsulfonyl) imide: comparison to other ionic liquids. Acc Chem Res 40(11):1208–1216

Anugwom I, Eta V, Virtanen P, Mäki-Arvela P, Hedenström M, Yibo Ma, Hummel M, Sixta H, Mikkola J-P (2014) Towards optimal selective fractionation for Nordic woody biomass using novel amine–organic superbase derived switchable ionic liquids (SILs). Biomass Bioenerg 70:373–381

Aristizábal V (2016) Integrated production of different types of bioenergy from oil palm through biorefinery concept. Waste Biomass Valorization 7(4):737–745

Ballivet-Tkatchenko D, Camy S, Condoret J-S (2005) Carbon dioxide, a solvent and synthon for green chemistry. Environmental chemistry. Springer, Berlin, Heidelberg, pp 541–5522005

Banchero M (2013) Supercritical fluid dyeing of synthetic and natural textiles–a review. Color Technol 129(1):2–17

Bansal P (2012) A review–status of CO_2 as a low temperature refrigerant: fundamentals and R&D opportunities. Appl Therm Eng 41:18–29

Beckman EJ (2003) Oxidation reactions in CO2: academic exercise or future green processes? Environ Sci Technol 37(23):5289–5296

Beckman EJ (2004) Supercritical and near-critical CO_2 in green chemical synthesis and processing. J Supercrit Fluids 28(2–3):121–191

Braithwaite DC, Sundaram M, Rasanayagam V (2014) Method and system for treating food items with an additive and solid carbon dioxide. U.S. Patent 8,691,308. Issued April 8, 2014; Champeau M, Thomassin J-M, Tassaing T, Jerome C (2015) Drug loading of sutures by supercritical CO_2 impregnation: effect of polymer/drug interactions and thermal transitions. Macromol Mater Eng 300 (6):596–610

Brunner G (2000) Fractionation of fats with supercritical carbon dioxide. Eur J Lipid Sci Technol 102(3):240–245

Campbell ML, Apodaca DL, Yates MZ, McCleskey TM, Birnbaum ER (2001) Metal extraction from heterogeneous surfaces using carbon dioxide microemulsions. Langmuir 17(18):5458–5463

Cao L, Chen L, Chen X, Zuo L, Li Z (2006) Synthesis of smart core–shell polymer in supercritical carbon dioxide. Polymer 47(13):45

Champeau M, Thomassin J-M, Tassaing T, Jérôme C (2015) Drug loading of polymer implants by supercritical CO_2 assisted impregnation: A review. J Controlled Release 209:248–259

Cooper AI (2000) Polymer synthesis and processing using supercritical carbon dioxide. J Mater Chem 10(2):207–234

Davies OR, Lewis AL, Whitaker MJ, Tai H, Shakesheff KM, Howdle SM (2008) Applications of supercritical CO_2 in the fabrication of polymer systems for drug delivery and tissue engineering. Adv Drug Deliv Rev 60(3):373–387

DeSimone JM (2002) Practical approaches to green solvents. Science 297(5582):799–803

DeSimone JM, Guan Z, Elsbernd CS (1992) Synthesis of fluoropolymers in supercritical carbon dioxide. Science 257(5072):945–947

Diep NQ, Sakanishi K, Nakagoshi N, Fujimoto S, Minowa T, Tran XD (2012) Biorefinery: concepts, current status, and development trends. Int J Biomass Renew 2(1):1–8

Du L, Kelly JY, Roberts GW, DeSimone JM (2009) Fluoropolymer synthesis in supercritical carbon dioxide. J Supercrit Fluids 47 (3):447–457

Duarte ARC, Mano JF, Reis RL (2009) Supercritical fluids in biomedical and tissue engineering applications: a review. Int Mater Rev 54(4):214–222

Eckert CA, Knutson BL, Debenedetti PG (1996) Supercritical fluids as solvents for chemical and materials processing. Nature 383 (6598):313

Enick RM, Olsen DK, Ammer JR, Schuller W (2012) Mobility and conformance control for CO2 EOR via thickeners, foams, and gels —a literature review of 40 years of research and pilot tests. In: SPE improved oil recovery symposium. Society of Petroleum Engineers

Eren HA, Yiğit İ, Eren S, Avinc O (2020) Sustainable textile processing with zero water utilization using super critical carbon dioxide technology. Sustainability in the textile and apparel industries. Springer, Cham, pp 179–1962020

F'Oliakoff M, George MW, Howdle SM, Bagratashvili VN, Han B-X, Yan H-K (1999) Supercritical fluids: clean solvents for green chemistry. Chin J Chem 17(3):212–222

Garcia-Gonzalez L, Geeraerd AH, Spilimbergo S, Elst K, Van Ginneken L, Debevere J, Van Impe JF, Devlieghere F (2007) High pressure carbon dioxide inactivation of microorganisms in foods: the past, the present and the future. Int J Food Microbiol 117(1):1–28

Giddings JC (1969) L., McLaren, MN Myers and RA Keller. Science 162:67

Gracia I, Rodríguez JF, de Lucas A, Fernandez-Ronco MP, García MT (2011) Optimization of supercritical CO_2 process for the concentration of tocopherol, carotenoids and chlorophylls from residual olive husk. J Supercrit Fluids 59:72–77

Guan Z, Combes JR, Elsbernd CS, DeSimone JM (1993) Synthesis of fluoropolymers in supercritical carbon dioxide. In: No. CONF-930304—American Chemical Society, Washington, DC (United States)

Hansen CM (1967) The three dimensional solubility parameter. Danish Tech: Copenhagen 14

Hobbs HR, Thomas NR (2007) Biocatalysis in supercritical fluids, in fluorous solvents, and under solvent-free conditions. Chem Rev 107 (6):2786–2820

Hyatt JA (1984) Liquid and supercritical carbon dioxide as organic solvents. J Org Chem 49(26):5097–5101

Hâncu D, Beckman EJ (2001) Generation of hydrogen peroxide directly from H_2 and O_2 using CO_2 as the solvent. Green Chem 3 (2):80–86

Ivanovic J, Milovanovic S, Zizovic I (2016) Utilization of supercritical CO_2 as a processing aid in setting functionality of starch-based materials. Starch-Stärke 68(9–10):821–833

Kainz S, Brinkmann A, Leitner W, Pfaltz A (1999) Iridium-catalyzed enantioselective hydrogenation of imines in supercritical carbon dioxide. J Am Chem Soc 121(27):6421–6429

Kawanami H, Ikushima Y (2000) Chemical fixation of carbon dioxide to styrene carbonate under supercritical conditions with DMF in the absence of any additional catalysts. Chem Commun 21:2089–2090

Kazarian SG (2000) Polymer processing with supercritical fluids. Polym Sci Ser CC/C Vysokomolekuliarnye Soedineniia 42(1):78–101

Kazarian SG, Martirosyan GG (2002) Spectroscopy of polymer/drug formulations processed with supercritical fluids: in situ ATR–IR and Raman study of impregnation of ibuprofen into PVP. Int J Pharm 232(1–2):81–90

Kendall JL, Canelas DA, Young JL, DeSimone JM (1999a) Polymerizations in supercritical carbon dioxide. Chem Rev 99(2):543–564

Kendall JL, Canelas DA, Young JL, DeSimone JM (1999b) Chem Rev 99:543

Krukonis VJ (1994) Supercritical fluid extraction: principles and practice. Butterworth-Heinemann

Lee D-S, Gloyna EF, Li L (1990) Efficiency of H2O2 and O2 in supercritical water oxidation of 2, 4-dichlorophenol and acetic acid. J Supercrit Fluids 3(4):249–255

Linthorst JA (2010) An overview: origins and development of green chemistry. Found Chem 12(1):55–68

Marriott R, Sin, E (2012) Supercritical CO_2 as an Environmentally Benign Medium for Biorefinery. Role Green Chem Biomass Proc Convers 181–204

De Matos MBC, Piedade AP, Alvarez-Lorenzo C, Concheiro A, Braga MEM, De Sousa HC (2013) Dexamethasone-loaded poly (ε-caprolactone)/silica nanoparticles composites prepared by supercritical CO2 foaming/mixing and deposition. Int J Pharm 456 (2):269–281

Maul JJ, Ostrowski PJ, Ublacker GA, Linclau B, Curran DP (1999) Benzotrifluoride and derivatives: useful solvents for organic synthesis and fluorous synthesis. In: Modern solvents in organic synthesis. Springer, Berlin, Heidelberg, pp 79–105

McFann GJ, Johnston KP, Howdle SM (1994) Solubilization in nonionic reverse micelles in carbon dioxide. AIChE J 40(3):543–555

McHugh MA, Krukonis VJ (1994) Supercritical fluid extraction: principles and practice, 2nd edn

Mendes RL, Fernandes HL, Coelho JP, Reis EC, Cabral JMS, Novais JM, Palavra AF (1995) Supercritical CO_2 extraction of carotenoids and other lipids from Chlorella vulgaris. Food Chem 53 (1):99–103

Michels A, Michels C (1935) Isotherms of CO_2 between 0° and 150° and pressures from 16 to 250 atm (Amagat densities 18–206). In:

Proceedings of the Royal Society of London. Series A-mathematical and physical sciences, vol 153, no 878, pp 201–214
Michels A, Michels C, Wouters H (1935) Isotherms of CO2 between 70 and 3000 atmospheres (amagat densities between 200 and 600). In: Proceedings of the Royal Society of London. Series A-mathematical and physical sciences, vol 153, no 878, pp 214–224
Modell M (1982) Processing methods for the oxidation of organics in supercritical water. U.S. Patent 4,338,199. Issued July 6, 1982
Mohammed F (2012) Application of carbon dioxide towards the development of smart materials, green reaction schemes and metallic nanoparticle synthesis
Nalawade SP, Picchioni F, Janssen LPBM (2006) Supercritical carbon dioxide as a green solvent for processing polymers; Das S, Tyagi AK, Kaur H (2000) Cancer modulation by glucosinolates: a review. Curr Sci 1665–1671
Nilsson WB, Gauglitz EJ, Hudson JK (1989) Supercritical fluid fractionation of fish oil esters using incremental pressure programming and a temperature gradient. J Am Oil Chem Soc 66(11):1596–1600
Ohde H, Hunt F, Wai CM (2001) Synthesis of silver and copper nanoparticles in a water-in-supercritical-carbon dioxide microemulsion. Chem Mater 13(11):4130–4135
Ohde M, Ohde H, Wai CM (2002) Catalytic hydrogenation of arenes with rhodium nanoparticles in a water-in-supercritical CO_2 microemulsion. Chem Commun 20:2388–2389
Ohde H, Rodriguez JM, Ye X-R, Wai CM (2000) Synthesizing silver halide nanoparticles in supercritical carbon dioxide utilizing a water-in-CO_2 microemulsion. Chem Commun 23:2353–2354
Payne SM, Kerton FM (2010) Solubility of bio-sourced feedstocks in 'green' solvents. Green Chem 12(9):1648–1653
Peach J, Eastoe J (2014) Supercritical carbon dioxide: a solvent like no other. Beilstein J Org Chem 10(1):1878–1895
Perretti G, Motori A, Bravi E, Favati F, Montanari L, Fantozzi P (2007) Supercritical carbon dioxide fractionation of fish oil fatty acid ethyl esters. J Supercrit Fluids 40(3):349–353
Persichilli M, Kacludis A, Zdankiewicz E, Held T (2012) Supercritical CO2 power cycle developments and commercialization: why sCO2 can displace steam ste. Power-Gen India Central Asia 2012:19–21
Poliakoff M, Fitzpatrick JM, Farren TR, Anastas PT (2002) Green chemistry: science and politics of change. Science 297(5582):807–810
Ramaswamy S, Huang H-J, Ramarao BV (eds) (2013) Separation and purification technologies in biorefineries. John Wiley & Sons
Reid RC, Prausnitz JM, Poling BE (1987) The properties of gases and liquids
Rothenberg G (2017) Catalysis: concepts and green applications. John Wiley & Sons
Saari H, Sjolander S, Zanganeh K, Pearson B, Shafeen A (2011) Supercritical CO_2 advanced Brayton cycle design. In: Presentation, Carleton University, supercritical CO_2 power cycle symposium
Saito M (2013) History of supercritical fluid chromatography: instrumental development. J Biosci Bioeng115, no. 6 (2013):590–599; Roller DHD, Thilorier M (1952) Thilorier and the first solidification of a "permanent" gas (1835). Isis 43(2):109–113
Sako T (2002) Supercritical fluids: molecular interactions, physical properties, and new applications. Springer science & business media; Beckman EJ (2004) Supercritical and near-critical CO_2 in green chemical synthesis and processing. J Supercrit Fluids 28(2–3):121–191
Saraf MK, Gerard S, Wojcinski LM, Charpentier PA, DeSimone JM, Roberts GW (2002) Continuous precipitation polymerization of vinylidene fluoride in supercritical carbon dioxide: formation of polymers with bimodal molecular weight distributions. Macromolecules 35(21):7976–7985
Sarbu T, Styranec T, Beckman EJ (2000) Non-fluorous polymers with very high solubility in supercritical CO_2 down to low pressures. Nature 405(6783):165–168
Sheldon RA (2005) Green solvents for sustainable organic synthesis: state of the art. Green Chem 7(5):267–278
Sherman J, Chin B, Huibers PD, Garcia-Valls R, Hatton TA (1998) Solvent replacement for green processing. Environ Health Perspect 106(1):253–271
Soh L, Curry J, Beckman EJ, Zimmerman JB (2014) Effect of system conditions for biodiesel production via transesterification using carbon dioxide–methanol mixtures in the presence of a heterogeneous catalyst. ACS Sustain Chem Eng 2(3):387–395
Soh L, Zimmerman J (2011) Biodiesel production: the potential of algal lipids extracted with supercritical carbon dioxide. Green Chem 13 (6):1422–1429
Soh L (2014) Carbon dioxide solvent applications in a biorefinery. In Green technologies for the environment. American Chemical Society, pp 9–35
Steiner RW (1993) Carbon dioxide's expanding role. Chem Eng 100 (3):114
Subramaniam B, Rajewski RA, Snavely K (1997) Pharmaceutical processing with supercritical carbon dioxide. J Pharm Sci 86 (8):885–890; Oakes RS, Clifford AA, Rayner CM (2001) The use of supercritical fluids in synthetic organic chemistry. J Chem Soc, Perkin Trans 1(9):917–941
Tomasko DL, Li H, Liu D, Han X, Wingert MJ, Lee LJ, Koelling KW (2003) A review of CO_2 applications in the processing of polymers. Ind Eng Chem Res 42(25):6431–6456
Tsang CY, Street WB (1981) Phase equilibria in the H_2/CO_2 system at temperatures from 220 to 290 K and pressures to 172 MPa. Chem Eng Sci 36(6):993–1000
Vieville C, Yoo JW, Pelet S, Mouloungui Z (1998) Synthesis of glycerol carbonate by direct carbonatation of glycerol in supercritical CO2 in the presence of zeolites and ion exchange resins. Catal Lett 56(4):245–247
Wissinger RG, Paulaitis ME (1987) Swelling and sorption in polymer–CO_2 mixtures at elevated pressures. J Polym Sci, Part b: Polym Phys 25(12):2497–2510
Wong S, Bioletti R (2002) Carbon dioxide separation technologies. Alberta Research Council
Wood CD, Cooper AI, DeSimone JM (2004) Green synthesis of polymers using supercritical carbon dioxide. Curr Opin Solid State Mater Sci 8(5):325–331
Yeo S-D, Kiran E (2005) Formation of polymer particles with supercritical fluids: a review. J Supercrit Fluids 34(3):287–308
Yoshida M, Hara N, Okuyama S (2000) Catalytic production of urethanes from amines and alkyl halides in supercritical carbon dioxide. Chem Commun 2:151–152

State-Of-The-Art Overview of CO_2 Conversions

Grazia Leonzio

Abstract

Due to the environmental problems about climate change and global warming, carbon dioxide is transformed and valorized from an emission to a raw material of many chemical production processes. This contributes transforming the linear economy into a circular one, considering the principles of reduction, reuse, recovery, recycle. However, the potential carbon dioxide uptake is lower compared to the worldwide CO_2 emissions, estimated of 37 Gt actually. In the literature, many utilization options are suggested and are evaluated according to the 3E performance criteria (engineering–economic–environmental), including nine key indicators. An economic estimation and an analysis about the market size of the main carbon dioxide-based products are here suggested. Moreover, with particular attention, in this work, different routes for carbon dioxide utilization are reviewed: chemicals, fuels, concrete building materials, horticulture and microalgae production as well as mineral carbonation, oil and methane recovery and some direct uses. Cycling, closed and open pathways can be realized. Carbon dioxide utilization pathways ensuring a promising development in the next years are suggested.

1 Introduction

Carbon dioxide (CO_2), being a waste emission, can be valorized as a feedstock and used in different pathways. Another important aspect is the fact that CO_2, through its utilization, can introduce renewable energies into the supply chain (Ampelli et al. 2015). Then, CO_2 utilization has a relevant role. In the literature, there are different pathways for CO_2 utilization that are mainly investigated and discussed (Hepburn et al. 2019):

- Chemicals
- Fuels
- Concrete building materials
- Mineral carbonation
- Oil and methane recovery
- Horticulture and microalgae production
- Direct uses

Each pathway is characterized by a defined potential development, economic perspective, use of energy and amount of CO_2, time of sequestration and environmental impact (Ampelli et al. 2015). These aspects are also enclosed in the definition of technology readiness level (TRL), shown for the most important CO_2-based compounds in Fig. 1, for which the time frame to deployment was assigned with an itemized literature analysis (Chauvy et al. 2019).

Data about CO_2 potential utilization for the main products are reported in Table 1, where CO_2 uptake potential (the amount of moles in a product according to the stoichiometry) is suggested. It is evident that the overall potential CO_2 uptake is smaller than anthropogenic CO_2 emissions in the world (actually about 37 Gt).

On the other hand, an economic estimation and market size of these products obtained from CO_2 are reported in Fig. 2.

Two typologies of CO_2-based products can be suggested: high unit price compounds with low market volume and low unit price products with significant market value. More specific production costs are suggested in the literature for CO_2-based products. In particular, the production cost for polymers is 1440 \$/ton, for methanol (CH_3OH) it is 510 \$/ton, for methane (CH_4) it is 200–250 \$/ton, for dimethyl ether it is 660 \$/ton, for urea it is 370–450 \$/ton, for calcium

G. Leonzio (✉)
Department of Industrial and Information Engineering and Economics, University of L'Aquila, Via Giovanni Gronchi 18, 67100 L'Aquila, Italy
e-mail: grazia.leonzio@graduate.univaq.it

Inamuddin et al. (eds.), *Carbon Dioxide Utilization to Sustainable Energy and Fuels*, Advances in Science, Technology & Innovation, https://doi.org/10.1007/978-3-030-72877-9_18

Table 1 CO_2 utilization potential

CO_2-based compound	Production (Mton/year)	CO_2 uptake potential (Mton/year)
Methane	1100–1500	3000–4000
Urea	180	132.30
Calcium carbonate	113.9	50.002
Ethanol	80	152.880
Methanol	65	89.245
Sodium carbonate	62	25.730
Microalgae	35	63.000
Formaldehyde	21	30.450
Magnesium carbonate	20.5	5.350
Polyurethane	15	4.500
Dimethyl ether	11.4	21.785
Acetic acid	10.25	7.513
Acrylic acid	5.85	3.574
Polycarbonates	5	0.865
Dimethyl carbonate	1.6	2.346
Formic acid	1	0.956
Ethylene carbonate	0.2	0.099
Propylene carbonate	0.2	0.086
Salicylic acid	0.17	0.054

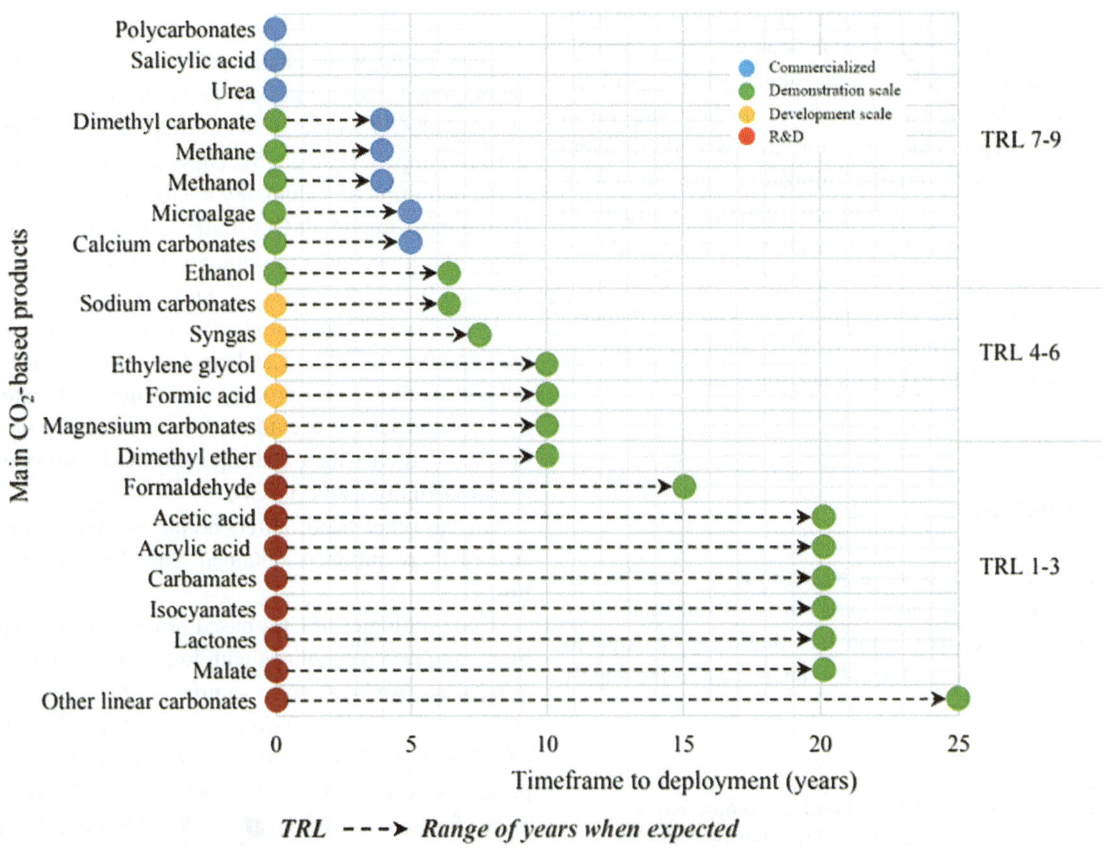

Fig. 1 TRL of main CO_2-based products (Chauvy et al. 2019)

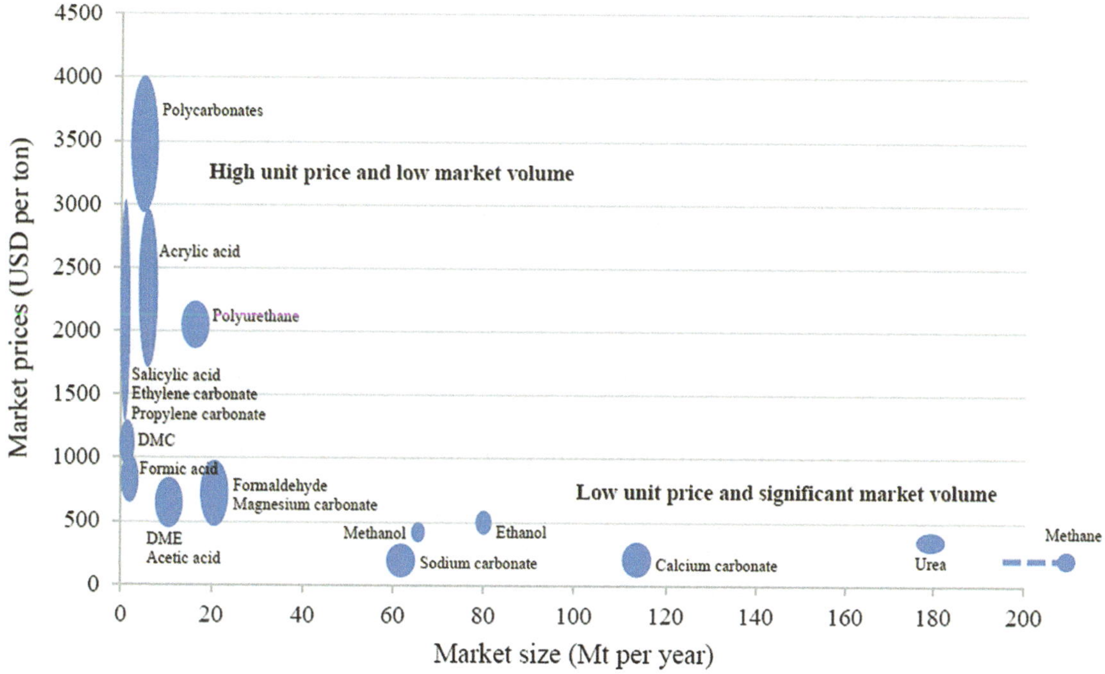

Fig. 2 Market sizes and prices of CO_2-based products (Chauvy et al. 2019)

carbonate it is 30–350 $/ton, for ethanol ($C_2H_5OH$) it is 480–530 $/ton, for formaldehyde is 490–1000 $/ton, for concrete curing it is 56 $/ton, and for dimethyl carbonate ($(CH_3O)_2CO$) it is 1000 $/ton (Hepburn et al. 2019; Tcvetkov et al. 2019).

A semiquantitative analysis is carried out by Chauvy et al. (2019) to measure 3E performance criteria (engineering performance, economic performance, environmental, health and safety performance) of main products obtained from CO_2. Figure 3 shows this analysis, while Fig. 4 shows the weighted overall results for 3E performances.

Methanol, polycarbonates, methane and microalgae have higher values. After the above overview and general information of CO_2 utilization, in the following sections of this chapter, each utilization route, listed before at the beginning, is discussed and reviewed. These discussed pathways are conventional. In the literature, are suggested other non-conventional CO_2 utilizations, as (i) bioenergy with carbon capture and storage (BECCS), (ii) enhanced weathering, (iii) forestry techniques, including afforestation/reforestation, forest management and wood products, (iv) land management via soil carbon sequestration techniques and (v) biochar (Hepburn et al. 2019). However, these last are not investigated in this work.

2 Carbon Dioxide Conversion to Chemicals

This route is regarding the catalytic chemical CO_2 transformation into chemical compounds, as urea (Pérez-Fortes et al. 2014) polyether carbonate polyols (Langanke 2014) for polymer production (polyurethane), polycarbonate and dimethyl carbonate. Urea is obtained from ammonia (NH_3), from the Haber–Bosch process and CO_2, as described by these reactions (see Eqs. 1–2):

$$2 \cdot NH_3 + CO_2 \leftrightarrow NH_2CO_2NH_4 \quad \Delta H = 157.5 \text{ kJ/mol} \tag{1}$$

$$NH_2CO_2NH_4 \leftrightarrow NH_2CONH_2 + H_2O \quad \Delta H = -26.44 \text{ kJ/mol} \tag{2}$$

with ammonium carbamate as an intermediate step. Urea is non-toxic and can be used as fertilizer or for polymer synthesis (melamine and urea–formaldehyde resins) (Mikulčić et al. 2019). Actually, urea production is the largest scale process for CO_2 utilization: 140 $MtCO_2$/year is used to produce 200 Mt/year of urea (Jarvis and Samsatli 2018). For this route, the value of TRL is 9 (Chauvy et al. 2019). A SWOT analysis is suggested by Chauvy et al.

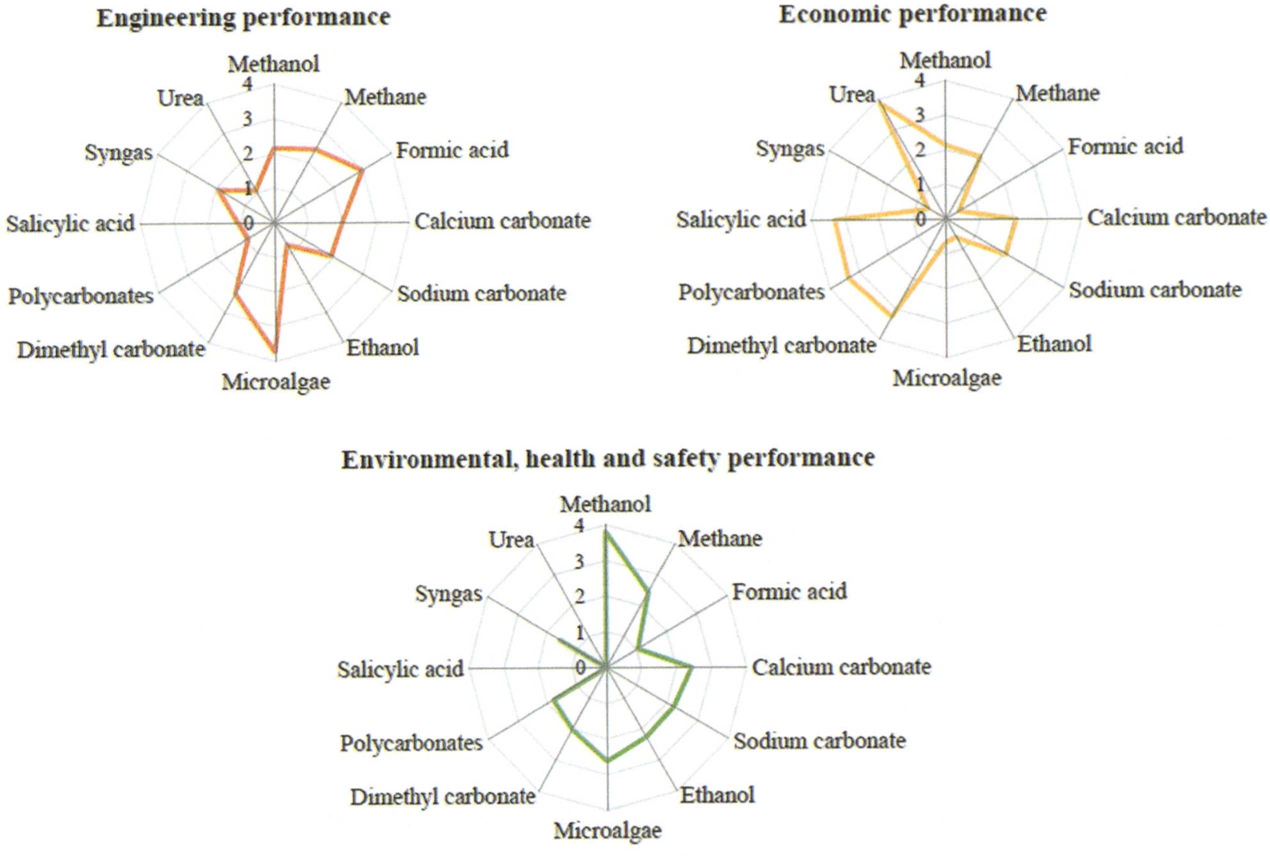

Fig. 3 Evaluation of the 3E performance criteria for CO_2-based products (Chauvy et al. 2019)

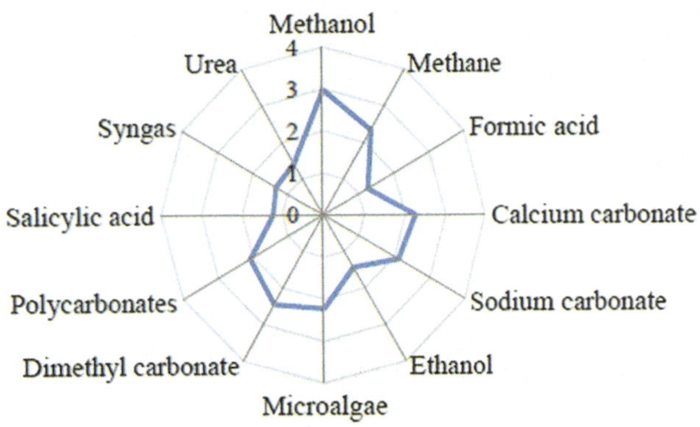

Fig. 4 Total weighted scores of CO_2-based products (Chauvy et al. 2019)

(2019) regarding this utilization route. Mature technology and an economically viable are the points of strengths. The points of weaknesses are the following: volatility in the price and demand of urea and ammonia, and the high capital costs of CO_2 (currently). The opportunity is the aim of decreasing CO_2 capture plant capital and operational costs. Threats are: international regulations, the fluctuation of prices and the demand of ammonia and urea.

The use of CO_2 for polymers is raising. Polyols (usually polyether and polyester) are petroleum derived products and these catalytically react with isocyanates to produce polyurethanes (Ramos, et al. 2016). However, polyether carbonate polyols can be produced in a more environmental way with the reaction of CO_2 and epoxides (alkylene oxide) (Artz et al. 2018; Muller et al. 2016). Actually, the company Bayer (Covestro) is studying to produce polyols for polyurethanes and the value of TRL is 8–9 (Chauvy et al. 2019; Fernández-Dacosta et al. 2017). The product is a novel polyol containing about 20% CO_2. The mechanical properties of polyurethanes are comparable with those obtained through a traditional way. The applications of polyurethane are in coatings, adhesives, sealants, elastomers and foams, heart valves and cardiovascular catheters and others (Grignard et al. 2019).

Other commercialized products from CO_2 are polycarbonates and salicylic acid.

Polycarbonates are largely produced through the reaction (melt polycondensation) between bisphenol A (aromatic polycarbonates) or diols (aliphatic polycarbonates) and phosgene or diphenyl carbonate (Singh 2014). Aromatic polycarbonates with good mechanical properties are used as plastics in automotive, electrical and electronic industries and construction. Aliphatic polycarbonates show poor mechanical properties are used in packaging, as a binder in ceramics, etc. (Grignard et al. 2019). Due to their biodegradability and biocompatibility, these polymers are used in the biomedical sector. In an alternative route, (mostly aliphatic) polycarbonates are produced from the reaction between CO_2 and epoxides but, with a different catalyst compared to the previous route, that leads to polyurethane production (Liu and Wang 2017). Generally, in polycarbonate reaction, the combination between many Lewis acid catalysts as organometallic complexes and a nucleophile has been considered (Grondin et al. 2019). In particular, recent reviews have been considered metal-based catalysts (zinc, aluminum, chromium, cobalt, magnesium, iron, titanium, copper, ytterbium, etc.) with a variety of ligands (Grondin et al. 2019; Taherimehr and Pescarmona 2014). The efficiency of this green reaction, as the physicochemical properties of the obtained product, is affected by: the nature of the epoxide, catalyst, reaction temperature, CO_2 pressure, solvent, reaction time and presence of impurities (Taherimehr and Pescarmona 2014). The TRL value for polycarbonate production from CO_2 is 9 (Chauvy et al. 2019).

Salicylic acid is produced through the Kolbe–Schmitt reaction (or Kolbe process): sodium phenoxide and CO_2 react to form salicylic acid, used for aspirin production (Bazzanella and Krämer 2019).

Another chemical compound that is obtained from CO_2 dimethyl carbonate, non-toxic and biodegradable (Marciniak et al. 2019). Generally, it is used in electrolytic solutions of lithium ion batteries, as an additive in fuels, as a methylating agent and for polycarbonates production (Kumar et al. 2017; Marin et al. 2016). Also, dimethyl carbonate can be used for glycerol carbonate production, pharmaceuticals, adhesives, detergents and biolubricants (Algoufi et al. 2017). Conventional routes for this product are the following: methoxycarbonylation with CO/oxygen (O_2)/CH_3OH, the alcoholysis of urea and the transesterification of cyclic carbonate with methanol (Zhao et al. 2017). However, dimethyl carbonate can be obtained through different ways from CO_2: (i) direct production from methanol and CO_2; (ii) production from methanol, CO_2 and epoxides; and (iii) production from CO_2 and ortho-ester or acetals (Rafiee et al. 2018). Each of them is characterized by a particular catalyst, reaction conditions and reactor type. The TRL value for this CO_2 utilization pathway is 8–9 (Chauvy et al. 2019).

At the research stage, there are other chemicals, such as carboxylic acids (acetic acid), carbamates (linear and cyclic), formaldehyde and isocyanates. In addition, CO_2 could also be used directly or indirectly to produce olefins such as ethene and propylene that are raw materials for the subsequent synthesis of polymers. However, catalysts for CO_2-to-olefins reactions actually have low yields and low selectivities.

In this pathway, the expected storage time of CO_2 depends on the produced chemical (days/decades), while the likelihood of release during storage is high (Hepburn et al. 2019).

3 Carbon Dioxide Conversion to Fuels

According to this route, a catalytic hydrogenation reaction is applied to transform CO_2 into fuels (methane, methanol, ethanol, dimethyl ether, syngas, hydrocarbons, etc.). Methanol, methane and syngas are the most important products. In fact, methanol, among fuel liquids, and methane, among gas fuels, have the highest value of EXc, as the carbon fuel exergy content per mole of carbon and then the maximum reversible work that it can ensure. A higher value of this parameter suggests that the carbon demand to store a unit of exergy is lower. Al-musleh et al. (2014) suggest that the EXc for methanol and methane is, respectively, 693 MJ/kmol C and 806 MJ/kmol C.

Syngas is suggested because it is the raw material of different fuels (hydrocarbons, methanol, dimethyl ether). However, methanol is preferred over methane for different reasons: the higher maturity of technology, the liquid phase that is easy to store and transport, the versatility to transform it into other chemicals (for gas there are different constraints in terms of storage, distribution and use respect to a liquid), the production process made of a reduced number of steps and a simpler separation (Ampelli et al. 2015). In addition, methanol synthesis is more efficient, because it takes place at milder conditions and requires less hydrogen (H_2).

Methanol is a relevant raw material, due to its utilization for energy and chemical uses (Bertau et al. 2014). Methanol may be converted to ethylene and propylene (Speybroeck et al. 2014). The conventional raw material for methanol production is the syngas, according to the following reactions (Leonzio 2018) (see Eqs. 3–5):

$$CO + 2 \cdot H_2 \leftrightarrow CH_3OH \quad \Delta H = -90.8 \text{ kJ/mol} \tag{3}$$

$$CO_2 + H_2 \leftrightarrow CO + H_2O \quad \Delta H = 41.6 \text{ kJ/mol} \tag{4}$$

$$CO_2 + 3 \cdot H_2 \leftrightarrow CH_3OH + H_2O \quad \Delta H = -49.2 \text{kJ/mol} \tag{5}$$

However, when methanol is produced via CO_2 hydrogenation (with a stoichiometric H_2/CO_2 ratio of 3), only independent reactions are considered (see Eqs. 4–5) (Leonzio 2018).

Table 2 Analysis of the open literature for methanol production

Syngas hydrogenation						
Reference	**Type of reactor**	**Pressure (bar)**	**Temperature (K)**	**Catalyst**	**Methanol production**	
Luu et al. (2016)	Packed bed	50-100	503-538	Cu/ZnO/Al_2O_3	776 ton/h	
Zhang et al. (2017)	Packed bed	80	523	Cu		
Zhang et al. (2017)	Packed bed	80	523	Cu	83.54 kmol/h	
Storch et al. (2016)		100	525-538			
Gai et al. (2016)		40	483		5.6 kmol/day	
Martin and Grossmann (2017)		50-100	473-573		207 Mgal/year	
Iaquaniello et al. (2017)			493		105000 t/year	
Specht et al. (1999)					2.2 t/h	
Co-electrolysis of CO_2 and H_2O						
References	**Type of reactor**	**Pressure (bar)**	**Temperature (K)**	**Catalyst**	**Methanol production**	**Type of electrolysis**
Al-Kalbani et al. (2016)	Packed bed		538-131		1500 ton/day	SOEC
Rivera-Tinoco et al. (2016)	Packed bed	80	533	Cu/ZnO/Al_2O_3		PEM/SOEC
CO_2 hydrogenation						
References	**Type of reactor**	**Pressure (bar)**	**Temperature (K)**	**Catalyst**	**Methanol production**	**Hydrogen source**
Kiss et al. (2016)	Packed bed	50-100	473-573	Cu/Zn/Al/Zr	100 kt/year	Chlor-alkali
Atsonios et al. (2016)	Membrane reactor	65	523	Cu-Zn-Al	94.38 ton/year	KOH water electrolysis
Peres Fortes et al. (2014)	Packed bed	76	483	Cu/ZnO/Al_2O_3		PEM
Bellotti et al. (2017)		50-100	523-573	Cu/ZnO/Al_2O_3	97 kg/h	PEM
Harp et al. (2015)		50-100	523	Zn	452 t/day	Steelwork
Electrocatalytic reaction						
References	**Type reactor**	**Pressure (bar)**	**Temperature (K)**	**Catalyst**	**Methanol production**	
Gai et al. (2016)		80	403		5.6 kmol/day	

In addition to the hydrogenation reaction, the co-electrolysis CO_2 and H_2O and the electrocatalytic reduction of CO_2 are investigated to produce methanol, as shown in Table 2.

The TRL value for hydrogenation reaction is of 8–9 (the "George Olah" plant in Iceland is close to a commercial plant with a capacity of about 5 Mton/year, while a pilot plant with a capacity of about 100 ton/year was built in Japan, Osaka, by Mitsui Chemicals Inc., in 2008), while other routes are at research and development scale (TRL of 1–3) (Chauvy et al. 2019). Chauvy et al. (2019) suggest an interesting SWOT analysis for the route producing methanol via CO_2 hydrogenation. Strengths for methanol production via CO_2 hydrogenation are the following: a big market, it is a raw material for synthetic hydrocarbons and it is a liquid under ambient conditions, so that it is easy to handle. Weaknesses for this route are, instead, the following: the use of hydrogen that should be obtained by renewable energies increasing the cost, the lack of a suitable catalyst, it is toxic and flammable so that it must be handled properly. On the other hand, opportunities are: the replacement of fossil fuels, it is a fuel and can be used to produce hydrocarbons, the increase of consumption due to the emerging application technologies. Threats are the following: the evolution of CO_2 and H_2 price, the possibility to have "catastrophic" events in all phases of the methanol value chain and the biomethanol produced from other feedstocks. Several reactors can be considered. For gas-phase reactors, isothermal or adiabatic technologies are suggested. Also, liquid-phase reactors and membrane reactors can be considered (Leonzio 2018). Adiabatic reactors are proposed by Imperial Chemical Industries, Casale, Toyo Engineering Corporation (2015a), Haldor Topsoe, Kellogg (Leonzio 2018) and Rahimpour (2008). Overall, these types of reactors consist of adiabatic fixed beds with intermediate refrigeration to control temperature. Isothermal reactors are proposed by Linde (2015b), Lurgi (Haid and Koss 2001), Mitsubishi Gas Chemical (MGC) and Mitsubishi Heavy Industry (MHI) (Leonzio 2018), Casale (Bozzano and Manenti 2016). These are shell and tube reactors with a catalyst inside (Lurgi, MGC, MHI) or outside tubes (radial steam-raising converter and tube-cooled converter). Liquid reactors, with respect to other systems, ensure better control of temperature due to a large heat capacity of a liquid to absorb the generated heat (Xu et al. 2009). Air products proposes a slurry liquid-phase technology (LPMEOH™) (Leonzio 2018): The catalyst is suspended in an inert mineral oil, improving heat and mass transfer at lower costs than tubular fixed bed reactors. The technology has a high conversion with a low recycle. Membrane reactors allow the removal of a product (methanol or water or both) improving efficiency (Gallucci et al. 2004). Other membrane reactors for methanol synthesis are analyzed by Parvasi et al. (2009), Rahimpour and Ghader (2003), Farsi and Jahanmiri (2011), Barbieri et al. (2002). In addition to these types of reactors, a gas-phase fluidized-bed reactor is proposed: the catalyst is fluidized by the fresh gas, ensuring a good conversion, less pressure drop, the elimination of diffusion limitations, good heat transfer capability and a more compact design (Abashar 2004).

Methane is obtained by CO_2 and H_2, according to the Sabatier reaction (see Eq. 6) (Catarina Faria et al. 2018):

$$CO_2 + 4 \cdot H_2 \leftrightarrow CH_4 + 2 \cdot H_2O \quad \Delta H = -165 \text{kJ/mol} \quad (6)$$

To be more precise, the methanation reaction is the combination of the carbon monoxide (CO) methanation and reverse water gas shift (RWGS) reaction (Bassano et al. 2019) (see Eqs. 7–8):

$$CO + 3 \cdot H_2 \leftrightarrow CH_4 + H_2O \quad \Delta H = -206 \text{kJ/mol} \quad (7)$$

$$CO_2 + H_2 \leftrightarrow CO + H_2O \quad \Delta H = 41 \text{ kJ/mol} \quad (8)$$

The H_2/CO_2 ratio for methane production should be 4 (Leonzio 2018). The obtained methane is called substitute natural gas (SNG), with properties close to that of natural gas (Riccia et al. 2019). When H_2 is obtained by renewable energies (solar or wind energy), via H_2O electrolysis the process is known as power-to-gas (Lewandowska-Bernat and Desideri 2017). The energy efficiency from renewable energies to gas is about 30–40% (Sauer et al. 2012). The TRL value for this hydrogenation reaction is 7, and tests are limited to the demonstration plant. A lower value of TRL is considered for the other routes of methane production: 4 for CO_2 electrochemical reduction, 3–4 for CO_2 microbial conversion and 3 for CO_2 photo-electrochemical reduction (Chauvy et al. 2019). Existing projects, on the catalytic hydrogenation of CO_2 in a power-to-gas process, are present in Germany, Switzerland, USA, Austria, Japan, the Netherlands, Poland, France, Canada and Thailand (Bailera et al. 2017; Chwoła et al. 2020). Plants with high installed power are in Germany (30.7 MW_{el}) Denmark (2.53 MW_{el}), Canada and the USA (both about 0.45 MW_{el}) (Thema et al. 2019).

Water electrolysis systems can be alkaline electrolysis (AEL), polymer electrolyte membrane electrolysis (PEM) and solid oxide electrolysis cell (SOEC). The first is the most mature technology, while the last is still at the pre-commercial scale. Specific electrical energy consumption of AEL, PEM and SOEC is, respectively, of 4.2–4.8, 4.4–5 and 3 kWh/Nm^3, while capital costs are 1000 €/kWel–1200 €/kWe, 1800–2300 €/kWe and above 2000 €/kWe (Schmidt et al. 2017; Buttler and Spliethoff 2018). Stages of fixed bed reactors with refrigeration, fluidized-bed reactors and three-phase reactors may be used for methanation reactions (Rönsch et al. 2016).

Most of the studies present in the literature are about the process and system analysis of a power-to-gas process, in addition to a simple analysis of methanation reaction (Bassano et al. 2019). Blanco et al. (2018) analyze the potential application of this system in Europe with two different scenarios: a reduction of 85 and 90% of emission by 2050 compared to the 1990 level. They find that the power-to-gas can be taken into account only at lower costs. Brunner and Thomas (2014) consider an analysis of a power-to-gas system for Germany, while an analysis of this system with large-scale renewable electricity production is considered in Vandewalle et al. (2015). Dickinson et al. (2010) study a power-to-gas process where hydrogen is produced through an electrolyzer using geothermal energy while carbon dioxide is captured in natural gas power plants in Australia. Kötter et al. (2015) consider using 100% of renewable energy for a power-to-gas system in Germany. They find that in this case, the leveled cost of electricity (LCOE) of 11 ct/kWhel is lower than that estimated by using batteries to store energies. An economic analysis is developed by Gassner and Maréchal (2012), Buchholz et al. (2014), Tsuparia et al. (2016), Petersa et al. (2019) finding a methane cost in the range of € 3.51–3.88 per kg. An economic analysis for Germany is carried out by Leonzio (2017), suggesting the need for economic incentives to have a profitable system. Balan et al. (2016) consider a techno-economic analysis for a power-to-gas in Romania, considering different values for the energy price. Considering a discounted cash flow analysis, it is found that the process requires economic incentives. Economic incentives are also required by Blanco and Faaij (2018) for this synthesis. In this context, an economic optimization is considered by Gorre et al. (2020): a reduction up to 17% in synthetic natural gas production is achieved in optimal conditions.

From this analysis, it results that there is a great interest on increase the integration of renewable energies and that this technology is characterized by high costs; then, economic incentives are required. A SWOT analysis for methanation reaction is also suggested by Chauvy et al. (2019). A strength is the fact that methane has a big market. Opportunities are the following: a reaction highly exothermic with the possibility to reuse heat from conversion, the produced methane may be transported using the current gas grid and infrastructure, the possibility to fit renewable methane and renewable hydrogen economies. Weaknesses are: the use of H_2 that should be produced by renewable energies at high costs, the reaction is highly exothermic and should be controlled, methane is flammable and it is in competition with natural gas. Threats are the following: the evolution of CO_2 and H_2 price, and the competition with biogas from fermentation.

Generally, syngas can be obtained from coal, petroleum coke, natural gas, biomass and even from organic wastes (Tao et al. 2011). However, syngas may be obtained from CO_2 in several routes, as reported in Table 3. Overall, it is possible to have, according to temperature, thermal and non-thermal (dielectric barrier discharge, glow discharge, corona discharge, plasma jets) plasma. Reactions of CO_2 reforming of CH_4 to syngas are involved; however, these techniques are not commercialized actually (Tao et al. 2011).

Li et al. (2009) analyze operating parameters of a glow discharge plasma reactor, finding CH_4 and CO_2 conversion of 60.97 and 49.91%, respectively, at CH_4/CO_2 rate of 4/6 and an input power of 69.85 W. Ozkan et al. (2015) study the dielectric barrier discharge (DBD) to produce syngas by CO_2 and CH_4: They find the negative effect of power on carbon dioxide and methane conversion. A plasma jet is analyzed by Ni et al. (2011): At the optimal experimental conditions, a value of energy efficiency of 74.63% is achieved which is higher compared to other plasma processes. The good performances of a thermal plasma reactor are verified by Yanpeng et al. (2014): Thermodynamic and experimental studies show that the CO_2/CH_4 volume ratio and the total feed flow rate are important parameters in the reforming process. Then, for thermal and non-thermal plasma reactors, operating parameters are analyzed in the existing literature.

Another technique for syngas production using CO_2 and that has not reached the commercialized level is the methane dry reforming. The value of TRL is, in fact, 6, and a major

Table 3 Investigated technologies to produce syngas from CO_2

Glow discharges
Li et al. (2009)
Dielectric barrier discharges
Wang et al. (2011)
Pham et al. (2011)
Ozkan et al. (2015)
Song et al. (2018)
Nguyen et al. (2019)
Plasma jets
Ni et al. (2011)
Rutberg et al. (2015)
Thermal plasma
Yanpeng et al. (2014)
Methane dry reforming
Gokon et al. (2011)
Dou et al. (2019)
Wang et al. (2019)
Microwave-assisted dry reforming
Fidalgo et al. (2008)
Co-electrolysis of H_2O and CO_2
Pardal et al. (2017)
Liu et al. (2016)
Stoots et al. (2009)
Delacourt et al. (2008)
Kleiminger et al. (2015)
How and Xie (2019)
Zhou et al. (2019)
Acid Gas To Syngas (AG2STM)
Bassani et al. (2016)

disadvantage is the formation and deposition of soot causing the deactivation of catalyst (Chauvy et al. 2019; Mustafa et al. 2020). It is an endothermic reaction that requires a lot of energy and high temperature (900–1200 K) (Leonzio 2018) (see Eq. 9):

$$CO_2 + CH_4 \leftrightarrow 2CO + 2 \cdot H_2 \quad \Delta H = 247.3 \text{ kJ/mol} \quad (9)$$

Most of the literature analysis is about the searching and analysis of an appropriate catalyst (Gokon 2011; Wang et al. 2019; Dou et al. 2019).

Methane dry reforming can be developed also using microwaves, under the microwave-assisted dry reforming of methane (Leonzio 2018). A comparison with the traditional dry reforming reaction is considered in the work of Fidalgo et al. (2008), showing higher CH_4 and CO_2 conversion for the microwave heating. A SWOT analysis for methane dry reforming is suggested by Chauvy et al. (2019) through strengths, weaknesses, threats and opportunities. The points of strength are the following: a huge market, two greenhouse gases consumed (CO_2 and CH_4), a cheap and an abundant CH_4 available, the crucial intermediate resource of many large-scale chemical building blocks and petrochemical processes. The points of weaknesses are the following: the competition with syngas from steam reforming, the lack of availability of suitable catalyst (fast catalyst deactivation and carbon deposition on catalyst), the energy consuming, the lack of information as syngas is usually not considered as a final product. The opportunity is the research of a catalyst, while the threat is the competition with syngas from biogas and biomass.

Another technology to produce syngas is the co-electrolysis of CO_2 and H_2O. Reactions in the cathode are (see Eqs. 10–11) (Li et al. 2013):

$$CO_2 + 2e^- \rightarrow CO + O^{2-} \quad (10)$$

$$H_2O + 2e^- \rightarrow H_2 + O^{2-} \quad (11)$$

At the anode, this reaction occurs (see Eq. 12) (Li et al. 2013):

$$2 \cdot O^{2-} \rightarrow O_2 + 4 \cdot e^- \quad (12)$$

Literature works demonstrate the feasibility of this technique at a large scale and analyze different configurations. Liu and Poon (2016) analyze the separate electrolysis of CO_2 and H_2O and the co-electrolysis of these molecules. Results show that simultaneous electrolysis in a single electrolyzer is possible as well as industrial production. Stoots et al. (2009) study experimentally the co-electrolysis of H_2O and CO_2 finding high yields. An electrolyzer system using also an ionic liquid (1-butyl-3-methylimidazolium triflate) is suggested by Pardal et al. (2017) in order to have high efficiencies and low energy demand. Kleiminger et al. (2015) use a microtubular solid oxide electrolyzer with yttria-stabilized zirconia (YSZ) electrolyte, Ni-YSZ cermet cathode and strontium (II)-doped lanthanum manganite (LSM) oxygen-evolving anode, for the co-electrolysis of H_2O and CO_2 to produce syngas. The value of TRL for this technology is 4–5 (Chauvy et al. 2019).

Bassani et al. (2016) propose a novel Acid Gas To Syngas (AG2STM) technology consisting of the gasification of coal or biomass while H_2S and CO_2 are transformed into syngas, using a regenerative thermal reactor.

Other researches are about the production of syngas through solar cells converting carbon dioxide from air directly into synthesis gas, utilizing solar energy (Leonzio 2018).

In this pathway for fuel production, the expected storage time of CO_2 depends on the produced fuel (weeks/months), while the likelihood of release during storage is high (Hepburn et al. 2019).

Figure 5 shows main fuels that may be obtained from CO_2, through different reactions.

4 Carbon Dioxide Conversion to Concrete Building Materials

Natural and accelerated curing carbonation can occur and these have been studied (Galan et al. 2010; Monkman and Shao 2006). Natural curing carbonation occurs in the air: calcium hydroxide and calcium silicate hydrates react with the atmospheric CO_2 to produce calcium carbonate ($CaCO_3$), improving mechanical properties and reducing the subsequent drying shrinkage of concrete product (Baojian et al. 2013). However, it is a slow process and it is a problem for steel-reinforced concrete structure because the carbonation reduces the concrete pH, improving the corrosion of reinforcing steel. In the other options, CO_2 is injected into the curing vessel at room temperature; then, CO_2 diffusing into the fresh concrete under low pressure, reacts with compounds like $3CaO{\cdot}SiO_2$, $2CaO{\cdot}SiO_2$ to produce $CaCO_3$ and calcium silicate hydrates, according to the following reactions (see Eqs. 13–14) (Xuan et al. 2018; Khan et al. 2018):

$$2(3CaO \cdot SiO_2) + 3CO_2 + 3H_2O \rightarrow 3CaO \cdot 2SiO_2 \cdot 3H_2O + 3CaCO_3 \quad (13)$$

$$2(2CaO \cdot SiO_2) + CO_2 + 3H_2O \rightarrow 3CaO \cdot 2SiO_2 \cdot 3H_2O + CaCO_3 \quad (14)$$

With an initial hydration products as calcium hydroxide (CH) and calcium silicate hydrate (C-S–H) can also be carbonated, according to the following reactions (see Eqs. 15–16) (He et al. 2019):

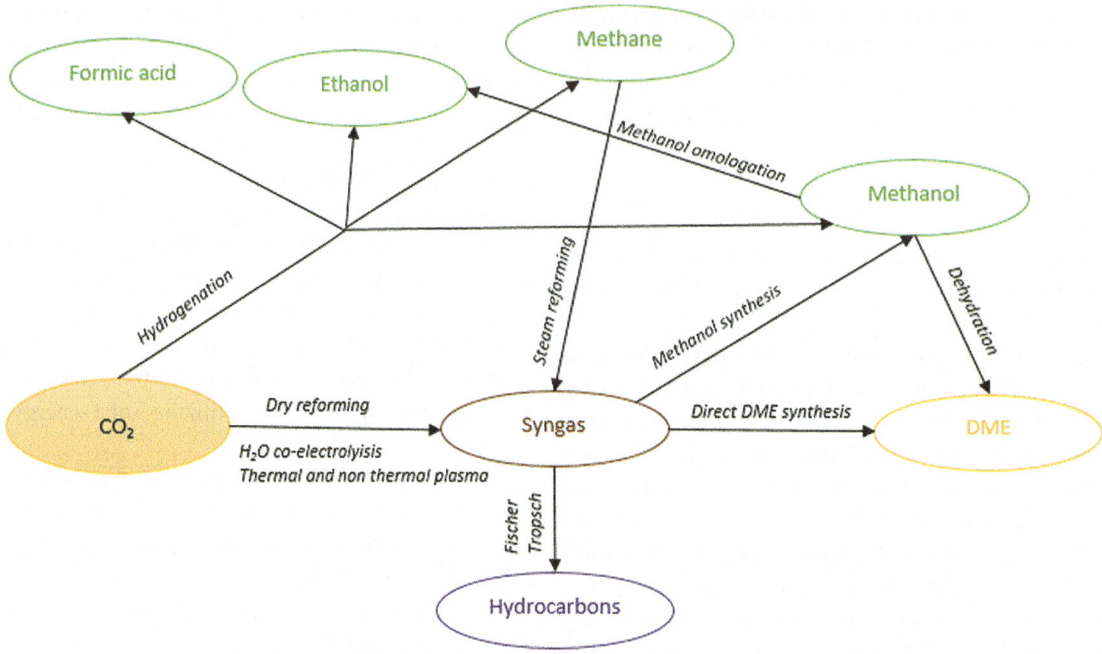

Fig. 5 Principal fuels obtained from CO_2 (Leonzio 2019)

$$Ca(OH)_2 + CO_2 \rightarrow CaCO_3 + H_2O \quad (15)$$

$$3CaO \cdot 2SiO_2 \cdot 3H_2O + 3CO_2 \rightarrow 3CaCO_3 + 2SiO_2 + 3H_2O \quad (16)$$

Overall, gaseous CO_2, after its dissolution into H_2O, is converted into a solid calcium carbonate during an accelerated curing process. For applications without reinforcing steel, the carbonated concrete products have better performances in terms of compressive strength, abrasion resistance, durability, stable dimensions, due to the near-complete depletion of calcium hydroxide (Baojian et al. 2013; Shi-Cong et al. 2014). Costs are also reduced due to a lower cement content. However, it is most suitable for concrete products, such as blocks and cement boards.

Also, these properties may be obtained in few hours contrary to natural carbonation (El-Hassan and Shao 2014; Zhan et al. 2016), and in any case, this process depends on CO_2 concentration, pressure, time of exposure, relative humidity, etc. (Shi et al. 2013). Khan et al. (2018) find that in the accelerated curing carbonation, concrete achieves in only 4 h the strength that should be achieved in 28 days, with natural curing carbonation. Xuan et al. (2018) find that important parameters to improve the maturity and strength of concrete are: a high CO_2 concentration, a fast gas flow rate and moderate relative humidity. On the other hand, Monkman and Shao (2006) report that with accelerated concrete curing the durability of the concrete is not compromised, but better performances (compressive strength, linear shrinkage, etc.) are obtained. The value of TRL for concrete curing is 7–8 (Alberici 2017).

In addition, concrete can be produced from CO_2, through the red mud or "bauxite residue," obtained by bauxite treatment during alumina production (two tons of red mud are obtained by one ton of bauxite, according to the bauxite characteristics and processing parameters) (Klauber and Gräfe 2009). Red mud is mainly made up of Al_2O_3, Fe_2O_3, SiO_2, TiO_2, CaO, Na_2O; then, it has an alkaline character with a high pH (10.5–12.5) (Patricio et al. 2017a; Ribeiro and Morelli 2011). If it is carbonated with CO_2, the pH is reduced and it can be used to produce concrete reducing the cement content (Sutar et al. 2014). Different works are about its application in concrete production. Ribeiro et al. (2012) suggest that red mud in this production route can improve the corrosion resistance, if with a composition in the range between 20 and 30% wt. Better mechanical properties than conventional concrete are obtained by Nikbin et al. (2018). Liu and Poon (2016) replace fly ash with red mud in concrete: Mechanical properties are improved; on the other hand, drying shrinkage is decreased. Rathod et al. (2014) suggest an optimal composition of red mud in the concrete of 25% wt. Other studies are proposed out by Sai (2017), Colella et al. (2007), Feng and Peng (2005), Hauri (2006), Dindi et al. (2019). Overall, literature works analyze the production process finding better operating conditions for this pathway. This pathway has a value of TRL of 9 (Patricio et al. 2017a).

In concrete production, the expected storage time of CO_2 may be of centuries, while the likelihood of release during storage is low (Hepburn et al. 2019).

5 Carbon Dioxide for Mineral Carbonation

A review about processes of CO_2 mineral carbonation has been proposed by Sanna et al. (2014). Mineral carbonation reaction can be in situ or ex situ (Chang et al. 2017a). The first case occurs in geological storages, when CO_2 reacts with alkaline minerals and it is transformed into a mineral carbonate. The second reaction, on the other hand, occurs chemically in an industrial plant, through a reaction between CO_2 and alkaline earth metals mostly calcium (Ca) or magnesium (Mg), obtained from nature via silicate minerals or from industrial by-products or waste materials (Gerdemann et al. 2007; Pan et al. 2012). This case occurs particularly where CO_2 storage is not allowed. In this case, a precipitated mineral carbonate is produced. In particular, for both carbonations, the reaction is between the metal oxide bearing the mineral and CO_2, as follows (see Eq. 17):

$$\text{Metal oxide} + CO_2 \rightarrow \text{Mineral carbonate} + \text{Heat} \quad (17)$$

The overall reaction for carbonation is exothermic and spontaneous ($\Delta G < 0$). On the other hand, the reaction rate of this reaction is very slow and should be improved (Zheng et al. 2017).

In addition to $CaCO_3$ and magnesium carbonate ($MgCO_3$), sodium carbonate ($NaCO_3$) and sodium bicarbonate ($NaHCO_3$) may be produced. The TRL value for the process producing $CaCO_3$, $MgCO_3$, $NaCO_3$ and $NaHCO_3$ is, respectively, 7, 3–4, 6 and 8–9 (Chauvy et al. 2019).

Compared to the in situ reaction, the ex situ reaction allows to valorize also industrial hazardous wastes, containing the considered metals and to use the obtained mineral carbonate products for different aims. In fact, for example, $CaCO_3$ may be used as adhesives, sealants, in the food and pharmaceutical sectors, in the paint, rubber and paper industry, for construction materials (Eloneva et al. 2008). In addition, while in the in situ reaction CO_2 is just injected into geological formations containing alkaline metals, the ex situ mineral carbonation involves different processes like mining, grinding and/or pre-treatment processes to ensure Ca or Mg bearing mineral feedstock. The different applications of mineral carbonates are related to their physicochemical characteristics such as particle size, shape, density, color, brightness which can be governed through operating parameters such as pH, temperature, concentration, additives, stirring and reaction time (Chang et al. 2017b).

Literature works are mainly considering steel slag as raw material to produce precipitated calcium carbonate. In the work of Zappa (2014), calcium is extracted from steel slag by using a solution of ammonium chloride (NH_4Cl). The author reports that temperature, calcium concentration, NH_4Cl solvent concentration, CO_2 flow rate and agitation speed are significant on the mineralization and quality of the precipitated calcium carbonate. In particular, a lower dimension can be obtained reducing temperature, calcium and NH_4Cl concentration, CO_2 flow rate and increasing the agitation speed. An economic analysis of the system is also developed: The production cost for the precipitated calcium carbonate is 65 €/ton. A techno-economic analysis of KIST process to produce calcium carbonate from steel slag is carried out by Lee et al. (2020): Results suggest the cost is 483 USD/ton $CaCO_3$; then, the process is feasible. Said et al. (2013) consider a parametric analysis of this process. They find that grinding the materials of steel slag to a smaller size can improve the efficiency and chemical conversion rates, while the smallest solid-to-liquid ratio ensures the maximum calcium extraction efficiency. A feasibility study to produce $CaCO_3$ from basic oxygen furnace slag is proposed by Kim et al. (2020). It is obvious that the most of literature works are regarding the analysis of the mineralization process, finding the costs and the influence of some operating parameters. In addition, researchers find that the use of an industrial waste or by-products with calcium or magnesium allows to decrease energy consumption and local pollution (Zheng et al. 2017).

A SWOT analysis for mineral carbonation of CO_2 is suggested by Chauvy et al. (2019). The strengths of this analysis are the following: the permanence of CO_2 storage, no feed quality requirements for CO_2 which can contain sulfur dioxide, particulate matter, mercury and other metals. Weaknesses are: the use of a large amount of minerals (as the precipitated calcium carbonate), direct competition with raw materials and building products usually used, high energy requirements of mining, transportation and preparation of minerals. Threats are the following points: the competition with cement, the price of the technology affected by changes, the environmental issues potentially generated by the storage of large amounts. On the other hand, the opportunity is that the environmental benefit leads to a commercial benefit especially in the region where carbon trading schemes exist.

As in concrete production, the expected storage time of CO_2 in mineral carbonates may be of centuries in a stable, inert and solid form, while the likelihood of release during storage is low due to the stability of mineral carbonates (Hepburn et al. 2019; Zevenhoven et al. 2006).

6 Carbon Dioxide for Oil and Methane Recovery

CO_2 may be utilized for oil and gas (methane) recovery, respectively, in CO_2 enhanced oil recovery (CO_2-EOR) and CO_2-enhanced coalbed methane (CO_2-ECBM) technologies. This allows to store CO_2, but at the same time, CO_2 is used as an injection fluid to recover oil and methane from respective reservoirs. In CO_2-enhanced coalbed methane, CO_2 is used to extracts CH_4 from the coalbed: With the injected gas, CH_4 is desorbed from the coalbed, varying the adsorption rate between gases and micropores while CO_2 is adsorbed and then stored (Cho et al. 2019). In fact, coalbed methane reservoirs are naturally fractured rocks, with cleats and matrix, which micropores have a lot amount of methane adsorbed on the walls (Verma and Sirvaiya 2016). This is explained by the different properties of the considered gases (Sayyafzadeh et al. 2015). In fact, CO_2 is smaller than CH_4 and it is also linear, facilitating the inlet of CO_2 in small pores and dislocating the pre-adsorbed CH_4 out of micropore surfaces. In addition, CO_2 has an interaction enthalpy better than hydrocarbons, increasing CO_2 solubility in coals compared to CH_4. An overview about this technology and the investigations that should be done is suggested by Li and Fang (2014) and Lau et al. (2017). Other studies present in the literature are about the analysis of geological and physical characteristics of a coalbed methane, underlining the process performances, as CO_2 injection and CH_4 separation. Mazzotti et al. (2009) show the best performances and efficiencies obtained by using CO_2 and not another gas (e.g., nitrogen or a mixture of nitrogen and carbon dioxide), due to the greater affinity of coal towards CO_2. Guan et al. (2018) analyze the adsorption capacity of CH_4 and CO_2 for an Illinois coal at several temperatures and find that these are correlated. In particular, the adsorption capacity of these gases decreases at higher temperatures, until a critical temperature value, where this dependence is vanished. Liu et al. (2013) propose a mathematical model to evaluate the wellhead and bottom hole pressure, considered as important parameters. In Sinayuc et al. (2011), the effect of some parameters, as the layout and number of wells, the composition of injected gas and permeability on CH_4 production are evaluated. In a more accurate analysis, Dai et al. (2017) find that the reservoir thickness, permeability and porosity are important to manage the process, with a positive effect on CO_2 injection rate and CH_4 production rate. The effect of injection pressure and initial reservoirs temperature on the process is analyzed by Fana et al. (2018) through a simulation in COMSOL Multiphysics. The authors find that the amount of stored CO_2 and extracted CH_4 decrease at a higher reservoir temperature and increase at a higher pressure. The positive effect of pressure is also documented in the research of Yin et al. (2017). The optimal CO_2 composition for gas recovery is found by Sayyafzadeh and Keshavarz (2016), based on the economic calculation of net present value. Studies for micropilot tests are carried out by Wong et al. (2007) and Ye et al. (2007). In recent years, also liquid CO_2 is used for this technology: Higher pressure and volume is able to crack the coal seams and improve the gas extraction (Chen et al. 2017; Hu et al. 2018; Wen et al. 2020).

The TRL value for this utilization option is 7 (demonstration scale) (Bui 2018).

In CO_2-EOR, as already mentioned, CO_2 is used to extract oil from reservoirs. This technology is based on chemical and physical mechanisms and involved during the interaction of CO_2 with rocks and fluids that are present in the reservoir. In particular, CO_2 has excellent dissolution properties, and above the minimum miscibility pressure, it is miscible with the in situ oil, dragging the oil out of pores (Farajzadeha et al. 2020). In addition to the USA, China and countries in the Middle East have been interested in CO_2-EOR (Azzolina et al. 2016). In 2014, 136 active CO_2-EOR projects were present in the USA (Koottungal 2014).

Studies present in the literature are mainly about the optimization of the process, techno-economic and environmental analysis. Generally, the co-optimization of CO_2 storage and oil recovery is considered by Kamali and Cinar (2014), Ampomah et al. (2016), Wang et al. (2018), Chen and Pawar (2019). In AlMazrouei et al. (2017), two different optimizations are suggested for a CO_2-EOR in California: One is minimizing the costs, and the other one is maximizing the revenues. The second one is preferable according to the market dynamics. Wei et al. (2015) perform a techno-economic analysis for an oil field in China, as well as Jiang et al. (2019). The net present value of the process is maximized by Kwak and Kim (2017), while Kemp and Kasim (2013) find that high CO_2 costs can provide negative values of net present value. Many other literature works regard a techno-economic analysis, suggesting the potential of this technology and its capacity to reduce CO_2 emissions (Davidson et al. 2011; Rubin et al. 2013; King et al. 2011; Mendelevitch 2014). Environmental analyses are also developed. Jaramillo et al. (2009) propone a model to calculate the cumulative CO_2 emissions over the life cycle of the considered process, including CO_2 capture from a power plant, CO_2 transportation, oil exploration and transportation, crude oil refining and its combustion. A detailed life cycle assessment (LCA) analysis for a CO_2-EOR system is developed by Cooney et al. (2015): They find that it is more environmentally beneficial to use anthropogenic CO_2 sources than natural CO_2. Azzolina et al. (2016) in their study find that the CO_2-EOR process produces oil with an emission factor of about 300 $kgCO_{2\text{-}eq}$/bbl, lower than the conventional process (500 $kgCO_{2\text{-}eq}$/bbl). In this study, CO_2 is captured from a coal-fired power plant. Greenhouse gas emission reduction per unit of produced oil is obtained through this kind of system in an offshore EOR in the North Sea, in the work of Stewart and Haszeldine (2015). Overall, the environmental benefits of this technology are reported in the literature. For this pathway, the value of TRL is 9; then, it is at a commercial scale (Bui 2018). In both discussed pathways, CO_2 is stored as geological sequestration; then, for millennia and the likelihood of release during storage, it is low (Hepburn et al. 2019).

7 Carbon Dioxide for Horticulture and Microalgae Production

CO_2 may be used for horticulture and microalgae production by using greenhouses with CO_2 enrichment. CO_2 enrichment is a technique used to enhance photosynthesis, improving yields and incomes (Chalabi et al. 2002). It is important to underline that these benefits are present only for plants with C3 photosynthesis pathway and not for plants with C4 pathway (Dion et al. 2011). C3 pathway includes cereals, all legumes, nearly all fruits, roots and tubers. C3 is also the pathway for sugar beet, fiber crops and oil crops, microalgae and trees. On the other hand, C4 pathway includes corn (maize), sorghum, millet and sugarcane. Then, C3 plants are greenhouse crops, even if have a lower efficiency to fix CO_2 at ambient concentrations (about 388 ppm) (Dion et al. 2011). However, for these crops, a higher CO_2 concentration determines a higher photosynthetic rate, creating more carbohydrates that translate into increased biomass. Generally, this positive effect for greenhouse crops is obtained for CO_2 concentrations in the range of 700 and 1000 ppm, increasing the yield from 21 to 61% in dry mass (Jaffrin et al. 2003). Also, photosynthesis can improve also up to 50% (Patricio et al. 2017a). The positive effect of CO_2 on the yield of rice, wheat, soybeans and grain has been studied (Dong et al. 2020; Senghor et al. 2017).

Also, these positive effects strongly depend on different greenhouse environmental parameters (temperature, relative humidity, nutrients and irrigation schedule) and different researches optimized the level of CO_2 injection according to these parameters (Klaring et al. 2007; Edwards 2008; Rachmilevitch et al. 2004). The economic benefits of CO_2 enrichment technology are reported by Marchi et al. (2018). It is evident that for horticulture production, the most of the literature works want to underline the better efficiencies obtained at a higher CO_2 concentration. Few works are about the economic analysis of this pathway.

Interesting considerations are reported in the literature for microalgae cultivation that can be used as feedstocks for biofuels, such as biodiesel (Hallenbeck et al. 2014; Choia et al. 2014), pyro-oil (Pisal and Lele 2005), colorants, vitamins (Chauvy et al. 2019) and other biochemicals (Pradhan et al. 2015) production. Algae can be converted into biofuels through several methods: transesterification, gasification, AD, direct combustion, fermentation and pyrolysis (Anwar et al. 2020).

Compared to other C3 plants, microalgae can tolerate higher ranges of temperature, pH value and CO_2 composition (Olaizola 2003); moreover, these are able to fix CO_2 at a rate several times higher than plants showing their high photosynthetic efficiencies (Bhola et al. 2014). Generally, microalgae have been received the interest of researchers due to the high CO_2-fixation efficiencies (up to 10%, compared with 1–4% for other biomass) and capacity to produce several products (Hepburn et al. 2019). From the literature analysis, it is evident that microalgae may be cultivated either in open raceway ponds or in photo-bioreactors used to enhance the photosynthetic process (Pradhan et al. 2015). Several works suggest how to increase the efficiency of these reactors (Zittelli et al. 2013; Ugwu et al. 2008; Zhang 2015).

A SWOT analysis for microalgae production is suggested by Chauvy et al. (2019). Strengths, weaknesses, threats and opportunities are evaluated. The point of strengths are: the use of solar energy, the use of various point source emissions, do not required high CO_2 purity, because NO_x and SO_x can be used as nutrients, the sewage wastewater can be used as the source of nutrients, do not require fertile land or food crops, the grow fast consuming CO_2. Weaknesses are the following: for open systems are a large specific area and an easy contamination of the culture by bacteria, for closed systems are high operative and capital costs, for both algal species depending on the environment and require small distance for transport and storage of CO_2 feed. Opportunity points are the following: an integrated system using wastewater and waste CO_2, an integrated biorefinery based on microalgae, a potential of algal oil in transportation fuel, livestock feed, agricultural fertilizer, oleo-chemicals, pharmaceutical and nutraceuticals markets. A threat is that modifying microalgae genetically could generate market and societal rejection.

The TRL value for horticulture and microalgae production is 9 and 8–9, respectively (Tcvetkov et al. 2019; Patricio et al. 2017a). The expected storage time of CO_2 depends on the product and can be of weeks or months, while the likelihood of release during storage is high (Hepburn et al. 2019).

8 Carbon Dioxide for Direct Use

CO_2 may be utilized in several processes of food industry, as an inert to avoid food deterioration and in packaging applications. In addition, CO_2 may be utilized in beverage carbonation (Linde 2019), in coffee decaffeination as an extraction solvent at supercritical conditions (Horticulture 2019) and wine making, preventing oxidation of the wine during maturation (Global CCS Institute 2011). In the food industry, CO_2 can be used also for the selective extraction of bioactive compounds (essential oils, vitamin E, etc.), to extract antioxidants from fruits and vegetable sources (tomato, rosemary, grape and aloe vera) and for the extraction of edible oils (Muthuraj and Mekonnen 2018). CO_2 is used also in the metal industry or basic oxygen furnaces for dust suppression (Leeson et al. 2017).

CO_2 can be also used in pulp and paper processing, for pH reduction during pulp washing operations (Linde 2012), wastewater treatment (after the reverse osmosis process to re-mineralize water or to control the pH) (Girdon et al. 2006), printed circuit board manufacture (Pieri et al. 2018) and in power generation as working fluid (Global CCS Institute 2011). Moreover, CO_2 may be utilized in pneumatic applications, in industrial fire protection systems (Pieri et al. 2018), in refrigeration units as the working fluid, avoiding the use of more toxic refrigerant gases (Leeson et al. 2017).

9 Conclusions

The interest in CO_2 utilization is increasing around the world, being a complementary option to CO_2 storage, for CO_2 emissions reduction. However, CO_2 utilization uptake potential is only a few percent of the anthropogenic emissions, even if it has the potential to produce revenues. Then, the use of CO_2 provides valid support to storage and other abatement methods.

Different CO_2 utilization pathways can be considered, and each of them is characterized by a value of TRL and particular performances, as economic, engineering and environmental (3E performance criteria). Generally, CO_2 can be used to produce chemicals, fuels, concrete building materials, microalgae, but CO_2 can be used also for mineral carbonation, oil and methane recovery and in the horticulture sector. In this chapter, these conventional routes are investigated. Different studies have been developed mainly for techno-economic analysis and results are quite encouraging, but further studies are required to make these technologies commercially viable.

A cycling or closed pathway can be ensured during the valorization of this emission. In the first case, the net CO_2 removal from the atmosphere is not provided, but CO_2 is only reduced. A cycling pathway is present when CO_2 is used to produce chemicals, fuels, microalgae. In the second case, a near permanent CO_2 storage and mineralization is present: CO_2 is used for oil and methane recovery and concrete building materials. On the other hand, an open pathway can be also present, but this last option about non-conventional use of CO_2 is not investigated in this chapter.

Overall, as suggested in the literature by Chauvy et al. (2019), CO_2 utilization options that show a high level of maturity, suitable for implementation in the near future include methanol, dimethyl carbonate, methane, calcium carbonates, microalgae and polycarbonates.

Due to the slow nature of the innovation process and the urgency of solving the climate change problem, these options should be considered. However, a life cycle analysis should be taken into account for them and policy should provide incentives for the route that is climate-beneficial.

References

AHDB Horticulture (2019) Available at: https://horticulture.ahdb.org.uk/sources-co2

Abashar MEE (2004) Coupling of steam and dry reforming of methane in catalytic fluidized bed membrane reactors. Int J Hydrogen Energy 29(8):799–808

Al-Kalbani H, Xuan J, García S, Wang H (2016) Comparative energetic assessment of methanol production from CO2: chemical versus electrochemical process. Appl Energy 165:1–13

Al-musleh EI, Mallapragada DS, Agrawal R (2014) Continuous power supply from a baseload renewable power plant. Appl Energy 122:83–93

AlMazrouei M, Asad O, Zahra MA, Mezher T, Tsai IT (2017) CO2-enhanced oil recovery system optimization for contract-based versus integrated operations. Energy Procedia 105:4357–4362

Alberici S et al (2017) Assessing the potential of CO2 utilization in the UK. Report Available at: https://assets.publishing.service.gov.uk/government/uploads/system/uploads/attachment_data/file/799293/SISUK17099AssessingCO2_utilisationUK_ReportFinal_260517v2__1_.pdf

Algoufi YT, Kabir G, Hameed BH (2017) Synthesis of glycerol carbonate from biodiesel by-product glycerol over calcined dolomite. J Taiwan Inst Chem Eng 70:179–187

Ampelli C, Perathoner S, Centi G (2015) CO_2 utilization: an enabling element to move to a resource- and energy-efficient chemical and fuel production. Phil Trans R Soc A 373:20140177

Ampomah W, Balch RS, Grigg RB, McPherson B, Will RA, Lee SY, Dai Z, Pan F (2016) Co-optimization of CO_2-EOR and storage processes in mature oil reservoirs. Greenh Gases Sci Technol 7:128–142

Anwar MN, Fayyaz A, Sohail NF, Khokhar MF, Baqar M, Yasar A, Rasool K, Nazir A, Raja MUF, Rehan M, Aghbashlo M, Tabatabaei M, Nizami AS (2020) CO2 utilization: Turning greenhouse gas into fuels and valuable products. J Environ Manag 260:110059

Artz J, Muller TE, Thenert K, Kleinekorte J, Meys R, Sternberg A, Bardow A (2018) Sustainable conversion of carbon dioxide: an integrated review of catalysis and life cycle assessment. Chem Rev 118:434–504

Atsonios K, Panopoulos KD, Kakaras E (2016) Investigation of technical and economic aspects for methanol production through CO2 hydrogenation. Int J Hydrogen Energy 41:2202–2214

Azzolina NA, Peck WD, Hamling JA, Gorecki CD, Ayash SC, Doll TE et al (2016) How green is my oil? A detailed look at greenhouse gas accounting for CO2-enhanced oil recovery (CO2-EOR) sites. Int J Greenhouse Gas Control 51:369–379

Bailera M, Lisbona P, Romeo LM, Espatolero S (2017) Power-to-gas projects review: Lab, pilot and demo plants for storing renewable energy and CO2. Renew Sustain Energy Rev 69:292–312

Balan OM, Buga MR, Bildea CS (2016) Conceptual design, performance and economic evaluation of carbon dioxide methanation plant. Rev Chim 67(11):2237–2242

Baojian Z, Chisun P, Caijun S (2013) CO2 curing for improving the properties of concrete blocks containing recycled aggregates. Cement Concr Compos 42:1–8

Barbieri G, Marigliano G, Golemme G, Drioli E (2002) Simulation of CO2 hydrogenation with CH3OH removal in a zeolite membrane reactor. Chem Eng J 85:53–59

Bassani A, Pirola C, Maggio E, Pettinau A, Frau C, Bozzano G, Pierucci S, Ranzi E, Manenti F (2016) Acid Gas to Syngas (AG2STM) technology applied to solid fuel gasification: cutting H2S and CO2 emissions by improving syngas production. Appl Energy 184:1284–1291

Bassano C, Deiana P, Lietti L, Visconti CG (2019) P2G movable modular plant operation on synthetic methane production from CO2 and hydrogen from renewables sources. Fuel 253:1071–1079

Bazzanella A, Krämer D (2019) Technologies for sustainability and climate protection—chemical processes and use of CO2. Available at https://dechema.de/en/energyandclimate/_/CO2_Buch_engl.pdf

Bellotti D, Rivarolo M, Magistri L, Massardo AF (2017) Feasibility study of methanol production plant from hydrogen and captured carbon dioxide. J. CO2 Util 21:132–138

Bertau M, Offermanns H, Plass L, Schmidt F, Wernicke HJ (2014) Methanol: the basic chemical and energy feedstock of the future. Springer-Verlag, Berlin Heidelberg

Bhola V, Swalaha F, Kumar RR, Singh M, Bux F (2014) Overview of the potential of microalgae for CO2 sequestration. Int J Environ Sci Technol 11:2103–2118

Blanco H, Faaij A (2018) A review at the role of storage in energy systems with a focus on Power-to-Gas and long-term storage. Renew Sustain Energy Rev 81:1049–1086

Blanco H, Nijs W, Ruf J, Faaij A (2018) Potential of power-to-methane in the EU energy transition to a low carbon system using cost optimization. Appl Energy 232:323–340

Bozzano G, Manenti F (2016) Efficient methanol synthesis: perspectives, technologies and optimization strategies. Prog Energy Combust Sci 56:71–105

Brunner C, Thomas A (2014) Arbeitspaket 6: Gasnetzanalysen und Wirtschaftlichkeitsbetrachtung. Wasser-Praxis 65:59–65

Buchholz OS, van der Ham AGJ, Veneman R, Brilman DWF, Kersten SRA (2014) Power-to-gas: storing surplus electrical energy. A Des Study, Energy Procedia 63:7993–8009

Bui M et al (2018) Carbon capture and storage (CCS): the way forward. Energy Environ Sci 11:1062–1176

Buttler A, Spliethoff H (2018) Current status of water electrolysis for energy storage, grid balancing and sector coupling via Power-to-Gas and power-to-liquids: a review. Renew Sustain Energy Rev 82:2440–2454

Catarina Faria A, Miguel CV, Madeira LM (2018) Thermodynamic analysis of the CO_2 methanation reaction with in situ water removal for biogas upgrading. J CO2 Util 26:271–280

Chalabi ZS, Biro A, Bailey BJ, Aikman DP, Cockshull KE (2002) Optimal control strategies for carbon dioxide enrichment in greenhouse tomato cropsdpart 1: using pure carbon dioxide. Biosyst Eng 81:421

Chang R, Choi D, Kim MH, Park Y (2017b) Tuning crystal polymorphisms and structural investigation of precipitated calcium carbonates for CO2 mineralization. ACS Sustain Chem Eng 5:1659–1667

Chang R, Kim S, Lee S, Choi S, Kim M, Park Y (2017a) Calcium carbonate precipitation for CO2 storage and utilization: a review of the carbonate crystallization and polymorphism. Front Energy Res 5 (17):1–12

Chauvy R, Meunier N, Thomas D, De Weireld G (2019) Selecting emerging CO_2 utilization products for short- to mid-term deployment. Appl Energy 236:662–680

Chen B, Pawar RJ (2019) Capacity assessment and co-optimization of CO2 storage and enhanced oil recovery in residual oil zones. J Petrol Sci Eng 182:106342

Chen H, Wang Z, Chen X, Chen X, Wang L (2017) Increasing permeability of coal seams using the phase energy of liquid carbon dioxide. J CO2 Util 19:112–119

Cho S, Kim S, Kim J (2019) Life-cycle energy, cost, and CO2 emission of CO2-enhanced coalbed methane (ECBM) recovery framework. J Nat Gas Sci Eng 70:102953

Choia JH, Woob HC, Suha DJ (2014) Pyrolysis of seaweeds for bio-oil and bio-char production. Chem Eng Trans 37:121–126

Chwoła T, Spietz T, Więcław-Solny L, Tatarczuk A, Krótki A, Dobra S, Wilk A, Tchórz J, Stec M, Zde J (2020) Pilot plant initial results for the methanation process using CO2 from amine scrubbing at the Łaziska power plant in Poland. Fuel 263:116804

Colella C, Cejka J, Van Bekkum H, Corma A, Schueth F (2007) Introduction to zeolite science and practice. Elsevier, Amsterdam, pp 999–1035

Cooney G, Littlefield J, Marriott J, Skone TJ (2015) Evaluating the climate benefits of CO2-enhanced oil recovery using life cycle analysis. Environ Sci Technol 49(12):7491–7500

Dai Z, Viswanathan H, Xiao T, Middleton R, Pan F, Ampomah W, Yang C, Zhoud Y, Jia W, Lee SY, Cather M, Balch R, McPherson B (2017) CO2 Sequestration and enhanced oil recovery at depleted oil/gas reservoirs. Energy Procedia 114:6957–6967

Davidson CL, Dahowski RT, Dooley JJ (2011) A quantitative comparison of the cost of employing EOR-coupled CCS supplemented with secondary DSF storage for two large CO2 point sources. Energy Procedia 4:2361–2368

Delacourt C, Ridgway PL, Kerr JB, Newman J (2008) Design of an electrochemical cell making syngas (CO + H2) from CO2 and H2O reduction at room temperature. J Electrochem Soc 155(1):B42–B44

Dickinson RR, Battye DL, Linton VM, Ashman PJ, Nathan GJ (2010) Alternative carriers for remote renewable energy sources using existing CNG infrastructure. Int J Hydrogen Energy 35(3):1321–1329

Dindi A, Quang DV, Vega LF, Nashef E, Abu-Zahr MRM (2019) Applications of fly ash for CO2 capture, utilization, and storage. J CO_2 Util 29:82–102

Dion LM, Lefsrud M, Orsat V (2011) Review of CO2 recovery methods from the exhaust gas of biomass heating systems for safe enrichment in greenhouses. Biomass Bioenerg 35:3422–3432

Dong J, Gruda N, Li X, Tang Y, Zhang P, Duan Z (2020) Sustainable vegetable production under changing climate: the impact of elevated CO2 on yield of vegetables and the interactions with environments-a review. J Clean Prod 253:119920

Dou J, Zhang R, Hao X, Bao Z, Wu T, Wang B, Yu F (2019) Sandwiched SiO2@Ni@ZrO2 as a coke resistant nanocatalyst for dry reforming of methane. Appl Catal B 254(2019):612–623

Edwards DR (2008) Towards a plant-based method of guiding CO2 enrichment in greenhouse tomato. Plant Science Department. University of British Columbia, Vancouver

El-Hassan H, Shao YX (2014) Carbon storage through concrete block carbonation curing. J Clean Energy Technol 2(3):287–291

Eloneva S, Teir S, Salminen J, Fogelholm C-J, Zevenhoven R (2008) Fixation of CO2 by carbonating calcium derived from blast furnace slag. Energy 33:1461–1467

Fana Y, Deng C, Zhanga X, Lid F, Wanga X, Qiao L (2018) Numerical study of CO2-enhanced coalbed methane recovery. Int J Greenhousess Gas Control 76:12–23

Farajzadeha R, Eftekhari AA, Dafnomilisa G, Lake W, Bruining J (2020) On the sustainability of CO2 storage through CO2—enhanced oil recovery. Appl Energy 261:114467

Farsi M, Jahanmiri A (2011) Application of water vapor and hydrogen perm-selective membranes in an industrial fixed-bed reactor for large scale methanol production. Chem Eng Res Des 89(12):2728–2735

Feng N, Peng G (2005) Applications of natural zeolite to construction and building materials in China. Constr Build Mater 19(8):579–584

Fernández-Dacosta C, Van Der Spek M, Hung CR, Oregionni GD, Skagestad R, Parihar P et al (2017) Prospective techno-economic and environmental assessment of carbon capture at a refinery and CO2 utilisation in polyol synthesis. J CO2 Util 21:405–422
Fidalgo F, Domínguez A, Pis JJ, Menéndez JA (2008) Microwave-assisted dry reforming of methane. Int J Hydrogen Energy 33(16):4337–4344
Gai S, Yub J, Yuc H, Eagle J, Zhao H, Lucas J, Doroodchi E, Moghtaderi B (2016) Process simulation of a near-zero-carbon-emission power plant using CO2 as the renewable energy storage medium. Int J Greenh Gas Control 47:240–249
Galan I, Andrade C, Mora P, Sanjuan MA (2010) Sequestration of CO2 by concrete carbonation. Environ Sci Technol 44(8):3181–3186
Gallucci F, Paturzo L, Basile A (2004) An experimental study of CO_2 hydrogenation into methanol involving a zeolite membrane reactor. Chem Eng Process 43:1029–1036
Gassner M, Maréchal M (2012) Thermo-economic optimisation of the polygeneration of synthetic natural gas (SNG), power and heat from lignocellulosic biomass by gasification and methanation. Energy Environ Sci 5(2):5768
Gerdemann SJ, O'Connor WK, Dahlin DC, Penner LR, Rush H (2007) Ex situ aqueous mineral carbonation. Environ Sci Technol 41:2587–2593
Girdon P, Gloger C, Gonzalez D, Henneqiun J, Krinninger K, de Lorenzi L, Wilyman P (2006) Minimum Specifications for Food Gas Applications; European Industrial Gases Association AISBL, Brussels, Belgium
Global CCS Institute (2011) Accelerating the uptake of CCS: Industrial Use of Captured Carbon Dioxide; Parsons Brickerhoff: New York, NY, USA
Gokon N et al (2011) Kinetics of methane reforming over Ru/γ-Al2O3-catalyzed metallic foam at 650–900 °C for solar receiver-absorbers. Int J Hydrogen Energy 36(1):203–215
Gorre J, Ruossa F, Karjunenb H, Schaffertc J, Tynjälä T (2020) Cost benefits of optimizing hydrogen storage and methanation capacities for Power-to-Gas plants in dynamic operation. Appl Energy 257:113967
Grignard B, Gennen S, Jerome C, Kleij AW, Detrembleur C (2019) Advances in the use of CO2 as a renewable feedstock for the synthesis of polymers. Chem Soc Rev 48:4466–4514
Grondin J, Aupetit C, Tassaing T (2019) a rational investigation of the lewis acid-promoted coupling of carbon dioxide with cyclohexene oxide: towards CO2-sourced polycyclohexene carbonate under solvent- and cocatalyst-free conditions. J Carbon Res 5:39
Guan C, Liu S, Li C, Wang Y, Zhao Y (2018) The temperature effect on the methane and CO_2 adsorption capacities of Illinois coal. Fuel 211:241–250
Haid J, Koss U (2001) Lurgi's mega-methanol technology opens the door for a new era in down-stream applications. Stud Surf Sci Catal 136:399–404
Hallenbeck PC, Leite GB, Abdelaziz AEM (2014) Exploring the diversity of microalgal physiology for applications in waste water treatment and biofuel production. Algal Res 6:111–118
Harp G, Harp KC, Bergins TG, Buddenberg T, Drach I, Koytsoumpa EI, Sigurbjornsson O (2015) Application of power to methanol technology to integrated steelworks for profitability, conversion efficiency, and CO2 reduction. Available at: http://www.mefco2.eu/pdf/2.%20Application%20of%20Power%20to%20Methanol%20Technology%20to%20Integrated%20Steelworks%20for%20Profitability,%20Conversion%20Efficiency,%20and%20CO2.pdf
Hauri F (2006) Natural zeolite from southern Germany: applications in concrete. In: Bowman RS, Delap SE (Eds) Proceedings of 7th international conference on the occurrence, properties, and utilization of natural zeolites. Socorro, pp 130–131
He Z, Jia Z, Wang S, Mahoutian M, Shao Y (2019) Maximizing CO2 sequestration in cement-bonded fiberboards through carbonation curing. Constr Build Mater 213:51–60
Hepburn C, Adlen E, Beddington J, Carter EA, Fuss S, Dowell NM, Minx JC, Smith P, Williams CK (2019) The technological and economic prospects for CO_2 utilization and removal. Nature 575:87–97
Hou S, Xie K (2019) Enhancing the performance of high-temperature H2O/CO2 co-electrolysis process on the solid oxide Sr2Fe1.6-Mo0.5O6-d-SDC/ LSGM/Sr2Fe1.5Mo0.5O6-d-SDC cell. Electrochim Acta 301:63–68
Hu GZ, He WR, Sun M (2018) Enhancing coal seam gas using liquid CO2 phase-transition blasting with crossmeasure borehole. J Nat Gas Sci Eng 60:164–173
Iaquaniello G, Centi G, Salladini A, Palo E, Perathoner S, Spadaccini L (2017) Wasteto-methanol: process and economics assessment. Bioresour Technol 243:611–619
Jaffrin A, Bentounes N, Joan AM, Makhlouf S (2003) Landfill biogas for heating greenhouses and providing carbon dioxide supplement for plant growth. Biosyst Eng 86:113
Jaramillo P, Griffin WM, McCoy ST (2009) Life cycle inventory of CO2 in an enhanced oil recovery system
Jarvis SM, Samsatli S (2018) Technologies and infrastructures underpinning future CO2 value chains: a comprehensive review and comparative analysis. Renew Sustain Energy Rev 85:46–68
Jiang J, Rui Z, Hazlett R, Lu J (2019) An integrated technical-economic model for evaluating CO2 enhanced oil recovery development. Appl Energy 247:190–211
Kamali F, Cinar Y (2014) Co-Optimizing Enhanced Oil Recovery and CO_2 Storage by Simultaneous Water and CO_2 Injection. Energy Explor Exploit 32(2):281–300
Kemp AG, Kasim S (2013) The economics of CO2-EOR cluster developments in the UK Central North Sea. Energy Policy 62:1344–1355
Khan MT, Saud KR, Irfan M, KA, Ibrahim (2018) Curing of Concrete by Carbon Dioxide. Int Res J Eng Technol 5(4):4410–4414
Kim SH, Jeong S, Chung H, Nam K (2020) Mechanism for alkaline leachate reduction through calcium carbonate precipitation on basic oxygen furnace slag by different carbonate sources: application of NaHCO3 and CO2 gas. Waste Manage 103:122–127
King C, Coleman S, Cohen S, Gülen G (2011) The economics of an integrated CO2 capture and sequestration system: Texas Gulf Coast case study. Energy Procedia 4:2588–2595
Kiss AA, Pragt JJ, Vos HJ, Bargeman G, de Groot MT (2016) Novel efficient process for methanol synthesis by CO2 hydrogenation. Chem Eng J 284:260–269
Klaring HP, Hauschild C, Heißner A, Bar-Yosef B (2007) Modelbased control of CO2 concentration in greenhouses at ambient levels increases cucumber yield. Agr Forest Meteor 143:208
Klauber C, Gräfe M (2009) Power, G. Review of Bauxite Residue "Re-use" Options. CSIRO Document DMR-3609, 2009. Available online: http://www.asiapacificpartnership.org/pdf/Aluminium/6th_aluminium_tf_meeting/Review_of_Bauxite_Residue_Re-use_Options_Aug09_sec.pdf
Kleiminger L, Li T, Li K, Kelsall GH (2015) Syngas (CO-H2) production using high temperature micro-tubular solid oxide electrolysers. Electrochim Acta 179:565–577
Koottungal L (2014) 2014 worldwide EOR survey. Oil Gas J (April)
Kumar P, With P, Srivastava VC, Gläser R, Mishra IM (2017) Efficient ceria-zirconium oxide catalyst for carbon dioxide conversions: characterization, catalytic activity and thermodynamic study. J Alloy Compd 696:718
Kwak DH, Kim JK (2017) Techno-economic evaluation of CO2 enhanced oil recovery (EOR) with the optimization of CO2 supply. Int J Greenhouse Gas Control 58:169–184

Kötter E, Schneider L, Sehnke F, Ohnmeiss K, Schröer R (2015) Sensitivities of powerto-gas within an optimised energy system. Energy Procedia 73:190–199

Langanke J et al (2014) Carbon dioxide (CO_2) as sustainable feedstock for polyurethane production. Green Chem 16:1865–1870

Lau HC, Li H, Huang S (2017) Challenges and opportunities of coalbed methane development in China. Energy Fuels 31(5):4588–4602

Lee J, Ryu KH, Ha HY, Jung K-D, Lee JH (2020) Techno-economic and environmental evaluation of nano calcium carbonate production utilizing the steel slag. J CO_2Utilize 37:113–121

Leeson D, Mac Dowell N, Shah N, Petit C, Fennell PS (2017) A Techno-economic analysis and systematic review of carbon capture and storage (CCS) applied to the iron and steel, cement, oil refining and pulp and paper industries, as well as other high purity source. Int J Greenh Gas Control 61:71–84

Leonzio G (2017) Design and feasibility analysis of a Power-to-Gas plant in Germany. J Clean Prod 162:609–623

Leonzio G (2018) State of art and perspectives about the production of methanol, dimethyl ether and syngas by carbon dioxide hydrogenation. J CO2 Util 27:326–354

Leonzio G (2019) Carbon capture utilization and storage supply chain: analysis, modeling and optimization. Sustain Agric Rev 37:37–72

Lewandowska-Bernat A, Desideri U (2017) Opportunities of Power-to-Gas technology. Energy Procedia 105:4569–4574

Li X, Fang Z (2014) Current status and technical challenges of CO_2 storage in coal seams and enhanced coalbed methane recovery: an overview. Int J Coal Sci Technol 1(1):93–102

Li D, Li X, Bai M, Tao X, Shang S, Dai X, Yin Y (2009) CO2 reforming of CH4 by atmospheric pressure glow discharge plasma: a high conversion ability. Int J Hydrogen Energy 34(1): 308–313

Li W, Wang H, Shi Y, Cai N (2013) Performance and methane production characteristics of H_2O-CO_2 co-electrolysis in solid-oxide electrolysis cells. Int J Hydrogen Energy 38:11104–11109

Linde (2012) Gas applications for the pulp and paper industry. Linde North America Inc., New York, NY, USA

Linde (2019) Available at: http://www.linde-gas.com/en/processes/freezing_and_cooling/metal_cooling/index.html

Linde Engineering (2015). Available at: https://www.linde-engineering.com/en/process_plants/hydrogen_and_synthesis_gas_plants/gas_generation/isothermal_reactor/index.html

Liu M, Bai B, Li X (2013) A unified formula for determination of wellhead pressure and bottom-hole pressure. Energy Procedia 37:3291–3298

Liu RX, Poon CS (2016) Utilization of red mud derived from bauxite in self-compacting concrete. J Clean Prod 112:384–391

Liu S, Wang X (2017) Polymers from carbon dioxide: polycarbonates, polyurethanes. Curr Opin Green Sustain Chem 3:61–66

Luu MT, Milani D, Abbas A (2016) Analysis of CO2 utilization for methanol synthesis integrated with enhanced gas recovery. J Clean Prod 112:3540–3554

Marchi B, Zanonia S, Pasetti M (2018) Industrial symbiosis for greener horticulture practices: the CO2 enrichment from energy intensive industrial processes. Procedia CIRP 69:562–567

Marciniak AA, Alves OC, Appel LG, Mota CJA (2019) Synthesis of dimethyl carbonate from CO2 and methanol over CeO2: role of copper as dopant and the use of methyl trichloroacetate as dehydrating agent. J Catal 371(2019):88–95

Marin CM, Li L, Bhalkikar A, Doyle JE, Zeng XC, Cheung CH (2016) Kinetic and mechanistic investigations of the direct synthesis of dimethyl carbonate from carbon dioxide over ceria nanorod catalysts. J Catal 340:295–301

Martín M, Grossmann IE (2017) Towards zero CO2 emissions in the production of methanol from switchgrass. CO2 to methanol. Comput Chem Eng 105:308–3016

Mazzotti M, Pini R, Storti G (2009) Enhanced coalbed methane recovery. J Supercrit Fluids 47(3):619–627

Mendelevitch R (2014) The role of CO2-EOR for the development of a CCTS infrastructure in the North Sea Region A techno-economic model and applications. Int J Greenhouse Gas Control 20:132–159

Mikulčić H, Skovc IR, Dominkovićd DF, Alwie SRW, Manane ZA, Tanf R, Duićb N, Mohamade SNH, Wang X (2019) Flexible Carbon Capture and Utilization technologies in future energysystems and the utilization pathways of captured CO_2. Renew Sustain Energy Rev 114:109338

Monkman S, Shao YX (2006) Assessing the carbonation behavior of cementitious materials. J Mater Civ Eng 18(6):768–776

Muller et al (2016) Polyether carbonate polyol production method. Patent US 9.273,183 B2

Mustafa A, Lougou BG, Shuai Y, Wang Z, Tan H (2020) Current technology development for CO2 utilization into solar fuels and chemicals: a review. J Energy Chem 49:96–123

Muthuraj R, Mekonnen T (2018) Recent progress in carbon dioxide (CO2) as feedstock for sustainable materials development: Co-polymers and polymer blends. Polymer 145:348–373

Nguyen DB, Trinh QH, Hossain MM, Lee WG, Mok YS (2019) Enhancement of plasma-assisted catalytic CO2 reforming of CH4 to syngas by avoiding outside air discharges from ground electrode. Int J Hydrogen Energy, In press

Ni G, Lan Y, Cheng C, Meng Y, Wang X (2011) Reforming of methane and carbon dioxide by DC water plasma at atmospheric pressure. Int J Hydrogen Energy 36(20):12869–12876

Nikbin IM, Aliaghazadeh M, Charkhtab Sb, Fathollahpour A (2018) Environmental impacts and mechanical properties of lightweight concrete containing bauxite residue (red mud). J Clean Prod 172:2683–2694

Olaizola O (2003) Microalgal removal of CO2 from flue gases: changes in medium pH and flue gas composition do not appear to affect the photochemical yield of microalgal cultures. Biotechnol Bioprocess Eng 8:360–367

Ozkan A, Dufour T, Arnoult G, De Keyzer P, Bogaerts A, Reniers F (2015) CO2–CH4 conversion and syngas formation at atmospheric pressure using a multi-electrode dielectric barrier discharge. J CO2 Util 9:74–81

Pan SY, Chang EE, Chiang PC (2012) CO2 capture by accelerated carbonation of alkaline wastes: a review on its principles and applications. Aerosol Air Qual Res 12:770–791

Pardal T, Messias S, Sousa M, Reis Machado AS, Rangel CM, Nunesc D, Pinto JV, Martins R, Nunes da Ponte M (2017) Syngas production by electrochemical CO2 reduction in an ionic liquid based-electrolyte. J CO2 Util 18:62–72

Parvasi P, Khosravanipour Mostafazadeh A, Rahimpour MR (2009) Dynamic modeling and optimization of a novel methanol synthesis loop with hydrogen permselective membrane reactor. Int J Hydrogen Energy 34(9):3717–3733

Patricio J, Angelis-Dimakis A, Castillo-Castillo A, Kalmykova Y, Rosado L (2017) Region prioritization for the development of carbon capture and utilization technologies. J CO2 Util 17:50–59

Petersa R, Baltruweita M, Grubea T, Can Samsuna R, Stolten D (2019) A techno economic analysis of the Power-to-Gas route. J CO2 Util 34:616–634

Pham M, Goujard V, Tatibouet JM, Batiot-Dupeyrat C (2011) Activation of methane and carbon dioxide in a dielectric-barrier discharge-plasma reactor to produce hydrocarbons—Influence of La_2O_3/γ-Al_2O_3 catalyst. Catal Today 171(2011):67–71

Pieri T, Nikitas A, Castillo-Castillo A, Angelis-Dimakis A (2018) Holistic assessment of carbon capture and utilization value chains. Environments 5(108):1–17

Pisal DS, Lele SS (2005) Carotenoid production from microalgae, Dunalliella salina, Indian. J Biotechnol 4:476–483

Pradhan L, Bhattacharjee V, Mitra R, Bhattacharya I, Chowdhury R (2015) Biosequestration of CO2 using power plant algae (Rhizoclonium hieroglyphicum JUCHE2) in a Flat Plate Photobio-Bubble-Reactor—Experimental and modeling. Chem Eng J 275:381–390

Pérez-Fortes M, Bocin-Dumitriu A, Tzimas E (2014) Techno-economic assessment of carbon utilisation potential in Europe. Chem Eng Trans 39:1453–1458

Rachmilevitch S, Cousins AB, Bloom AJ (2004) Nitrate assimilation in plant shoots depends on photorespiration. In: Proceedings of the national academy of sciences of the United States of America, vol 101, pp 11506

Rafiee A, Khalilpour KR, Milani D, Panahie M (2018) Trends in CO2 conversion and utilization: a review from process systems perspective. J Environ Chem Eng 6:5771–5794

Rahimpour MR (2008) A two-stage catalyst bed concept for conversion of carbon dioxide into methanol. Fuel Process Technol 89:556–566

Rahimpour MR, Ghader S (2003) Theoretical investigation of a Pd-membrane reactor for methanol synthesis. Chem Eng Technol 26(8):902–907

Ramos G et al (2016) A thermoplastic polyurethane based on polyether carbonate polyol. Patent WO 2016/120399 Al

Rathod RR, Suryawanshi NT, Memade PD (2014) Evaluation of the properties of Red Mud Concrete. IOSR J Mech Civ Eng (IOSR-JMCE) 1:31–34. ISSN: 2278-1684

Ribeiro DV, Labrincha JA, Morelli MR (2012) Effect of the addition of red mud on the corrosion parameters of reinforced concrete. Cem Concr Res 42:124–133

Ribeiro DV, Morelli MR (2011) Use of red mud as addition for Portland cement mortars. Available at.: https://inis.iaea.org/collection/NCLCollectionStore/_Public/47/081/47081549.pdf

Riccia A, Truda L, Palma V (2019) Study of the role of chemical support and structured carrier on the CO2 methanation reaction. Chem Eng J 377:120461

Rivera-Tinoco R, Farran M, Bouallou C, Aupretre F, Valentin S, Millet P, Ngameni JR (2016) Investigation of power-to-methanol processes coupling electrolytic hydrogen production and catalytic CO2 reduction. Int J Hydrogen Energy 41:4546–4559

Rubin ES, Short C, Booras G, Davison J, Ekstrom C, Matuszewski M, McCoy S (2013) A proposed methodology for CO2 capture and storage cost estimates. Int J Greenh Gas Control 17:488–503

Rutberg PG, Kuznetsov VA, Popov VE, Popov DS, Surov AV, Subbotin DI, Bratsev AN (2015) Conversion of methane by CO2 +H2O+CH4 plasma. Appl Energy 148:159–168

Rönsch S, Schneider J, Matthischke S, Schlüter M, Götz M, Lefebvre J et al (2016) Review on methanation—from fundamentals to current projects. Fuel 166:276–296

Sai PS (2017) Strength properties of concrete by using red mud as a replacement of cement with hydrated lime. Int J Civ Eng Technol (IJCIET) 8(3):38–49

Said et al (2013)

Sanna A, Uibu M, Caramanna G, Kuusik R, Maroto-Valer MM (2014) A review of mineral carbonation technologies to sequester CO2. Chem Soc Rev 43:8049–8080

Sauer DU, Fuchs G, Lunz B, Leuthold M (2012) Technology Overview on Electricity Storage—Overview on the potential and on the deployment perspectives of electricity storage technologies.

Sayyafzadeh M, Keshavarz A (2016) Optimisation of gas mixture injection for enhanced coalbed methane recovery using a parallel genetic algorithm. J Nat Gas Sci Eng 33:942–953

Sayyafzadeh M, Keshavarz A, Alias ARM, Dong KA, Manser M (2015) Investigation of varying-composition gas injection for coalbed methane recovery enhancement: a simulation-based study. J Nat Gas Sci Eng 27:1205–1212

Schmidt O, Gambhir A, Staffell I, Hawkes A, Nelson J, Few S (2017) Future cost and performance of water electrolysis: an expert elicitation study. Int J Hydrogen Energy 42(52):30470–30492

Senghor A, Dioha RMN, Müllerb C, Youm I (2017) Cereal crops for biogas production: a review of possible impact of elevated CO2. Renew Sustain Energy Rev 71:548–554

Shi CS, Liu M, He PP, Ou ZH (2013) Factors affecting kinetics of CO2 curing of concrete. J Sustain Cement-Based Mater 1(1–2):24–33

Shi-Cong K, Bao-jian Z, Chi-Sun P (2014) Use of a CO2 curing step to improve the properties of concrete prepared with recycled aggregates. Cement Concr Compos 45:22

Sinayuc C, Shi JQ, Imrie EC, Syed SA, Korre A, Durucan S (2011) Implementation of horizontal well CBM/ECBM technology and the assessment of effective CO2 storage capacity in a Scottish coalfield. Energy Procedia 4:2150–2156

Singh A (2014) Encyclopedia of polymeric nanomaterials. In: Kobayashi S, Mullen K (eds), Springer Berlin Heidelberg, Berlin, Heidelberg, pp 1–5

Song C, Liu Q, Ji N, Deng S, Zhao J, Li Y, Song Y, Lid H (2018) Alternative pathways for efficient CO_2 capture by hybrid processes —A review. Renew Sustain Energy Rev 82:215–231

Specht M, Bandi A, Baumgart F, Murray CN, Gretz J, Eliasson B, Riemer PWF, Wokaun A (1999) Synthesis of methanol from biomass/CO2 resources, greenhouse gas control technologies, vol 723. Pergamon, Amsterdam

Speybroeck VV, De Wispelaere K, Van der Mynsbrugge J, Vandichel M, Hemelsoet K, Waroquier M (2014) First principle chemical kinetics in zeolites: the methanol-to-olefin process as a case study. Chem Soc Rev 43:7326–7357

Stewart RJ, Haszeldine RS (2015) Can producing oil store carbon? Greenhouse Gas footprint of CO2-EOR, offshore North Sea. Environ Sci Technol 49(9):5788–5795

Stoots CM, O'Brien JE, Herring JS, Hartvigsen JJ (2009) Syngas production via high-temperature co-electrolysis of steam and carbon dioxide. J Fuel Cell Sci Technol 6:011014

Storch HV, Roeb M, Stadler H, Sattler C, Bardow A, Hoffschmidt B (2016) On the assessment of renewable industrial processes: case study for solar co-production of methanol and power. Appl Energy 183:121–132

Sutar H, Mishra SC, Sahoo SK, Chakraverty AP, Maharana HS (2014) Progress of red mud utilization: an overview. Am Chem Sci J 4 (3):255–279

Taherimehr M, Pescarmona PP (2014) Green polycarbonates prepared by the copolymerization of CO2 with epoxides. J Appl Polym Sci 41141:1–17

Tao X, Bai M, Li X, Long H, Shang S, Yin Y, Dai X (2011) CH4-CO2 reforming by plasma e challenges and opportunities. Prog Energy Combust Sci 37:113–124

Tcvetkov P, Cherepovitsyn A, Fedoseev S (2019) The changing role of CO2 in the transition to a circular economy: review of carbon sequestration projects. Sustainability 11:5834

Thema M, Bauer F, Sterner M (2019) Power-to-gas: electrolysis and methanation status review. Renew Sustain Energy Rev 112:775–787

Toyo Engineering (2015) Available at: http://www.toyo-eng.com/jp/en/products/petrochmical/methanol

Tsuparia E, Kärki J, Vakkilainen E (2016) Economic feasibility of Power-to-Gas integrated with biomass fired CHP plant. J Energy Storage 5:62–69

Ugwu CU, Aoyagi H, Uchiyama H (2008) Photobioreactors for mass cultivation of algae. Bioresour Techol 99:4021–4028

Vandewalle J, Bruninx K, W, D (2015) Effects of large-scale Power-to-Gas conversion on the power, gas and carbon sectors and their interactions. Energy Convers Manage 94:28–39

Verma AK, Sirvaiya A (2016) Comparative analysis of intelligent models for prediction of Langmuir constants for CO_2 adsorption of Gondwana coals in India. Geomech Geophy Geo-Energy Geo-Res 2 (2):97–109
Wang Q, Shi H, Yan B, Jin Y, Cheng Y (2011) Steam enhanced carbon dioxide reforming of methane in DBD plasma reactor. Int J Hydrogen Energy 36(14):8301–8306
Wang X, van't Veld K, Marcy P, Huzurbazar S, Alvarado V (2018) Economic co-optimization of oil recovery and CO2 sequestration. Appl Energy 222:132–14
Wang F, Wang Y, Zhang L, Zhu J, Han B, Fan W, Xu L, Yu H, Cai W, Li Z, Deng Z, Shi W (2019) Performance enhancement of methane dry reforming reaction for syngas production over Ir/Ce0.9La0.1O2-nanorods catalysts. Catalysis today, In proof
Wei N, Li X, Dahowski RT, Davidson CL, Liu S, Zha Y (2015) Economic evaluation on CO2- EOR of onshore oil fields in China. Int J Greenhouse Gas Control 37:170–181
Wen H, Cheng X, Chen J, Zhang C, Yu Z, Li Z, Fan S, Wei G, Cheng B (2020) Micro-pilot test for optimized pre-extraction boreholes and enhanced coalbed methane recovery by injection of liquid carbon dioxide in the Sangshuping coal mine. In press, Process Safety and Environmental Protection
Wong S, Law D, Deng X, Robinson J, Kadatz B, Gunter WD et al (2007) Enhanced coalbed methane and CO2 storage in anthracitic coals—micro-pilot test at south Qinshui, Shanxi, China. Int J Greenhouse Gas Control 1(2):215–222
Xu B, Yang R, Meng F, Reubroycharoen P, Vitidsant T, Zhang Y, Yoneyama Y, Tsubaki N (2009) A new method of low temperature methanol synthesis. Catal Surv Asia 13:147–163
Xuan D, Zhan B, Sun Poon C (2018) A maturity approach to estimate compressive strength development of CO2-cured concrete blocks. Cement Concr Compos 85:153–160
Yanpeng S, Yong N, Angshan W, Dengxiang J, Fengwen Y, Jianbing J (2014) Carbon dioxide reforming of methane to syngas by thermal plasma. Plasma Sci Technol 14(3):252–256
Ye JP, Feng SL, Fan ZQ, Wang GQ, Gunter WD, Sam W et al (2007) Micro-pilot test for enhanced coalbed methane recovery by injecting carbon dioxide in south part of Qinshui Basin. Acta Petrolei Sinica 28(4):77–80
Yin G, Deng B, Li M, Zhang D, Wang W, Li W, Shang D (2017) Impact of injection pressure on CO2–enhanced coalbed methane recovery considering mass transfer between coal fracture and matrix. Fuel 196:288–297
Zappa W (2014) Pilot-scale experimental work on the production of precipitated calcium carbonate (PCC) from steel slag for CO2 fixation. Master Thesis, Aalto University. Department of Energy Technology
Zevenhoven R, Eloneva S, Teir S (2006) Chemical fixation of CO2 in carbonates: routes to valuable products and long-term storage. Catal Today 115(1–4):73–79
Zhan BJ, Xuan DX, Poon CS, Shi CJ (2016) Effect of curing parameters on CO2 curing of concrete blocks containing recycled aggregates. Cem Concr Comp 71:122–130
Zhang C, Jun KW, Gao R, Kwak G, Park HG (2017) Carbon dioxide utilization in a gas-to-methanol process combined with CO2/steam-mixed reforming: technoeconomic analysis. Fuel 190:303–311
Zhang X (2015) Microalgae removal of CO2 from flue gas. Clean Coal Technol Res Rep
Zhao T, Hu X, Wu D, Li R, Yang G, Wu Y (2017) Direct synthesis of dimethylcarbonate from carbon dioxide and methanol at room temperature using imidazolium hydrogen carbonate ionic liquid as recyclable catalyst and dehydrant. Chemsuschem 10:2046–2052
Zheng Y, Zhang W, Li Y, Chen J, Yu B, Wang J, Zhang L, Zhang J (2017) Energy related CO2 conversion and utilization: advanced materials/nanomaterials, reaction mechanisms and technologies. Nano Energy 40:512–539
Zhou J, Ma Z, Zhang L, Liu C, Pu J, Chen X, Zheng Y, Chan SH (2019) Study of CO2 and H2O direct co-electrolysis in an electrolyte-supported solid oxide electrolysis cell by aqueous tape casting technique. Int J Hydrogen Energy 44:28939–28946
Zittelli GC, Biondi N, Rodolfi L, Tredici MR (2013) Photobioreactors for mass production of microalgae. In: Richmond A, Amos Q (eds) Handbook of microalgal culture: applied phycology and biotechnology, 2nd edn. John Wiley & Sons Ltd., Oxford, UK, pp 225–266

The manufacturer's authorised representative in the EU is Springer Nature Customer Service Centre GmbH, Europaplatz 3, 69115 Heidelberg, Germany. If you have any concerns regarding our products, please contact ProductSafety@springernature.com

Printed and bound by CPI Group (UK) Ltd, Croydon, CR0 4YY
22/09/2026
02221072-0002